Applied Statistics (Continued)

GOULDEN · Methods of Statistical Analysis, *Second Edition*
HALD · Statistical Tables and Formulas
HALD · Statistical Theory with Engineering Applications
HANSEN, HURWITZ, and MADOW · Sample Survey Methods and Theory, Volume I
HOEL · Elementary Statistics
JOHNSON and LEONE · Statistics and Experimental Design: In Engineering and the Physical Sciences, Volumes I and II
KEMPTHORNE · An Introduction to Genetic Statistics
MEYER · Symposium on Monte Carlo Methods
RICE · Control Charts
ROMIG · 50–100 Binomial Tables
SARHAN and GREENBERG · Contributions to Order Statistics
TIPPETT · Technological Applications of Statistics
WILLIAMS · Regression Analysis
WOLD and JURÉEN · Demand Analysis
YOUDEN · Statistical Methods for Chemists

Books of Related Interest

ALLEN and ELY · International Trade Statistics
ARLEY and BUCH · Introduction to the Theory of Probability and Statistics
CHERNOFF and MOSES · Elementary Decision Theory
HAUSER and LEONARD · Government Statistics for Business Use, *Second Edition*
STEPHAN and McCARTHY · Sampling Opinions—An Analysis of Survey Procedures

A WILEY PUBLICATION IN APPLIED STATISTICS

Statistical Theory with Engineering Applications

A. HALD
Professor of Statistics
University of Copenhagen

New York · John Wiley & Sons, Inc.
New York · London · Sydney

sixth printing, may, 1965

PRINTED IN THE UNITED STATES OF AMERICA

Preface

THE PRESENT BOOK IS A SLIGHTLY EXPANDED VERSION OF MY DANISH textbook, published in 1948, which was written on the basis of lectures given to members of the Danish Society of Civil Engineers.

It is the aim of this book to provide a fairly elementary mathematical treatment of statistical methods of importance to the engineer in his daily work.

The choice of the statistical methods considered has been much influenced by the works of the English statistician Professor R. A. Fisher. The main stress has been laid on the normal distribution and the tests of significance connected with this distribution, since these tests have proved to be of great practical value. The problems of statistical quality control have been dealt with according to the point of view of Dr. W. A. Shewhart. Certain subjects, for example the theory of growth curves, time series, and stochastic processes, have only been outlined to draw attention to the problems. This is due partly to the fact that these subjects are very extensive and a detailed treatment of them would cause this book to exceed its bounds, and partly to the fact that the statistical methods involved are not as yet fully developed.

Statistics cannot be properly understood without mathematics. The mathematical techniques employed here, however, are reasonably elementary, and only a knowledge of standard differential and integral calculus is assumed. Admittedly, this results in somewhat longer proofs of certain theorems than would be required if more advanced mathematics had been employed, but the point aimed at was as said above to provide proofs requiring the minimum of mathematical knowledge.

The full significance of a theorem can be grasped only when its applications are studied. This point of view underlies the design of the book, and the theory is therefore illustrated by a great number of examples, mainly from my own experience in engineering and industrial practice. However, similar applications could have been found within physics, biology, sociology, economics, and many other fields.

I am indebted to many friends for their help in the writing of this book.

Dr. G. Fagerholt read the manuscript of the Danish edition, and the resulting discussions have to a considerable extent set their stamp upon it. Dr. Niels Arley read the proofs and suggested many valuable corrections. Mr. Victor Hansen, Mechanical Engineer, helped me with the planning of the figures and made all the drawings.

Dr. G. Seidelin translated the Danish text, and Mr. J. E. Kerrich, of the University of the Witwatersrand, South Africa, read through the translation and gave me valuable suggestions for the revision of certain parts of the original text.

A survey of the notation used and the most important formulas together with tables of the distributions employed in the tests of significance and some other auxiliary tables have been published as a separate volume.

A. Hald

University of Copenhagen
August, 1951

Contents

References to tables numbered with Roman numerals are to A. Hald: *Statistical Tables and Formulas*, John Wiley & Sons, New York, 1952.

CHAPTER 1.

FUNDAMENTAL CALCULUS OF PROBABILITIES

1.1. The Concept of Probability.

The theory of statistics is a branch of applied mathematics aiming at a mathematical description and analysis of observations as a basis for the prediction of certain events under given conditions. As the theory of statistics is based on the theory of probability a short account of the fundamental calculus of probabilities is given in the first chapter.

Many processes are influenced by interacting factors, which for fundamental or practical reasons cannot be analyzed completely. This, for instance, applies to industrial production processes where the properties of the final product depend initially on the properties of the raw materials and the manufacturing process itself, and where these two factors are further influenced by a large number of less important factors, such as, for instance, contamination of the raw materials, variations in temperature and humidity, and irregularities in the working of the machines. *Repetition* of the manufacturing process under as uniform conditions as possible leads to products of varying quality as a certain number of the above-mentioned factors are not completely under control. For example, in the manufacture of incandescent lamps, the length of life of the individual lamps will vary even though a uniform quality has been aimed at. Similar variations are found in the capacity of condensers, the diameter of rivets, the strength of linen thread, etc. It therefore follows *that a single observation cannot be reproduced, but experience shows that a set of observations, resulting from repetition of the process, presents certain characteristic features which can be reproduced.* In statistical terminology this is expressed by saying that the observations present "random" variations, and it is the purpose of statistics to describe and analyze such random variations. Any observed quantity which varies from one repetition of a given process to another is called a *random* or *stochastic variable* (στοχάζεσθαι = to aim, to guess).

Similarly, any observed quantity associated with a set of n repetitions of a given process, which can vary from one set to another, is a stochastic variable.

As an example of fundamental importance, consider a production process where any article manufactured will be either accepted as satisfactory or rejected as defective. Among n such articles a certain number a will be defective. The number a, and hence the proportion $h = \frac{a}{n}$ will vary somewhat from one set of n articles to the next. Both a and h are examples of stochastic variables.

The above may be generalized as follows:

In a process in which the occurrence of an "event" depends on chance, the frequency with which the event occurs in a given number of *repetitions* is a stochastic variable. If the process is repeated n times, and the event occurs a_1 times, we define *the frequency* h_1 *of the event* as the ratio

$$h_1 = \frac{a_1}{n} \qquad (0 \leq h_1 \leq 1).$$

Sometimes a_1 is called the *absolute* and h_1 the *relative* frequency. If nothing is said to the contrary, the word frequency will in the following pages denote *relative* frequency. If the process is again repeated n times, we get a new frequency

$$h_2 = \frac{a_2}{n},$$

where h_2 differs at random from h_1; and if we proceed with such repetitions, we get a series of frequencies

$$h_1, h_2, h_3, \ldots,$$

which vary at random about "a cluster point". The random variation of

TABLE 1.1.

Number of items per group		
25	250	2500
Number of defective items per group		
1	12	157
4	14	152
0	17	157
0	11	136
1	22	152
1	9	135
2	15	143
0	14	160
1	21	149
1	8	153

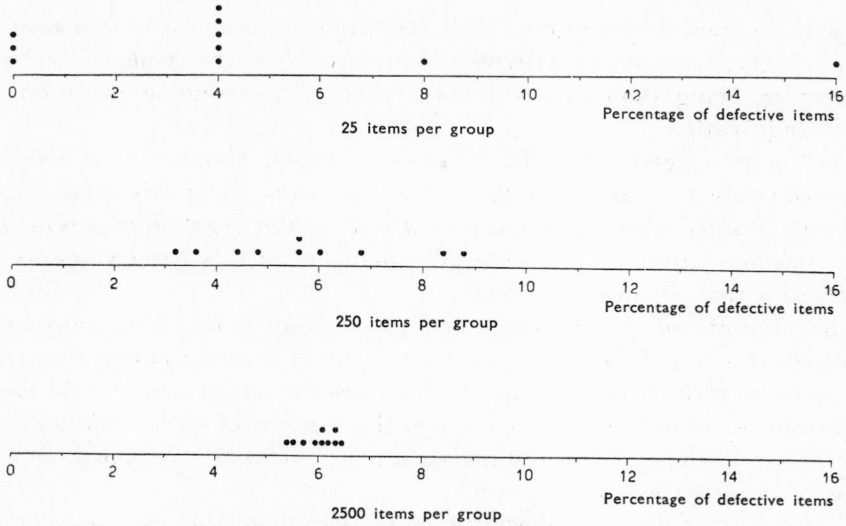

FIG. 1.1. Variation in percentage of defective items.

TABLE 1.2.

Number of items per group		
25	250	2500
Percentage of defective items		
4	4·8	6·28
16	5·6	6·08
0	6·8	6·28
0	4·4	5·44
4	8·8	6·08
4	3·6	5·40
8	6·0	5·72
0	5·6	6·40
4	8·4	5·96
4	3·2	6·12

the frequencies depends on the value of n as shown in the following example.

As an example of what can happen when an industrial product is manufactured under stable conditions, Table 1.1 gives the number of defective items in 10 groups of 25 items, 10 groups of 250 items, and 10 groups of 2500 items, respectively, and in Table 1.2 the corresponding percentages have been tabulated.

In the groups with 25 items each the percentage of defective items varies at random between 0 and 16, in the groups with 250 items it varies between

3·2 and 8·8, and in the groups with 2500 items between 5·40 and 6·40. In Fig. 1.1 these percentages have been plotted on the same scale, each percentage being shown as a dot above the corresponding point on the percentage scale.

The figure suggests that the frequencies cluster about a fixed value of approximately 6%, and that *the random variation about this value is considerably smaller when the frequencies are calculated from large groups than when calculated from small groups.* We shall define this fixed value as the *probability* that an item is defective.

This stability of the frequency of a given event is found in many other fields where *a well-defined process is the object of repeated observations.* For a *mathematical* description of any such experience it will therefore be useful to introduce, as in the above example, the concept of *the probability of an event,* i. e., *a number between* 0 *and* 1, *which denotes the theoretical, or ideal value of the frequency of the event.*

A central problem is to decide how to obtain reliable estimates of the probability of an event, based on the observed frequencies. As the solution of this problem requires some knowledge of the calculus of probabilities it will be discussed later.

The calculus of probabilities bears the same relationship to the real world as do other branches of applied mathematics, e. g., geometry and mechanics. By idealizing experience, certain fundamental theoretical concepts and rules are introduced and by deduction from these definitions and axioms the theory is developed.

In § 1.3 the axioms of the calculus of probabilities are introduced. If this calculus is to furnish us with an adequate description of the real world, the axioms must be chosen in such a manner that *the rules for operating with probabilities are the same as those used for frequencies.* Before introducing the axioms of the calculus of probabilities, we therefore first investigate the fundamental properties of frequencies in § 1.2.

1.2. Some Fundamental Properties of Frequencies.

When 6805 pieces of moulded vulcanite made from a resinous powder were sorted out according to two criteria, porosity and dimension, 473 pieces were rejected on account of defects in their dimensions. Detailed examination of the 473 defective pieces showed that 142 were porous. Inspection of the remaining $6805 - 473 = 6332$ pieces resulted in the rejection of a further 1233 pieces because they too were porous. Tables 1.3 and 1.4 have been calculated from these four numbers.

The two frequencies in the vertical margin, 7·0 and 93·0%, are called *the marginal distribution* according to dimensional defects. Correspondingly

TABLE 1.3.

Classification of 6805 moulded pieces of vulcanite.

	Porous	Non-porous	Total
With defective dimensions	142	331	473
Without defective dimensions..	1233	5099	6332
Total...	1375	5430	6805

TABLE 1.4.

Classification of 6805 moulded pieces of vulcanite.
Frequencies in per cent.

	Porous	Non-porous	Total
With defective dimensions	2·1	4·9	7·0
Without defective dimensions..	18·1	74·9	93·0
Total...	20·2	79·8	100·0

the frequencies in the horizontal margin, 20·2 and 79·8%, are called the marginal distribution according to porosity. *Any marginal frequency is equal to the sum of the frequencies in the corresponding row (or column) of the table.*

The total frequency of defective pieces

$$2{\cdot}1+4{\cdot}9+18{\cdot}1 = 25{\cdot}1\%,$$

cannot be computed from the marginal frequencies alone, as some of the pieces are rejected for both reasons, but it can be expressed in terms of the marginal frequencies and the frequency of the simultaneous occurrence of both reasons for rejection since

$$7{\cdot}0+20{\cdot}2-2{\cdot}1 = 25{\cdot}1\%,$$

i. e., *the total frequency of defective pieces is equal to the sum of the marginal frequencies minus the frequency of the simultaneous occurrence of both reasons for rejection.*

If we compute the frequency of dimensional defects among the 1375 pieces rejected on account of porosity, we get 142/1375 = 10·3%. This frequency of dimensional defects, given that the pieces are porous, is called the *conditional frequency* of dimensional defects, given porosity. Correspondingly we find the frequency of dimensional defects among the non-porous pieces to be 6·1%, see Table 1.5.

TABLE 1.5.

Conditional distributions of dimensional defects.

	Conditional distributions		Marginal distribution
	Porous	Non-porous	
With defective dimensions	10·3	6·1	7·0
Without defective dimensions .	89·7	93·9	93·0
Total...	100·0	100·0	100·0

The frequency of dimensional defects is somewhat larger among the porous than among the non-porous pieces, which indicates dependence between dimensional defects and porosity.

The frequency of pieces rejected for both reasons may be computed from the frequency of porous pieces by multiplying that frequency by the frequency of dimensional defects among the porous pieces, i. e.,

$$\frac{20{\cdot}2}{100} \times \frac{10{\cdot}3}{100} = 2{\cdot}1\%\,.$$

The conditional distributions according to porosity may be computed in a similar manner, as shown in Table 1.6. The frequency of porosity is considerably greater among pieces with defective dimensions than among pieces without defective dimensions.

As above, *the frequency of pieces rejected for both reasons* may be expressed as the *product of a marginal and a conditional frequency*:

$$\frac{7{\cdot}0}{100} \times \frac{30{\cdot}0}{100} = 2{\cdot}1\%.$$

TABLE 1.6.

Conditional distributions according to porosity.

		Porous	Non-porous	Total
Conditional distributions	With defective dimensions	30·0	70·0	100·0
	Without defective dimensions .	19·5	80·5	100·0
Marginal distribution		20·2	79·8	100·0

The above analysis can be generalized as follows: Let U and V denote two events, and $\overline{U}$ and $\overline{V}$ the corresponding *complementary* events, i. e., $\overline{U}$ denotes the event that U does not occur, and similarly for $\overline{V}$. The results of n observations can then be tabulated as shown in Table 1.7, where a_1 indicates the absolute frequency of the simultaneous occurrence of U and V, a_2 the absolute frequency of the occurrence of U when V fails to occur, etc.

TABLE 1.7.

Distribution of n observations according to two events and the corresponding combinations of events.

Event	V	$\overline{V}$	Total
U	a_1	a_2	a_1+a_2
$\overline{U}$	a_3	a_4	a_3+a_4
Total	a_1+a_3	a_2+a_4	n

Furthermore, let UV denote *the compound event* that *both U and V occur* and $U\dot{+}V$ the event that *at least one of the two events occurs*, i. e., one of the events $U\overline{V}$, $\overline{U}V$, or UV occurs. If U and V are *mutually exclusive*, i. e., the occurrence of one of them precludes the occurrence of the other, then the event $U\dot{+}V$ is equal to the event that *either U or V* occurs and is denoted by $U+V$. The three symbols UV, $U+V$, and $U\dot{+}V$ may therefore be read as "*both U and V*", "*either U or V*", and "*U and/or V*", respectively. These concepts and system of notation may easily be extended to more than two events.

The definitions of the symbols imply for instance the following relations:

$$U\dot{+}V = U\overline{V}+\overline{U}V+UV$$

and—if U and V are mutually exclusive—

$$U+V = U\overline{V}+\overline{U}V.$$

The corresponding frequencies are written $H\{U\}$, $H\{UV\}$, $H\{U\dot{+}V\}$, etc. Expressing these frequencies by means of the symbols in Table 1.7 we get

$$H\{U\} = \frac{a_1+a_2}{n},$$

$$H\{UV\} = \frac{a_1}{n},$$

and

$$H\{U\dot{+}V\} = \frac{a_1+a_2+a_3}{n}.$$

Dividing all numbers in Table 1.7 by n we get Table 1.8, which corresponds to Table 1.4 of the example above.

TABLE 1.8.

The distribution of n observations according to two events and the corresponding combinations of events. Frequencies.

Event	V	$\bar{V}$	Marginal distribution
U	$H\{UV\}$	$H\{U\bar{V}\}$	$H\{U\}$
$\bar{U}$	$H\{\bar{U}V\}$	$H\{\bar{U}\bar{V}\}$	$H\{\bar{U}\}$
Marginal distribution	$H\{V\}$	$H\{\bar{V}\}$	1

Tables 1.7 and 1.8 immediately give us the following relations:

$$H\{U\}+H\{\bar{U}\} = \frac{a_1+a_2}{n}+\frac{a_3+a_4}{n} = 1, \tag{1.2.1}$$

i. e., *the frequency of an event plus the frequency of the complementary event is equal to* 1, and

$$H\{U\} = \frac{a_1+a_2}{n} = \frac{a_1}{n}+\frac{a_2}{n} = H\{UV\}+H\{U\bar{V}\}, \tag{1.2.2}$$

i. e., *any marginal frequency is equal to the sum of the frequencies in the corresponding row (or column) of the table.*

The frequency of at least one of the events U and V may be written

$$\frac{a_1+a_2+a_3}{n} = \frac{a_1+a_2}{n}+\frac{a_1+a_3}{n}-\frac{a_1}{n},$$

or

$$\boxed{H\{U\dot{+}V\} = H\{U\}+H\{V\}-H\{UV\},} \tag{1.2.3}$$

which means that *the frequency of the occurrence of at least one of the events is equal to the sum of the marginal frequencies minus the frequency of the compound event.*

$H\{U\dot{+}V\}$ may also be expressed by $H\{\bar{U}\bar{V}\}$:

$$H\{U\dot{+}V\} = \frac{a_1+a_2+a_3}{n} = \frac{n-a_4}{n} = 1-H\{\bar{U}\bar{V}\}, \tag{1.2.4}$$

i. e., *the frequency of the occurrence of at least one of the events is equal to* 1 *minus the frequency of the occurrence of none of the events* ("none" is complementary to "at least one").

Consider the a_1+a_2 observations where the event U has occurred. These observations constitute a *subset* of the total set of observations. The frequency of the event V in this subset is denoted by $H\{V|U\}$ and is called *the conditional frequency of V, given U.*

Division of the first row of figures in Table 1.7 by the marginal number gives us

$$H\{V|U\} = \frac{a_1}{a_1+a_2}$$

and

$$H\{\overline{V}|U\} = \frac{a_2}{a_1+a_2}.$$

Correspondingly

$$H\{U|V\} = \frac{a_1}{a_1+a_3}$$

and

$$H\{\overline{U}|V\} = \frac{a_3}{a_1+a_3}.$$

The frequency of the compound event UV may then be written

$$\frac{a_1}{n} = \frac{a_1+a_2}{n} \cdot \frac{a_1}{a_1+a_2}$$

or

$$\frac{a_1}{n} = \frac{a_1+a_3}{n} \cdot \frac{a_1}{a_1+a_3},$$

i. e.,

$$\boxed{\begin{aligned} H\{UV\} &= H\{U\} \cdot H\{V|U\} \\ &= H\{V\} \cdot H\{U|V\}, \end{aligned}} \qquad (1.2.5)$$

which means that *the frequency of the compound event equals the frequency of one event multiplied by the conditional frequency of the other event, given that the first event occurs.*

1.3. Definitions and Axioms of the Calculus of Probabilities.

As mentioned in § 1.1, the definitions and axioms of the calculus of probabilities are chosen in such a manner that the rules for operating with probabilities are the same as those used for frequencies.

The calculus of probabilities may be based on the following five definitions and axioms:

I. A probability is *a real number between* 0 *and* 1, both limits included.

II. The probability of *a certain event*, i. e., an event which occurs at every observation, is 1.

III. The probability of *an impossible event*, i. e., an event which cannot occur at any observation, is 0.

IV. *The addition formula.*

The probability that *at least one of two events occurs* is equal to the sum of the probabilities of each event minus the probability of the compound event.

V. *The multiplication formula.*

The probability that *two events occur together* is equal to the probability of one event multiplied by the conditional probability of the other, given that the first occurs.

The concept of probability was introduced in § 1.1 as the ideal value of the relative frequency of an event in a set of observed data, generated by repetition of the observation under uniform conditions. Thus, the definition of the probability of an event requires the definition or description of both the event itself and the process observed under the given conditions. For instance, the fraction of defective items in a set of n items of an industrial product manufactured under stable conditions depends not only on the definition of a defective item but also on the manner in which the n items have been selected from the total production of items in the plant. We will probably find a difference between the fraction of defectives in a random sample of n items from the total production composed of items from many machines and the fraction defectives in a random sample of n items from a single machine.

It is often convenient to interpret the given set of observations as a random sample drawn from a *population* of all conceivable observations that might have been made under the given conditions. In some cases this population is *finite* and consists of N elements, as, for instance, when a random sample of n items is taken from a group of N items (a lot) submitted for inspection for the purpose of accepting or rejecting the lot on the basis of the quality (fraction defectives) of the n items in the sample. In other cases it is practical to regard the population as *indefinitely large.* A random sample of n items from the production of a given machine on a given day may be conceived of as a random sample of n items from the hypothetical infinite population of items that might have been produced under the given conditions. Thus the probability of an event may be spoken of as the frequency of the event in the population. The concept of population will be discussed further in § 5.4.

In practical work probability statements derive their meaning from the interpretation of probability as a relative frequency. When we apply the theory of probability we assume that *in the long run an event will occur with a relative frequency which is practically equal to its probability.* For instance, an event U with a *small* probability will occur only in a small number of cases in a long series of observations. Consequently, if we perform only one observation we do not *expect* the event U to occur or in other words we consider it *practically certain* that the event U will not occur.

Later on, see for instance § 9.4, we shall derive probability statements regarding the acceptance or rejection of a hypothesis. We may, for instance, want to accept or reject a lot of produced items, submitted for inspection, on the basis of the fraction of defective items found in a random sample. We make the hypothesis that the fraction defectives in the lot is less than or equal to a specified percentage and calculate the probability of the observed event under the assumption that the hypothesis is true. If this probability is small, for instance less than 1%, we usually reject the hypothesis because it leads to a small probability for an event which has actually occurred in a single observation.

It has been stated in Definition III that the probability of an impossible event is zero. The reverse statement, however, is not true. As an example consider a roulette wheel with a pointer and assume that it is equally likely for the wheel to stop in any position. The probability that the pointer falls within a given interval on the scale of the wheel is consequently equal to the ratio between the length of that interval and the length of the whole scale. Letting the interval converge to zero we find that the probability of the pointer pointing at a single point specified in advance is zero. Thus any single point specified in advance has the probability zero but none of these points are impossible events. In most cases where the stochastic variable is continuous we get similar results. This is not in contradiction with the frequency interpretation of probability because for any preassigned number of observations we only expect a small proportion to fall within any small interval and the smaller the interval the smaller becomes the expected frequency.

The object of the calculus of probabilities is the development of rules which make it possible to calculate the probabilities of combinations of events from given initial events and probabilities. Like the rules of other calculi these rules are derived from the definitions and axioms by purely formal reasoning. The calculus of probabilities, however, plays an important rôle in testing whether our system of axioms has been suitably chosen, for in most cases not the axioms but more complex probability statements are compared with frequency statements derived from experiments and observations.

In §§ 1.4 and 1.5 it is shown that the five axioms lead to rules for operating with probabilities analogous to the rules for frequencies developed in § 1.2. In Chapter 2 these rules are used to derive the probabilities of combinations of events frequently occurring in practical work.

1.4. The Addition Formula.

Let $P\{U\}$ denote the probability that U will occur. (Braces are always used in connection with the term P, as U is not necessarily quantitatively expressed). Table 1.9 gives the probabilities of the possible combinations of U, V, $\overline{U}$, and $\overline{V}$.

TABLE 1.9.
Probabilities of two events and the corresponding combinations of events.

Event	V	$\overline{V}$	Marginal probability
U	$P\{UV\}$	$P\{U\overline{V}\}$	$P\{U\}$
$\overline{U}$	$P\{\overline{U}V\}$	$P\{\overline{U}\overline{V}\}$	$P\{\overline{U}\}$
Marginal probability	$P\{V\}$	$P\{\overline{V}\}$	1

On basis of the axioms in § 1.3, we will now show that the probabilities in Table 1.9 will fulfill relations corresponding to those of the frequencies in Table 1.8.

According to the addition formula, we have

$$\boxed{P\{U\dot{+}V\} = P\{U\}+P\{V\}-P\{UV\}.} \tag{1.4.1}$$

If the events U and V are *mutually exclusive*, i. e., they cannot occur together, the event UV is an impossible event, and $P\{UV\}$ is 0. In that case the addition formula reduces to

$$\boxed{P\{U+V\} = P\{U\}+P\{V\},} \tag{1.4.2}$$

i. e., *the probability of the occurrence of either one or the other of two mutually exclusive events is equal to the sum of the probabilities of the occurrence of each event separately.* (The "either—or" rule).

The event $U+\overline{U}$ is certain, as either U or $\overline{U}$ must occur at any observation. Therefore, $P\{U+\overline{U}\}$ is 1. As U and $\overline{U}$ are mutually exclusive, we find

$$P\{U+\overline{U}\} = P\{U\}+P\{\overline{U}\},$$

i. e.,

$$\boxed{P\{U\}+P\{\overline{U}\} = 1,} \tag{1.4.3}$$

or *the sum of the probability of an event and the probability of the complementary event is* 1, corresponding to (1.2.1). This means that the sum of the probabilities in the marginal distributions is 1, see Table 1.9.

The necessary and sufficient condition for the occurrence of U is the occurrence of either of the events UV or $U\overline{V}$, i. e.,

$$U = UV + U\overline{V} .$$

Using the "either—or" rule, we get

$$\boxed{P\{U\} = P\{UV\} + P\{U\overline{V}\} .} \qquad (1.4.4)$$

We thus see that we get *the marginal probabilities* (as we got the marginal frequencies of Table 1.8) *by adding the probabilities in the corresponding row or column.*

The probability that *at least* one of two events will occur is often easier to calculate from the complementary event, *none* of the events will occur, than from the addition formula. Table 1.9 and (1.4.1) give us

$$\boxed{P\{U \dot{+} V\} = 1 - P\{\overline{U}\,\overline{V}\} .} \qquad (1.4.5)$$

This result may be generalized directly: *The probability that at least one of k events will occur is equal to* 1 *minus the probability that none of the k events will occur.*

The addition formula can be generalized to more than two events. Introducing the event W together with U and V we get

$$U \dot{+} V \dot{+} W = U \dot{+} (V \dot{+} W)$$

and consequently

$$\begin{aligned} P\{U \dot{+} V \dot{+} W\} &= P\{U \dot{+} (V \dot{+} W)\} \\ &= P\{U\} + P\{V \dot{+} W\} - P\{U(V \dot{+} W)\} \end{aligned} \qquad (1.4.6)$$

according to the addition formula for two events.

As

$$U(V \dot{+} W) = UV \dot{+} UW$$

we get

$$\begin{aligned} P\{U(V \dot{+} W)\} &= P\{UV \dot{+} UW\} \\ &= P\{UV\} + P\{UW\} - P\{UVW\} . \end{aligned} \qquad (1.4.7)$$

Inserting (1.4.7) in (1.4.6) we finally get

$$\begin{aligned} P\{U \dot{+} V \dot{+} W\} &= P\{U\} + P\{V\} + P\{W\} \\ &\quad - P\{UV\} - P\{UW\} - P\{VW\} + P\{UVW\} . \end{aligned} \qquad (1.4.8)$$

This formula may be illustrated by means of tables analogous to Tables 1.7, 1.8, and 1.9.

Further generalization gives

$$
\begin{aligned}
P\{U_1 \dotplus U_2 \dotplus \ldots \dotplus U_m\} = {} & P\{U_1\}+P\{U_2\}+\ldots+P\{U_m\} \\
& -P\{U_1U_2\}-P\{U_1U_3\}-\ldots-P\{U_{m-1}U_m\} \\
& +P\{U_1U_2U_3\}+P\{U_1U_2U_4\}+\ldots+P\{U_{m-2}U_{m-1}U_m\} \\
& \cdots\cdots\cdots\cdots\cdots\cdots \\
& (-1)^{m-1}P\{U_1U_2\ldots U_m\}\,.
\end{aligned}
\tag{1.4.9}
$$

The general "either—or" rule becomes

$$P\{U_1+U_2+\ldots+U_m\} = P\{U_1\}+P\{U_2\}+\ldots+P\{U_m\}\,, \tag{1.4.10}$$

if

$$P\{U_iU_j\} = 0 \quad \text{for} \quad i \neq j\,, \tag{1.4.11}$$

i. e., *the probability that one out of m events, any two of which are mutually exclusive, occurs is equal to the sum of the probabilities of the occurrence of each event separately.*

If the result of an observation is that one or other of m mutually exclusive events $U_1, \ldots, U_m$ must occur, and if the probability of occurrence is the same for each event, being equal to P, say, then

$$P\{U_1\} = \ldots = P\{U_m\} = P\,.$$

Further, since one of the events must occur,

$$P\{U_1+\ldots+U_m\} = 1\,,$$

and, by the "either—or" rule,

$$P\{U_1+\ldots+U_m\} = P\{U_1\}+\ldots+P\{U_m\}\,;$$

hence,

$$mP = 1$$

or

$$P = \frac{1}{m}\,.$$

If we denote the first k of the m *equally possible* events as *favourable* events, we find according to the "either—or" rule, that the probability of a favourable event is

$$P\{U_1+\ldots+U_k\} = \frac{k}{m}\,. \tag{1.4.12}$$

This result which is derived from our axioms was the starting point in the classical calculus of probabilities. Since the first applications of this calculus were to simple games of chance, e. g., gambling with dice, it was natural to *define the probability of a favourable event as the number of ways in which a favourable event can occur divided by the number of all possible results of an observation, it being assumed that all results are "equally possible"*. If in throwing a die the chances are the same for all 6 faces, the probability of a particular face is $\frac{1}{6}$. The probability that the number thrown is even is $\frac{3}{6} = \frac{1}{2}$, i. e., 3 favourable cases divided by 6 equally possible cases. Many probability problems of games of chance may be solved by simple enumerations of favourable and equally possible cases using the theory of permutations and combinations.

1.5. The Multiplication Formula.

The conditional probability of V, given U, is written $P\{V|U\}$. According to the multiplication formula, we have:

$$\boxed{P\{UV\} = P\{U\} \cdot P\{V|U\}} \tag{1.5.1}$$

and

$$\boxed{P\{UV\} = P\{V\} \cdot P\{U|V\}.} \tag{1.5.2}$$

For $P\{U\}$ and $P\{V\}$ different from zero we get

$$P\{V|U\} = \frac{P\{UV\}}{P\{U\}}$$

and

$$P\{U|V\} = \frac{P\{UV\}}{P\{V\}}.$$

If (1.5.1) and the corresponding relations with the conditional probabilities, given U and $\bar{U}$, are entered in Table 1.9, we get

TABLE 1.10.

Event	V	$\bar{V}$	Total
U	$P\{U\} \cdot P\{V\|U\}$	$P\{U\} \cdot P\{\bar{V}\|U\}$	$P\{U\}$
$\bar{U}$	$P\{\bar{U}\} \cdot P\{V\|\bar{U}\}$	$P\{\bar{U}\} \cdot P\{\bar{V}\|\bar{U}\}$	$P\{\bar{U}\}$
Total	$P\{V\}$	$P\{\bar{V}\}$	1

Dividing the two rows by the marginal probabilities, we get the conditional distributions, given U and $\bar{U}$, as in Table 1.11.

TABLE 1.11.

The conditional distributions, given U and $\overline{U}$.

	Event	V	$\overline{V}$	Total
Conditional distributions	U	$P\{V\|U\}$	$P\{\overline{V}\|U\}$	1
	$\overline{U}$	$P\{V\|\overline{U}\}$	$P\{\overline{V}\|\overline{U}\}$	1
Marginal distribution		$P\{V\}$	$P\{\overline{V}\}$	1

A corresponding table may be calculated for the conditional distributions of U, given V and $\overline{V}$, respectively.

If

$$P\{V|U\} \neq P\{V|\overline{U}\}, \tag{1.5.3}$$

U and V are said to be *stochastically dependent*. The corresponding relation for frequencies indicates that the frequency of the event V among the observations where the event U occurs is different from the frequency of V among observations where U does not occur, see for example Table 1.6 which shows that porosity occurs more often among pieces with defective dimensions than among pieces without defective dimensions.

If

$$P\{V|U\} = P\{V|\overline{U}\}, \tag{1.5.4}$$

V is said to be *stochastically independent of U as the probability of the occurrence of V is the same whether U occurs or not.*

The relation (1.5.4) implies

$$\boxed{P\{V|U\} = P\{V|\overline{U}\} = P\{V\},} \tag{1.5.5}$$

as

$$\begin{aligned} P\{V\} &= P\{UV\}+P\{\overline{U}V\} \\ &= P\{U\}P\{V|U\}+P\{\overline{U}\}P\{V|\overline{U}\} \\ &= P\{V|U\}(P\{U\}+P\{\overline{U}\}) \\ &= P\{V|U\}. \end{aligned}$$

From (1.5.1) and (1.5.5) we get

$$P\{UV\} = P\{U\} \cdot P\{V\},$$

which together with (1.5.2) gives us

$$P\{U\} = P\{U|V\},$$

whence

$$P\{U\} = P\{U|V\} = P\{U|\overline{V}\}.$$

Thus, the relation

$$P\{V|U\} = P\{V|\overline{U}\}$$

implies

$$P\{U|V\} = P\{U|\overline{V}\},$$

i. e., if V is stochastically independent of U then U is stochastically independent of V. U and V are therefore said to be mutually stochastically independent. In an analogous manner it may be proved that either of the events U and $\overline{U}$ is independent of either of the events V and $\overline{V}$.

The theorem corresponding to (1.5.5) reads: *If U and V are stochastically independent, the conditional distributions are equal to the corresponding marginal distributions.*

In this case the multiplication formula takes the form

$$\boxed{P\{UV\} = P\{U\} \cdot P\{V\},} \tag{1.5.6}$$

i. e., *the probability that two stochastically independent events occur together is equal to the product of the probabilities of the occurrence of each event separately.*

The multiplication formula can be generalized as follows:

$$\boxed{P\{U_1U_2\ldots U_m\} = P\{U_1\}P\{U_2|U_1\}\ldots P\{U_m|U_1U_2\ldots U_{m-1}\}\,.} \tag{1.5.7}$$

In the special case where each event is stochastically independent of any combination of the others we get

$$\boxed{P\{U_1U_2\ldots U_m\} = P\{U_1\}P\{U_2\}\ldots P\{U_m\}\,.} \tag{1.5.8}$$

Example 1.1. In the assembling of a machine two different parts are used originating from two independent processes. Let the probability that the parts are defective be P_1 and P_2, respectively. By means of the multiplication formula we get the following results: The probability that both parts in the final product are defective is $P_1 \cdot P_2$, and the probability that both are without defects is $(1-P_1) \cdot (1-P_2)$. The probability that part No. 1 is defective and part No. 2 without defects is $P_1(1-P_2)$, and the probability that No. 1 is without defects and No. 2 defective is $(1-P_1) \cdot P_2$. As one of these four events must necessarily occur and as the events are mutually exclusive, the sum of the four probabilities must be 1:

$$P_1P_2+(1-P_1)(1-P_2)+P_1(1-P_2)+(1-P_1)P_2 = 1\,.$$

Example 1.2. Suppose that a number of lots of N items each are submitted for inspection and that a random sample of n items from each lot is inspected. If a sample contains no defective items then the lot is accepted; if the sample contains one or more defective items then the lot is rejected. What is the probability that a lot containing M defective items will be accepted?

The probability of no defectives in the sample is

$$P\{U_1U_2\ldots U_n\} = P\{U_1\} \cdot P\{U_2|U_1\}\ldots P\{U_n|U_1U_2\ldots U_{n-1}\},$$

where U_i denotes that the ith randomly chosen item is not defective.

In accordance with (1.4.12) we get

$$P\{U_1\} = \frac{N-M}{N} = 1-\frac{M}{N},$$

since the lot contains M defective and $N-M$ nondefective items.

After the first item has been chosen and has turned out to be not defective, the lot now contains M defective items and $N-M-1$ nondefective items so that

$$P\{U_2|U_1\} = \frac{N-M-1}{N-1} = 1-\frac{M}{N-1}.$$

For the ith item we get

$$P\{U_i|U_1U_2\ldots U_{i-1}\} = \frac{N-M-(i-1)}{N-(i-1)} = 1-\frac{M}{N-i+1}.$$

Thus, the probability that a sample contains no defective items is equal to

$$P\{U_1U_2\ldots U_n\} = \left(1-\frac{M}{N}\right)\left(1-\frac{M}{N-1}\right)\ldots\left(1-\frac{M}{N-n+1}\right).$$

For the practical application of this formula we have to determine n. Usually we can state a number M_0 such that we want to reject lots with a number of defectives $M \geq M_0$. It is, however, impossible to make sure of the rejection of *all* lots with $M \geq M_0$ by means of *sampling* inspection, so we must specify another number P_0 giving the (small) probability of accepting a lot with M_0 defectives. For the determination of n we then have the following relation:

$$P\{U_1\ldots U_n\} = P_0 \quad \text{for} \quad M = M_0,$$

i. e.,

$$\left(1-\frac{M_0}{N}\right)\left(1-\frac{M_0}{N-1}\right)\ldots\left(1-\frac{M_0}{N-n+1}\right) = P_0,$$

which gives us $n = n(M_0, P_0) = n_0$.

The probability of accepting a lot with M defectives is then

$$P_M = \left(1-\frac{M}{N}\right)\left(1-\frac{M}{N-1}\right)\ldots\left(1-\frac{M}{N-n_0+1}\right).$$

This probability is a decreasing function of M, which for $M = 0$ gives $P = 1$, for $M = M_0$, $P_{M_0} = P_0$ and for $M \geq N-n_0+1$, $P = 0$. Thus, of the good lots (M small) submitted for inspection a large proportion will in the long run be accepted and a small proportion rejected, and of the bad lots (M large) a small proportion will be accepted and a large proportion rejected.

Example 1.3. At a telephone exchange, the probability that there will be a call within a time element dt is assumed to be equal to λdt, and the calls are assumed to be stochastically independent. What is the probability that no calls will occur within a time interval of length t?

The probability that there will not be any calls within a time element dt is $1-\lambda dt$. Dividing the interval t into n equal parts of duration $\Delta t = \frac{t}{n}$ the required probability may be found as the probability that there will not be any calls during all n intervals for $n \to \infty$, i. e., as

$$(1-\lambda\Delta t)^n = \left(1-\frac{\lambda t}{n}\right)^n \text{ for } n \to \infty.$$

Introducing

$$x = -\frac{n}{\lambda t}$$

we have

$$\left(1-\frac{\lambda t}{n}\right)^n = \left(1+\frac{1}{x}\right)^{-x\lambda t},$$

which leads to

$$\lim_{n\to\infty}(1-\lambda\Delta t)^n = e^{-\lambda t},$$

since $x \to -\infty$ as $n \to \infty$ and

$$\left(1+\frac{1}{x}\right)^x \to e\,.$$

1.6. Stochastic and Causal Dependence.

In laboratory investigations of the relationship between two factors, stochastic dependence often indicates a corresponding causal dependence, it being possible to eliminate all disturbing factors during the experiments. As an example we may take an investigation[1]) of the relationship between the amount of copper present in butter and the taste of the butter. As far as possible all butter samples were identical except for the quantity of copper added. Fig. 1.2 shows a distinct relationship between copper content and assessment of taste, a high copper content giving a high value for taste, i. e., the taste is "oily."

---

[1]) S. HARTMANN: *De oxydative Virkninger i Smør og Mælk*, Dissertation, Copenhagen, 1944, p. 66.

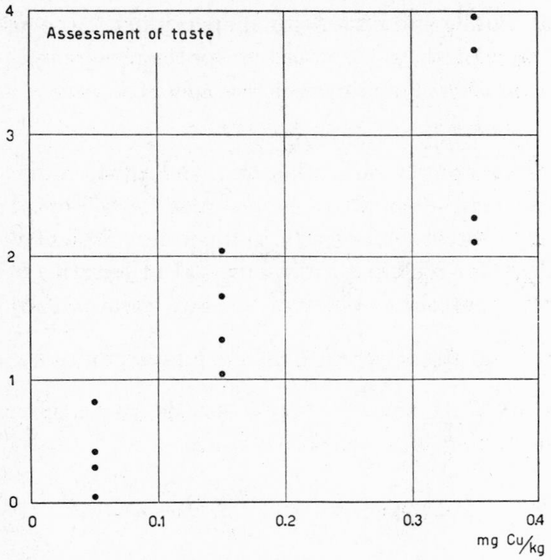

FIG. 1.2. Relationship between copper content and assessment of taste in butter.

An investigation of the same problem was also attempted by means of assessment of taste and analysis of copper content in 295 samples of butter chosen at random from different dairies.[1]) The results are seen in Table 1.12.

TABLE 1.12.
Relationship between the copper content and assessment of taste in butter.

mg Cu per kg butter		Assessment of taste: V 0·0—1·0	$\overline{V}$ 1·1—	Total
U	0·01—0·06	80	106	186
$\overline{U}$	0·07—	45	64	109
Total...		125	170	295

The *conditional frequencies*

$$H\{V|U\} = \frac{80}{186} = 43{\cdot}0\%$$

and

$$H\{V|\overline{U}\} = \frac{45}{109} = 41{\cdot}3\%$$

[1]) *Statens Smørbedømmelser*, Aarsberetning for 1941, Table 11, p. 57.

are of practically the same magnitude. In this investigation, therefore, copper content and taste appear to be stochastically independent. The relationship found in the laboratory experiments does not show up in this investigation, because other and equally decisive factors, which could not be controlled outside the laboratory, also influence the taste of the butter.

Just as stochastic independence may conceal a causal relationship, two events may present themselves as stochastically dependent, though they are causally independent. If the events U and V are stochastically and causally independent, but each one is separately dependent on a third event W, U and V will often appear to be stochastically dependent, if the relationship with W is overlooked. On examination of a large number of observations relating to the casting of pipes in a steel-works[1]), there appeared to be a close relationship between the melting time and the percentage of rejected pipes. It was impossible to give any causal explanation of this stochastic relationship, but all the same it was possible to reduce the percentage of rejected pipes by limiting the melting time. Several years later it was discovered that a long melting time always coincided with the use of raw materials of a special composition. This special kind of raw material caused both a long melting time and a large rejection percentage, although these two events themselves were mutually independent. Thus the concepts of stochastic and causal dependence must be carefully differentiated.

Bernard Shaw in his characteristic manner illuminates this point in the following quotation[2]) from the section "Statistical Illusions" of the preface to *The Doctor's Dilemma*: "Thus it is easy to prove that the wearing of tall hats and the carrying of umbrellas enlarges the chest, prolongs life, and confers comparative immunity from disease; for the statistics shew that the classes which use these articles are bigger, healthier, and live longer than the class which never dreams of possessing such things. It does not take much perspicacity to see that what really makes this difference is not the tall hat and the umbrella, but the wealth and nourishment of which they are evidence, and that a gold watch or membership of a club in Pall Mall might be proved in the same way to have the like sovereign virtues."

1.7. Notes and References.

The calculus of probabilities was developed during the seventeenth century in connection with the solution of problems of games of chance. The classical definition of probability given in § 1.4 took form in a correspondence between Pascal and Fermat in 1654. On basis of this definition an extensive mathematical theory was developed. Important contributions were made

[1]) K. Daeves: *Praktische Grosszahl-Forschung*, VDI-Verlag, Berlin, 1933, p. 63.

[2]) Quotation from Bernard Shaw: *The Doctor's Dilemma*, Brentano's, New York, 1918, p. LXIV, by permission of The Public Trustee and The Society of Authors, London.

by J. Bernoulli in *Ars Conjectandi*, 1713, and by P. S. de Laplace, who in *Théorie analytique des probabilités*, 1812, gave a very extensive and systematic exposition of the results and methods of the calculus of probabilities.

The use of the calculus of probabilities was very soon extended to other fields, e. g., problems of economics and sociology, where the classical definition of probability had no sense. A new definition was therefore formed on the basis of the observed frequencies, the probability of an event being defined as the limiting value of the frequency of the event. The exact formulation of the definition, however, has given rise to great difficulties. One of the best-known advocates of the above definition is R. v. Mises, who in *Probability, Statistics and Truth*, William Hodge and Company, London, 1939, has given a popular exposition of his opinions.

The construction of a calculus of probabilities from a set of axioms, as done in § 1.3, is analogous to the way in which other branches of applied mathematics have been built up. The first systematic exposition of this method and its consequences was given by A. Kolmogoroff: *Foundations of the Theory of Probability*, Chelsea Publishing Company, New York, 1950, originally published in German in 1933. An elementary exposition of the modern mathematical concept of probability may be found in P. R. Halmos: *The Foundations of Probability*, Amer. Math. Monthly, 51, 1944, 493–510.

The basic principles of the calculus of probabilities and its interpretation are discussed in *Les fondements du calcul des probabilités*, Actualités scientifiques et industrielles, No. 735, Hermann & Cie, Paris, 1938. It contains the following contributions: W. Feller: *Sur les axiomatiques du calcul des probabilités et leurs relations avec les expériences*, pp. 7–21; M. Fréchet: *Exposé et discussion de quelques recherches récentes sur les fondements du calcul des probabilités*, pp. 23–55; R. de Misès: *Quelques remarques sur les fondements du calcul des probabilités*, pp. 57–66; and J. F. Steffensen: *Fréquence et probabilités*, pp. 67–78. A short exposition of the main viewpoints presented in these papers is given by M. Fréchet: *The Diverse Definitions of Probability*, Journ. of Unified Science, 8, 1939, 7–23.

An elementary exposition of the different concepts of probability from a philosophical point of view can be found in E. Nagel: *Principles of the Theory of Probability*, International Encyclopedia of Unified Science, Vol. 1, No. 6, Chicago, 1939.

A simple introduction to the calculus of probabilities has been given by J. E. Kerrich: *An Experimental Introduction to the Theory of Probability*, Ejnar Munksgaard, Copenhagen, 1946.

CHAPTER 2.

SOME FUNDAMENTAL APPLICATIONS OF THE CALCULUS OF PROBABILITIES

2.1. The Binomial Distribution.

Suppose that a possible outcome of an observation is the event U, and that $P\{U\} = \theta$. We will assume that the probability of U is equal to θ at every repetition of the observation, irrespective of the outcomes of the previous observations, i. e., at each observation the result is assumed to be stochastically independent of the previous results.

Then, when two observations are made, it follows from the multiplication formula, that the probability that the event U will not occur on either occasion is $(1-\theta)^2$, and the probability that the event will occur on both occasions is θ^2. Symbolically, we have

$$P\{\overline{U}_1\overline{U}_2\} = P\{\overline{U}\}\,P\{\overline{U}\} = (1-\theta)^2$$

and

$$P\{U_1U_2\} = P\{U\}\,P\{U\} = \theta^2 ,$$

where U_i denotes that U occurs at the ith observation.

The third possible result, that the event U occurs only on one of the two occasions, may arise in two different ways: U may occur at the first observation but not at the second—the probability of this result is $\theta(1-\theta)$—or U may fail to occur at the first observation but occurs at the second—the probability of this result being $(1-\theta)\theta$. As these two results are exclusive, the probability that the event U will occur exactly once is, according to the addition formula: $\theta(1-\theta)+(1-\theta)\theta = 2(1-\theta)\theta$. Symbolically we have

$$P\{U_1\overline{U}_2+\overline{U}_1U_2\} = P\{U_1\overline{U}_2\}+P\{\overline{U}_1U_2\}$$
$$= 2P\{\overline{U}\}P\{U\} = 2(1-\theta)\theta .$$

Denote *the probability that the event U will occur x times when 2 observations are made* by $p_2\{x\}$. Thus, the probability that U will occur 0, 1, or 2 times, respectively, is

$$p_2\{0\} = (1-\theta)^2 ,$$
$$p_2\{1\} = 2(1-\theta)\theta ,$$
$$p_2\{2\} = \theta^2 .$$

As the event U must occur either 0, 1, or 2 times in 2 observations, the sum of the corresponding probabilities must be 1, which may be used to check the results derived:

$$p_2\{0\}+p_2\{1\}+p_2\{2\} = (1-\theta)^2+2(1-\theta)\theta+\theta^2 = \big((1-\theta)+\theta\big)^2 = 1\,.$$

Upon 3 repetitions, the event U may occur 0, 1, 2, or 3 times. The corresponding probabilities are

$$\begin{aligned}
p_3\{0\} &= P\{\overline{U}_1\overline{U}_2\overline{U}_3\}\\
&= (1-\theta)^3\,,\\
p_3\{1\} &= P\{U_1\overline{U}_2\overline{U}_3+\overline{U}_1U_2\overline{U}_3+\overline{U}_1\overline{U}_2U_3\}\\
&= P\{U_1\overline{U}_2\overline{U}_3\}+P\{\overline{U}_1U_2\overline{U}_3\}+P\{\overline{U}_1\overline{U}_2U_3\}\\
&= 3(1-\theta)^2\,\theta\,,\\
p_3\{2\} &= P\{U_1U_2\overline{U}_3+U_1\overline{U}_2U_3+\overline{U}_1U_2U_3\}\\
&= P\{U_1U_2\overline{U}_3\}+P\{U_1\overline{U}_2U_3\}+P\{\overline{U}_1U_2U_3\}\\
&= 3(1-\theta)\,\theta^2\,,\\
p_3\{3\} &= P\{U_1U_2U_3\}\\
&= \theta^3\,.
\end{aligned}$$

As a check we may calculate

$$\begin{aligned}
&p_3\{0\}+p_3\{1\}+p_3\{2\}+p_3\{3\}\\
&= (1-\theta)^3+3(1-\theta)^2\theta+3(1-\theta)\theta^2+\theta^3\\
&= \big((1-\theta)+\theta\big)^3 = 1\,.
\end{aligned}$$

This result shows that in 3 observations we get the probabilities of the 4 possible results by expanding the expression $\big((1-\theta)+\theta\big)^3$. In the following it will be shown that the probabilities of the $n+1$ different possible results of n repetitions arise by expanding the expression $\big((1-\theta)+\theta\big)^n$ in powers of θ.

The probability, $p_n\{x\}$, that the event U will occur exactly x times on n repetitions, is, according to the "either—or" rule a sum of probabilities of the form

$$P\{U_1U_2\overline{U}_3U_4\overline{U}_5\,\ldots\,U_{n-1}\overline{U}_n\}\,,$$

U appearing x times and $\overline{U}$ $n-x$ times. According to the multiplication formula these probabilities all take on the value $\theta^x(1-\theta)^{n-x}$.

The number of terms in the sum is

$$\binom{n}{x} = \frac{n!}{x!(n-x)!} \tag{2.1.1}$$

where $n! = 1\cdot 2\cdot 3\ldots n$, this expression denoting the number of permutations of x U's and $(n-x)$ $\overline{U}$'s. (According to definition $0!$ is equal to 1, $\binom{n}{0}$ thus being equal to 1). The required probability is therefore

$$\boxed{p_n\{x\} = \binom{n}{x} \theta^x (1-\theta)^{n-x}} \qquad (0 \leq x \leq n)\,. \qquad (2.1.2)$$

As a check we calculate the sum of the probabilities of the $n+1$ different possible results:

$$\sum_{x=0}^{n} p_n\{x\} = (1-\theta)^n + n(1-\theta)^{n-1}\theta + \ldots + n(1-\theta)\theta^{n-1} + \theta^n$$
$$= ((1-\theta)+\theta)^n = 1\,.$$

As the probabilities can be obtained by the expansion of $((1-\theta)+\theta)^n$, this series of probabilities is called *the binomial distribution.*

The binomial distribution was derived by J. Bernoulli in *Ars Conjectandi*, 1713. The binomial coefficients, $\binom{n}{x}$, were already dealt with by Pascal, who arranged them in a table, Pascal's triangle:

Table 2.1.
Pascal's triangle.

n	The binomial coefficients															
0	1															
1	1	1														
2	1	2	1													
3	1	3	3	1												
4	1	4	6	4	1											
5	1	5	10	10	5	1										
6	1	6	15	20	15	6	1									
7	1	7	21	35	35	21	7	1								
8	1	8	28	56	70	56	28	8	1							
9	1	9	36	84	126	126	84	36	9	1						
10	1	10	45	120	210	252	210	120	45	10	1					
11	1	11	55	165	330	462	462	330	165	55	11	1				
12	1	12	66	220	495	792	924	792	495	220	66	12	1			
13	1	13	78	286	715	1287	1716	1716	1287	715	286	78	13	1		
14	1	14	91	364	1001	2002	3003	3432	3003	2002	1001	364	91	14	1	
15	1	15	105	455	1365	3003	5005	6435	6435	5005	3003	1365	455	105	15	1

The triangle is based on the relationship

$$\binom{n}{x} + \binom{n}{x+1} = \binom{n+1}{x+1}, \qquad (2.1.3)$$

e. g.,

$$\binom{8}{3}+\binom{8}{4} = 56+70 = 126 = \binom{9}{4}.$$

The probability that the event U will occur exactly 3 times in 9 repetitions may be read directly from this table, as

$$p_9\{3\} = 84\,\theta^3(1-\theta)^6\,.$$

The value of $p_n\{x\}$ for a single value of x may be computed from the formula

$$\log p_n\{x\} = \log\binom{n}{x}+x\log\theta+(n-x)\log(1-\theta)\,. \qquad (2.1.4)$$

Table XIV[1]) gives $\log\binom{n}{x}$ for values of n from $n = 1$ to $n = 100$, and Table XIII gives $\log n!$ for $n = 1$ to $n = 1000$. Methods for computing a series of values of $p_n\{x\}$ are discussed in Chapter 21.

According to the addition formula, the probability that the event U will occur at most x times in n repetitions, called *the cumulative probability*, $P_n\{x\}$, is equal to the sum of the first $x+1$ terms of the binomial distribution

$$P_n\{x\} = \sum_{\nu=0}^{x}\binom{n}{\nu}\theta^\nu(1-\theta)^{n-\nu}\,. \qquad (2.1.5)$$

The binomial distribution is dealt with in detail in Chapter 21. At present we shall only give a couple of examples of its application.

Example 2.1. A bag contains a number of red and white balls in the proportion θ to $(1-\theta)$. One ball is drawn from the bag in such a manner that all balls have the same probability of being drawn. According to (1.4.12) the probability that the ball drawn is red, is θ. The ball is *replaced* in the bag, and the drawing is repeated. *If we make n drawings and replace the ball each time,* the probability that exactly x of the balls drawn are red will according to (2.1.2) be

$$p_n\{x\} = \binom{n}{x}\theta^x(1-\theta)^{n-x}.$$

We have tried to fulfil the assumption that the events are stochastically independent by replacing the ball drawn and mixing all the balls carefully before the next drawing.

If the ball is *not replaced* after each drawing, the results become stochastically dependent. If, e. g., the bag contains 5 red and 5 white balls, the probability of drawing a red ball, $P\{R_1\}$, is $\frac{1}{2}$ in the first drawing. If, at the first drawing we get a red ball and do not replace it, the probability,

[1]) The tables numbered with Roman numerals are to be found in A. Hald: *Statistical Tables and Formulas*, John Wiley & Sons, New York, 1952.

$P\{R_2|R_1\}$, of drawing a red ball next time is $\frac{4}{9}$. If, on the other hand, the ball we get at the first drawing is white, and we do not replace it, the probability, $P\{R_2|\overline{R}_1\}$, of getting a red ball in the second drawing is $\frac{5}{9}$. Thus, we see that the probability of the result R in the second drawing depends on the result of the first drawing, i. e., the events are stochastically dependent, cf. (1.5.3).

If the ball is replaced before the next drawing, we have

$$P\{R_2|R_1\} = P\{R_2|\overline{R}_1\} = \tfrac{1}{2},$$

i. e., the results are stochastically independent, cf. (1.5.4).

Using the multiplication formula, we can write the probability $P\{R_1R_2\}$ of getting a red ball in two drawings

$$P\{R_1R_2\} = P\{R_1\} \cdot P\{R_2|R_1\}.$$

When we *replace* the balls, we have

$$P\{R_1R_2\} = \tfrac{1}{2} \cdot \tfrac{1}{2} = \tfrac{1}{4},$$

and when we *do not replace* them

$$P\{R_1R_2\} = \tfrac{1}{2} \cdot \tfrac{4}{9} = \tfrac{2}{9}.$$

The probabilities at the third drawing are, if we do not replace the balls

$$\begin{aligned} P\{R_3|R_1R_2\} &= \tfrac{3}{8} \\ P\{R_3|R_1\overline{R}_2\} &= \tfrac{4}{8} \\ P\{R_3|\overline{R}_1R_2\} &= \tfrac{4}{8} \\ P\{R_3|\overline{R}_1\overline{R}_2\} &= \tfrac{5}{8}, \end{aligned}$$

whereas if we replace them, all these probabilities are $\frac{1}{2}$. The probability of getting a red ball in every drawing is

$$P\{R_1R_2R_3\} = P\{R_1\}\ P\{R_2|R_1\}P\{R_3|R_1R_2\}.$$

By substitution, we get

$$P\{R_1R_2R_3\} = \tfrac{5}{10} \cdot \tfrac{4}{9} \cdot \tfrac{3}{8} = \tfrac{1}{12}$$

and

$$P\{R_1R_2R_3\} = (\tfrac{1}{2})^3 = \tfrac{1}{8},$$

respectively, i. e., the probabilities are very different in the two cases.

If the bag contains 5000 red balls and 5000 white balls, and the balls are not replaced after drawing, we get

$$P\{R_2|R_1\} = \tfrac{4999}{9999}$$

and

$$P\{R_2|\overline{R}_1\} = \tfrac{5000}{9999}.$$

For the third drawing we have

$$P\{R_3|R_1R_2\} = \tfrac{4998}{9998},$$

$$P\{R_3|R_1\overline{R}_2\} = \tfrac{4999}{9998},$$

$$P\{R_3|\overline{R}_1R_2\} = \tfrac{4999}{9998},$$

and

$$P\{R_3|\overline{R}_1\overline{R}_2\} = \tfrac{5000}{9998}.$$

For all practical purposes, these 6 probabilities are of the same magnitude and equal to $\frac{1}{2}$.

The probability that all three drawings will give red balls is

$$P\{R_1R_2R_3\} = \tfrac{5000}{10000} \cdot \tfrac{4999}{9999} \cdot \tfrac{4998}{9998} = 0{\cdot}12496 .$$

The value of this probability is practically the same as $(\frac{1}{2})^3 = 0{\cdot}125$, which is the probability of the corresponding result of 3 drawings if the balls are replaced. Correspondingly, we find that the other possible results of 3 drawings are also practically the same whether we replace the ball or not.

Therefore, instead of speaking about drawings *with* replacement, we often imagine that we are drawing balls from a bag containing an *infinite* number of red and white balls in the proportion θ to $1-\theta$. If we now make a *finite* number of drawings, the probability of drawing a red ball will still be θ, even though we have not replaced the balls.

Example 2.2. Consider a production process where in the long run 2% of the items produced are defective and the defective items occur independently of each other. What is the probability that 0, 1, 2,... items, respectively, are defective in a lot of 100 items?

The probability, θ, of a defective item is 0·02. The probability that a lot of 100 items contains exactly x defectives is, according to (2.1.2)

$$p_{100}\{x\} = \binom{100}{x} 0{\cdot}02^x 0{\cdot}98^{100-x}.$$

If we substitute $x = 0, 1, 2,\ldots,$ we get the results given in Table 2.2. A graphical representation of $p_{100}\{x\}$ is given in Fig. 2.1.

The probability of finding exactly 2 defective items among 100 items is 0·2734. One defective item will occur with almost the same probability, i. e., 0·2707, whilst 0 and 3 defective items will occur somewhat less frequently. The probability of getting 10 or more defective items is practically zero.

The expected percentage of defective items occurring in the production process, $\theta = 2\%$, may be controlled by regularly collecting samples of the product and determining the number of defectives among 100 items. Table 2.2 shows that the percentage of defectives thus observed must be expected to lie between 0 and 4, the probability that the percentage of defectives

TABLE 2.2.

A binomial distribution.

$$p_{100}\{x\}=\binom{100}{x}0{\cdot}02^x 0{\cdot}98^{100-x}$$

Number of defective items	Probability	Cumulative probability
x	$p_{100}\{x\}$	$\Sigma p\{x\}$
0	0·1326	0·1326
1	0·2707	0·4033
2	0·2734	0·6767
3	0·1823	0·8590
4	0·0902	0·9492
5	0·0353	0·9845
6	0·0114	0·9959
7	0·0031	0·9990
8	0·0007	0·9997
9	0·0002	0·9999

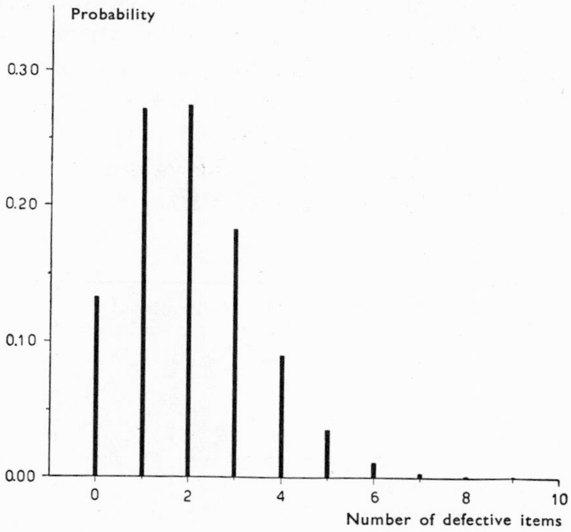

FIG. 2.1. A binomial distribution. Number of defective items in a lot of 100 items.

will be 4 or less being about 0·95. The probability of finding 5 or 6% defectives is 0·047, whilst the probability of finding 7% or more is so small that the occurrence of such percentages would justify a more detailed examination of the production process to find out if new factors entering the process have caused this extremely high percentage. Thus the calculus of probabilities furnishes us with a means of judging the significance of deviations from the "expected" percentage of defectives.

Example 2.3. Let x denote the tensile strength of a metal rod and let us assume that initially all rods employed in a certain production process have the same tensile strength, x_0. During the first "phase" of the process all rods undergo for instance a heat treatment, which aims at changing the tensile strength of each rod from x_0 to x_1. Because of "chance variations" in the treatment each rod does not acquire exactly the desired strength, but the results obtained vary "at random" about x_1. The rods are then subjected to the second phase of the process, which aims at altering x from x_1 to x_2, the result being a random variation about x_2, etc.

Three simple mathematical models of such processes are given below.

A. We will assume that the deviations from the quality aimed at can only take on the values $+\varepsilon$ and $-\varepsilon$, and that the probability of each of these deviations is $\frac{1}{2}$. We assume further that the successive deviations are stochastically independent.

During the first phase of the process, the tensile strength, x_0, is altered by an amount $\Delta x_0-\varepsilon$ or $\Delta x_0+\varepsilon$, so that after the first phase x is equal to either $x_0+\Delta x_0-\varepsilon=x_1-\varepsilon$ or $x_0+\Delta x_0+\varepsilon=x_1+\varepsilon$. The corresponding probabilities are each $\frac{1}{2}$. During the second phase the tensile strength is altered by an amount $\Delta x_1-\varepsilon$ or $\Delta x_1+\varepsilon$; therefore, at the end of this phase x has one or other of the values $(x_1-\varepsilon)+(\Delta x_1-\varepsilon) = x_2-2\varepsilon$, $(x_1-\varepsilon)+(\Delta x_1+\varepsilon) = x_2$, $(x_1+\varepsilon)+(\Delta x_1-\varepsilon)=x_2$, or $(x_1+\varepsilon)+(\Delta x_1+\varepsilon) = x_2+2\varepsilon$. The corresponding probabilities are $(\frac{1}{2})^2 = \frac{1}{4}$, but as two of the values are identical, we have the following distribution of the values of x:

Values of x	Probability
$x_2-2\varepsilon$	$\frac{1}{4}$
x_2	$\frac{1}{2}$
$x_2+2\varepsilon$	$\frac{1}{4}$
Total	1

After n such phases we get the distribution given in Table 2.3.

The difference between two successive possible values of x is constant and equal to 2ε. The probabilities are generated by expanding

$$(\tfrac{1}{2}+\tfrac{1}{2})^n .$$

The distribution is symmetrical with a strongly pronounced concentration

TABLE 2.3.

Values of x	Probability
$x_n - n\varepsilon$	$(\frac{1}{2})^n$
$x_n - (n-2)\varepsilon$	$\binom{n}{1}(\frac{1}{2})^n$
$x_n - (n-4)\varepsilon$	$\binom{n}{2}(\frac{1}{2})^n$
$\vdots$	$\vdots$
$x_n + n\varepsilon$	$(\frac{1}{2})^n$

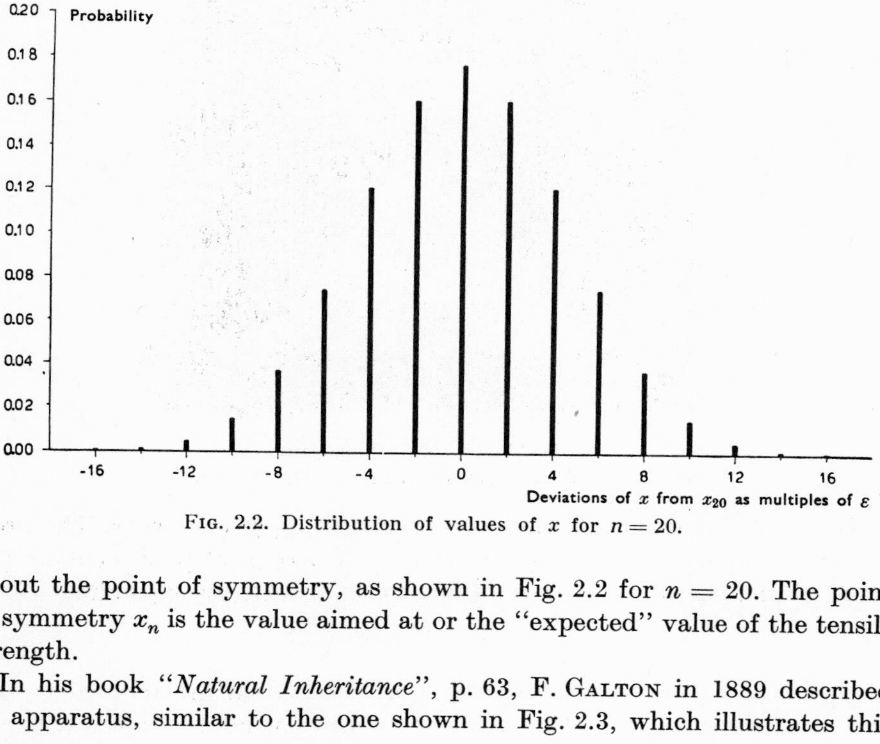

FIG. 2.2. Distribution of values of x for $n = 20$.

about the point of symmetry, as shown in Fig. 2.2 for $n = 20$. The point of symmetry x_n is the value aimed at or the "expected" value of the tensile strength.

In his book "*Natural Inheritance*", p. 63, F. GALTON in 1889 described an apparatus, similar to the one shown in Fig. 2.3, which illustrates this process.

The apparatus consists of a board with nails in rows, the nails of a given row being placed below the midpoints of the intervals in the row above. Steel balls are poured into the apparatus through a funnel, and the balls will then be "influenced" by the nails in such a manner that they take up positions deviating from the point vertically below the funnel. The distribution of the balls in Fig. 2.3 according to position is of the same type as the theoretical distribution in Fig. 2.2.

FIG. 2.3. GALTON'S Apparatus.
(From G. WEBER: *Belysningsteknik*, Vol. 2, p. 132.)

The above assumptions may be generalized in several ways. The probability of the deviation $+\varepsilon$ may be equal to θ instead of $\frac{1}{2}$. This entails that the probabilities of the above values are expressed by expanding

$$((1-\theta)+\theta)^n$$

in powers of θ. For $\theta \neq \frac{1}{2}$ this distribution is skew, but for large values of n, the skewness is hardly noticeable. With a given value of θ and for $n \to \infty$ the distribution tends to become symmetrical, approximating to the so-called *normal distribution*, as shown in Table 21.4 and Fig. 21.2 for $n = 100$ and $\theta = 0{\cdot}1$. This limiting process is demonstrated in § 21.5.

So far we have assumed that the deviations were of the magnitude $+\varepsilon$ and $-\varepsilon$ at every phase of the process. The distribution of measures x of the quality in question, however, also tends to become "normal", even if the size of the deviations is not the same at every phase of the process, as long as all deviations are of the *same order of magnitude.*

TABLE 2.4.

Values of x	Probability
$x_n(1-\varepsilon)^n$	$(\frac{1}{2})^n$
$x_n(1-\varepsilon)^{n-1}(1+\varepsilon)$	$\binom{n}{1}(\frac{1}{2})^n$
$x_n(1-\varepsilon)^{n-2}(1+\varepsilon)^2$	$\binom{n}{2}(\frac{1}{2})^n$
$\vdots$	$\vdots$
$x_n(1+\varepsilon)^n$	$(\frac{1}{2})^n$

Our assumptions may be generalized even further, as in § 8.2.

B. We assume that the *relative* deviations at each phase of the process take on the values $+\varepsilon$ and $-\varepsilon$, their probability being $\frac{1}{2}$, and that successive deviations are stochastically independent.

After the tensile strength has been altered in the first phase of the process, x_0 has become either $x_0(1+\Delta x_0)(1-\varepsilon)=x_1(1-\varepsilon)$ or $x_0(1+\Delta x_0)(1+\varepsilon)=x_1(1+\varepsilon)$. After the second phase we have $x_2(1-\varepsilon)^2$, $x_2(1-\varepsilon)(1+\varepsilon)$ or $x_2(1+\varepsilon)^2$. The corresponding probabilities are $\frac{1}{4}$, $\frac{1}{2}$, $\frac{1}{4}$. After n phases we get the distribution given in Table 2.4. The ratio between two successive terms is constant and equal to $\dfrac{1+\varepsilon}{1-\varepsilon}$.

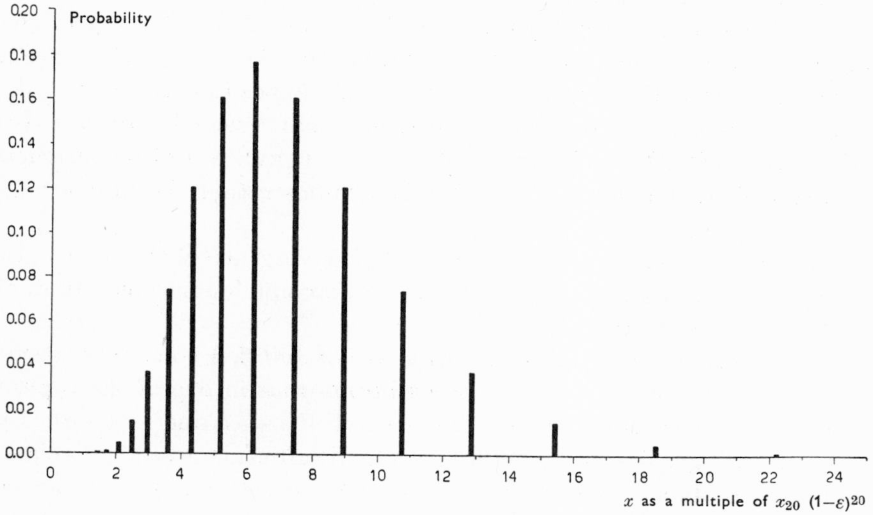

FIG. 2.4. Distribution of values of x for $n=20$.

Fig. 2.4 gives the distribution for $n = 20$ and $\dfrac{1+\varepsilon}{1-\varepsilon} = 1{\cdot}2$. As the ratio between two successive terms is constant, the difference between their logarithms is also constant. If we write

$$\log\,(x_n(1-\varepsilon)^n) = y_n$$

and

$$\log \frac{1+\varepsilon}{1-\varepsilon} = \delta\,,$$

the distribution in Table 2.4 takes the same form as the distribution in Table 2.3, as shown in Table 2.5.

TABLE 2.5.

Logarithm of x	Probability
y_n	$(\frac{1}{2})^n$
$y_n+\delta$	$\binom{n}{1}(\frac{1}{2})^n$
$y_n+2\delta$	$\binom{n}{2}(\frac{1}{2})^n$
$\vdots$	$\vdots$
$y_n+n\delta$	$(\frac{1}{2})^n$

Thus, *the logarithm of the variable is approximately "normally" distributed.* We may generalize the assumptions in the same manner as discussed in A.

These ideas have been further developed in: J. C. KAPTEYN: *Skew Frequency Curves*, 1903, and J. C. KAPTEYN and M. J. VAN UVEN: *Skew Frequency Curves*, Groningen, 1916. A popular exposition is given in J. C. KAPTEYN: *Skew Frequency Curves*, Recueil des Travaux botaniques Néerlandais, Vol. XIII, Livr. II, 1916, 105–157. KAPTEYN has also constructed an apparatus analogous to GALTON's for demonstrating the derivation of skew distributions.

C. As mentioned in A a condition for the convergence of the distribution of x to a normal distribution is that all the possible deviations are of the same order of magnitude.

As an example[1]) where this condition is not fulfilled consider a process similar to the one in A with the one exception that in one of the n phases the possible deviations are not $+\varepsilon$ and $-\varepsilon$ but $+(1+\alpha)\varepsilon$ and $-(1+\alpha)\varepsilon$. By a reasoning analogous to that in A we get the distribution given in Table 2.6.

This distribution contains two components corresponding to the two terms $+\alpha\varepsilon$ and $-\alpha\varepsilon$. It may be conceived of as a composition of two

[1]) This example is based on a similar example mentioned by W. A. SHEWHART (unpublished).

TABLE 2.6.

Values of x	Probability	Values of x	Probability
$x_n-n\varepsilon-\alpha\varepsilon$	$(\frac{1}{2})^n$		
$x_n-(n-2)\varepsilon-\alpha\varepsilon$	$\binom{n-1}{1}(\frac{1}{2})^n$	$x_n-(n-2)\varepsilon+\alpha\varepsilon$	$(\frac{1}{2})^n$
$x_n-(n-4)\varepsilon-\alpha\varepsilon$	$\binom{n-1}{2}(\frac{1}{2})^n$	$x_n-(n-4)\varepsilon+\alpha\varepsilon$	$\binom{n-1}{1}(\frac{1}{2})^n$
$\vdots$	$\vdots$	$\vdots$	$\vdots$
$x_n+(n-2)\varepsilon-\alpha\varepsilon$	$(\frac{1}{2})^n$	$x_n+(n-2)\varepsilon+\alpha\varepsilon$	$\binom{n-1}{n-2}(\frac{1}{2})^n$
		$x_n+n\varepsilon+\alpha\varepsilon$	$(\frac{1}{2})^n$

binomial distributions, each with the weight $\frac{1}{2}$, of the type given in Table 2.3 with $n-1$ instead of n, where the means of the two distributions are displaced $+\alpha\varepsilon$ and $-\alpha\varepsilon$, respectively, from x_n. For $\alpha = 0$ we get the distribution in Table 2.3. For $\alpha > n-1$ the two components are completely separated since

$$x_n+(n-2)\varepsilon-\alpha\varepsilon < x_n-(n-2)\varepsilon+\alpha\varepsilon \quad \text{for} \quad \alpha \geq n-1 \,.$$

In this case the size of the deviations in the one phase is of the same order of magnitude as the sum of the deviations of all other phases.

For $0 < \alpha < n-1$ we get the intermediate cases. As α increases from 0 to $n-1$, the distribution becomes more and more flat and finally passes into a bimodal distribution, as shown in Fig. 2.5 for $n = 20$ and $\alpha = 8$.

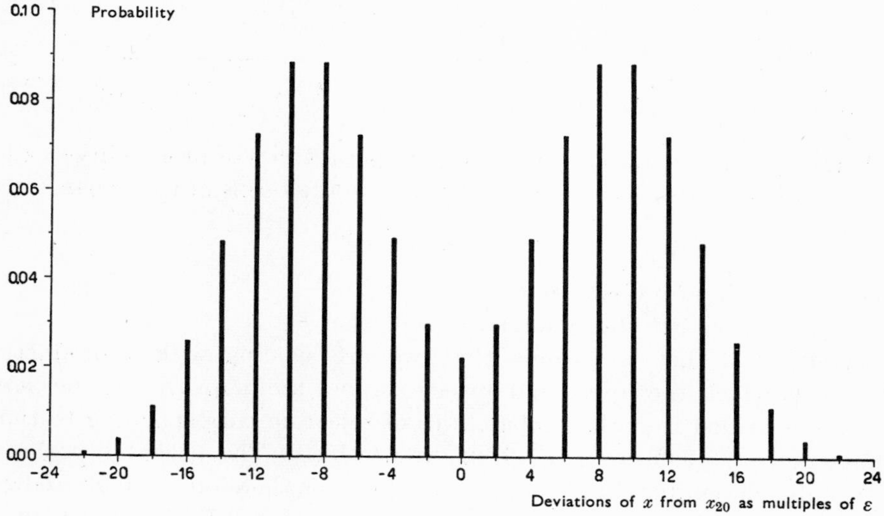

FIG. 2.5. Distribution of values of x for $n=20$ and $\alpha=8$.

In accordance with this theoretical background for the generation of a bimodal distribution we often in practice expect that an observed bimodal distribution has been generated by a process where at least one phase or cause produces a dominating effect. It is often possible to find and remove such a cause of variation and in this way reduce the variability of the quality of the manufactured product.

2.2. The Multinomial Distribution.

Let the outcome of an observation be one of k different results, $U_1, U_2, \ldots, U_k$, which are mutually exclusive. The probability of the result U_i, $P\{U_i\}$, is denoted by θ_i, and as the event $U_1+U_2+\ldots+U_k$ is certain, $\theta_1+\theta_2+\ldots+\theta_k = 1$. Let us make n stochastically independent observations. What is the probability that U_1 will occur x_1 times, U_2 x_2 times,... and U_k x_k times, where $x_1+x_2+\ldots+x_k = n$?

This result may happen in many different ways, corresponding to the number of different permutations of x_1 U_1's, x_2 U_2's, ..., x_k U_k's. According to the multiplication formula, the probability of each of these results is

$$\theta_1^{x_1}\theta_2^{x_2}\ldots\theta_k^{x_k}.$$

The number of different permutations is

$$\frac{n!}{x_1!\,x_2!\ldots x_k!},$$

therefore according to the addition formula the required probability is

$$p_n\{x_1, x_2, \ldots, x_k\} = \frac{n!}{x_1!\,x_2!\cdots x_k!}\,\theta_1^{x_1}\theta_2^{x_2}\cdots\theta_k^{x_k}, \tag{2.2.1}$$

where $0 \le x_i \le n$, and $\sum_{i=1}^{k} x_i = n$.

If the x's take on all possible values, we get a series of probabilities called the *multinomial distribution*, because they are the terms of the expansion of

$$(\theta_1+\theta_2+\ldots+\theta_k)^n.$$

The sum of these probabilities is 1.

For $k = 2$ we get the binomial distribution.

Example 2.4. Let us assume that we are dealing with a production process in which the upper and lower tolerance limits for the product have been so specified that the probabilities of rejection corresponding to these limits are $\theta_1 = 0{\cdot}06$ and $\theta_2 = 0{\cdot}02$, respectively. The probability that a randomly chosen item lies between these limits is therefore $1-0{\cdot}06-0{\cdot}02= 0{\cdot}92$. In a sample of 100 items a number of items will be rejected because they exceed the specified limits. According to (2.2.1) the probability that x_1 and x_2 items, respectively, will be rejected is

$$p_{100}\{x_1, x_2, x_3\} = \frac{100!}{x_1!\, x_2!\, x_3!} 0{\cdot}06^{x_1} 0{\cdot}02^{x_2} 0{\cdot}92^{x_3},$$

where $x_3 = 100 - x_1 - x_2$.

Table 2.7 gives these probabilities as a function of x_1 and x_2.

TABLE 2.7.

A multinomial distribution.

$$p_{100}\{x_1, x_2, x_3\} = \frac{100!}{x_1!\, x_2!\, x_3!} 0{\cdot}06^{x_1} 0{\cdot}02^{x_2} 0{\cdot}92^{x_3}. \qquad (x_3 = 100 - x_1 - x_2).$$

x_1 \ x_2	0	1	2	3	4	5	6	7	8	Marginal distribution
0	·0002	·0005	·0006	·0004	·0002	·0001	·0000	·0000	·0000	·0020
1	·0016	·0034	·0036	·0025	·0013	·0005	·0002	·0001	·0000	·0132
2	·0050	·0107	·0113	·0079	·0041	·0017	·0006	·0002	·0000	·0415
3	·0107	·0226	·0236	·0163	·0083	·0034	·0011	·0003	·0001	·0864
4	·0170	·0354	·0366	·0249	·0126	·0050	·0017	·0005	·0001	·1338
5	·0213	·0439	·0448	·0302	·0151	·0060	·0020	·0005	·0001	·1639
6	·0219	·0448	·0453	·0302	·0149	·0059	·0019	·0005	·0001	·1655
7	·0192	·0389	·0389	·0256	·0125	·0048	·0016	·0004	·0001	·1420
8	·0146	·0291	·0288	·0188	·0091	·0035	·0011	·0003	·0001	·1054
9	·0097	·0192	·0188	·0121	·0058	·0022	·0007	·0002	·0000	·0687
10	·0058	·0113	·0109	·0069	·0033	·0012	·0004	·0001	·0000	·0399
11	·0031	·0060	·0057	·0036	·0017	·0006	·0002	·0001	·0000	·0210
12	·0015	·0029	·0027	·0017	·0008	·0003	·0001	·0000	·0000	·0100
13	·0007	·0012	·0012	·0007	·0003	·0001	·0000	·0000	·0000	·0042
14	·0003	·0005	·0005	·0003	·0001	·0001	·0000	·0000	·0000	·0018
15	·0001	·0002	·0002	·0001	·0001	·0000	·0000	·0000	·0000	·0007
Marginal distribution	·1327	·2706	·2735	·1822	·0902	·0354	·0116	·0032	·0006	1·0000

The expected number of items above the upper tolerance limit is $100 \times 0{\cdot}06 = 6$, and the expected number of items lying below the lower tolerance limit is $100 \times 0{\cdot}02 = 2$. The probability that $(x_1, x_2) = (6,2)$ is 0·0453, which is the maximum value of $p_{100}\{x_1, x_2, x_3\}$. According to the addition formula, the probability that exactly 6 units are too large, irrespective of the number of units that are too small, is

$$\sum_{x_2=0}^{94} p_{100}\{6, x_2, 94 - x_2\}.$$

This gives us the marginal probability for $x_1 = 6$:

$$\sum_{x_2=0}^{94} p_{100}\{6, x_2, 94 - x_2\} = 0{\cdot}0219 + 0{\cdot}0448 + \ldots + 0{\cdot}0001 = 0{\cdot}1655\,.$$

Alternatively, it follows from the binomial distribution that the marginal probability of the result x_1 must be

$$\binom{100}{x_1} 0{\cdot}06^{x_1} 0{\cdot}94^{100-x_1} .$$

Correspondingly, the marginal probability of the result x_2 is

$$\binom{100}{x_2} 0{\cdot}02^{x_2} 0{\cdot}98^{100-x_2} .$$

Therefore, the bottom row in Table 2.7 must—apart from errors due to rounding off—be equal to the terms of the binomial distribution in Table 2.2.

The probability that among 100 items 10 at most are too large and 5 at most are too small, is equal to the sum of the probabilities in Table 2.7 for values of x_1 and x_2 which satisfy the inequalities

$$0 \leq x_1 \leq 10$$

and

$$0 \leq x_2 \leq 5 .$$

This gives a probability of 0·9473.

The probability of obtaining at most 5 defective items among 100 is found by adding the probabilities of Table 2.7 in the domain which is defined by the inequality $0 \leq x_1 + x_2 \leq 5$, i. e., the triangle in the upper lefthand corner. This probability is equal to 0·1799. We can see directly that the probabilities of x defective items in all—whether large or small—is

$$\binom{100}{x} 0{\cdot}08^{x}\, 0{\cdot}92^{100-x} ,$$

the probability of one defective item being $0{\cdot}06 + 0{\cdot}02 = 0{\cdot}08$. The probability of at most 5 defective items can therefore also be computed as

$$\sum_{x=0}^{5} \binom{100}{x} 0{\cdot}08^{x} 0{\cdot}92^{100-x} .$$

2.3. Pascal's Distribution.

Let θ be the probability that the event U will be the outcome of an observation. The observation is repeated until the event occurs for the first time. What is the probability that we shall have to repeat the observation just x times? The observations are assumed to be stochastically independent.

If the observation must be carried out x times, the event cannot have occurred the first $(x-1)$ times, but must have occurred the xth time. The required probability is therefore $P\{\overline{U}_1 \overline{U}_2 \ldots \overline{U}_{x-1} U_x\}$. According to the multiplication formula, the probability, $p\{x\}$, of this result is

$$\boxed{p\{x\} = (1-\theta)^{x-1}\,\theta} \qquad (x \geq 1). \qquad (2.3.1)$$

As a check we may calculate

$$\sum_{x=1}^{\infty} p\{x\} = \theta \sum_{x=1}^{\infty} (1-\theta)^{x-1} = \theta \frac{1}{1-(1-\theta)} = 1.$$

The probability that the observation must be repeated at most x times for the event to occur is, according to the addition formula,

$$P\{x\} = p\{1\}+p\{2\}+\ldots+p\{x\} = \theta \sum_{\nu=1}^{x} (1-\theta)^{\nu-1} = 1-(1-\theta)^x. \quad (2.3.2)$$

This result may also be obtained by reasoning as follows: If we have to repeat the observation more than x times, it is because the event has not occurred in the first x observations, which has the probability $(1-\theta)^x$, and the probability of the complementary event is therefore $1-(1-\theta)^x$.

The problem may be generalized as follows: The observation is repeated until the event U occurs for the ath time. What is the probability that the observation must be repeated exactly x times?

If we have to repeat the observation exactly x times, U must have occurred $(a-1)$ times in the first $(x-1)$ observations and then U must occur at the xth observation. According to the binomial distribution the probability that the event will occur $(a-1)$ times in $(x-1)$ observations is

$$\binom{x-1}{a-1} \theta^{a-1}(1-\theta)^{x-a}.$$

By multiplying this probability with the probability θ that the event will occur at the xth observation, we get the required probability

$$p\{x\} = \binom{x-1}{a-1} (1-\theta)^{x-a}\,\theta^a \qquad (x \geq a)$$

or

$$\boxed{p\{x\} = \left(\frac{\theta}{1-\theta}\right)^a \binom{x-1}{a-1} (1-\theta)^x} \qquad (x \geq a). \qquad (2.3.3)$$

For $a = 1$, we get the result derived above.

If we substitute $x = a, a+1, \ldots$ we get a series of probabilities called PASCAL's *distribution*, as the formula (2.3.3) was derived by PASCAL when he was solving a problem involving games of chance.

The *cumulative probability*

$$P\{x\} = \sum_{\nu=a}^{x} p\{\nu\}$$

denotes the probability that an observation must be repeated at most x times in order to make the event occur a times. The cumulative probability is most easily calculated from the complementary event. If this complementary event, i. e., more than x repetitions of the observation are necessary to make the event occur a times, is to occur, the event can only have occurred $(a-1)$ times in the first x observations. According to (2.1.2) the probability that an event will occur exactly μ times in x repetitions is

$$\binom{x}{\mu} \theta^{\mu}(1-\theta)^{x-\mu},$$

and therefore

$$1-\sum_{\nu=a}^{x} p\{\nu\} = \sum_{\mu=0}^{a-1} \binom{x}{\mu} \theta^{\mu}(1-\theta)^{x-\mu},$$

and

$$P\{x\} = \sum_{\nu=a}^{x} p\{\nu\} = 1-\sum_{\mu=0}^{a-1} \binom{x}{\mu} \theta^{\mu}(1-\theta)^{x-\mu}, \tag{2.3.4}$$

which we also get by direct summation of (2.3.3).

Example 2.5. In a production process 2% of the items produced are defective and the defects occur independently of one another. The probability that in a sample of x items none will be defective is therefore $0{\cdot}98^x$.

Table 2.8 shows this probability as a function of x. The probability that among 200 items none will be defective is 0·018.

TABLE 2.8.

x	$0{\cdot}98^x$
10	0·8171
20	0·6676
50	0·3642
100	0·1326
200	0·0176

2.4. The Hypergeometric Distribution.

Let us assume that we have a population of N elements, M of which belong to class U, and that we draw n elements without replacements in such a manner that all N elements have the same chance of being drawn, i. e., we draw a random sample of n elements from the population. What is the probability that among our n elements exactly x will belong to the class U?

This problem is most easily solved by using the classical definition of probability. n elements may be drawn from N elements in $\binom{N}{n}$ different ways, the probability being the same for each way, as the probability of being drawn is the same for all N elements. We get a favourable case if x of the n elements are drawn from the M elements belonging to class U, and $n-x$ from the $N-M$ elements that do not belong to class U. This may happen in $\binom{M}{x}$ and $\binom{N-M}{n-x}$ different ways, respectively, and each of the $\binom{M}{x}$ ways may be combined with the $\binom{N-M}{n-x}$ ways, the total number of favourable cases being equal to the product of these two. Therefore, the required probability is

$$\boxed{p\{x\} = \frac{\binom{M}{x}\binom{N-M}{n-x}}{\binom{N}{n}}.} \tag{2.4.1}$$

The range of variation of our variables is defined by the inequalities $(0 \leq x \leq n)$, $(0 \leq x \leq M)$ and $(0 \leq n-x \leq N-M)$.

These probabilities form what we call the *hypergeometric distribution* on account of their connection with the hypergeometric function.

Example 2.6. From a bag containing M red and $N-M$ white balls n balls are drawn *without replacements*. The n balls may be drawn simultaneously or successively, but at any rate in such a manner that all balls have the same probability of being drawn. The probability that exactly x out of the n balls will be red is

$$p\{x\} = \frac{\binom{M}{x}\binom{N-M}{n-x}}{\binom{N}{n}}.$$

For $M = N-M = 5000$, $n = 3$ and $x = 3$, we get

$$p\{3\} = \frac{\binom{5000}{3}\binom{5000}{0}}{\binom{10000}{3}}$$

$$= \frac{5000 \cdot 4999 \cdot 4998}{10000 \cdot 9999 \cdot 9998} = 0{\cdot}12496,$$

i. e., the same result as on p. 28.

If the ratio $\frac{M}{N}$ of red balls in the bag is called θ, (2.4.1) may be written

$$p\{x\} = \frac{\dfrac{M(M-1)\cdots(M-x+1)}{x!}\cdot\dfrac{(N-M)(N-M-1)\cdots(N-M-n+x+1)}{(n-x)!}}{\dfrac{N(N-1)\cdots(N-n+1)}{n!}}$$

$$= \binom{n}{x}\frac{M\cdots(M-x+1)(N-M)\cdots(N-M-n+x+1)}{N(N-1)\cdots(N-n+1)}$$

$$= \binom{n}{x}\frac{\theta\cdots\left(\theta-\dfrac{x-1}{N}\right)(1-\theta)\cdots\left(1-\theta-\dfrac{n-x-1}{N}\right)}{1\left(1-\dfrac{1}{N}\right)\cdots\left(1-\dfrac{n-1}{N}\right)}.$$

For $N \to \infty$, we get

$$p\{x\} \to \binom{n}{x}\theta^x(1-\theta)^{n-x}. \tag{2.4.2}$$

Thus we see that the hypergeometric distribution converges to the binomial distribution for $N \to \infty$, which proves the statement concerning drawings from an infinite population made on p. 28.

Example 2.7. We have, say, 1000 items 20 (2%) of which are defective. If we take a random sample of 100 items, the probability that x items will be defective is

$$p\{x\} = \frac{\dbinom{20}{x}\dbinom{980}{100-x}}{\dbinom{1000}{100}} \qquad (0 \leq x \leq 20).$$

According to Table 2.9, the most probable result is that our random sample will include 2 defective items. The result $x = 1$, however, has almost the same probability. More than 5 defective items will occur comparatively seldom in a random sample of 100 items.

If we take a sample of 100 items from a group of 1000 items without knowing the percentage defectives, and this sample, e. g., contains 8 defectives, Table 2.9 shows that it is reasonable to assume that the percentage of defective items among the 1000 items is larger than 2, or that the sample is not a random sample.

The hypergeometric distribution of Table 2.9 differs but little from the binomial of Table 2.2. Thus, for $N = 1000$ we get almost the same distribution as for $N \to \infty$.

TABLE 2.9.

A hypergeometric distribution.

$$p\{x\}=\frac{\binom{20}{x}\binom{980}{100-x}}{\binom{1000}{100}}.$$

x	$p\{x\}$	$\Sigma p\{x\}$
0	0·1190	0·1190
1	0·2701	0·3891
2	0·2881	0·6772
3	0·1918	0·8690
4	0·0895	0·9585
5	0·0311	0·9896
6	0·0083	0·9979
7	0·0018	0·9997
8	0·0003	1·0000

CHAPTER 3.

GRAPHICAL AND TABULAR REPRESENTATION OF OBSERVATIONS

EMPIRICAL DISTRIBUTION OF A SINGLE CONTINUOUS VARIABLE

3.1. Dot Diagram and Enumeration.

The observations are generally listed in the order in which they are obtained. This comprises our *primary data*. For example, Table 3.1 gives the measurements of the diameters of 200 rivet heads.

In order to obtain a clearer impression of the data, the measurements are ranked according to magnitude in the following manner: On a piece of squared paper (ruled, say, in millimetres) a suitable scale is marked off on a horizontal baseline. The results are then marked in as dots above the corresponding numbers on the line, in the order in which they have been listed, each dot representing an observation. Such a *dot diagram* shows us how the results are distributed according to magnitude and forms a basis for tabulating the distribution. Fig. 3.1 shows the dot diagram corresponding to Table 3.1.

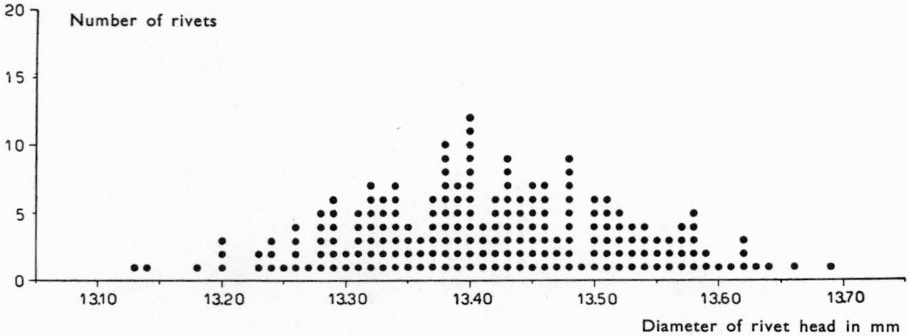

FIG. 3.1. *Dot diagram.* Distribution of the diameters of 200 rivet heads measured in millimetres.

The accuracy of the dot diagram should be *checked*. This may, for instance, be done by enumerating the observational data and registering the results in a *table*. For each observation a vertical mark is drawn in the table. Every fifth mark is placed across the previous four, which facilitates the final enumeration. (On account of printing difficulties this has not been done in Table 3.2). Table 3.2 shows the enumeration of the data in Table 3.1.

TABLE 3.1.
Primary data.

Diameters of 200 rivet heads in millimetres							
13·39	13·43	13·54	13·64	13·40	13·55	13·40	13·26
13·42	13·50	13·32	13·31	13·28	13·52	13·46	13·63
13·38	13·44	13·52	13·53	13·37	13·33	13·24	13·13
13·53	13·53	13·39	13·57	13·51	13·34	13·39	13·47
13·51	13·48	13·62	13·58	13·57	13·33	13·51	13·40
13·30	13·48	13·40	13·57	13·51	13·40	13·52	13·56
13·40	13·34	13·23	13·37	13·48	13·48	13·62	13·35
13·40	13·36	13·45	13·48	13·29	13·58	13·44	13·56
13·28	13·59	13·47	13·46	13·62	13·54	13·20	13·38
13·43	13·35	13·56	13·51	13·47	13·40	13·29	13·20
13·46	13·44	13·42	13·29	13·41	13·39	13·50	13·48
13·53	13·34	13·45	13·42	13·29	13·38	13·45	13·50
13·55	13·33	13·32	13·69	13·46	13·32	13·32	13·48
13·29	13·25	13·44	13·60	13·43	13·51	13·43	13·38
13·24	13·28	13·58	13·31	13·31	13·45	13·43	13·44
13·34	13·49	13·50	13·38	13·48	13·43	13·37	13·29
13·54	13·33	13·36	13·46	13·23	13·44	13·38	13·27
13·66	13·26	13·40	13·52	13·59	13·48	13·46	13·40
13·43	13·26	13·50	13·38	13·43	13·34	13·41	13·24
13·42	13·55	13·37	13·41	13·38	13·14	13·42	13·52
13·38	13·54	13·30	13·18	13·32	13·46	13·39	13·35
13·34	13·37	13·50	13·61	13·42	13·32	13·35	13·40
13·57	13·31	13·40	13·36	13·28	13·58	13·58	13·38
13·26	13·37	13·28	13·39	13·32	13·20	13·43	13·34
13·33	13·33	13·31	13·45	13·39	13·45	13·41	13·45

The accuracy of the dot diagram or the enumeration table may also be checked by summation of the observations, partly from the primary data and partly from the dot diagram or the table. In order to be able to trace possible errors, enumeration and summation should be made on groups of not more than 100 observations. Checking by summation is, of course, not quite foolproof, as errors in opposite directions counterbalance one another.

Observations may also be enumerated by the aid of *cards*, each observation being entered on a card and the cards then being sorted according to the magnitude of the results.

TABLE 3.2.

Enumeration of the distribution of the diameters of 200 rivet heads in millimetres.

mm	Number of rivets		mm	Number of rivets	
13·10			13·40	IIIII IIIII II	12
13·11			13·41	IIII	4
13·12			13·42	IIIII I	6
13·13	I	1	13·43	IIIII IIII	9
13·14	I	1	13·44	IIIII I	6
13·15			13·45	IIIII II	7
13·16			13·46	IIIII II	7
13·17			13·47	III	3
13·18	I	1	13·48	IIIII IIII	9
13·19			13·49	I	1
13·20	III	3	13·50	IIIII I	6
13·21			13·51	IIIII I	6
13·22			13·52	IIIII	5
13·23	II	2	13·53	IIII	4
13·24	III	3	13·54	IIII	4
13·25	I	1	13·55	III	3
13·26	IIII	4	13·56	III	3
13·27	I	1	13·57	IIII	4
13·28	IIIII	5	13·58	IIIII	5
13·29	IIIII I	6	13·59	II	2
13·30	II	2	13·60	I	1
13·31	IIIII	5	13·61	I	1
13·32	IIIII II	7	13·62	III	3
13·33	IIIII I	6	13·63	I	1
13·34	IIIII II	7	13·64	I	1
13·35	IIII	4	13·65		
13·36	III	3	13·66	I	1
13·37	IIIII I	6	13·67		
13·38	IIIII IIIII	10	13·68		
13·39	IIIII II	7	13·69	I	1
					Total: 200

3.2. Tabulation and Grouping.

Fig. 3.1 and Table 3.2 give a clearer description of the distribution of the observations than the original list of 200 unranked numbers. Clarity may, however, be further facilitated by grouping the results in classes.

When *grouping or classifying*, we divide a suitable interval which includes all n observational results, $x_1, x_2, \ldots, x_n$, into m intervals, *the class intervals*.

The *lengths of the class intervals* are denoted by $\Delta t_1, \Delta t_2, \ldots, \Delta t_m$, and the *midpoints* of the m class intervals are termed $t_1, t_2, \ldots, t_m$. The phrases "length of the class interval" and "midpoint of the class interval" will be abbreviated to "class length" and "class midpoint", respectively.

The *number of observations*, a_j, in the jth class interval indicates the number of observations satisfying the inequality

$$t_j - \frac{\Delta t_j}{2} < x \leq t_j + \frac{\Delta t_j}{2}. \tag{3.2.1}$$

The distribution now appears as shown in Table 3.3, where

$$h_j = \frac{a_j}{n} \tag{3.2.2}$$

denotes the *frequency of observations in the jth class interval* and

$$H_j = \sum_{\nu=1}^{j} h_\nu \tag{3.2.3}$$

gives the corresponding *cumulative frequency*, i. e., the frequency of the observations *smaller than or equal to the class limit* $t_j + \frac{\Delta t_j}{2}$.

TABLE 3.3.

Grouped distribution.

Class midpoint	Class length	Number of observations	Frequency	Cumulative frequency
t_1	Δt_1	a_1	h_1	H_1
t_2	Δt_2	a_2	h_2	H_2
$\vdots$	$\vdots$	$\vdots$	$\vdots$	$\vdots$
t_m	Δt_m	a_m	h_m	H_m
		n	1	

Table 3.4 gives the grouped distribution of the diameters of the rivet heads. The lengths of all the class intervals are 0·05 mm. The class interval with midpoint 13·32 mm thus has the limits 13·32±0·025 mm, i. e., the class boundaries are 13·295 and 13·345 mm, respectively.

In official statistical tables the class intervals are often of varying size. This particularly refers to the statistics of economic data, where the distributions are usually skew with a pronounced condensation of small values, see for example Table 3.5, which gives the distribution of Danish estates classified according to taxable land values in 1936. (*Statistisk Aarbog*, 1943, Table 44). In Table 3.5 the class intervals are denoted by their lower limit

TABLE 3.4.

Distribution of the diameters of 200 rivet heads in millimetres.
Class length: 0·05 mm.

Class midpoint mm	Number of rivets	Cumulative number	Frequency in per cent	Cumulative frequency in per cent
13·12	2	2	1·0	1·0
13·17	1	3	0·5	1·5
13·22	8	11	4·0	5·5
13·27	17	28	8·5	14·0
13·32	27	55	13·5	27·5
13·37	30	85	15·0	42·5
13·42	37	122	18·5	61·0
13·47	27	149	13·5	74·5
13·52	25	174	12·5	87·0
13·57	17	191	8·5	95·5
13·62	7	198	3·5	99·0
13·67	2	200	1·0	100 0
Total:	200		100·0	

TABLE 3.5.

Distribution of Danish estates according to taxable land values in 1936.

Taxable land value in 1000 kr.	Number of estates	Class length in 1000 kr.	Number of estates per 1000 kr. class length	Cumulative number of estates	Frequency of estates in per cent	Frequency in per cent per 1000 kr. class length	Cumulative frequency in per cent
0—	3770	1	3770	3770	1·84	1·84	1·84
1—	15266	1	15266	19036	7·47	7·47	9·31
2—	19543	1	19543	38579	9·56	9·56	18·87
3—	19753	1	19753	58332	9·67	9·67	28·54
4—	19245	1	19245	77577	9·42	9·42	37·96
5—	16333	1	16333	93910	7·99	7·99	45·95
6—	23901	2	11951	117811	11·70	5·85	57·65
8—	14399	2	7200	132210	7·05	3·52	64·70
10—	10436	2	5218	142646	5·11	2·55	69·81
12—	12420	3	4140	155066	6·08	2·03	75·89
15—	14802	5	2960	169868	7·24	1·45	83·13
20—	9907	5	1981	179775	4·85	0·97	87·98
25—	7299	5	1460	187074	3·57	0·71	91·55
30—	9239	10	924	196313	4·52	0·45	96·07
40—	3614	10	361	199927	1·77	0·18	97·84
50—	1601	10	160	201528	0·78	0·08	98·62
60—	1255	20	63	202783	0·61	0·03	99·23
80—	518	20	26	203301	0·25	0·01	99·48
100—	1049	?	?	204350	0·51	?	99·99
Total:	204350				99·99		

in 1000 Danish crowns (kr.). The term "2—" denotes the class interval $2000 \leq x < 3000$ kr. and "12—" denotes $12000 \leq x < 15000$ kr.

On account of the increasing size of the class intervals, the second and sixth columns of the table may give a wrong impression of the distribution; two columns have therefore been added, giving the number and the frequency of estates per unit of class interval (1000 kr.).

3.3. The Histogram and the Frequency Polygon.

The grouped distribution in Table 3.3 may be represented graphically in a (t, h)-coordinate system as the "step-function"

$$h(t) = \frac{h_j}{\Delta t_j} \quad \text{for} \quad t_j - \frac{\Delta t_j}{2} < t \leq t_j + \frac{\Delta t_j}{2}. \tag{3.3.1}$$

Thus, not the frequency but the *frequency divided by the class length* is used as ordinate. In this manner the frequency, h_j, of observations in the jth class interval is represented by a rectangular *area* of magnitude h_j, the base line being Δt_j and the height $\frac{h_j}{\Delta t_j}$. If the class lengths are the same for all class intervals, division by Δt is usually omitted and h_j or a_j are used as ordinates, as in Fig. 3.2, which shows the graphical representation of the distribution given in Table 3.4. Such a figure is called a *histogram* (ἱστός =

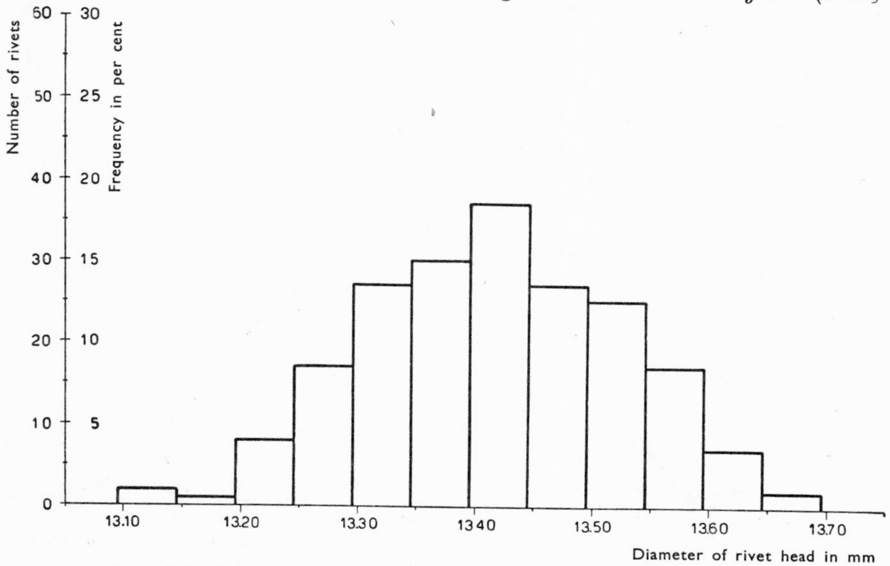

FIG. 3.2. *Histogram.* Grouped distribution of the diameters of 200 rivet heads in millimetres.
Length of class interval: 0·05 mm.
Ordinate: Number or frequency of rivets in each class interval.

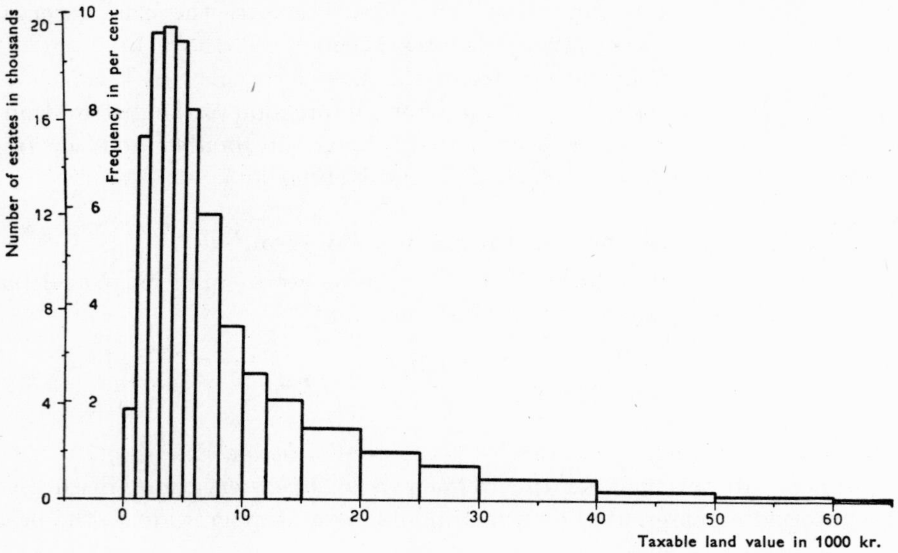

FIG. 3.3. *Histogram.* Distribution of Danish estates according to taxable land value in 1936.
Ordinate: Number or frequency of estates in each class interval per 1000 kr. class length.

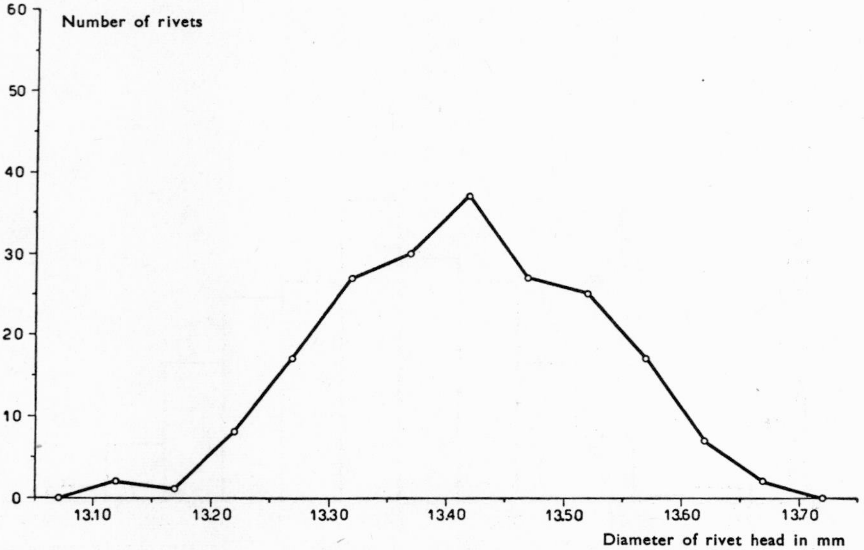

FIG. 3.4. *Frequency polygon.*
Grouped distribution of the diameters of 200 rivet heads in millimetres.
Length of class interval: 0·05 mm.
Ordinate: Number of rivets in each class interval.

ship's mast, cellular tissue). The property that observed frequencies are represented by *areas* proves to be very useful.

Fig. 3.3 gives the histogram corresponding to Table 3.5. The ordinate used here is the frequency divided by the class length, see columns four and seven of the table, the class length not being the same for all class intervals.

Sometimes another graphical representation, *the frequency polygon*, is used. It is drawn by connecting the points $\left(t_j, \frac{h_j}{\Delta t_j}\right)$ by straight lines. Fig. 3.4 shows the frequency polygon corresponding to the distribution in Table 3.4. As the class limits are equidistant, the frequencies have not been divided by the class lengths.

The histogram should, however, be preferred to the frequency polygon, because the area under the polygon has no simple interpretation in terms of frequencies.

3.4. The Length of the Class Interval.

Any grouping is arbitrary, as the boundaries of the class intervals may be chosen at will. When representing the distribution of our observations graphically we should aim at choosing our class intervals in such a manner that the characteristic features of the distribution are emphasized and chance variations are obscured.

Figs. 3.5–3.8 show the effect of varying the class length upon the

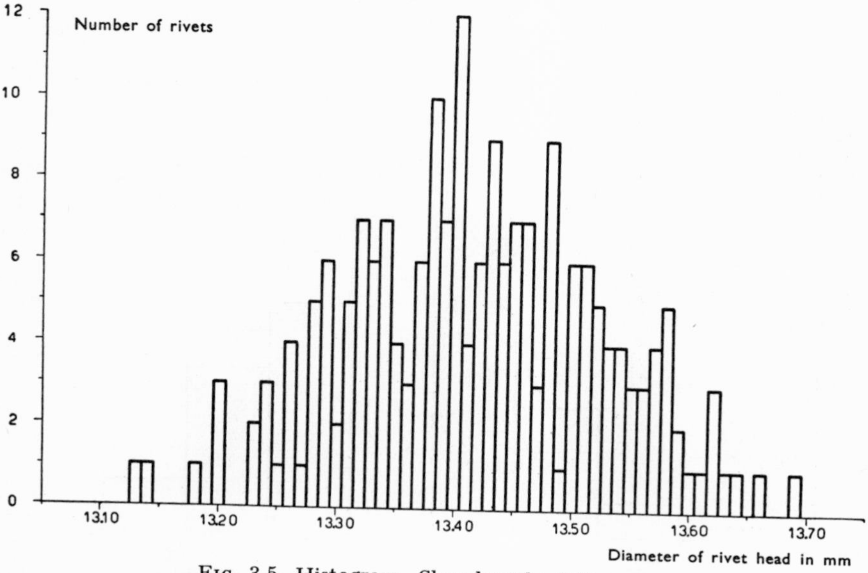

FIG. 3.5. Histogram. Class length: 0·01 mm.
Ordinate: Number of rivets in each class interval.

appearance of the histogram. Frequencies, and not frequencies divided by class lengths, have been used as ordinates; therefore, the unit of the ordinate is inversely proportional to the class length, 1 sq. cm representing the same number of observations in all figures. If the class lengths are small, chance variations dominate because each interval includes only a small

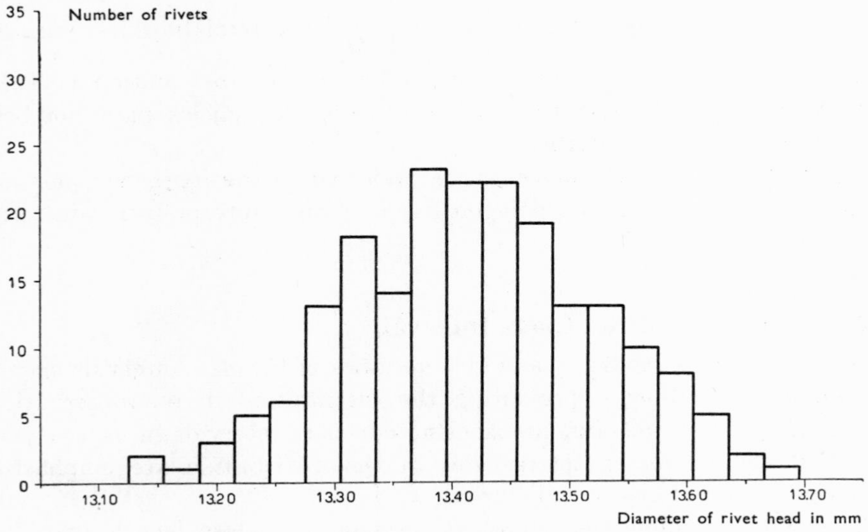

FIG. 3.6. Histogram. Class length: 0·03 mm.

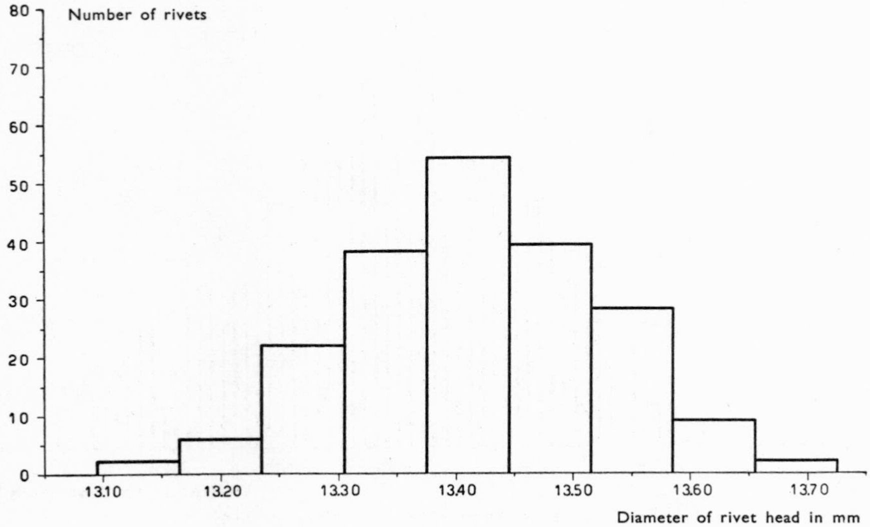

FIG. 3.7. Histogram. Class length: 0·07 mm.

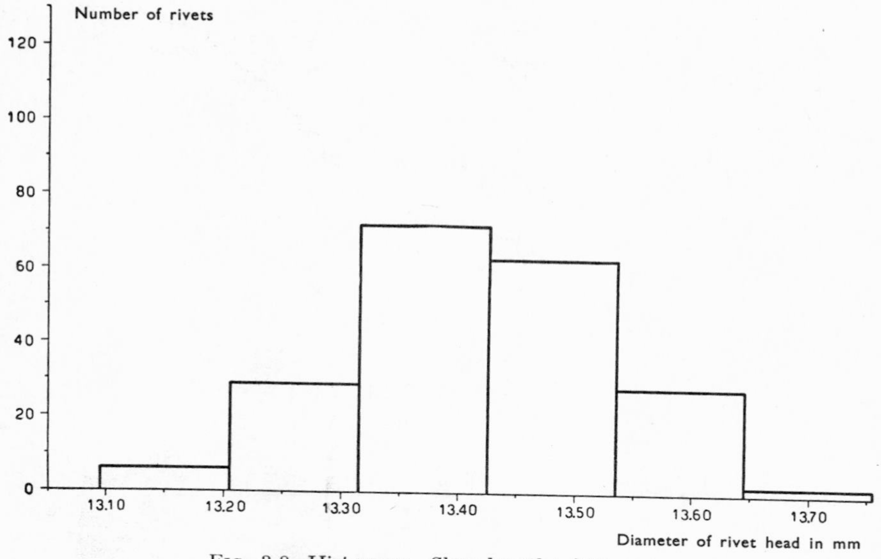

FIG. 3.8. Histogram. Class length: 0·11 mm.

number of observations; if the class length is large, the characteristics of the distribution are obscured.

If the grouped distribution is to form the basis of computations the class intervals should be small and of equal length, see § 4.3 and § 4.4.

3.5. The Cumulative Frequency Polygon.

Let $x_1, x_2, \ldots, x_n$ denote n observed results, taken in the "random" order in which the observations have been made, and let $x_{(1)}, x_{(2)}, \ldots, x_{(n)}$ denote the same n results ranked in order of magnitude, cf. Tables 3.1 and 3.2. The frequency $H\{x\}$ of observations that are smaller than or equal to x, called the *cumulative frequency*, is

$$H\{x\} = \begin{cases} 0 & \text{for} \quad x < x_{(1)}\,, \\ \dfrac{i}{n} & \text{for} \quad x_{(i)} \le x < x_{(i+1)},\ i = 1, 2, \ldots, n-1\,, \\ 1 & \text{for} \quad x \ge x_{(n)}\,. \end{cases} \qquad (3.5.1)$$

$H\{x\}$ is represented as a step curve, called the *cumulative frequency polygon*, which increases from 0 to 1 in "jumps" of $\frac{1}{n}$ for $x = x_{(1)}, x_{(2)}, \ldots, x_{(n)}$. If several observations take on the same value, the "jump" is a multiple of $\frac{1}{n}$. Fig. 3.9 shows the cumulative frequency polygon for the distribution in Table 3.2.

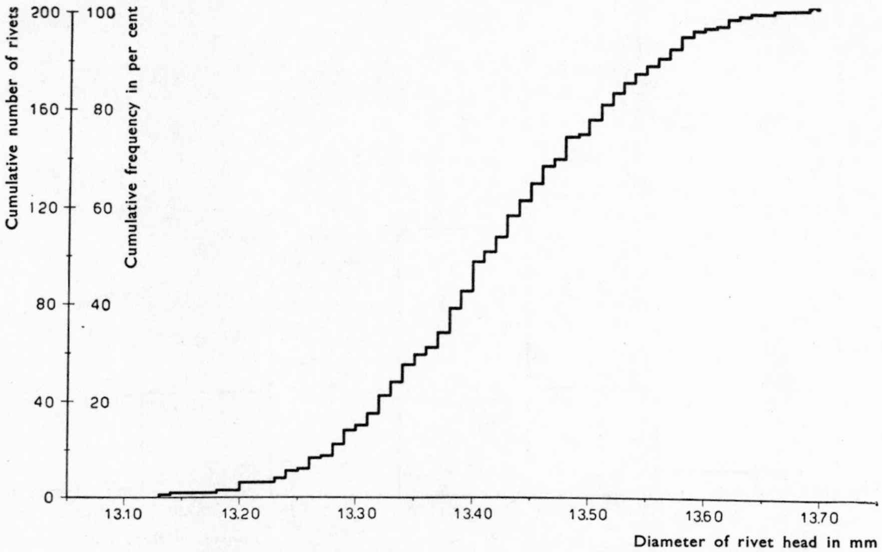

FIG. 3.9. *Cumulative frequency polygon of an ungrouped distribution.* Distribution of the diameters of 200 rivet heads in millimetres.

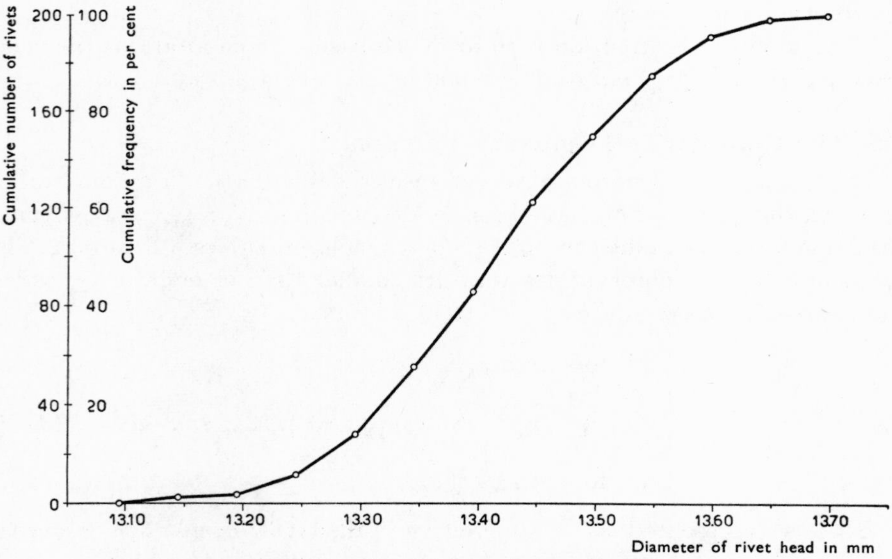

FIG. 3.10. *Cumulative frequency polygon of a grouped distribution.* Distribution of the diameters of 200 rivet heads in millimetres.

When a *grouped* distribution is represented as a cumulative frequency polygon, the class limits $t_j + \frac{\Delta t_j}{2}$ are used as abscissas and the corresponding

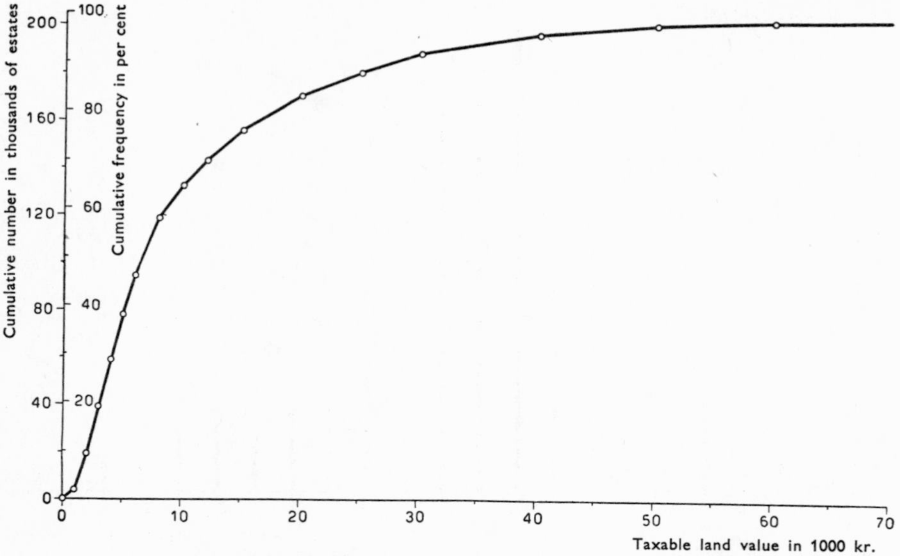

FIG. 3.11. *Cumulative frequency polygon of a grouped distribution.* Distribution of Danish estates according to taxable land value in 1936.

cumulative numbers or frequencies, H_j, as ordinates. The points are connected by straight lines. Figs. 3.10 and 3.11 give the cumulative frequency polygons corresponding to the distributions in Tables 3.4 and 3.5.

The cumulative frequency polygon is far less sensitive to variations in class lengths than either the histogram or the frequency polygon. The four cumulative frequency polygons corresponding to the histograms in Figs. 3.5–3.8 give practically the same impression of the distribution.

EMPIRICAL DISTRIBUTION OF A SINGLE DISCONTINUOUS VARIABLE

3.6. Enumeration and Graphical Representation.

In the above examples, the variables are either *continuous* or may be regarded as continuous for all practical purposes, i. e., they are capable of taking on *any* value within a certain range.

Sometimes, however, the variable is *discontinuous*. For instance consider the number of defective items in lots of 250 mass produced articles. Here, the variable is discontinuous, as it can only take on the values 0, 1, 2. . . . Enumeration of the observational results is done in the same way as for a continuous variable. Table 3.6 shows the result of enumerating the number of lots with given numbers of rejected items among 100 lots of 250 items each.

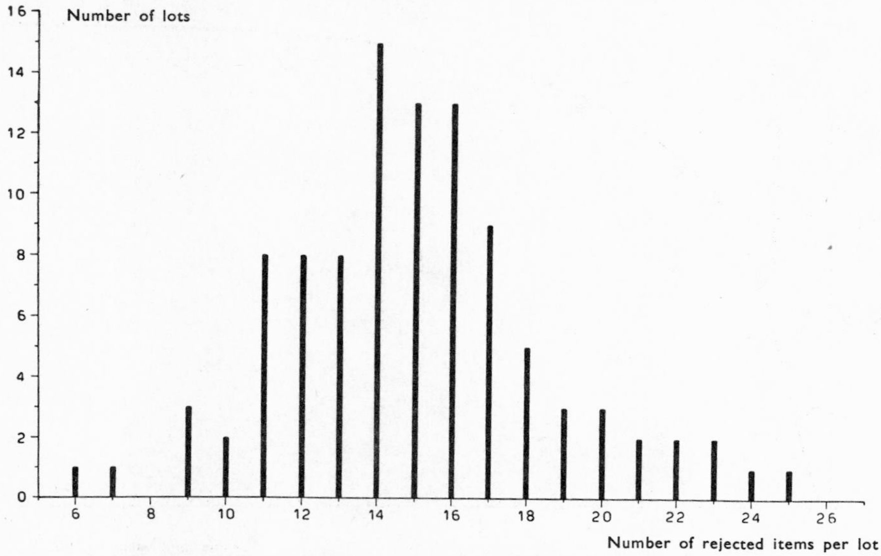

FIG. 3.12. Distribution of 100 lots of 250 items according to number of rejected items per lot.

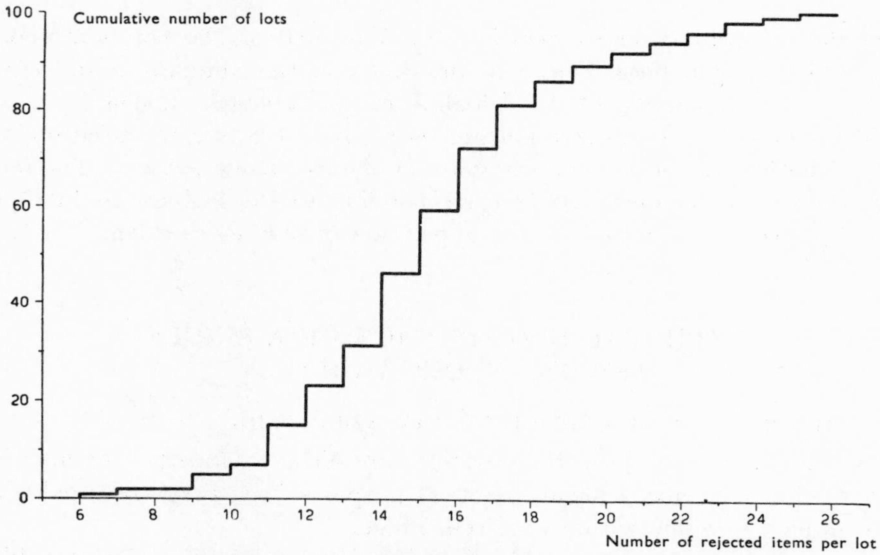

FIG. 3.13. Cumulative distribution of 100 lots of 250 items according to number of rejected items per lot.

This distribution is represented graphically by plotting the number of rejected items in each lot as abscissa and the corresponding number of lots as ordinate, as in Fig. 3.12. For these ordinates we may substitute rectangles of the same height and with unit-base, which gives us a histogram of the

TABLE 3.6.
Distribution of 100 lots of 250 items according to number of rejected items per lot.

Number of rejected items per lot	Number of lots	Cumulative number of lots
6	1	1
7	1	2
8	0	2
9	3	5
10	2	7
11	8	15
12	8	23
13	8	31
14	15	46
15	13	59
16	13	72
17	9	81
18	5	86
19	3	89
20	3	92
21	2	94
22	2	96
23	2	98
24	1	99
25	1	100

same kind as for continuous variables. The representation used in Fig. 3.12 should, however, be preferred, as the variable can only take isolated values, and this is clearly indicated in the figure.

The cumulative frequency polygon for discontinuous variables can be represented by isolated ordinates or by a step curve, the height of the steps denoting the number of results corresponding to the abscissa, as in Fig. 3.13.

MULTI-DIMENSIONAL DISTRIBUTIONS

3.7. Dot Diagram.

If our observations form number-pairs, e. g., water content and calorific value of coal, the corresponding distribution is called two-dimensional. In a two-dimensional dot diagram each observation is plotted as a point with the observed values of the variables as coordinates, as in Fig. 3.14, which shows the plotting of water content against calorific value for 74 samples of coal.

The dot diagram furnishes us with a good description of the relation

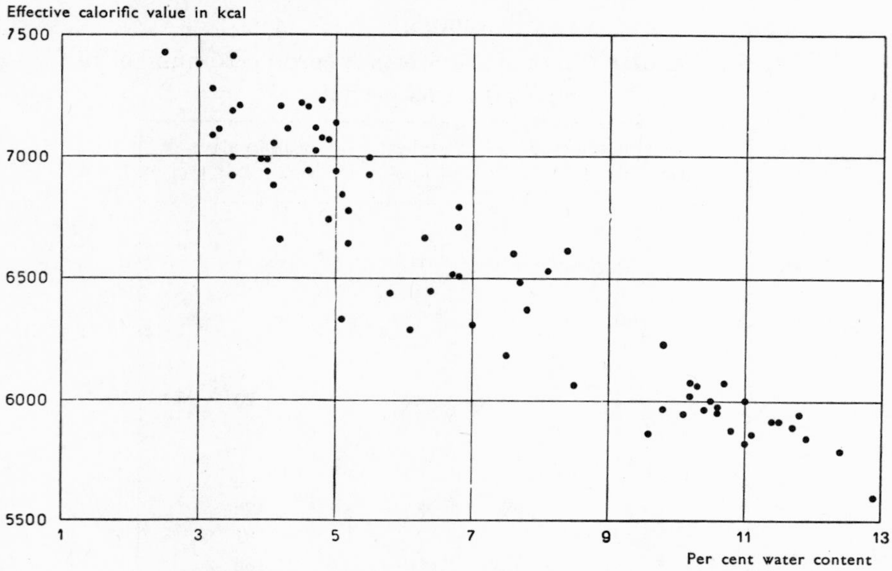

FIG. 3.14. Distribution of per cent water content and calorific value in kcal for 74 samples of coal.

between the variables and is a useful tool for a first analysis of two-dimensional distributions.

3.8. Punched Cards.

In two-dimensional distributions the observations may be plotted in dot diagrams or recorded in tables or registered on cards, as described for one-dimensional distributions. For multi-dimensional distributions, punched cards may also be used for enumeration. Instead of writing the results on the cards, they may be recorded by the aid of holes punched in the cards. There are both manual and mechanical systems of punched cards.

The *manual* system is based on cards with a row of circular holes placed along each edge without any hole being open to the edge, see Fig. 3.16, p. 61. An observation is marked by notching a piece off the card corresponding to a hole, thereby opening the hole to the edge; if we wish, for example, to group the observed values in k classes, each class is allotted a hole on the cards. Using the primary data, each value is registered by notching the hole which represents the value in question. The cards are then collected in a heap, the square corner ensuring that all cards turn the same way, and a sorting needle (a knitting needle) is inserted in hole number 1; the cards which fall down give us the number of observations in class number 1. In this manner the cards are sorted into k groups. It is easy to check whether the cards have been correctly sorted, as all cards belonging to the same group have had their edge notched in the same position. In order to avoid copying and notching errors, the primary data may be marked on the cards in two phases: First the holes to be notched are marked directly according to the primary data with a pencil; the markings are then checked by comparison with the primary data and then the cards are notched according to the pencil marks with a suitable tool similar to an ordinary ticket punch.

When we are dealing with multi-dimensional distributions each variable can usually only monopolize a small number of holes. If there is not room enough to give each class interval a separate hole, a better use is made of the holes if we give the holes numbers according to a coding system and record the observed values by notching several holes for each, the sum of

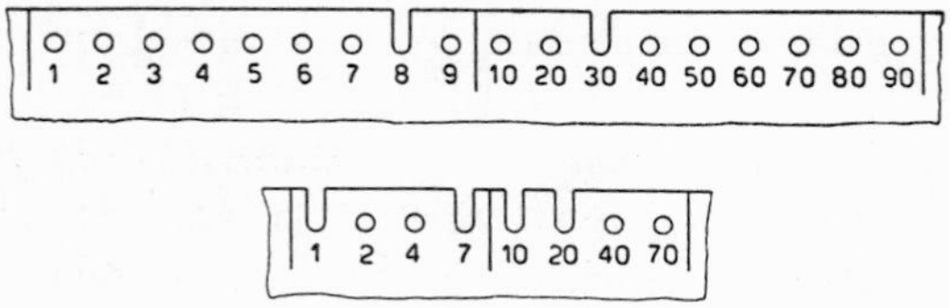

FIG. 3.15. Two examples of coding systems.

the corresponding numbers being equal to the value to be recorded. Fig. 3.15 shows two examples of such coding systems, which may be used for recording every integer between 0 and 99. In both illustrations the value 38 has been registered.

The first illustration needs no further explanation. In the second system any integer from 1 to 9 may be recorded by notching not more than two holes, and any integer from 1 to 99 by notching not more than four holes. If none of the holes of a card have been notched, the card corresponds to the value 0. Utilising the holes of the cards in this complicated manner makes notching and sorting cumbersome, and the coding system must be chosen with great care for each new problem.

A. T. McKAY has in *A New Method of Handling Statistical Data*, Suppl. Journ. Roy. Stat. Soc., 1, 1934, 62–75, given a coding system based on the numbers 1, 2, 4, 8, 16,... (the scale two system). With k holes denoted by these numbers every integer between 1 and $1+2+4+\ldots+2^{k-1} = 2^k-1$ may be represented in one way and one way only. With four holes numbered 1, 2, 4, and 8, 15 classes may be recorded, and if a fifth hole numbered 16 is added, the number of classes that may be recorded increases to 31, which is more than is generally needed. The numbers 1, 2, 4, 8 may also be used to represent all integers from 1 to 99 as in the two systems illustrated in Fig. 3.15. In the scale two system the number 7 is represented by $1+2+4$, i. e., 3 holes, whilst in the system 1, 2, 4, 7 we need at most 2 holes for the numbers from 1 to 9; on the other hand the number 8 needs only one hole in the scale two system but two holes in the 1, 2, 4, 7 system. The advantages of the scale two system first become evident when we are dealing with more advanced operations than simple enumeration. McKay has demonstrated how sums and sums of squares and products may be obtained by simple computations utilizing only the number of cards in suitably chosen groups.

15 groups represented by 4 holes with values 1, 2, 4, 8 should be sorted as follows:

1. 4 sorting needles are inserted in the four holes. The cards that fall away represent class interval number 15 (1+2+4+8).
2. The sorting needle is removed from hole 1. The cards now falling represent class interval number 14 (2+4+8).
3. A sorting needle is again inserted in hole 1, whilst that in hole 2 is removed. The falling cards represent class interval number 13 (1+4+8), etc.

H. A. C. Todd has in *A Note on Systematic Coding for Card Sorting Systems*, Suppl. Journ. Roy. Stat. Soc., 7, 1941, 151–54, proposed the coding system 0, 1, 2, 4, 7, which allows the ten numbers 0, 1, 2, . . . , 9 to be recorded uniquely by combination of *two holes for every integer*, the combination (4, 7) being used to represent 0.

Fig. 3.16 shows a punched card from a steel works on which the following data have been registered:

Furnace No.	10
Charge No.	4718
Yielding point	34·5 kg/sq.mm
Tensile strength	61·7 kg/sq.mm
Elongation	12·4 %
Carbon	0·46 %
Silicon	0·27 %
Manganese	0·65 %
Sulphur	0·05 %
Phosphorus	0·04 %

When sorting the cards, they are first grouped according to the magnitude of one of the variables. Then each of these groups is divided according to the magnitude of another variable, etc.

When we are dealing with a large mass of data with many categories of information, the mutual relationship of which is to be worked out, *mechanical* systems of punched cards are used. Fig. 3.17 is an illustration of such a card. Each column of the card includes the values 0, 1, 2, . . . , 9. Combination of two columns gives all integers 0–99, etc. The primary data are recorded on the cards with a machine which punches the cards according to the pre-fixed code. The cards are also sorted and counted by machine. Besides machines

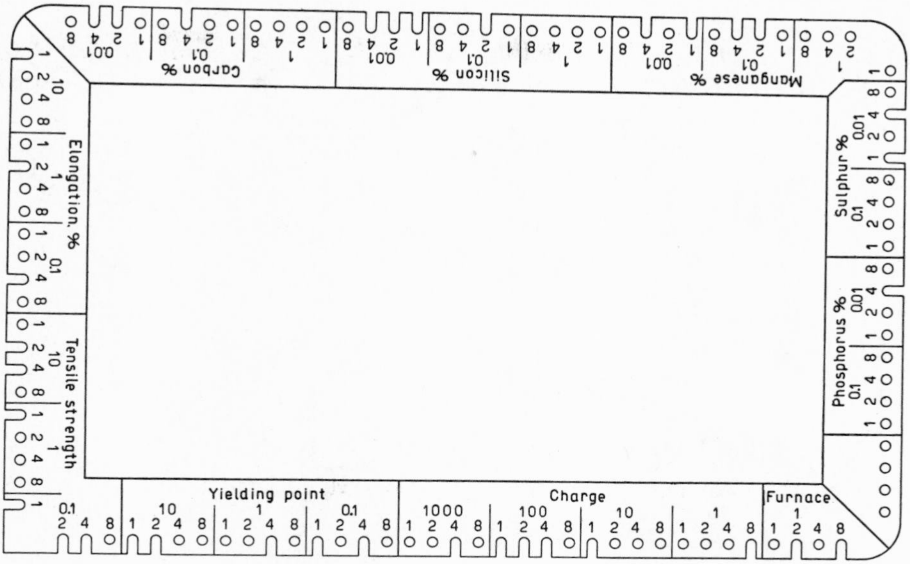

FIG. 3.16. Punched card for manual enumeration.

for sorting and counting, calculating machines have also been constructed which will add and multiply the recordings on the cards and print the results.

Fig. 3.17 shows a punched card from the laboratory of a cement plant. The following 30 categories of information have been registered:

1. Number of sample of cement   796/45
2. Sieve-residue on 0·09 mm sieve   6·3 %
3. Flour   49 %
4. Specific surface   1125 sq.cm/grm
5. Water to normal consistency   26·5 %

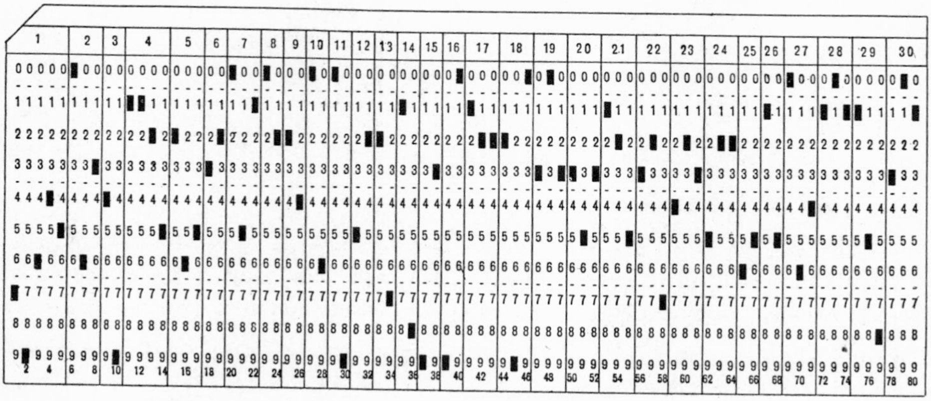

FIG. 3.17. Punched card for mechanical sorting (Hollerith system). This card is typical of similar systems developed by International Business Machines, Remington Rand, Powers, and Bull.

Setting time

6.	Initial setting time	3·2 hours
7.	Final setting time	5·1 hours
8.	Expansion according to Chatelier	2 mm

Chemical analysis

9.	$CaSO_4$	2·4 %
10.	C_4AF	6 »
11.	C_3A	9 »
12.	C_3S	52 »
13.	C_2S	27 »
14.	Free lime	1·8 »
15.	Lime saturation	93 »
16.	Combined lime	90 »

Strength of mortar

	Tensile strength	
17.	1 day	12·2 kg/sq.cm
18.	3 days	29·0 »
19.	7 days	30·3 »
20.	28 days	35·3 »
	Compressive strength	
21.	1 day	125 kg/sq.cm
22.	3 days	327 »
23.	7 days	423 »
24.	28 days	522 »

Strength of concrete

25.	Water-cement ratio	0·65
26.	Slump test	15 cm
27.	1 day	64 kg/sq.cm
28.	3 days	101 »
29.	7 days	158 »
30.	28 days	301 »

3.9. Tables and Histograms.

Table 2.7, p. 37, is an example of the tabulation of a theoretical, discontinuous, two-dimensional distribution. Table 3.7 shows an example of an empirical, (continuous), two-dimensional distribution[1]). The variables are

[1]) C. von Schéele, G. Svensson och J. Rasmusson: *Om Bestämning av Potatisens Stärkelse och Torrsubstanshalt med Tillhjälp av dess specifika Vikt*, Nordisk Jordbrugsforskning, 1935, p. 24.

TABLE 3.7.

Distribution of 560 samples of potatoes according to specific gravity and starch content.
Class lengths: 0·004 grm/cc and 1·0% starch.

Starch content in per cent	Specific gravity																				Total
	1·064	1·068	1·072	1·076	1·080	1·084	1·088	1·092	1·096	1·100	1·104	1·108	1·112	1·116	1·120	1·124	1·128	1·132	1·136	1·140	
9·5	1																				1
10·5	1	1																			2
11·5			5	1																	6
12·5		1	5	2	3	1															12
13·5			1	2	7	9															19
14·5					6	9	11														26
15·5						3	19	18	6												46
16·5							11	30	43	10											94
17·5							1	11	33	54	6	1									106
18·5								2	4	39	43	22									110
19·5										2	22	35	16	1							76
20·5											1	6	22	9	1						39
21·5														11	3						14
22·5														1	1	4	1				7
23·5																1					1
24·5																					0
25·5																				1	1
Total	2	2	11	5	16	22	42	61	86	105	72	64	38	22	5	5	1	0	0	1	560

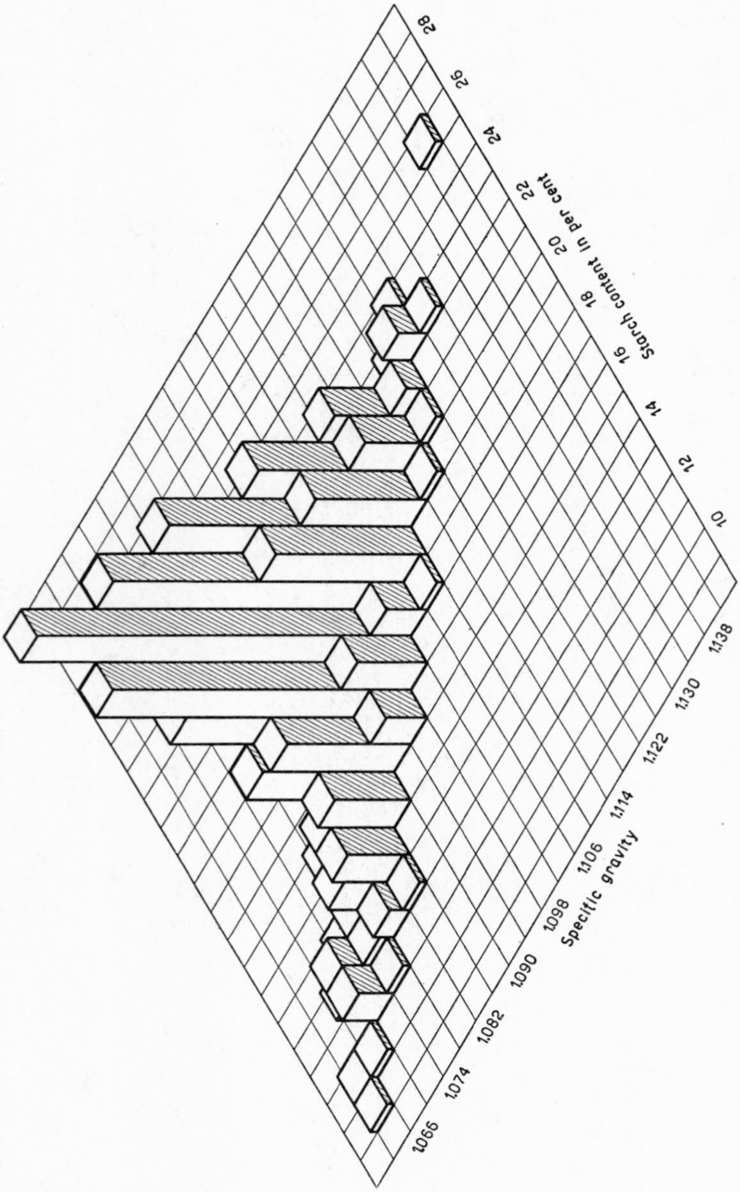

FIG. 3.18. Histogram. The distribution of 560 samples of potatoes according to specific gravity and starch content.

starch content (in per cent) and specific gravity of potatoes for industrial use.

Such a table is called a *correlation table.* It is produced by grouping the two variables, as in § 3.2, and enumerating the observations in each of the corresponding two-dimensional cells. The distribution may also be represented by a histogram, the class intervals of the variables being plotted along the axes of a base plane and the number of observations in each rectangular cell being represented by the *volume* of a regular prism standing on the cell, see Fig. 3.18. As all the rectangles (representing the class intervals) in this example are of equal size, the height of each prism indicates the number of observations in the corresponding cell of the table.

CHAPTER 4.

DEFINITIONS AND FUNDAMENTAL PROPERTIES OF EMPIRICAL DISTRIBUTIONS

4.1. Statement of the Problem.

One of the main objects of statistics is to give a mathematical description of observed data in such a manner that the observed phenomena and the method of observation are characterized by a few numbers. The method of approach is to attempt a formulation of a mathematical expression for the distribution, the so-called theoretical distribution, and to determine the "constants" of this distribution from the observations in such a manner that all essential information is concentrated in these constants and the functional form of the distribution.

In much the same way as an examination of the properties of frequencies forms a natural introduction to the basic theory of probability so a clue as to how the mathematical theory of distributions may be developed is found by examining the empirical distributions and certain descriptive "constants" associated with them. In the remainder of this chapter, therefore, a number of important empirical quantities are introduced, e. g., fractile, mean, variance and covariance. The significance of these *empirical or sample* quantities becomes clear in Chapter 5 and later chapters, where corresponding *theoretical or population* concepts are discussed.

4.2. Fractiles.

Fig. 4.1 gives the cumulative frequency polygon for the distribution of 500 rivets according to size of diameter of their heads given in Table 4.1.

Corresponding to a given abscissa in Fig. 4.1 we find an ordinate which gives us the *fraction* of the total number of observations less than or equal to the given value of the abscissa.

Conversely, corresponding to a given ordinate we find an abscissa which gives the limit below which we have that fraction of the observations indicated by the ordinate. This abscissa will be termed the *fractile* corresponding to the given fraction.

By graphical interpolation we read from Fig. 4.1 that the 50% fractile

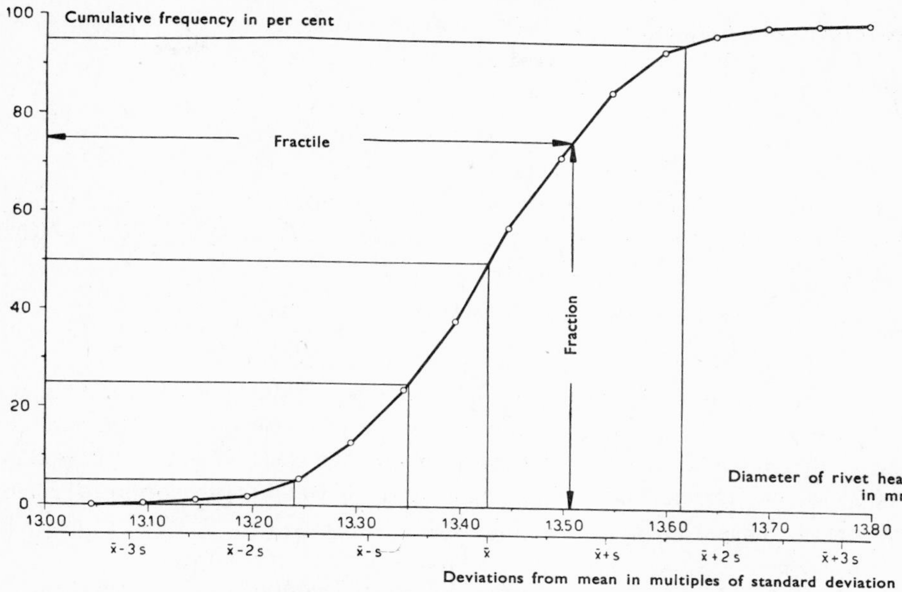

FIG. 4.1. *Definition of fractiles.*

Distribution of 500 rivets according to diameter of heads in millimetres.

is 13·43 mm, i. e., half the observations are smaller than or equal to 13·43 mm. The 5% and 95% fractiles are 13·24 and 13·62 mm. These fractiles characterize the distribution as follows: 5% of the observations are smaller than or equal to 13·24 mm, 5% larger than 13·62 mm; this interval is bisected by the value 13·43 mm, and 45% of our observed results are situated in the interval $13{\cdot}24 < x \leq 13{\cdot}43$, and 45% in the interval $13{\cdot}43 < x \leq 13{\cdot}62$ mm. If we include the 25% and 75% fractiles we have a still better description of our distribution, 25% corresponding to 13·35 mm and 75% to 13·51 mm.

In Chapters 6 and 7 it is shown how the equation for a theoretical cumulative distribution curve is derived on basis of the corresponding cumulative frequency polygon.

Some of the fractiles have been given special names. The 50% fractile is called the *median*, because half the observations are below and half above this value. Correspondingly the 25% and 75% fractiles are called *quartiles*, the 10, 20, . . ., 90% fractiles *deciles*, and the 1, 2, . . ., 99% fractiles *percentiles*. From here on, however, the term fractile will be used in most cases, followed by the corresponding cumulative frequency in per cent. Some authors use percentile or *quantile* as synonymous with fractile.

4.3. Definition and Computation of the Mean.

The *mean* (arithmetic mean) or average of the observations is defined as the sum of the observed values divided by their number. Let us call the n

observed values $x_1, x_2, \ldots, x_n$; the mean, $\bar{x}$, is then

$$\boxed{\bar{x} = \frac{x_1+x_2+\ldots+x_n}{n} = \frac{1}{n}\sum_{i=1}^{n} x_i.} \tag{4.3.1}$$

The following figures give the weight in grammes of 9 pieces of copper wire of equal length: 18·457, 18·434, 18·444, 18·461, 18·453, 18·447, 18·452, 18·440, 18·443. The mean is

$$\bar{x} = \frac{18{\cdot}457+18{\cdot}434+\ldots+18{\cdot}443}{9} = \frac{166{\cdot}031}{9} = 18{\cdot}4479.$$

Computation is facilitated if we introduce a suitably chosen number, the "*computing origin*", and read the observed values as the sum of this number and a difference. It is then necessary to operate only with the differences. For example, taking the computing origin at 18·450, the computation of the mean takes the form

$$\bar{x} = 18{\cdot}450+\frac{0{\cdot}007-0{\cdot}016-\ldots-0{\cdot}007}{9} = 18{\cdot}450-\frac{0{\cdot}019}{9} = 18{\cdot}4479.$$

This mode of expression implies that any observed value, x_i, may be written

$$\boxed{x_i = a+v_i,} \tag{4.3.2}$$

the sum of the observations being

$$\sum_{i=1}^{n} x_i = na+\sum_{i=1}^{n} v_i$$

and the mean

$$\boxed{\bar{x} = a+\frac{\sum v_i}{n} = a+\bar{v}.} \tag{4.3.3}$$

Direct computation of the mean can thus be replaced by the computation of the n differences $v_i = x_i-a$ and of $\bar{v}$ and $a+\bar{v}$. If a calculating machine is not used, this method often saves time. With a calculating machine, the method is useful only if a is so simple that the differences can be computed mentally as the summation proceeds on the machine. In the above example, 18·400 might be used as the origin when employing a calculating machine.

If the data are numerous, the computation can be further simplified by using grouped data.

The first two columns of Table 4.1 give the grouped distribution of 500 rivets according to size of diameter. For example, the number of diameters between 13·345 and 13·395 mm is 69. The sum of these diameters must lie

TABLE 4.1.

Computation of the mean of a grouped distribution.

Distribution of 500 rivets according to diameter of heads in millimetres.
Class length: 0·05 mm.

Class midpoint	Number of rivets	$t = 13{\cdot}07 + 0{\cdot}05\, w'$		$t = 13{\cdot}42 + 0{\cdot}05\, w''$	
t	a	w'	$w' \cdot a$	w''	$w'' \cdot a$
13·07	1	0	0	—7	—7
13·12	4	1	4	—6	—24
13·17	4	2	8	—5	—20
13·22	18	3	54	—4	—72
13·27	38	4	152	—3	—114
13·32	56	5	280	—2	—112
13·37	69	6	414	—1	—69
13·42	96	7	672	0	0
13·47	72	8	576	1	72
13·52	68	9	612	2	136
13·57	41	10	410	3	123
13·62	18	11	198	4	72
13·67	12	12	144	5	60
13·72	2	13	26	6	12
13·77	1	14	14	7	7
Sum	500		3564		64
Mean			7·128		0·128

$\bar{t} = 13{\cdot}07 + 0{\cdot}05 \times 7{\cdot}128 = 13{\cdot}4264$

$\bar{t} = 13{\cdot}42 + 0{\cdot}05 \times 0{\cdot}128 = 13{\cdot}4264$

between $69 \times 13{\cdot}345$ and $69 \times 13{\cdot}395$ mm, and as an approximation to this sum we use $69 \times 13{\cdot}37$ mm, i. e., the number of observations multiplied by the midpoint of the class interval. Computation from the primary data gives the sum 922·63 mm instead of $69 \times 13{\cdot}37 = 922{\cdot}53$ mm, which is only a small difference. Moreover, for class intervals situated at either side of the mean, the differences will usually take on opposite signs; therefore, a good approximation to the sum of the 500 diameters will be *the sum of the class midpoints multiplied by the number of observations in the respective classes.*

In the present example this means that the 500 additions can be replaced by 15 multiplications and additions giving

$$\bar{x} \simeq \frac{1 \times 13{\cdot}07 + 4 \times 13{\cdot}12 + \ldots + 2 \times 13{\cdot}72 + 1 \times 13{\cdot}77}{500}$$

$$= \frac{6713{\cdot}20}{500} = 13{\cdot}4264 \text{ mm},$$

where $\simeq$ is read: Is approximately equal to.

Computation from the primary data gives

$$\bar{x} = \frac{6713{\cdot}00}{500} = 13{\cdot}4260 \text{ mm},$$

so the approximation is very good in this case.

The computation can be further abbreviated by choosing a computing origin. If we choose, e. g., 13·07 as origin, the above expression takes the form

$$\bar{x} \simeq 13{\cdot}07 + \frac{1\times 0{\cdot}00 + 4\times 0{\cdot}05 + \ldots + 2\times 0{\cdot}65 + 1\times 0{\cdot}70}{500}.$$

In the product sum of the numerator the class interval, 0·05, is a common factor and we have

$$\bar{x} \simeq 13{\cdot}07 + 0{\cdot}05\,\frac{1\times 0 + 4\times 1 + \ldots + 2\times 13 + 1\times 14}{500}$$

$$= 13{\cdot}07 + 0{\cdot}05\,\frac{3564}{500} = 13{\cdot}4264 \text{ mm}.$$

Thus, when employing these short cut methods, the class interval can be taken as the *"computing unit"*. By choosing a suitable computing origin and unit, the original 500 additions can be replaced by a few multiplications and additions involving quite small integers, as shown in Table 4.1.

The above methods can be expressed algebraically as follows:

The n observed values, $x_1, x_2, \ldots, x_n$, are grouped in m classes with midpoints $t_1, t_2, \ldots, t_m$ and the same class length Δt, see Table 4.2, in which a_j denotes the number of observations in the class interval with the midpoint t_j. The sum of the a_j observed values in the jth class interval is approximately

TABLE 4.2.
Grouped distribution.
Class length: Δt.

Class midpoint	Number of observations
t_1	a_1
t_2	a_2
$\vdots$	$\vdots$
t_m	a_m
Total...	n

equal to the product $a_j \times t_j$. The sum of the observations may now be expressed by the t's, as

$$\sum_{i=1}^{n} x_i \simeq \sum_{j=1}^{m} t_j a_j .$$

Division by n gives

$$\frac{\sum x_i}{n} \simeq \frac{\sum t_j a_j}{n}$$

or

$$\boxed{\bar{x} \simeq \bar{t} = \frac{\sum_{j=1}^{m} t_j a_j}{\sum_{j=1}^{m} a_j}.} \tag{4.3.4}$$

If *the class intervals are small* and *the number of observations is large*, the difference between $\bar{x}$ and $\bar{t}$ is of no importance.

Introducing

$$h_j = \frac{a_j}{n} \qquad \left(\sum_{j=1}^{m} h_j = 1\right), \tag{4.3.5}$$

which gives the frequency of the observations in the jth class interval, $\bar{t}$ takes the form

$$\bar{t} = \sum t_j \frac{a_j}{n} = \sum_{j=1}^{m} t_j h_j .$$

Thus, in the computation of the mean every t value is associated with a "weight" h, which denotes the frequency of the observations represented by t. An expression such as

$$\sum t_j a_j$$

or

$$\sum t_j h_j$$

is therefore called "the weighted sum" of the t values with weights a_j and h_j, respectively, and

$$\bar{t} = \frac{\sum t_j a_j}{\sum a_j} = \frac{\sum t_j h_j}{\sum h_j} = \sum t_j h_j \tag{4.3.6}$$

is called the "*weighted mean*" of the t-values with weights a and h, respectively.

As in the above example,

$$\sum t_j a_j$$

is most easily computed by writing

$$\boxed{t_j = \alpha + \beta w_j .} \tag{4.3.7}$$

Multiplication by a_j gives

$$t_j a_j = \alpha a_j + \beta w_j a_j$$

and the summation now results in

$$\sum t_j a_j = \alpha \sum a_j + \beta \sum w_j a_j$$
$$= \alpha n + \beta \sum w_j a_j .$$

Division by n gives the mean

$$\bar{t} = \alpha + \beta \frac{\sum w_j a_j}{n}$$

or

$$\boxed{\bar{t} = \alpha + \beta \bar{w} .} \qquad (4.3.8)$$

Usually α is chosen as one of the t values and β is chosen equal to the class length, the w's becoming integers. In the above example t has been chosen as

$$t_j = 13{\cdot}07 + 0{\cdot}05 w_j ,$$

w taking on the values 0, 1, 2,..., 14, and

$$\sum w_j a_j = 1 \times 0 + 4 \times 1 + \ldots + 2 \times 13 + 1 \times 14 = 3564 ,$$

see Table 4.1.

As computing origin it is sometimes convenient to choose a value near $\bar{t}$, since the w's then take on small integral values. If we choose $\alpha = 13{\cdot}42$ and $\beta = 0{\cdot}05$, i. e.,

$$t_j = 13{\cdot}42 + 0{\cdot}05 w_j ,$$

we have

$$\bar{x} \simeq \bar{t} = 13{\cdot}42 + 0{\cdot}05 \frac{1 \times (-7) + 4 \times (-6) + \ldots + 2 \times 6 + 1 \times 7}{500}$$
$$= 13{\cdot}42 + 0{\cdot}05 \frac{64}{500}$$
$$= 13{\cdot}4264 ,$$

as shown in columns 5 and 6 of Table 4.1.

4.4. Definition and Computation of the Variance.

The mean gives us some information about the "magnitude" of the observations. The deviations from the mean, $x_1 - \bar{x}$, $x_2 - \bar{x}, \ldots$, $x_n - \bar{x}$, describe the *scattering of the observations* about this mean. From the said n deviations we may calculate *the variance*, s^2, defined by

$$s^2 = \frac{\sum_{i=1}^{n}(x_i - \bar{x})^2}{n-1}, \tag{4.4.1}$$

which under certain conditions provides us with the best possible description of the variation of the observations contained in a single number. This point is discussed further in § 4.5. The positive value of the square root of the variance is called the *standard deviation s*.

It is usually laborious to calculate the sum of the squares of all deviations

$$\sum_{i=1}^{n}(x_i - \bar{x})^2$$

directly from the deviations from the mean. Squaring each term of the above sum gives

$$\sum_{i=1}^{n}(x_i - \bar{x})^2 = \sum(x_i^2 - 2\bar{x}x_i + \bar{x}^2) = \sum x_i^2 - 2\bar{x}\sum x_i + n\bar{x}^2$$
$$= \sum x_i^2 - 2\bar{x}n\bar{x} + n\bar{x}^2 = \sum x_i^2 - n\bar{x}^2.$$

In order to avoid errors due to rounding, the last term should be calculated as

$$n\bar{x}^2 = \frac{(n\bar{x})^2}{n} = \frac{\left(\sum x\right)^2}{n},$$

so that

$$\sum_{i=1}^{n}(x_i - \bar{x})^2 = \sum_{i=1}^{n} x_i^2 - \frac{1}{n}\left(\sum_{i=1}^{n} x_i\right)^2. \tag{4.4.2}$$

If we introduce the terms

$$S_x = \sum_{i=1}^{n} x_i \tag{4.4.3}$$

$$SS_x = \sum_{i=1}^{n} x_i^2 \tag{4.4.4}$$

and

$$SSD_x = \sum_{i=1}^{n}(x_i - \bar{x})^2 \tag{4.4.5}$$

where S, SS, and SSD are abbreviations for sum, sum of squares, and sum of squares of deviations from the mean, (4.4.2) may be written

$$SSD = SS - \frac{S^2}{n}, \tag{4.4.6}$$

and we have

$$s^2 = \frac{SSD}{n-1}. \qquad (4.4.7)$$

If we use the formula (4.4.6) it is unnecessary to compute the deviations, and the influence of errors due to "rounding off" can be easily judged.

If the observed values are large numbers, computation of the variance may be facilitated by a procedure similar to that used for calculating $\bar{x}$.

If we put

$$x_i = a + v_i$$

we have

$$x_i - \bar{x} = v_i - \bar{v},$$

i. e., the *deviations* from the mean are independent of the computing origin. It then follows that

$$\sum_{i=1}^{n} (x_i - \bar{x})^2 = \sum_{i=1}^{n} (v_i - \bar{v})^2$$

and therefore

$$SSD_x = SSD_v = SS_v - \frac{S_v^2}{n} \qquad (4.4.8)$$

and

$$s_x^2 = s_v^2 . \qquad (4.4.9)$$

We may often choose the computing origin in such a manner that the computation of $x_i - a$ is simple, and SS_v becomes much easier to compute than SS_x. In Table 4.3, column 2, SSD_x has been computed for the 9 weighings of copper wire (column 3 is explained on p. 75).

TABLE 4.3.

Computation of the mean and the variance of an ungrouped distribution.
Distribution of 9 pieces of copper wire according to weight in grammes.

x	$v = x - 18{\cdot}450$	$2v = 2(x - 18{\cdot}450)$
18·457	0·007	0·014
18·434	—0·016	—0·032
18·444	—0·006	—0·012
18·461	0·011	0·022
18·453	0·003	0·006
18·447	—0·003	—0·006
18·452	0·002	0·004
18·440	—0·010	—0·020
18·443	—0·007	—0·014
S	—0·019	—0·038
SS	0·000633	0·002532
S^2/n	0·0000401	0·0001604
SSD	0·0005929	0·0023716
s^2	0·00007411	0·00029645
s	0·00861	0·01722
S/n	—0·0021	—0·0042
$\bar{x}$	18·4479	

If a calculating machine is used, SS is determined without recording the separate values of the squares. The computing origin is usually chosen so that the differences can be computed mentally as they are squared on the machine. In our present example, for instance, 18·400 can be used as the computing origin; the numbers to be squared are, it is true, somewhat larger than if we choose 18·450, but this is counterbalanced by the fact that the differences need not be recorded.

All computations should be checked. This may be done by repeating the computations with another computing origin and/or unit. It is important that the figures employed in this second computation differ from those of the first, in order to avoid errors due to psychological factors (memorizing). Variation of the computing origin will not influence the value of s^2, as shown in (4.4.9). On the other hand, if the computing unit is varied, this leads to a proportional variation in s. If we introduce

$$x - \alpha = v = \beta w$$

we have

$$S_v = \beta S_w ,$$

$$SS_v = \beta^2 SS_w ,$$

$$SSD_v = \beta^2 SSD_w ,$$

and

$$\boxed{s_x^2 = s_v^2 = \beta^2 s_w^2 .} \tag{4.4.10}$$

In Table 4.3 this check has been worked out in column 3 for $\beta = \frac{1}{2}$.

If the observations are numerous, S and SS are checked by groups of, e. g., 20 observations at a time, which makes it easy to locate possible errors.

For grouped distributions we have

$$SSD_x = \sum_{i=1}^{n} (x_i - \bar{x})^2 \simeq \sum_{j=1}^{m} (t_j - \bar{t})^2 a_j ,$$

the a_j observations in the class interval $t_j \pm \dfrac{\Delta t}{2}$ approximately contributing $(t_j - \bar{t})^2 \times a_j$ to the sum of squares.

If we write

$$\boxed{t_j = \alpha + \beta w_j ,}$$

we have

$$\bar{t} = \alpha + \beta \bar{w}$$

and

$$t_j - \bar{t} = \beta (w_j - \bar{w}) ,$$

i. e., the t and w deviations from their means are proportional, the factor being the computing unit.

Substitution results in

$$\sum_{j=1}^{m}(t_j-\bar{t})^2 a_j = \beta^2 \sum (w_j-\bar{w})^2 a_j$$

$$= \beta^2 \left(\sum w_j^2 a_j - \frac{\left(\sum w_j a_j\right)^2}{n} \right),$$

or

$$SSD_x \simeq SSD_t = \beta^2 SSD_w = \beta^2 \left(SS_w - \frac{S_w^2}{n} \right) \tag{4.4.11}$$

and

$$\boxed{s_x^2 \simeq s_t^2 = \beta^2 s_w^2 .} \tag{4.4.12}$$

Table 4.4 shows the computation of the standard deviation for the distribution of the 500 rivet heads according to diameter.

The last column of the table shows the check. The computation includes the following three product sums

$$S_w = \sum_{j=1}^{m} w_j a_j \ , \quad SS_w = \sum_{j=1}^{m} w_j^2 a_j \, , \text{ and } \ SS_{w+1} = \sum_{j=1}^{m} (w_j+1)^2 a_j \, ,$$

which are computed on the machine without recording the separate products. The following relation shows how the values have been checked

$$SS_{w+1} = SS_w + 2S_w + n \, . \tag{4.4.13}$$

The value of SSD may be checked from

$$SSD_w = SSD_{w+1} \, . \tag{4.4.14}$$

The correction for SS_{w+1} we get from

$$S_{w+1} = S_w + n \, ,$$

so that

$$SSD_{w+1} = SS_{w+1} - \frac{S_{w+1}^2}{n} \, .$$

The variance

$$s^2 = \frac{1}{n-1} \sum_{i=1}^{n} (x_i - \bar{x})^2 \simeq \frac{1}{n-1} \sum_{j=1}^{m} (t_j - \bar{t})^2 a_j$$

may be written

$$s^2 = \frac{n}{n-1} \frac{1}{n} \sum_{i=1}^{n} (x_i - \bar{x})^2 \simeq \frac{n}{n-1} \sum_{j=1}^{m} (t_j - \bar{t})^2 h_j \, , \tag{4.4.15}$$

i. e., the variance is equal to $\frac{n}{n-1}$ multiplied by the weighted mean of the squared deviations, the weights being the frequencies of the deviations. For large values of n, $\frac{n}{n-1}$ is practically equal to 1.

TABLE 4.4.

Computation of mean and variance for a grouped distribution.
Distribution of 500 rivets according to diameter of their heads in millimetres. Class length: 0·05 mm.

Class midpoint t	Number of rivets a	$t = 13{\cdot}42 + 0{\cdot}05\, w$		
		w	w^2	$(w+1)^2$
13·07	1	—7	49	36
13·12	4	—6	36	25
13·17	4	—5	25	16
13·22	18	—4	16	9
13·27	38	—3	9	4
13·32	56	—2	4	1
13·37	69	—1	1	0
13·42	96	0	0	1
13·47	72	1	1	4
13·52	68	2	4	9
13·57	41	3	9	16
13·62	18	4	16	25
13·67	12	5	25	36
13·72	2	6	36	49
13·77	1	7	49	64
S	500	64		
SS			2638	3266
S^2/n			8	636
SSD			2630	2630
s_w^2			5·271	
s_w			2·296	
S/n		0·128		

$\bar{t} = 13{\cdot}42 + 0{\cdot}05 \times 0{\cdot}128 = 13{\cdot}426$ mm

$s_t = 0{\cdot}05 \times 2{\cdot}296 = 0{\cdot}115$ mm

As a measure of dispersion we sometimes use the *coefficient of variation* which is defined as the standard deviation divided by the mean, i.e.,

$$c_x = \frac{s_x}{\bar{x}}. \tag{4.4.16}$$

4.5. The Normal Distribution.

Fig. 4.2 represents the cumulative frequency polygon for the distribution of 650 condensers according to capacity. The capacity should be 2 μF but varies from 1·85 to 2·33 μF. The mean and the standard deviation are 2·086 and 0·079 μF, respectively, see Table 4.5.

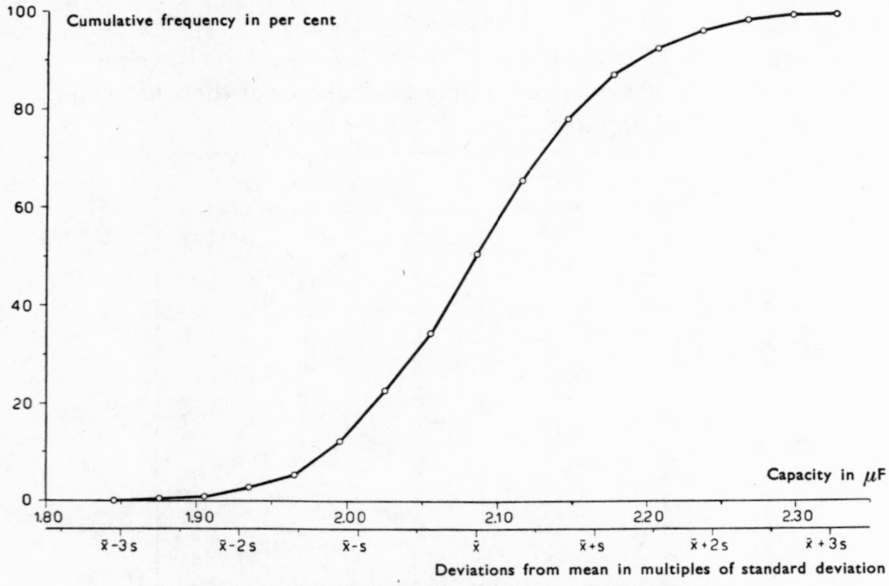

FIG. 4.2. Distribution of 650 condensers according to capacity.

If we compare Figs. 4.1 and 4.2 we find that the shape of the two cumulative frequency polygons is practically the same. A more detailed comparison may be made by computing $\bar{x}-3s$, $\bar{x}-2x$, $\bar{x}-s$, $\bar{x}-0{\cdot}5s$, $\bar{x}$, $\bar{x}+0{\cdot}5s$, $\bar{x}+s$, $\bar{x}+2s$, $\bar{x}+3s$ for both distributions and reading the corresponding cumulative frequencies from the cumulative frequency polygons, see Figs. 4.1 and 4.2 and Table 4.6.

The two columns of cumulative frequencies are practically identical for each value of $\bar{x}+us$; we find that the value of the cumulative frequency depends on u, but is independent of $\bar{x}$ and s. Experience shows that this agreement can be found again and again for distributions of all kinds of variables. This type of distribution, "the normal distribution", has been formulated mathematically, and its properties are discussed in Chapter 6. The normal distribution is a mathematical abstraction in the same sense as is the concept of probability. At present, only the relation between the "theoretical" cumulative frequencies (probabilities) and the corresponding multiple, u, of the standard deviation, is given in Table 4.6. It is clear that the agreement between the normal distribution and the other two distributions is excellent. Fig. 4.3 shows the cumulative normal distribution curve.

Introduction of the normal distribution lends far greater importance to the mean and standard deviation than may be directly inferred from their definition. The mean and the standard deviation determine the curve of the normal distribution in the same manner as the constants a and b in the equation

TABLE 4.5.

Computation of mean and standard deviation.

Distribution of 650 condensers according to capacity.

Class length: 0·03 μF.

Class midpoint μF	Number of condensers	$t = 2{\cdot}07 + 0{\cdot}03\, w$		
		w	w^2	$(w+1)^2$
1·86	2	—7	49	36
1·89	3	—6	36	25
1·92	13	—5	25	16
1·95	16	—4	16	9
1·98	45	—3	9	4
2·01	68	—2	4	1
2·04	77	—1	1	0
2·07	103	0	0	1
2·10	99	1	1	4
2·13	81	2	4	9
2·16	61	3	9	16
2·19	35	4	16	25
2·22	24	5	25	36
2·25	14	6	36	49
2·28	8	7	49	64
2·31	1	8	64	81
S	650	343		
SS			4633	5969
S^2/n			181	1517
SSD			4452	4452
s_w^2			6·860	
s_w			2·619	
S/n		0·528		

$$\bar{t} = 2{\cdot}07 + 0{\cdot}03 \times 0{\cdot}528 = 2{\cdot}086\,\mu F$$

$$s_t = 0{\cdot}03 \times 2{\cdot}619 = 0{\cdot}0786\,\mu F$$

$$y = a + bx$$

determine the corresponding straight line. When we know that the diameters of our 500 rivet heads are practically normally distributed with mean 13·426 mm and standard deviation 0·115 mm, it is possible to determine the corresponding normal distribution, as each value of $13{\cdot}426 + u \times 0{\cdot}115$ mm corresponds to a certain cumulative probability.

Table 4.6 shows that the distribution of the 500 rivet heads is essentially characterized by two values, the mean and the standard deviation, and by the type of the distribution.

TABLE 4.6.

Relation between deviations from mean, given in multiples of standard deviation, and cumulative frequencies.

Diameter of rivet heads: $\bar{x} = 13{\cdot}426$ mm $s = 0{\cdot}115$ mm

Capacity of condensers: $\bar{x} = 2{\cdot}086\ \mu F$ $s = 0{\cdot}079\ \mu F$

$\bar{x}+us$	Distribution of 500 rivets according to diameter of their heads in mm		Distribution of 650 condensers according to capacity in μF		Normal distribution	
	Diameter in mm	Cumulative frequency in per cent	Capacity in μF	Cumulative frequency in per cent	Deviations in multiples of standard deviation	Cumulative probability in per cent
$\bar{x}-3s$	13·081	0·2	1·849	0·0	—3	0·1
$\bar{x}-2s$	13·196	1·8	1·928	2·5	—2	2·3
$\bar{x}-s$	13·311	16·0	2·007	15·5	—1	15·9
$\bar{x}-0{\cdot}5s$	13·368	30·0	2·047	30·5	—0·5	30·9
$\bar{x}$	13·426	50·0	2·086	50·0	0	50·0
$\bar{x}+0{\cdot}5s$	13·484	68·5	2·125	69·0	0·5	69·1
$\bar{x}+s$	13·541	84·0	2·165	84·0	1	84·1
$\bar{x}+2s$	13·656	97·5	2·244	97·0	2	97·7
$\bar{x}+3s$	13·771	99·9	2·323	100·0	3	99·9

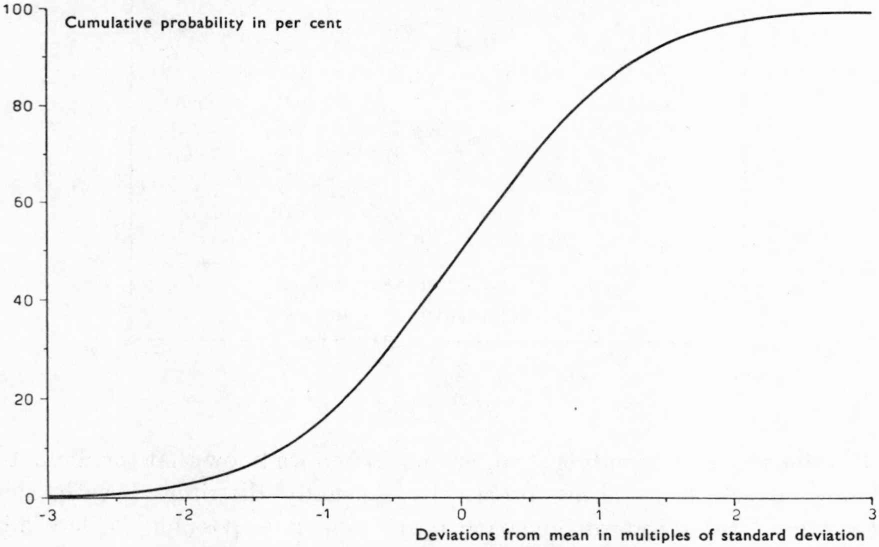

FIG. 4.3. The cumulative normal distribution curve.

4.6. Marginal Means and Variances.

Table 4.7 shows the relationship, *correlation*, between the specific gravity and the starch content of potatoes, see Table 3.7 and Fig. 3.18, pp. 63–64.

By summing the rows and the columns, respectively, we get the marginal distributions according to specific gravity and to starch content. The means and variances of these distributions are called *the marginal means and variances.*

If we introduce computing origins and units according to the equations

$$t = 1{\cdot}100 + 0{\cdot}004\, v,$$

$$u = 17{\cdot}5 + w,$$

where t and u denote the midpoints of the class intervals for the specific gravity and the starch content, respectively, the marginal means and variances may be computed according to the rules given in § 4.3 and § 4.4. The results are shown in the small table of Table 4.7. The average specific gravity is 1·0988 grm/cc and the average starch content 17·546%; the corresponding standard deviations are 0·01056 grm/cc and 2·221%, respectively.

4.7. Conditional Means and Variances.

The marginal distribution of specific gravities shows that the specific gravity of 64 samples lies between 1·106 and 1·110 grm/cc. The column corresponding to this class of specific gravities gives the distribution of these 64 samples according to starch content. Table 4.8 shows this conditional distribution as well as the determination of its mean and standard deviation.

The mean and standard deviation of the other conditional distributions may be computed in similar manner, see Table 4.7. $S_{w|v}$ denotes the sum of the w's in the conditional distribution of w's, given v, and the other symbols have corresponding meanings, see Tables 4.7 and 4.8. (The remaining computations are explained in § 4.8).

The average starch content, $\bar{y}_x$, increases gradually as the specific gravity increases, as shown in Fig. 4.4. The points vary at random about a straight line, *the regression line.* (The term "regression" was originally used by F. Galton in a statistical examination of human inheritance; he used "regression" to denote a certain hereditary relationship. From this work the word has passed into statistical terminology, and it now denotes the statistical method employed by him). Thus, the average starch content may be considered to be a linear function of the specific gravity. The starch content of the separate samples varies about this line with a standard deviation which is practically constant, as shown by the values of $s_{y|x}$ in Table 4.7.

A corresponding analysis may be made of the conditional distributions of the specific gravities for given starch content. The average specific gravity is practically a linear function of the starch content, and the standard

TABLE 4.7. Computation of conditional mean

Starch content in per cent (y)	Specific gravity (x)	1·064	1·068	1·072	1·076	1·080	1·084	1·088	1·092	1·096	1·100	1·104	1·108	1·11
	w \ v	—9	—8	—7	—6	—5	—4	—3	—2	—1	0	1	2	3
9·5	—8	1												
10·5	—7	1	1											
11·5	—6			5	1									
12·5	—5		1	5	2	3	1							
13·5	—4			1	2	7	9							
14·5	—3					6	9	11						
15·5	—2						3	19	18	6				
16·5	—1							11	30	43	10			
17·5	0							1	11	33	54	6	1	
18·5	1								2	4	39	43	22	
19·5	2										2	22	35	16
20·5	3											1	6	22
21·5	4													
22·5	5													
23·5	6													
24·5	7													
25·5	8													
Total	n_v	2	2	11	5	16	22	42	61	86	105	72	64	38
$u = 17{\cdot}5 + w$	$S_{w\|v}$	—15	—12	—59	—24	—61	—74	—82	—64	—51	33	90	110	98
	$SS_{w\|v}$	113	74	321	118	241	262	186	104	71	57	140	216	262
	S^2/n_v	112·50	72·00	316·45	115·20	232·56	248·91	160·10	67·15	30·24	10·37	112·50	189·06	252·7
	$SSD_{w\|v}$	0·50	2·00	4·55	2·80	8·44	13·09	25·90	36·85	40·76	46·63	27·50	26·94	9·2
	$s^2_{w\|v}$	0·500	2·000	0·455	0·700	0·563	0·623	0·632	0·614	0·480	0·448	0·387	0·428	0·2
	$s_{w\|v} \simeq s_{y\|x}$	0·707	1·414	0·675	0·837	0·750	0·789	0·795	0·784	0·693	0·669	0·622	0·654	0·5
	$\bar{w}_v$	—7·50	—6·00	—5·36	—4·80	—3·81	—3·36	—1·95	—1·05	—0·59	0·31	1·25	1·72	2·5
	$\bar{y}_x \simeq \bar{u}_t$	10·00	11·50	12·14	12·70	13·69	14·14	15·55	16·45	16·91	17·81	18·75	19·22	20·0

and variances and the covariance.

1·116	1·120	1·124	1·128	1·132	1·136	1·140	Total	$t = 1{\cdot}100 + 0{\cdot}004\,v$								
4	5	6	7	8	9	10	n_w	$S_{v\mid w}$	$SS_{v\mid w}$	S^2/n_w	$SSD_{v\mid w}$	$s^2_{v\mid w}$	$s_{v\mid w}$	$\bar{v}_w$	$\bar{x}_y \simeq \bar{t}_u$	$s_{x\mid y}$
							1	—9	81	81·00	0·00	—	—	—9·00	1·0640	—
							2	—17	145	144·50	0·50	0·500	0·707	—8·50	1·0660	0·00283
							6	—41	281	280·17	0·83	0·166	0·407	—6·83	1·0727	0·00163
							12	—74	472	456·33	15·67	1·425	1·194	—6·17	1·0753	0·00478
							19	—90	440	426·32	13·68	0·760	0·872	—4·74	1·0810	0·00349
							26	—99	393	376·96	16·04	0·642	0·801	—3·81	1·0848	0·00320
							46	—111	297	267·85	29·15	0·648	0·805	—2·41	1·0904	0·00322
							94	—136	262	196·77	65·23	0·701	0·837	—1·45	1·0942	0·00335
							106	—50	96	23·58	72·42	0·690	0·831	—0·47	1·0981	0·00332
							110	79	143	56·74	86·26	0·791	0·889	0·72	1·1029	0·00356
1							76	144	322	272·84	49·16	0·655	0·809	1·89	1·1076	0·00324
9	1						39	120	392	369·23	22·77	0·599	0·774	3·08	1·1123	0·00310
11	3						14	59	251	248·64	2·36	0·182	0·427	4·21	1·1168	0·00171
1	1	4	1				7	40	234	228·57	5·43	0·905	0·951	5·71	1·1228	0·00380
		1					1	6	36	36·00	0·00	—	—	6·00	1·1240	—
							0	0	—	—	—	—	—	—	—	—
						1	1	10	100	100·00	0·00	—	—	10·00	1·1400	—
22	5	5	1	0	0	1	560	—169	3945	3565·50	379·50					

78	20	26	5	—	—	8	26
86	82	136	25	—	—	64	2758
76·55	80·00	135·20	25·00	—	—	64·00	2500·53
9·45	2·00	0·80	0·00	—	—	0·00	257·47
0·450	0·500	0·200	—	—	—	—	
0·671	0·707	0·447	—	—	—	—	
3·55	4·00	5·20	5·00	—	—	8·00	
21·05	21·50	22·70	22·50	—	—	25·50	

3101 $= SP_{vw}$

—7·8 $= S_v \cdot S_w/n$

3108·8 $= SPD_{vw}$

5·5614 $= s_{vw}$

0·02225 $= s_{xy}$

0·949 $= r_{xy}$

Marginal means and variances

	v	w
S	—169	26
SS	3945	2758
S^2/n	51·0	1·2
SSD	3894·0	2756·8
s^2	6·9660	4·9317
s	2·639	2·221
S/n	—0·30	0·046

$\bar{x} \simeq \bar{t} = 1{\cdot}100 - 0{\cdot}004 \times 0{\cdot}30 = 1{\cdot}0988$

$s_x \simeq s_t = 0{\cdot}004 \times 2{\cdot}639 = 0{\cdot}01056$

$\bar{y} \simeq \bar{u} = 17{\cdot}5 + 0{\cdot}046 = 17{\cdot}546$

$s_y \simeq s_u = 2{\cdot}221$

TABLE 4.8.

Conditional distribution according to starch content.
Specific gravity: 1·106–1·110.

Starch content in per cent	Number of samples	$u = 17{\cdot}5 + w$	
		w	w^2
17·5	1	0	0
18·5	22	1	1
19·5	35	2	4
20·5	6	3	9
S	64	110	
SS		216	
S^2/n		189·06	
SSD		26·94	
s^2		0·428	
s		0·654	
S/n		1·72	

Mean = 17·5 + 1·72 = 19·22 % starch
Standard deviation = 0·654 % starch

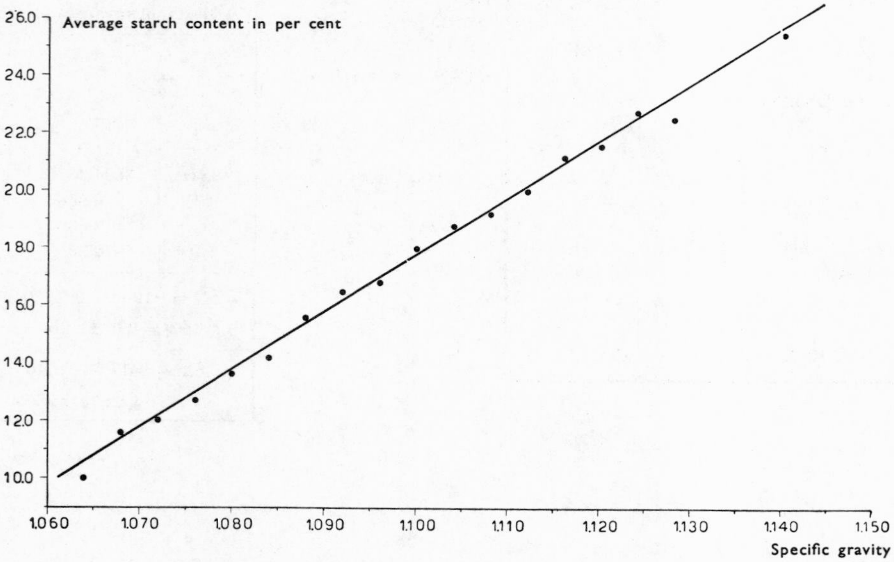

FIG. 4.4. The dependence of average starch content on specific gravity.

deviations of the conditional distributions, $s_{x|y}$, is independent of y, see Table 4.7.

The methods for checking the computations described in § 4.3 and § 4.4 may here be replaced by other methods, based on the relations between the conditional and the marginal distributions. $S_{w|v}$ means the sum of the w's, given v. If $S_{w|v}$ is summed for all values of v, we get the sum of all the w values, i. e., the marginal sum S_w. Corresponding relations apply to SS; hence we may form the following check formulas:

$$\sum_v S_{w|v} = S_w \quad \text{and} \quad \sum_w S_{v|w} = S_v$$

and

$$\sum_v SS_{w|v} = SS_w \quad \text{and} \quad \sum_w SS_{v|w} = SS_v\,,$$

see Table 4.7.

4.8. Definition and Computation of the Covariance.

The relationship between two variables may under certain conditions (normal correlation) be described by means of a single number, which is a symmetrical function of the observations. This number is called the *covariance* and is defined by

$$s_{xy} = \frac{1}{n-1}\sum_{i=1}^{n}(x_i-\bar{x})(y_i-\bar{y}), \tag{4.8.1}$$

where $(x_1, y_1), (x_2, y_2), \ldots (x_n, y_n)$ denote the observed values of x and y.

If large values of one variable usually occur together with large values of the other and similarly small values also occur together as in Table 4.7, the covariance of these two variables is positive, the majority of the deviations $x_i-\bar{x}$ and $y_i-\bar{y}$ as well as the largest of these deviations having the same sign; their products will therefore usually be positive; we say that the variables are *positively correlated.* If large values of one variable are most frequently associated with small values of the other, and vice versa, corresponding deviations will most often be of opposite sign, and the covariance will be negative. The variables are then said to be *negatively correlated.*

The sum of the products of the deviations from the mean may be transformed in much the same manner as the corresponding sum of squares:

$$\sum_{i=1}^{n}(x_i-\bar{x})(y_i-\bar{y}) = \sum_{i=1}^{n}x_iy_i - n\bar{x}\bar{y}\,.$$

For computations we use the form

$$\sum_{i=1}^{n}(x_i-\bar{x})(y_i-\bar{y}) = \sum_{i=1}^{n}x_iy_i - \frac{1}{n}\sum_{i=1}^{n}x_i\sum_{i=1}^{n}y_i \tag{4.8.2}$$

or

$$SPD_{xy} = SP_{xy} - \frac{S_x S_y}{n}, \tag{4.8.3}$$

where *SPD*, the sum of the products of the deviations, is defined by

$$SPD_{xy} = \sum_{i=1}^{n} (x_i - \bar{x})(y_i - \bar{y}), \tag{4.8.4}$$

and *SP*, the sum of the products, by

$$SP_{xy} = \sum_{i=1}^{n} x_i y_i. \tag{4.8.5}$$

These formulas are analogous to (4.4.2) and (4.4.6).

If we introduce computing origins and units by means of the following equations

$$x_i = \alpha + \beta v_i,$$

$$y_i = \gamma + \varkappa w_i,$$

we have

$$(x_i - \bar{x})(y_i - \bar{y}) = \beta\varkappa(v_i - \bar{v})(w_i - \bar{w}),$$

i. e.,

$$SPD_{xy} = \beta\varkappa SPD_{vw} = \beta\varkappa \left(SP_{vw} - \frac{S_v S_w}{n} \right)$$

and

$$s_{xy} = \beta\varkappa s_{vw} \tag{4.8.6}$$

analogous to (4.4.10).

For checking our computations, we introduce

$$z_i = x_i + y_i \tag{4.8.7}$$

and S_z, SS_z, and SSD_z are then computed. The following relations are used for the checks:

$$S_z = S_x + S_y, \tag{4.8.8}$$

$$SS_z = SS_x + SS_y + 2SP_{xy}, \tag{4.8.9}$$

$$SSD_z = SSD_x + SSD_y + 2SPD_{xy}, \tag{4.8.10}$$

and

$$s_z^2 = s_x^2 + s_y^2 + 2s_{xy}. \tag{4.8.11}$$

Instead of $x+y$, another linear function of x and y, e. g., $x-y$, may be used. The function chosen should always be such that the computation of z, s_z, etc., is not too laborious.

In grouped distributions, where the range of variations for x and y are divided into k and l intervals of equal length, respectively, with midpoints $t_1, t_2, \ldots, t_k$ for x and $u_1, u_2, \ldots, u_l$ for y, SPD_{xy} is approximately equal to

$$SPD_{tu} = \sum_{\nu=1}^{k} \sum_{\mu=1}^{l} (t_\nu - \bar{t})(u_\mu - \bar{u}) a_{\nu\mu}, \tag{4.8.12}$$

where $a_{\nu\mu}$ denotes the number of (x, y)-values in the cells defined by the inequalities

$$t_\nu - \frac{\Delta t}{2} < x \leq t_\nu + \frac{\Delta t}{2},$$

$$u_\mu - \frac{\Delta u}{2} < y \leq u_\mu + \frac{\Delta u}{2}.$$

The sum of the $a_{\nu\mu}$ products of the form

$$(x_i - \bar{x})(y_i - \bar{y})$$

is thus replaced by

$$(t_\nu - \bar{t})(u_\mu - \bar{u}) a_{\nu\mu},$$

i. e., the observed results are replaced by the midpoints of the class intervals. If we put

$$t_\nu = \alpha + \beta v_\nu$$

$$u_\mu = \gamma + \varkappa w_\mu,$$

we get

$$SPD_{tu} = \beta\varkappa SPD_{vw} = \beta\varkappa \left(SP_{vw} - \frac{S_v S_w}{n} \right), \tag{4.8.13}$$

analogous to (4.4.11).

The computation of

$$SP_{vw} = \sum_{\nu=1}^{k} \sum_{\mu=1}^{l} v_\nu w_\mu a_{\nu\mu}$$

may, e. g., be carried out by first computing the weighted sums

$$S_{w|v_\nu} = \sum_{\mu=1}^{l} w_\mu a_{\nu\mu}$$

for all values of v_ν, and then the weighted sum

$$SP_{vw} = \sum_{\nu=1}^{k} v_\nu S_{w|v_\nu}.$$

As a check, the computation may now be made by taking v and w in the reverse order, i. e.,

$$S_{v|w_\mu} = \sum_{\nu=1}^{k} v_\nu a_{\nu\mu}$$

and

$$SP_{vw} = \sum_{\mu=1}^{l} w_\mu S_{v|w_\mu}.$$

In Table 4.7 the computations have been carried out in both ways:

$$SP_{vw} = (-15) \cdot (-9) + (-12) \cdot (-8) + \ldots + 8 \cdot 10 = 3101$$

and

$$SP_{vw} = (-9) \cdot (-8) + (-17) \cdot (-7) + \ldots + 10 \cdot 8 = 3101 \,.$$

Then

$$\frac{S_v S_w}{n} = \frac{-169 \cdot 26}{560} = -7{\cdot}8$$

is computed, so that

$$SPD_{vw} = 3101 - (-7{\cdot}8) = 3108{\cdot}8$$

and

$$s_{vw} = \frac{3108{\cdot}8}{559} = 5{\cdot}5614 \,,$$

which gives

$$s_{xy} \simeq s_{tu} = 0{\cdot}004 \times 1{\cdot}0 \times 5{\cdot}5614 = 0{\cdot}02225 \,.$$

A check may also be made as shown in Table 4.9. From Table 4.7 the distribution of $w-v$ is computed by summation of the numbers $(a_{\nu\mu})$ along the oblique lines. The S, SS, SSD and s^2 values of this distribution are

TABLE 4.9.

Check on computation of covariance.

$w-v$	$(w-v)^2$	a
—2	4	5
—1	1	77
0	0	252
1	1	175
2	4	46
3	9	5
S 195		560
SS 501		
S^2/n 67·9		
SSD 433·1		
s^2 0·7748		

$$S_{w-v} = S_w - S_v = 26 - (-169) = 195$$

$$SS_{w-v} = SS_w + SS_v - 2SP_{wv}$$
$$= 2758 + 3945 - 2 \times 3101 = 501$$

$$SSD_{w-v} = SSD_w + SSD_v - 2SPD_{wv}$$
$$= 2756{\cdot}8 + 3894{\cdot}0 - 2 \times 3108{\cdot}8 = 433{\cdot}2$$

$$s^2_{w-v} = s^2_w + s^2_v - 2s_{wv}$$
$$= 4{\cdot}9317 + 6{\cdot}9660 - 2 \times 5{\cdot}5614 = 0{\cdot}7749$$

computed, and the check is performed as shown in Table 4.9, by means of a formula analogous to (4.8.11).

The covariance

$$s_{xy} = \frac{1}{n-1}\sum_{i=1}^{n}(x_i-\bar{x})(y_i-\bar{y}) \simeq \frac{1}{n-1}\sum_{\nu=1}^{k}\sum_{\mu=1}^{l}(t_\nu-\bar{t})(u_\mu-\bar{u})a_{\nu\mu}$$

may be written

$$s_{xy} = \frac{n}{n-1}\cdot\frac{1}{n}\sum_{i=1}^{n}(x_i-\bar{x})(y_i-\bar{y}) \simeq \frac{n}{n-1}\sum_{\nu=1}^{k}\sum_{\mu=1}^{l}(t_\nu-\bar{t})(u_\mu-\bar{u})h_{\nu\mu}, \quad (4.8.14)$$

where

$$h_{\nu\mu} = \frac{a_{\nu\mu}}{n}$$

indicates the frequency of (x, y)-values in the corresponding cell. Thus, we see that the covariance is equal to the weighted mean of the products of the deviations, multiplied by $\frac{n}{n-1}$.

The covariance is not suitable as a measure of the relationship between the two variables because its value depends on the units in which the variables are measured. Dividing the covariance by the product of the standard deviations we get the *coefficient of correlation*

$$\boxed{r_{xy} = \frac{s_{xy}}{s_x s_y} = \frac{SPD_{xy}}{\sqrt{SSD_x \cdot SSD_y}}} \quad (4.8.15)$$

which is independent of the units of measurement. From (4.8.1) it follows that

$$r_{xy} = \frac{1}{n-1}\sum_{i=1}^{n}\frac{x_i-\bar{x}}{s_x}\,\frac{y_i-\bar{y}}{s_y}, \quad (4.8.16)$$

i. e., the coefficient of correlation is equal to the covariance of the two transformed variables $\frac{x_i-\bar{x}}{s_x}$ and $\frac{y_i-\bar{y}}{s_y}$.

From Table 4.7 we get

$$r_{xy} = \frac{3108{\cdot}8}{\sqrt{3894{\cdot}0 \times 2756{\cdot}8}} = 0{\cdot}949.$$

It can be shown, see § 19.7, that

$$-1 \leq r_{xy} \leq 1\,. \quad (4.8.17)$$

The significance of the coefficient of correlation is discussed in Chapter 19.

CHAPTER 5.

DEFINITIONS AND FUNDAMENTAL PROPERTIES OF THEORETICAL DISTRIBUTIONS

ONE-DIMENSIONAL DISTRIBUTIONS

5.1. The Cumulative Distribution Function.

In accordance with the notation introduced in § 1.4 the probability that a stochastic variable x will take on a value less than or equal to a specified number x_a is written $P\{x \le x_a\}$. This probability, which is a function of x_a, will be written more briefly as $P\{x_a\}$, i. e.,

$$\boxed{P\{x \le x_a\} = P\{x_a\}\,.} \tag{5.1.1}$$

The probability that $x = x_a$ will be written $P\{x = x_a\}$.

In general *we write* $P\{x\}$ *for the probability that the stochastic variable in question is less than or equal to the real number* x. Usually we denote the stochastic variable and the real variable x by the same letter if no confusion can arise. $P\{x\}$ is called *the cumulative distribution function* of the variable.

The probability that the stochastic variable x *takes on a value in the interval* $x_a < x \le x_b$ is written $P\{x_a < x \le x_b\}$. According to the addition formula of the probability calculus, we have

$$P\{x_b\} = P\{x_a\} + P\{x_a < x \le x_b\} \tag{5.1.2}$$

or

$$\boxed{P\{x_a < x \le x_b\} = P\{x_b\} - P\{x_a\}\,.} \tag{5.1.3}$$

From (5.1.2) it follows that

$$P\{x_a\} \le P\{x_b\} \quad \text{for} \quad x_a \le x_b\,. \tag{5.1.4}$$

Thus, for increasing values of x, the cumulative distribution function never decreases in value. As the event $-\infty \le x \le \infty$ is certain, we have

$$P\{\infty\} = 1\,. \tag{5.1.5}$$

It is assumed that the value of the cumulative distribution function at a point x_a is equal to the limiting value of $P\{x\}$ when x approaches x_a from *the right.*

5.2. Continuous Distributions.

If the cumulative distribution function is continuous everywhere and possesses a continuous derivative, except possibly at certain points, of which at most a finite number is contained in any finite interval, *then the stochastic variable and its distribution are called continuous.*

The derivative

$$\boxed{\frac{dP\{x\}}{dx} = p\{x\}} \tag{5.2.1}$$

is called the *distribution function* of x. Integration of (5.2.1) leads to

$$P\{x\} = \int_{-\infty}^{x} p\{t\}dt\,, \tag{5.2.2}$$

or, as we frequently write it,

$$\boxed{P\{x\} = \int_{-\infty}^{x} p\{x\}dx\,,} \tag{5.2.3}$$

the limit of integration and the variable of integration being denoted by the same letter.

From (5.1.5) we have

$$\int_{-\infty}^{\infty} p\{x\}dx = 1\,. \tag{5.2.4}$$

The distribution function is represented by a curve, the *distribution curve*,

$$y = p\{x\}\,, \tag{5.2.5}$$

which together with the x-axis encloses an area of unit size. The probability that the stochastic variable x will take on a value between two given numbers, x_a and x_b, is represented by the *area* enclosed by the curve, the x-axis and the ordinates through x_a and x_b, as shown in Fig. 5.1. This area is equal to

$$P\{x_a < x \leq x_b\} = P\{x_b\} - P\{x_a\} = \int_{x_a}^{x_b} p\{x\}dx\,. \tag{5.2.6}$$

Thus, the probability that the variable falls within the interval $(x, x+\Delta x)$ is

$$\int_{x}^{x+\Delta x} p\{x\}dx\,, \tag{5.2.7}$$

which, according to the mean value theorem of the integral calculus, is equal to $p\{x'\}\Delta x$, where $x < x' < x+\Delta x$. We therefore write

$$dP\{x\} = p\{x\}dx\,, \tag{5.2.8}$$

which is called the *probability element* and indicates *the probability that the variable falls within the interval* $(x, x+dx)$.

The probability that the variable will take on a given value x_a is equal to 0, as

$$P\{x = x_a\} = \lim_{x \to x_a} (P\{x_a\} - P\{x\}) = 0\,. \tag{5.2.9}$$

This, however, does not mean that $x = x_a$ is an impossible event—we cannot reverse axiom III, § 1.3—but it is a mathematical model of the fact that for any preassigned number n of observations we only expect a small proportion h to fall within any small class interval and, the smaller the class

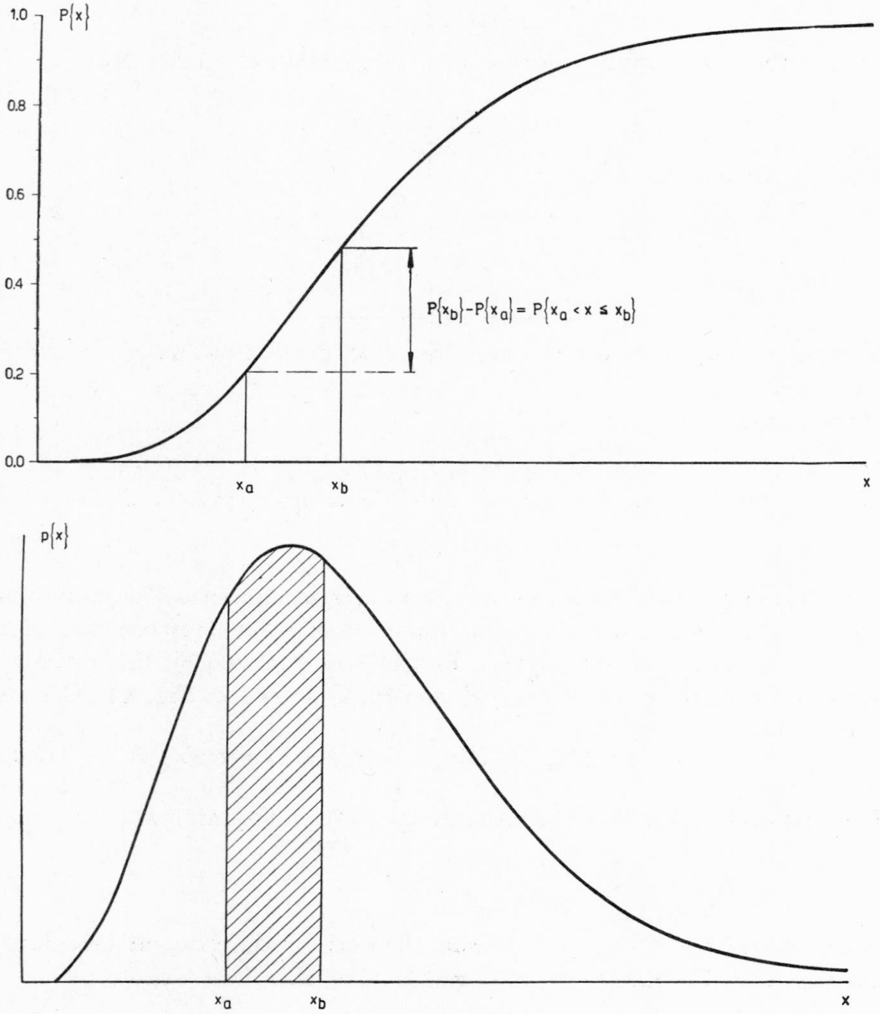

FIG. 5.1. Continuous cumulative distribution curve, $y = P\{x\}$, and the corresponding distribution curve, $y = p\{x\}$. The hatched area represents the probability $P\{x_a < x \leq x_b\}$.

interval, the smaller the expected value of h becomes. For continuous distributions we may therefore write $P\{x_a < x < x_b\}$ instead of $P\{x_a < x \leq x_b\}$.

Consequently, a continuous distribution function $p\{x\}$ must not itself be interpreted as a probability, but only as a probability density, whereas the probability element $p\{x\}dx$ is a probability.

The cumulative distribution function is represented by a curve, the *cumulative distribution curve*,

$$y = P\{x\}, \tag{5.2.10}$$

which is never decreasing from 0 to 1, and the *ordinate* of which gives the *area* to the left of the corresponding ordinate of the distribution curve, see Fig. 5.1.

For empirical distributions, the graphical representations called the histogram and the cumulative frequency polygon correspond to the distribution curve and the cumulative distribution curve, respectively.

Example 5.1. *The rectangular distribution.*

As a simple example of a continuous distribution consider the *rectangular distribution*, the distribution function and cumulative distribution function of which are

$$p\{x\} = \frac{1}{\alpha} \qquad \left(-\frac{\alpha}{2} \leq x \leq \frac{\alpha}{2}\right) \tag{5.2.11}$$

and

$$P\{x\} = \frac{1}{2} + \frac{x}{\alpha} \qquad \left(-\frac{\alpha}{2} \leq x \leq \frac{\alpha}{2}\right). \tag{5.2.12}$$

The distribution curve is a straight line parallel to the abscissa axis at a distance $\frac{1}{\alpha}$ and the cumulative distribution curve is a straight line with slope $\frac{1}{\alpha}$ and passing through the point $(0, \frac{1}{2})$, as shown in Fig. 5.3, p.101, for $\alpha = \pi$.

If a roulette wheel has the circumference α, the distribution of the results should be rectangular if the wheel is unbiased, i. e., in the long run we expect the results to scatter evenly along the periphery and the probability of a result taking on a value in the interval dx to be $\frac{dx}{\alpha}$.

Example 5.2. *The normal distribution.*

The *normal distribution* is defined by

$$p\{x\} = \frac{1}{\sqrt{2\pi}\,\sigma} e^{-\frac{(x-\xi)^2}{2\sigma^2}} \qquad (-\infty < x < \infty). \tag{5.2.13}$$

The distribution curve and the cumulative distribution curve have been plotted in Fig. 6.1, p.122, for $\xi = 0$ and $\sigma = 1$. This distribution is of fundamental importance and will be dealt with in detail in Chapter 6.

Example 5.3. *The distribution of the mth smallest observation.*

Let $x_1, x_2, \ldots, x_n$ denote n stochastically independent observations with the same distribution function, and let $x_{(1)}, x_{(2)}, \ldots, x_{(n)}$ denote the same values, *ranked according to order of magnitude*. The necessary and sufficient condition that *the largest observed value*, $x_{(n)}$, is less than or equal to x, is that all the values are less than or equal to x. According to the multiplication formula we have

$$P\{x_{(n)} \leq x\} = P\{x_1 \leq x\} \cdot P\{x_2 \leq x\} \ldots P\{x_n \leq x\} = (P\{x\})^n . \qquad (5.2.14)$$

Differentiation with respect to x gives us the distribution function, $f(x)$, of the largest observation as

$$f(x) = n(P\{x\})^{n-1}p\{x\} . \qquad (5.2.15)$$

This formula may also be derived by direct reasoning, the probability $f(x)dx$ that the largest observed value falls within the interval $(x, x+dx)$ being equal to the probability that $n-1$ values are less than or equal to x and that one value lies between x and $x+dx$.

Correspondingly, we may derive the distribution of the mth smallest observation $x_{(m)}$. The necessary and sufficient condition that $x_{(m)} \leq x$ is that at least m of the n values are less than or equal to x. The probability of the latter event may be obtained as follows:

According to the binomial distribution the probability that exactly ν of the n values are less than or equal to x is

$$\binom{n}{\nu} (P\{x\})^{\nu}(1-P\{x\})^{n-\nu}.$$

The probability that at least m observations are less than or equal to x is then according to the addition formula

$$P\{x_{(m)} \leq x\} = \sum_{\nu=m}^{n} \binom{n}{\nu} (P\{x\})^{\nu}(1-P\{x\})^{n-\nu}. \qquad (5.2.16)$$

Differentiation of (5.2.16) with respect to x and reduction of the expression found leads to the distribution $f(x)$ of our variable:

$$f(x) = \frac{n!}{(m-1)!\ (n-m)!} (P\{x\})^{m-1}(1-P\{x\})^{n-m}p\{x\} . \qquad (5.2.17)$$

This result may also be derived directly by applying the multinomial distribution; the probability that a result is less than or equal to x is $P\{x\}$, the probability of a result lying in the interval $(x, x+dx)$ is $p\{x\}dx$, and the probability of a result larger than x is $1-P\{x\}$. According to (2.2.1) the probability of $m-1$ values $\leq x$, one value "equal to x" and $n-m$ values $> x$, is

$$\frac{n!}{(m-1)!\,1!\,(n-m)!}(P\{x\})^{m-1}p\{x\}\,dx\,(1-P\{x\})^{n-m},$$

which agrees with (5.2.17).

5.3. Discontinuous Distributions.

If the cumulative distribution function is a step function, i. e., there exist certain values of x at which $P\{x\}$ makes positive jumps, and $P\{x\}$ is constant between two consecutive values of x, *then the stochastic variable and its distribution are called discontinuous.*

The variable takes on the values

$$\ldots < x_{-2} < x_{-1} < x_0 < x_1 < x_2 < \ldots$$

corresponding to the points of discontinuity of the cumulative distribution function with probabilities

$$\ldots\ p\{x_{-2}\},\ p\{x_{-1}\},\ p\{x_0\},\ p\{x_1\},\ p\{x_2\},\ldots\ .$$

equal to the jumps of $P\{x\}$.

The function

$$y = p\{x\} \qquad (x = \ldots, x_{-2}, x_{-1}, x_0, x_1, x_2, \ldots) \tag{5.3.1}$$

is called *the distribution function* of x. According to the addition formula we have

$$\boxed{P\{x_\nu\} = \sum_{i=-\infty}^{\nu} p\{x_i\}} \qquad (\nu = \ldots, -2, -1, 0, 1, 2, \ldots), \tag{5.3.2}$$

and from (5.1.5)

$$\sum_{i=-\infty}^{\infty} p\{x_i\} = 1. \tag{5.3.3}$$

Thus, the cumulative distribution function is obtained by summation of the distribution function. If the cumulative distribution function is given, *the distribution function is obtained by forming the differences of the cumulative distribution function*:

$$\boxed{p\{x_\nu\} = P\{x_\nu\} - P\{x_{\nu-1}\}.} \tag{5.3.4}$$

It will be seen that

$$P\{x_a < x \leq x_b\} = P\{x_b\} - P\{x_a\}$$
$$= \sum_{i=a+1}^{i=b} p\{x_i\}. \tag{5.3.5}$$

Fig. 5.2 shows the graphical representation of a distribution function and

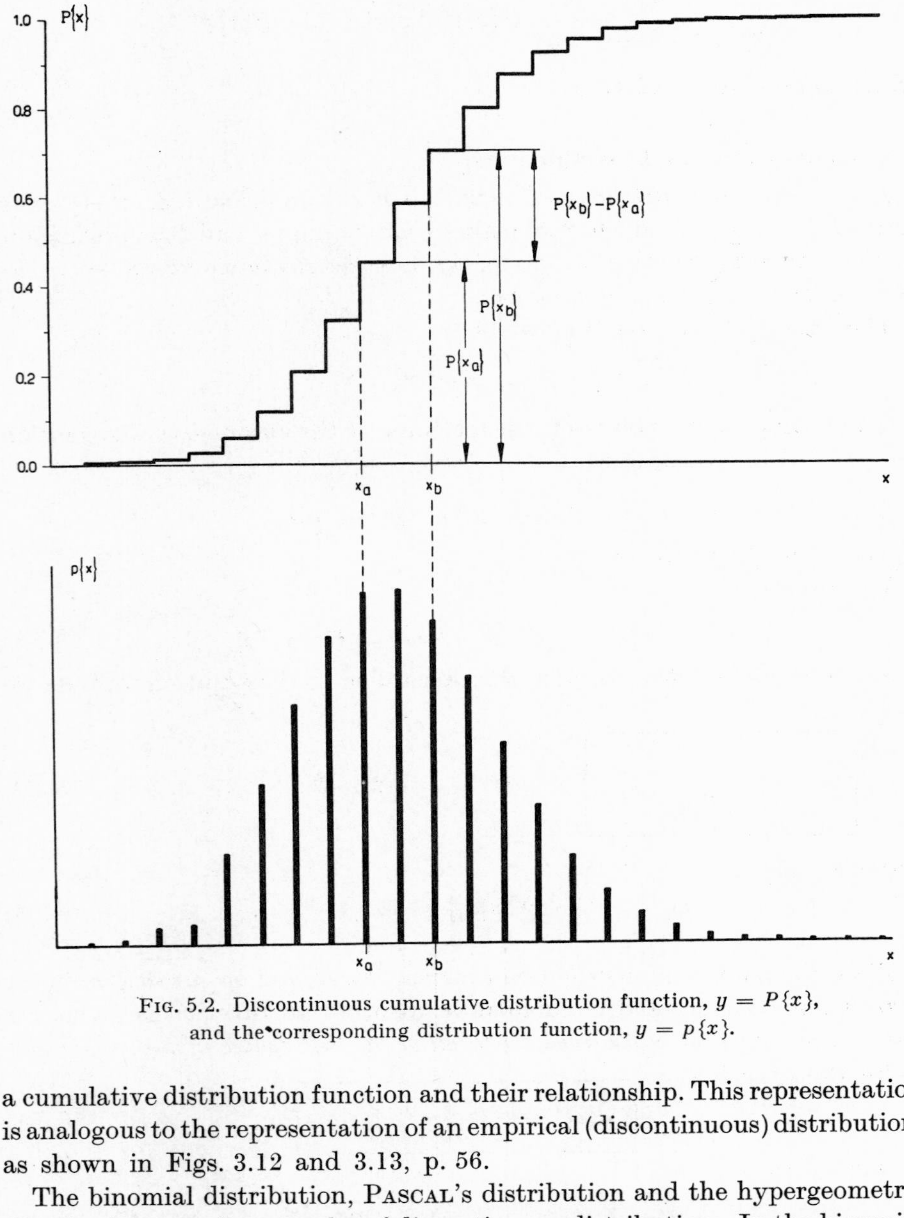

FIG. 5.2. Discontinuous cumulative distribution function, $y = P\{x\}$, and the corresponding distribution function, $y = p\{x\}$.

a cumulative distribution function and their relationship. This representation is analogous to the representation of an empirical (discontinuous) distribution, as shown in Figs. 3.12 and 3.13, p. 56.

The binomial distribution, PASCAL's distribution and the hypergeometric distribution are all examples of discontinuous distributions. In the binomial and the hypergeometric distributions, the variable can take on the values $x = 0,1,\ldots, n$ whilst in PASCAL's distribution it can take on an infinite number of values, $x = a, a+1,\ldots,$ cf. (2.1.2), (2.4.1) and (2.3.3).

5.4. The Concept of a Population.

The relation between *empirical* and *theoretical* distributions is often elucidated by interpreting a set of observations as a *sample* drawn at *random* from a *population*.

We distinguish between two types of populations: (1) populations containing a *finite* number of elements, and (2) populations containing an *infinite* number of elements.

The drawing of n balls without replacement from a bag containing N balls is a simple example of sampling from a finite population, see Example 2.1 and § 2.4. In general a finite population consists of N elements associated with certain numbers, $x_1, x_2, \ldots, x_N$. Drawing an element at *random* from this population means that the drawing is performed in such a way that each of the N elements has the same probability, $p = \frac{1}{N}$, of being drawn, i. e., by repetition of this process we expect all the N elements to occur equally frequently in the long run and to occur in a random order. What is meant by "random order" may be illustrated by considering the results of usual games of chance, see also Chapter 13.

If a certain number of the elements of the population, $N(x)$ say, are associated with the same value of the variable x, then the probability of drawing such an element is equal to $N(x)/N$ according to (1.4.12), i. e., equal to the frequency of this value of the variable in the population.

As the drawing without replacement of an element from a finite population affects the composition of the population, the results of successive drawings are not stochastically independent even if each drawing is random, see Example 2.1. The definition of *a random sample of n elements* is therefore perhaps most easily understood by considering all possible different samples of n elements drawn without replacement from the given population, i. e., after each sample has been drawn we replace the n elements and draw a new sample in the same manner. From the theory of combinations we know that we can select $N^{(n)} = N(N-1)\ldots(N-n+1)$ different samples of n elements from the given population. (If we disregard the *order* of the elements within the samples there are only $\frac{N^{(n)}}{n!} = \binom{N}{n}$ different samples). In accordance with the above definition of the random selection of a single element from a population we define a random sample of n elements as a sample selected among the $N^{(n)}$ different possible samples in such a manner that every sample has the same probability, $p = \frac{1}{N^{(n)}}$, of being chosen.

In practice it is often difficult to secure a random selection of elements from a population. It is not sufficient that the selection is haphazard, but we must be sure that the method of selection and the values of the variable in the population are independent. If the population can be enumerated, however, it is possible to select a random sample by means of a table of *random numbers*, see Table XIX or M. G. KENDALL and B. BABINGTON SMITH: *Random Sampling Numbers*, Tracts for Computers, No. 24, Cambridge, 1939. These numbers are constructed by means of some sort of game of chance where each of the integers from 0 to 9 as far as possible has the same probability of occurrence.

Very often it is convenient to speak of a set of observations as a *random* sample from the *hypothetical infinite* number of observations which might have arisen under the given circumstances.

Using this terminology, we look upon the observed values resulting from a series of throws with a die as a set of observations, chosen at random, from the infinite number of observations which might be obtained under similar conditions. Likewise, the 200 diameters of rivet heads given in Table 3.1 are considered as a sample of diameters, chosen at random, from the infinite number of diameters of rivet heads which might be generated by repetition of the given production process under the same conditions.

Correspondingly, *we interpret the probability of an event as the frequency with which the event occurs in the population* and a theoretical distribution function is said to give the frequency with which the values of the variable occur in the population just as an empirical distribution gives the frequency with which the values of the variable occur in the sample. Thus, in a population with a continuous distribution the frequency of values of the variable falling within the interval $(x, x+dx)$ is $p\{x\}dx$, and in a population with a discontinuous distribution the frequency of values of the variable equal to x_i is $p\{x_i\}$.

The *drawing of an element at random* from an infinite population means that the probability of drawing an element associated with values of the variable less than or equal to a specified value x is equal to the corresponding value of the cumulative distribution function $P\{x\}$, i. e., equal to the frequency of such values in the population.

As the *drawing of a finite number of elements* from an infinite population will not affect the composition of the population, the results of successive random drawings are *stochastically independent*. (This may be illustrated in practice by drawings from a finite population with replacement, see Example 2.1). It follows that the probability of getting a sample where the first element takes on a value less than or equal to x_1, the second element a value less than or equal to x_2 and so forth by random sampling is equal to

$$P\{x_1\}P\{x_2\}\ldots P\{x_n\},$$

according to the multiplication formula (1.5.8).

The first step to be taken when we are to analyze a set of observations is to investigate whether the observations may be regarded as a *random* sample from a stable population, i. e., whether the observations occur in a random order and are produced under the same essential conditions, see Chapter 13.

A discussion of both practical and theoretical aspects of sampling and randomness may be found in E. S. PEARSON: *Sampling Problems in Industry*, Suppl. Journ. Roy. Stat. Soc., 1, 1934, 107–136, M. G. KENDALL and B. BABINGTON SMITH: *Randomness and Random Sampling Numbers*, Journ. Roy. Stat. Soc., 101, 1938, 147—167, and W. A. SHEWHART: *Statistical Method from the Viewpoint of Quality Control*, The Department of Agriculture, Washington, 1939.

The second step is to formulate a hypothesis for *the type of the distribution* of the population from which the sample is supposed to be randomly drawn, i. e., we specify the *form* of the theoretical distribution function but not the constants or *parameters* entering into this function.

The third step in our analysis is therefore to compute *empirical values of the parameters*, or as we generally say, to determine *estimates of the parameters*, based on the sample.

The following paragraphs of this chapter give some general properties of theoretical distributions, and in Chapters 6 and 7 a special system of distribution functions which has proved useful in practical work is introduced.

5.5. Transformation of Distributions.

Let x be a stochastic variable and $y = \varphi(x)$ a function of x which determines y uniquely. The variable y is then a stochastic variable, the distribution of which can be derived from the distribution of x.

If $y = \varphi(x)$ is *an increasing function*, the inequalities $x \leq x_0$ and $y = \varphi(x) \leq \varphi(x_0) = y_0$ will always be satisfied simultaneously so that

$$P\{x \leq x_0\} = P\{y \leq y_0\}$$

or

$$P\{x_0\} = P\{y_0\}, \tag{5.5.1}$$

i. e., *the two cumulative distribution functions take on the same values for corresponding values of the variables.*

In order not to mix up the two cumulative distribution functions we will introduce the terms $P\{x\} = F(x)$ and $P\{y\} = G(y)$; we can then write (5.5.1) as

$$G(y) = F(x) \quad \text{for} \quad y = \varphi(x), \tag{5.5.2}$$

or

$$G(y) = [F(x)]_{x=\psi(y)} = F(\psi(y)), \tag{5.5.3}$$

where $x = \psi(y)$ denotes the inverse function of $y = \varphi(x)$.

For continuous distributions we get the distribution function $g(y)$ of y by differentiating $G(y)$:

$$\frac{dG(y)}{dy} = g(y) = \frac{dF(x)}{dx} \cdot \frac{dx}{dy} = \left[f(x) \cdot \frac{dx}{dy}\right]_{x=\psi(y)} \tag{5.5.4}$$

or

$$g(y) = f(\psi(y))\psi'(y)\,, \tag{5.5.5}$$

where $f(x)$ denotes the distribution function of x.

If, for example, $y = x^2$ and $p\{x\} = f(x)$ for $x \geq 0$ and $p\{x\} = 0$ for $x < 0$ we get

$$g(y) = \left[f(x)\frac{1}{2x}\right]_{x=\sqrt{y}} = \frac{1}{2\sqrt{y}}\,[f(x)]_{x=\sqrt{y}} \text{ for } y \geq 0, \text{ and } g(y) = 0 \text{ for } y<0.$$

If $y = \varphi(x)$ is a decreasing function, we get

$$P\{x \leq x_0\} = P\{y \geq y_0\}$$

or

$$G(y) = 1-F(x)\,. \tag{5.5.6}$$

Differentiation gives

$$g(y) = -f(x) \cdot \frac{dx}{dy}. \tag{5.5.7}$$

This formula and (5.5.4) may be expressed by the single formula

$$g(y) = f(x)\left|\frac{dx}{dy}\right| \quad \text{for} \quad y = \varphi(x) \tag{5.5.8}$$

or

$$p\{y\}\,|dy| = p\{x\}\,|dx| \quad \text{for} \quad y = \varphi(x)\,. \tag{5.5.9}$$

As in the above the two p's denote different functions; the symbol $p\{\ \}$ does not here denote a special function, but stands for the "distribution function of".

For discontinuous distributions, we get the distribution of y by forming the differences of $G(y)$.

Example 5.4. Fig. 5.3 shows a roulette wheel on which the observed value, the angle x, is read off a semicircle with a scale from $-\frac{\pi}{2}$ to $\frac{\pi}{2}$. These values form a rectangular distribution

$$f(x) = \frac{1}{\pi} \qquad \left(-\frac{\pi}{2} \leq x \leq \frac{\pi}{2}\right), \tag{5.5.10}$$

the probability of a result in the interval $(x, x+dx)$ being $\frac{dx}{\pi}$, cf. Example 5.1. The cumulative distribution function is

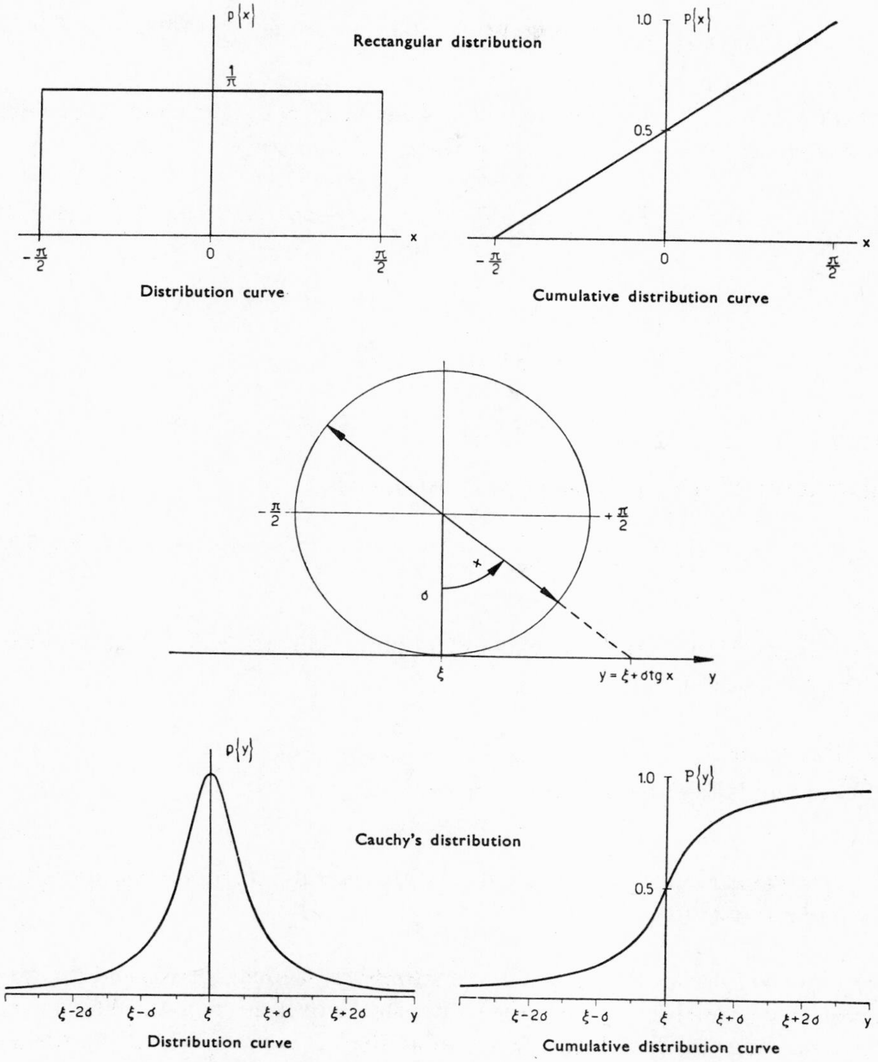

FIG. 5.3. Transformation of a rectangular distribution to CAUCHY's distribution.

$$F(x) = \frac{1}{2} + \frac{x}{\pi} \qquad \left(-\frac{\pi}{2} \leq x \leq \frac{\pi}{2}\right). \tag{5.5.11}$$

The distribution function and the cumulative distribution function have been plotted in Fig. 5.3.

Corresponding to every result x we find a value y on the horizontal tangent to the circle. The tangent bears an equidistant scale, and the point at which it touches the circle is called ξ, so that

$$y = \xi + \sigma \tan x \qquad \left(-\frac{\pi}{2} \leq x \leq \frac{\pi}{2}\right), \tag{5.5.12}$$

the radius of the circle being σ.

If the value on the tangent is to be smaller than or equal to a certain value y, the angle of the pointer must not exceed the corresponding

$$x = \arctan \frac{y-\xi}{\sigma} \qquad (-\infty < y < \infty)\,. \tag{5.5.13}$$

This leads to (cf. (5.5.3))

$$G(y) = \left[\frac{1}{2} + \frac{x}{\pi}\right]_{x = \arctan \frac{y-\xi}{\sigma}}$$

$$= \frac{1}{2} + \frac{1}{\pi} \arctan \frac{y-\xi}{\sigma} \qquad (-\infty < y < \infty)\,. \tag{5.5.14}$$

Differentiation of $G(y)$ with respect to y gives

$$g(y) = \frac{1}{\sigma\pi} \cdot \frac{1}{1+\left(\dfrac{y-\xi}{\sigma}\right)^2} \qquad (-\infty < y < \infty)\,. \tag{5.5.15}$$

This distribution, which is called CAUCHY's distribution, is represented graphically in Fig. 5.3.

5.6. Fractiles.

The cumulative distribution function

$$P = F(x) \tag{5.6.1}$$

gives the dependence of the cumulative probability, P, upon the variable x. The *inverse function*

$$x = G(P) \tag{5.6.2}$$

gives the values of the variable, the *fractiles*, which correspond to given cumulative probabilities; compare with the definition in § 4.2. The fractile corresponding to the cumulative probability P is called the *P-fractile* and is written x_P. We thus have the following identity

$$\boxed{P\{x \leq x_P\} = P\{x_P\} = P\,,} \tag{5.6.3}$$

i. e., *the probability that the variable is smaller than or equal to the P-fractile is equal to P.*

For continuous distributions x_P thus forms the solution of the equation

$$\int_{-\infty}^{x_P} p\{x\}dx = P\,. \tag{5.6.4}$$

For discontinuous distributions, x_P can take on only the values $\ldots, x_{-2}, x_{-1}, x_0, x_1, x_2, \ldots$, corresponding to values of P equal to

$$P\{x_\nu\} = \sum_{i=-\infty}^{\nu} p\{x_i\} \qquad (\nu = \ldots, -2, -1, 0, 1, 2, \ldots) .$$

If $y = \varphi(x)$ *is an increasing function of* x, we have

$$P\{x \le x_P\} = P\{y \le \varphi(x_P)\} = P ,$$

i. e.,

$$\boxed{y_P = \varphi(x_P) .} \tag{5.6.5}$$

This result shows an important property of the fractiles, namely that *the fractiles of a transformed variable are given by the transformed fractiles of the original variable,* provided that the transformation is performed by an increasing function.

For graphical analysis of empirical and theoretical distributions the cumulative distribution polygon or curve is often plotted on special graph paper, the P-scale of which is replaced by a G(P)-scale, so that the cumulative distribution curve is represented by a straight line, if the distribution in question really has the cumulative distribution function $P = F(x)$, see for example § 6.6.

5.7. The Mean of a Stochastic Variable.

The mean, $\mathfrak{M}\{x\}$, *of a stochastic variable* x with continuous or discontinuous distribution function is defined by

$$\boxed{\mathfrak{M}\{x\} = \frac{\int_{-\infty}^{\infty} x p\{x\} dx}{\int_{-\infty}^{\infty} p\{x\} dx} = \int_{-\infty}^{\infty} x p\{x\} dx = \xi} \tag{5.7.1}$$

and

$$\boxed{\mathfrak{M}\{x\} = \frac{\sum_{i=-\infty}^{\infty} x_i p\{x_i\}}{\sum_{i=-\infty}^{\infty} p\{x_i\}} = \sum_{i=-\infty}^{\infty} x_i p\{x_i\} = \xi ,} \tag{5.7.2}$$

respectively.

As $p\{x\}dx$ and $p\{x\}$, respectively, denote the frequency of "the value x" in the population, $\mathfrak{M}\{x\}$ may be interpreted as the mean of the infinite number of observations in the population; compare with the corresponding formulas (4.3.1) and (4.3.6) for empirical distributions. (The definition of the mean corresponds to the definition of the centre of gravity in mechanics). Therefore $\mathcal{M}\{x\}$ is also called the *population mean* to distinguish it from the *sample mean* $\bar{x}$. Often the words "population" and "sample" may be omitted, because the context makes it clear which mean is discussed. (Another notation for $\mathcal{M}\{x\}$ is $E(x)$, which is read "the mathematical expectation of x").

The symbol $\mathcal{M}\{\ \}$ is read "the mean of" and does *not* denote a function of the variable in question. The mean of x is not a function of x, but *a function of the parameters of the distribution of* x, as shown in the following examples.

Example 5.5. For the rectangular distribution

$$p\{x\} = \frac{1}{\alpha} \qquad \left(-\frac{\alpha}{2} \leq x \leq \frac{\alpha}{2}\right)$$

we have

$$\mathcal{M}\{x\} = \int_{-\frac{\alpha}{2}}^{\frac{\alpha}{2}} x \cdot \frac{1}{\alpha}\, dx = 0\,, \tag{5.7.3}$$

i. e., the mean value is equal to the point of symmetry of the distribution.

Example 5.6. For the binomial distribution, we get

$$\mathcal{M}\{x\} = \sum_{x=0}^{n} x \binom{n}{x} \theta^x (1-\theta)^{n-x}$$

$$= n\theta \sum_{x=1}^{n} \binom{n-1}{x-1} \theta^{x-1}(1-\theta)^{n-x}\,,$$

since

$$x\binom{n}{x} = x\frac{n!}{x!\,(n-x)!} = n\frac{(n-1)!}{(x-1)!\,(n-x)!} = n\binom{n-1}{x-1}.$$

The sum becomes

$$\sum_{x=1}^{n} \binom{n-1}{x-1} \theta^{x-1}(1-\theta)^{n-x} = \sum_{\nu=0}^{n-1} \binom{n-1}{\nu} \theta^{\nu}(1-\theta)^{n-\nu-1}$$

$$= (\theta+(1-\theta))^{n-1} = 1\,,$$

so that

$$\mathcal{M}\{x\} = n\theta\,. \tag{5.7.4}$$

Thus, the mean of x is equal to the number of observations multiplied by the probability θ of the occurrence of U in a single observation; this is the same as the number of times we "a priori" would expect U to occur on the average. The mean $n\theta$ is therefore also called *the expected number* of times U occurs in n observations.

5.8. The Mean of a Function of a Stochastic Variable.

Let $y = \varphi(x)$ be a monotonic function of x. In

$$\mathcal{M}\{y\} = \int_{-\infty}^{\infty} yp\{y\}dy$$

we substitute $y = \varphi(x)$ and (5.5.9), which gives

$$\boxed{\mathcal{M}\{\varphi(x)\} = \int_{-\infty}^{\infty} \varphi(x)p\{x\}dx .} \tag{5.8.1}$$

Thus, the mean of $y = \varphi(x)$ may be calculated directly from the distribution function of x.

For discontinuous distributions we have the corresponding formula

$$\mathcal{M}\{y\} = \sum_{i=-\infty}^{\infty} y_i p\{y_i\}$$

which may be written

$$\boxed{\mathcal{M}\{\varphi(x)\} = \sum_{i=-\infty}^{\infty} \varphi(x_i)p\{x_i\} ,} \tag{5.8.2}$$

since

$$p\{y_i\} = p\{x_i\} \quad \text{for} \quad y_i = \varphi(x_i) .$$

Let us consider the case where $\varphi(x) = (x-\xi)^r$ for $r = 2, 3, \ldots$. This leads to

$$\mu_r = \mathcal{M}\{(x-\xi)^r\} = \int_{-\infty}^{\infty} (x-\xi)^r p\{x\}\, dx \tag{5.8.3}$$

which is called *the moment of order r* of the distribution. The second order moment, the variance, is especially important and is treated in the next section.

If y is a linear function of x,

$$y = \alpha+\beta x ,$$

we get

$$\begin{aligned}\mathcal{M}\{y\} &= \int_{-\infty}^{\infty} (\alpha+\beta x)p\{x\}dx \\ &= \alpha\int_{-\infty}^{\infty} p\{x\}dx+\beta\int_{-\infty}^{\infty} xp\{x\}dx \\ &= \alpha+\beta\, \mathcal{M}\{x\} ,\end{aligned}$$

or

$$\boxed{\mathcal{M}\{\alpha+\beta x\} = \alpha+\beta\, \mathcal{M}\{x\} ,} \tag{5.8.4}$$

a result which is analogous to (4.3.8).

If $y = \varphi(x)$ *is approximately linear* over practically the whole range of the variation of x, it follows from TAYLOR'S formula that

$$y = \varphi(x) \simeq \varphi(\xi) + (x-\xi)\varphi'(\xi) .$$

Inserting this in (5.8.4) we get

$$\boxed{\mathcal{M}\{\varphi(x)\} \simeq \varphi(\mathcal{M}\{x\}) ,} \tag{5.8.5}$$

since

$$\mathcal{M}\{x-\xi\} = 0 .$$

The results (5.8.4) and (5.8.5) show that *the mean of a transformed variable is not given by the transformed value of the mean of the original variable unless the transformation is linear*, i. e., $\mathcal{M}\{\varphi(x)\} \neq \varphi(\mathcal{M}\{x\})$ unless $\varphi(x) = \alpha+\beta x$. The approximation formula (5.8.5), however, has important practical applications as shown in § 9.9.

If we consider the sum of two functions of the same variable

$$y = f_1(x)+f_2(x) ,$$

we get

$$\mathcal{M}\{y\} = \int_{-\infty}^{\infty} (f_1(x)+f_2(x))\, p\{x\}dx$$

$$= \int_{-\infty}^{\infty} f_1(x)\, p\{x\}dx + \int_{-\infty}^{\infty} f_2(x)\, p\{x\}dx ,$$

so that

$$\boxed{\mathcal{M}\{f_1(x)+f_2(x)\} = \mathcal{M}\{f_1(x)\}+\mathcal{M}\{f_2(x)\} .} \tag{5.8.6}$$

Analogous relations apply to discontinuous distributions.

5.9. The Variance of a Stochastic Variable.

The variance $\mathcal{V}\{x\}$ of a stochastic variable x with mean $\xi = \mathcal{M}\{x\}$ is defined by

$$\boxed{\mathcal{V}\{x\} = \mathcal{M}\{(x-\xi)^2\} ,} \tag{5.9.1}$$

so for continuous distributions

$$\mathcal{V}\{x\} = \int_{-\infty}^{\infty} (x-\xi)^2 p\{x\}dx \tag{5.9.2}$$

and for discontinuous distributions

$$\mathcal{V}\{x\} = \sum_{i=-\infty}^{\infty} (x_i-\xi)^2 p\{x_i\} . \tag{5.9.3}$$

Thus, the population variance is defined as the mean of the squared deviations from the population mean; compare with the corresponding expressions for the sample variance, (4.4.1) and (4.4.15). (The definition of the variance corresponds to the definition of the moment of inertia about the centre of gravity).

The positive value of the square root of the variance is called the *standard deviation* and is written σ or $\sigma\{x\}$. The variance may also be written σ^2 or $\sigma^2\{x\}$.

In most cases *Greek letters* are used for *parameters or population values* and corresponding *Roman letters* for their *estimates or sample values*, such as σ^2 and s^2.

The symbol $\mathcal{V}\{\ \}$ is read "the variance of" and—like $\mathcal{M}\{\ \}$—it is not a function of the variable but a function of the parameters of the distribution, as shown in Examples 5.7 and 5.8.

Squaring $(x-\xi)$ and using (5.8.6) we get

$$\begin{aligned}\mathcal{V}\{x\} &= \mathcal{M}\{x^2-2\xi x+\xi^2\}\\ &= \mathcal{M}\{x^2\}-2\xi\mathcal{M}\{x\}+\xi^2\ ,\end{aligned}$$

i. e.,

$$\boxed{\mathcal{V}\{x\} = \mathcal{M}\{x^2\}-\mathcal{M}^2\{x\}\ ,} \tag{5.9.4}$$

which is analogous to (4.4.2) divided by n. This formula is often used for calculating the variance of continuous distributions. For discontinuous distributions, it is often more convenient to calculate the variance from

$$\mathcal{V}\{x\} = \mathcal{M}\{x(x-1)\}-\mathcal{M}\{x\}(\mathcal{M}\{x\}-1)\ , \tag{5.9.5}$$

which is easily verified.

In the special case where y is a linear function of x

$$y = \alpha+\beta x\ ,$$

the deviation from the mean is equal to

$$\begin{aligned}y-\mathcal{M}\{y\} &= \alpha+\beta x-(\alpha+\beta\xi)\\ &= \beta(x-\xi)\ ,\end{aligned}$$

so that

$$\mathcal{M}\{(y-\mathcal{M}\{y\})^2\} = \beta^2\mathcal{M}\{(x-\xi)^2\}$$

or

$$\boxed{\mathcal{V}\{\alpha+\beta x\} = \beta^2\mathcal{V}\{x\}\ ,} \tag{5.9.6}$$

which is analogous to (4.4.10). Thus, *the standard deviation* is independent of the computing origin and is proportional to the computing unit.

If $y = \varphi(x)$ *is approximately linear* over practically the whole range of the variation of x, it follows from TAYLOR's formula that

$$y = \varphi(x) \simeq \varphi(\xi)+(x-\xi)\varphi'(\xi)\ ,$$

so that (5.9.6) leads to

$$\boxed{\mathcal{V}\{\varphi(x)\} \simeq (\varphi'(\xi))^2 \mathcal{V}\{x\}.} \tag{5.9.7}$$

If the variable is measured from the mean, using the standard deviation as unit, we get a new variable

$$\boxed{u = \frac{x - \mathcal{M}\{x\}}{\sigma\{x\}},} \tag{5.9.8}$$

which is called *the standardized variable.* It is easily seen that a *standardized variable has mean* 0 *and variance* 1, i. e.,

$$\mathcal{M}\{u\} = 0 \tag{5.9.9}$$

and

$$\mathcal{V}\{u\} = \mathcal{M}\{u^2\} = 1\,. \tag{5.9.10}$$

The *coefficient of variation* is defined as

$$\boxed{\mathcal{C}\,\{x\} = \frac{\sigma\{x\}}{\mathcal{M}\{x\}}} \tag{5.9.11}$$

or corresponding to $c = s/\bar{x}$ as

$$\gamma = \frac{\sigma}{\xi}. \tag{5.9.12}$$

Example 5.7. For the rectangular distribution we have

$$\mathcal{M}\{x^2\} = \int_{-\frac{\alpha}{2}}^{\frac{\alpha}{2}} x^2 \frac{1}{\alpha}\, dx = \frac{\alpha^2}{12},$$

and since $\mathcal{M}\{x\} = 0$,

$$\mathcal{V}\{x\} = \frac{\alpha^2}{12}. \tag{5.9.13}$$

Example 5.8. For the binomial distribution, we get

$$\begin{aligned} \mathcal{M}\{x(x-1)\} &= \sum_{x=0}^{n} x(x-1)\binom{n}{x}\theta^x(1-\theta)^{n-x} \\ &= n(n-1)\theta^2 \sum_{x=2}^{n} \binom{n-2}{x-2}\theta^{x-2}(1-\theta)^{n-x} \\ &= n(n-1)\theta^2 \cdot (\theta+(1-\theta))^{n-2} \\ &= n(n-1)\theta^2\,, \end{aligned}$$

whence from (5.9.5)

$$\begin{aligned} \mathcal{V}\{x\} &= n(n-1)\theta^2 - n\theta(n\theta-1) \\ &= n\theta(1-\theta)\,. \end{aligned} \tag{5.9.14}$$

If we take the frequency $\frac{x}{n}$ as our variable, it follows from (5.8.4) and (5.9.6) that

$$m\left\{\frac{x}{n}\right\} = \frac{1}{n}\cdot n\theta = \theta \tag{5.9.15}$$

and

$$v\left\{\frac{x}{n}\right\} = \frac{1}{n^2}\cdot n\theta(1-\theta) = \frac{\theta(1-\theta)}{n}. \tag{5.9.16}$$

5.10. Tchebycheff's Theorem.

From the distribution function of a stochastic variable can be derived the mean and the standard deviation, $m\{x\} = \xi$ and $\sigma\{x\} = \sigma$, as defined by the above formulas. TCHEBYCHEFF's theorem gives some information about an inverse problem: *To what extent do the mean and the standard deviation characterize the distribution when we do not know the mathematical form of the distribution function?* TCHEBYCHEFF's theorem says that *the probability of obtaining a value of the standardized variable which is numerically less than or equal to a specified number α is larger than $1-\frac{1}{\alpha^2}$*, i. e.,

$$P\{|u| \le \alpha\} = P\left\{\left|\frac{x-\xi}{\sigma}\right| \le \alpha\right\} > 1-\frac{1}{\alpha^2} \tag{5.10.1}$$

or

$$\boxed{P\{|x-\xi| \le \alpha\sigma\} > 1-\frac{1}{\alpha^2}.} \tag{5.10.2}$$

The required probability, P, can be expressed as

$$P\{|x-\xi| \le \alpha\sigma\} = \int_{|x-\xi|\le\alpha\sigma} p\{x\}\,dx = \int_{\xi-\alpha\sigma}^{\xi+\alpha\sigma} p\{x\}\,dx\,. \tag{5.10.3}$$

To find a lower bound for this probability we partition the integral defining σ^2 in the following way

$$\sigma^2 = \int_{|x-\xi|\le\alpha\sigma} (x-\xi)^2 p\{x\}dx + \int_{|x-\xi|>\alpha\sigma} (x-\xi)^2 p\{x\}dx \tag{5.10.4}$$

and use the inequalities

$$\int_{|x-\xi|\le\alpha\sigma} (x-\xi)^2 p\{x\}dx \ge 0 \tag{5.10.5}$$

and

$$\int_{|x-\xi|>\alpha\sigma} (x-\xi)^2 p\{x\}dx > \alpha^2\sigma^2 \int_{|x-\xi|>\alpha\sigma} p\{x\}dx = \alpha^2\sigma^2(1-P)\,. \tag{5.10.6}$$

Inserting (5.10.5) and (5.10.6) into (5.10.4) we get

$$\sigma^2 > \alpha^2\sigma^2(1-P)$$

from which follows

$$P > 1-\frac{1}{\alpha^2}.$$

A similar proof may be given for discontinuous distributions.

The following table gives some examples of values of $1-\frac{1}{\alpha^2}$.

TABLE 5.1.

$P\{\lvert x-\xi\rvert \leq \alpha\sigma\} > 1-\frac{1}{\alpha^2}$			
α	$1-\frac{1}{\alpha^2}$	α	$1-\frac{1}{\alpha^2}$
1·0	0·000	1·41	0·50
1·5	0·556	1·58	0·60
2·0	0·750	1·83	0·70
2·5	0·840	2·24	0·80
3·0	0·889	3·16	0·90
3·5	0·918	4·47	0·95
4·0	0·938	10·00	0·99
4·5	0·951		
5·0	0·960		

TWO-DIMENSIONAL DISTRIBUTIONS

5.11. The Cumulative Distribution Function.

A pair of stochastic variables (x, y) are represented graphically as a point in a rectangular coordinate system. The probability that a point (x, y) belongs to the region in the plane defined by the inequalities $x \leq x_a$ and $y \leq y_a$ is written

$$\boxed{P\{x \leq x_a, y \leq y_a\} = P\{x_a, y_a\}.} \tag{5.11.1}$$

The *cumulative distribution function*, $P\{x, y\}$, fully describes the distribution of the variables, $P\{x, y\}$ *denoting the probability that the variables will take on values smaller than or equal to* (x, y).

According to the addition formula the probability that (x, y) lies within the rectangular domain defined by $x_a < x \leq x_b$ and $y_a < y \leq y_b$ is

$$P\{x_a < x \leq x_b, y_a < y \leq y_b\} = P\{x_b, y_b\}-P\{x_a, y_b\}-P\{x_b, y_a\}+P\{x_a, y_a\}.$$

From the addition formula and axiom II, § 1.3, it follows that

1. $P\{x, y\}$ never decreases for increasing values of x and fixed y.
2. $P\{x, y\}$ never decreases for increasing values of y and fixed x.
3. $P\{x_a, y_a\} \leq P\{x_b, y_b\}$ for $x_a < x_b$ and $y_a < y_b$, and $P\{\infty, \infty\} = 1$.

5.12. Continuous Distributions.

If a *function*, $p\{x, y\} \geq 0$, exists and is *continuous*, except possibly at certain points on at most a finite number of curves, and

$$P\{x, y\} = \int_{-\infty}^{x} \int_{-\infty}^{y} p\{x, y\} dx\, dy, \tag{5.12.1}$$

then the distribution is called *continuous*. The function $p\{x, y\}$ is called the *distribution function*. The definition (5.12.1) implies that the probability element

$$dP\{x, y\} = p\{x, y\} dx\, dy \tag{5.12.2}$$

denotes the probability that the variables will take on values in the corresponding rectangle with sides (dx, dy).

Analogous to (5.2.4) we get

$$\int_{-\infty}^{\infty} \int_{-\infty}^{\infty} p\{x, y\} dx\, dy = 1 . \tag{5.12.3}$$

The distribution function is represented by a surface

$$z = p\{x, y\} ,$$

and the probability that (x, y) lies within a certain domain of the (x, y) plane is represented by the volume bounded by the surface with the corresponding area of the (x, y) plane as base, cf. the histogram of the empirical distribution in Fig. 3.18. p. 64.

According to the addition formula, *the marginal distribution function*, $p\{x\}$, is

$$p\{x\} = \int_{-\infty}^{\infty} p\{x, y\} dy . \tag{5.12.4}$$

The probability element $p\{x\}dx$ denotes the probability that the first variable will take on a value between x and $x+dx$, irrespective of the value of the other variable. Correspondingly, the marginal distribution function of y is

$$p\{y\} = \int_{-\infty}^{\infty} p\{x, y\} dx . \tag{5.12.5}$$

The *conditional distribution functions* are written $p\{x|y\}$ and $p\{y|x\}$. According to the multiplication formula, we have

$$p\{x, y\}\,dx\,dy = p\{x\}\,dx \cdot p\{y|x\}\,dy = p\{y\}\,dy \cdot p\{x|y\}\,dx\,,$$

i. e.,

$$\boxed{p\{x, y\} = p\{x\} \cdot p\{y|x\} = p\{y\} \cdot p\{x|y\}\,.} \tag{5.12.6}$$

If $p\{x\} \neq 0$ and $p\{y\} \neq 0$, we get

$$p\{x|y\} = \frac{p\{x, y\}}{p\{y\}} \tag{5.12.7}$$

and

$$p\{y|x\} = \frac{p\{x, y\}}{p\{x\}}. \tag{5.12.8}$$

If

$$p\{x|y\} = p\{x\}$$

for all values of y, (5.12.6) leads to

$$p\{y|x\} = p\{y\}\,,$$

i. e., x and y are *stochastically independent*. If this is the case, the two-dimensional distribution is completely determined by the marginal distributions, since

$$p\{x, y\} = p\{x\} \cdot p\{y\}\,. \tag{5.12.9}$$

Example 5.9. *The two-dimensional normal distribution.*

Let x and y be normally distributed with distribution functions

$$p\{x\} = \frac{1}{\sqrt{2\pi}\,\sigma_x} e^{-\frac{(x-\xi)^2}{2\sigma_x^2}} \qquad (-\infty < x < \infty)$$

and

$$p\{y\} = \frac{1}{\sqrt{2\pi}\,\sigma_y} e^{-\frac{(y-\eta)^2}{2\sigma_y^2}} \qquad (-\infty < y < \infty).$$

If x and y are stochastically independent, it follows from (5.12.9) that

$$p\{x, y\} = \frac{1}{2\pi\,\sigma_x\sigma_y} e^{-\frac{1}{2}\left(\left(\frac{x-\xi}{\sigma_x}\right)^2 + \left(\frac{y-\eta}{\sigma_y}\right)^2\right)}. \tag{5.12.10}$$

This distribution is a special case of the two-dimensional normal distribution

$$p\{x, y\} = \frac{1}{2\pi\,\sigma_x\sigma_y\sqrt{1-\varrho^2}} e^{-\frac{1}{2(1-\varrho^2)}\left(\left(\frac{x-\xi}{\sigma_x}\right)^2 - 2\varrho\frac{x-\xi}{\sigma_x}\frac{y-\eta}{\sigma_y} + \left(\frac{y-\eta}{\sigma_y}\right)^2\right)}. \tag{5.12.11}$$

If we put $\varrho = 0$, we get (5.12.10). The two-dimensional normal distribution is dealt with in Chapter 19.

5.13. Discontinuous Distributions.

Let the possible values of the variable (x, y) be of the form (x_i, y_j) where

$$\ldots < x_{-2} < x_{-1} < x_0 < x_1 < x_2 < \ldots$$

$$\ldots < y_{-2} < y_{-1} < y_0 < y_1 < y_2 < \ldots$$

The discontinuous distribution function $p\{x_i, y_j\}$ *denotes the probability that the variables will take on the values* (x_i, y_j).

By means of the addition formula we get *the cumulative distribution function*

$$P\{x_i, y_j\} = \sum_{\nu=-\infty}^{i} \sum_{\mu=-\infty}^{j} p\{x_\nu, y_\mu\} \tag{5.13.1}$$

and

$$\sum_{\nu=-\infty}^{\infty} \sum_{\mu=-\infty}^{\infty} p\{x_\nu, y_\mu\} = 1\,. \tag{5.13.2}$$

The distribution of (x_1, x_2) in Example 2.4, p. 37, is an example of a two-dimensional, discontinuous distribution.

Using the addition formula we get the *marginal distribution functions*

$$\boxed{p\{x_i\} = \sum_{j=-\infty}^{\infty} p\{x_i, y_j\}\,,} \tag{5.13.3}$$

and

$$\boxed{p\{y_j\} = \sum_{i=-\infty}^{\infty} p\{x_i, y_j\}\,.} \tag{5.13.4}$$

The *conditional distribution functions* are written $p\{x_i|y_j\}$ and $p\{y_j|x_i\}$. Using the multiplication formula, we find

$$\boxed{p\{x_i, y_j\} = p\{x_i\} \cdot p\{y_j|x_i\} = p\{y_j\}\, p\{x_i|y_j\}\,,} \tag{5.13.5}$$

and hence

$$p\{x_i|y_j\} = \frac{p\{x_i, y_j\}}{p\{y_j\}}$$

and

$$p\{y_j|x_i\} = \frac{p\{x_i, y_j\}}{p\{x_i\}}$$

assuming that $p\{x_i\} \neq 0$ and $p\{y_j\} \neq 0$.

If

$$p\{x_i|y_j\} = p\{x_i\}$$

for all values of j, (5.13.5) leads to

$$p\{y_j|x_i\} = p\{y_j\}\,,$$

i. e., x and y are stochastically independent. In this case we have

$$p\{x_i, y_j\} = p\{x_i\} \cdot p\{y_j\}\,, \tag{5.13.6}$$

the two-dimensional distribution being completely determined by the two marginal distributions.

5.14. Marginal Means and Variances.

The following definitions and formulas hold for continuous distributions. The corresponding results for discontinuous distributions are obtained by substituting sums for integrals.

The marginal means and variances are defined as the means and variances of the marginal distributions:

$$\mathcal{M}\{x\} = \int_{-\infty}^{\infty} x p\{x\}\,dx = \xi\,, \tag{5.14.1}$$

$$\mathcal{M}\{y\} = \int_{-\infty}^{\infty} y p\{y\}\,dy = \eta\,, \tag{5.14.2}$$

$$\mathcal{V}\{x\} = \int_{-\infty}^{\infty} (x-\xi)^2 p\{x\}\,dx = \sigma^2\{x\}\,, \tag{5.14.3}$$

and

$$\mathcal{V}\{y\} = \int_{-\infty}^{\infty} (y-\eta)^2 p\{y\}\,dy = \sigma^2\{y\}\,. \tag{5.14.4}$$

5.15. Conditional Means and Variances.

The conditional means are defined as the means of the conditional distributions:

$$\boxed{\mathcal{M}\{x|y\} = \int_{-\infty}^{\infty} x p\{x|y\}\,dx} \tag{5.15.1}$$

and

$$\boxed{\mathcal{M}\{y|x\} = \int_{-\infty}^{\infty} y p\{y|x\}\,dy,} \tag{5.15.2}$$

cf. the corresponding empirical definitions in § 4.7. To each value of y corresponds a mean value of x, i. e., *$\mathcal{M}\{x|y\}$ is a function of y*. This function is called *the regression of x on y*. An example of an empirical regression is given on p. 84.

Correspondingly, the conditional variances are defined as

$$\boxed{\mathcal{V}\{x|y\} = \int_{-\infty}^{\infty} (x-\mathcal{M}\{x|y\})^2 p\{x|y\}\,dx} \tag{5.15.3}$$

and

$$\mathcal{V}\{y|x\} = \int_{-\infty}^{\infty} (y - \mathcal{M}\{y|x\})^2 p\{y|x\} dy . \tag{5.15.4}$$

Thus, $\mathcal{V}\{x|y\}$ is the variance of x in the conditional distribution $p\{x|y\}$ and therefore, like $\mathcal{M}\{x|y\}$, it is a function of y.

The formulas valid for marginal means and variances are also valid for the conditional means and variances, e. g.,

$$\mathcal{V}\{x|y\} = \mathcal{M}\{x^2|y\} - (\mathcal{M}\{x|y\})^2 ,$$

cf. (5.9.4).

If the variables are stochastically independent, i. e., $p\{x|y\} = p\{x\}$ and $p\{y|x\} = p\{y\}$, *then the conditional means and variances are equal to the corresponding marginal values.*

5.16. The Covariance.

The function

$$z = \varphi(x, y)$$

is a stochastic variable, the distribution of which may be derived from the two-dimensional distribution of (x, y) in the same manner as the distribution of $y = \varphi(x)$ is derived from the distribution of x in § 5.5. Correspondingly it may be shown that[1])

$$\mathcal{M}\{z\} = \int_{-\infty}^{\infty} \int_{-\infty}^{\infty} \varphi(x, y)\, p\{x, y\}\, dx\, dy , \tag{5.16.1}$$

cf. (5.8.1).

If we introduce $z_1 = \varphi_1(x, y)$ and $z_2 = \varphi_2(x, y)$ and put

$$z = \varphi(x, y) = \varphi_1(x, y) + \varphi_2(x, y)$$

then

$$\mathcal{M}\{z\} = \int_{-\infty}^{\infty} \int_{-\infty}^{\infty} (\varphi_1(x, y) + \varphi_2(x, y))\, p\{x, y\}\, dx\, dy$$

$$= \int_{-\infty}^{\infty} \int_{-\infty}^{\infty} \varphi_1(x, y)\, p\{x, y\}\, dx\, dy + \int_{-\infty}^{\infty} \int_{-\infty}^{\infty} \varphi_2(x, y)\, p\{x, y\}\, dx\, dy ,$$

i. e.,

$$\mathcal{M}\{z_1 + z_2\} = \mathcal{M}\{z_1\} + \mathcal{M}\{z_2\} , \tag{5.16.2}$$

which means that *the mean of the sum of two variables is equal to the sum of the means of the variables.*

---

[1]) See, e. g., H. CRAMÉR: *Random Variables and Probability Distributions*, Cambridge, 1937, pp. 19—20.

The *covariance*, $\mathcal{V}(x, y)$, is defined by

$$\boxed{\mathcal{V}\{x, y\} = \mathcal{M}\{(x-\xi)(y-\eta)\},} \tag{5.16.3}$$

where $\xi = \mathcal{M}\{x\}$ and $\eta = \mathcal{M}\{y\}$, cf. the definition of the empirical covariance (4.8.1) and (4.8.14).

If we introduce

$$(x-\xi)(y-\eta) = xy-\xi y-\eta x+\xi\eta ,$$

(5.16.2) leads to

$$\boxed{\mathcal{V}\{x, y\} = \mathcal{M}\{xy\}-\mathcal{M}\{x\}\mathcal{M}\{y\},} \tag{5.16.4}$$

analogous to (4.8.2) divided by n.

Let us put

$$x = \alpha+\beta v$$

$$y = \gamma+\varkappa w ,$$

and we get

$$(x-\mathcal{M}\{x\})(y-\mathcal{M}\{y\}) = \beta\varkappa(v-\mathcal{M}\{v\})(w-\mathcal{M}\{w\})$$

hence

$$\mathcal{V}\{x, y\} = \beta\varkappa\mathcal{V}\{v, w\} , \tag{5.16.5}$$

which is analogous to (4.8.6).

If we choose the special case

$$x = \mathcal{M}\{x\}+\sigma\{x\}\cdot v$$

$$y = \mathcal{M}\{y\}+\sigma\{y\}\cdot w ,$$

we have

$$\mathcal{V}\{x, y\} = \sigma\{x\}\sigma\{y\}\,\mathcal{V}\{v, w\}$$

or

$$\mathcal{V}\left\{\frac{x-\xi}{\sigma\{x\}}, \frac{y-\eta}{\sigma\{y\}}\right\} = \frac{\mathcal{V}\{x, y\}}{\sqrt{\mathcal{V}\{x\}\mathcal{V}\{y\}}} . \tag{5.16.6}$$

This covariance of the two standardized variables is called the correlation coefficient of (x, y) *and is usually written* $\varrho = \varrho_{xy}$, compare with (4.8.15).

From (5.16.4) we get

$$\begin{aligned}\mathcal{M}\{xy\} &= \mathcal{M}\{x\}\mathcal{M}\{y\}+\mathcal{V}\{x, y\}\\ &= \mathcal{M}\{x\}\mathcal{M}\{y\}+\varrho_{xy}\sigma_x\sigma_y ,\end{aligned} \tag{5.16.7}$$

which means that *the mean of the product of two stochastic variables is different from the product of the means unless the coefficient of correlation* (*or the covariance*) *is zero.*

This condition, at any rate, is fulfilled for two stochastically independent variables since

$$\mathcal{V}\{x, y\} = \int_{-\infty}^{\infty}\int_{-\infty}^{\infty}(x-\xi)(y-\eta)\,p\{x\}\,p\{y\}\,dx\,dy$$

$$= \int_{-\infty}^{\infty}(x-\xi)\,p\{x\}\,dx \int_{-\infty}^{\infty}(y-\eta)\,p\{y\}\,dy = 0\,,$$

as each of the two integrals has the value 0. In this case we get

$$\mathcal{M}\{xy\} = \mathcal{M}\{x\}\,\mathcal{M}\{y\}\,, \tag{5.16.8}$$

i. e., *the mean of the product of two stochastically independent variables is equal to the product of their means.*

The covariance (or correlation coefficient) is specially important in the case of the two-dimensional normal distribution, as this distribution has 5 *parameters, which are equal to the marginal means and variances and the covariance. In this special case, the above theorem may be reversed*; i. e., $\varrho_{xy} = 0$ *implies that the variables are stochastically independent*, see Example 5.9 and § 19.5.

5.17. The Mean and Variance of a Linear Function of Stochastic Variables.

In the particular case when

$$z = x+y$$

(5.16.2) leads to

$$\mathcal{M}\{z\} = \mathcal{M}\{x\}+\mathcal{M}\{y\} \tag{5.17.1}$$

or

$$\zeta = \mathcal{M}\{z\} = \xi+\eta\,.$$

The variance of z is the mean value of

$$(z-\zeta)^2 = \big((x+y)-(\xi+\eta)\big)^2$$

$$= (x-\xi)^2+(y-\eta)^2+2(x-\xi)(y-\eta)\,,$$

and hence

$$\mathcal{V}\{z\} = \mathcal{M}\{(x-\xi)^2\}+\mathcal{M}\{(y-\eta)^2\}+2\mathcal{M}\{(x-\xi)(y-\eta)\}$$

or

$$\mathcal{V}\{z\} = \mathcal{V}\{x\}+\mathcal{V}\{y\}+2\mathcal{V}\{x, y\}\,, \tag{5.17.2}$$

cf. the corresponding empirical relations (4.8.11).

These results may be directly generalized. Let us put

$$z = \alpha_0+\alpha_1x_1+\alpha_2x_2+\ldots+\alpha_kx_k\,,$$

where $x_1, x_2, \ldots, x_k$ are k stochastic variables, and $\alpha_0, \alpha_1, \ldots, \alpha_k$ are constants. This leads to

$$\boxed{\mathcal{M}\{\alpha_0+\alpha_1x_1+\ldots+\alpha_kx_k\} = \alpha_0+\alpha_1\mathcal{M}\{x_1\}+\ldots+\alpha_k\mathcal{M}\{x_k\}} \tag{5.17.3}$$

and

$$\begin{aligned} \mathcal{V}\{\alpha_0+\alpha_1x_1+\ldots+\alpha_kx_k\} = \alpha_1^2\mathcal{V}\{x_1\}+\ldots+\alpha_k^2\mathcal{V}\{x_k\} \\ +2\alpha_1\alpha_2\mathcal{V}\{x_1, x_2\}+2\alpha_1\alpha_3\mathcal{V}\{x_1, x_3\}+\ldots \\ +2\alpha_{k-1}\alpha_k\mathcal{V}\{x_{k-1}, x_k\}. \end{aligned} \tag{5.17.4}$$

For

$$\mathcal{V}\{x_i, x_j\} = 0 \quad \text{for} \quad i \neq j ,$$

which at any rate is true *if any two variables are stochastically independent* we get

$$\mathcal{V}\{\alpha_0+\alpha_1x_1+\ldots+\alpha_kx_k\} = \alpha_1^2\mathcal{V}\{x_1\}+\ldots+\alpha_k^2\mathcal{V}\{x_k\} . \tag{5.17.5}$$

If $z = \varphi(x_1,\ldots, x_k)$ is approximately linear over practically the whole range of the variation of $(x_1, \ldots, x_k)$, TAYLOR's formula leads to

$$z = \varphi(x_1, \ldots, x_k) \simeq \varphi(\xi_1, \ldots, \xi_k)+(x_1-\xi_1)\frac{\partial\varphi}{\partial x_1}+\ldots+(x_k-\xi_k)\frac{\partial\varphi}{\partial x_k},$$

where

$$\frac{\partial\varphi}{\partial x_i}=\left[\frac{\partial\varphi(x_1, \ldots, x_k)}{\partial x_i}\right]_{(x_1, \ldots, x_k) = (\xi_1, \ldots, \xi_k)} .$$

Using (5.17.3) and (5.17.4), we get

$$\mathcal{M}\{\varphi(x_1, \ldots, x_k)\} \simeq \varphi(\xi_1, \ldots, \xi_k) \tag{5.17.6}$$

and

$$\begin{aligned} \mathcal{V}\{\varphi(x_1, \ldots, x_k)\} \simeq \left(\frac{\partial\varphi}{\partial x_1}\right)^2 \mathcal{V}\{x_1\}+\ldots+\left(\frac{\partial\varphi}{\partial x_k}\right)^2 \mathcal{V}\{x_k\} \\ +2\frac{\partial\varphi}{\partial x_1}\frac{\partial\varphi}{\partial x_2}\mathcal{V}\{x_1, x_2\}+\ldots+2\frac{\partial\varphi}{\partial x_{k-1}}\frac{\partial\varphi}{\partial x_k}\mathcal{V}\{x_{k-1}, x_k\}, \end{aligned} \tag{5.17.7}$$

see § 9.9 regarding applications of this result.

CHAPTER 6.

THE NORMAL DISTRIBUTION

6.1. Definitions.

The *normal distribution function* is defined by

$$p\{x\} = \frac{1}{\sqrt{2\pi}\,\sigma} e^{-\frac{(x-\xi)^2}{2\sigma^2}} \qquad (-\infty < x < \infty). \qquad (6.1.1)$$

Thus, the probability that the variable is less than or equal to x is

$$P\{x\} = \frac{1}{\sqrt{2\pi}\,\sigma} \int_{-\infty}^{x} e^{-\frac{(t-\xi)^2}{2\sigma^2}} dt. \qquad (6.1.2)$$

The constants ξ and σ^2 are called the mean and the variance or *the parameters of the distribution.*

A variable with distribution function (6.1.1) is called *normally distributed with parameters* (ξ, σ^2).

According to (5.2.4) the constant $\frac{1}{\sqrt{2\pi}\,\sigma}$ has been determined so that the probability of the variable falling within the interval $(-\infty < x < \infty)$ is equal to 1, i. e.,

$$\frac{1}{\sqrt{2\pi}\,\sigma} \int_{-\infty}^{\infty} e^{-\frac{(t-\xi)^2}{2\sigma^2}} dt = 1. \qquad (6.1.3)$$

Proof. Instead of proving directly that

$$I = \frac{1}{\sqrt{2\pi}\,\sigma} \int_{-\infty}^{\infty} e^{-\frac{(t-\xi)^2}{2\sigma^2}} dt = 1,$$

it is simpler to prove that $I^2 = 1$. If we introduce a new variable

$$u = \frac{t-\xi}{\sigma},$$

we get

$$I = \frac{1}{\sqrt{2\pi}} \int_{-\infty}^{\infty} e^{-\frac{u^2}{2}} du,$$

and

$$I^2 = \frac{1}{\sqrt{2\pi}}\int_{-\infty}^{\infty} e^{-\frac{u^2}{2}}\, du \cdot \frac{1}{\sqrt{2\pi}}\int_{-\infty}^{\infty} e^{-\frac{v^2}{2}}\, dv$$

$$= \frac{1}{2\pi}\int_{-\infty}^{\infty}\int_{-\infty}^{\infty} e^{-\frac{1}{2}(u^2+v^2)}\, du\, dv\,.$$

In this integral we introduce new variables r and φ (polar coordinates) from the equations

$$u = r\cos\varphi$$

$$v = r\sin\varphi\,,$$

so that

$$u^2+v^2 = r^2$$

and the element of area corresponding to $du\, dv$ is $r\, dr\, d\varphi$ [1]). This leads to the result

$$I^2 = \frac{1}{2\pi}\int_{r=0}^{\infty}\int_{\varphi=0}^{2\pi} e^{-\frac{r^2}{2}}\, r\, dr\, d\varphi$$

$$= \int_0^{\infty} r e^{-\frac{r^2}{2}}\, dr \cdot \frac{1}{2\pi}\int_0^{2\pi} d\varphi$$

$$= \left[-e^{-\frac{r^2}{2}}\right]_{r=0}^{\infty} = 1\,.$$

6.2. The Parameters of the Distribution.

The parameter ξ is equal to the mean of the variable, since

$$\mathcal{M}\{x\} = \frac{1}{\sqrt{2\pi}\,\sigma}\int_{-\infty}^{\infty} x\, e^{-\frac{(x-\xi)^2}{2\sigma^2}}\, dx$$

$$= \xi + \frac{1}{\sqrt{2\pi}\,\sigma}\int_{-\infty}^{\infty} (x-\xi) e^{-\frac{(x-\xi)^2}{2\sigma^2}}\, dx = \xi\,. \tag{6.2.1}$$

The integral disappears, as the negative and positive values of $(x-\xi)$ occur with equal weights.

The parameter σ^2 is equal to the variance of the variable:

$$\mathcal{V}\{x\} = \mathcal{M}\{(x-\xi)^2\} = \int_{-\infty}^{\infty} (x-\xi)^2\, p\{x\}\, dx = \sigma^2\,. \tag{6.2.2}$$

Proof. Making the transformation

$$u = \frac{x-\xi}{\sigma}\,,$$

[1]) See section on volume integration in any standard textbook on calculus.

we get

$$\mathcal{V}\{x\} = \frac{1}{\sqrt{2\pi}\,\sigma}\int_{-\infty}^{\infty}(x-\xi)^2 e^{-\frac{(x-\xi)^2}{2\sigma^2}}dx$$

$$= \frac{\sigma^2}{\sqrt{2\pi}}\int_{-\infty}^{\infty}u^2 e^{-\frac{u^2}{2}}du$$

$$= \frac{\sigma^2}{\sqrt{2\pi}}\int_{-\infty}^{\infty}u\,d\left(-e^{-\frac{u^2}{2}}\right)$$

$$= \frac{\sigma^2}{\sqrt{2\pi}}\left[-ue^{-\frac{u^2}{2}}\right]_{-\infty}^{\infty} + \frac{\sigma^2}{\sqrt{2\pi}}\int_{-\infty}^{\infty}e^{-\frac{u^2}{2}}du = \sigma^2,$$

since

$$ue^{-\frac{u^2}{2}} \to 0 \quad \text{for} \quad |u| \to \infty,$$

and

$$\frac{1}{\sqrt{2\pi}}\int_{-\infty}^{\infty}e^{-\frac{u^2}{2}}du = 1$$

as follows from (6.1.3) for $\xi = 0$ and $\sigma = 1$.

6.3. The Standardized Normal Distribution. The u-Distribution.

The variable

$$\boxed{u = \frac{x-\xi}{\sigma}} \tag{6.3.1}$$

is a standardized variable, see (5.9.8). We have

$$\mathcal{M}\{u\} = 0, \tag{6.3.2}$$

and

$$\mathcal{V}\{u\} = 1. \tag{6.3.3}$$

According to (5.5.9) the distribution function of u is

$$p\{u\} = \left[p\{x\}\frac{dx}{du}\right]_{x=\xi+u\sigma}$$

$$= \frac{1}{\sqrt{2\pi}}e^{-\frac{u^2}{2}},$$

i. e., the variable u is *normally distributed with mean* 0 *and variance* 1.

The function

$$\boxed{\varphi(u) = \frac{1}{\sqrt{2\pi}}e^{-\frac{u^2}{2}}} \tag{6.3.4}$$

is called *the standardized normal distribution function.* A variable which is distributed according to (6.3.4) will be denoted by the letter u.

The standardized distribution function is illustrated in Fig. 6.1. The curve is symmetrical about $u = 0$, since

$$\boxed{\varphi(-u) = \varphi(u)\,.} \tag{6.3.5}$$

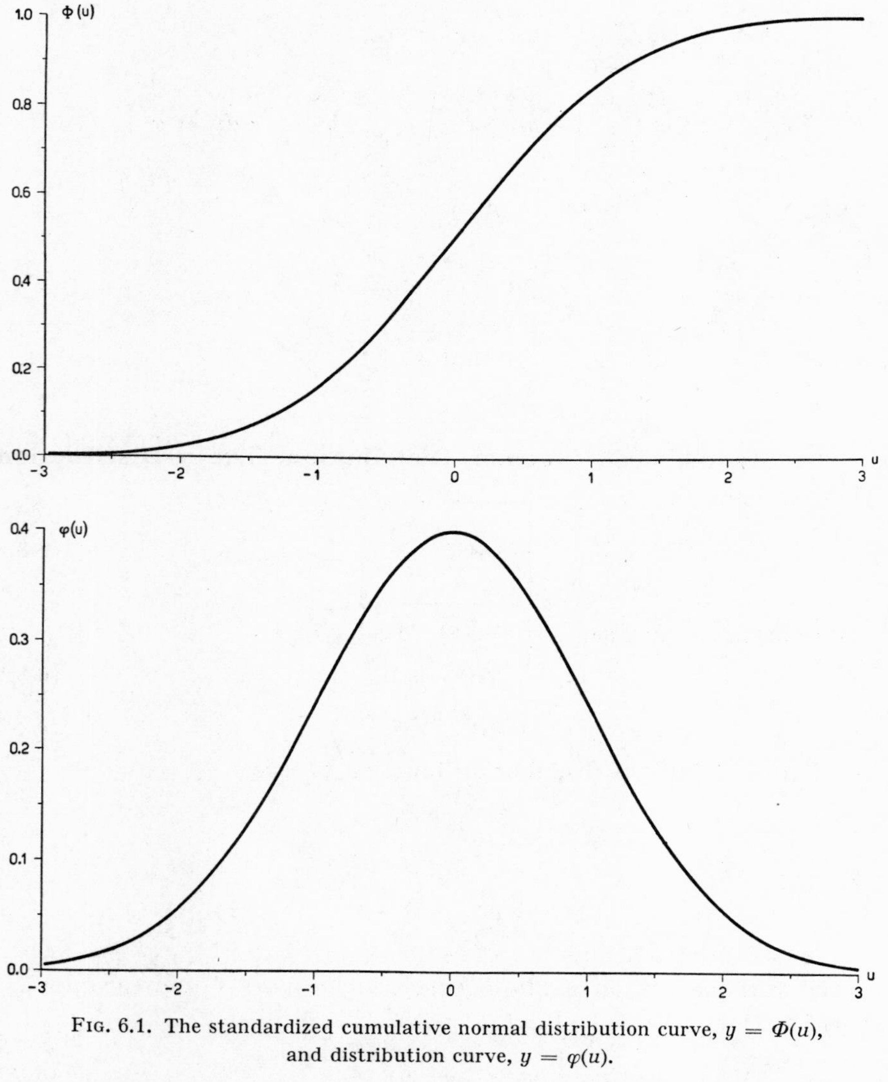

FIG. 6.1. The standardized cumulative normal distribution curve, $y = \Phi(u)$, and distribution curve, $y = \varphi(u)$.

The slope of the tangent is obtained by differentiating (6.3.4):

$$\varphi'(u) = \frac{1}{\sqrt{2\pi}} e^{-\frac{u^2}{2}} (-u) = -u\varphi(u)\,. \tag{6.3.6}$$

As $\varphi(u) > 0$, the slope of the tangent is positive for $u < 0$, 0 for $u = 0$, and negative for $u > 0$. Thus, the distribution function has a maximum for $u = 0$, and the maximum ordinate is

$$\varphi(0) = \frac{1}{\sqrt{2\pi}} = 0{\cdot}3989.$$

Differentiation of (6.3.6) leads to

$$\varphi''(u) = (u^2 - 1)\varphi(u) , \tag{6.3.7}$$

which is 0 for $u = \pm 1$. Thus, the distribution curve has points of inflection for these values of u, see Fig. 6.1.

In Table I the function $\varphi(u)$ has been tabulated to four significant figures.

The standardized cumulative distribution function, $\Phi(u)$, is equal to

$$\boxed{\Phi(u) = \frac{1}{\sqrt{2\pi}} \int_{-\infty}^{u} e^{-\frac{u^2}{2}} du.} \tag{6.3.8}$$

This function is illustrated in Fig. 6.1. The derivatives of the first and second order are

$$\Phi'(u) = \varphi(u) \tag{6.3.9}$$

and

$$\Phi''(u) = \varphi'(u) = -u\varphi(u) , \tag{6.3.10}$$

cf. (6.3.6). Thus, the cumulative distribution curve increases monotonically from 0 to 1 with a point of inflection at $u = 0$.

TABLE 6.1.

The standardized normal distribution.
The cumulative distribution function.

u	$\Phi(u)$	u	$\Phi(u)$
— 3·29	0·0005	0·00	0·50
— 3·09	0·001	0·25	0·60
— 2·58	0·005	0·52	0·70
— 2·33	0·01	0·84	0·80
— 1·96	0·025	1·28	0·90
— 1·64	0·05	1·64	0·95
— 1·28	0·10	1·96	0·975
— 0·84	0·20	2·33	0·99
— 0·52	0·30	2·58	0·995
— 0·25	0·40	3·09	0·999
— 0·00	0·50	3·29	0·9995

From the symmetry of the distribution function it follows that

$$\Phi(u)+\Phi(-u) = 1\,. \tag{6.3.11}$$

The cumulative distribution function has been tabulated to four significant figures in Table II. Table 6.1 shows some characteristic values of the function.

6.4. The Distribution Function.

The distribution function $p\{x\}$ may be expressed by the standardized distribution function $\varphi(u)$, since

$$p\{x\} = \frac{1}{\sigma}\varphi(u) \quad \text{for} \quad u = \frac{x-\xi}{\sigma}. \tag{6.4.1}$$

When transforming the distribution curve $y = \varphi(u)$ to the curve $y = p\{x\}$, the abscissas are first multiplied by σ and the ordinates divided by σ, and then the curve is displaced to the point of symmetry $x = \xi$, as shown in Fig. 6.2.

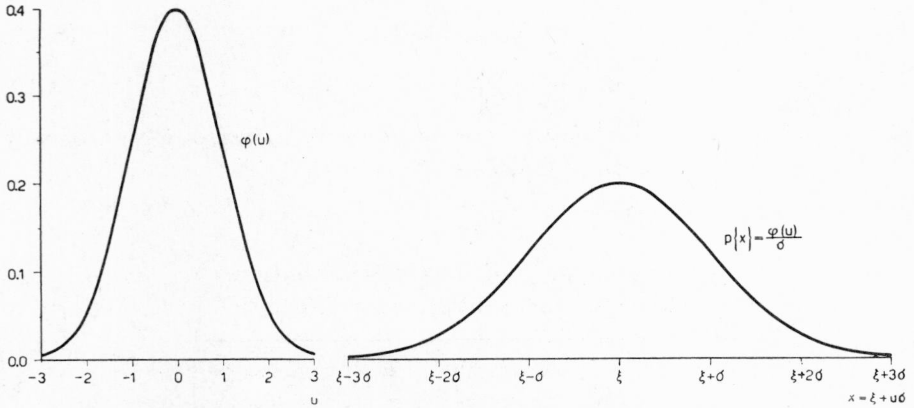

FIG. 6.2. Transformation of the standardized distribution curve $y = \varphi(u)$ to the distribution curve $y = p\{x\}$ with mean ξ and standard deviation σ.

The derivatives of $p\{x\}$ with respect to x are, cf. (6.3.6) and (6.3.7),

$$p'\{x\} = -\frac{1}{\sigma^2}u\varphi(u) \tag{6.4.2}$$

and

$$p''\{x\} = \frac{1}{\sigma^3}(u^2-1)\varphi(u)\,. \tag{6.4.3}$$

Thus, the maximum of the distribution curve corresponds to $x = \xi$, and the points of inflection to $x = \xi-\sigma$ and $x = \xi+\sigma$.

Fig. 6.2 also shows how the shape of the distribution curve depends upon the standard deviation. The larger the standard deviation, the flatter the curve.

For a linear function, $y = \alpha+\beta x$, the use of (6.3.1) leads to

$$y = \alpha+\beta\xi+\beta\sigma u \, , \tag{6.4.4}$$

i. e., y is normally distributed with parameters $(\alpha+\beta\xi, \beta^2\sigma^2)$.

6.5. The Cumulative Distribution Function.

The cumulative distribution function $P\{x\}$ can be expressed in terms of the standardized cumulative distribution function $\Phi(u)$ as

$$P\{x\} = \frac{1}{\sqrt{2\pi}\,\sigma}\int_{-\infty}^{x} e^{-\frac{(t-\xi)^2}{2\sigma^2}}\,dt$$

$$= \frac{1}{\sqrt{2\pi}}\int_{-\infty}^{\frac{x-\xi}{\sigma}} e^{-\frac{u^2}{2}}\,du$$

$$= \Phi\left(\frac{x-\xi}{\sigma}\right),$$

or

$$\boxed{P\{x\} = \Phi(u) \quad \text{for} \quad u = \frac{x-\xi}{\sigma}.} \tag{6.5.1}$$

Fig. 6.3 shows the transformation of the curve $y = \Phi(u)$ to $y = P\{x\}$, as given by (6.5.1). When transforming we first multiply the abscissas by σ, and then displace the curve to the point of symmetry $x = \xi$. The larger the standard deviation, the flatter the curve.

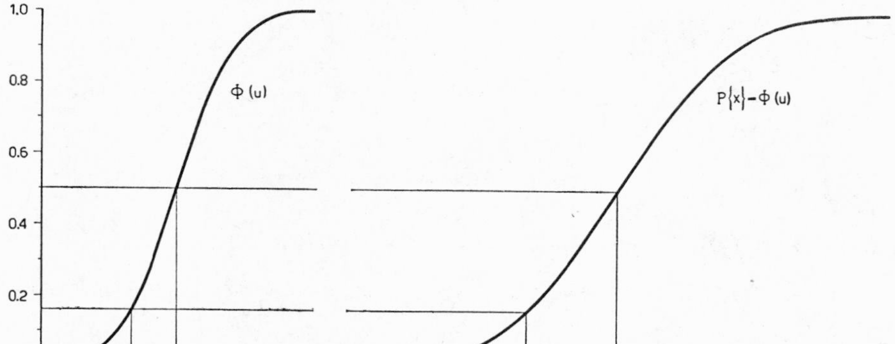

FIG. 6.3. Transformation of the standardized cumulative distribution curve $y = \Phi(u)$ to the cumulative distribution curve $y = P\{x\}$ with mean ξ and standard deviation σ.

The parameters (ξ, σ) *may be derived from two given values of the cumulative distribution function*, (x_1, P_1) and (x_2, P_2), for, as illustrated in Fig. 6.3, we have the following linear equations

$$\xi + u_1\sigma = x_1 \tag{6.5.2}$$

$$\xi + u_2\sigma = x_2 , \tag{6.5.3}$$

where u_1 and u_2 are determined from the equations $\Phi(u_1) = P_1$ and $\Phi(u_2) = P_2$. This leads to

$$\sigma = \frac{x_1 - x_2}{u_1 - u_2} \tag{6.5.4}$$

and

$$\xi = x_1 - u_1\sigma = x_2 - u_2\sigma . \tag{6.5.5}$$

The result is particularly simple for $P_1 = 0{\cdot}500$ and $P_2 = 0{\cdot}159$, which give $u_1 = 0$ and $u_2 = -1$, so that $\xi = x_1$ and $\sigma = x_1 - x_2$, as shown in Fig. 6.3.

From the symmetry of the distribution curve about $x = \xi$ we see that

$$\boxed{P\{x_1\} + P\{x_2\} = 1 \quad \text{for} \quad \xi - x_1 = x_2 - \xi .} \tag{6.5.6}$$

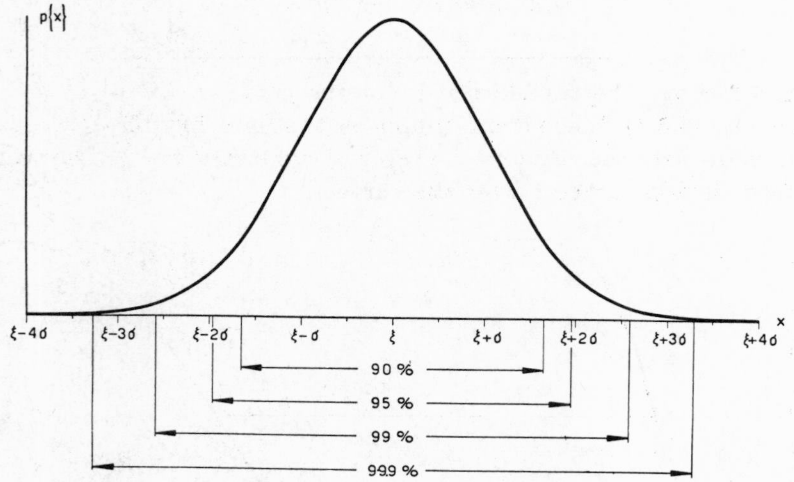

FIG. 6.4. Normal distribution. The interval $(\xi - \alpha\sigma, \xi + \alpha\sigma)$ is given corresponding to $P\{\xi - \alpha\sigma < x < \xi + \alpha\sigma\} = 0{\cdot}90, 0{\cdot}95, 0{\cdot}99, 0{\cdot}999$.

The probability that x deviates more than α times the standard deviation from the mean is

$$\begin{aligned} P\{|x-\xi| > \alpha\sigma\} &= P\{x < \xi-\alpha\sigma\}+P\{x > \xi+\alpha\sigma\} \\ &= P\{x < \xi-\alpha\sigma\}+(1-P\{x < \xi+\alpha\sigma\}) \\ &= \Phi(-\alpha)+(1-\Phi(\alpha)) \\ &= 2(1-\Phi(\alpha))\,. \end{aligned} \tag{6.5.7}$$

Table 6.1 shows that

$$P\{|x-\xi| > 1{\cdot}96\sigma\} = 2\times 0{\cdot}025 = 0{\cdot}05$$

and

$$P\{|x-\xi| > 2{\cdot}58\sigma\} = 2\times 0{\cdot}005 = 0{\cdot}01;$$

cf. Fig. 6.4. Observations which deviate more than 1·96 times the standard deviation from the mean are therefore comparatively rare, the probability of obtaining such a result being 5%.

This result may be compared with Table 5.1 which shows the corresponding probabilities according to TCHEBYCHEFF's theorem. If we know only the mean, ξ, and the standard deviation, σ, of a distribution we can only say that at least 99% of the distribution falls within the limits $\xi \pm 10\sigma$. However, if we get the further information that the distribution is normal we know that exactly 99% of the distribution falls within the limits $\xi \pm 2{\cdot}58\sigma$.

6.6. The Fractiles.

We may derive the fractiles, u_P, of the standardized distribution from the equation

$$\boxed{\Phi(u_P) = P} \qquad (0 \leq P \leq 1)\,. \tag{6.6.1}$$

u_P may be read from the cumulative distribution curve in Fig. 6.1 as the abscissa corresponding to a given ordinate, P, as in Fig. 4.1. u_P increases from $-\infty$ to ∞ as P increases from 0 to 1; for $P = 0{\cdot}5$ we have $u_P = 0$.

In computations, it is often inconvenient that u_P takes on negative as well as positive values. By choosing a suitable computing origin this difficulty can be avoided, the u_P-values being replaced by values of $u_P+\alpha$. As u very seldom takes on values below -5, BLISS[1]) has suggested giving α the value 5, and he has tabulated u_P+5, cf. Table III. BLISS calls the values of $u+5$ *probits*, an abbreviation of "*probability units*"; we thus have

$$\boxed{\text{Fractile } +5 = \text{Probit.}} \tag{6.6.2}$$

Fig. 6.5 shows u_P and u_P+5 as a function of P. The asymptotic behaviour of the curve for $P \to 0$ and $P \to 1$ implies that a variation of, say, 0·01

---

[1]) C. I. BLISS: *The Calculation of the Dosage-Mortality Curve,* The Annals of Applied Biology, 22, 1935, 134—167.

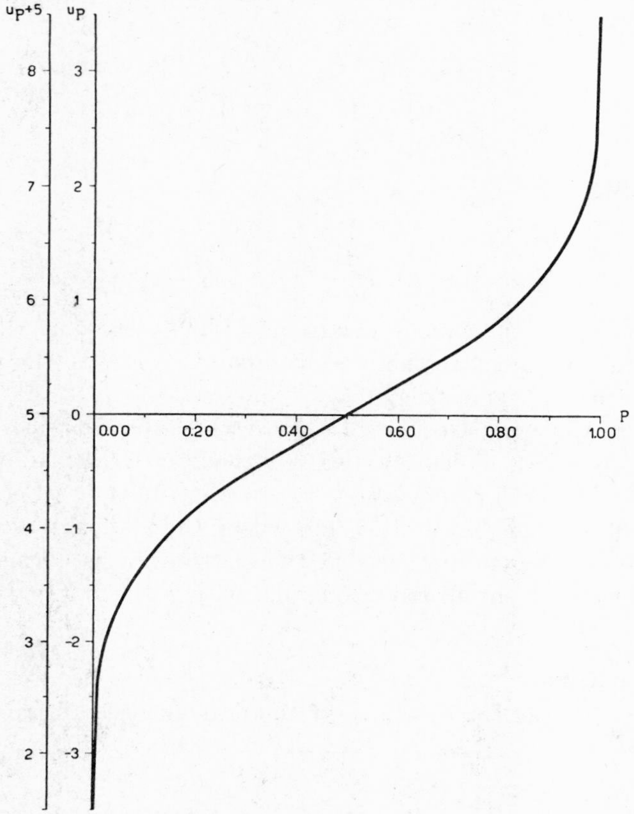

FIG. 6.5. The fractiles of the normal distribution.

in P leads to a considerably larger change in u_P when the P values are close to 0 or 1 than if they are near 0·5. For example, we find from Table III that

$$(u_{.995}+5)-(u_{.985}+5) = 7{\cdot}576-7{\cdot}170 = 0{\cdot}406$$

and

$$(u_{.505}+5)-(u_{.495}+5) = 5{\cdot}013-4{\cdot}987 = 0{\cdot}026\,.$$

The symmetry of the normal distribution leads to

$$u_P+u_{1-P} = 0 \tag{6.6.3}$$

or

$$(u_P+5)+(u_{1-P}+5) = 10\,. \tag{6.6.4}$$

As an example we find from Table III that

$$(u_{.25}+5)+(u_{.75}+5) = 4{\cdot}326+5{\cdot}674 = 10{\cdot}000\,.$$

The fractiles of a variable with mean ξ and standard deviation σ may be derived from the equation

$$P\{x_P\} = P\,.$$

As

$$P\{x_P\} = \Phi\left(\frac{x_P-\xi}{\sigma}\right), \tag{6.6.5}$$

we get

$$\boxed{u_P = \frac{x_P-\xi}{\sigma}} \tag{6.6.6}$$

or

$$x_P = \xi+u_P\sigma\,, \tag{6.6.7}$$

see Fig. 6.3.

This result shows that x is a linear function of the corresponding u in the equation $P\{x\} = \Phi(u)$. *If, using rectangular coordinates, the x values are plotted as abscissas and the u values as ordinates, we get a straight line through the point $(\xi, 0)$ with slope $\frac{1}{\sigma}$.*

Suppose that k points, $(x_1, P_1), \ldots, (x_k, P_k)$, of a cumulative distribution curve are given and that we do not know the mathematical form of the corresponding function. We may then investigate whether the k points can be interpreted as points on a cumulative *normal* distribution curve by looking up the fractiles u_P or the probits u_P+5 in Table III and plotting the k values of (x, u_P) on ordinary graph paper. If the k points lie on a straight line, then they may be interpreted as points on a cumulative normal distribution curve.

Instead of plotting (x, u_P) on ordinary graph paper, we may plot (x, P) directly on a special graph paper, *probability paper*, which has the property

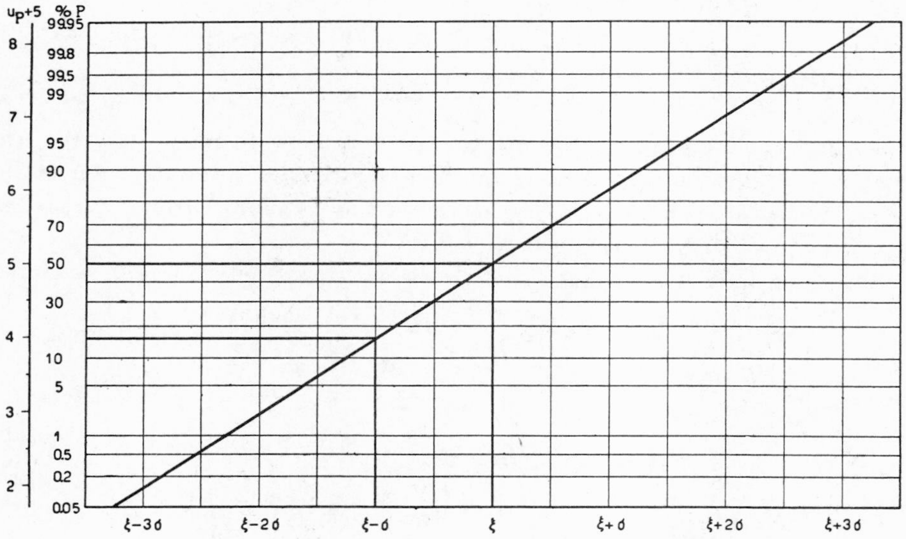

FIG. 6.6. Probability paper.

that *the graph of a cumulative normal distribution is a straight line.* This is accomplished by using a special ruling of the ordinate axis, namely a scale where instead of the u-values the corresponding values of $P = \Phi(u)$ are marked. This paper has the advantage that the (x, u_P)-values are plotted directly from the (x, P)-values without using the probit table.

Fig. 6.6 shows this *probability paper* with the line (6.6.6) plotted on it. If we read two values on this line (x_1, u_1) and (x_2, u_2), we get two equations from which ξ and σ may be computed, see (6.5.4) and (6.5.5). For $P = 50\%$, we get $x = \xi$, and for $P = 15{\cdot}9\%$, $x = \xi - \sigma$, see Fig. 6.6.

Probability paper is used in the same manner as other special graph papers, for example logarithmic paper. If we plot our observed values on the paper, we can then decide whether it is reasonable to describe their relationship by a function of the type which is represented by a straight line on the paper.

6.7. The Fractile Diagram.

A. Ungrouped observations.

Let $x_{(1)}, x_{(2)}, \ldots, x_{(n)}$ denote n observations, ranked according to magnitude. The *cumulative frequency* is computed for each value of x as in § 3.5:

$$H\{x\} = \begin{cases} 0 & \text{for} \quad x < x_{(1)} \, , \\ \dfrac{i}{n} & \text{for} \quad x_{(i)} \leq x < x_{(i+1)} \text{ and } i = 1, 2, \ldots, n-1 \, , \\ 1 & \text{for} \quad x \geq x_{(n)} \, . \end{cases} \tag{6.7.1}$$

$H\{x\}$ is the *observed value of the cumulative probability*, $P\{x\}$, which is written in symbols:

$$H\{x\} \gtrsim P\{x\} \quad \text{or} \quad P\{x\} \gtrsim H\{x\} \, ,$$

the sign $\gtrsim$ being read "*is an estimate of*" or "*has the estimate*". We are dealing neither with an ordinary equation nor with an approximation in the usual mathematical sense. $H\{x\}$ deviates "at random" from $P\{x\}$. The significance of this is discussed in § 6.7 C.

Assuming that x is *normally distributed* with parameters (ξ, σ^2), we have

$$H\{x\} \gtrsim \Phi\left(\frac{x-\xi}{\sigma}\right), \tag{6.7.2}$$

which may also be written

$$u_{H\{x\}} \gtrsim \frac{x-\xi}{\sigma}. \tag{6.7.3}$$

According to the definition of $H\{x\}$, the function $u_{H\{x\}}$ is a step-function which may be written as

$$u_{H\{x\}} = \begin{cases} -\infty & \text{for} \quad x < x_{(1)}\,, \\ u_{i/n} & \text{for} \quad x_{(i)} \leq x < x_{(i+1)} \text{ and } i = 1, 2, \ldots, n-1\,, \\ \infty & \text{for} \quad x \geq x_{(n)}\,. \end{cases} \tag{6.7.4}$$

The values of $u_{i/n}+5$ are obtained from Table III.

Thus, the step-function

$$u = u_{H\{x\}} \tag{6.7.5}$$

forms an estimate of the straight line

$$u = u_{P\{x\}} = \frac{x-\xi}{\sigma}, \tag{6.7.6}$$

in the same way as the cumulative frequency polygon $y = H\{x\}$ forms an estimate of the cumulative distribution curve $y = P\{x\}$. The graphical representation of the function $u_{H\{x\}}$ is called the *fractile diagram.*

The plotting of the function $u = u_{H\{x\}}$ in an (x, u) coordinate-system is equivalent to plotting the cumulative frequency polygon $y = H\{x\}$ on probability paper. This has the advantage that we do not have to look up probits in Table III, which saves a great deal of time, see Example 6.1.

In practical work the step-curve is usually not drawn in full, but is indicated by n of its points. At the point $x = x_{(i)}$, $H\{x\}$ jumps from $\frac{i-1}{n}$ to $\frac{i}{n}$. An obvious substitution for the step-curve is therefore given by the n points $\left(x_{(i)}, \frac{i-\frac{1}{2}}{n}\right)$ which are situated at the midpoints of the vertical parts of the step-curve. These points are plotted on probability paper, i. e., the points $\left(x_{(i)}, u_{\frac{i-1/2}{n}}\right)$ are plotted in an (x, u) coordinate-system; see the last two columns of Table 6.2 and Fig. 6.9. This procedure has the practical advantages that it is easier to plot n points than to draw a step-curve, and that possible systematic deviations from a straight line are more easily detected from a dot diagram than from a step-curve.

If the parameters are estimated from a graph and are used for further computations, it is important that the points on the diagram do not lead to a line with systematic errors. IPSEN and JERNE[1]) have pointed out that in principle the above method implies a systematic error, but for $n > 25$ the error is insignificant.

Example 6.1. In the manufacture of linen thread samples were taken at regular intervals for testing the breaking strength of the thread. Table 6.2

[1]) JOHS. IPSEN and N. K. JERNE: *Graphical Evaluation of the Distribution of Small Experimental Series,* Acta pathologica, 21, 1944, 343–361.

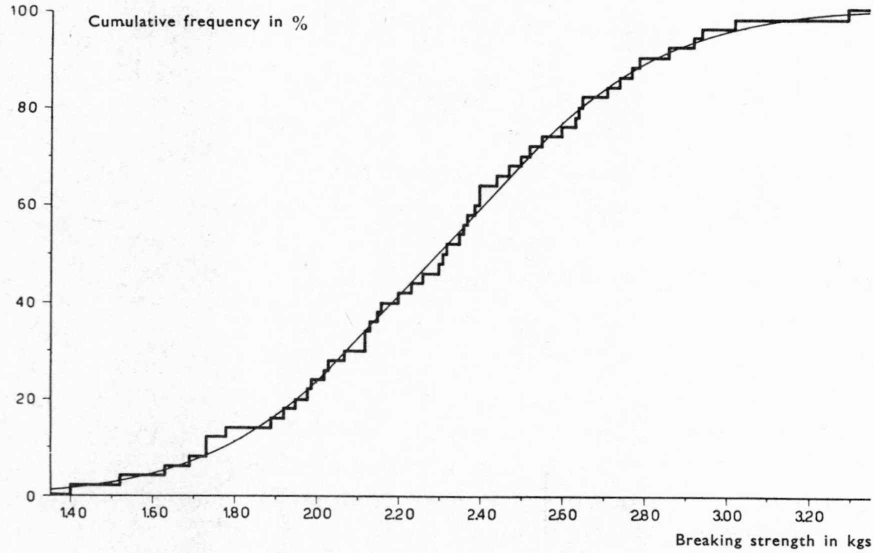

FIG. 6.7. *Cumulative frequency polygon of an ungrouped distribution.*
The distribution of 50 samples of linen thread according to breaking strength.

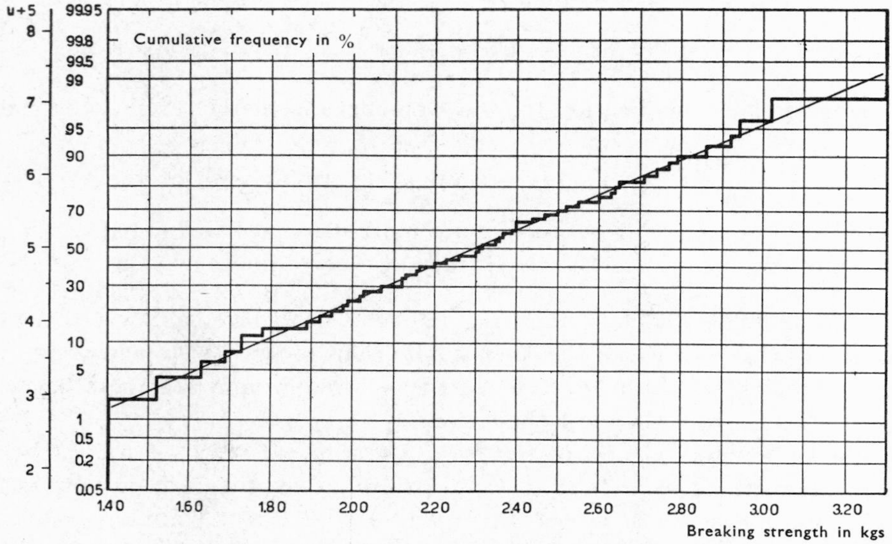

FIG. 6.8. *Fractile diagram of an ungrouped distribution. Step-curve.*
The distribution of 50 samples of linen thread according to breaking strength.

gives the breaking strength of 50 samples, ranked according to increasing values of the breaking strength. The third and fourth columns of the table give the cumulative frequencies and the corresponding probits. Fig. 6.7 shows the cumulative frequency polygon $(x, H\{x\})$, and Fig. 6.8 its trans-

TABLE 6.2.

Computation of cumulative frequencies and probits of an ungrouped distribution.

The breaking strength of 50 samples of linen thread in kgs.

No.	Breaking strength in kgs	Cumulative frequency in %	Probit	Cumulative frequency in %	Probit	No.	Breaking strength in kgs	Cumulative frequency in %	Probit	Cumulative frequency in %	Probit
i	$x_{(i)}$	$100\frac{i}{n}$	$u+5$	$100\frac{i-\frac{1}{2}}{n}$	$u+5$	i	$x_{(i)}$	$100\frac{i}{n}$	$u+5$	$100\frac{i-\frac{1}{2}}{n}$	$u+5$
1	1·40	2	2·95	1	2·67	26	2·32	52	5·05	51	5·03
2	1·52	4	3·25	3	3·12	27	2·35	54	5·10	53	5·08
3	1·63	6	3·45	5	3·36	28	2·36	56	5·15	55	5·13
4	1·69	8	3·60	7	3·52	29	2·37	58	5·20	57	5·18
5	1·73	10	3·72	9	3·66	30	2·39	60	5·25	59	5·23
6	1·73	12	3·83	11	3·77	31	2·40	62	5·31	61	5·28
7	1·78	14	3·92	13	3·87	32	2·40	64	5·36	63	5·33
8	1·89	16	4·01	15	3·96	33	2·44	66	5·41	65	5·39
9	1·92	18	4·09	17	4·05	34	2·47	68	5·47	67	5·44
10	1·95	20	4·16	19	4·12	35	2·50	70	5·52	69	5·50
11	1·98	22	4·23	21	4·19	36	2·52	72	5·58	71	5·55
12	1·99	24	4·29	23	4·26	37	2·55	74	5·64	73	5·61
13	2·02	26	4·36	25	4·33	38	2·60	76	5·71	75	5·67
14	2·03	28	4·42	27	4·39	39	2·63	78	5·77	77	5·74
15	2·07	30	4·48	29	4·45	40	2·64	80	5·84	79	5·81
16	2·12	32	4·53	31	4·50	41	2·65	82	5·92	81	5·88
17	2·12	34	4·59	33	4·56	42	2·71	84	5·99	83	5·95
18	2·13	36	4·64	35	4·62	43	2·74	86	6·08	85	6·04
19	2·15	38	4·70	37	4·67	44	2·77	88	6·18	87	6·13
20	2·16	40	4·75	39	4·72	45	2·79	90	6·28	89	6·23
21	2·20	42	4·80	41	4·77	46	2·86	92	6·41	91	6·34
22	2·23	44	4·85	43	4·82	47	2·92	94	6·56	93	6·48
23	2·26	46	4·90	45	4·87	48	2·94	96	6·75	95	6·65
24	2·30	48	4·95	47	4·93	49	3·02	98	7·05	97	6·88
25	2·31	50	5·00	49	4·98	50	3·30	100	∞	99	7·33

formation to a step-curve, $(x, u_{H\{x\}})$. Based on the step-curve, a straight line has been drawn, which gives the *graphical estimate* of the line

$$u = \frac{x-\xi}{\sigma}.$$

Since the step-curve appears to deviate only at random from the straight line, the theoretical distribution is assumed to be normal.

If we read the abscissa corresponding to the ordinate 50% from the straight line, we get

$$2{\cdot}30 \text{ kgs} \approx \xi .$$

The abscissa corresponding to 15·9% is determined as 1·90 kgs, which gives the graphical estimate of σ

$$2{\cdot}30 - 1{\cdot}90 = 0{\cdot}40 \text{ kgs} \approx \sigma .$$

The cumulative distribution curve of Fig. 6.7 has been plotted by reading suitable values from the straight line in Fig. 6.8.

The last two columns of Table 6.2 give the "adjusted" cumulative

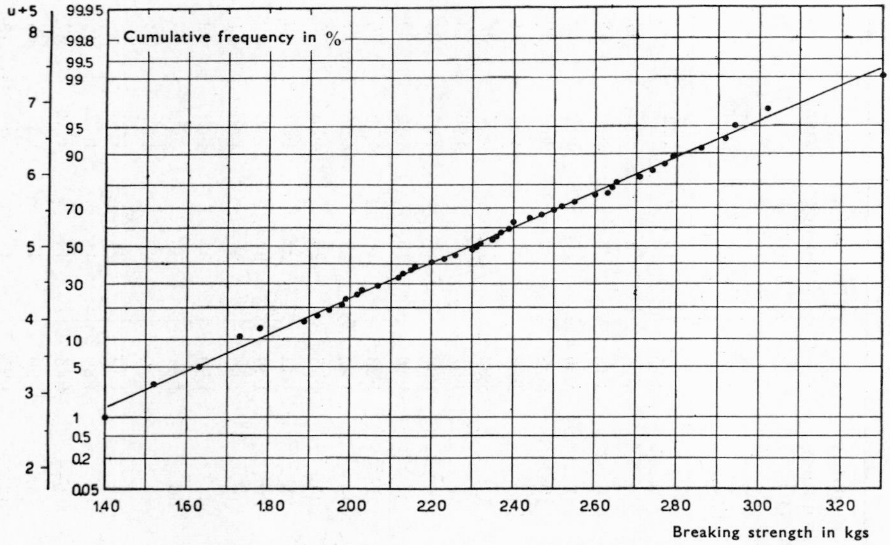

FIG. 6.9. *Fractile diagram of an ungrouped distribution. Dot diagram.*
The distribution of 50 samples of linen thread according to breaking strength.

frequencies and the corresponding probits. From these numbers the points given in Fig. 6.9 have been plotted. The straight line here is identical with that of Fig. 6.8.

B. Grouped observations.

If the observed values are grouped, the class limits $t_j+\frac{\Delta t_j}{2}$ are plotted as abscissas and the corresponding probits, $u_{H_j}+5$, as ordinates. This is equivalent to plotting the cumulative frequency polygon on probability paper.

TABLE 6.3.

Computation of cumulative frequencies and probits of a grouped distribution.

The distribution of 500 rivets according to diameter of heads in millimetres. Class length: 0·05 mm.

Upper class limit	Number of rivets	Cumulative frequency in %	Probit
$t_j+\frac{\Delta t}{2}$	a_j	$H_j=\frac{\Sigma a_j}{n}$	$u_{H_j}+5$
13·095	1	0·2	2·12
13·145	4	1·0	2·67
13·195	4	1·8	2·90
13·245	18	5·4	3·39
13·295	38	13·0	3·87
13·345	56	24·2	4·30
13·395	69	38·0	4·70
13·445	96	57·2	5·18
13·495	72	71·6	5·57
13·545	68	85·2	6·05
13·595	41	93·4	6·51
13·645	18	97·0	6·88
13·695	12	99·4	7·51
13·745	2	99·8	7·88
13·795	1	100·0	∞
Total:	500		

Example 6.2. Table 6.3 gives the probit values corresponding to the upper class limits of the distribution of 500 rivets according to diameter of heads given in Table 4.4. Fig. 6.10 shows the cumulative frequency polygon of Fig. 4.1, plotted on probability paper. The points are situated practically on a straight line.

From this diagram we read off the fractiles corresponding to the cumulative frequencies 50% and 15·9%. The first fractile is an estimate of ξ and the difference between the two is an estimate of σ as explained on p. 130. Reading the fractiles from the class limit scale and transforming the results

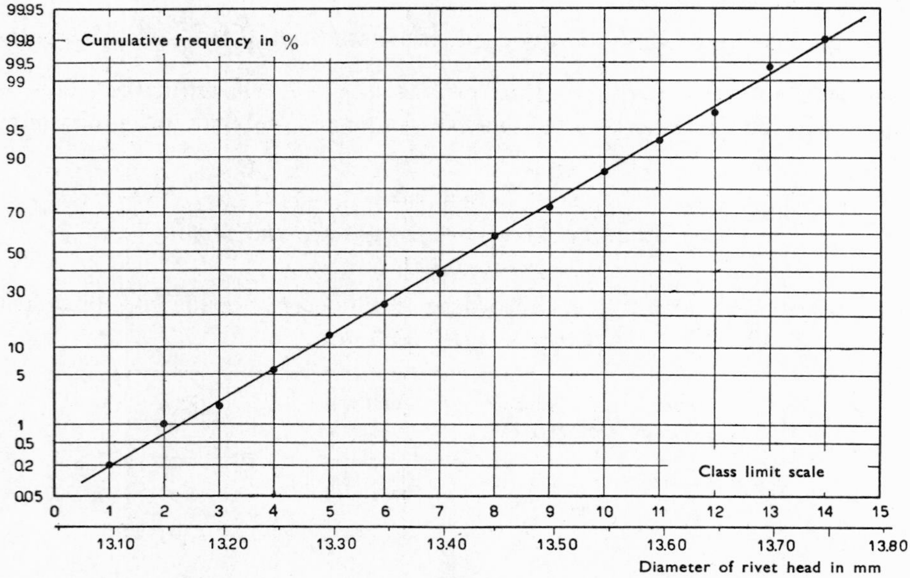

FIG. 6.10. *Fractile diagram of a grouped distribution.* The distribution of 500 rivets according to diameter of heads in millimetres. Class length: 0·05 mm.

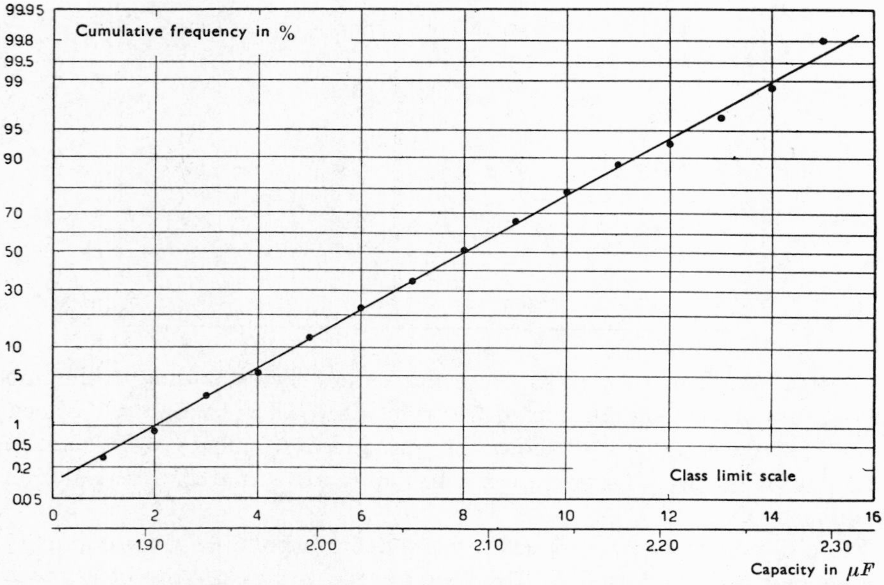

FIG. 6.11. *Fractile diagram of a grouped distribution.* The distribution of 650 condensers according to capacity in μF. Class length: 0·03 μF.

TABLE 6.4.

Computation of cumulative frequencies and probits of a grouped distribution.

The distribution of 650 condensers according to capacity in μF.
Class length: 0·03 μF.

Upper class limit	Number of condensers	Cumulative frequency in %	Probit
$t_j+\frac{\Delta t}{2}$	a_j	$H_j = \frac{\Sigma a_j}{n}$	$u_{H_j}+5$
1·875	2	0·31	2·26
1·905	3	0·77	2·58
1·935	13	2·8	3·09
1·965	16	5·2	3·37
1·995	45	12·2	3·84
2·025	68	22·6	4·25
2·055	77	34·5	4·60
2·085	103	50·3	5·01
2·115	99	65·5	5·40
2·145	81	78·0	5·77
2·175	61	87·4	6·15
2·205	35	92·8	6·46
2·235	24	96·5	6·81
2·265	14	98·62	7·20
2·295	8	99·85	7·97
2·325	1	100·00	∞
Total:	650		

back to millimetres, we get

$$13{\cdot}045+0{\cdot}05\times 7{\cdot}6 = 13{\cdot}425 \text{ mm} \approx \xi$$

and

$$0{\cdot}05(7{\cdot}6-5{\cdot}3) = 0{\cdot}115 \text{ mm} \approx \sigma .$$

These values are in good agreement with the mean, $\bar{x} = 13{\cdot}426$ mm, and the standard deviation, $s = 0{\cdot}115$ mm, as computed in Table 4.4.

Example 6.3. Table 6.4 gives the probit values corresponding to the class limits for the distribution of 650 condensers according to capacity, cf. Table 4.5. Fig. 6.11 illustrates the cumulative frequency polygon of Fig. 4.2, plotted on probability paper. The values (2·085, 50%) and (2·010, 15·9%) are read off the straight line, which gives

$$2{\cdot}085\ \mu F \approx \xi$$

and

$$2{\cdot}085-2{\cdot}010 = 0{\cdot}075\ \mu F \approx \sigma ,$$

which are in good agreement with $\bar{x} = 2{\cdot}086\ \mu F$ and $s = 0{\cdot}0786\ \mu F$, as computed in Table 4.5.

C. The random variation about the theoretical straight line.

If the theoretical distribution is normal, the points of the fractile diagram vary at random about a straight line. The variance of the fractile u_H corresponding to the cumulative frequency H is found in the following way.

According to (5.9.7) we have

$$\mathcal{V}\{u_H\} \simeq \mathcal{V}\{H\} \left(\frac{du}{dP}\right)^2 \tag{6.7.7}$$

where

$$H = \frac{i}{n} \approx P \tag{6.7.8}$$

and

$$P = \Phi(u)\,. \tag{6.7.9}$$

Further

$$\frac{dP}{du} = \varphi(u) \tag{6.7.10}$$

and according to (5.9.16)

$$\mathcal{V}\{H\} = \frac{P(1-P)}{n}$$

$$= \frac{\Phi(u_P)\,\Phi(-u_P)}{n}\,. \tag{6.7.11}$$

Substitution of these values in (6.7.7) leads to

$$\mathcal{V}\{u_H\} \simeq \frac{1}{n}\,\frac{\Phi(u_P)\,\Phi(-u_P)}{(\varphi(u_P))^2}\,, \tag{6.7.12}$$

i. e., the variance is inversely proportional to the number of observations. The quantity $n\mathcal{V}\{u_H\}$ depends only on u_P, and has a minimum at $u_P = 0$, where $P = 50\%$. $n\mathcal{V}\{u_H\}$ is symmetrical about $u = 0$ and increases as the numerical value of $P-0{\cdot}5$ increases.

Assuming that u_H is normally distributed about u_P with standard deviation $\sigma\{u_H\} = \sqrt{\mathcal{V}\{u_H\}}$, and that the observed values, $u_{H_1}, u_{H_2}, \ldots$ are stochastically independent, we may, from the theoretical values, calculate limits between which the observations should lie with a certain probability, for example

$$P\{|u_H - u_P| < 1{\cdot}96\sigma\{u_H\}\} = 95\%\,,$$

i. e., u_H must be expected to fall within the interval $u_P \pm 1{\cdot}96\sigma\{u_H\}$ on about 95% of occasions (95% limits). For $n = 100$ this interval has been represented in Fig. 6.12. Thus, the "permissible" magnitude of random deviations from the theoretical line are considerably larger for P-values close to 0 and 100% than for values about 50%. This in turn means that

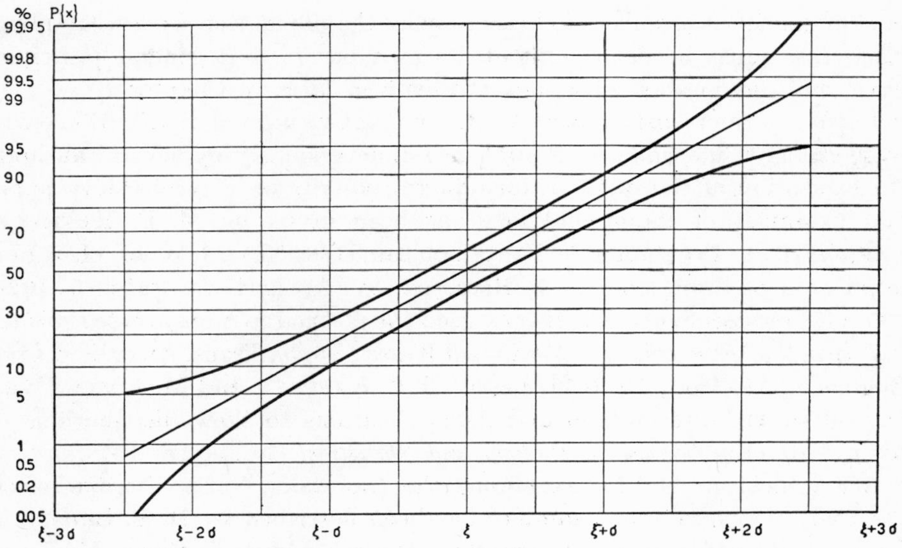

FIG. 6.12. 95% limits of the random variation about the theoretical straight line for $n = 100$.

when determining the straight line which gives the graphical estimate of the theoretical straight line, the extreme points are less important than the mid-points.

The above assumptions are, however, not strictly fulfilled. u_H is only approximately normally distributed about u_P with the said variance. For small values of n and for P-values near 0 and 100% the formula gives only a rough idea of the range of variation of u_H. Further, the fractiles are not stochastically independent, each cumulative frequency depending on the preceding values. It is therefore incorrect to expect the observed points to vary independently within these limits. If a fractile happens to be near the upper limit, the next one will probably also be comparatively large, because the corresponding cumulative frequency includes the former frequency. Owing to this "dependence" the points often seem to wind round the straight line in a systematic manner.

In a small sample, $n < 20$ say, the permissible random variation of the points in the fractile diagram is so great that it is difficult to be confident that systematic deviations from a straight line exist. If the data include *many small samples*, systematic deviations from the straight line can, however, often be recognized, as deviations which in a single diagram have been interpreted as random must be considered systematic when they are found repeatedly in several diagrams, see § 7.3.

When we are satisfied that the u_P-curve could not be a straight line, then we have shown that the distribution is not normal. This of course complicates our fundamental problem of characterizing the population we are sampling. Each case must be dealt with on its own merits, but general principles governing our method of attack are evolved later in Chapter 7.

Papers on the transformation of the cumulative normal distribution curve to a straight line have been published independently by several authors. An exposition of the method, including a description of probability paper and examples of its application, has been given by M. P. HENRY in *Probabilité du Tir*, Cours de Fontainebleau, 1894, fasc. 1, p. 96; cf. a new impression of these lectures in Mémorial de l'Artillerie française, 5, 1926, 295–447. Independently, A. HAZEN used the method in *Storage to be Provided in Impounding Reservoirs for Municipal Water Supply*, Trans. Amer. Soc. Civil Engineers, 77, 1914, 1549–50. Lastly, J. C. KAPTEYN and M. J. VAN UVEN have described the method and its applications to skew distributions in *Skew Frequency Curves in Biology and Statistics*, Groningen, 1916.

A graphical method for examining the "normality" of a distribution on the basis of several small samples has been described by M. E. LHOSTE in *Étude d'un procédé rationnel permettant de vérifier la loi de Gauss*, Mémorial de l'Artillerie française, 4, 1925, 245–251; see also a paper by M. J. HAAG in Mémorial de l'Artillerie française, 4, 1925, 1027–30, E. S. PEARSON and C. CH. SEKAR: *The Efficiency of Statistical Tools and a Criterion for the Rejection of Outlying Observations*, Biometrika, 28, 1936, 308–320, and N. ARLEY: *On the Distribution of Relative Errors from a Normal Population of Errors*, Det kgl. Danske Videnskabernes Selskab, Math.-fysiske Meddelelser, 18, 3, Copenhagen, 1940.

Example 6.4. In Example 6.2 u_H corresponding to the class limit 13·495 mm is equal to 0·57. The estimate of the theoretical straight line in Fig. 6.10 gives a corresponding u-value of 0·61. The deviation between the observed value and our estimate of the theoretical value is thus 0·04. An estimate of the variance of the observed value is obtained by substituting $u_P \approx 0{\cdot}61$ into (6.7.12) which gives

$$\mathcal{V}\{u_H\} \simeq \frac{1}{500} \frac{0{\cdot}729 \times 0{\cdot}271}{0{\cdot}331^2} = 0{\cdot}00361$$

and hence

$$\sigma\{u_H\} \simeq 0{\cdot}060 .$$

Therefore, for the u-value in question it is permissible that the difference $|u_H - u_P|$ takes on values up to $2\sigma\{u_H\} = 0{\cdot}12$.

6.8. Fitting a Normal Distribution to Observed Data.

The parameters of the normal distribution corresponding to the sample of diameters of rivet heads in Table 4.4 are unknown, and therefore we cannot compute the theoretical distribution. We may obtain an *estimate* of the distribution by substituting $\bar{x}$ and s for ξ and σ in the formulas for the distribution function and the cumulative distribution function. This is termed "fitting" a normal distribution to observed data.

The probability of an observation less than or equal to the jth class limit is

$$P\left\{x \leq t_j + \frac{\Delta t}{2}\right\} = \Phi\left(\frac{t_j + \frac{\Delta t}{2} - \xi}{\sigma}\right), \qquad (6.8.1)$$

and the probability of an observation belonging to the jth class is

$$p_j = P\left\{t_j - \frac{\Delta t}{2} < x \leq t_j + \frac{\Delta t}{2}\right\} = \Phi\left(\frac{t_j + \frac{\Delta t}{2} - \xi}{\sigma}\right) - \Phi\left(\frac{t_j - \frac{\Delta t}{2} - \xi}{\sigma}\right). \qquad (6.8.2)$$

If ξ and σ are replaced by $\bar{x}$ and s, we obtain an estimate of these probabilities, and by multiplying by 500 we get estimates of the theoretical number of rivets belonging to each class interval as shown in Table 6.5, column 5. The corresponding observed numbers are to be found in Table 4.4.

TABLE 6.5.

Fitting a normal distribution to observed data.

The distribution of 500 rivets according to diameter of their heads in millimetres.

$\bar{x} = 13.426$ mm. $s = 0.115$ mm. Class length: 0·05 mm.

Class limit $t+\frac{\Delta t}{2}$	$u = \frac{t+\frac{\Delta t}{2}-13{\cdot}426}{0{\cdot}115}$	$\Phi(u)$	$500\,\Phi(u)$	Differences	$\varphi(u)$	$500\times0{\cdot}05\,\frac{\varphi(u)}{0{\cdot}115}$
				0.3		
13·045	−3·31	0·0005	0·3		0·0017	0·4
				0.7		
13·095	−2·88	0·0020	1·0		0·0063	1·4
				2.7		
13·145	−2·44	0·0073	3·7		0·0203	4·4
				7.4		
13·195	−2·01	0·0222	11·1		0·0529	11·5
				18.0		
13·245	−1·57	0·0582	29·1		0·1163	25·3
				34.5		
13·295	−1·14	0·1271	63·6		0·2083	45·3
				57.4		
13·345	−0·70	0·2420	121·0		0·3123	67·9
				75.8		
13·395	−0·27	0·3936	196·8		0·3847	83·6
				87.0		
13·445	0·17	0·5675	283·8		0·3932	85·5
				79.1		
13·495	0·60	0·7257	362·9		0·3332	72·4
				61.4		
13·545	1·03	0·8485	424·3		0·2347	51·0
				40.3		
13·595	1·47	0·9292	464·6		0·1354	29·4
				21.1		
13·645	1·90	0·9713	485·7		0·0656	14·3
				9.5		
13·695	2·34	0·9904	495·2		0·0258	5·6
				3.4		
13·745	2·77	0·9972	498·6		0·0086	1·9
				1.1		
13·795	3·21	0·9993	499·7		0·0023	0·5
				0.3		

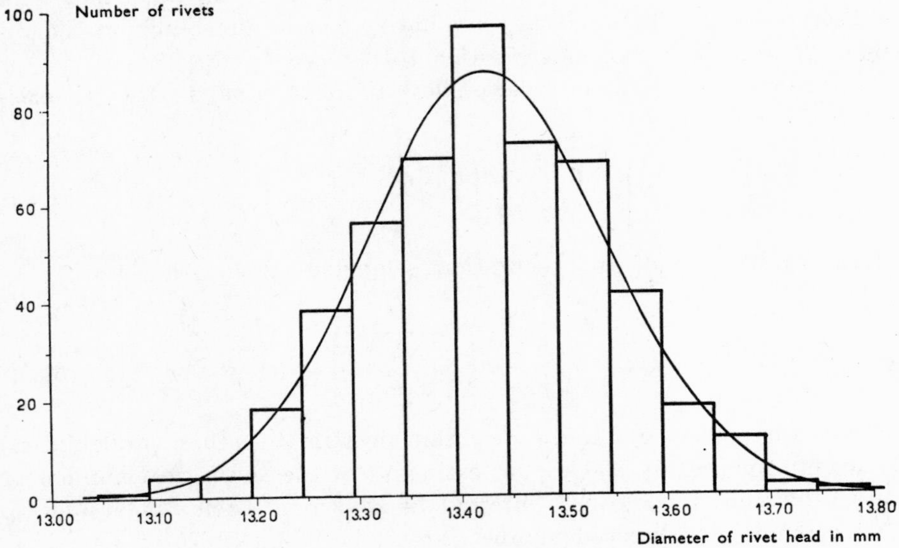

FIG. 6.13. *Histogram and fitted distribution curve.*
The distribution of 500 rivet heads according to diameter in millimetres.
Class length: 0·05 mm.

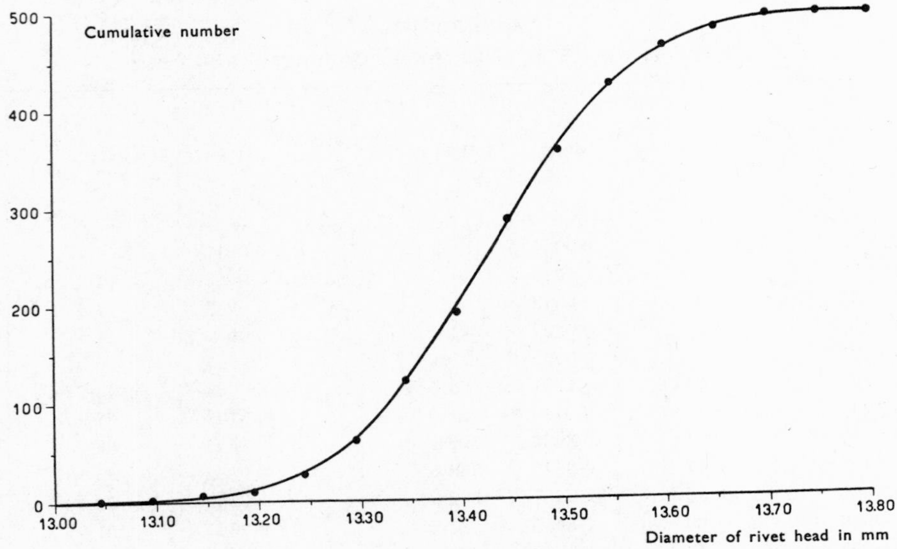

FIG. 6.14. *Cumulative frequency polygon and fitted cumulative distribution curve.*
(The points of the cumulative frequency polygon have not been connected by straight lines).
The distribution of 500 rivet heads according to diameter in millimetres.
Class length: 0·05 mm.

When computing the distribution curve corresponding to the histogram of Fig. 6.13, the fact that the histogram represents the number of rivets, a_j, in each class interval, and not the relative frequency divided by the class length, $h_j/\triangle t$, must be taken into consideration. As mentioned in § 3.3 and § 5.2 the two functions $h(t) = h_j/\triangle t$ and $p\{x\}$ are equivalent in the sense that frequencies and probabilities, respectively, are represented by *areas* enclosed by the corresponding curves. The distribution curve corresponding to a histogram with ordinates h_j or a_j instead of $h_j/\triangle t$ will therefore have the equation $\triangle t\, p\{x\}$ or $n\triangle t\, p\{x\}$, respectively. The equation of the distribution curve corresponding to the histogram of Fig. 6.13 is thus

$$\begin{aligned} y &= 500\, \Delta t p\{x\} \\ &= 500\, \Delta t \frac{1}{\sigma} \varphi(u) \quad \text{for} \quad u = \frac{x-\xi}{\sigma}. \end{aligned}$$

In Table 6.5 the ordinates of this curve have been computed for $x = t_j + \dfrac{\Delta t}{2}$ $(j = 1, \ldots, k)$, using $\bar{x}$ and s instead of ξ and σ. The curve is shown in Fig. 6.13 together with the histogram.

The general formula for a normal distribution curve corresponding to a histogram with class intervals of length Δt and the observed numbers, a_j, as ordinates is

$$y = \frac{n\Delta t}{\sigma} \varphi(u) \quad \text{for} \quad u = \frac{x-\xi}{\sigma}, \tag{6.8.3}$$

where n denotes the number of observations in the sample.

In Fig. 6.14 the points denote the cumulative frequency polygon $\left(t_j + \dfrac{\Delta t}{2}, H_j\right)$, cf. Fig. 4.1, and the curve is the cumulative distribution curve computed in Table 6.5.

From the diagrams and the table it would appear that we have (1) chosen a plausible type of distribution, and (2) successfully estimated its parameters, but this is just a subjective opinion and may well be wrong. We require an objective test of the goodness of fit. Such a test is found in § 23.2.

6.9. The Truncated Normal Distribution.

Suppose that a normally distributed population is given and that for some reason we are not able to sample the whole population but only a part of it. This happens, for example, when the elements of a population are sorted according to some criterion and our sample is taken after the sorting has been performed; we are then *sampling an incomplete population*. The population may, for instance, consist of manufactured items, which are normally distributed with respect to a certain property, x. Before selling the product, the items are inspected and all items for which $x < x_0$ are rejected. When the consumer takes a random sample of the product he is only able to sample the incomplete or *truncated* population consisting of all elements for which $x \geq x_0$.

Another situation arises when we are able to *sample the complete population but the individual values of observations below* (*or above*) *a given value are not specified*. Suppose, for example, that the distribution of lifetimes of incandescent lamps is normal. To estimate this distribution we start an experiment where 100 lamps, say, are kept burning. After a certain time, for example 1000 hours, we end the experiment and our observations now consist of two groups: (1) lamps for which we do not know the individual lifetimes but only that they all exceed 1000 hours, and (2) lamps for which we know the individual lifetimes (which are less than 1000 hours). Observations of this type are found in many other cases where the observed quantity is time and some of the observed individuals have a reaction time longer than the period of observation. A distribution (population) of this type is called a *censored* distribution because the obtainable information in a sense has been censored, either by nature or by ourselves.

In the following we shall discuss the one-sided truncated and the one-sided censored distributions under the assumption that the point of truncation is known. Further, some remarks are given concerning other cases of truncation.

A. The one-sided truncated normal distribution.

Let a normally distributed population with parameters (ξ, σ^2) be given. The corresponding one-sided truncated population is formed by removing all elements which take on a value below (or above) a certain limit, *the point of truncation*. In the following we assume that *the point of truncation is known and is chosen as origin* for the variable; i. e., the underlying distribution of the variable x is a normal distribution with parameters (ξ, σ^2) and the truncation is performed by limiting x to values larger than or equal to zero. (In practice the point of truncation is usually not zero but a certain value x_0. We then introduce the new variable $X = x - x_0$, which has the desired properties).

From this it follows that *the standardized point of truncation* is

$$\zeta = \frac{0-\xi}{\sigma} = -\frac{\xi}{\sigma}, \tag{6.9.1}$$

so that a proportion $\Phi(\zeta)$, called *the degree of truncation*, has been excluded from the original population.

The cumulative distribution function and the distribution function of the one-sided truncated population is then

$$P\{x\} = \frac{\Phi\left(\dfrac{x-\xi}{\sigma}\right) - \Phi(\zeta)}{1-\Phi(\zeta)} \qquad (x \geq 0) \tag{6.9.2}$$

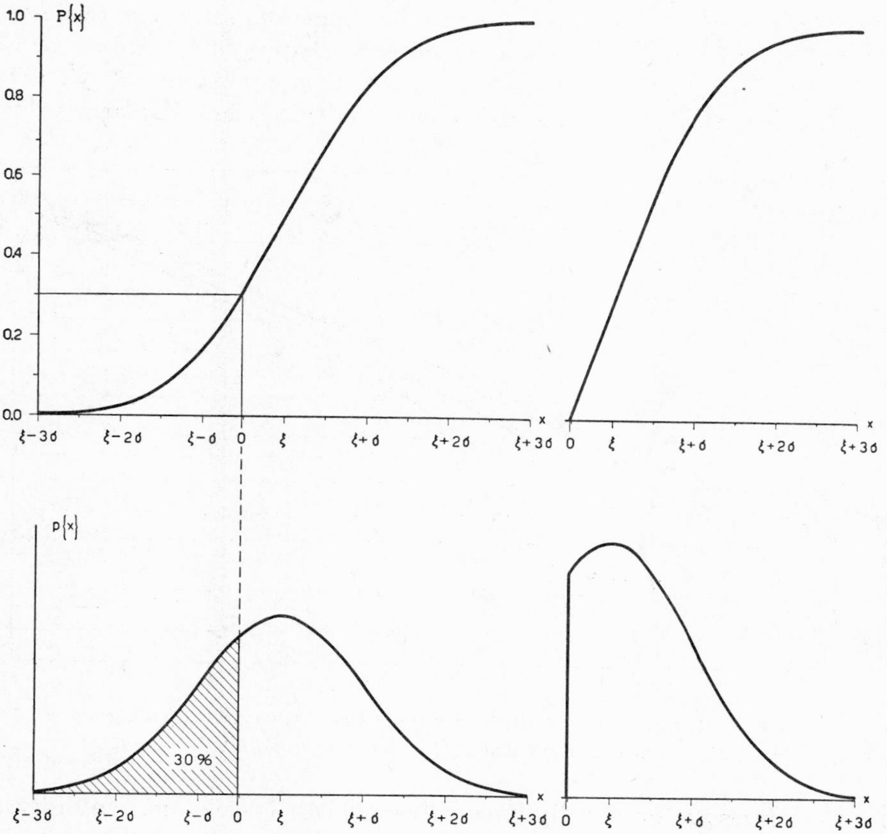

FIG. 6.15. Cumulative distribution curve and distribution curve for a normal distribution and a truncated normal distribution with 30 per cent truncation.

and

$$p\{x\} = \frac{1}{1-\Phi(\zeta)}\frac{1}{\sigma}\varphi\left(\frac{x-\xi}{\sigma}\right) \qquad (x \geq 0). \qquad (6.9.3)$$

The relation between the distribution function and the cumulative distribution function of a normal and a truncated normal population with 30% truncation is shown in Fig. 6.15. Table III shows that when $\Phi(\zeta)=0{\cdot}30$, $\zeta=4{\cdot}58-5{\cdot}00=-0{\cdot}52$, the origin thus being $\xi-0{\cdot}52\sigma$. The cumulative distribution function and the distribution function have the equations

$$P\{x\} = \frac{1}{0{\cdot}7}\left(\Phi\left(\frac{x-\xi}{\sigma}\right)-0{\cdot}3\right) \qquad (x \geq \xi-0{\cdot}52\sigma)$$

and

$$p\{x\} = \frac{1}{0{\cdot}7}\frac{1}{\sigma}\varphi\left(\frac{x-\xi}{\sigma}\right) \qquad (x \geq \xi-0{\cdot}52\sigma).$$

In Fig. 6.16 the cumulative distribution functions of three truncated distributions with the same ξ and σ and with degrees of truncation of 10, 30 and 50%, respectively, have been plotted on probability paper. The curves have vertical asymptotes at the points of truncation. For large values

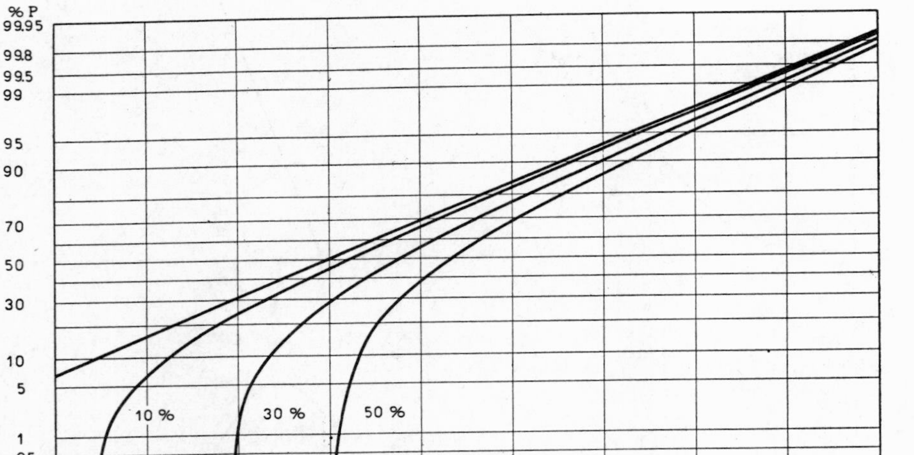

FIG. 6.16. The cumulative distribution functions of three truncated normal distributions plotted on probability paper.

of x the curves are practically linear. From this it is clear that, in order to determine a truncation less than 10–20%, the number of observations must be very large, as otherwise, for small values of x, the deviation of the fractiles from the straight line will fall within the permissible limits of random variation.

If the theoretical degree of truncation, $\Phi(\zeta)$, is known, the cumulative distribution function, $P\{x\}$, may be transformed into

$$P_1\{x\} = (1-\Phi(\zeta))P\{x\}+\Phi(\zeta) = \Phi\left(\frac{x-\xi}{\sigma}\right) \quad (x \geq 0). \tag{6.9.4}$$

If the corresponding curve is plotted on probability paper, we obtain the straight line

$$u = \frac{x-\xi}{\sigma} \quad (x \geq 0), \tag{6.9.5}$$

i. e., that part of the straight line corresponding to the normal distribution with the parameters (ξ, σ^2) which starts at the transformed point of truncation, $(0, \zeta)$.

When analyzing an observed distribution, the degree of truncation may be guessed at from the fractile diagram, and utilizing this estimated value, G, we may compute

$$H_1\{x\} = (1-G)H\{x\}+G\,. \tag{6.9.6}$$

If the observed distribution is truncated normal and the degree of truncation has been estimated fairly correctly, the transformed cumulative frequencies $H_1\{x\}$ will, when plotted on probability paper, give us an estimate of the straight line (6.9.5), and we can then read an estimate of ξ and σ from the diagram. By trial and error an analysis of this kind will often be successful.

R. A. FISHER[1]) has given a *method for estimating the parameters when the point of truncation is known*. Let $x_1, x_2, \ldots, x_n$ denote n observed values with the point of truncation as origin. We first compute

$$y = \frac{n\sum_{i=1}^{n} x_i^2}{2\left(\sum_{i=1}^{n} x_i\right)^2} = \frac{nSS}{2S^2}, \tag{6.9.7}$$

and from Table IX we then get *an estimate of the standardized point of truncation* as

$$z = f(y) \approx \zeta\,. \tag{6.9.8}$$

Using Table IX, we further compute

$$s = \frac{\sum_{i=1}^{n} x_i}{n} g(z) \approx \sigma \tag{6.9.9}$$

---

[1]) R.A. FISHER: *The Truncated Normal Distribution*, British Association for the Advancement of Science, Mathematical Tables, I, 1931, pp. XXXIII–XXXIV.

as an estimate of the parameter σ, and

$$\bar{x} = -zs \approx \xi \tag{6.9.10}$$

as an estimate of the parameter ξ.

Example 6.5. As an example of the numerical treatment of observations from a normal distribution which is truncated at a known point we will estimate ξ and σ from the distribution of rivet heads according to diameter, imagining that all rivets with a diameter less than or equal to 13·395 mm have been excluded, see Table 6.6.

TABLE 6.6.

Computation of estimates of the parameters of a normal distribution which is truncated at a known point.

Distribution of 310 rivets according to diameter of heads in millimetres.
Class length: 0·05 mm.
Computing origin = point of truncation = 13·395 mm.

Class midpoint	Number of rivets	$t_j = 13{\cdot}395 + 0{\cdot}025\, w_j$	
t_j	a_j	w_j	w_j^2
13·42	96	1	1
13·47	72	3	9
13·52	68	5	25
13·57	41	7	49
13·62	18	9	81
13·67	12	11	121
13·72	2	13	169
13·77	1	15	225
S	310	1274	
SS		7926	
S/n		4·110	

$$y = \frac{nSS}{2S^2} = \frac{310 \times 7926}{2 \times 1274^2} = 0{\cdot}7569.$$

$z = f(0{\cdot}7569) = -0{\cdot}310$, see Table IX.

$s = \frac{S}{n}\, 0{\cdot}025 \times g(z) = 4{\cdot}110 \times 0{\cdot}025 \times 1{\cdot}085 = 0{\cdot}111$, see Table IX.

$-zs = 0{\cdot}310 \times 0{\cdot}111 = 0{\cdot}034.$

$\bar{t} = 13{\cdot}395 + 0{\cdot}034 = 13{\cdot}429.$

$\Phi(-0{\cdot}310) = 37{\cdot}8\%.$

In this table the grouping of the data is fairly coarse, and therefore the corresponding fractile diagram does not show the asymptotic course of the

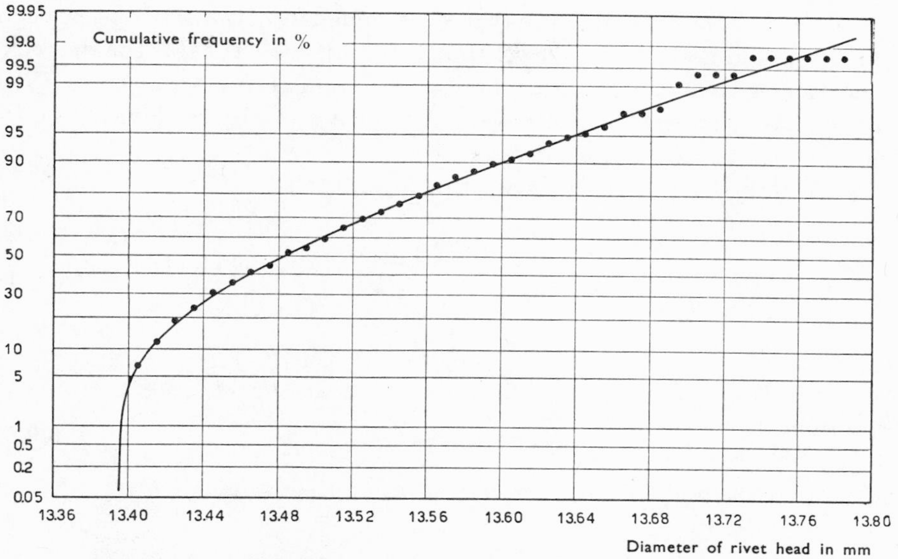

FIG. 6.17. *Fractile diagram for a normal distribution truncated at a known point.* Distribution of 310 rivets according to diameter of heads in millimetres. Class length: 0·01 mm.

transformed cumulative frequency polygon. The fractile diagram given in Fig. 6.17 was therefore plotted from the ungrouped distribution.

The computations in Table 6.6 give 13·429 mm as an estimate of ξ, 0·111 mm as an estimate of σ, and 37·8% as an estimate of the degree of truncation. The corresponding estimates computed from the total 500 original observations are 13·426 mm, 0·115 mm, and 38%, respectively, cf. Table 4.4.

B. The one-sided censored normal distribution.

Let n observations from a normally distributed population be divided into two groups, a observations being less than or equal to 0 (the point of truncation), and $n-a$ observations greater than 0. We assume that the a observations are not further specified, while the individual values of the $n-a$ observations, $x_1, x_2, \ldots, x_{n-a}$, are known. The distribution in Table 6.6 complies with these conditions if it is supplemented by the information that 190 rivets were rejected because their diameter was less than or equal to 13·395 mm.

The observed values may be interpreted as $n-a$ observations from a truncated normal distribution, where an estimate of the degree of truncation, $\Phi(\zeta)$, is given as $h = \dfrac{a}{n}$. Therefore, the fractile diagram shows a series of points which vary at random about the straight line (6.9.5), the first point being $(0, u_h) \approx (0, \zeta)$.

Estimates of the parameters may be determined from the $n-a$ observations by the method described under A, but if so our knowledge of the degree of truncation is not utilized. A system of equations[1]) analogous to (6.9.7)—(6.9.10) has therefore been derived, in which the observed value of the degree of truncation is included.

From the $n-a$ observations we first compute

$$y = \frac{(n-a)\sum_{i=1}^{n-a} x_i^2}{2\left(\sum_{i=1}^{n-a} x_i\right)^2}, \tag{6.9.11}$$

and from Table X we then obtain *an estimate of the standardized point of truncation* as

$$z = f(h, y) \approx \zeta . \tag{6.9.12}$$

Next *an estimate of the parameter* σ is computed from

$$s = \frac{\sum_{i=1}^{n-a} x_i}{n-a} g(h, z) \approx \sigma \tag{6.9.13}$$

with the aid of Table X, and an estimate of the parameter ξ from

$$\bar{x} = -zs \approx \xi . \tag{6.9.14}$$

Example 6.6. We use the distribution in Table 6.6 with the additional infortion that 190 rivet heads had a diameter smaller than or equal to 13·395 mm. Table 6.7 gives the computations.

A comparison of the three sets of estimates obtained is given in §§ 9.8 and 11.11.

C. One-sided truncation with unknown point of truncation.

In this case the distribution has three parameters: ξ, σ, and the point of truncation. Estimation of these parameters by formulas analogous to the above is very laborious, as tables which might facilitate the work have not yet been computed.

Graphical analysis of the observations may, however, often lead to an estimate of the point of truncation, and then the observations may be dealt with as described under A and B, the estimated point of truncation being taken as equal to the theoretical point. This procedure is not correct, and the uncertainty of the estimates of ξ and σ therefore cannot be found by

---

[1]) C. I. Bliss: *The Calculation of the Time-Mortality Curve.* Appendix by W. L. Stevens: *The Truncated Normal Distribution,* Annals of Applied Biology, 24, 1937, 815–52.

A. Hald: *Maximum Likelihood Estimation of the Parameters of a Normal Distribution Which is Truncated at a Known Point,* Skandinavisk Aktuarietidskrift, 1949, 119—134.

TABLE 6.7.

Computation of estimates of the parameters of a censored normal distribution.
Distribution of 500 rivets according to diameter of heads in millimetres.
Class length: 0·05 mm. Degree of truncation: $^{190}/_{500} = 0{\cdot}380$.
Computing origin = point of truncation = 13·395 mm.

Class midpoint	Number of rivets	$t_j = 13{\cdot}395 + 0{\cdot}025\, w_j$	
t_j	a_j	w_j	w_j^2
13·42	96	1	1
13·47	72	3	9
13·52	68	5	25
13·57	41	7	49
13·62	18	9	81
13·67	12	11	121
13·72	2	13	169
13·77	1	15	225
S	310	1274	
SS		7926	
$S/(n-a)$		4·110	

$$y = \frac{(n-a)SS}{2S^2} = \frac{310 \times 7926}{2 \times 1274^2} = 0{\cdot}757$$

$z = f(0{\cdot}380,\ 0{\cdot}757) = -0{\cdot}306$, see Table X.

$$s = 4{\cdot}110 \times 0{\cdot}025 \times g(0{\cdot}380,\ -0{\cdot}306)$$
$$= 4{\cdot}110 \times 0{\cdot}025 \times 1{\cdot}085 = 0{\cdot}111, \text{ see Table X.}$$

$$\bar{x} = 13{\cdot}395 - zs = 13{\cdot}395 + 0{\cdot}306 \times 0{\cdot}111 = 13{\cdot}429.$$

using the expressions given in § 9.8 and § 11.11, which expressions can only be regarded as giving the lower limits of the uncertainty.

D. Double-sided truncation.

When the truncation is double-sided, formulas equivalent to (6.9.7)–(6.9.14) may be derived. If the cumulative distribution function is plotted on probability paper, the resulting curve tends asymptotically towards both 0% ($-\infty$) and 100% ($+\infty$) as x tends to the upper and lower point of truncation, respectively.

Here, as for C, estimation of the parameters is complicated, and adequate tables are lacking.

When we have a large sample, the two degrees of truncation may be guessed at from the fractile diagram, and then the parameters may be estimated from a new fractile diagram, utilizing adjusted, cumulative frequencies computed from a formula analogous to (6.9.6).

6.10. Heterogeneous Distributions.

Imagine that a *heterogeneous population* is formed by combining two given populations in a given proportion. The distribution of this third population may be derived from the two given distributions and the proportion mentioned. This distribution is termed *a heterogeneous distribution.*

In a plant where the product is manufactured by several machines of the same type, the distribution of the quality of the items produced will often be heterogeneous, on account of individual differences between the machines. One machine may, for example, be more worn than another, and differences between the operators of the machines and variations in the quality of the raw materials may also result in heterogeneous distributions.

If a population is composed of two normal populations with parameters (ξ_1, σ_1^2) and (ξ_2, σ_2^2), combined in the ratio α_1/α_2, $(\alpha_1+\alpha_2 = 1)$, we get a heterogeneous population with the distribution function:

$$p\{x\} = \frac{\alpha_1}{\sigma_1}\varphi\left(\frac{x-\xi_1}{\sigma_1}\right) + \frac{\alpha_2}{\sigma_2}\varphi\left(\frac{x-\xi_2}{\sigma_2}\right). \qquad (6.10.1)$$

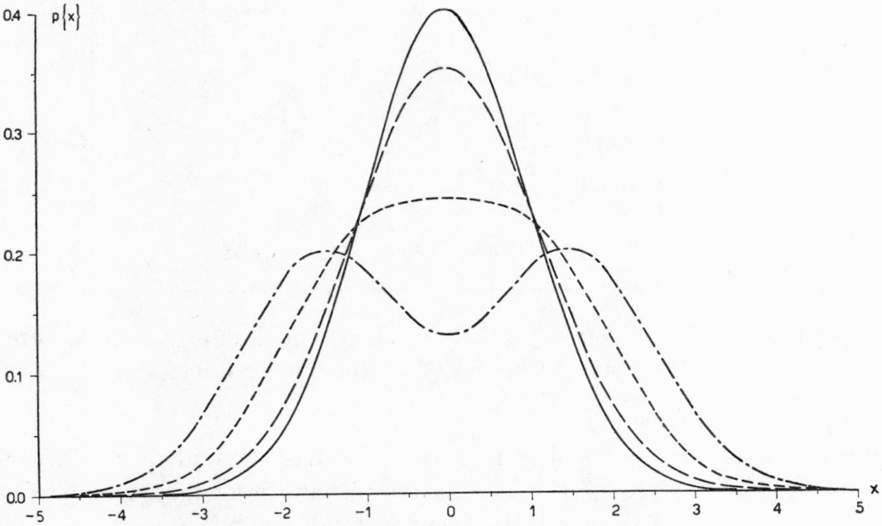

FIG. 6.18. *Heterogeneous distributions.*

The distribution curve for $p\{x\} = \frac{1}{2}\varphi\left(x-\frac{\delta}{2}\right)+\frac{1}{2}\varphi\left(x+\frac{\delta}{2}\right)$ for $\delta = 0, 1, 2$ and 3.

Fig. 6.18 shows a heterogeneous population composed of two normal populations with standard deviations 1 combined in the ratio 1/1. The distance between the two means is 1, 2, and 3 times the standard deviation, respectively. The equation of the distribution curve is

$$p\{x\} = \frac{1}{2}\varphi\left(x-\frac{\delta}{2}\right)+\frac{1}{2}\varphi\left(x+\frac{\delta}{2}\right), \tag{6.10.2}$$

where $\delta = 1, 2, 3$. When the distance between the two means is 3 times the standard deviation, the form of the curve is bimodal, showing the heterogeneity very clearly. With a distance of twice the standard deviation between the means the distribution curve is flat, while for $\delta = 1$ the heterogeneous distribution curve does not differ conspicuously from the normal curve.

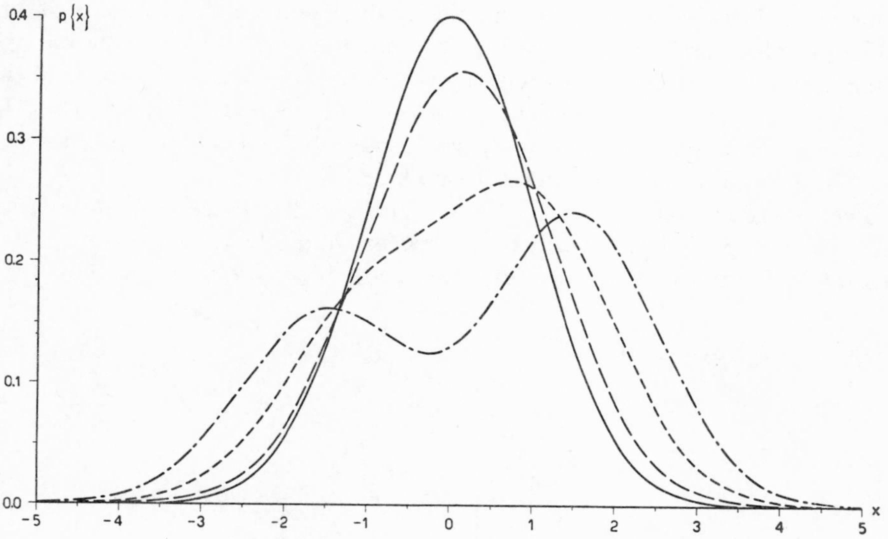

FIG. 6.19. *Heterogeneous distributions.*

The distribution curve for $p\{x\} = 0{\cdot}6\varphi\left(x-\frac{\delta}{2}\right)+0{\cdot}4\varphi\left(x+\frac{\delta}{2}\right)$ for $\delta = 0, 1, 2$ and 3.

Fig. 6.19 shows the distribution curve corresponding to

$$p\{x\} = 0{\cdot}6\varphi\left(x-\frac{\delta}{2}\right)+0{\cdot}4\varphi\left(x+\frac{\delta}{2}\right) \tag{6.10.3}$$

for $\delta = 0, 1, 2$ and 3. For $\delta = 3$ the distribution curve is bimodal, and for $\delta = 2$ it is clearly skew.

A heterogeneous distribution in the first place raises the question: *What has caused the heterogeneity*? It is often possible, when considering what we know about the process which leads to the observed values, to dissect the observations into two or more groups each corresponding to a normal distribution. Such a dissection should never be purely formal, but *a justified reason must be demonstrated for making the dissection.* The fundamental attitude towards the dissection of a heterogeneous distribution must necessarily be: *Is it possible to dissect the observed values into two or more groups on the basis of other criteria than the property considered?*

For industrial products we often take the following factors into account when dissecting heterogeneous distributions: Machines, attendants (day and night shift), the properties of the raw material, and time, which again may represent the result of several influences, such as wear, routine, etc.

Heterogeneity is often caused by *a false formulation of the problem*, which leads to a one-dimensional investigation instead of a two- or multi-dimensional approach. If one criterion is stochastically dependent on a second criterion, and only the distribution according to the first criterion is observed, the distribution will often become heterogeneous on account of variations of the second criterion. In such cases, however, the second criterion must be included in the analysis, which then becomes an analysis of variance or a regression analysis, see Chapters 16 and 18.

The purely formal dissection of heterogeneous distributions by numerical methods is very complicated[1]). If the data are numerous it is, however, often possible to make a graphical dissection, as shown in Example 6.8.

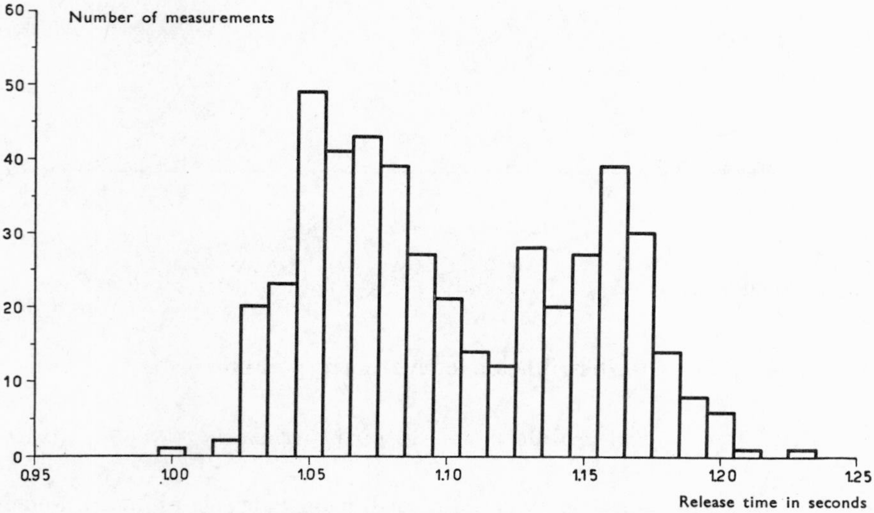

FIG. 6.20. Histogram of the distribution of 466 determinations of the release time of a maximum time relay.

---

[1]) C. BURRAU: *The Half-Invariants of Two Typical Laws of Errors, with an Application to the Problem of Dissecting a Frequency Curve into Components*, Skandinavisk Aktuarietidskrift, 17, 1934, 1–6.

B. STRÖMGREN: *Tables and Diagrams for Dissecting a Frequency Curve into Components by the Half-Invariant Method*, Skandinavisk Aktuarietidskrift, 17, 1934, 7–54.

TABLE 6.8.
The distribution of 466 measurements of release time of a relay.
Class length: 0·01 sec.

Release time in seconds	Number of measurements	Release time in seconds	Number of measurements
1·00	1	1·12	12
1·01	0	1·13	28
1·02	2	1·14	20
1·03	20	1·15	27
1·04	23	1·16	39
1·05	49	1·17	30
1·06	41	1·18	14
1·07	43	1·19	8
1·08	39	1·20	6
1·09	27	1·21	1
1·10	21	1·22	0
1·11	14	1·23	1
		Total:	466

Example 6.7. Table 6.8 gives the results of 466 measurements of the release time of a maximum time relay (measured in seconds). Fig. 6.20 shows the corresponding histogram, which indicates that the distribution is composed of two presumably normal distributions in about the ratio 1/1 with the same standard deviation and a difference between the means of about 0·08 second. A more detailed examination of the relay led to the following explanation of the heterogeneity: When the current through the coil of the relay is closed, a worm, which is connected with an induction break disc, meshes with a toothed sector, which causes the release key to close. The above difference of 0·08 second corresponds to two positions, separated from each other by one tooth. If the relative position of the worm and the toothed sector has not been adjusted with extreme accuracy, which of these two positions is taken up depends on the position at which the induction break disc has stopped at the end of the previous operation.

The relay was adjusted, and a new distribution was observed in which by far the greater number of the measurements belonged to the distribution with the larger mean. Further adjustment of the relay would no doubt have removed the heterogeneity completely[1]).

Example 6.8. Analysis of samples of peat from a bog resulted in the ash content values given in the first two columns of Table 6.9. The table shows

[1]) K. DANØ: *Tidsaftrapning ved Overstrømsbeskyttelse*, Ingeniøren, 53, 1944, E. 13–19.

TABLE 6.9.

Dissection of a heterogeneous distribution.

Distribution of 430 samples of peat according to ash content.

Class length: 0·50% ash.

Ash percentage of dry substance. Class midpoint	Observations		Component I		Component II	
	Number of analyses	Cumulative frequency in %	Number of analyses	Cumulative frequency in %	Number of analyses	Cumulative frequency in %
0·25	1	0·23			1	1·16
0·75	1	0·47			1	2·33
1·25	2	0·93			2	4·7
1·75	5	2·09			5	10·5
2·25	12	4·9			12	24·4
2·75	18	9·1	1	0·29	17	44·2
3·25	20	13·7	2	0·87	18	65·1
3·75	19	18·1	3	1·74	16	83·7
4·25	16	21·9	6	3·5	10	95·3
4·75	14	25·1	11	6·7	3	98·84
5·25	20	29·8	19	12·2	1	100·00
5·75	25	35·6	25	19·5		
6·25	35	43·7	35	29·7		
6·75	43	53·7	43	42·2		
7·25	48	64·9	48	56·1		
7·75	45	75·3	45	69·2		
8·25	35	83·5	35	79·4		
8·75	26	89·5	26	86·9		
9·25	17	93·5	17	91·9		
9·75	13	96·5	13	95·6		
10·25	9	98·60	9	98·26		
10·75	4	99·53	4	99·42		
11·25	2	100·00	2	100·00		
Total:	430		344		86	

that the distribution is bimodal and that the abscissas of the maximum values are approximately 3·25 and 7·25% ash. In Fig. 6.21 the class midpoints have been plotted on semi-logarithmic paper.

For a normal distribution we have:

$$a_j \approx n \int_{t_j - \frac{\Delta t}{2}}^{t_j + \frac{\Delta t}{2}} p\{x\}\, dx$$

$$\simeq n \Delta t\, p\{t_j\}$$

$$= \frac{n \Delta t}{\sqrt{2\pi}\, \sigma} e^{-\frac{(t_j - \xi)^2}{2\sigma^2}} \qquad (6.10.4)$$

if Δt is sufficiently small. Hence

$$\log a_j \approx \log\left(\frac{n\Delta t}{\sqrt{2\pi}\,\sigma}\right) - \frac{M}{2\sigma^2}(t_j-\xi)^2 \qquad (M = 0{\cdot}4343)\,, \qquad (6.10.5)$$

from which it is seen that the points $(t_j, \log a_j)$ are distributed at random about a parabola with the axis of symmetry $t = \xi$.

If the frequency polygon (t_j, a_j) of a heterogeneous distribution is plotted on semi-logarithmic paper, it should be possible to dissect it into a sum of parabolas, corresponding to the components of the heterogeneous distribution.

In the present case it seems obvious that the descending limb of the curve on the right is not influenced by the component with the smaller mean value. If this part is rotated about the presumed axis of symmetry, $t = 7{\cdot}25\%$,

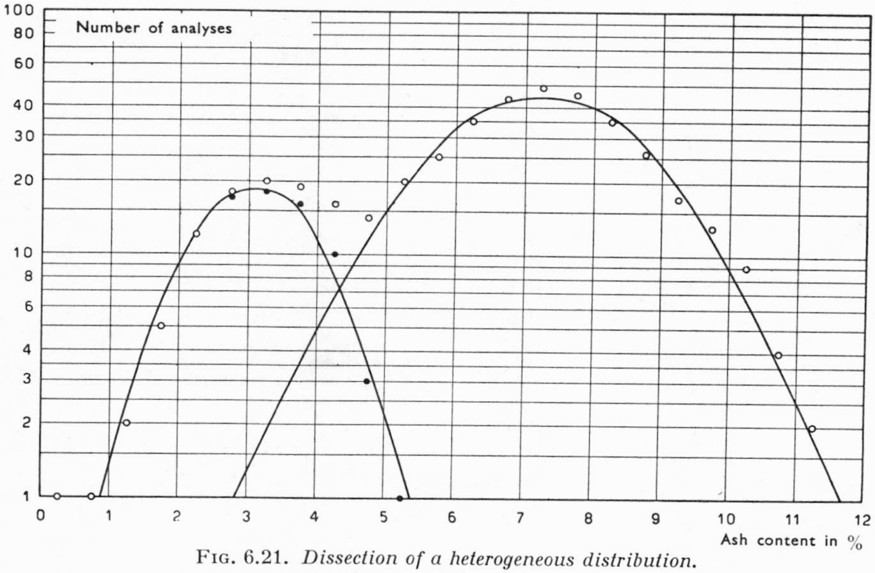

FIG. 6.21. *Dissection of a heterogeneous distribution.*
Distribution of 430 samples of peat according to ash content in per cent.

we obtain a parabola corresponding to the larger component of the heterogeneous distribution. The number of observations belonging to this component is now estimated from the curve. These estimates have been tabulated in the fourth column of Table 6.9; 344 analyses or 80% of the observations seem to belong to this distribution. In order to check that the curve we have drawn is a parabola, the cumulative frequencies are computed and plotted on probability paper, see Table 6.9 and Fig. 6.22. The figure suggests that the distribution is normal, and as estimates of the mean and the standard deviation we get $7{\cdot}2\%$ and $1{\cdot}5\%$ ash, respectively.

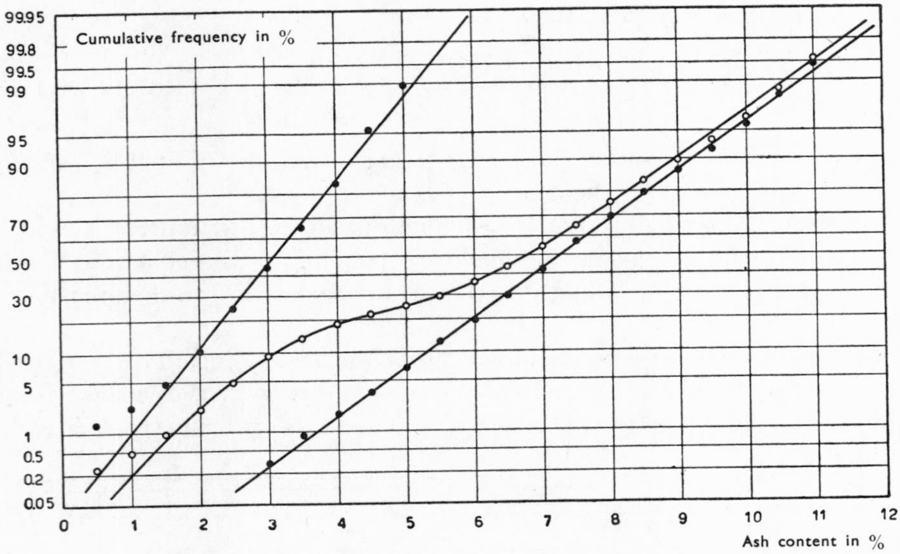

FIG. 6.22. *Dissection of a heterogeneous distribution.* Fractile diagram of the distribution of 430 samples of peat according to ash content.

By subtraction, the number of observations belonging to the smaller component of the total distribution is determined for each class interval, see column 6, Table 6.9, and the corresponding points are plotted in Fig. 6.21. The figure shows that a parabola may also be adjusted to these points, and the corresponding fractile diagram is shown in Fig. 6.22. Estimates of the mean and standard deviation of this smaller component are read off as 3·1% and 0·8% ash, respectively.

Thus, the bog seems to contain two kinds of peat. By taking samples at different places in the bog, it was ascertained that areas with peat of two such different qualities could be delimited.

CHAPTER 7.

SKEW DISTRIBUTIONS

7.1. Transformation of Skew Distributions.

Experience shows that only a comparatively small number of the distributions met with in practical life can be described by the normal distribution. Distributions influenced by economic, psychological and biological factors are generally skew. The points (x, u_H) in the fractile diagram of a skew distribution are scattered about a theoretical curve, the equation of which may be written

$$u = f(x) , \tag{7.1.1}$$

where x is the stochastic variable. This means that the theoretical cumulative distribution function has the equation

$$\boxed{P\{x\} = \Phi\big(f(x)\big) ,} \tag{7.1.2}$$

and hence

$$dP\{x\} = p\{x\}\,dx = \varphi\big(f(x)\big)\,df(x) \tag{7.1.3}$$

or

$$\boxed{p\{x\} = \frac{1}{\sqrt{2\pi}} e^{-\frac{(f(x))^2}{2}} f'(x) .} \tag{7.1.4}$$

Thus, the variable x itself is not normally distributed, but *the function* $f(x)$ *is normally distributed.*

Theoretically it is always possible to determine a function $f(x)$ which will transform the skew distribution into a normal one. In practical work this is of interest mainly because in numerous cases it is possible to choose a function of the type

$$f(x) = \frac{g(x) - g(\xi)}{\sigma} , \tag{7.1.5}$$

where $g(x)$ *does not include unknown parameters.* In § 7.2 we will deal with the transformation function $g(x) = \log x$, while in § 7.3 examples are given of several other transformations. Thus, the function $g(x)$ is normally distributed about a constant which we call $g(\xi)$ with standard deviation σ.

Usually it is possible to determine the function $g(x)$ from the fractile diagram. As a check *a new fractile diagram is plotted, using the transformed observations as abscissas.* If the function $g(x)$ has been chosen correctly, the points of the diagram scatter at random about the straight line

$$u = \frac{g(x) - g(\xi)}{\sigma} . \tag{7.1.6}$$

Estimates of $g(\xi)$ and σ may be read off the fractile diagram as described in § 6.7.

Estimates of the parameters $g(\xi)$ and σ^2 are computed from the observations by transforming each observed value x_i to $g(x_i)$, and then using the following formulas to determine the mean and the variance:

$$\overline{g(x)} = \frac{1}{n} \sum_{i=1}^{n} g(x_i) \approx g(\xi) \tag{7.1.7}$$

and

$$s^2 = \frac{1}{n-1} \sum_{i=1}^{n} \left(g(x_i) - \overline{g(x)}\right)^2 \approx \sigma^2 . \tag{7.1.8}$$

Examples of transformations of skew distributions are given in the following paragraphs.

7.2. The Logarithmic Normal Distribution.

A variable is said to have *a logarithmic normal distribution if the logarithm of the variable is normally distributed.*

The cumulative distribution function is defined from the equation

$$\boxed{P\{x\} = \Phi(u), \quad u = \frac{\log x - \log \xi}{\sigma}} \quad (0 < x < \infty) . \tag{7.2.1}$$

Differentiation of the cumulative distribution function leads to

$$\begin{aligned} p\{x\} dx &= \varphi(u) du \\ &= \frac{1}{\sqrt{2\pi}\,\sigma} e^{-\frac{(\log x - \log \xi)^2}{2\sigma^2}} d \log x \end{aligned} \tag{7.2.2}$$

or

$$\boxed{p\{x\} = \frac{M}{\sigma x} \varphi(u), \quad u = \frac{\log x - \log \xi}{\sigma},} \tag{7.2.3}$$

where $M = \log e = 0{\cdot}4343$ and $0 < x < \infty$.

From the definition we have

$$\mathcal{M}\{\log x\} = \log \xi \tag{7.2.4}$$

and

$$\mathcal{V}\{\log x\} = \mathcal{M}\{(\log x - \log \xi)^2\} = \sigma^2 \,. \tag{7.2.5}$$

Thus, the symbol ξ does *not* denote the mean of the variable x, but ξ is defined by $\log \xi$ being the mean of $\log x$. As $\log \xi$ is the 50% fractile, the median, in the distribution of $\log x$, ξ is equal to the median in the distribution of x, as shown in (5.6.5).

The mean of x is equal to

$$\mathcal{M}\{x\} = \xi \times 10^{\frac{\sigma^2}{2M}}, \tag{7.2.6}$$

and the abscissa of the maximum value of the distribution curve, *the mode*, is equal to $\xi \times 10^{-\frac{\sigma^2}{M}}$, as shown in the proof given below.

The mode, the median and the mean of x are thus determined by the equations:

the mode: $\log x = \log \xi - 2{\cdot}3026\ \sigma^2$ (7.2.7)

the median: $\log x = \log \xi$ (7.2.8)

the mean: $\log x = \log \xi + 1{\cdot}1513\ \sigma^2$. (7.2.9)

The mode is less than the median, which is again less than the mean, and the greater the standard deviation, the greater the difference between these three characteristics.

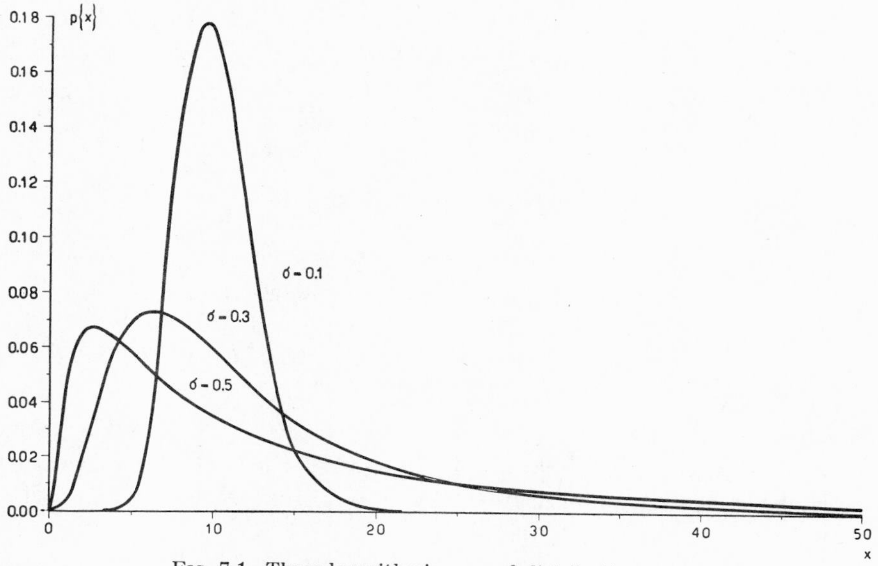

FIG. 7.1. Three logarithmic normal distribution curves. $\mathcal{M}\{\log x\} = 1{\cdot}0$ and $\sigma\{\log x\} = 0{\cdot}1,\ 0{\cdot}3$, and $0{\cdot}5$.

Fig. 7.1 shows three logarithmic normal distribution curves with

$$\mathcal{M}\{\log x\} = \log \xi = 1{\cdot}0$$

and

$$\sigma\{\log x\} = \sigma = \begin{cases} 0{\cdot}1 \\ 0{\cdot}3 \\ 0{\cdot}5 . \end{cases}$$

In all three distributions the median is equal to 10. The corresponding cumulative distribution curves have been plotted in Fig. 7.2. The curves intersect at the point (10, 50%). For $\sigma = 0{\cdot}5$ and $0{\cdot}3$ the distributions are

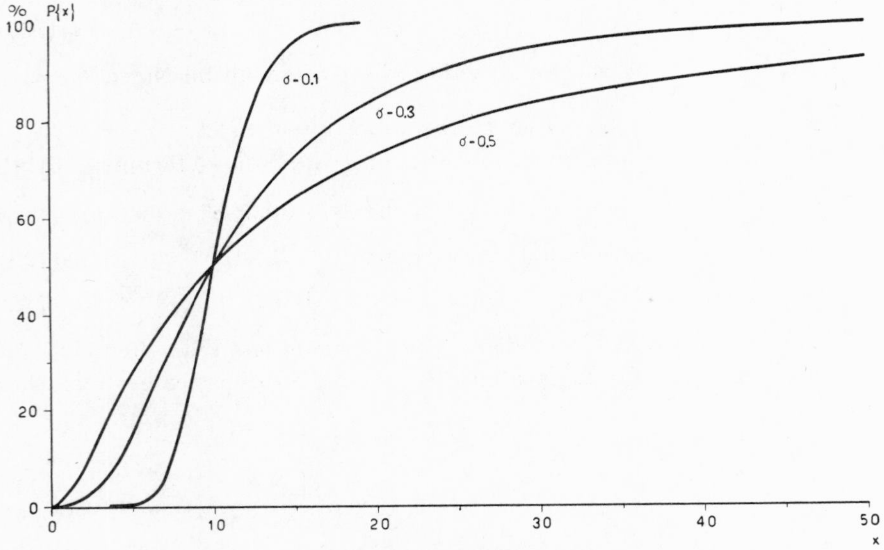

FIG. 7.2. Three cumulative logarithmic normal distribution curves. $\mathcal{M}\{\log x\} = 1{\cdot}0$ and $\sigma\{\log x\} = 0{\cdot}1,\ 0{\cdot}3$, and $0{\cdot}5$.

markedly skew, while for $\sigma = 0{\cdot}1$ the distribution does not differ noticeably from a normal distribution.

In Fig. 7.3 the cumulative distribution curves have been plotted on probability paper. The three curves are "logarithmic curves" with the equation

$$u = \frac{\log x - 1{\cdot}0}{\sigma},$$

for $\sigma = 0{\cdot}1$, $0{\cdot}3$, and $0{\cdot}5$. Fig. 7.4 shows the same curves on *logarithmic probability paper*, i. e., a special graph paper with a logarithmic scale as the abscissa and a normal probability scale as the ordinate. Here the cumulative distribution curves form straight lines.

Figs. 7.1–7.3 show that the skewness of the logarithmic normal distribution depends on the standard deviation $\sigma\{\log x\}$. The larger the standard deviation,

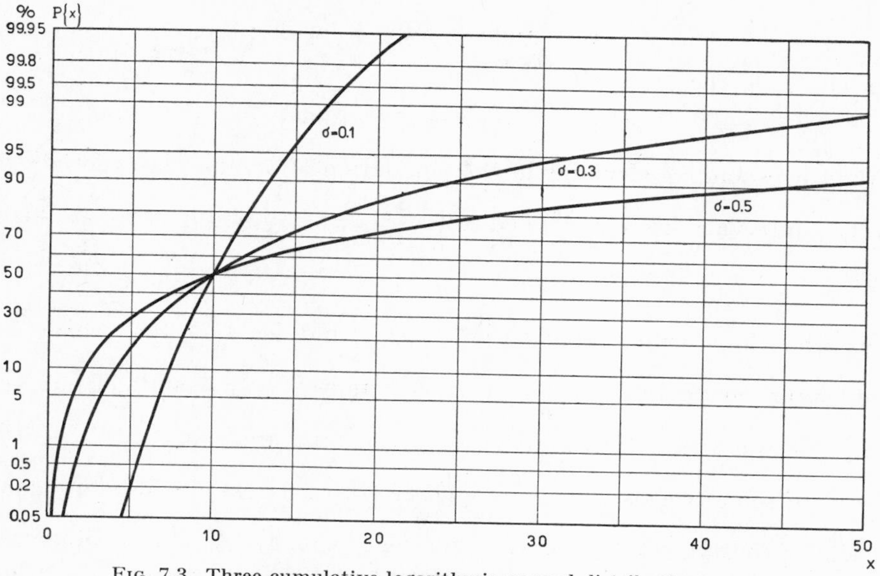

FIG. 7.3. Three cumulative logarithmic normal distribution curves plotted on probability paper.
$m\{\log x\} = 1{\cdot}0$ and $\sigma\{\log x\} = 0{\cdot}1,\ 0{\cdot}3,$ and $0{\cdot}5$.

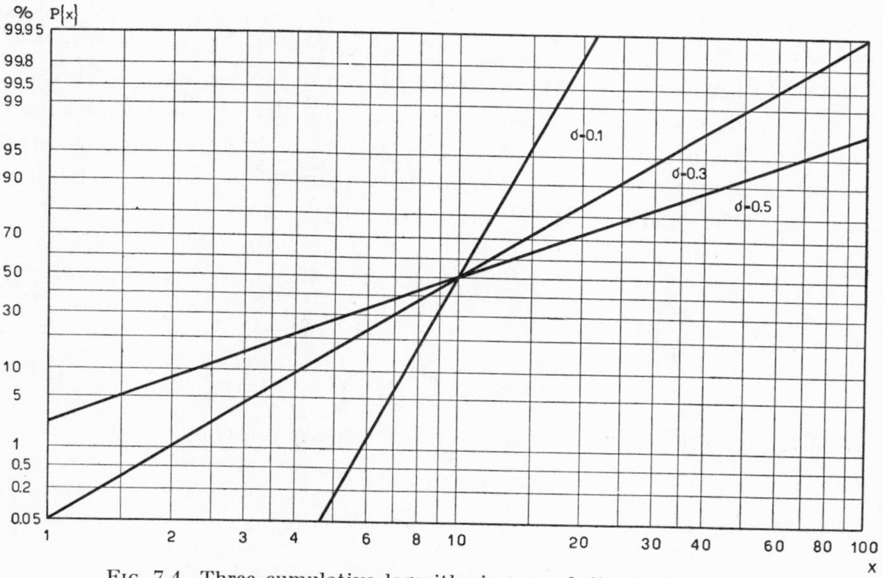

FIG. 7.4. Three cumulative logarithmic normal distribution curves plotted on logarithmic probability paper.
$m\{\log x\} = 1{\cdot}0$ and $\sigma\{\log x\} = 0{\cdot}1,\ 0{\cdot}3,$ and $0{\cdot}5$.

the more pronounced the skewness. In the normally distributed population of $\log x$-values $47{\cdot}5\%$ of the elements belong to the interval

$$(\log \xi - 1{\cdot}96\sigma < \log x < \log \xi)$$

and 47·5% to the interval

$$(\log \xi < \log x < \log \xi + 1{\cdot}96\sigma)\ .$$

If the corresponding elements in the population of x-values are considered, the two intervals are transformed to $\left(\frac{\xi}{1+\alpha} < x < \xi\right)$ and $(\xi < x < \xi(1+\alpha))$, where

$$10^{1{\cdot}96\sigma} = 1+\alpha\ .$$

Thus, the three values $\log \xi - 1{\cdot}96\sigma$, $\log \xi$, and $\log \xi + 1{\cdot}96\sigma$, which are equidistant, are transformed into three values $\frac{\xi}{1+\alpha}$, ξ, and $\xi(1+\alpha)$, the ratio between the distances now being constant and equal to $1+\alpha$, the lengths of the intervals being $\frac{\alpha\xi}{1+\alpha}$ and $\alpha\xi$. If σ is large, α is also large, and the size of the two intervals is very different; i. e., 47·5% of the population are squeezed together in a small interval to the left of the median, while another 47·5% are scattered over a comparatively large interval to the right of the median. If σ is small and α consequently small, the two intervals are almost of the same size and the distribution of x is practically normal. This agrees with the fact that, for small values of σ, the function $y = \log x$ is practically linear in the interval $(\log \xi - 1{\cdot}96\sigma < \log x < \log \xi + 1{\cdot}96\sigma)$. In this case the logarithmic transformation is practically identical with a linear transformation, so that both $\log x$ and x can be regarded as normally distributed. In practical work we generally reckon that both $\log x$ and x can be regarded as normally distributed as long as

$$c_x = \frac{s_x}{\bar{x}} < \frac{1}{3}\,. \qquad (7.2.10)$$

From (5.9.7) we have

$$\sigma\{\log x\} \simeq M\,\frac{\sigma\{x\}}{\mathcal{M}\{x\}}, \qquad (7.2.11)$$

if $y = \log x$ is approximately linear in the neighbourhood of $x = \mathcal{M}\{x\}$, and as

$$\frac{s_x}{\bar{x}} \approx \frac{\sigma\{x\}}{\mathcal{M}\{x\}}, \qquad (7.2.12)$$

the above rule approximately gives $s_{\log x} < \frac{M}{3} = 0{\cdot}14$.

Fig. 7.3 shows that even for $\sigma = 0{\cdot}1$ the part of the logarithmic curve corresponding to the interval $(0{\cdot}0005 < P < 0{\cdot}9995)$ has a definite curvature. For a corresponding empirical distribution, however, a curvature of this order of magnitude will hardly be demonstrable on account of the random variation of the points about the theoretical curve.

Derivation of the mode and the mean.

The abscissa of the maximum of the distribution, the mode, is determined by differentiating the distribution function. (7.2.3) and (6.3.6) lead to

$$p'\{x\} = -\frac{M}{\sigma x^2}\varphi(u) - \frac{M^2}{\sigma^2 x^2}u\varphi(u)$$

$$= -\frac{M}{\sigma x^2}\varphi(u)\left(1+\frac{M}{\sigma}u\right).$$

Thus, the mode is determined by the equation

$$1+\frac{M}{\sigma}u = 1+\frac{M}{\sigma}\frac{\log x - \log \xi}{\sigma} = 0,$$

which leads to

$$\log x = \log \xi - \frac{\sigma^2}{M}$$

and

$$x = \xi \times 10^{-\frac{\sigma^2}{M}}.$$

The mean of x is defined as

$$m\{x\} = \frac{M}{\sigma}\int_0^\infty x\frac{1}{x}\varphi(u)\,dx, \quad u = \frac{\log x - \log \xi}{\sigma}.$$

If u is introduced as a new variable, we get

$$\frac{dx}{du} = \frac{\sigma}{M}\xi e^{\frac{\sigma u}{M}}$$

and

$$m\{x\} = \frac{\xi}{\sqrt{2\pi}}\int_{-\infty}^\infty e^{-\frac{u^2}{2}+\frac{\sigma u}{M}}du$$

$$= \xi e^{\frac{\sigma^2}{2M^2}}\frac{1}{\sqrt{2\pi}}\int_{-\infty}^\infty e^{-\frac{1}{2}\left(u-\frac{\sigma}{M}\right)^2}du$$

$$= \xi \times 10^{\frac{\sigma^2}{2M}},$$

since

$$\frac{1}{\sqrt{2\pi}}\int_{-\infty}^\infty e^{-\frac{1}{2}\left(u-\frac{\sigma}{M}\right)^2}du = 1.$$

Example 7.1. Table 7.1 gives the distribution of 750 consumers of electricity according to the time their equipment has been used at maximum load, during a certain period, and Fig. 7.5 shows the corresponding histogram. The distribution is skew, and in the fractile diagram, Fig. 7.6, which has been plotted from the class limits and the cumulative frequencies, the points scatter about a curved line, which rises comparatively steeply for

TABLE 7.1.
Distribution of 750 consumers of electricity according to time of using their equipment.
Class length: 100 hours.

Time in units of 100 hours	Number of consumers	Cumulative frequency in %	Time in units of 100 hours	Number of consumers	Cumulative frequency in %
0·5	2	0·27	20·5	1	98·13
1·5	15	2·27	21·5	3	98·53
2·5	44	8·1	22·5	1	98·67
3·5	83	19·2	23·5	2	98·93
4·5	108	33·6	24·5	0	,,
5·5	110	48·3	25·5	1	99·07
6·5	83	59·3	26·5	0	,,
7·5	75	69·3	27·5	1	99·20
8·5	49	75·9	28·5	1	99·33
9·5	34	80·4	29·5	2	99·60
10·5	27	84·0	30·5	0	,,
11·5	21	86·8	31·5	1	99·73
12·5	24	90·0	32·5	0	,,
13·5	13	91·7	33·5	0	,,
14·5	13	93·5	34·5	0	,,
15·5	19	96·0	35·5	0	,,
16·5	8	97·1	36·5	1	99·87
17·5	3	97·5	37·5	0	,,
18·5	2	97·73	38·5	0	,,
19·5	2	98·00	39·5	1	100·00
			Total:	750	

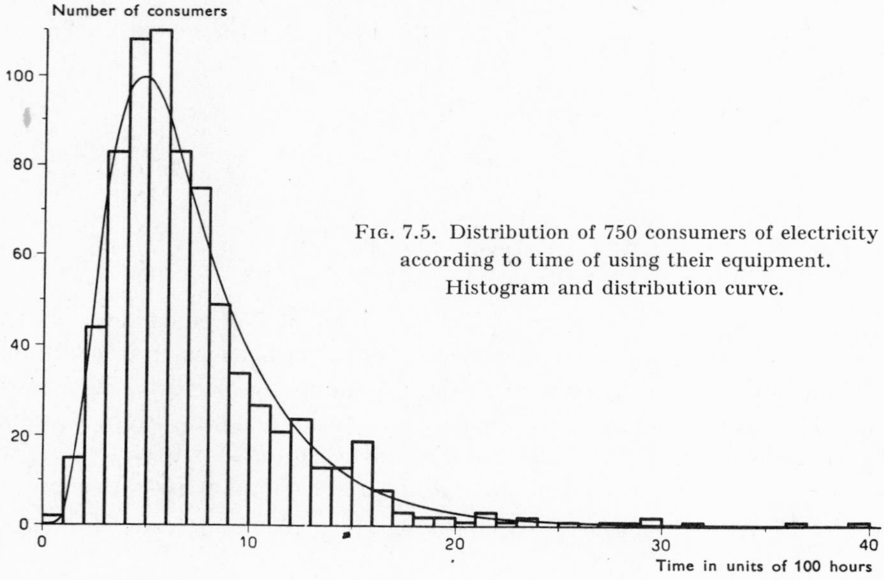

FIG. 7.5. Distribution of 750 consumers of electricity according to time of using their equipment. Histogram and distribution curve.

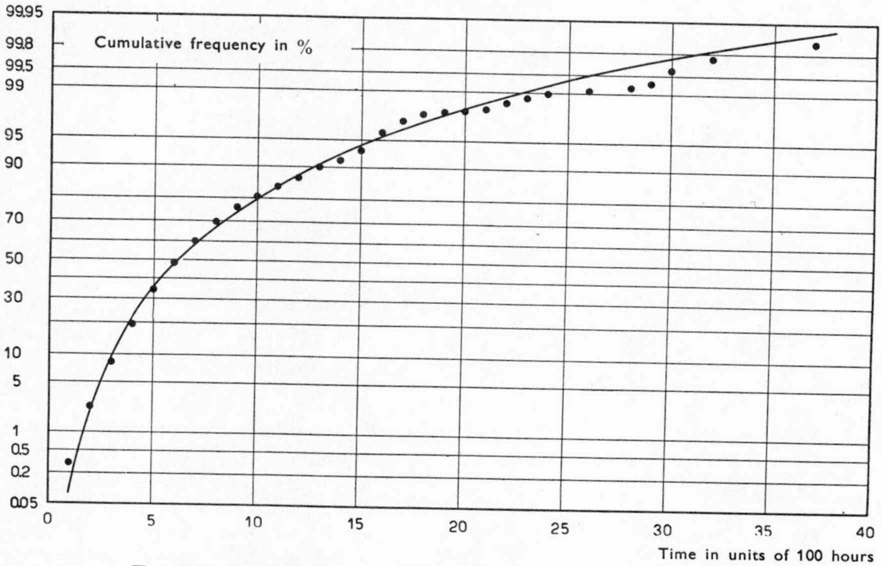

FIG. 7.6. Fractile diagram for the distribution of 750 consumers of electricity according to time of using their equipment.

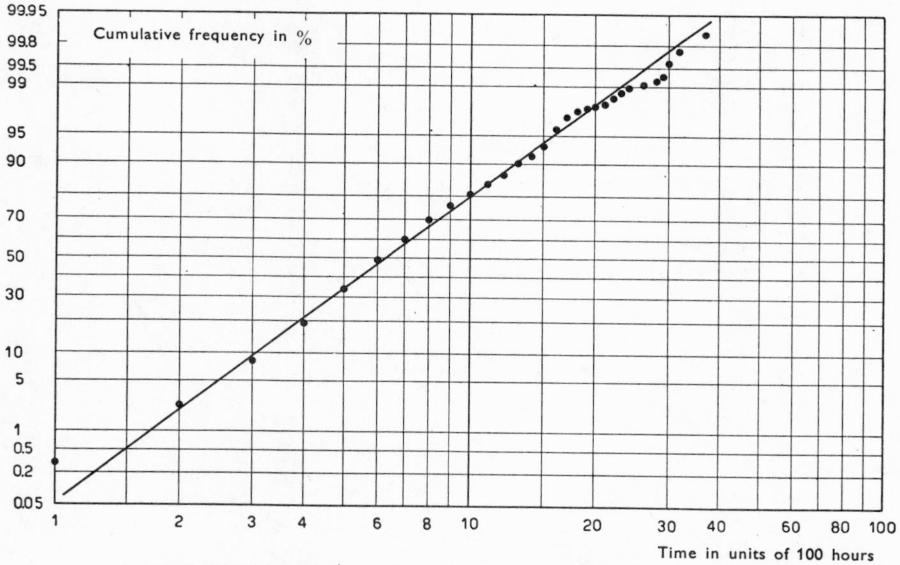

FIG. 7.7. Fractile diagram for the distribution of 750 consumers of electricity according to the logarithm of the consumption time.

small values of the variable. As this curve resembles a logarithmic curve, a new fractile diagram is tentatively plotted on logarithmic probability paper, see Fig. 7.7. Here the points are grouped about a straight line and we may therefore consider the logarithm of the consumption time to be normally distributed.

Two points are read off Fig. 7.7, e. g. (640 hours, 50%) and (360 hours, 15·9%). This gives us as a graphical estimate of the parameters,

$$\log 640 = 2{\cdot}806 \approx \log \xi$$

and

$$\log 640 - \log 360 = 0{\cdot}250 \approx \sigma\ .$$

TABLE 7.2.

Computation of mean and variance.

Distribution of 750 consumers of electricity according to the logarithm of the consumption time in hours.

Class length: 0·100.

Class midpoint t	Number of consumers a	$t = 1{\cdot}850+0{\cdot}100\,w$		
		w	w^2	$(w+1)^2$
1·850	1	0	0	1
1·950	1	1	1	4
2·050	0	2	4	9
2·150	7	3	9	16
2·250	8	4	16	25
2·350	15	5	25	36
2·450	39	6	36	49
2·550	73	7	49	64
2·650	109	8	64	81
2·750	141	9	81	100
2·850	123	10	100	121
2·950	86	11	121	144
3·050	65	12	144	169
3·150	52	13	169	196
3·250	14	14	196	225
3·350	8	15	225	256
3·450	5	16	256	289
3·550	3	17	289	324
	750			
S		7094		
SS		71490		86428
S^2/n		67100		82038
SSD		4390		4390
s_w^2		5·861		
s_w		2·421		
$\overline{w}$		9·459		

$$\bar{t} = 1{\cdot}850+0{\cdot}100\times 9{\cdot}459 = 2{\cdot}7959$$

$$s_t = 0{\cdot}100\times 2{\cdot}421 = 0{\cdot}2421$$

The computation of the mean and the standard deviation is carried out according to the scheme shown in Table 4.4, the logarithm of the 750 consumption times being suitably grouped, see Table 7.2. The number of consumers in each class interval is enumerated from the primary data by noting the antilogarithms of the class limits chosen and then counting the number of consumers with consumption time between these limits. Fig. 7.8 shows the histogram corresponding to the distribution in Table 7.2.

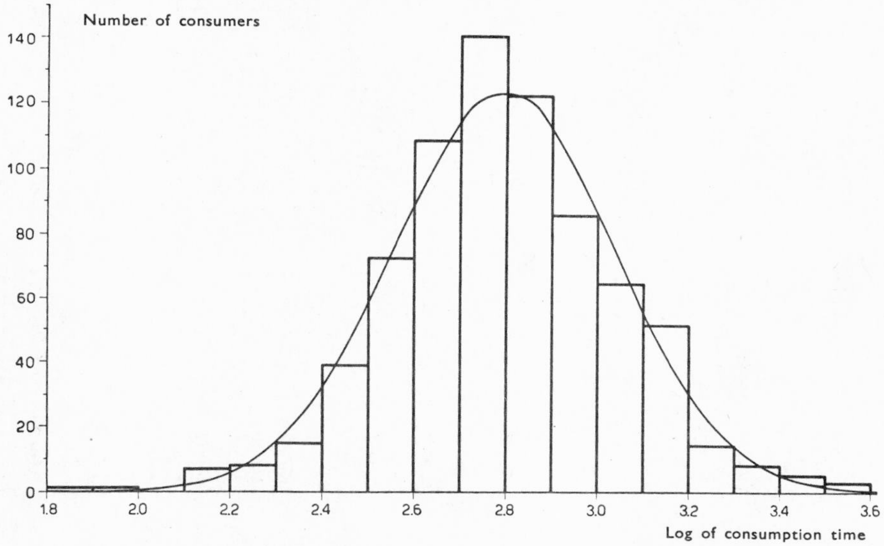

FIG. 7.8. Distribution of 750 consumers of electricity according to the logarithm of the consumption time. Histogram and distribution curve.

The mean of the logarithms of the consumption time is 2·7959, and the standard deviation 0·2421. These numbers are in good agreement with the graphical estimates given previously.

The distribution curve drawn in Fig. 7.8 has been obtained according to the scheme given in Table 6.5, the ordinate, z, being equal to

$$z = 750 \times 0{\cdot}100 \frac{\varphi(u)}{0{\cdot}2421},$$

where

$$u = \frac{y - 2{\cdot}7959}{0{\cdot}2421},$$

and y denotes the abscissa, i. e., the logarithm of the consumption time.

Table 7.3 shows the fitting of the logarithmic normal distribution to the observed data. The equation of the cumulative distribution function is

$$P\{x\} = \Phi(u), \quad u = \frac{\log x - 0{\cdot}7959}{0{\cdot}2421},$$

TABLE 7.3.

Fitting a logarithmic normal distribution.

Distribution of 750 consumers of electricity according to time of using their equipment (unit: 100 hours).

$$u = \frac{\log\left(t+\frac{\Delta t}{2}\right) - 0{\cdot}7959}{0{\cdot}2421}.$$

Class limits $t+\frac{\Delta t}{2}$	$\log\left(t+\frac{\Delta t}{2}\right)$	u	$\Phi(u)$	$750\ \Phi(u)$	Cumulative number of consumers	$\varphi(u)$	$\frac{\varphi(u)}{t+\frac{\Delta t}{2}}$	$\frac{1345{\cdot}4\ \varphi(u)}{t+\frac{\Delta t}{2}}$
1	0·0000	−3·29	0·0005	0·4	2	0·00178	0·00178	2·4
2	0·3010	−2·04	0·0207	15·5	17	0·04880	0·02440	32·8
3	0·4771	−1·32	0·0934	70·1	61	0·1669	0·05563	74·8
4	0·6021	−0·80	0·2119	158·9	144	0·2897	0·07243	97·4
5	0·6990	−0·40	0·3446	258·5	252	0·3683	0·07366	99·1
6	0·7782	−0·07	0·4721	354·1	362	0·3980	0·06633	89·2
7	0·8451	0·20	0·5793	434·5	445	0·3910	0·05586	75·2
8	0·9031	0·44	0·6700	502·5	520	0·3621	0·04526	60·9
9	0·9543	0·65	0·7422	556·7	569	0·3230	0·03589	48·3
10	1·0000	0·84	0·7995	599·6	603	0·2803	0·02803	37·7
12	1·0792	1·17	0·8790	659·3	651	0·2012	0·01677	22·6
14	1·1462	1·45	0·9265	694·9	688	0·1394	0·00996	13·4
16	1·2042	1·69	0·9545	715·9	720	0·0957	0·00598	8·0
18	1·2553	1·90	0·9713	728·5	731	0·0656	0·00364	4·9
20	1·3010	2·09	0·9817	736·3	735	0·0449	0·00225	3·0
25	1·3979	2·49	0·9936	745·2	742	0·0180	0·00072	1·0
30	1·4771	2·81	0·9975	748·1	747	0·0077	0·00026	0·3
40	1·6021	3·33	0·9996	749·7	750	0·0016	0·00004	0·1

x denoting time in units of 100 hours. The distribution curve corresponding to the histogram in Fig. 7.5 has the equation

$$y = 750 \frac{0{\cdot}4343}{0{\cdot}2421} \frac{\varphi(u)}{x}$$

$$= 1345{\cdot}4 \frac{\varphi(u)}{x}, \quad u = \frac{\log x - 0{\cdot}7959}{0{\cdot}2421},$$

see (7.2.3) and (6.8.3). (The class interval, Δt, is here equal to 1).

According to (7.2.7)—(7.2.9) the mode, the median and the mean of the logarithmic normal distribution may be estimated from the following equations:

$$\log x = 2{\cdot}7959 - 2{\cdot}3026 \times 0{\cdot}05861 = 2{\cdot}6609$$

$$\log x = 2{\cdot}7959$$

$$\log x = 2{\cdot}7959 + 1{\cdot}1513 \times 0{\cdot}05861 = 2{\cdot}8634\,,$$

which lead to

$$\text{mode} \approx 458 \text{ hours}$$

$$\text{median} \approx 625 \text{ hours}$$

$$\text{mean} \approx 730 \text{ hours.}$$

Thus, the consumption period with the greatest frequency is about 460 hours, and half of the consumers use their equipment at maximum load for more than 625 hours. The average time of using the equipment is about 730 hours. (If the mean is computed directly from Table 7.1 we get 734 hours).

Hence, the distribution of our 750 consumers according to the time they use their equipment at maximum load may be characterized by two numbers: the mean and the standard deviation of the logarithms of their consumption times. This is of great value when we wish to compare this distribution with similar distributions pertaining to other groups of consumers or other periods of the year, it now being possible to make the comparison by using means and standard deviations only.

Example 7.2. Table 7.4 gives the results of two experiments in which the reaction of flies was investigated after they had been subjected to a nerve poison for 30 and 60 seconds, respectively. The paralytic effect of the poison is defined by the time—called reaction time—which elapses from the moment the fly is brought into contact with the poison until the paralysis is so extensive that the fly can no longer stand, but lies on its back.

Table 7.4 shows that the distributions are markedly skew, which fact may be verified from fractile diagrams. Fig. 7.9 shows the fractile diagram for the logarithms of the reaction times. These points are apparently scattered at random about a straight line, and we therefore assume the distribution

TABLE 7.4.
Distribution of 16 and 15 flies according to reaction time.

Contact time: 30 seconds		Contact time: 60 seconds	
Reaction time in minutes	Cumulative frequency in %	Reaction time in minutes	Cumulative frequency in %
3	3·1	2	3·3
5	9·4	5	10·0
5	15·6	5	16·7
7	21·9	7	23·3
9	28·1	8	30·0
9	34·4	9	36·7
10	40·6	14	43·3
12	46·9	18	50·0
20	53·1	24	56·7
24	59·4	26	63·3
24	65·6	26	70·0
34	71·9	34	76·7
43	78·1	37	83·3
46	84·4	42	90·0
58	90·6	90	96·7
140	96·9		

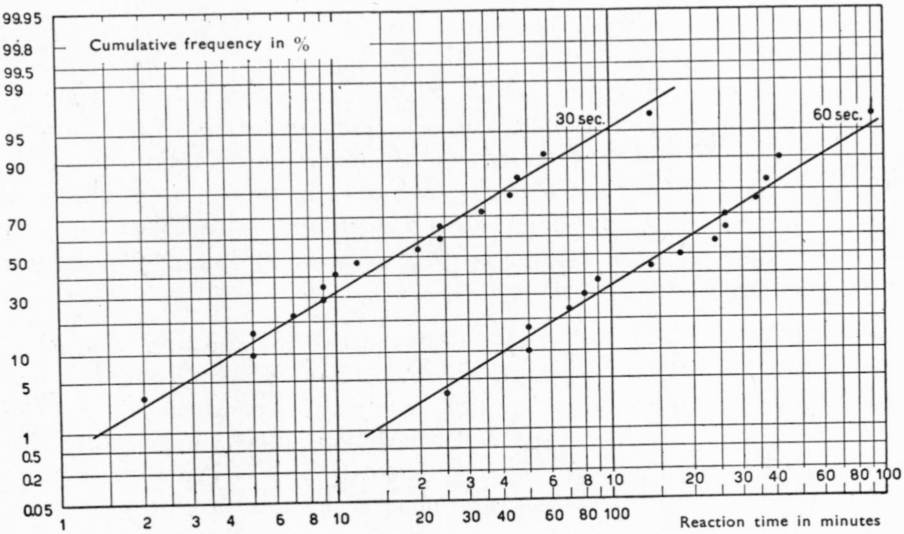

FIG. 7.9. Fractile diagram for the distributions of two groups of flies according to reaction time. (Contact time: 30 and 60 seconds, respectively). Abscissa: The logarithm of the reaction time in minutes.

TABLE 7.5.

Computation of the mean and the standard deviation of the logarithms of the reaction times.

The distributions of 16 and 15 flies according to reaction time.

	Contact time: 30 seconds	Contact time: 60 seconds
	Logarithm of reaction time	Logarithm of reaction time
	0·477	0·301
	0·699	0·699
	0·699	0·699
	0·845	0·845
	0·954	0·903
	0·954	0·954
	1·000	1·146
	1·079	1·255
	1·301	1·380
	1·380	1·415
	1·380	1·415
	1·532	1·532
	1·634	1·568
	1·663	1·623
	1·763	1·954
	2·146	—
S	19·506	17·689
SS	26·900664	23·562437
S^2/n	23·780252	20·860049
SSD	3·120412	2·702388
s^2	0·2080	0·1930
s	0·456	0·439
S/n	1·219	1·179

to be logarithmic normal. (The cumulative frequencies in Table 7.4 have been calculated as $\frac{i-\frac{1}{2}}{n}$, cf. § 6.7). Analogous results were found in a considerable number of experiments in which the time of contact with the poison, the poison compound, and the fly species were varied.

Table 7.5 shows the computation of the means and standard deviations of the logarithms of the reaction times. The two means and standard deviations are 1·219 and 1·179, and 0·456 and 0·439, respectively. The mean and the standard deviation are smaller when the contact time is 60 seconds than when it is 30 seconds. This difference in means and standard deviations may be due to the difference in contact time, but it might also be caused by random variation, i. e., the "error" of the experiment. To distinguish between these possibilities we need to know the reliability of our results. Methods for dealing with this important problem are given in Chapter 9 et seq.

7.3. Other Transformations.

If different transformations of the variable are attempted, it is often possible to find a function of the observational results which may be regarded as normally distributed. The choice of the transformation function is to some extent arbitrary, particularly if the determination of the function has to be based on a single, comparatively short series of observations. In practical work, however, we often have *several empirical distributions of the same phenomenon*, corresponding to the variation of some factor, the magnitude of which will influence the parameters of the distribution, while *the type of the distribution* may be considered independent of the magnitude of this factor. For example, consider a number of sample pieces of steel, each with the same carbon content, distributed according to tensile strength. For other carbon percentages we obtain similar distributions, the "levels" of which depend upon the carbon percentages. When handling such observational data, we attempt to transform all the distributions to *normal distributions* by using *the same transformation function*. In the further analysis, the mean and standard deviation of the transformed observations are computed for each distribution, and we examine how these quantities depend on the factor in question (the carbon percentage).

If we have only one series of observations at our disposal, we run the risk that the part of the transformation function covered is so short that we can not properly distinguish its characteristics. If, however, *several sets of observations with considerably different means and variances* are available, we may be able from the corresponding diagrams to get a fuller picture of the form of the transformation function. As examples of other transformation functions of the same type as $g(x) = \log x$, we may mention $g(x) = \frac{1}{x}$ and $g(x) = \sqrt{x}$.

The advantage of having the transformation function in a form not containing parameters which have to be estimated from the sample lies in the fact that the distribution is then described fully by means of two parameters, the mean and the variance of the transformed variable. Hence, further analysis of the distribution can be carried out in terms of standard theory concerning estimates of these parameters as given in the following chapters.

If we employ a more general type of transformation function, such as

$g(x) = \log(x+\alpha)$, $g(x) = \frac{1}{x+\alpha}$ or $g(x) = \sqrt{x+\alpha}$, we introduce an additional parameter into the distribution which has to be estimated, unless its value has been given a priori.

The theoretically correct method of estimating α lies beyond the scope of this book. For practical purposes, however, it is often possible by trial and error, using a fractile diagram, to find a value of α which leads to a transformation such that the transformed observations appear to be normally distributed. Accepting this value of α as the theoretical value of the parameter we then proceed as in the case discussed above. It is important that the determination of α be based on several sets of observations with different means and variances, and that all these distributions lead to the same value of α, so that the transformation is of a general nature.

The transformation functions $g(x) = \log(x-\alpha)$, $(x > \alpha)$, and $g(x) = -\log(\alpha-x)$, $(x < \alpha)$ have been discussed in J. C. KAPTEYN and M. J. VAN UVEN: *Skew Frequency Curves*, 1916, pp. 46–50. A generalization

$$g(x) = \log(\alpha_2 - x) - \log(x - \alpha_1), \quad (\alpha_1 < x < \alpha_2),$$

has been dealt with in M. J. VAN UVEN: *Logarithmic Frequency Distributions*, Kon. Akad. v. Wetenschappen, 19, 1917, 533–46.

An example of an analysis of several empirical distributions, for which

$$g(x) = \frac{1}{\alpha - x}, \quad (x < \alpha),$$

is given by TH. MADSEN and G. RASCH: *On Immunization of Rabbit Groups*, Acta Pathol. et Microbiol. Scand., Suppl. 37, 1938, 369–80.

A more complicated transformation function

$$g(x) = \log\left(1 + \left(\frac{\alpha}{x-\beta}\right)^{\gamma}\right)$$

has been dealt with by J. IPSEN: *Contribution to the Theory of Biological Standardization*, Copenhagen, 1941, p. 98 and ff.

If the coefficient of variation of a set of observations is small, and the variable itself can be regarded as normally distributed, several functions of the variable may also be normally distributed, for all practical purposes, since the functions will be approximately linear in the comparatively small range of variation. In such cases the transformation leading to the greatest simplicity in the further handling of the data is chosen. The following chapters will show that it is often essential for the statistical analysis that *the standard deviation is the same for the different sets of observations to be compared.* When this is the case, the comparison of the distributions may be reduced to a comparison of the means. Equality of the standard deviations may often be obtained by choosing a suitable transformation.

Let a series of k distributions be given, the parameters being (ξ_1, σ_1), $(\xi_2, \sigma_2), \ldots, (\xi_k, \sigma_k)$, and suppose that a relation

$$\sigma = h(\xi) \tag{7.3.1}$$

between the mean and the standard deviation of the variable x exists. In each distribution x is replaced by $g(x)$. According to (5.9.7) we have

$$\sigma_i\{g(x)\} \simeq \sigma_i\{x\} g'(\xi_i) \,, \tag{7.3.2}$$

and from (7.3.1)

$$\sigma_i\{g(x)\} \simeq h(\xi_i) g'(\xi_i) \,. \tag{7.3.3}$$

If we choose the transformation function, $g(x)$, in such a manner that the standard deviation $\sigma\{g(x)\}$ is the same for all distributions, we have

$$h(\xi) g'(\xi) = c \,,$$

where c is a constant, or

$$g'(\xi) = \frac{c}{h(\xi)} .$$

Thus, the transformation function is determined by the equation[1])

$$g(x) = c \int \frac{dx}{h(x)} . \tag{7.3.4}$$

The practical technique is as follows: From the k empirical distributions we calculate $(\bar{x}_1, s_1), \ldots, (\bar{x}_k, s_k)$, and these values are plotted in a $(\bar{x}, s)$ coordinate system. From this diagram we obtain an estimate of the corresponding theoretical relation $\sigma = h(\xi)$, and then the transformation function is determined from (7.3.4). As a simple example we may mention $\sigma = \gamma\xi$, i. e., the standard deviation is proportional to the mean; this leads to $g(x) = \log x$.

The following examples of transformations of skew distributions illustrate a series of simple transformation functions. The examples have been chosen from many different fields, as so far a sufficient number of good examples pertaining to industrial problems is not available.

Example 7.3. The first two columns of Table 7.6[2]) give the distribution of 924 English convicts according to pulse rate per minute. The first column gives the upper class limit, and the third column the corresponding cumulative frequency. If the latter quantities $\left(t_j + \frac{\Delta t}{2}, H_j\right)$ are plotted on prob-

[1]) A more complete proof can be found in J. H. CURTISS: *On Transformations Used in the Analysis of Variance*, Ann. Math. Stat., 14, 1943, 107–122.

[2]) See WHITING, M. H.: *A Study in Criminal Anthropometry*, Biometrika, 11, 1915, 1–37.

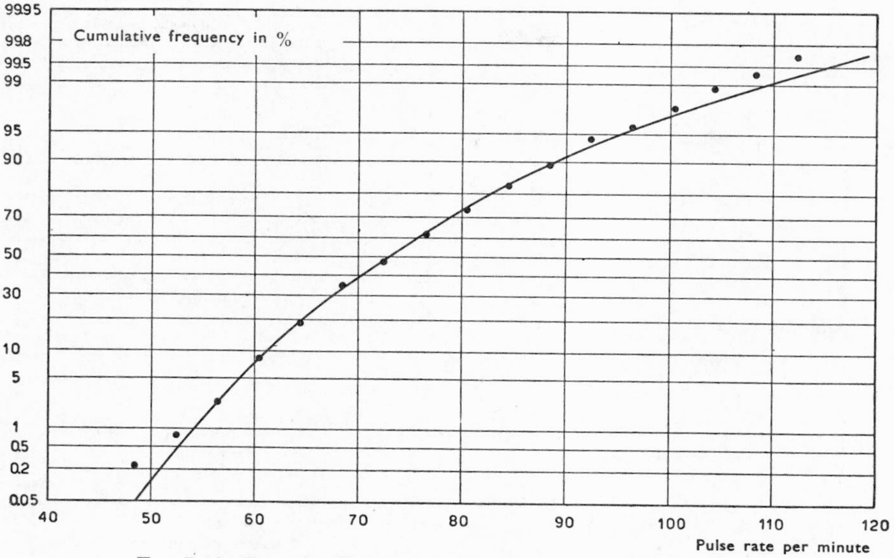

Fig. 7.10. Fractile diagram for the distribution of 924 English convicts according to pulse rate per minute.

Table 7.6.

The distribution of 924 English convicts according to pulse rate per minute. (The figures in the first column denote the upper class limit).

Pulse rate per minute	Number of convicts	Cumulative frequency in %	Duration of pulse wave in seconds	Cumulative frequency in %
48·5	2	0·22	1·237	99·78
52·5	5	0·76	1·143	99·24
56·5	17	2·6	1·062	97·4
60·5	57	8·8	0·992	91·2
64·5	90	18·5	0·930	81·5
68·5	150	34·7	0·876	65·3
72·5	120	47·7	0·828	52·3
76·5	131	61·9	0·784	38·1
80·5	109	73·7	0·745	26·3
84·5	86	83·0	0·710	17·0
88·5	62	89·7	0·678	10·3
92·5	42	94·3	0·649	5·7
96·5	15	95·9	0·622	4·1
100·5	18	97·84	0·597	2·16
104·5	9	98·81	0·574	1·19
108·5	5	99·35	0·553	0·65
112·5	3	99·68	0·533	0·32
116·5	3	100·00	0·515	0·00
Total:	924			

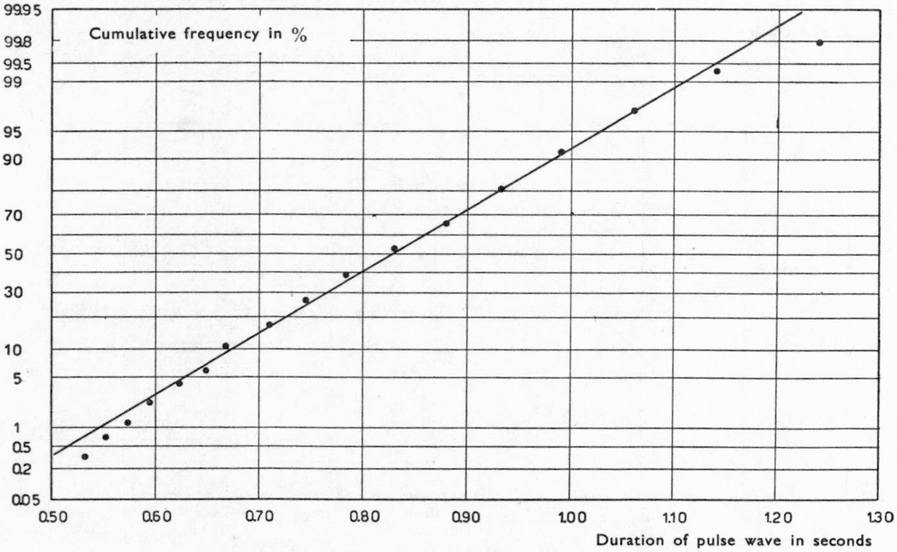

FIG. 7.11. Fractile diagram for the distribution of 924 English convicts according to the duration of the pulse wave in seconds.

TABLE 7.7.
Distribution according to age of previously unmarried females, who were married in Copenhagen during the period 1926–1930.

Age of females in years	Number	Cumulative frequency in %	Upper class limit minus 16 years
—17	275	1·04	2
18—19	2630	11·0	4
20—21	5030	30·0	6
22—24	8201	61·1	9
25—29	7426	89·2	14
30—34	1872	96·3	19
35—39	587	98·53	24
40—44	210	99·32	29
45—49	101	99·70	34
50—54	48	99·89	39
55—59	16	99·95	44
60—64	9	99·98	49
65—	5	100·00	—
Total:	26410		

ability paper, we get a set of points which scatter at random about a curved line as shown in Fig. 7.10.

It seems fairly obvious to try whether the duration of the pulse wave, i. e., the reciprocal value of the pulse rate, can be regarded as normally distributed. The last two columns of Table 7.6 give the class limits, $60/t_j+\frac{\Delta t}{2}$, and the cumulative frequencies, $100(1-H_j)$, of this distribution. The fractile diagram, Fig. 7.11, shows that the deviations of these points from a straight line are comparatively small, but that nevertheless a slight curvature still remains. Thus, in this case $g(x) = \frac{1}{x}$ appears to be more nearly normally distributed than x itself.

Example 7.4. The distribution, according to age, of previously unmarried females, who were married in Copenhagen[1]) during the period 1926–1930, is given in Table 7.7. The distribution is very skew with a pronounced accumulation between 20 and 30 years. According to law the lower limit is

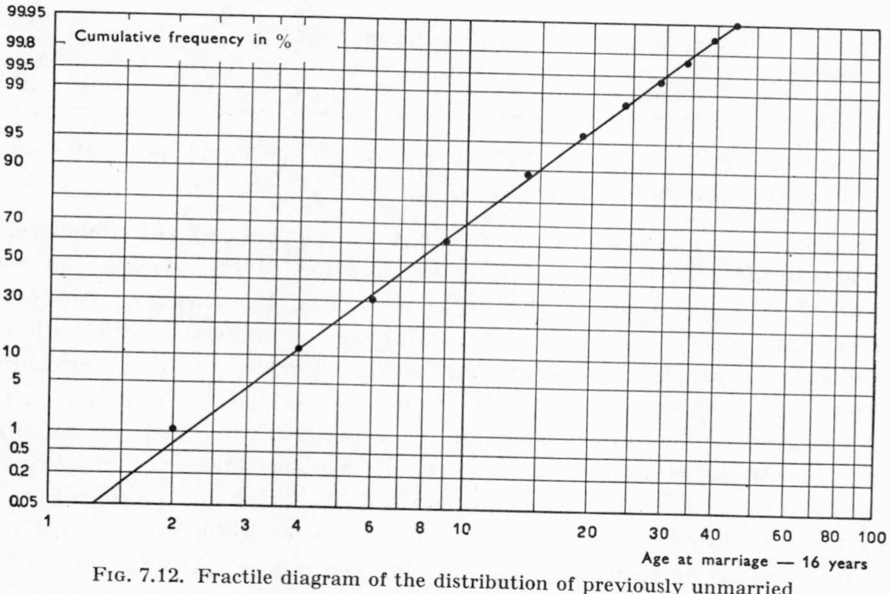

FIG. 7.12. Fractile diagram of the distribution of previously unmarried females according to age at marriage.

16 years. A fractile diagram indicates that the distribution may be regarded as logarithmic normal with origin at $x = 16$ years. If the transformation

$$g(x) = \log (x-16)$$

---

[1]) *Ægteskaber, Fødte og Døde i Aarene 1926–30,* Statistisk Tabelværk, Femte Række, Litra A, Nr. 19 (1934), Table II D, p. 11.

TABLE 7.8.
Distribution of males according to the duration of stomach or duodenal ulcer (lethal cases).

x = duration of disease in months. Upper class limit				Age at onset of disease (years)					
				15—19	20—29	30—39	40—49	50—59	60—69
				Cumulative frequencies					
x	$\log x$	$\sqrt{x}$	$\log(x+12)$	%	%	%	%	%	%
0·5	−0·30	0·71	1·10	7·1	7·7	5·3	9·8	5·3	10·5
3·5	0·54	1·87	1·19	7·1	15·4	12·3	17·1	18·4	26·3
10·5	1·02	3·24	1·35	14·3	15·4	19·3	19·5	28·9	42·1
22·5	1·35	4·74	1·54	14·3	28·2	24·6	31·7	39·5	63·2
60·0	1·78	7·75	1·86	42·9	51·3	50·9	58·5	81·6	92·1
120·0	2·08	10·95	2·12	64·3	71·8	80·7	78·0	94·7	100·0
180·0	2·26	13·42	2·28	64·3	79·5	91·2	85·4	97·4	
270·0	2·43	16·43	2·45	78·6	87·2	98·2	97·6	100·0	
390·0	2·59	19·75	2·60	92·9	94·9	100·0	100·0		
576·0	2·76	24·00	2·77	100·0	100·0				
Number of observations:				14	39	57	41	38	38

is tried, a fractile diagram is obtained in which the points are scattered at random about a straight line; see Fig. 7.12.

Example 7.5. In Table 7.8 the distribution of lethal cases of stomach and duodenal ulcer in males according to duration of the disease has been grouped in 6 age-groups according to the age of the patient at the onset of the disease. The first column of the table gives the duration of the disease in months (upper class limits), and the last six columns give the corresponding cumulative frequencies. The table shows at once that the duration of the disease decreases with increasing age of the patient. In order to characterize the association between duration of the disease and age at the onset of the disease further, it is necessary to know something about the distributions according to duration of the disease within each age-group. For these distributions 6 fractile diagrams may be plotted. The points scatter about curved lines which, broadly speaking, are of similar shape and resemble logarithmic curves. Six fractile diagrams were therefore plotted with $\log x$ as abscissas, see Table 7.8, column 2, and Fig. 7.13. This transformation seems to be "overdone", the curvature of the lines now being opposite to that of the "original" curves. If we try $\sqrt{x}$, see the third column of Table 7.8, the result is better, but not quite satisfactory, as a certain amount of curvature is still apparent as shown in Fig. 7.14. Log $(x+a)$ was then tried for various values of a, and after several attempts it was found that $\log(x+12)$ may be

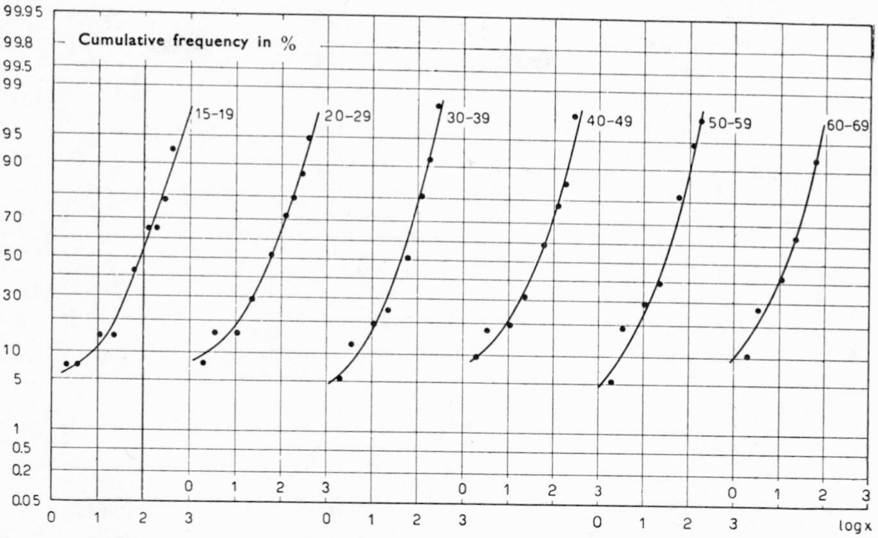

FIG. 7.13. Fractile diagram for the distribution of lethal cases of ulcer according to duration of the disease.
Abscissa: The logarithm of the duration of the disease in months.

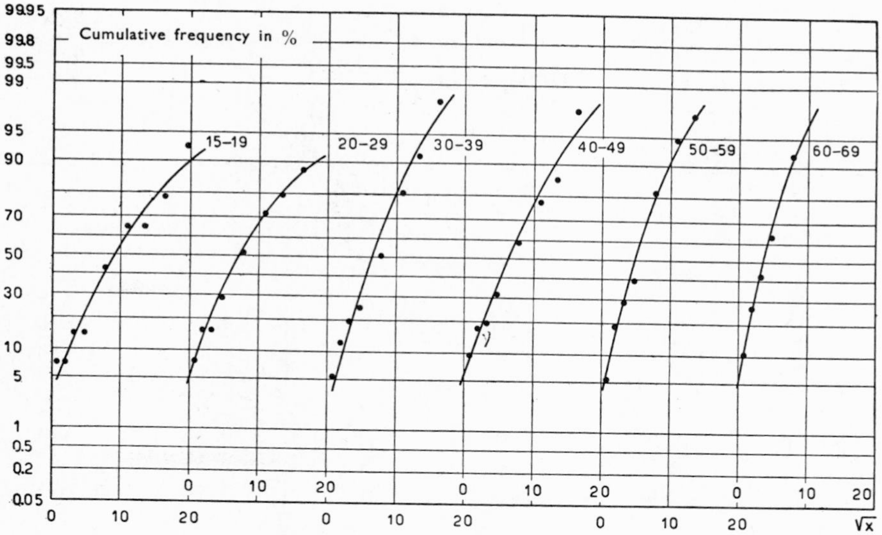

FIG. 7.14. Fractile diagram for the distribution of lethal cases of ulcer according to duration of the disease.
Abscissa: The square root of the duration of the disease in months.

considered normally distributed, see Fig. 7.15. The same transformation holds good for the corresponding data for females.

The data were then revised, each observed value x_i being transformed to

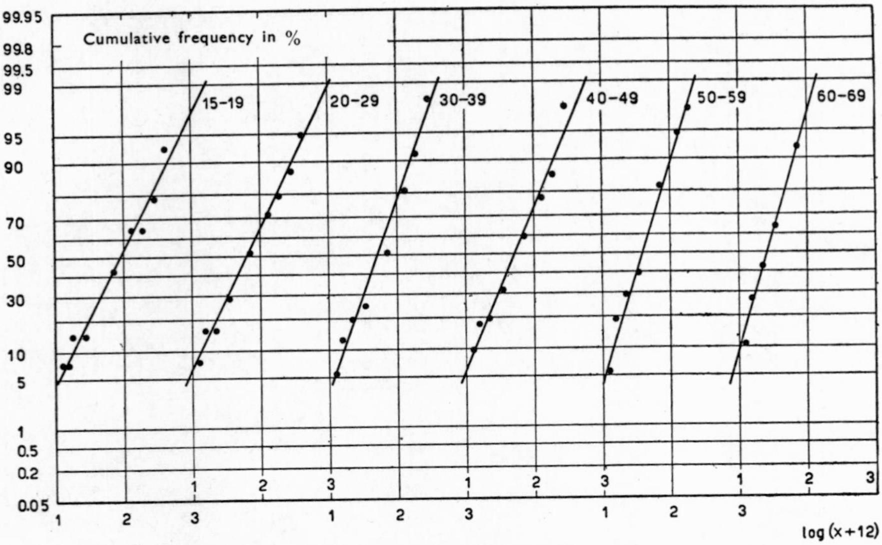

FIG. 7.15. Fractile diagram for the distribution of lethal cases of ulcer according to duration of the disease.
Abscissa: The logarithm of the duration of the disease plus 12 months.

TABLE 7.9.
Distribution of sand according to size of the particles after grinding in a tube mill.
x = size of particles in μ (10^{-3} mm).
$H\{x\}$ = cumulative frequency in %.

Time of grinding					
16 minutes		64 minutes		256 minutes	
x	$H\{x\}$	x	$H\{x\}$	x	$H\{x\}$
1·17	0·70	0·286	0·63	0·172	1·15
1·25	0·7	0·337	0·65	0·457	4·68
2·20	1·43	1·25	4·7	1·08	13·0
2·64	1·9	1·40	5·20	1·24	13·5
3·81	2·92	2·57	9·9	2·57	25·1
3·83	2·8	3·83	14·5	3·83	34·7
7·08	4·9	7·08	24·2	7·06	54·3
12·4	9·2	12·4	37·5	12·4	73·5
21·7	15·8	21·7	55·3	21·7	90·7
86	48·6	86	98·3	86	100·0
143	73·8	143	99·9		
181	89·3	181	99·9		
257	98·2	257	100·0		
419	99·9				
515	100·0				

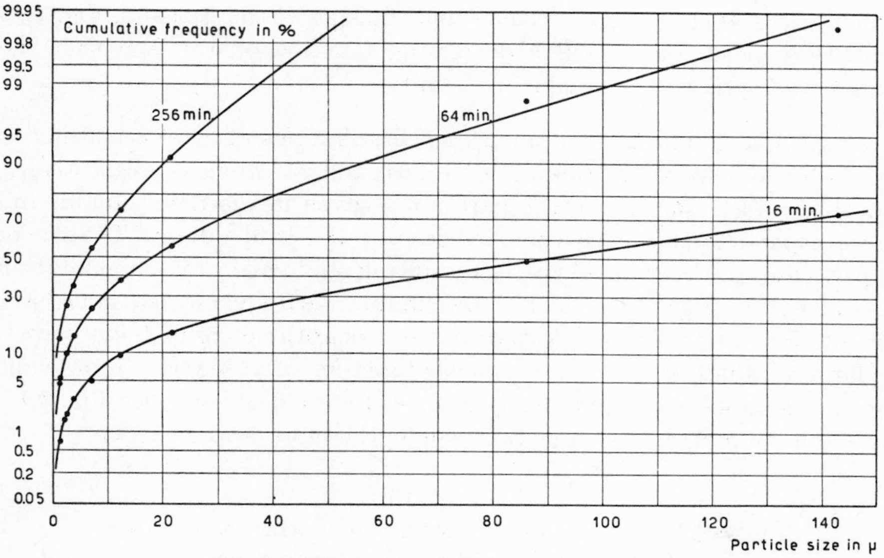

FIG. 7.16. Fractile diagram for the distribution of particles of sand according to size after 16, 64, and 256 minutes grinding in a tube mill.

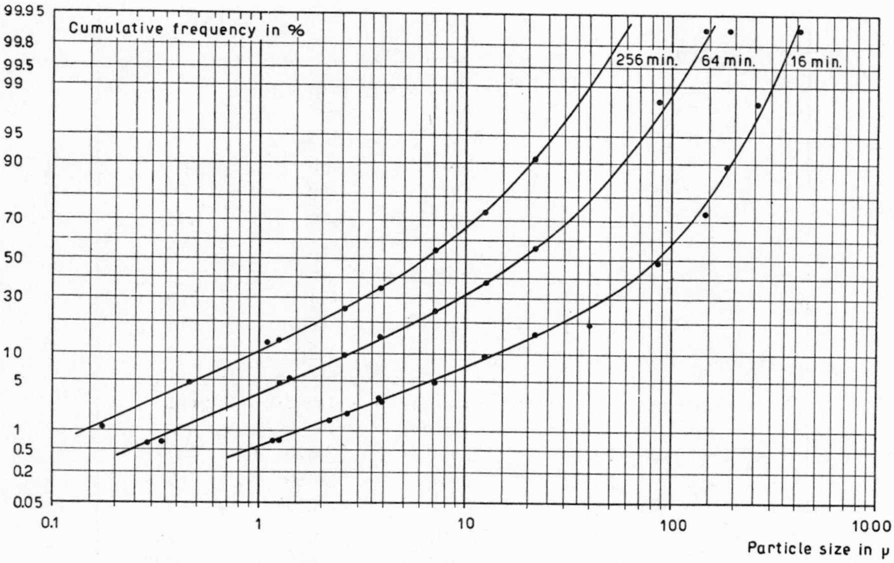

FIG. 7.17. Fractile diagram for the distribution of particles of sand according to size after 16, 64, and 256 minutes grinding in a tube mill. Abscissa: Logarithm of particle size in μ.

log (x_i+12), and the mean and standard deviation of the transformed values were computed. The association between the duration of the disease and the age of the patient could now be characterized by the relationship between these means and standard deviations and the ages of the patients. Fig. 7.15 shows that mean duration is a decreasing function of age, and that the standard deviations are roughly constant.

Example 7.6. Table 7.9 contains three distributions of sand according to size of the particles after grinding for 16, 64, and 256 minutes, respectively, in a tube mill[1]). The size of the particles is given in the first column, and the corresponding cumulative frequencies in the second column. The size of the particles was determined partly by sieving, and partly by sedimentation experiments. In Figs. 7.16 and 7.17 the cumulative frequency polygons have been plotted on probability paper and on logarithmic probability paper. For large x-values, u_H is almost a linear function of x, see Fig. 7.16, while for small x-values, u_H is practically a linear function of $\log x$, see Fig. 7.17. A possible transformation function would therefore appear to be

TABLE 7.10.
Compressive strength of 4×5 cement-mortar cubes after curing for 1, 3, 7, and 28 days, respectively.

	Curing time, days			
	1	3	7	28
	Compressive strength, kg/cm²			
	134	309	478	590
	137	299	489	608
	128	316	472	597
	136	310	498	582
	135	302	485	619
$\bar{x}$	134·0	307·2	484·4	599·2
s	3·54	6·76	10·01	14·62
$100s/\bar{x}$	2·6	2·2	2·1	2·4
	Logarithm of compressive strength in kg/cm²			
	2·127	2·490	2·679	2·771
	2·137	2·476	2·689	2·784
	2·107	2·500	2·674	2·776
	2·134	2·491	2·697	2·765
	2·130	2·480	2·686	2·792
$\overline{\log x}$	2·127	2·487	2·685	2·778
$s_{\log x}$	0·0118	0·0095	0·0089	0·0106

[1]) G. FAGERHOLT: *Particle Size Distribution of Products Ground in Tube Mill*, Copenhagen, 1945, Table 35.

$$u = \alpha \log x + \beta x + \gamma,$$

where the three parameters must be determined from the experimental data.

The experimental technique employed gives direct measures of $H\{x\}$, so that the $H\{x\}$ values given in Table 7.9 are *stochastically independent*, in contrast to the cumulative frequencies of our previous examples where the values of $H\{x\}$ were determined by summing the frequencies of observations less than or equal to x. To estimate the parameters of a distribution from this type of data we therefore require a different technique from that discussed in this chapter and Chapter 6. This technique, the regression analysis, is given in Chapter 20.

Example 7.7. Table 7.10 gives the distribution of 4×5 cement-mortar cubes according to compressive strength after curing for 1, 3, 7, and 28 days, respectively. The table shows that the mean compressive strength increases with the curing time, and that the standard deviation increases in the same proportion, the ratio $s/\bar{x}$ being practically constant. According to what was stated on p. 176, we might therefore expect the standard deviations of the distributions of the logarithms of the compressive strengths to be constant, and Table 7.10 shows that these four standard deviations are practically identical. The average value of $s/\bar{x}$ is 0·023, and we should therefore expect $s_{\log x} = 0{\cdot}434 \times 0{\cdot}023 = 0{\cdot}010$, which agrees with the values found in Table 7.10.

When determining the relation between the compressive strength and the curing time we can thus assume that the logarithms of the compressive strengths are normally distributed with a standard deviation which is independent of the curing time. The association between the compressive strength and the curing time may therefore be expressed by the relation between the logarithmic mean and the curing time.

7.4. Notes and References.

The idea of transforming skew distributions was set forth by F. Y. Edgeworth in *On the Representation of Statistics by Mathematical Formulæ*, Journ. Roy. Stat. Soc., 61, 1898, 670–700, and independently by J. C. Kapteyn in *Skew Frequency Curves*, 1903.

Edgeworth has published a large number of papers on his form of the method, based on the concept that the observed variable x is a polynomium in the normally distributed variable u, cf. *Untried Methods of Representing Frequency*, Journ. Roy. Stat. Soc., 87, 1924, 571–94, which includes references to his previously published papers.

In J. C. Kapteyn and M. J. van Uven: *Skew Frequency Curves*, 1916, the derivation of the skew distributions has been improved as compared with the exposition of 1903, and the interpretation of empirically determined transformation functions is discussed. KAPTEYN has stressed this point strongly; he is not content with being able to *describe* the distribution of the observational data, but he endeavours to find the *reason* why the distribution takes the form in question, see § 8.3.

Neither of these authors discusses the unreliability of the estimates of the parameters, which, however, plays a decisive part in the further treatment of the data, e. g., when several empirical distributions are to be compared in order to test various hypotheses with regard to the theoretical values. In view of these problems, the application of the EDGEWORTH-KAPTEYN principle given in this book is based on *transformation functions which do not contain unknown parameters,* so that the mean and the standard deviation of the transformed data form estimates of the corresponding theoretical values which characterize the population exhaustively. The following chapters deal with the distributions of the mean and the variance.

An empirical distribution may be described in several other ways than that described above. In *Skew Variation in Homogeneous Material,* Phil. Trans. Roy. Soc., A, 186, 1895, 343–414; 197, 1901, 443–459; and 216, 1916, 429–457, K. PEARSON has given a system of distribution functions, resulting from the integration of the differential equation

$$\frac{d\log y}{dx} = \frac{x+\alpha}{\beta_0+\beta_1 x+\beta_2 x^2}, \tag{7.4.1}$$

where $y = p\{x\}$ denotes the distribution function. This system includes 12 types of distribution functions. We may mention

$$\text{Type I: } p\{x\} = c \cdot (x-\alpha)^{\nu}(\beta-x)^{\mu} \quad (\alpha \leq x \leq \beta) \tag{7.4.2}$$

and

$$\text{Type III: } p\{x\} = c \cdot x^{\nu-1}e^{-\mu x} \quad (x \geq 0)\ . \tag{7.4.3}$$

A discussion of this system of distribution functions has been given by W. P. ELDERTON: *Frequency Curves and Correlation,* Cambridge, 3d Ed., 1938.

A third system of distributions is based on expansions in series of the following type

$$p\{x\} = \varphi(u)\sum_{i=0}^{\infty}\alpha_i H_i(u)\ , \quad u = \frac{x-\xi}{\sigma}, \tag{7.4.4}$$

$H_i(u)$ denoting a special kind of polynomials in u (HERMITE's polynomials). This system is connected with the names H. BRUNS: *Warscheinlichkeitsrechnung und Kollektivmasslehre,* 1906, C. V. L. CHARLIER: *Researches into the Theory of Probability,* Meddelanden från Lunds astronomiske Observatorium, 1906, and F. Y. EDGEWORTH: *The Law of Error,* Camb. Phil. Trans., 120, 1904, 36–65 and 113–141. In *On the Composition of Elementary Errors,* Skandinavisk Aktuarietidskrift, 11, 1928, 13–74 and 141–180, H. CRAMÉR has discussed the properties of this expansion and has given examples of its use.

7.5. Summary.

An important step in a statistical analysis of a given set of observations is *the mathematical formulation of the type of distribution of the population.* The distribution may sometimes be derived from what we know about the process that generates the observational data, but—at any rate as far as continuous distributions are concerned—this is an exception. As a rule, a distribution type applying to the population must be chosen on the basis of the empirical distribution.

The distribution of the population, is, however, not determined uniquely by an empirical distribution. It is possible to find a "sufficiently good" description of an empirical distribution by different mathematical formulas, for instance by using formulas from the systems of distribution functions mentioned in § 7.4. Even though these systems differ considerably from a mathematical point of view, they will often, when utilized for a given set of observational data, lead to distribution functions the graphical representations of which are practically identical.

The advantages and disadvantages of the three systems have been much discussed; see, e. g., K. PEARSON: "*Das Fehlergesetz und seine Verallgemeinerungen durch Fechner und Pearson*". *A Rejoinder*, Biometrika, 4, 1905, 169–212, and F. Y. EDGEWORTH: *Untried Methods of Representing Frequency*, Journ. Roy. Stat. Soc. 87, 1924, 571–94.

When choosing between the three systems, not only the extent of the computations involved should be considered, but also the possibilities of a further analysis of the data. This latter point is of the greatest importance, and at the present stage of development of theoretical statistics, the most important "tests" presuppose that the distribution of the observations is normal or has been transformed to a normal distribution. To this may be added that the normal distribution is characterized by two parameters, the mean and the variance, and that the estimates of these two quantities are easily computed.

The following exposition in Chapters 9–20 is therefore based on the normal distribution and the transformation of skew distributions to normal ones, according to the principles laid down in § 7.3. Experience shows that a considerable number of the distributions met with in practical work may be dealt with in a satisfactory manner by these means.

CHAPTER 8.

SOME LIMIT THEOREMS AND SAMPLING DISTRIBUTIONS

8.1. Introduction.

Even if the main part of this book is based on the assumption that the observations themselves or a known function of the observations can be interpreted as a random sample drawn from a normally distributed population, we shall in the present chapter, without complete proofs[1]), give some fundamental theorems applicable under very general assumptions regarding the distribution of the underlying population. Most of these theorems are *limit theorems* and lead to the normal distribution for large numbers of observations. These theorems will make it easier to understand why the normal distribution so often appears as the limiting distribution for many other distributions met with in the later chapters.

8.2. The Central Limit Theorem.

Let $y_1, y_2, \ldots$ denote *independent stochastic variables*, where y_i has the mean η_i and the variance σ_i^2. From these variables we derive the new variables $z_1, z_2, \ldots$, where

$$z_n = \sum_{i=1}^{n} y_i, \tag{8.2.1}$$

and according to (5.17.3) and (5.17.5)

$$\zeta_n = \mathcal{M}\{z_n\} = \sum_{i=1}^{n} \eta_i \tag{8.2.2}$$

and

$$\varkappa_n^2 = \mathcal{V}\{z_n\} = \sum_{i=1}^{n} \sigma_i^2 . \tag{8.2.3}$$

This process may be illustrated in the following way. From the population corresponding to the variable y_1 we draw elements, $y_{11}, y_{12}, \ldots$, at random, and similarly for the other variables. We then get the following table of elements drawn at random from the populations.

---

[1]) Proofs may be found in H. CRAMÉR: *Mathematical Methods of Statistics*, Princeton, 1946, Chapters 17, 32, and 33.

TABLE 8.1.

Element No.	Population No. 1 2 ... i ... n	Sum
1	y_{11} y_{21} ... y_{i1} ... y_{n1}	$z_{n1} = \sum_{i=1}^{n} y_{i1}$
2	y_{12} y_{22} ... y_{i2} ... y_{n2}	$z_{n2} = \sum_{i=1}^{n} y_{i2}$
$\vdots$	$\vdots$ $\vdots$ $\vdots$ $\vdots$	$\vdots$
j	y_{1j} y_{2j} ... y_{ij} ... y_{nj}	$z_{nj} = \sum_{i=1}^{n} y_{ij}$
$\vdots$	$\vdots$ $\vdots$ $\vdots$ $\vdots$	$\vdots$
Mean	η_1 η_2 ... η_i ... η_n	$\mathcal{M}\{z_n\} = \sum_{i=1}^{n} \eta_i$
Variance	σ_1^2 σ_2^2 ... σ_i^2 ... σ_n^2	$\mathcal{V}\{z_n\} = \sum_{i=1}^{n} \sigma_i^2$

The generation of the distribution of z_n is seen in the margin of the table.

We do not assume that the distribution functions of the variables $y_1, y_2, \ldots$ are identical, but only that the means and variances exist. Under very general conditions it may then be proved that z_n *is approximately normally distributed for large n.* A *sufficient* condition is that two positive constants α and β exist such that for all i we have $\sigma_i^2 > \alpha$ and $\mathcal{M}\{|y_i - \eta_i|^3\} < \beta$.

This implies that *the distribution of the sum of n independent random variables tends to the normal distribution for $n \to \infty$ under fairly general conditions, two of which are that the variance of the sum tends to infinity, and the variance of each variable divided by the variance of the sum tends to zero.* This theorem is called *the central limit theorem.*

In the special case where the variables all have *the same distribution,* the central limit theorem is valid if only *the mean and the variance of this common distribution exist.*

A special case of the central limit theorem was derived by A. DE MOIVRE: *Approximatio ad Summam Terminorum Binomii* $\overline{a+b}^n$ *in Seriem Expansi,* 1733, cf. K. PEARSON: *Historical Note on the Normal Curve of Errors,* Biometrika, 16, 1924, 402–404. The general theorem was given by P. S. DE LAPLACE in 1812. Also C. F. GAUSS derived the normal distribution from various hypotheses, and this distribution is therefore also called the LAPLACE-GAUSS distribution. A complete proof of the central limit theorem under general conditions was first given by A. LIAPOUNOFF in 1901. A systematic discussion of this and related theorems in the calculus of probabilities has been published by A. KHINTCHINE: *Asymptotische Gesetze der Wahrscheinlichkeitsrechnung*, 1933, and by H. CRAMÉR: *Random Variables and Probability Distributions*, 1937.

Consider a manufacturing process divided into "phases" or "stages" as in Example 2.3. The final quality, z, of the product is built up of increments, y_i, produced at the successive stages. For a given item the actual increment, y_{ij}, at a given stage may deviate at random from the expected increment, η_i. If these deviations are independent of the size of the increment, if they are stochastically independent from one stage to the next, if the variance of any deviation is small compared with the variance of the quality of the final product, and finally if the number of stages is large, then we expect the distribution of this quality to be normal.

In practice it is seldom possible to analyze whether these assumptions hold. Hence, the central limit theorem is not of much value for predicting which processes can be expected to generate a normal distribution. Before asserting that a distribution is normal it is therefore always advisable to examine the distribution of the given observations, for instance by plotting the cumulative frequency polygon on probability paper.

In the *theory of errors*, which deals with the distribution of errors of measurements, experience has, however, shown that such errors can usually be considered normally distributed; therefore, when such measurements are found to have a non-normal distribution the question arises as to the fundamental causes of this abnormality, whether of a theoretical or a practical nature.

Example 8.1. In Example 2.3, part A, a simple example of the central limit theorem has been given in the case where all the variables have the same distribution, viz.

Value of variable (y_i)	Probability
$-\varepsilon$	$1-\theta$
$+\varepsilon$	θ

For this distribution we have

$$\eta_i = \mathcal{M}\{y_i\} = -\varepsilon(1-\theta)+\varepsilon\theta = \varepsilon(2\theta-1) ,$$

$$\mathcal{M}\{y_i^2\} = \varepsilon^2(1-\theta)+\varepsilon^2\theta = \varepsilon^2 ,$$

and

$$\sigma_i^2 = \mathcal{V}\{y_i\} = \varepsilon^2-\varepsilon^2(2\theta-1)^2 = 4\varepsilon^2\theta(1-\theta) .$$

Hence

$$\zeta_n = n\varepsilon(2\theta-1)$$

and

$$\varkappa_n^2 = 4n\varepsilon^2\theta(1-\theta) .$$

According to the central limit theorem the distribution of $\sum_{i=1}^{n} y_i$ converges toward the normal distribution as $n \to \infty$. Fig. 2.2 shows that for $n = 20$, the distribution of the variable in question is practically normal when $\theta = \frac{1}{2}$.

In Example 2.3, part C, we have $n-1$ components with the same distribution as above, and one component where the two possible values of the variable have been multiplied by $1+\alpha$. For this latter component, y_1 say, we get

$$\mathcal{M}\{y_1\} = (1+\alpha)\mathcal{M}\{y_i\} = (1+\alpha)\varepsilon(2\theta-1)$$

and

$$\mathcal{V}\{y_1\} = (1+\alpha)^2\mathcal{V}\{y_i\} = (1+\alpha)^2 4\varepsilon^2\theta(1-\theta)\,.$$

Hence

$$\zeta_n = \varepsilon(2\theta-1)(\alpha+n)$$

and

$$\varkappa_n^2 = 4\varepsilon^2\theta(1-\theta)[(1+\alpha)^2+n-1]\,.$$

Consequently, if α is large compared with n, the condition

$$\frac{\mathcal{V}\{y_1\}}{\mathcal{V}\{z_n\}} \to 0$$

is not fulfilled, and the distribution of $\sum_{i=1}^{n} y_i$ does not converge toward the normal distribution, cf. Fig. 2.5.

Example 8.2. Consider a discontinuous population where the variable, y, takes on the values $0,1,\ldots,9$ with equal probabilities, $p = 0{\cdot}1$. The mean and variance are

$$\mathcal{M}\{y\} = 0{\cdot}1\ \ (0+1+\ldots+9) = 4{\cdot}5$$

and

$$\mathcal{V}\{y\} = 0{\cdot}1\ \ (0^2+1^2+\ldots+9^2)-4{\cdot}5^2 = 8{\cdot}25\,.$$

The table of random numbers, Table XIX, contains 15000 observations from such a population.

According to the central limit theorem the sum of n observations is approximately normally distributed for large n. To illustrate this theorem we compute the sums of the first four random numbers of each row on the first four pages of Table XIX. In this way we get 200 sums, the first being $1+5+7+7 = 20$, the second $8+5+4+0 = 17$, etc. Similarly, we compute the sums of the first 16 random numbers of each row.

Interpreting the numbers in each column of Table XIX as a series of observations from the same population we see that the sums above correspond to the sums $z_{n1}, z_{n2},\ldots$ in Table 8.1 for $n = 4$ and $n = 16$, respectively, with the modification that the populations are all equal.

The distributions of the 200 sums are given in Tables 8.2 and 8.3. For comparison the corresponding expected numbers of sums are computed from the normal distribution with parameters $\zeta_4 = 4\times 4{\cdot}5 = 18$ and $\varkappa_4^2 =$

$4 \times 8{\cdot}25 = 33$ for sums of four observations, and $\zeta_{16} = 16 \times 4{\cdot}5 = 72$ and $\varkappa^2_{16} = 16 \times 8{\cdot}25 = 132$ for sums of 16 observations. It is seen that for $n = 4$ the sums are nearly normally distributed.

TABLE 8.2.
The distribution of 200 sums of 4 random numbers.

Sum	Observed distribution	Normal distribution	Theoretical distribution
≤ 4	0	1·88	1·40
5 – 6	2	2·68	2·80
7 – 8	3	5·34	5·70
9 – 10	7	9·12	10·04
11 – 12	21	14·68	15·26
13 – 14	16	20·48	20·40
15 – 16	20	25·30	24·50
17 – 18	34	27·70	26·60
19 – 20	25	26·82	25·86
21 – 22	22	22·46	22·64
23 – 24	28	17·70	17·90
25 – 26	10	11·96	12·60
27 – 28	8	7·16	7·70
29 – 30	4	3·80	4·08
≥ 31	0	2·80	2·52
Total	200	199·88	200·00

TABLE 8.3.
The distribution of 200 sums of 16 random numbers.

Sum	Observed distribution	Normal distribution	Theoretical distribution
≤ 44	0	1·68	1·57
45 – 49	3	3·32	3·38
50 – 54	14	7·86	7·90
55 – 59	14	14·72	15·08
60 – 64	19	23·98	23·80
65 – 69	28	31·02	31·20
70 – 74	39	34·84	34·14
75 – 79	33	31·02	31·20
80 – 84	25	23·98	23·80
85 – 89	12	14·72	15·08
90 – 94	10	7·86	7·90
95 – 99	2	3·32	3·38
≥ 100	1	1·68	1·57
Total	200	200·00	200·00

Besides the two normal distributions the two theoretical distributions of sums of 4 and 16 random numbers, respectively, have been computed and the results are given in Tables 8.2 and 8.3.

The theoretical distribution for sums of two random numbers may be derived by simple enumerations, since the sum can take on only the values 0,1, . . ., 18, and the probability of a given value is equal to $0{\cdot}1^2$ times the number of ways the given value may be obtained as a sum of two integers between 0 and 9. Having obtained the distribution of the sum of two random numbers we combine this distribution with itself to obtain the distribution of the sum of four random numbers, and so forth. It is seen from Table 8.2 that even for $n = 4$ the theoretical distribution does not deviate very much from the corresponding normal distribution.

TABLE 8.4.

Distribution of 140 differences between duplicate titrations of raw meal with a $CaCO_3$-content between 76 and 78%. Class length: 0·05% $CaCO_3$.

Difference % $CaCO_3$	Number of differences	Cumulative frequency in %
−0·25	1	0·71
−0·20	1	1·43
−0·15	5	5·0
−0·10	13	14·3
−0·05	22	30·0
0·00	55	69·3
0·05	15	80·0
0·10	13	89·3
0·15	7	94·3
0·20	6	98·57
0·25	0	98·57
0·30	2	100·00
Total	140	

Example 8.3. Table 8.4 shows the distribution of the differences between 140 duplicate titrations of raw meal, made in the laboratory of a cement factory. As two corresponding titrations should give the same percentage of $CaCO_3$, apart from errors of titration, the differences are expected to be approximately normally distributed about zero. Table 8.4 shows that the

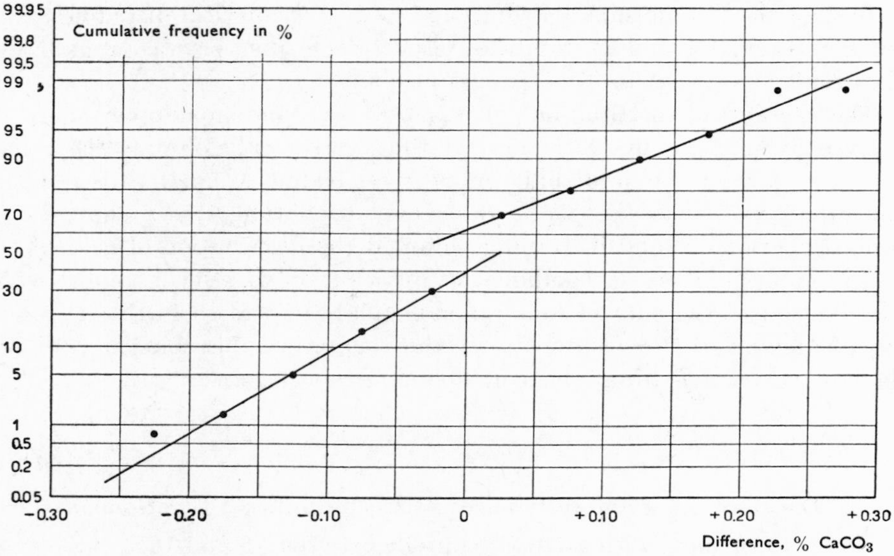

Fig. 8.1. Fractile diagram for the distribution of 140 differences between duplicate titrations of raw meal.

Table 8.5.

Distribution of 88 differences between duplicate titrations of raw meal with a $CaCO_3$-content between 76 and 78%. Class length: 0·10% $CaCO_3$.

Difference % $CaCO_3$	Number of differences	Cumulative frequency in %
−0·60	1	1·14
−0·50	0	1·14
−0·40	3	4·5
−0·30	6	11·4
−0·20	10	22·7
−0·10	15	39·8
0·00	15	56·8
0·10	14	72·7
0·20	13	87·5
0·30	4	92·0
0·40	4	96·6
0·50	1	97·73
0·60	1	98·86
0·70	1	100·00
Total	88	

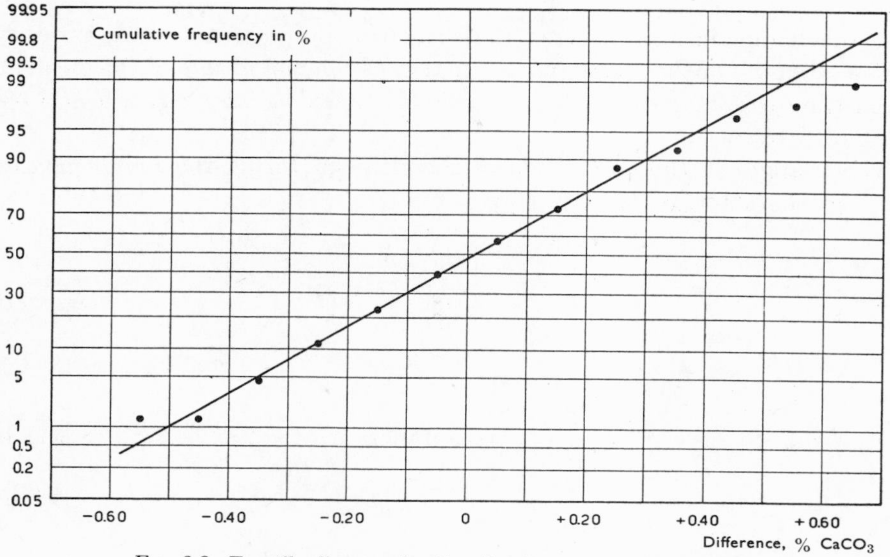

FIG. 8.2. Fractile diagram for the distribution of 88 differences between duplicate titrations of raw meal.

distribution of the 140 differences is almost symmetrical about zero with a pronounced accumulation at this point. If the cumulative frequencies are plotted on probability paper we get Fig. 8.1, which suggests that the distribution is not normal.

As the titration technique was satisfactory, the following explanation is suggested: The fact that the technician knows the result of the first titration influences his titration of the second sample, and therefore this determination is not independent of the first. This hypothesis was tested by means of 88 duplicate titrations, the correspondence of the samples now being unknown to the technician. These results are shown in Table 8.5 and Fig. 8.2. This distribution is practically normal and the range of variation is considerably larger than in the first set of observations. Thus, in order not to underestimate the uncertainty of the titration, and to make the second determination as reliable as the first, it is necessary that duplicate analyses be made independently.

8.3. Kapteyn's Derivation of Skew Distributions.

Let an element of magnitude (quality) ξ_0 be subjected to a process which successively alters the expected magnitude of ξ_0 to $\xi_1, \xi_2, \ldots$, corresponding to the different phases of the process. The change in magnitude at the ith phase, $\xi_i - \xi_{i-1}$, is assumed to depend on a "cause" acting with intensity η_i, and the magnitude of the element, ξ_{i-1}, in the following manner

$$\xi_i - \xi_{i-1} = \eta_i h(\xi_{i-1}) , \tag{8.3.1}$$

i. e., the change in magnitude of the element is proportional to the product of the intensity of the cause and a function of the magnitude of the element when the cause starts to act. η_i *is called the reaction intensity, and* $h(\xi)$ *the reaction function.*

The changes in the magnitude of the element during the first n phases may be characterized by the equations

$$\begin{aligned} \xi_1 &= \xi_0 + \eta_1 h(\xi_0) , \\ \xi_2 &= \xi_1 + \eta_2 h(\xi_1) , \\ &\vdots \\ \xi_n &= \xi_{n-1} + \eta_n h(\xi_{n-1}) . \end{aligned} \tag{8.3.2}$$

In order to determine $\xi_n - \xi_0$ as a function of $\eta_1, \eta_2, \ldots, \eta_n$, (8.3.1) is written

$$\eta_i = \frac{\xi_i - \xi_{i-1}}{h(\xi_{i-1})}, \tag{8.3.3}$$

and hence

$$\sum_{i=1}^{n} \eta_i = \sum_{i=1}^{n} \frac{\xi_i - \xi_{i-1}}{h(\xi_{i-1})}. \tag{8.3.4}$$

Assuming that the number of causes influencing the final result is large and the changes in magnitude at every stage comparatively small, we have

$$\sum_{i=1}^{n} \eta_i = \sum_{i=1}^{n} \frac{\xi_i - \xi_{i-1}}{h(\xi_{i-1})} \simeq \int_{\xi_0}^{\xi_n} \frac{dx}{h(x)}. \tag{8.3.5}$$

If we introduce

$$\zeta_n = \sum_{i=1}^{n} \eta_i \tag{8.3.6}$$

and

$$g(\xi) = \int_{\xi_0}^{\xi} \frac{dx}{h(x)}, \tag{8.3.7}$$

(8.3.5) may be written

$$\zeta_n = g(\xi_n) , \tag{8.3.8}$$

it now being possible to determine the size of the element at the end of the nth phase by solving (8.3.8) with respect to ξ_n.

In practical work it is not possible to keep the conditions of the process constant, and at each phase the reaction intensity will therefore deviate from the above stated theoretical values, and the changes in magnitude of the elements, partaking in the process, will vary. Assuming that the reaction intensity at the ith phase, y_i, is a stochastic variable with mean value η_i and variance σ_i^2, the changes in magnitude of a given element may be characterized by the following equations, equivalent to (8.3.2):

$$\begin{aligned} x_1 &= x_0 + y_1 h(x_0)\,, \quad x_0 = \xi_0\,, \\ x_2 &= x_1 + y_2 h(x_1)\,, \\ &\vdots \\ x_n &= x_{n-1} + y_n h(x_{n-1})\,. \end{aligned} \tag{8.3.9}$$

In analogy with (8.3.5) we get

$$\sum_{i=1}^{n} y_i = \sum_{i=1}^{n} \frac{x_i - x_{i-1}}{h(x_{i-1})} \simeq \int_{x_0}^{x_n} \frac{dx}{h(x)} \tag{8.3.10}$$

and

$$\sum_{i=1}^{n} y_i = z_n = g(x_n)\,. \tag{8.3.11}$$

According to the central limit theorem z_n will be normally distributed under certain general conditions when $n \to \infty$, the mean being ζ_n. (8.3.11) then implies that the elements will not be normally distributed according to size, but that *a function, $g(x)$, of the size will be normally distributed.*

For $h(x) = 1$, we find that $g(x) = x - x_0$, i. e., x is normally distributed. Thus, if the reaction function is constant, which means that the changes in magnitude are independent of the size already obtained when the causes start to act, then the distribution according to size will be normal, cf. § 8.2.

If the reaction function $h(x)$ is equal to x, we have

$$g(x) = \int_{x_0}^{x} \frac{dx}{x} = \ln x - \ln x_0\,, \tag{8.3.12}$$

i. e., $\ln x$ is normally distributed. Thus, if the change in magnitude corresponding to a given cause is proportional to the intensity of that cause and further to the size of the element, the distribution obtained will be logarithmic normal, compare with the simple case given in Example 2.3, *B*, p. 33.

The above simple explanation of the generation of skew distributions was given by KAPTEYN in 1903.

8.4. The Concept of a Sampling Distribution.

A given set of observations may be of interest in itself, for instance from a historical or administrative point of view. From a statistical viewpoint, however, *a set of observations is always interpreted as a random sample from a population*, as the purpose of a statistical analysis is to draw inferences about the population or about future samples from the population. Thus, the given sample is only one of a whole "*population of samples*", which might have been generated by repeated selections of random samples of the same size from the given population.

Consider a sample consisting of the 10 numbers resulting from 10 throws with a die. Repeating this process under similar conditions we get a new sample of 10 observations, and so forth. Likewise, the 200 diameters of rivet heads given in Table 3.1 is just one sample of diameters, and more samples, each consisting of 200 diameters of rivet heads, may be produced by repetition of the whole procedure by which the first sample came into existence.

For each sample we may compute the mean, the variance, the frequency of observations below a given value, etc. *These numbers vary from sample to sample in a random way* and the corresponding distributions are called *sampling distributions* or distributions of sample means, variances, frequencies, etc.

We distinguish between *empirical and theoretical sampling distributions* in the same manner as between empirical and theoretical distributions of the basic stochastic variable.

An empirical distribution of sample means may be obtained by taking for instance 125 samples of 4 rivets each and computing the 125 mean diameters, see Table 9.2 where this has actually been carried out. The corresponding theoretical distribution of sample means may be derived from the theoretical distribution of the given variable.

To obtain a sample mean less than or equal to a specified number, t, the observed values, $x_1, x_2, \ldots, x_n$, must satisfy the relation

$$x_1+x_2+\ldots+x_n \leq nt\,. \tag{8.4.1}$$

If we know the distribution of x, the probability of getting a set of sample values falling within the intervals (x_1, x_1+dx_1), $(x_2, x_2+dx_2), \ldots, (x_n, x_n+dx_n)$ is

$$p\{x_1\}p\{x_2\}\ldots p\{x_n\}dx_1dx_2\ldots dx_n\,, \tag{8.4.2}$$

according to the definition of random sampling, see § 5.4. Integrating this probability over the domain defined by (8.4.1) we get the probability that the sample mean takes on a value less than or equal to the given number t, i. e., the cumulative distribution function of the sample mean.

Other theoretical sampling distributions may be derived in a similar way but very often the resulting expressions are too complicated to be of any practical value. In the case of the normal distribution, however, the resulting distributions of sample means and variances are fairly simple as shown in Chapters 9 and 11.

Theoretical sampling distributions are of fundamental importance, because they permit us to predict the variations of sample means and variances. Further, the sampling distribution of the mean usually depends on the mean of the given population, so that by utilizing the sampling distribution we are able to draw inferences from the sample mean about the population mean.

In cases where we do not know the distribution of the sample mean completely, the mean and variance of this distribution may be of some value, since the mean and the variance to a certain extent characterize the distribution, see TCHEBYCHEFF's theorem in § 5.10. Further, for large n we may apply the central limit theorem, so that the sample mean and variance under certain conditions may be considered approximately normally distributed. In the following sections we therefore derive the theoretical mean and variance of the sample mean and variance.

8.5. The Sampling Distribution of the Mean.

Consider an infinite population of values of a variable x with mean ξ and variance σ^2, from which random samples of n elements are drawn. We denote the n elements of a sample by $x_1, x_2, \ldots, x_n$, where the values of each element are distributed independently of the other values and according to the given distribution function of x, i. e., $\mathcal{M}\{x_i\} = \xi$ and $\mathcal{V}\{x_i\} = \sigma^2$.

The sum and the mean of the n values,

$$S_x = x_1 + x_2 + \ldots + x_n \tag{8.5.1}$$

and

$$\bar{x} = \frac{1}{n}(x_1 + x_2 + \ldots + x_n)\,, \tag{8.5.2}$$

vary from sample to sample. According to (5.17.3) and (5.17.5) we get

$$\mathcal{M}\{S_x\} = \mathcal{M}\{x_1\} + \mathcal{M}\{x_2\} + \ldots + \mathcal{M}\{x_n\} = n\xi \tag{8.5.3}$$

and

$$\mathcal{V}\{S_x\} = \mathcal{V}\{x_1\} + \mathcal{V}\{x_2\} + \ldots + \mathcal{V}\{x_n\} = n\sigma^2 \tag{8.5.4}$$

and consequently

$$\boxed{\mathcal{M}\{\bar{x}\} = \xi} \tag{8.5.5}$$

and

$$\boxed{\mathcal{V}\{\bar{x}\} = \frac{\sigma^2}{n}}\,, \tag{8.5.6}$$

i. e., the sample means vary at random about the population mean with a standard deviation of $\sigma/\sqrt{n}$.

As a special case consider n independent observations where at each observation the probability is θ that a particular event U occurs, see § 2.1.

Let the variable x take on the value 1 if U occurs, and the value zero if U does not occur, i. e., the distribution of x is

$$p\{x = 0\} = 1-\theta ,$$

$$p\{x = 1\} = \theta . \tag{8.5.7}$$

It follows that

$$\mathcal{m}\{x\} = 0 \cdot (1-\theta)+1 \cdot \theta = \theta \tag{8.5.8}$$

and

$$\mathcal{v}\{x\} = (0-\theta)^2(1-\theta)+(1-\theta)^2\theta = \theta(1-\theta) . \tag{8.5.9}$$

The sum of n observations, S_x, then denotes the number of times the event U has occurred, and the mean

$$h = \frac{1}{n}(x_1+x_2+\ldots+x_n)$$

denotes the frequency of U in a sample of n observations. Consequently, the sampling distribution of S_x is *the binomial distribution* which has been derived in § 2.1. (In § 2.1 the notation x is used for the number of times the event U occurs).

According to (8.5.3)–(8.5.6) we get

$$\mathcal{m}\{S_x\} = n\theta , \tag{8.5.10}$$

$$\mathcal{v}\{S_x\} = n\theta(1-\theta) , \tag{8.5.11}$$

$$\mathcal{m}\{h\} = \theta , \tag{8.5.12}$$

and

$$\mathcal{v}\{h\} = \frac{\theta(1-\theta)}{n} . \tag{8.5.13}$$

These results have been derived directly from the binomial distribution in Examples 5.6 and 5.8.

It follows directly from the central limit theorem that *for large n the sample mean is approximately normally distributed*, provided that the population mean and variance are finite. This result may be useful in cases where we do not know the exact sampling distribution of the mean or where this distribution is too complicated to be handled in practice. However, the question always arises: *How large* must n be before the exact sampling distribution can be replaced by the normal distribution with sufficient accuracy? Very little can be said on this matter in general as the sampling distribution depends on the distribution of the population and the accuracy required depends on the purpose of the analysis.

It will be proved in § 9.1 that the sample mean is normally distributed for *any n*, if the population is normally distributed. Consequently, the sampling distribution of the mean does not deviate very much from the normal distribution, even if n is small, if the population is approximately normally distributed. It is therefore recommended to transform the given observations so that the distribution becomes as nearly normal as possible, even if we do not succeed in getting a "completely" normal distribution. In many cases it has been found that for $n > 30$ the sampling distribution of the mean is fairly normal and even for smaller values of n the normal distribution is often sufficiently accurate, see Example 8.2.

Consider now *a finite population* consisting of N elements, where the variable x takes on the values $x_1, x_2, \ldots, x_N$ with the same probability

$$p\{x = x_i\} = \frac{1}{N}. \tag{8.5.14}$$

Consequently we have

$$\mathcal{m}\{x\} = \frac{1}{N}\sum_{i=1}^{N} x_i = \xi \tag{8.5.15}$$

and

$$\mathcal{v}\{x\} = \frac{1}{N}\sum_{i=1}^{N} (x_i - \xi)^2 = \sigma^2 . \tag{8.5.16}$$

By random sampling from this population (without replacement) we get $N^{(n)}$ different samples as explained in § 5.4. For each sample we can compute the mean and the variance and then find the mean and variance of all sample means and variances.

The result is

$$\boxed{\mathcal{m}\{\bar{x}\} = \xi} \tag{8.5.17}$$

and

$$\boxed{\mathcal{v}\{\bar{x}\} = \frac{\sigma^2}{n}\frac{N-n}{N-1},} \tag{8.5.18}$$

i. e., the sample means vary at random about the population mean with a standard deviation of $\dfrac{\sigma}{\sqrt{n}}\sqrt{\dfrac{N-n}{N-1}}$, see § 17.2.

For increasing values of n the variance of $\bar{x}$ decreases, and for $n = N$, the variance becomes zero because the sample is identical with the whole population.

For a given value of n and increasing values of N we get

$$\frac{N-n}{N-1} \to 1 ,$$

i. e., the population becomes infinite and (8.5.18) coincides with (8.5.6).

The fraction

$$\alpha = \frac{n}{N} \tag{8.5.19}$$

is called *the sampling fraction*. Introducing $n = N\alpha$ into (8.5.18) we get

$$\mathcal{V}\{\bar{x}\} = \frac{\sigma^2}{n}(1-\alpha)\frac{N}{N-1} \tag{8.5.20}$$

or approximately

$$\mathcal{V}\{\bar{x}\} \simeq \frac{\sigma^2}{n}(1-\alpha). \tag{8.5.21}$$

In the special case where all values of x are either 1 or 0, i. e., the event U occurs or does not occur (see § 2.4 on the hypergeometric distribution), and the frequency of U in the population is θ, we get

$$\boxed{\mathcal{M}\{h\} = \theta} \tag{8.5.22}$$

and

$$\boxed{\mathcal{V}\{h\} = \frac{\theta(1-\theta)}{n}\frac{N-n}{N-1},} \tag{8.5.23}$$

where h denotes the frequency of the event U at n observations of x. (This result may also be derived by finding the mean and variance of the hypergeometric distribution).

By sampling from a finite population the central limit theorem does not apply. Nevertheless it is found that the sampling distribution of the mean is approximately normal for large values of n ($n > 30$) provided that the sampling fraction is small.

8.6. The Sampling Distribution of the Variance.

To find the mean of the sampling distribution of the variance

$$s^2 = \frac{1}{n-1}\sum_{i=1}^{n}(x_i-\bar{x})^2$$

we introduce

$$x_i-\bar{x} = (x_i-\xi)-(\bar{x}-\xi)$$

which leads to

$$(n-1)s^2 = \sum_{i=1}^{n}(x_i-\xi)^2-n(\bar{x}-\xi)^2 .$$

As

$$\mathcal{M}\{(x_i-\xi)^2\} = \mathcal{V}\{x_i\} = \sigma^2$$

and

$$\mathcal{M}\{(\bar{x}-\xi)^2\} = \mathcal{V}\{\bar{x}\} = \frac{\sigma^2}{n}$$

we get

$$\mathcal{M}\{(n-1)s^2\} = n\sigma^2 - n\frac{\sigma^2}{n} = (n-1)\sigma^2$$

or

$$\boxed{\mathcal{M}\{s^2\} = \sigma^2,} \tag{8.6.1}$$

i. e., the sample variances vary at random about the population variance.

The variance of the theoretical distribution of sample variances is defined as

$$\mathcal{V}\{s^2\} = \mathcal{M}\{(s^2-\sigma^2)^2\} \tag{8.6.2}$$

and requires the evaluation of $\mathcal{M}\{s^4\}$. The resulting variance is

$$\mathcal{V}\{s^2\} = \frac{\sigma^4}{n}\left[\frac{\mu_4}{\sigma^4} - \frac{n-3}{n-1}\right] \tag{8.6.3}$$

where

$$\mu_4 = \mathcal{M}\{(x-\xi)^4\}\,. \tag{8.6.4}$$

The sampling distribution of the variance will, according to the central limit theorem, converge to the normal distribution for increasing values of n provided that μ_4 is finite.

8.7. The Law of Large Numbers.

Combining the results of § 8.5 and TCHEBYCHEFF's theorem we can find a lower bound to the probability

$$P\{|\bar{x}-\xi| \leq \eta\}$$

where η is a specified (small) number. Substituting $\bar{x}$ for x and η for $\alpha\sigma_{\bar{x}}$ in (5.10.2) we get

$$P\{|\bar{x}-\xi| \leq \eta\} > 1 - \frac{\sigma_{\bar{x}}^2}{\eta^2} = 1 - \frac{1}{n}\frac{\sigma^2}{\eta^2}. \tag{8.7.1}$$

Since σ^2 and η^2 are given numbers, this lower bound converges to 1 as n increases. Thus, the probability that the mean of a random sample deviates at most by a given (small) amount from the population mean converges to 1 as the size of the sample increases, or in other words: *the sample mean converges in probability to the population mean.* This result is called *the law of large numbers.*

By similar reasoning on the basis of the mean and variance of the theoretical distributions of sample frequencies, h, and sample variances, s^2, it can be proved that the sample frequency, h, converges in probability to θ, and that the sample variance, s^2, converges in probability to the population variance, σ^2.

8.8. The Method of Maximum Likelihood.

Let $p\{x;\theta\}$ denote a distribution function of *known mathematical form*, containing *an unknown parameter* θ, and let $x_1, x_2, \ldots, x_n$ denote a random sample from the corresponding population. To *estimate* the unknown parameter we compute the value of a function of the n observations,

$$t = t(x_1, \ldots, x_n), \tag{8.8.1}$$

called a *statistic* or an *estimate* of θ. It is assumed that the function $t(x_1, \ldots, x_n)$ does not depend on θ. The properties of t may then be derived from its sampling distribution.

The main problem within the theory of estimation is to formulate the desired properties of estimates and to derive methods for finding estimates with these properties. The following remarks are based on the principles laid down by R. A. FISHER.

A statistic t is said to be a consistent estimate of θ if t converges in probability to θ as $n \to \infty$. From (8.7.1) it follows that the sample mean is a consistent estimate of the population mean if the population variance is finite. Even if the variance is not finite it can be proved that the sample mean converges in probability to the population mean provided that the population mean is finite (KHINTCHINE's theorem).

As an example, where the sample mean is an inconsistent estimate of a parameter, consider *Cauchy*'s distribution as given in (5.5.15). The distribution curve is symmetrical about $x = \xi$, but the population mean does not exist as the corresponding integral is not convergent. Further, it can be proved that the distribution of the sample mean is itself a *Cauchy*-distribution with the *same* parameters as the parent distribution. Thus, as an estimate of ξ, the sample mean is not better than any single observation. However, the median is a consistent estimate of ξ.

Usually, several consistent estimates of a parameter exist and some of these are asymptotically normally distributed, i. e., their sampling distributions tend to the normal distribution as $n \to \infty$. Comparing the variances of these estimates *we term the estimate with the smallest variance an asymptotically efficient estimate.* The *efficiency* of an estimate is defined as the ratio between this minimum variance and the variance of the estimate in question.

Consider, for example, the mean and the median of n observations from a normally distributed population. It can be proved that both estimates are asymptotically normally distributed with means equal to ξ, the population mean, and variances $\frac{\sigma^2}{n}$ and $\frac{\sigma^2}{n}\frac{\pi}{2}$, respectively. Thus, the variance of the mean is less than the variance of the median. Further, it can be shown that the mean is an efficient estimate of ξ. It follows that the efficiency of the median is $\frac{2}{\pi} = 0{\cdot}64$, i. e., the variance of the mean of 64 observations is equal to the variance of the median of 100 observations.

R. A. FISHER has given a method, *the method of maximum likelihood*, which under certain general conditions leads to efficient estimates.

We define the *likelihood function* L of a random sample as

$$L(x_1, \ldots, x_n; \theta) = p\{x_1; \theta\} \ldots p\{x_n; \theta\}\,. \tag{8.8.2}$$

According to the method of maximum likelihood we choose as an estimate of θ the value which maximizes L for the given values of $(x_1, \ldots, x_n)$. As it is often more convenient to work with $\log L$ than with L itself, the required value of θ is usually found by solving the *likelihood equation*

$$\frac{\partial \log L}{\partial \theta} = \sum_{i=1}^{n} \frac{\partial \log p\{x_i; \theta\}}{\partial \theta} = 0 \tag{8.8.3}$$

with respect to θ. *The solution t_n of the likelihood equation is an asymptotically efficient estimate of θ and*

$$v\{t_n\} = \frac{1}{nm\left\{\left(\frac{\partial \log p\{x; \theta\}}{\partial \theta}\right)^2\right\}}\,. \tag{8.8.4}$$

The denominator is calculated as

$$m\left\{\left(\frac{\partial \log p\{x; \theta\}}{\partial \theta}\right)^2\right\} = \int_{-\infty}^{\infty} \left(\frac{\partial \log p\{x; \theta\}}{\partial \theta}\right)^2 p\{x; \theta\}\,dx$$

$$= \int_{-\infty}^{\infty} \frac{1}{p\{x; \theta\}} \left(\frac{\partial p\{x; \theta\}}{\partial \theta}\right)^2 dx \tag{8.8.5}$$

in the continuous case and as

$$m\left\{\left(\frac{\partial \log p\{x; \theta\}}{\partial \theta}\right)^2\right\} = \sum_{i=-\infty}^{\infty} \left(\frac{\partial \log p\{x_i; \theta\}}{\partial \theta}\right)^2 p\{x_i; \theta\}$$

$$= \sum_{i=-\infty}^{\infty} \frac{1}{p\{x_i; \theta\}} \left(\frac{\partial p\{x_i; \theta\}}{\partial \theta}\right)^2 \tag{8.8.6}$$

in the discontinuous case.

This theorem can be generalized to two or more unknown parameters in the following way. Let the distribution function $p\{x; \theta_1, \theta_2\}$ contain two unknown parameters so that the likelihood function takes the form

$$L = p\{x_1; \theta_1, \theta_2\} \ldots p\{x_n; \theta_1, \theta_2\}. \tag{8.8.7}$$

The likelihood equations are

$$\frac{\partial \log L}{\partial \theta_1} = \sum_{i=1}^{n} \frac{\partial \log p\{x_i; \theta_1, \theta_2\}}{\partial \theta_1} = 0 \tag{8.8.8}$$

and

$$\frac{\partial \log L}{\partial \theta_2} = \sum_{i=1}^{n} \frac{\partial \log p\{x_i; \theta_1, \theta_2\}}{\partial \theta_2} = 0. \tag{8.8.9}$$

The solutions, t_1 and t_2, are joint efficient estimates of the unknown parameters, i. e., the sampling distribution of (t_1, t_2) tends to a two-dimensional normal distribution as $n \to \infty$ with a greater "concentration" than the distribution of any other pair of asymptotically normally distributed estimates.

The variances and the coefficient of correlation are

$$\mathcal{V}\{t_1\} = \frac{1}{nm\left\{\left(\dfrac{\partial \log p}{\partial \theta_1}\right)^2\right\}} \frac{1}{1-\varrho^2}, \tag{8.8.10}$$

$$\mathcal{V}\{t_2\} = \frac{1}{nm\left\{\left(\dfrac{\partial \log p}{\partial \theta_2}\right)^2\right\}} \frac{1}{1-\varrho^2}, \tag{8.8.11}$$

and

$$\varrho = \varrho\{t_1, t_2\} = \frac{- m\left\{\dfrac{\partial \log p}{\partial \theta_1} \dfrac{\partial \log p}{\partial \theta_2}\right\}}{\sqrt{m\left\{\left(\dfrac{\partial \log p}{\partial \theta_1}\right)^2\right\} m\left\{\left(\dfrac{\partial \log p}{\partial \theta_2}\right)^2\right\}}}. \tag{8.8.12}$$

If

$$m\left\{\frac{\partial \log p}{\partial \theta_1} \frac{\partial \log p}{\partial \theta_2}\right\} = 0, \tag{8.8.13}$$

then the two estimates are stochastically independent for large n.

In the case where the given distribution function is normal

$$p\{x; \xi, \sigma^2\} = \frac{1}{\sqrt{2\pi}\,\sigma} e^{-\frac{(x-\xi)^2}{2\sigma^2}} \tag{8.8.14}$$

we get

$$L = (\sqrt{2\pi}\,\sigma)^{-n} e^{-\frac{1}{2\sigma^2} \sum_{i=1}^{n} (x_i-\xi)^2} \tag{8.8.15}$$

and

$$\log L = -n \log \sqrt{2\pi} - n \log \sigma - \frac{1}{2\sigma^2} \sum_{i=1}^{n} (x_i-\xi)^2 \tag{8.8.16}$$

from which it follows that maximizing L or $\log L$ with respect to ξ is the same as minimizing the sum of squares $\sum(x_i-\xi)^2$. Thus, *the method of least squares*, which we shall use several times in the following chapters, is contained in the method of maximum likelihood when the observations are normally distributed.

Example 8.4. *The binomial distribution.* Consider the distribution

$$p\{x = 0\} = 1-\theta$$

and

$$p\{x = 1\} = \theta .$$

Let the outcome of n observations be that the result $x = 1$ is obtained a times and the result $x = 0$ is obtained $n-a$ times, so that

$$L = \theta^a(1-\theta)^{n-a} .$$

The likelihood equation is

$$\frac{\partial \log L}{\partial \theta} = \frac{a}{\theta} - \frac{n-a}{1-\theta} = 0 ,$$

which has the solution

$$\theta \approx \frac{a}{n} = h ,$$

i. e., the frequency h is an asymptotically efficient estimate of θ.

To find the variance we calculate

$$\frac{\partial p}{\partial \theta} = \begin{cases} -1 & \text{for} \quad x = 0 \\ 1 & \text{for} \quad x = 1 . \end{cases}$$

Introducing this result into (8.8.6) we get

$$m\left\{\left(\frac{\partial \log p\{x;\, \theta\}}{\partial \theta}\right)^2\right\} = \frac{1}{1-\theta} + \frac{1}{\theta} = \frac{1}{\theta(1-\theta)}$$

which leads to

$$v\{h\} = \frac{\theta(1-\theta)}{n}$$

for large n.

In Example 5.8 we found that $m\{h\} = \theta$ and $v\{h\} = \dfrac{\theta(1-\theta)}{n}$ for any n. The above result shows that the sampling distribution of h is approximately normal for large n with the said mean and variance and that no "better" estimate of θ can be found.

Example 8.5. For *the normal distribution* we get

$$\frac{\partial \log p}{\partial \xi} = \frac{x-\xi}{\sigma^2}$$

and
$$\frac{\partial \log p}{\partial(\sigma^2)} = -\frac{1}{2\sigma^2}\left(1-\left(\frac{x-\xi}{\sigma}\right)^2\right),$$

which lead to the following likelihood equations

$$\frac{\partial \log L}{\partial \xi} = \frac{1}{\sigma^2}\sum_{i=1}^{n}(x_i-\xi) = 0$$

and
$$\frac{\partial \log L}{\partial(\sigma^2)} = -\frac{1}{2\sigma^2}\left(n-\frac{1}{\sigma^2}\sum_{i=1}^{n}(x_i-\xi)^2\right) = 0.$$

The solution is
$$\xi \approx \bar{x} = \frac{1}{n}\sum_{i=1}^{n} x_i$$

and
$$\sigma^2 \approx \frac{1}{n}\sum_{i=1}^{n}(x_i-\bar{x})^2 = \frac{n-1}{n}s^2 .$$

Further
$$m\left\{\left(\frac{\partial \log p}{\partial \xi}\right)^2\right\} = \frac{1}{\sigma^4} m\{(x-\xi)^2\} = \frac{\sigma^2}{\sigma^4} = \frac{1}{\sigma^2}$$

and
$$m\left\{\left(\frac{\partial \log p}{\partial(\sigma^2)}\right)^2\right\} = \frac{1}{4\sigma^4} m\left\{1-2\left(\frac{x-\xi}{\sigma}\right)^2+\left(\frac{x-\xi}{\sigma}\right)^4\right\}$$
$$= \frac{1}{4\sigma^4}(1-2+3) = \frac{1}{2\sigma^4}$$

since $m\{u^2\} = 1$ and $m\{u^4\} = 3$, which can be proved by integration by parts.

Finally
$$m\left\{\frac{\partial \log p}{\partial \xi}\,\frac{\partial \log p}{\partial(\sigma^2)}\right\} = -\frac{1}{2\sigma^3} m\left\{\frac{x-\xi}{\sigma}-\left(\frac{x-\xi}{\sigma}\right)^3\right\} = 0$$

since
$$m\{u\} = m\{u^3\} = 0$$

on account of the symmetry of the normal distribution.

Thus, $\bar{x}$ and $\frac{n-1}{n}s^2$ are asymptotically efficient estimates of ξ and σ^2 with variances $\frac{\sigma^2}{n}$ and $\frac{2\sigma^4}{n}$, respectively, and $\varrho = 0$.

In Chapters 9 and 11 the exact sampling distributions of $\bar{x}$ and s^2 will be derived.

8.9. Unbiased and Sufficient Estimates.

An estimate, t_n, is said to be an unbiased estimate of the parameter θ if for every n

$$m\{t_n\} = \theta\,, \tag{8.9.1}$$

i. e., by drawing random samples of size n from the given population we get a population of t_n's whose mean is θ.

Most of the estimates we shall use in the following are unbiased. For example we have

$$\mathcal{m}\{s^2\} = \sigma^2$$

for every n, as shown in § 8.6. Consequently,

$$\mathcal{m}\left\{\frac{1}{n}\sum_{i=1}^{n}(\bar{x}_i - \bar{x})^2\right\} = \mathcal{m}\left\{\frac{n-1}{n}s^2\right\} = \frac{n-1}{n}\sigma^2 \tag{8.9.2}$$

so that $\frac{n-1}{n} s^2$ is a biased estimate of σ^2. For $n \to \infty$ both estimates converge in probability to σ^2 as $\frac{n-1}{n} \to 1$, i. e., they are both consistent, but for small n we prefer the unbiased estimate s^2.

It has been proved by H. CRAMÉR[1]) that under certain general conditions there exists a lower bound to the variance of unbiased estimates, namely,

$$\mathcal{v}\{t_n\} \geq \frac{1}{n\mathcal{m}\left\{\left(\frac{\partial \log p\{x;\theta\}}{\partial\theta}\right)^2\right\}}. \tag{8.9.3}$$

Consider an estimate t_n of θ and its sampling distribution $p\{t_n;\theta\}$. *If the probability density of the sample may be written as*

$$p\{x_1;\theta\}\ldots p\{x_n;\theta\} = p\{t_n;\theta\}f(x_1,\ldots,x_n) \tag{8.9.4}$$

where $f(x_1,\ldots,x_n)$ *does not contain* θ, *then the estimate* t_n *is said to be sufficient.*

The significance of this definition is that *all the information regarding* θ *is contained in* t_n. Thus, when we know a sufficient estimate of θ no other estimate calculated from the same sample can supply *further* information concerning θ.

It can be shown that, when a sufficient estimate exists, the maximum likelihood method will lead to this estimate. It follows that $\bar{x}$ and s^2 are sufficient estimates of the parameters of a normal distribution, which may also be shown directly from the definition (8.9.4) by means of the exact sampling distributions of $\bar{x}$ and s^2.

As mentioned in the introduction to this chapter the following chapters are based on the assumption that the observations are a random sample from a normally distributed population. As the exact sampling distributions of various statistics under this assumption are known (and will be derived in these chapters) the limit theorems might seem of little interest apart from giving an explanation as to why these sampling distributions are

---

[1]) H. CRAMÉR: *Mathematical Methods of Statistics*, Princeton, 1946, § 32.3.

asymptotically normal. However, in practice we are never certain that our population is exactly normally distributed, and even if we transform the observations we can only hope that the resulting distribution is not far from normal. The true sampling distribution of the statistics used will therefore deviate from the "exact" sampling distributions derived from the normal distribution. What gives us confidence to use the "exact" distributions in practice, when the distribution of the population does not deviate much from the normal distribution, is the fact that the above limit theorems show that the "exact" and the true sampling distributions are all asymptotically normal, and furthermore that only small differences have been found between them for small n in several cases where experimental sampling from non-normal populations has been carried out.

8.10. Stochastic Dependence and Independence.

A boiler of a steam engine is tested for 100 minutes, the rate of steam production being measured every 30 seconds. The measurements—expressed in

TABLE 8.6.

200 consecutive measurements of the rate of steam production in a boiler. Unit: Tons per hour.

32·9	34·8	33·0	34·9	34·3	34·2	34·8	35·4
32·8	34·8	33·1	34·8	34·2	34·2	34·7	35·2
32·8	34·1	33·0	35·3	34·2	33·5	33·9	35·4
33·0	34·1	33·1	35·7	34·3	32·8	33·5	35·5
33·2	33·9	33·4	35·7	34·3	33·4	33·7	35·3
33·5	34·0	33·6	34·7	34·3	34·1	33·4	34·9
33·8	33·8	33·9	35·0	34·2	34·3	33·5	34·8
34·7	34·0	34·2	35·2	34·1	34·5	34·3	34·4
34·4	33·8	34·2	35·4	34·1	34·5	34·2	34·3
34·5	33·8	34·2	35·3	34·1	34·6	34·0	34·2
34·7	33·8	35·2	35·0	33·8	35·3	33·4	33·9
34·7	33·7	35·8	34·2	33·8	35·3	33·6	32·7
34·6	33·6	35·8	34·2	34·7	35·5	33·7	33·1
34·7	34·4	35·3	34·2	34·9	35·5	34·4	34·3
34·8	34·7	34·5	34·2	35·0	35·4	34·6	33·9
34·9	35·1	33·5	33·3	35·4	34·5	34·6	34·2
34·9	35·0	32·9	32·3	35·8	34·6	34·7	34·4
34·8	34·8	33·1	32·2	36·0	34·6	34·5	34·8
35·3	34·7	33·3	32·4	36·6	34·5	34·5	34·9
35·7	34·7	33·3	32·4	36·4	33·9	34·6	34·8
36·0	34·6	33·3	34·1	36·0	33·9	34·5	34·8
36·1	34·2	33·5	34·1	35·3	34·7	34·6	34·7
35·7	34·4	33·8	34·2	35·2	34·9	35·1	34·7
34·8	33·7	34·5	34·1	35·3	34·8	35·0	35·0
34·8	33·3	34·5	34·0	35·2	35·0	35·0	35·4

TABLE 8.7.

Distribution of 200 measurements of the rate of steam production in a boiler.

Unit: Tons per hour. Class length: 0·2 t/h.

t/h	Number of measurements	t/h	Number of measurements
32·25	2	34·65	23
32·45	2	34·85	21
32·65	1	35·05	10
32·85	5	35·25	14
33·05	7	35·45	9
33·25	6	35·65	4
33·45	10	35·85	3
33·65	7	36·05	4
33·85	15	36·25	0
34·05	13	36·45	1
34·25	25	36·65	1
34·45	17		
		Total:	200

tons per hour—have been entered chronologically in Table 8.6. If the observations are ranked according to order of magnitude, the distribution in Table 8.7 is obtained. The corresponding fractile diagram suggests that the distribution is normal.

Fig. 8.3 shows the measurements in chronological order. The points do not vary at random, but show cyclical movements. The probability that an observation will be above a certain point, say the mean of the 200 measurements, is not the same for every measurement, but depends on its "previous history". If, for example, the two preceding values were above the mean, the probability that the measure in question will also be above the mean is considerably larger than $\frac{1}{2}$, while this probability is considerably smaller than $\frac{1}{2}$ if the preceding values were below the mean. Thus, the *observations are not stochastically independent.*

The interdependence of the observations in Fig. 8.3 may also be seen from the fact that the values of the mean and the standard deviation of a component part of the observations depend on which part is considered. Thus, the standard deviation of the 15 determinations numbered 70–84 is very large compared with that of the numbers 96–110.

A different state of affairs exists in Fig. 8.4 which shows the measurements of the 200 diameters of rivet heads first mentioned on p. 45 also arranged in chronological order. In this case, the structure of the diagram appears uneven or random instead of cyclical.

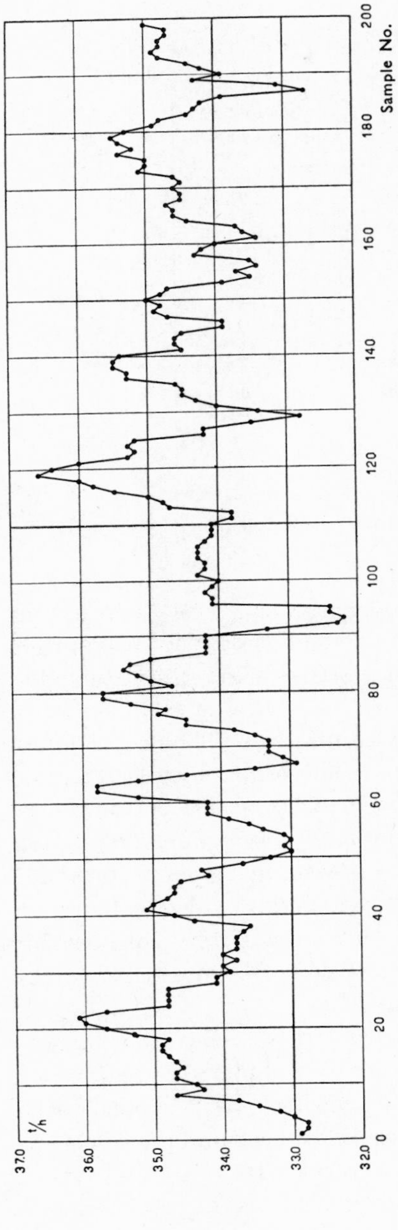

FIG. 8.3. 200 measurements of the rate of steam production in a boiler. Unit: Tons per hour.

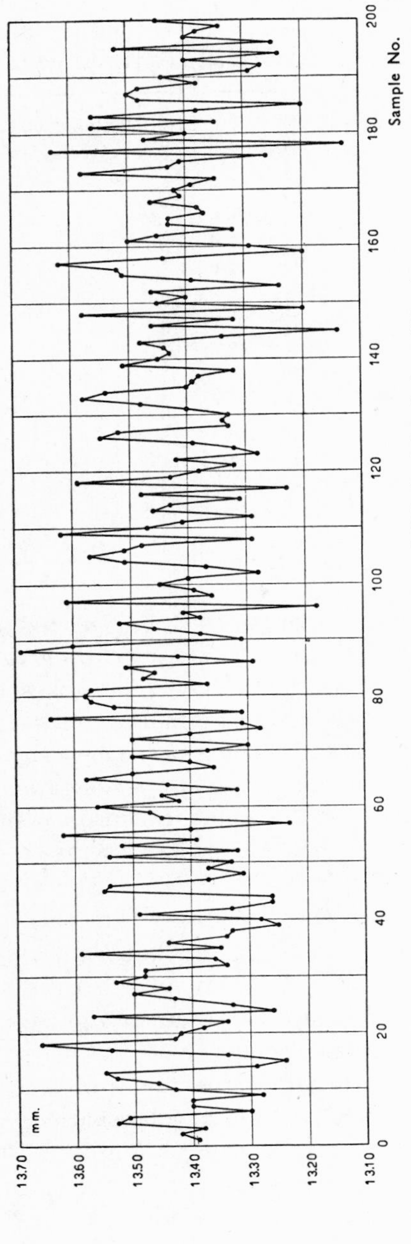

FIG. 8.4. Measurements of the diameters of 200 rivet heads in millimetres.

If we consider only the *distributions* of the two sets of 200 measurements, and disregard the order of the measurements, they appear to be of the same type and may therefore both be *described* by the normal distribution. As shown above, however, further analysis reveals definite differences in the structure of the data.

In the following chapters it is *assumed that the observations are stochastically independent*, and, therefore, the theorems derived should not be applied to observations such as those given in Table 8.6. Some tests for stochastical independence will be given in Chapter 13.

CHAPTER 9.

THE DISTRIBUTION OF THE MEAN

9.1. The Addition Theorem for the Normal Distribution.

Let two variables, x_1 and x_2, with means ξ_1 and ξ_2, and variances σ_1^2 and σ_2^2, be *stochastically independent*. According to (5.17.3) and (5.17.5) the mean and the variance of the variable $x = x_1+x_2$ are

$$\mathcal{M}\{x\} = \xi = \xi_1+\xi_2$$

and

$$\mathcal{V}\{x\} = \sigma^2 = \sigma_1^2+\sigma_2^2 .$$

If the variables x_1 and x_2 are normally distributed, their sum $x = x_1+x_2$ will also be normally distributed. The proof is given below.

To each of the variables x_1 and x_2 corresponds a normally distributed population with the above stated parameters. An element in the population of the variable $x = x_1+x_2$ is derived by random choice of one element from each of the two given populations and addition of these two elements. If, for instance, the standard deviations of the two populations are the same, the standard deviation of the x-population will be equal to this standard deviation multiplied by $\sqrt{2}$.

The above theorem may be generalized as follows: Let the k variables $x_1, x_2, \ldots, x_k$ be *stochastically independent and normally distributed* with parameters $(\xi_1, \sigma_1^2), (\xi_2, \sigma_2^2), \ldots, (\xi_k, \sigma_k^2)$. *The variable*

$$x = \alpha_0+\alpha_1 x_1+\ldots+\alpha_k x_k \tag{9.1.1}$$

will then be normally distributed with parameters (ξ, σ^2), *given by*

$$\xi = \alpha_0+\alpha_1\xi_1+\ldots+\alpha_k\xi_k \tag{9.1.2}$$

and

$$\sigma^2 = \alpha_1^2\sigma_1^2+\ldots+\alpha_k^2\sigma_k^2 . \tag{9.1.3}$$

This theorem is called *the addition theorem for the normal distribution.*

The above theorem may be compared with the central limit theorem, according to which x will be normally distributed for $k \to \infty$ if only the distribution functions of the component parts satisfy some very general conditions, cf. § 8.2. The addition theorem for the normal distribution says that, if the component parts are normally distributed, x will be normally distributed not only for $k \to \infty$ but for any value of k.

The distribution of the sum of n elements from a normally distributed population with parameters (ξ, σ^2) forms a special case of the theorem. The sum of the n values is normally distributed about $n\xi$ with variance $n\sigma^2$, and thus the sample mean is normally distributed about the population mean, ξ, with variance σ^2/n.

Proof. The theorem is first proved for $\xi_1 = \xi_2 = 0$, the distribution functions for x_1 and x_2 being

$$p\{x_i\} = \frac{1}{\sigma_i}\varphi\left(\frac{x_i}{\sigma_i}\right) \qquad (i = 1, 2)\ .$$

According to the multiplication formula, the probability that x_1 belongs to the interval (x_1, x_1+dx_1), and at the same time $x_1+x_2 < x$, i. e., $x_2 < x-x_1$, is

$$p\{x_1\}dx_1 P\{x_2 < x-x_1\} = \frac{1}{\sigma_1}\varphi\left(\frac{x_1}{\sigma_1}\right)dx_1\,\Phi\left(\frac{x-x_1}{\sigma_2}\right).$$

According to the addition formula we obtain the probability that $x_1+x_2<x$, irrespective of the value of x_1, by summing the above probabilities for all values of x_1, i. e.,

$$P\{x_1+x_2 < x\} = P\{x\} = \frac{1}{\sigma_1}\int_{-\infty}^{\infty}\varphi\left(\frac{x_1}{\sigma_1}\right)\Phi\left(\frac{x-x_1}{\sigma_2}\right)dx_1\ . \qquad (9.1.4)$$

Differentiating (9.1.4) with respect to x, we get the distribution function

$$p\{x\} = \frac{1}{\sigma_1\sigma_2}\int_{-\infty}^{\infty}\varphi\left(\frac{x_1}{\sigma_1}\right)\varphi\left(\frac{x-x_1}{\sigma_2}\right)dx_1$$

$$= \frac{1}{2\pi\sigma_1\sigma_2}\int_{-\infty}^{\infty} e^{-\frac{x_1^2}{2\sigma_1^2}}e^{-\frac{(x-x_1)^2}{2\sigma_2^2}}dx_1\ . \qquad (9.1.5)$$

The exponent may be written as the sum of two terms, one of which does not include x_1, since

$$\frac{x_1^2}{\sigma_1^2}+\frac{(x-x_1)^2}{\sigma_2^2} = x_1^2\left(\frac{1}{\sigma_1^2}+\frac{1}{\sigma_2^2}\right)-\frac{2x_1x}{\sigma_2^2}+\frac{x^2}{\sigma_2^2}$$

$$= \frac{\sigma_1^2+\sigma_2^2}{\sigma_1^2\sigma_2^2}\left(x_1-x\frac{\sigma_1^2}{\sigma_1^2+\sigma_2^2}\right)^2+\frac{x^2}{\sigma_1^2+\sigma_2^2}$$

$$= \frac{\sigma^2}{\sigma_1^2\sigma_2^2}\left(x_1-x\frac{\sigma_1^2}{\sigma^2}\right)^2+\frac{x^2}{\sigma^2},$$

where $\sigma_1^2+\sigma_2^2 = \sigma^2$. Introducing this term in (9.1.5), we obtain

$$p\{x\} = \frac{1}{2\pi\sigma_1\sigma_2} e^{-\frac{x^2}{2\sigma^2}} \int_{-\infty}^{\infty} e^{-\frac{\sigma^2}{2\sigma_1^2\sigma_2^2}\left(x_1-x\frac{\sigma_1^2}{\sigma^2}\right)^2} dx_1 .$$

If we further introduce

$$u = \frac{\sigma}{\sigma_1\sigma_2}\left(x_1-x\frac{\sigma_1^2}{\sigma^2}\right)$$

as a new variable, we get

$$p\{x\} = \frac{1}{2\pi\sigma_1\sigma_2} e^{-\frac{x^2}{2\sigma^2}} \frac{\sigma_1\sigma_2}{\sigma} \int_{-\infty}^{\infty} e^{-\frac{1}{2}u^2} du$$

$$= \frac{1}{\sqrt{2\pi}\,\sigma} e^{-\frac{x^2}{2\sigma^2}}, \tag{9.1.6}$$

because, according to (6.1.3),

$$\frac{1}{\sqrt{2\pi}} \int_{-\infty}^{\infty} e^{-\frac{1}{2}u^2} du = 1.$$

Thus, the variable $x = x_1+x_2$ is normally distributed about 0 with variance $\sigma^2 = \sigma_1^2+\sigma_2^2$.

If x_1 and x_2 are normally distributed about ξ_1 and ξ_2, then $x_1-\xi_1$ and $x_2-\xi_2$ are normally distributed about 0, and we may apply the above theorem to these variables. This implies that $x = x_1+x_2$ is normally distributed about $\xi = \xi_1+\xi_2$ with variance $\sigma^2 = \sigma_1^2+\sigma_2^2$. The theorem may be readily generalized as suggested above, since a linear function of a normally distributed variable is itself normally distributed, cf. (6.4.4).

Example 9.1. Let us assume that a machine packs a certain product in doses of weight x_1, where x_1 is normally distributed about $\xi_1 = 25{\cdot}0$ grams with standard deviation $\sigma_1 = 0{\cdot}4$ gram and further that the weight of the empty packet, x_2, is normally distributed about $\xi_2 = 5{\cdot}0$ grams with standard deviation $\sigma_2 = 0{\cdot}2$ gram. According to the addition theorem the weight of the filled packet, x_1+x_2, will be normally distributed about $\xi_1+\xi_2 = 25{\cdot}0+5{\cdot}0 = 30{\cdot}0$ grams with standard deviation $\sqrt{\sigma_1^2+\sigma_2^2} = \sqrt{0{\cdot}4^2+0{\cdot}2^2} = \sqrt{0{\cdot}20} = 0{\cdot}45$ gram. Thus, 95% of the packets will weigh between $30{\cdot}0-1{\cdot}96\times0{\cdot}45 = 29{\cdot}1$ grams and $30{\cdot}0+1{\cdot}96\times0{\cdot}45 = 30{\cdot}9$ grams.

The factory delivers the packed product in cartons with 10 packets in each. The weight of the cartons, x_3, is normally distributed about $\xi_3 = 30{\cdot}0$ grams with standard deviation $\sigma_3 = 1{\cdot}0$ gram. According to the addition theorem the weight of the cartons with contents will be normally distributed about $\xi = 10(\xi_1+\xi_2)+\xi_3 = 10\times30+30 = 330$ grams with the variance $\sigma^2 = 10(\sigma_1^2+\sigma_2^2)+\sigma_3^2 = 10\times0{\cdot}20+1{\cdot}0 = 3{\cdot}0$ which gives a standard deviation $\sigma = 1{\cdot}7$ grams, i. e., 95% of the packed cartons will weigh between $330{\cdot}0-1{\cdot}96\times1{\cdot}7 = 326{\cdot}7$ grams and $330{\cdot}0+1{\cdot}96\times1{\cdot}7 = 333{\cdot}3$ grams.

Example 9.2. Let a characteristic of the quality of an industrial product be normally distributed with mean ξ and standard deviation σ_Q. If a measurement is performed on a randomly chosen item, the value observed will not be the "true" quality, x_1, but will be $x = x_1+x_2$, where x_2 stands for the error of measurement. Let us assume that the method used for measuring is such that the errors (x_2) are normally distributed about 0 with standard deviation σ_M. When measuring a large number of items we therefore get a set of x-values, which are normally distributed about ξ with the total standard deviation

$$\sigma_T = \sqrt{\sigma_Q^2+\sigma_M^2} = \sigma_Q \sqrt{1+\left(\frac{\sigma_M}{\sigma_Q}\right)^2}. \tag{9.1.7}$$

On account of the uncertainty of the method of measurement, the observed values have a standard deviation that is greater than that of the true quality characteristic in question. The ratio between the two standard deviations is

$$\frac{\sigma_T}{\sigma_Q} = \sqrt{1+\left(\frac{\sigma_M}{\sigma_Q}\right)^2},$$

see Table 9.1.

TABLE 9.1.

$\dfrac{\sigma_M}{\sigma_Q}$	$\dfrac{\sigma_T}{\sigma_Q} = \sqrt{1+\left(\dfrac{\sigma_M}{\sigma_Q}\right)^2}$
0·00	1·00
0·20	1·02
0·25	1·03
0·33	1·05
0·50	1·12
0·75	1·25
1·00	1·41

If, for instance, the standard deviation, σ_M, of the method of measurement equals $^1/_3$ of the standard deviation of the quality, σ_Q, the total standard deviation, σ_T, will be about 5% larger than σ_Q.

Example 9.3. Let (x_{11}, x_{12}), (x_{21}, x_{22}), ..., (x_{k1}, x_{k2}) be a series of duplicate determinations, two corresponding values (x_{i1}, x_{i2}) having the same mean, ξ_i, and the same variance, σ_i^2, and further let all the variances be equal, $\sigma_1^2 = \ldots = \sigma_k^2 = \sigma^2$, which means that the uncertainty of the method of measurement is the same for all duplicate values, irrespective of the value of the means.

The mean

$$\bar{x}_i = \tfrac{1}{2}(x_{i1}+x_{i2})$$

will then be normally distributed about ξ_i with variance $\frac{\sigma^2}{2}$, and the differences

$$d_i = x_{i1}-x_{i2}$$

will be normally distributed about 0 with variance $2\sigma^2$.

Example 9.4. A relay is adjusted to be released when a certain action has lasted for ξ_1 seconds. The release period, however, is not always exactly ξ_1 seconds, but is normally distributed about ξ_1 with variance σ_1^2. Another relay—in series with the first—is adjusted to be released after $\xi_2 > \xi_1$ seconds, the release period being normally distributed about ξ_2 with variance σ_2^2. On the one hand, it is important that the difference between ξ_1 and ξ_2 is as small as possible, while on the other hand overlapping must not occur too often, i. e., relay No. 2 should very seldom be released before relay No. 1. How large should the difference be between ξ_1 and ξ_2 in order that relay No. 2 is released before No. 1 on only one out of every 1000 occasions?[1])

If the release periods are denoted x_1 and x_2, ξ_1 and ξ_2 must take such values that $P\{x_2 < x_1\} = 0{\cdot}001$, which is the same as $P\{x_1-x_2 > 0\} = 0{\cdot}001$. According to the addition theorem $d = x_1-x_2$ is normally distributed about $\delta = \xi_1-\xi_2$ with variance $\sigma^2 = \sigma_1^2+\sigma_2^2$. Hence

$$\begin{aligned} P\{x_1-x_2 > 0\} &= P\{d > 0\} \\ &= 1-P\{d < 0\} \\ &= 1-\Phi\left(\frac{0-\delta}{\sigma}\right) = \Phi\left(\frac{\delta}{\sigma}\right). \end{aligned}$$

According to Table III, the equation

$$\Phi\left(\frac{\delta}{\sigma}\right) = 0{\cdot}001$$

has the root

$$\frac{\delta}{\sigma} = -3{\cdot}09\,.$$

Thus, the equation $P\{x_2 < x_1\} = 0{\cdot}001$ is equivalent to $\delta = -3{\cdot}09\sigma$, i.e.,

$$\xi_2-\xi_1 = 3{\cdot}09\sqrt{\sigma_1^2+\sigma_2^2}\,.$$

If the difference between the two means of the release periods is larger than or equal to $3{\cdot}09\sqrt{\sigma_1^2+\sigma_2^2}$, overlapping can be expected on at most one out of every 1000 occasions.

For $\xi_1 = 1{\cdot}00$ sec and $\sigma_1 = \sigma_2 = 0{\cdot}06$ sec, we find

$$\xi_2 = 1{\cdot}00+3{\cdot}09\times 0{\cdot}06\sqrt{2} = 1{\cdot}26\text{ sec}\,.$$

---

[1]) K. Danø: *Tidsaftrapning ved Overstrømsbeskyttelse*, Ingeniøren, 53, 1944, E. 13—19.

9.2. The Distribution of the Mean.

As mentioned above, the addition theorem for the normal distribution leads to the following theorem regarding the mean: *The mean of n stochastically independent observations from a normally distributed population with parameters* (ξ, σ^2) *is normally distributed with parameters* $(\xi, \sigma^2/n)$.

It follows that

$$\boxed{u = \frac{\bar{x}-\xi}{\sigma/\sqrt{n}},} \tag{9.2.1}$$

where

$$\bar{x} = \frac{1}{n}(x_1+x_2+\ldots+x_n),$$

is normally distributed about 0 with standard deviation 1, and that the distribution function and the cumulative distribution function of $\bar{x}$ will be

$$p\{\bar{x}\} = \frac{\sqrt{n}}{\sqrt{2\pi}\,\sigma} e^{-\frac{n(\bar{x}-\xi)^2}{2\sigma^2}} \tag{9.2.2}$$

and

$$P\{\bar{x}\} = \Phi\left(\frac{\sqrt{n}(\bar{x}-\xi)}{\sigma}\right), \tag{9.2.3}$$

respectively.

Writing

$$x_i-\xi = (x_i-\bar{x})+(\bar{x}-\xi)$$

and

$$\sum_{i=1}^{n}(x_i-\xi)^2 = \sum_{i=1}^{n}(x_i-\bar{x})^2 + n(\bar{x}-\xi)^2, \tag{9.2.4}$$

it is seen that the distribution of the mean satisfies the relation (8.9.4). Thus, *the sample mean is an unbiased and sufficient estimate of the population mean when the population is normally distributed.*

The standard deviation of the mean—or, as it is also called, the *standard error* of the mean—is equal to $\frac{1}{\sqrt{n}}$ times the standard deviation of the observations and no other estimate of ξ exists with a smaller standard error.

Example 9.5. The primary data relating to the distribution of the diameters of 500 rivet heads, given in Table 4.1, may also be read as 125 groups, each including 4 observations from the same population, or as 31 groups including 16 observations each. Table 9.2 gives the distributions of the

TABLE 9.2.

The distribution of the means of the diameters of 4 and 16 rivet heads, respectively.

Mean of 4 diameters in mm	Number of means	Mean of 16 diameters in mm	Number of means
13·26—13·28	1	13·37—13·38	2
13·28—13·30	1	13·38—13·39	1
13·30—13·32	1	13·39—13·40	1
13·32—13·34	5	13·40—13·41	5
13·34—13·36	3	13·41—13·42	1
13·36—13·38	13	13·42—13·43	7
13·38—13·40	12	13·43—13·44	5
13·40—13·42	22	13·44—13·45	6
13·42—13·44	19	13·45—13·46	0
13·44—13·46	19	13·46—13·47	2
13·46—13·48	14	13·47—13·48	1
13·48—13·50	6		
13·50—13·52	3		
13·52—13·54	3		
13·54—13·56	3		
	125		31

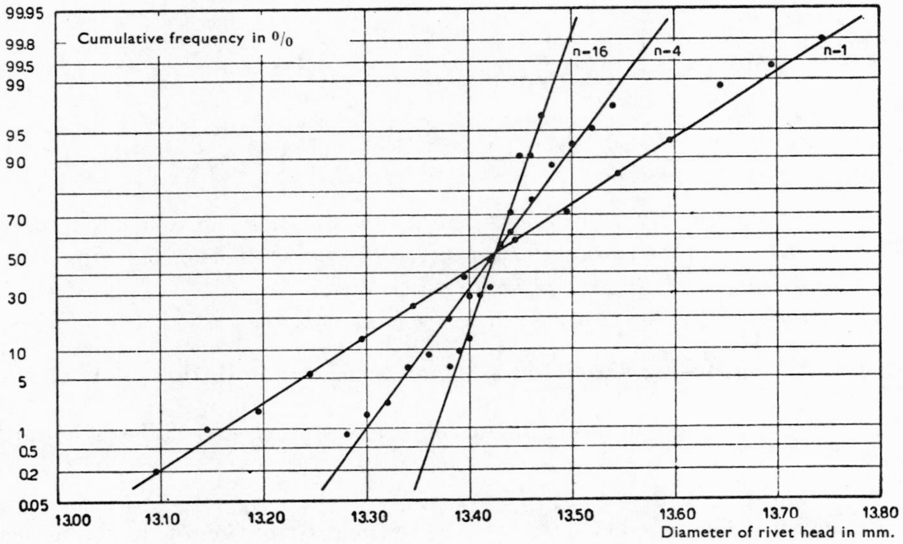

FIG. 9.1. Fractile diagram of the distribution of 500 rivets according to diameter of heads in millimetres, together with diagrams of the distributions of the means of the diameters of 4 and 16 rivet heads, respectively.

corresponding means, and Fig. 9.1 the corresponding fractile diagrams together with the fractile diagram of the grouped, original observations, cf. Fig. 6.10. The diagram shows that (1) it is permissible to regard both the means and the original observations as normally distributed, (2) all three distributions seem to have the same theoretical mean—the three lines intersect at the point (13·425 mm, 50%), and (3) the standard deviations are in the proportion $1:\sqrt{4}:\sqrt{16} = 1:2:4$, which is evident from the slope of the lines.

9.3. The Fractiles.

As

$$u = \frac{\bar{x}-\xi}{\sigma/\sqrt{n}}$$

is normally distributed about 0 with standard deviation 1, the fractiles for $\bar{x}$ are given directly by the relation

$$\boxed{P\left\{\frac{\bar{x}-\xi}{\sigma/\sqrt{n}} < u_P\right\} = P,} \qquad (9.3.1)$$

where P is a given probability and u_P denotes the P-fractile for u, as explained in § 6.6. Solving the inequality with respect to $\bar{x}$, we have

$$P\left\{\bar{x} < \xi + u_P\frac{\sigma}{\sqrt{n}}\right\} = P, \qquad (9.3.2)$$

i. e., the P-fractile for $\bar{x}$ is equal to

$$\boxed{\bar{x}_P = \xi + u_P\frac{\sigma}{\sqrt{n}}.} \qquad (9.3.3)$$

If ξ and σ are known, the fractiles for $\bar{x}$ may easily be calculated with the aid of the table of u_P. Table III, for instance, gives

$$\bar{x}_{.95} = \xi + 1{\cdot}64\frac{\sigma}{\sqrt{n}}.$$

An upper and a lower limit for the variation of the sample mean corresponding to given probabilities may be obtained directly from the fractiles, as

$$P\{\bar{x}_{P_1} < \bar{x} < \bar{x}_{P_2}\} = P_2 - P_1$$

or

$$P\left\{\xi+u_{P_1}\frac{\sigma}{\sqrt{n}} < \bar{x} < \xi+u_{P_2}\frac{\sigma}{\sqrt{n}}\right\} = P_2-P_1\,. \tag{9.3.4}$$

These limits may be calculated from given values of ξ, σ, n, P_1 and P_2. Putting $P_1 = 2{\cdot}5\%$ and $P_2 = 97{\cdot}5\%$, we obtain

$$P\left\{\xi-1{\cdot}96\frac{\sigma}{\sqrt{n}} < \bar{x} < \xi+1{\cdot}96\frac{\sigma}{\sqrt{n}}\right\} = 95\%\,. \tag{9.3.5}$$

This indicates that approximately 95 % of a large number of means must be expected to lie within the interval $\xi \pm 1{\cdot}96\frac{\sigma}{\sqrt{n}}$, when the means have been computed from samples of n observations each, drawn from a normally distributed population with parameters (ξ, σ^2).

This result may be used for controlling the quality of an industrial product, *if* this quality under the usual production conditions is normally distributed about ξ with standard deviation σ. In practical work we very seldom know the true values of the parameters, but they may be estimated from observations made during an earlier production period; based on this experience values of ξ and σ may be chosen as the manufacturing standard to which future production is required to adhere. If for instance n control measurements are made every day, their mean should belong to the interval stated in (9.3.5) with a probability of 95%. In practical work a control is carried out by plotting the means on a *control chart*, i. e., the number of the sample is plotted as abscissa and the corresponding mean as ordinate, cf. Fig. 9.2. The preassigned mean ξ and the control limits $\xi \pm 1{\cdot}96\,\frac{\sigma}{\sqrt{n}}$ are marked by lines parallel to the abscissa axis. As long as the quality of the product adheres to the fixed manufacturing standard, the

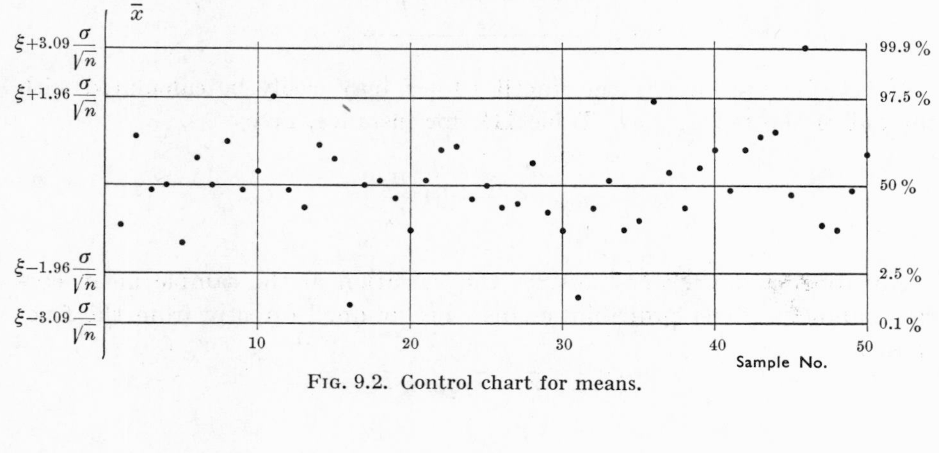

FIG. 9.2. Control chart for means.

points must vary at random about the centre-line and only 2·5% of the points are allowed to occur below or above the control lines, respectively. Further control of the variation of the means may be obtained by drawing more control lines, e. g., the control limits $\xi \pm 3{\cdot}09 \frac{\sigma}{\sqrt{n}}$, corresponding to a probability of 99·8%.

Systematic deviations from the standard will usually soon be detected. For instance an increase in the population mean will lead to more than 2·5% of the sample means being situated above the upper control line, and at the same time the frequency of the values below the lower limit will decrease. If the population standard deviation increases while the mean remains constant, the frequency of points outside both control limits will increase. If the population mean takes to oscillating about the pre-assigned mean, the points in the control chart will present corresponding oscillations.

Systematic deviations from the standard indicate that the production process is not functioning as it should; an investigation of the raw materials, the machines and the manner in which the machines are worked is therefore necessary in order to find the causes of the change.

In practical work, if a mean falls outside the 99·8% limits it is taken as an indication of a change in the production process, and the appearance of such a point in the diagram should at once be followed by an investigation of the production process. If the production process adheres to the standard and we use this rule, then we will look for trouble in the production process without any trouble being there for two out of every 1000 samples. If, however, there has been a change in the population mean, then the probability of finding a point outside the control limits will be greater than 0·2% and the larger the change in the population mean the greater the probability of getting warned through the control chart. The consequences of applying such a rule will be more fully discussed in the next paragraph.

In the above the chosen limits are symmetrical about the population mean. Sometimes a one-sided delimitation is useful, e. g., if the object of the control is to eliminate products in which the quality measured is below a certain limit.

It is not possible from purely statistical considerations to state general rules for the choice of control limits. As the number of values of P_2 and P_1 which satisfy the equation

$$P\left\{\xi + u_{P_1}\frac{\sigma}{\sqrt{n}} < \bar{x} < \xi + u_{P_2}\frac{\sigma}{\sqrt{n}}\right\} = P_2 - P_1 = P$$

are infinite, the number of control limits corresponding to a given probability

P is also infinite. Therefore, in order to reach a unique determination of the limits, the problem itself (i. e., the aim of the control) must suggest some further restriction than the size of P alone, e. g., that delimitation must be unilateral, or bilateral and symmetrical.

9.4. The u-Test of Significance.

In the discussion of the errors of a method of measurement it is of major importance that a distinction is made between the *random* errors and the *systematic* errors of the method in question. When the *same* object is measured repeatedly we usually get a series of measurements which due to a large number of uncontrollable factors of small importance vary at random about a certain value in such a way that this random variation, the random errors, can usually be looked upon as normally distributed about zero with a standard deviation characterizing the uncertainty of the method. In contrast to the random errors systematic errors are usually due to one or more isolated factors which cause a displacement of the measurements in one direction, the result being that the observations are distributed about a certain value, ξ, differing from the theoretical or "true" value, ξ_0, of the object. The systematic error is consequently equal to $\delta = \xi - \xi_0$.

When examining the possible systematic error of a method of measurement, it is sometimes possible to design the experiment in such a manner that the theoretical value, ξ_0, of the measured quantity is known. This is for instance the case in chemical analyses of a substance where the quantities of the components are known.

In the following we will assume that the uncertainty of the method is known, i. e., we know the theoretical value of the standard deviation, σ. (Chapter 15 deals with the corresponding problem where σ is unknown but an estimate, s, of σ is known). The mean $\bar{x}$ of n measurements of the same object is calculated and the question is now to decide whether $\bar{x}$ differs *significantly* from ξ_0, i. e., whether $\bar{x}$ deviates so much from ξ_0 that it is unreasonable to assume that the population mean, $\xi = \mathcal{M}\{x\}$, is equal to the true value, ξ_0, of the object.

In general, when n *observations from a normally distributed population with known variance* σ^2 *and unknown mean* ξ *are giwen we may test the hypothesis that* ξ *has a specified value* ξ_0. The procedure used for investigating this problem is a prototype for all tests of significance.

Assuming the test hypothesis to be true, the mean of n observations will be normally distributed about ξ_0 with standard deviation $\dfrac{\sigma}{\sqrt{n}}$, which means that

$$u = \frac{\bar{x} - \xi_0}{\sigma/\sqrt{n}} \tag{9.4.1}$$

is normally distributed about 0 with standard deviation 1. If we substitute the observed $\bar{x}$, we find a value of u which enables us to answer the question whether the test hypothesis seems reasonable or not. It will be recognized that, if the numerical value of u is large, it is reasonable to reject our test hypothesis, while this is not the case if $|u|$ is small. We must, however, choose some limit for $|u|$, so that the test hypothesis will be rejected if $|u|$ exceeds this limit, the *significance limit.* If for instance $|u| = 1{\cdot}96$ is chosen as limit and the test hypothesis is rejected every time $|u| > 1{\cdot}96$, *then the hypothesis will be rejected in* 5% *of the cases where it is true and accepted in* 95% *of such cases.*

Just as the above rule does not lead to acceptance of the test hypothesis in every case where it is true, neither will the hypothesis be rejected in every case where it is false. This is due to the fact that the test usually does not reveal small deviations between the "true theory" and the test hypothesis. In the above example this means that a systematic error which is small in comparison with $\frac{\sigma}{\sqrt{n}}$ will usually not be discovered. In such a case, *the test hypothesis will usually be accepted as working hypothesis* until so many observations have been gathered that the presence of a systematic error can be demonstrated.

To analyze the test in mathematical terms two new concepts are introduced: *the critical region* and *the power of the test.* If

$$\left|\frac{\bar{x}-\xi_0}{\sigma/\sqrt{n}}\right| < 1{\cdot}96 \tag{9.4.2}$$

is chosen as criterion for accepting the test hypothesis $\mathcal{M}\{x\} = \xi_0$, the corresponding region for the sample mean

$$\xi_0 - 1{\cdot}96\frac{\sigma}{\sqrt{n}} < \bar{x} < \xi_0 + 1{\cdot}96\frac{\sigma}{\sqrt{n}}$$

is called *the region of acceptance.* The region defined by

$$\left|\frac{\bar{x}-\xi_0}{\sigma/\sqrt{n}}\right| > 1{\cdot}96 \tag{9.4.3}$$

or

$$\bar{x} < \xi_0 - 1{\cdot}96\frac{\sigma}{\sqrt{n}} \text{ and } \bar{x} > \xi_0 + 1{\cdot}96\frac{\sigma}{\sqrt{n}}$$

is called *the region of rejection* or *the critical region.*

The probability of $\bar{x}$ falling within the critical region calculated under the assumption that $\mathcal{M}\{x\} = \xi_0$ is 5%.

Usually the first step in constructing a test of significance is to choose a *level of significance*, a small positive number α as for instance $\alpha = 0{\cdot}05$, and then to determine a critical region such that the probability of rejecting the test hypothesis when it is true is equal to α. *If* the test hypothesis is true, i.e., $\mathcal{M}\{x\} = \xi_0$, then the test will lead to the wrong decision, the rejection of the test hypothesis, on $100\alpha\%$ of occasions and to the right decision, the acceptance of the test hypothesis, on $100(1-\alpha)\%$ of occasions.

If, however, the test hypothesis is false, i. e., $\mathcal{M}\{x\} = \xi_1 \neq \xi_0$, then the probability of $\bar{x}$ falling within the critical region will depend on ξ_1. Denoting this probability by

$$\pi(\xi_1) = P\left\{\left|\frac{\bar{x}-\xi_0}{\sigma/\sqrt{n}}\right| > 1{\cdot}96; \mathcal{M}\{x\} = \xi_1\right\} \tag{9.4.4}$$

and writing

$$\frac{\bar{x}-\xi_0}{\sigma/\sqrt{n}} = \frac{\bar{x}-\xi_1}{\sigma/\sqrt{n}} + \lambda_1$$

where

$$\lambda_1 = \frac{\xi_1-\xi_0}{\sigma/\sqrt{n}}$$

we get

$$\begin{aligned}\pi(\xi_1) &= P\left\{\left|\frac{\bar{x}-\xi_1}{\sigma/\sqrt{n}}+\lambda_1\right| > 1{\cdot}96;\ \mathcal{M}\{x\} = \xi_1\right\}\\ &= P\{|u+\lambda_1| > 1{\cdot}96\}\\ &= \Phi(-1{\cdot}96-\lambda_1)+\Phi(-1{\cdot}96+\lambda_1)\end{aligned}$$

since $\frac{\bar{x}-\xi_1}{\sigma/\sqrt{n}}$ is normally distributed with parameters (0,1) under the assumption that $\mathcal{M}\{x\} = \xi_1$.

This probability is called *the power of the test with respect to the alternative hypothesis* $\mathcal{M}\{x\} = \xi_1 \neq \xi_0$. If $\mathcal{M}\{x\} = \xi_1$ then the test will lead to the right decision, i. e., rejection of the test hypothesis, with a probability given by the power of the test and to the wrong decision, i. e., acceptance of the test hypothesis, with a probability equal to 1 minus the power of the test.

Thus we see that wrong statements resulting from the application of a test may be classified as follows: (1) rejecting the test hypothesis when it is true; the probability of committing such an error (an error of the first kind) is denoted by α and is called the level of significance of the test; (2) failing to reject the test hypothesis when it is false; the probability of committing such an error (an error of the second kind) is denoted by β and is equal to 1 minus the power of the test.

Generally the *power function of a test gives the probability of rejecting the test hypothesis as a function of the unknown parameter*, i. e., *the power function gives the probability of making a correct decision* for all possible values of ξ different from ξ_0, and for $\xi = \xi_0$ it gives the probability of making a wrong decision.

If the set of alternative values is defined as $\xi \neq \xi_0$ we find the corresponding power function as

$$\pi(\xi) = P\left\{\left|\frac{\bar{x}-\xi_0}{\sigma/\sqrt{n}}\right| > 1{\cdot}96;\, \mathcal{M}\{x\} = \xi\right\} = \Phi(-1{\cdot}96-\lambda)+\Phi(-1{\cdot}96+\lambda) \quad (9.4.5)$$

where

$$\lambda = \frac{\xi-\xi_0}{\sigma/\sqrt{n}}. \quad (9.4.6)$$

The power function is an increasing function of $|\lambda|$, as shown in Table 9.3 and Fig. 9.3. If the distance between the population mean ξ and the hypothetical mean ξ_0 is small the power of the test is also small, i. e., the test will in many cases not detect the falsehood of the hypothesis tested. However, it will usually be of minor practical importance that the test fails to reject the test hypothesis when it is only slightly in error.

TABLE 9.3.

The power function of the two-tailed u-test at the 5% level of significance.

Reject the test hypothesis $\mathcal{M}\{x\} = \xi_0$ when $\left|\frac{\bar{x}-\xi_0}{\sigma/\sqrt{n}}\right| > 1{\cdot}96$

$\pi(\xi) = \Phi(-1{\cdot}96-\lambda)+\Phi(-1{\cdot}96+\lambda)$ where $\lambda = \frac{\xi-\xi_0}{\sigma/\sqrt{n}}$

$\lvert\lambda\rvert$	0·0	0·5	1·0	1·5	2·0	2·5	3·0	3·5	4·0	4·5	5·0
$100\pi(\xi)$	5·0	7·9	17·0	32·3	51·6	70·5	85·1	93·8	97·9	99·4	99·9

As an example consider a case where the systematic error is 1·5 times the standard error of the mean, i. e., $|\lambda| = 1{\cdot}5$. Table 9.3 shows that the power function in this case takes on the value 0·32. If we have a large number of samples each consisting of n observations from this population, the test hypothesis will be rejected for only 32% of these samples and incorrectly accepted for 68% of the samples. If $|\lambda| = 4{\cdot}0$, however, the corresponding figures are 98% and 2%.

The number of cases in which the test actually leads to the right conclusion depends on the values of the means in the populations from which the samples are actually drawn.

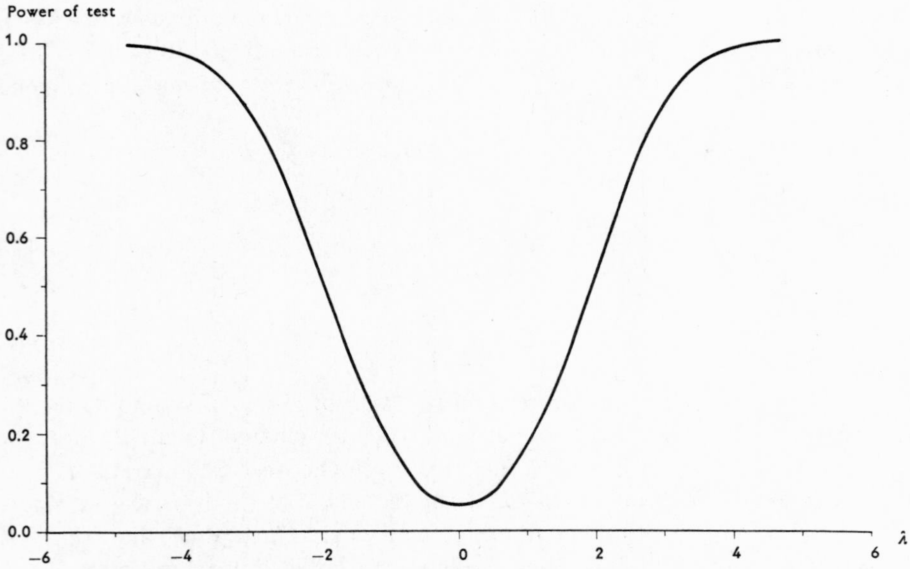

FIG. 9.3. The power curve for the two-tailed u-test at the 5 % level of significance.

$$\pi(\xi) = \Phi(-1{\cdot}96-\lambda)+\Phi(-1{\cdot}96+\lambda) \text{ for } \lambda = \frac{\xi-\xi_0}{\sigma/\sqrt{n}}.$$

The critical region at the α level of significance of the two-tailed u-test is determined by the inequality

$$\left|\frac{\bar{x}-\xi_0}{\sigma/\sqrt{n}}\right| > u_{1-\frac{\alpha}{2}} \tag{9.4.7}$$

since

$$P\left\{\left|\frac{\bar{x}-\xi_0}{\sigma/\sqrt{n}}\right| > u_{1-\frac{\alpha}{2}};\ \mathcal{M}\{x\} = \xi_0\right\} = \alpha\,. \tag{9.4.8}$$

The corresponding power function is

$$\pi(\xi) = P\left\{\left|\frac{\bar{x}-\xi_0}{\sigma/\sqrt{n}}\right| > u_{1-\frac{\alpha}{2}};\ \mathcal{M}\{x\} = \xi\right\}$$

$$= \Phi(u_{\frac{\alpha}{2}}-\lambda)+\Phi(u_{\frac{\alpha}{2}}+\lambda) \quad \text{for} \quad \lambda = \frac{\xi-\xi_0}{\sigma/\sqrt{n}}\,. \tag{9.4.9}$$

For $\alpha = 0{\cdot}05$ we get (9.4.3) and (9.4.5). The power function has been tabulated for $\alpha = 10\%$, 5%, 1%, and $0{\cdot}1\%$ in Table 9.4, and Fig. 9.4 shows the corresponding power curves plotted on probability paper. It is seen that a few values of $\pi(\xi)$ suffice for drawing a power curve when probability paper is used. The reason is that for large values of λ we get approximately

TABLE 9.4.

The power function for the two-tailed u-test at 4 levels of significance.

Reject the test hypothesis $\mathcal{M}\{x\} = \xi_0$ when $\left|\frac{\bar{x}-\xi_0}{\sigma/\sqrt{n}}\right| > u_{1-\frac{\alpha}{2}}$

$$\pi(\xi) = \Phi(u_{\frac{\alpha}{2}} - \lambda) + \Phi(u_{\frac{\alpha}{2}} + \lambda) \quad \text{for} \quad \lambda = \frac{\xi-\xi_0}{\sigma/\sqrt{n}}$$

Level of significance	$\alpha = 0{\cdot}1$	$\alpha = 0{\cdot}05$	$\alpha = 0{\cdot}01$	$\alpha = 0{\cdot}001$
$u_{1-\frac{\alpha}{2}}$	$u_{\cdot 95} = 1{\cdot}645$	$u_{\cdot 975} = 1{\cdot}960$	$u_{\cdot 995} = 2{\cdot}576$	$u_{\cdot 9995} = 3{\cdot}291$
$\lvert\lambda\rvert$	Power of test			
0·0	0·100	0·050	0·010	0·001
0·2	0·107	0·055	0·012	0·001
0·4	0·127	0·069	0·016	0·002
0·6	0·160	0·092	0·025	0·004
0·8	0·206	0·126	0·038	0·006
1·0	0·264	0·170	0·058	0·011
1.2	0·330	0·224	0·084	0·018
1·4	0·404	0·288	0·120	0·029
1·6	0·483	0·360	0·165	0·045
1·8	0·562	0·436	0·219	0·068
2·0	0·639	0·516	0·282	0·098
2·2	0·711	0·595	0·353	0·138
2·4	0·775	0·670	0·430	0·186
2·6	0·830	0·739	0·510	0·245
2·8	0·876	0·800	0·589	0·312
3·0	0·912	0·851	0·664	0·386
3·2	0·940	0·892	0·734	0·464
3·4	0·960	0·925	0·795	0·543
3·6	0·975	0·950	0·847	0·621
3·8	0·984	0·967	0·890	0·695
4·0	0·991	0·979	0·923	0·761
4·2	0·995	0·987	0·948	0·818
4·4	0·997	0·993	0·966	0·866
4·6	0·998	0·996	0·979	0·905
4·8	0·999	0·998	0·987	0·934
5·0	1·000	0·999	0·992	0·956
5·2		0·999	0·996	0·972
5·4		1·000	0·998	0·983
5·6			0·999	0·990
5·8			0·999	0·994
6·0			1·000	0·997
6·2				0·998
6·4				0·999
6·6				1·000

$$\pi(\xi) \simeq \Phi(u_{\frac{\alpha}{2}} + \lambda),$$

which is represented by a straight line on probability paper.

Fig. 9.4 shows that the power of the test increases with α, i. e., if we are willing to tolerate a large risk of rejecting the test hypothesis when it is true, then the critical region will be relatively large and a given power is obtained for a relatively small value of λ.

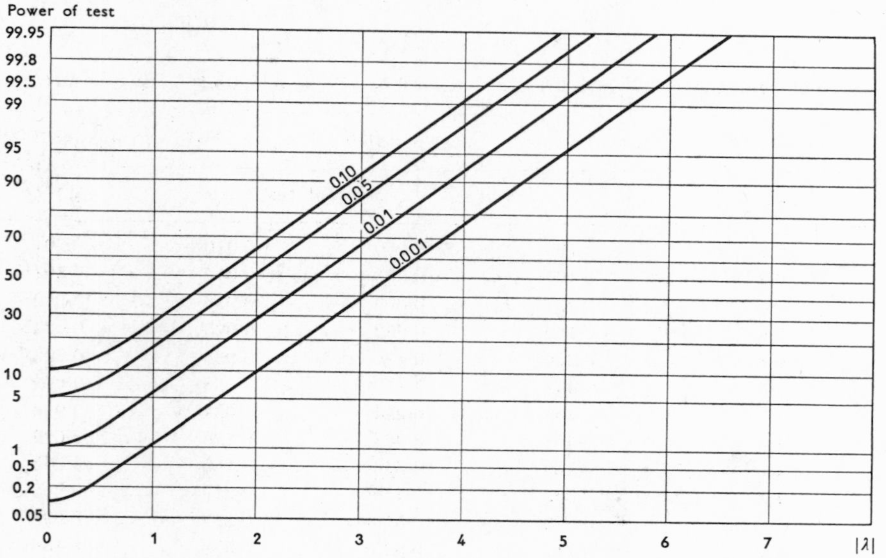

FIG. 9.4. Power curves for the two-tailed u-test at four levels of significance.

$$\pi(\xi) = \Phi(u_{\frac{\alpha}{2}} - \lambda) + \Phi(u_{\frac{\alpha}{2}} + \lambda) \quad \text{for} \quad \lambda = \frac{\xi - \xi_0}{\sigma/\sqrt{n}} \quad \text{and } \alpha = 0{\cdot}10,\ 0{\cdot}05,\ 0{\cdot}01, \text{ and } 0{\cdot}001\,.$$

In the above theory we have intuitively used the two tails of the normal distribution to define a critical region for the sample mean. Many other critical regions for the sample mean may be chosen and moreover other functions of the observations than the mean may be used to construct a test.

Suppose we have two different tests of the same hypothesis so designed that the probabilities of rejecting the test hypothesis when it is true take on the same (small) value, i. e., we are using *the same level of significance*, 5% say. The choice between the two tests is then made on the basis of a comparison of the corresponding power functions.

A general principle in the theory of testing statistical hypotheses, developed by J. NEYMAN and E. S. PEARSON, is to prefer the one test to the other if the one test for all possible alternative values of the unknown parameter has a greater power than the other, i. e., if it gives a greater probability of discovering the falsehood of the test hypothesis than the other. The one test is therefore called uniformly more powerful than the other. Whenever a test exists with a power function which is larger than the power function of any other test we shall prefer this *uniformly most powerful test* to all other tests. In many cases, however, uniformly most powerful tests do not exist.

Let us consider a test for the hypothesis $\xi = \xi_0$ on the basis of the sample median. For large n the median is approximately normally distributed about $\mathcal{M}\{x\}$ with variance $\frac{\sigma^2}{n}\frac{\pi}{2}$. Denoting the median by $\hat{x}$ we get a critical region analogous to (9.4.3) as

$$\left| \frac{\hat{x}-\xi_0}{\frac{\sigma}{\sqrt{n}}\sqrt{\frac{\pi}{2}}} \right| > 1{\cdot}96 \,. \tag{9.4.10}$$

Similarly the power function becomes

$$\pi_2(\xi) = \Phi\left(-1{\cdot}96-\lambda\sqrt{\frac{2}{\pi}}\right)+\Phi\left(-1{\cdot}96+\lambda\sqrt{\frac{2}{\pi}}\right)$$

where

$$\lambda = \frac{\xi-\xi_0}{\sigma/\sqrt{n}} \,.$$

Further, let us consider a one-tailed test for the hypothesis $\xi = \xi_0$ based on the sample mean. At the 5% level of significance we get as a critical region

$$\frac{\bar{x}-\xi_0}{\sigma/\sqrt{n}} > 1{\cdot}64 \tag{9.4.11}$$

and the power function becomes

$$\pi_3(\xi) = \Phi(-1{\cdot}64+\lambda) \,. \tag{9.4.12}$$

Fig. 9.5 shows these two power functions together with the power function $\pi_1(\xi)$ given by (9.4.5).

It is seen that the two-tailed test based on the mean is uniformly more powerful than the two-tailed test based on the median, i. e., $\pi_1(\xi) > \pi_2(\xi)$ for $\xi \neq \xi_0$. This result was to be expected since the mean is an efficient and the median an inefficient estimate of the population mean.

Comparing the curves corresponding to the two-tailed and the one-tailed test based on the mean it is observed that $\pi_3(\xi) > \pi_1(\xi)$ for $\xi > \xi_0$ and $\pi_3(\xi) < \pi_1(\xi)$ for $\xi < \xi_0$. It is quite common that the power curves corresponding to different tests intersect so that a particular test is best for certain values of the unknown parameter but not for other values.

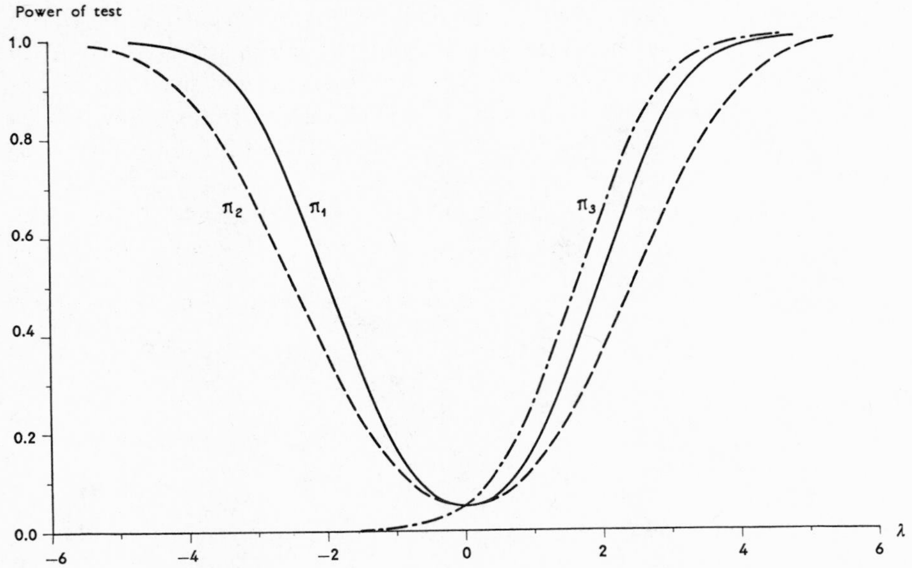

FIG. 9.5. The power curves corresponding to three tests of the same hypothesis. Level of significance: 5%.

Practical considerations in each case together with the power functions decide what test is to be preferred. In the above discussion we assumed that the alternative to the test hypothesis is that the population mean differs from ξ_0 in either direction, i. e., the alternative is $\xi \neq \xi_0$, without it a priori being possible to decide whether $\xi > \xi_0$ or $\xi < \xi_0$. If, however, we know a priori, i. e., independently of the observations in question, that the alternative to the test hypothesis is $\xi > \xi_0$, while $\xi < \xi_0$ is an impossibility or of no practical importance, then we prefer the one-tailed test to the two-tailed because it has the greater power for $\xi > \xi_0$. It has been proved that the one-tailed test is uniformly most powerful for testing the hypothesis $\xi = \xi_0$ against the alternative $\xi > \xi_0$. No uniformly most powerful test exists for testing the hypothesis $\xi = \xi_0$ against the alternative $\xi \neq \xi_0$ but the two-tailed test above is the uniformly most powerful among the tests with power functions having a minimum at $\xi = \xi_0$. (Such a test is called unbiased).

The level of significance must be chosen for each case in the light of practical considerations. Most commonly the conventional values 5% or 1% are used. In the two-tailed test, for instance, values of $|u|$ between 1·96 and 2·58 usually will give rise to doubt regarding the test hypothesis. As a rule a $|u|$-value larger than the 99% limit will be taken as an indication that it is practical to reject the test hypothesis, while values below the 95% limit very seldom lead to rejection of the hypothesis.

Thus, when using *the u-test at the* α *level of significance to test the hypothesis* $\xi = \xi_0$ *against the alternative* $\xi \neq \xi_0$ we proceed as follows:

1. The test hypothesis specifies the population mean $\mathcal{M}\{x\} = \xi_0$.
2. We calculate the quantity

$$u = \frac{\bar{x}-\xi_0}{\sigma/\sqrt{n}}$$

inserting the observed mean for $\bar{x}$.

3. If $|u| < u_{1-\frac{\alpha}{2}}$ the test hypothesis is for the present accepted as a working hypothesis.
4. If $|u| > u_{1-\frac{\alpha}{2}}$ the test hypothesis is rejected.

The above rules imply that

a. The test hypothesis will be accepted in $100(1-\alpha)\%$ of the cases in which it is true.

b. The test hypothesis will be rejected in $100\alpha\%$ of the cases in which it is true.

c. If the test hypothesis is false, it will all the same be accepted in a certain number of cases where the systematic error is small compared with the standard error of the mean. The power function (9.4.9) gives the probability of rejecting the test hypothesis as a function of the unknown parameter.

If, however, *the alternative to the test hypothesis is* $\xi > \xi_0$ we use

$$\frac{\bar{x}-\xi_0}{\sigma/\sqrt{n}} > u_{1-\alpha}$$

as the criterion for rejecting the test hypothesis at the α level of significance.

Correspondingly, if *the alternative is* $\xi < \xi_0$ then we use the criterion

$$\frac{\bar{x}-\xi_0}{\sigma/\sqrt{n}} < u_\alpha \,.$$

For the two last mentioned one-tailed tests the power function is

$$\pi(\xi) = \Phi(u_\alpha+\lambda) \quad \text{where} \quad \lambda = \frac{|\xi-\xi_0|}{\sigma/\sqrt{n}}\,, \tag{9.4.13}$$

cf. (9.4.12).

It must be remembered, however, that *statistical significance* is not always the same as *practical significance*. If, for example, we have a large number of observations we are perhaps able to show that a given production process does not correspond to the specified standard ξ_0 but that ξ deviates only by a *small amount* from ξ_0. Small values of $\xi-\xi_0$ may, however, be of no practical importance, for instance from an economic point of view, so even if the deviation of $\bar{x}$ from ξ_0 is statistically significant it need not be of any practical significance.

In the above discussion we have assumed that the number of observations n was given and the power function was considered as a function of ξ only. The steepness of the power curve depends, however, on n as shown in Fig. 9.6 for the one-tailed u-test with the power function

$$\pi(\xi) = \Phi\left(-1{\cdot}64+\frac{\xi-\xi_0}{\sigma/\sqrt{n}}\right).$$

The figure illustrates the fact that the probability of detecting the falsehood of the test hypothesis increases with the number of observations.

For $n \to \infty$ we get the ideal power function which takes on the value 1 for all $\xi > \xi_0$.

Two different samples of size n_1 and n_2, respectively, will give the same power if

$$(\xi_1-\xi_0)\sqrt{n_1} = (\xi_2-\xi_0)\sqrt{n_2}$$

or

$$\xi_2-\xi_0 = (\xi_1-\xi_0)\sqrt{\frac{n_1}{n_2}}. \tag{9.4.14}$$

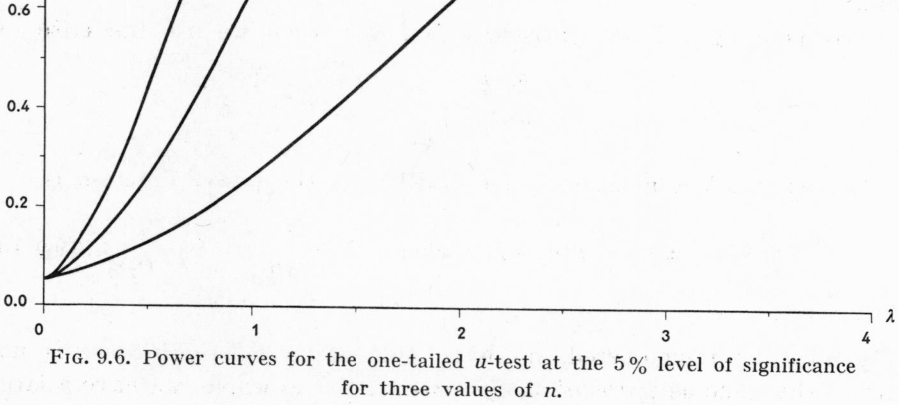

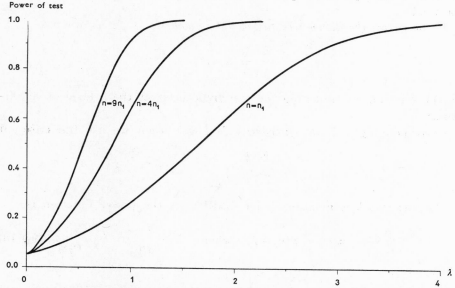

FIG. 9.6. Power curves for the one-tailed u-test at the 5% level of significance for three values of n.

The dependence of the power function on the number of observations may be used in the planning of experiments to determine *the number of observations necessary to obtain a given discriminating power*. Suppose that we want to test the hypothesis $\xi = \xi_0$ against the single alternative $\xi = \xi_1$, $(\xi_1 > \xi_0)$. How many observations are necessary if we want to be reasonably sure of accepting the right hypothesis? Choosing for instance $\alpha = 5\%$ as the level of significance for the test hypothesis $\xi = \xi_0$ we find the power of the u-test with respect to the alternative hypothesis as

$$\pi(\xi_1) = \Phi\left(-1{\cdot}64+\frac{\xi_1-\xi_0}{\sigma/\sqrt{n}}\right).$$

If we want the probability of making a wrong decision when $\xi = \xi_1$ to be the same as when $\xi = \xi_0$ we require the power function to take on the value 0·95. This leads to the equation

$$\pi(\xi_1) = \Phi\left(-1{\cdot}64+\frac{\xi_1-\xi_0}{\sigma/\sqrt{n}}\right) = 0{\cdot}95$$

or

$$-1{\cdot}64+\frac{\xi_1-\xi_0}{\sigma/\sqrt{n}} = u_{\cdot 95} = 1{\cdot}64\,,$$

from which we get

$$\sqrt{n} = 2\times 1{\cdot}64\times\frac{\sigma}{\xi_1-\xi_0}\,. \tag{9.4.15}$$

Thus, the square root of the number of observations is proportional to the standard deviation and inversely proportional to the distance between the two hypothetical means.

The critical region

$$\bar{x} > \xi_0+1{\cdot}64\frac{\sigma}{\sqrt{n}} \tag{9.4.16}$$

may by means of (9.4.15) be written as

$$\bar{x} > \tfrac{1}{2}(\xi_0+\xi_1)\,. \tag{9.4.17}$$

The required test is consequently carried out in the following way: By means of (9.4.15) we determine the number of observations n. If the mean of the n observations is less than $\frac{1}{2}(\xi_0+\xi_1)$ we accept the hypothesis $\xi = \xi_0$; otherwise we accept $\xi = \xi_1$. This rule will lead to the acceptance of $\xi = \xi_0$ in 95% of the cases where it is true, and similarly for $\xi = \xi_1$. If we are not satisfied with a probability of 95% of making a correct decision we have to increase the number of observations correspondingly.

If the consequences of making the two possible kinds of wrong decisions are not equally serious but the consequences of rejecting $\xi = \xi_0$ when it is true are considered more serious than the consequences of failing to reject $\xi = \xi_0$ when it is false, then we would want to choose α (the probability of making an error of the first kind) smaller than $\beta = 1-\pi(\xi_1)$ (the probability of making an error of the second kind).

For given values of α and β we get

$$\sqrt{n} = (u_{1-\beta}+u_{1-\alpha})\frac{\sigma}{\xi_1-\xi_0} \tag{9.4.18}$$

and the critical region

$$\bar{x} > \frac{u_{1-\beta}\xi_0+u_{1-\alpha}\xi_1}{u_{1-\beta}+u_{1-\alpha}} . \tag{9.4.19}$$

For $\alpha = 5\%$ and $\beta = 10\%$ we get $u_{\cdot 95} = 1{\cdot}64$, $u_{\cdot 90} = 1{\cdot}28$,

$$\sqrt{n} = 2{\cdot}92\frac{\sigma}{\xi_1-\xi_0} ,$$

and

$$\bar{x} > 0{\cdot}44\xi_0+0{\cdot}56\xi_1 .$$

The above formulas may also be applied to the following situation. Suppose that we want to be able to discriminate between populations with means falling on either side of the interval (ξ_0, ξ_1), i. e., if $\xi \leq \xi_0$ we want to be reasonably sure that the test leads to the right decision, and similarly for $\xi \geq \xi_1$. For values of ξ between ξ_0 and ξ_1, however, it is of minor practical importance which decision is taken. When we on the basis of practical considerations have chosen ξ_0 and ξ_1, and the corresponding (small) probabilities α and β of making incorrect decisions then (9.4.18) and (9.4.19) give us the number of observations and the critical region.

By means of the power function for the two-tailed test we may determine in a similar way the number of observations necessary for discriminating between the two alternatives. The critical region corresponding to the α level of significance is determined by the inequality (9.4.7) and the number of observations n is determined from the equation

$$\Phi(u_{\frac{\alpha}{2}}-\lambda)+\Phi(u_{\frac{\alpha}{2}}+\lambda) = 1-\beta \tag{9.4.20}$$

for

$$\lambda = \frac{\delta}{\sigma/\sqrt{n}} , \tag{9.4.21}$$

where $\delta > 0$ is chosen such that for $|\xi-\xi_0| < \delta$ it is of minor practical importance whether the test hypothesis is rejected or not, but for $|\xi-\xi_0| \geq \delta$ we want to be reasonably sure to detect the falsehood of the test hypothesis. Equation (9.4.20) does not have an explicit solution but must be solved by means of Table II.

In cases where α und β are small the main term of the equation is

$$\Phi(u_{\frac{\alpha}{2}}+\lambda) \simeq 1-\beta \tag{9.4.22}$$

so that an approximate solution is given by

$$\lambda \simeq u_{1-\beta}+u_{1-\frac{\alpha}{2}} \tag{9.4.23}$$

which leads to

$$\sqrt{n} \simeq (u_{1-\beta}+u_{1-\frac{\alpha}{2}})\,\frac{\sigma}{\delta}\,. \tag{9.4.24}$$

The above results may be useful for setting up a *sampling plan for the inspection of the quality of an industrial product.* Let us suppose that *the quality* of an item is found by measurement of a single quality characteristic x which is normally distributed and that the product is submitted for inspection in lots (populations) containing a large number of items so that the sample is only a small part, less than 10% say, of each lot. If the standard deviation of x is the same for all lots but the mean varies, then *the quality of a lot* is given by the mean, ξ, for instance so that small values of ξ denote good quality and large values of ξ denote bad quality. Usually it is not feasible or economical to inspect all the items of a lot to decide whether or not the lot is of acceptable quality but the decision must be based on a random sample from the lot. The producer (seller) will require a sampling plan which assures him that lots of good quality are nearly always accepted and the consumer (buyer) will require that lots of bad quality are nearly always rejected. These requirements lead to the specification of the four numbers ξ_0, ξ_1, α, and β, so that if $\xi \leq \xi_0$ the risk is less than α for rejecting the lot and if $\xi \geq \xi_1$ the risk is less than β for accepting the lot. The sample size, n, and the critical region are then determined from (9.4.18) and (9.4.19). The probability α is called the producer's (maximum) risk and β is called the consumer's (maximum) risk. Most commonly the values $\alpha = 0{\cdot}05$ and $\beta = 0{\cdot}10$ are used. The sampling plan then has the following form: A sample of n items is drawn at random from each lot and the mean is computed. If the mean falls within the critical region the lot is rejected; otherwise, the lot is accepted.

The sensitivity of a control chart to changes in the population mean may be studied by means of the power function. Let us suppose that changes in the production process influence only the mean but not the variance of the quality characteristic measured. Using the two-sided 95% limits, $\xi_0 \pm 1{\cdot}96\frac{\sigma}{\sqrt{n}}$, for the sample mean the corresponding power function is given by (9.4.5). As long as the process adheres to the standard only 2·5% of the points on the control chart are expected to fall outside each limit.

If ξ deviates from ξ_0 then the frequency of points outside the 95% limits will increase according to the value of the power function given in Table 9.3. An increase in the frequency of points above the upper limit from 2·5% to 25%, say, will be detected very soon after the change has taken place. The magnitude of the corresponding change in the population mean is determined from the equation

$$\Phi(-1{\cdot}96+\lambda) = 0{\cdot}25$$

which leads to

$$1{\cdot}96-\lambda = u_{\cdot 75} = 0{\cdot}67$$

or $\lambda = 1{\cdot}29$. Thus the increase in the population mean has to be $1{\cdot}29 \frac{\sigma}{\sqrt{n}}$ to give an expected frequency of 25% of points above the upper limit. At the same time the frequency of points below the lower limit decreases from 2·5% to practically zero since $\Phi(-1{\cdot}96-1{\cdot}29) = \Phi(-3{\cdot}25) = 0{\cdot}0006$.

The magnitude of the change in the population mean necessary to produce the above effect is seen to be inversely proportional to $\sqrt{n}$. If we want a change of given magnitude $\delta = |\xi-\xi_0|$ to result in an increase in the frequency of points outside the one limit from 2·5% to 25% then the sample size is determined from the equation

$$\delta = 1{\cdot}29\frac{\sigma}{\sqrt{n}}$$

or

$$\sqrt{n} = 1{\cdot}29\sigma/\delta\,.$$

The u-test may also be applied to *test the equality of two population means*. Suppose that a random sample is drawn from each of two normally distributed populations with parameters (ξ_1, σ_1^2) and (ξ_2, σ_2^2), respectively, and further that the two variances are known. We want to test the hypothesis that the two unknown population means are equal, i. e., $\xi_1 = \xi_2$. Introducing the new variable

$$d = \bar{x}_1-\bar{x}_2\,, \tag{9.4.25}$$

where $\bar{x}_1$ and $\bar{x}_2$ denote the two sample means, it follows from the addition theorem for the normal distribution that d is normally distributed with mean

$$\mathcal{M}\{d\} = \xi_1-\xi_2 = \delta \tag{9.4.26}$$

and variance

$$\mathcal{V}\{d\} = \frac{\sigma_1^2}{n_1}+\frac{\sigma_2^2}{n_2} = \sigma_d^2\,. \tag{9.4.27}$$

The hypothesis $\xi_1 = \xi_2$ or $\delta = 0$ may thus be tested by means of the u-test where

$$u = \frac{d}{\sigma_d} = \frac{\bar{x}_1 - \bar{x}_2}{\sigma_d} . \tag{9.4.28}$$

The theory of testing statistical hypotheses is not in its final form and at present many tests are to some extent based on practical conventions. A choice between different tests for the same hypothesis may in some cases be made by introducing a weight function specifying the relative importance of (or the loss caused by) the error committed by accepting the test hypothesis when an alternative hypothesis is true. The best test may then be found as the test which minimizes the expected loss. A general theory of statistical decision functions is at present under development, see the reference in § 9.10.

Example 9.6. Let us suppose that we want to test the hypothesis that the true mean diameter ξ_0 of rivets produced under the same conditions is 13·42 mm. We assume that the standard deviation σ is known, $\sigma = 0{\cdot}12$ mm, and that the diameters are normally distributed.

If we choose $\alpha = 0{\cdot}01$ as the level of significance and are willing to tolerate a risk of $\beta = 0{\cdot}05$ of not rejecting the test hypothesis when $|\xi - \xi_0| = 0{\cdot}02$ mm, then we get the following equations for the determination of n according to (9.4.20) and (9.4.21):

$$\Phi(-2{\cdot}58-\lambda)+\Phi(-2{\cdot}58+\lambda) = 0{\cdot}95$$

and

$$\lambda = \frac{0{\cdot}02}{0{\cdot}12}\sqrt{n} = \frac{1}{6}\sqrt{n} .$$

By means of Table II we find $\lambda = 4{\cdot}22$ which leads to

$$\sqrt{n} = 6\lambda = 25{\cdot}32$$

and $n = 642$.

Thus, the test may be performed in the following way: A random sample of 642 rivets is taken and the mean diameter $\bar{x}$ is calculated. The standard error of the mean is $0{\cdot}12/\sqrt{642} = 0{\cdot}00474$ mm. The critical region is defined as

$$|\bar{x}-\xi_0| > 2{\cdot}58\frac{\sigma}{\sqrt{n}} = 0{\cdot}012$$

or

$$\bar{x} < 13{\cdot}408 \text{ mm} \quad \text{and} \quad \bar{x} > 13{\cdot}432 \text{ mm} .$$

If the observed $\bar{x}$ falls within the critical region we reject the test hypothesis; otherwise, we accept it.

9.5. Confidence Limits for ξ.

In § 9.3 it is shown how with a preassigned probability we may predetermine the position of the sample mean from the parameters of the distribution of the population. On the other hand, the same relation may be used for obtaining limits for the value of ξ, the mean of the population, when we know the mean of n observations. (As before, we assume that σ is known. If σ is not known, the methods described in Chapter 15 must be used).

We start with the same relation as used for the determination of the fractiles

$$P\left\{\frac{\bar{x}-\xi}{\sigma/\sqrt{n}} < u_P\right\} = P\,.$$

If the inequality is solved with respect to ξ, we have

$$\boxed{P\left\{\bar{x}-u_P\frac{\sigma}{\sqrt{n}} < \xi\right\} = P\,.} \qquad (9.5.1)$$

Correspondingly, the relation

$$P\left\{u_{P_1} < \frac{\bar{x}-\xi}{\sigma/\sqrt{n}} < u_{P_2}\right\} = P_2-P_1$$

leads to

$$\boxed{P\left\{\bar{x}-u_{P_2}\frac{\sigma}{\sqrt{n}} < \xi < \bar{x}-u_{P_1}\frac{\sigma}{\sqrt{n}}\right\} = P_2-P_1\,.} \qquad (9.5.2)$$

Thus, for *every* $\mathcal{M}\{x\} = \xi$ *the probability that the inequality*

$$\bar{x}-u_{P_2}\frac{\sigma}{\sqrt{n}} < \xi < \bar{x}-u_{P_1}\frac{\sigma}{\sqrt{n}} \qquad (9.5.3)$$

will be satisfied is P_2-P_1.

If we put $P_1 = 2{\cdot}5\%$ and $P_2 = 97{\cdot}5\%$, we get

$$P\left\{\bar{x}-1{\cdot}96\frac{\sigma}{\sqrt{n}} < \xi < \bar{x}+1{\cdot}96\frac{\sigma}{\sqrt{n}}\right\} = 95\%, \qquad (9.5.4)$$

which says that from a long series of samples, each containing n observations drawn from a normally distributed population with parameters (ξ, σ^2), we may calculate a corresponding series of intervals $\left(\bar{x}\pm 1{\cdot}96\frac{\sigma}{\sqrt{n}}\right)$ and that *about* 95% *of these intervals must be expected to include the fixed quantity* ξ.

The intervals defined by (9.5.2) are called *confidence intervals* with a *confidence coefficient* of P_2-P_1. The corresponding limits are called *confidence limits* for ξ. Practical considerations of the same nature as mentioned in § 9.4 determine the level of the confidence coefficient and whether the confidence interval should be unilateral or bilateral. Very often the conventional values 95% and 99% are used as confidence coefficients.

It will be seen that *the limits for ξ vary from sample to sample as the mean $\bar{x}$ varies.* In the long run the proportion $P_2 - P_1$ of the number of confidence intervals will contain ξ. It may happen that a particular interval does not contain ξ, either $\bar{x} - u_{P_2}\frac{\sigma}{\sqrt{n}}$ being larger than ξ, or $\bar{x} - u_{P_1}\frac{\sigma}{\sqrt{n}}$ smaller than ξ; in the long run this will happen with frequencies $1 - P_2$ and P_1, respectively.

In practice we have only one sample from which to calculate the mean. The problem then is to determine the position of ξ from this single sample mean, $\bar{x}_0$, and the given standard deviation, σ. The method used is to calculate the limits of ξ using (9.5.3), e. g., by computing $\bar{x}_0 \pm 1{\cdot}96\frac{\sigma}{\sqrt{n}}$.

Fig. 9.7 illustrates the ideas underlying the above procedure. The figure demonstrates the $\bar{x}_0$-value found and five distribution curves of $\bar{x}$ with the different population means $\xi_1, \xi_2, \ldots, \xi_5$ and the same standard error

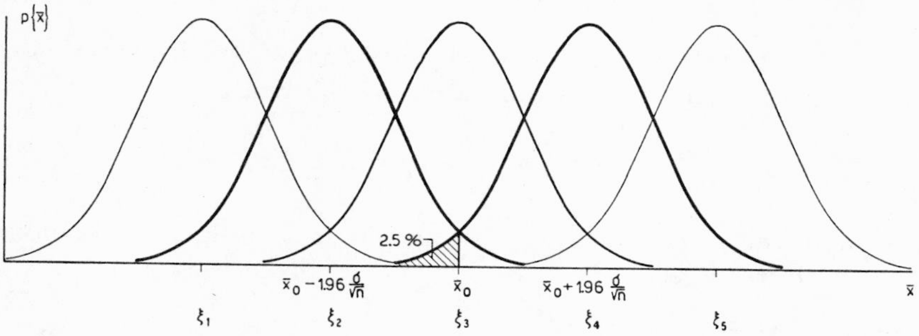

FIG. 9.7.

$\frac{\sigma}{\sqrt{n}}$. It seems unreasonable to suppose that the observed mean originates from the populations with means ξ_1 or ξ_5, as $\bar{x}_0$ deviates so much from the said population means that the probability of such a deviation or a more extreme deviation in the same direction is small. It may be seen directly that populations with mean values in the neighbourhood of ξ_3 will often "lead to" sample means close to $\bar{x}_0$. The position of ξ_2 has been chosen so that $\bar{x}_0$ is equal to the 97·5% fractile, $\xi_2 + 1{\cdot}96\frac{\sigma}{\sqrt{n}}$, in the distribution of $\bar{x}$'s derived from a population with mean ξ_2. In analogy to this, ξ_4 has been placed so that $\bar{x}_0$ is equal to the 2·5% fractile, $\xi_4 - 1{\cdot}96\frac{\sigma}{\sqrt{n}}$, in the distribution of $\bar{x}$'s with mean ξ_4, cf. the shaded area of Fig. 9·7. We consider it reasonable that ξ should be in the neighbourhood of $\bar{x}_0$, so that $\bar{x}_0$ does not occur as an extreme value. The equations

$$\xi_2 + 1{\cdot}96\frac{\sigma}{\sqrt{n}} = \bar{x}_0 = \xi_4 - 1{\cdot}96\frac{\sigma}{\sqrt{n}}$$

lead to *the lower confidence limit*

$$\xi_2 = \bar{x}_0 - 1{\cdot}96\frac{\sigma}{\sqrt{n}}$$

and to *the upper confidence limit*

$$\xi_4 = \bar{x}_0 + 1{\cdot}96\frac{\sigma}{\sqrt{n}},$$

cf. (9.5.4).

If we wish to accept a value of ξ which is greater than ξ_4, then we must admit that our $\bar{x}_0$ is a very unusual observation. We must have important additional information of a non-statistical nature before it seems reasonable to make such a judgment.

9.6. Summary.

The above discussion shows how the derivation of fractiles, tests of significance and confidence limits are based on the relation

$$P\left\{\frac{\bar{x}-\xi}{\sigma/\sqrt{n}} < u_P\right\} = P. \tag{9.6.1}$$

Throughout the discussion it is assumed that σ is known (σ unknown is dealt with in Chapter 15).

I. *Fractiles for $\bar{x}$. ξ known.*

The inequality in (9.6.1) is solved with respect to $\bar{x}$, which leads to

$$P\left\{\bar{x} < \xi + u_P\frac{\sigma}{\sqrt{n}}\right\} = P.$$

$\xi + u_P\frac{\sigma}{\sqrt{n}}$ gives the P-fractile for $\bar{x}$.

II. *Confidence limits for ξ. $\bar{x}$ observed.*

The inequality in (9.6.1) is solved with respect to ξ, which leads to

$$P\left\{\bar{x} - u_P\frac{\sigma}{\sqrt{n}} < \xi\right\} = P.$$

$\bar{x} - u_P\frac{\sigma}{\sqrt{n}}$ gives the unilateral P-confidence limit for ξ.

III. *Tests of significance.* $\bar{x}$ *observed and* ξ *postulated.*

The agreement between $\bar{x}$ and ξ is tested by calculating

$$\frac{\bar{x}-\xi}{\sigma/\sqrt{n}}$$

and comparing this quantity with u_P for a suitably chosen value of P.

In all the above relations we are dealing with *only one stochastic variable*, $\bar{x}$, while the other quantities, ξ, σ, u_P and n, are constants.

In all formulas in this chapter it is assumed that σ is known. This is true only in exceptional cases, and if σ is to be replaced by its estimate, it is necessary to modify the formulas. In the following chapters the distribution function of the sample variance will be discussed together with the consequences implied by employing s instead of σ.

9.7. The Weighted Mean.

Suppose k random samples are given and that the k population means are identical, as shown in Table 9.5. From each sample we may compute

TABLE 9.5.

Sample No.	Observations	Sample		Population	
		Mean	Variance	Mean	Variance
1	$x_{11}, x_{12}, \ldots, x_{1n_1}$	$\bar{x}_1$	s_1^2	ξ	σ_1^2
2	$x_{21}, x_{22}, \ldots, x_{2n_2}$	$\bar{x}_2$	s_2^2	ξ	σ_2^2
.	. . .	.	.	.	.
.	. . .	.	.	.	.
.	. . .	.	.	.	.
k	$x_{k1}, x_{k2}, \ldots, x_{kn_k}$	$\bar{x}_k$	s_k^2	ξ	σ_k^2

the mean and variance as the best estimates of the corresponding population quantities. Since the k population means are identical we want to combine the k sample means to a single estimate of ξ.

Considering the linear combination

$$\bar{x} = \sum_{i=1}^{k} \alpha_i \bar{x}_i \tag{9.7.1}$$

of the k means we will determine the unknown coefficients α_i so that $\bar{x}$ becomes an unbiased estimate of ξ with the least possible variance.

The mean and variance of $\bar{x}$ are

$$\mathcal{M}\{\bar{x}\} = \sum_{i=1}^{k} \alpha_i \mathcal{M}\{\bar{x}_i\} = \xi \sum_{i=1}^{k} \alpha_i \tag{9.7.2}$$

and

$$\mathcal{V}\{\bar{x}\} = \sum_{i=1}^{k} \alpha_i^2 \mathcal{V}\{\bar{x}_i\} = \sum_{i=1}^{k} \alpha_i^2 \frac{\sigma_i^2}{n_i} . \tag{9.7.3}$$

For $\bar{x}$ to be an unbiased estimate of ξ we must have

$$\sum_{i=1}^{k} \alpha_i = 1 . \tag{9.7.4}$$

Introducing $\alpha_k = 1-\alpha_1-\ldots-\alpha_{k-1}$ into $\mathcal{V}\{\bar{x}\}$ and differentiating with respect to α_i, $i = 1, \ldots, k-1$, to get the equations for determination of the values of α_i which minimize $\mathcal{V}\{\bar{x}\}$, we find

$$\frac{\partial \mathcal{V}}{\partial \alpha_i} = 2\alpha_i \frac{\sigma_i^2}{n_i} - 2\alpha_k \frac{\sigma_k^2}{n_k} = 0 , \quad i = 1, \ldots, k-1 , \tag{9.7.5}$$

or

$$\alpha_i = \alpha_k \frac{\sigma_k^2}{n_k} \frac{n_i}{\sigma_i^2} = \alpha_k \frac{w_i}{w_k} , \tag{9.7.6}$$

where

$$w_i = \frac{n_i}{\sigma_i^2} , \qquad i = 1, \ldots, k . \tag{9.7.7}$$

By means of (9.7.4) we obtain

$$\sum_{i=1}^{k} \alpha_i = \frac{\alpha_k}{w_k} \sum_{i=1}^{k} w_i = 1 \tag{9.7.8}$$

or

$$\alpha_k = w_k \Big/ \sum_{i=1}^{k} w_i . \tag{9.7.9}$$

From (9.7.6) follows

$$\alpha_i = w_i \Big/ \sum_{i=1}^{k} w_i \tag{9.7.10}$$

and finally

$$\boxed{\bar{x} = \sum w_i \bar{x}_i \Big/ \sum w_i = \sum \frac{n_i \bar{x}_i}{\sigma_i^2} \Big/ \sum \frac{n_i}{\sigma_i^2} .} \tag{9.7.11}$$

Hence, *the weights of the sample means are inversely proportional to their variances.*

The minimized variance of the combined estimate becomes

$$\boxed{\mathcal{V}\{\bar{x}\} = 1 \Big/ \sum_{i=1}^{k} w_i .} \tag{9.7.12}$$

In the special case where $\sigma_1^2 = \ldots = \sigma_k^2 = \sigma^2$ we get $w_i = n_i/\sigma^2$, $\alpha_i = n_i/\Sigma n_i$,

$$\bar{x} = \frac{\sum_{i=1}^{k} n_i \bar{x}_i}{\sum_{i=1}^{k} n_i} = \frac{\sum_{i=1}^{k}\sum_{\nu=1}^{n_i} x_{i\nu}}{\sum_{i=1}^{k} n_i} \tag{9.7.13}$$

the mean of all observations, and

$$\mathcal{V}\{\bar{x}\} = \sigma^2/\Sigma n_i \tag{9.7.14}$$

in agreement with the results previously found for a sample from a single population.

The same results will be obtained from the method of least squares giving as estimate of ξ the value which minimizes

$$\sum_{i=1}^{k}\sum_{\nu=1}^{n_i}\left(\frac{x_{i\nu}-\xi}{\sigma_i}\right)^2 \tag{9.7.15}$$

with respect to ξ. Putting the derivate of (9.7.15) with respect to ξ equal to zero we obtain the solution given by (9.7.11).

We will return to these problems in Chapter 16.

9.8. The Distribution of the Estimate of the Parameter ξ in a Truncated Normal Distribution.

The distribution function of the estimate $\bar{x}$ of the parameter ξ in a truncated normal distribution, cf. (6.9.10) and (6.9.14), is not known for finite values of n, but for $n \to \infty$ it converges toward the normal distribution function with mean ξ and variance

$$\mathcal{V}\{\bar{x}\} = \frac{\sigma^2}{n}\mu_{11}(\zeta), \tag{9.8.1}$$

where $\mu_{11}(\zeta)$ has been tabulated in Tables IX and X[1]). Thus, the variance depends on the degree of truncation. For $\zeta \to -\infty$, i. e., the degree of truncation $= 0$, the factor $\mu_{11}(\zeta)$ converges to 1, and for $\zeta \to \infty$, i. e., the degree of truncation $= 100\%$, $\mu_{11}(\zeta)$ converges to ∞.

Example 9.7. To find an estimate of the variance of $\bar{x}$ for the two distributions in Tables 6.6 and 6.7 we replace σ and ζ by their estimates s and z.

For the one-sided truncated distribution of 310 diameters of rivet heads we find

$$\mu_{11}(-0{\cdot}310) = 11{\cdot}9, \text{ see Table IX,}$$

---

1) The derivation of this formula may be found in A. Hald: *Maximum Likelihood Estimation of the Parameters of a Normal Distribution Which is Truncated at a Known Point*, Skandinavisk Aktuarietidskrift, 1949, 119—134.

so that
$$v\{\bar{x}\} \approx \frac{0\cdot111^2}{310} 11\cdot9 = 0\cdot0^3473$$
and
$$\sqrt{v\{\bar{x}\}} \approx 0\cdot022 \text{ mm}\,.$$

For the one-sided censored distribution of 500 diameters of rivet heads we find
$$\mu_{11}(-0\cdot306) = 1\cdot24, \text{ see Table X},$$
so that
$$v\{\bar{x}\} \approx \frac{0\cdot111^2}{500} 1\cdot24 = 0\cdot0^4306$$
and
$$\sqrt{v\{\bar{x}\}} \approx 0\cdot0055 \text{ mm}\,.$$

For comparison we find an estimate of the variance of the mean of the complete distribution of the 500 diameters of rivet heads, see Table 4.4, as
$$v\{\bar{x}\} \approx \frac{0\cdot115^2}{500} = 0\cdot0^4265$$
and
$$\sqrt{v\{\bar{x}\}} \approx 0\cdot0051 \text{ mm}\,.$$

The factor 11·9 shows that a degree of truncation of 38% gives rise to a large increase in the variance of the mean. If the distribution is only censored but not truncated the increase in the variance of the mean is relatively small.

9.9. An Approximate Formula for the Mean and Variance of Indirect Observations.

In many problems, particularly those dealing with the theory of errors, *the quantity required is a known function of the directly observed quantities.* In the simplest case the quantity required, y, is a function of one variable only, $y = \varphi(x)$ say. If, for instance, x is the diameter and y the area of the cross section of a circular metal wire, we have $y = \frac{\pi}{4} x^2$. After making n measurements of the diameter, $x_1, x_2, \ldots, x_n$, we may calculate the corresponding areas, $y_1, y_2, \ldots, y_n$, the mean $\bar{y}$, the variance s_y^2 and the estimated standard error of the mean $s_y/\sqrt{n}$. This is, however, a rather troublesome procedure and the question is now whether the transformation of each single observation can be avoided and the estimates required calculated from $\bar{x}$ and s_x.

If the mean and the variance of the y's are to be included in a more detailed statistical analysis as shown in the following chapters, each y-value should be calculated from the corresponding value of x, and the mean and variance calculated directly from the y-values, as for instance in Examples

7.2 and 7.7. If, however, we only require a comparatively rough estimate of $\mathcal{M}\{y\}$ and $\sigma\{y\}$ the approximation formulas given in this paragraph may be used.

If y is a linear function of x, and x is normally distributed, then y is also normally distributed, as shown in § 6.4. If y is not a linear function of x, but approximately so throughout practically the whole range of x, the theoretical distribution of y may be approximated to by a normal distribution with mean

$$\varphi(\xi) \simeq \mathcal{M}\{y\} \tag{9.9.1}$$

and standard deviation

$$|\varphi'(\xi)|\,\sigma\{x\} \simeq \sigma\{y\}\,, \tag{9.9.2}$$

cf. (5.8.5) and (5.9.7). As already stated, it is necessary that $y = \varphi(x)$ is linear throughout practically the whole range of x's variation, say from $\xi - 2{\cdot}58\sigma\{x\}$ to $\xi + 2{\cdot}58\sigma\{x\}$, so that only 1% of the x's take on values outside these limits. Whether this assumption is fulfilled may be partly checked by calculating $\varphi(\xi - 2{\cdot}58\sigma\{x\})$, $\varphi(\xi)$, and $\varphi(\xi + 2{\cdot}58\sigma\{x\})$, which quantities must satisfy the relation

$$\varphi(\xi) - \varphi(\xi - 2{\cdot}58\,\sigma\{x\}) \simeq \varphi(\xi + 2{\cdot}58\,\sigma\{x\}) - \varphi(\xi) \simeq 2{\cdot}58\,\sigma\{x\}\varphi'(\xi)\,. \tag{9.9.3}$$

In the theory of errors the coefficients of variation are often small, and the functions of such a nature that both x and y may be considered normally distributed, cf. §§ 7.2 and 7.3. We thus get the following estimates

$$\varphi(\bar{x}) \approx \mathcal{M}\{y\} \tag{9.9.4}$$

and

$$|\varphi'(\bar{x})|\,s_x \approx \sigma\{y\} \tag{9.9.5}$$

which may be used to find approximate confidence limits for $\mathcal{M}\{y\}$.

As a special case of the above formulas we may mention

$$y = \log_{10} x$$

which leads to

$$\sigma\{y\} \simeq 0{\cdot}4343\,\frac{\sigma\{x\}}{\mathcal{M}\{x\}}\,,$$

or

$$\boxed{\sigma\{\log_{10} x\} \simeq 0{\cdot}4343\,C\{x\}\,.} \tag{9.9.6}$$

We thus see that the standard deviation of $\log x$ is proportional to the coefficient of variation of x.

For $y = x^{\alpha}$ we get

$$\sigma\{y\} \simeq |\alpha\xi^{\alpha-1}|\,\sigma\{x\} = |\alpha\xi^{\alpha}|\,\frac{\sigma\{x\}}{\xi}$$

or, as $\xi^{\alpha} \simeq \mathcal{M}\{y\}$,

$$\boxed{C\{x^{\alpha}\} \simeq |\alpha|\,C\{x\}\,.} \tag{9.9.7}$$

In particular we find

$$C\left\{\frac{1}{x}\right\} \simeq C\{x\} \tag{9.9.8}$$

and

$$C\{\sqrt{x}\} \simeq \tfrac{1}{2} C\{x\} . \tag{9.9.9}$$

If y is a function of several normally distributed variables

$$y = \varphi(x_1, x_2, \ldots, x_k) ,$$

y itself will be approximately normally distributed if the function can be considered linear in all variables throughout practically the whole range of the variation of the variables. The mean and variance of y are given by (5.17.6) and (5.17.7).

In particular, if

$$y = \alpha_0 x_1^{\alpha_1} x_2^{\alpha_2} \ldots x_k^{\alpha_k} , \tag{9.9.10}$$

we have

$$\boxed{C^2\{y\} \simeq \sum_{i=1}^{k} \alpha_i^2 C^2\{x_i\} + 2 \sum_{i=1}^{k-1} \sum_{j=i+1}^{k} \alpha_i \alpha_j \varrho_{ij} C\{x_i\} C\{x_j\}} \tag{9.9.11}$$

where ϱ_{ij} denotes the coefficient of correlation for (x_i, x_j).

As two special cases of this formula we may mention

$$y = \alpha x_1 x_2 ,$$

$$C^2\{y\} \simeq C^2\{x_1\} + C^2\{x_2\} + 2\varrho_{12} C\{x_1\} C\{x_2\} , \tag{9.9.12}$$

and

$$y = \alpha x_1 x_2 / x_3 ,$$

$$\begin{aligned} C^2\{y\} \simeq & \; C^2\{x_1\} + C^2\{x_2\} + C^2\{x_3\} + 2\varrho_{12} C\{x_1\} C\{x_2\} \\ & - 2\varrho_{13} C\{x_1\} C\{x_3\} - 2\varrho_{23} C\{x_2\} C\{x_3\} . \end{aligned} \tag{9.9.13}$$

In the theory of errors we know in many cases that the variables are independent so that the coefficients of correlation are zero, as in the following examples.

Example 9.8. In the already cited investigation[1]) on the distribution of particle sizes, the average particle size, $\overline{K}$, of a certain fraction of the product was determined by weighing and counting by means of the formula

$$\overline{K} = \left(\frac{G}{\varrho N}\right)^{\frac{1}{3}} ,$$

G denoting the weight of the particles, N their number and ϱ their specific gravity.

This expression is of the form (9.9.10), and therefore, according to (9.9.11),

---

[1]) G. Fagerholt: *Particle Size Distribution of Products Ground in Tube Mill*, 1945, p. 65.

$$C^2\{\overline{K}\} = \tfrac{1}{9}(C^2\{G\}+C^2\{\varrho\}+C^2\{N\})$$

since the three variables are stochastically independent.

As the squares of the first two coefficients of variation, $C^2\{G\}$ and $C^2\{\varrho\}$, are negligible compared with $C^2\{N\}$, we have

$$C^2\{\overline{K}\} \simeq \tfrac{1}{9} C^2\{N\}$$

or

$$\frac{\sigma_{\overline{K}}}{\overline{K}} \simeq \tfrac{1}{3} \frac{\sigma_N}{N}.$$

Example 9.9. When the particle distributions were determined on the basis of sedimentation analyses, the particle size, k, could be computed from the formula[1])

$$k = 3{\cdot}42 \left(\frac{h\eta}{(\varrho_k-\varrho_f)\,gt}\right)^{\frac{1}{2}},$$

where h denotes the sedimentation height, η the viscosity, g the gravitational constant, t the sedimentation time, ϱ_k the specific gravity of the particles and ϱ_f the specific gravity of the suspension fluid.

Defining $d = \varrho_k - \varrho_f$ this expression may be thrown into the form (9.9.10) and we have according to (9.9.11) since the variables are stochastically independent

$$C^2\{k\} = \tfrac{1}{4}(C^2\{h\}+C^2\{\eta\}+C^2\{d\}+C^2\{g\}+C^2\{t\}),$$

where

$$C^2\{d\} = \frac{\sigma^2\{d\}}{m^2\{d\}} \simeq \frac{\sigma^2_{\varrho_k}+\sigma^2_{\varrho_f}}{(m\{\varrho_k\}-m\{\varrho_f\})^2}.$$

Under the conditions stated[1]), the estimates of the 5 coefficients of variation were 0·005, 0·010, 0·003, 0·000, and 0·020, respectively, so that

$$\begin{aligned} C^2\{k\} &= \tfrac{1}{4}(0{\cdot}000025+0{\cdot}000100+0{\cdot}000009+0{\cdot}000000+0{\cdot}000400) \\ &= 0{\cdot}00013\,. \end{aligned}$$

This leads to

$$C\{k\} = \frac{\sigma\{k\}}{m\{k\}} \simeq 0{\cdot}01\,.$$

Example 9.10. A raw material containing $x_1\%$ of a certain component is worked up, for instance by flotation, the treatment resulting in a concentrate containing $x_2\%$ of the component and a waste product with $x_3\%$ of the component. The efficiency of the process is controlled by calculating the two coefficients y and z defined as

$$y = \frac{x_1-x_3}{x_2-x_3},$$

[1]) G. FAGERHOLT: loc. cit. p. 83.

which gives the ratio between the weight of the concentrate and that of the raw material, and

$$z = \frac{x_2}{x_1} y = \frac{x_2(x_1 - x_3)}{x_1(x_2 - x_3)},$$

which denotes the ratio between the amount of the component present in the concentrate and that present in the raw material. This means that 100 grams raw material containing x_1 grams of the component yields $100y$ grams of concentrate, which contains the fraction z of the original x_1 grams of the component.

It is required to investigate what variation in y and z is to be expected under stable manufacturing conditions. The quantities x_1, x_2, and x_3 will vary on account of variations in the composition of the raw material, in the working of the flotation cells, in collecting the samples, in the chemical analyses, etc. We assume that x_1, x_2, and x_3 are stochastically independent. (This assumption must be checked by a correlation analysis, cf. Chapter 19).

For the determination of the variance of

$$y = \frac{x_1 - x_3}{x_2 - x_3}$$

we apply (5.17.7) directly. Differentiating

$$\ln y = \ln (x_1 - x_3) - \ln (x_2 - x_3)$$

we obtain

$$\frac{1}{y}\frac{\partial y}{\partial x_1} = \frac{1}{x_1 - x_3},$$

$$\frac{1}{y}\frac{\partial y}{\partial x_2} = \frac{-1}{x_2 - x_3},$$

and

$$\frac{1}{y}\frac{\partial y}{\partial x_3} = -\frac{1}{x_1 - x_3} + \frac{1}{x_2 - x_3}.$$

As $\eta = \mathcal{M}\{y\}$, (5.17.7) leads to

$$\sigma^2\{y\} \simeq \left(\frac{\eta}{\xi_1 - \xi_3}\right)^2 \sigma^2\{x_1\} + \left(\frac{\eta}{\xi_2 - \xi_3}\right)^2 \sigma^2\{x_2\} + \left(\frac{\eta}{\xi_2 - \xi_3} - \frac{\eta}{\xi_1 - \xi_3}\right)^2 \sigma^2\{x_3\},$$

which may be written

$$C^2\{y\} = \left(\frac{\sigma\{x_1\}}{\xi_1 - \xi_3}\right)^2 + \left(\frac{\sigma\{x_2\}}{\xi_2 - \xi_3}\right)^2 + \left(\frac{\sigma\{x_3\}}{\xi_2 - \xi_3} - \frac{\sigma\{x_3\}}{\xi_1 - \xi_3}\right)^2.$$

Correspondingly, we have

$$C^2\{z\} = \left(\frac{\xi_3\sigma\{x_1\}}{\xi_1(\xi_1-\xi_3)}\right)^2 + \left(\frac{\xi_3\sigma\{x_2\}}{\xi_2(\xi_2-\xi_3)}\right)^2 + \left(\frac{\sigma\{x_3\}}{\xi_2-\xi_3} - \frac{\sigma\{x_3\}}{\xi_1-\xi_3}\right)^2.$$

If we insert the following values

$$\xi_1 = 2{\cdot}5\%, \quad \sigma\{x_1\} = 0{\cdot}2\%,$$

$$\xi_2 = 17{\cdot}3\%, \quad \sigma\{x_2\} = 0{\cdot}1\%,$$

$$\xi_3 = 0{\cdot}5\%, \quad \sigma\{x_3\} = 0{\cdot}1\%,$$

we obtain

$$\mathcal{m}\{y\} \simeq \frac{2{\cdot}5-0{\cdot}5}{17{\cdot}3-0{\cdot}5} = 0{\cdot}119,$$

$$\mathcal{m}\{z\} \simeq \frac{17{\cdot}3}{2{\cdot}5}\,0{\cdot}119 = 0{\cdot}823,$$

$$C^2\{y\} = \left(\frac{0{\cdot}2}{2{\cdot}0}\right)^2 + \left(\frac{0{\cdot}1}{16{\cdot}8}\right)^2 + \left(\frac{0{\cdot}1}{16{\cdot}8} - \frac{0{\cdot}1}{2{\cdot}0}\right)^2 = 0{\cdot}0120,$$

and

$$C^2\{z\} = \left(\frac{0{\cdot}5\times 0{\cdot}2}{2{\cdot}5\times 2{\cdot}0}\right)^2 + \left(\frac{0{\cdot}5\times 0{\cdot}1}{17{\cdot}3\times 16{\cdot}8}\right)^2 + \left(\frac{0{\cdot}1}{16{\cdot}8} - \frac{0{\cdot}1}{2{\cdot}0}\right)^2 = 0{\cdot}0025.$$

The two coefficients of variation take on the values 11% and 5%, respectively.

In this process 100 grams of raw material yield 11·9 grams of concentrate, which contains 82.3 % of the original amount of the component in the raw material. Under stable manufacturing conditions, as characterized by the above means and variances, it is to be expected that y will vary around 0·119 with standard deviation $0{\cdot}119\times 0{\cdot}11 = 0{\cdot}013$, and z around 82.3 % with standard deviation $0{\cdot}823\times 0{\cdot}05\times 100 = 4.1\,\%$.

9.10. Notes and References.

Most of the tests of significance discussed in this book are due to R. A. FISHER and references to the appropriate papers are given in the following chapters.

The exposition of the ideas underlying the u-test in § 9.4 is based on a general theory for testing statistical hypotheses which has been developed by J. NEYMAN and E. S. PEARSON. The most important papers by these authors are: *On the Problem of the Most Efficient Tests of Statistical Hypotheses*, Phil. Trans. Roy. Soc. London, A, 231, 1933, 289–337; *Contributions to the Theory of Testing Statistical Hypotheses*, Statistical Research Memoirs, 1, 1936, 1–37, and 2, 1938, 25–57; and *Sufficient Statistics and Uniformly*

Most Powerful Tests of Statistical Hypotheses, Statistical Research Memoirs, 1, 1936, 113–137. An introduction to this theory may be found in J. NEYMAN: *Basic Ideas and Some Recent Results of the Theory of Testing Statistical Hypotheses*, Journ. Roy. Stat. Soc., 105, 1942, 292–327, and also in A. WALD: *On the Principles of Statistical Inference*, Notre Dame Mathematical Lectures, No. 1, 1942.

A more general theory has been developed by A. WALD in *Statistical Decision Functions*, J. Wiley and Sons, New York, 1950.

CHAPTER 10.

THE χ^2-DISTRIBUTION

10.1. Statement of the Problem.

If we have n stochastically independent observations, $x_1, x_2, \ldots, x_n$, from a normally distributed population with parameters (ξ, σ^2), an estimate s^2 of σ^2 is calculated from the formula:

$$s^2 = \frac{1}{n-1}\sum_{i=1}^{n}(x_i-\bar{x})^2 . \tag{10.1.1}$$

If the population mean ξ is known, which is very seldom the case, the estimate s^2 is replaced by

$$\hat{s}^2 = \frac{1}{n}\sum_{i=1}^{n}(x_i-\xi)^2 . \tag{10.1.2}$$

The n variables, $x_1-\xi, x_2-\xi, \ldots, x_n-\xi$, included in the above sum of squares, are stochastically independent and normally distributed with parameters $(0, \sigma^2)$. Standardization of these variables leads to

$$\hat{s}^2 = \sigma^2\frac{1}{n}\sum_{i=1}^{n}\left(\frac{x_i-\xi}{\sigma}\right)^2$$

$$= \sigma^2\frac{1}{n}\sum_{i=1}^{n}u_i^2 ,$$

where

$$u_i = \frac{x_i-\xi}{\sigma}.$$

Denoting the standardized sum of squares by χ^2,

$$\chi^2 = \sum_{i=1}^{n}u_i^2 , \tag{10.1.3}$$

$\hat{s}^2$ may be written

$$\hat{s}^2 = \sigma^2\frac{\chi^2}{n}. \tag{10.1.4}$$

If we consider a population of samples, each consisting of n stochastically independent observations from a normally distributed population with parameters (ξ, σ^2), and we calculate the quantity

$$\chi^2 = \sum_{i=1}^{n} \left(\frac{x_i - \xi}{\sigma}\right)^2 \tag{10.1.5}$$

for each sample, we obtain a population of χ^2-values whose distribution function, the χ^2-distribution, must be independent of ξ and σ, as χ^2 is a function of n standardized variables. Multiplying the χ^2-values by σ^2/n, we will obtain a population of $\hat{s}^2$-values according to (10.1.4).

Hence the χ^2-distribution depends only on n, and the distribution of $\hat{s}^2$ may be derived from the distribution of χ^2 by utilizing the transformation (10.1.4).

If s^2 is written in a manner similar to $\hat{s}^2$, we find

$$s^2 = \sigma^2 \frac{1}{n-1} \sum_{i=1}^{n} \left(\frac{x_i - \bar{x}}{\sigma}\right)^2$$

$$= \sigma^2 \frac{\chi_1^2}{n-1}, \tag{10.1.6}$$

where

$$\chi_1^2 = \sum_{i=1}^{n} \left(\frac{x_i - \bar{x}}{\sigma}\right)^2. \tag{10.1.7}$$

The n variables, $\dfrac{x_1 - \bar{x}}{\sigma}, \ldots, \dfrac{x_n - \bar{x}}{\sigma}$, included in χ_1^2, satisfy the relation

$$\sum_{i=1}^{n} \frac{x_i - \bar{x}}{\sigma} = 0, \tag{10.1.8}$$

and it will be shown that they are not stochastically independent but that χ_1^2 may be transformed to a sum of squares of $n-1$ stochastically independent, standardized variables, the distribution function of χ_1^2 thus being of the same type as that of χ^2, the only difference being that it has $n-1$ as a parameter instead of n, see § 10.6.

It is thus found that the distributions of $\hat{s}^2$ and s^2 may both be derived from the χ^2-distribution by simple transformations. These two distributions are dealt with separately in Chapter 11. It is also possible in a similar way to derive the distribution of estimates of σ^2 when the problems in question are more complicated, cf. Chapters 16 and 18. The theoretical basis for the solution of these questions is given in the following paragraphs.

10.2. The Distribution Function of χ^2.

Consider f stochastically independent variables, $u_1, u_2, \ldots, u_f$, which are normally distributed with parameters (0, 1). *The sum of squares of these* *variables is denoted by* χ^2

$$\boxed{\chi^2 = \sum_{i=1}^{f} u_i^2} \qquad (0 \leq \chi^2 < \infty), \qquad (10.2.1)$$

and f *is termed the number of degrees of freedom for* χ^2.

The distribution function of χ^2 depends solely upon f, as the variables of the sum of squares are *standardized*. The probability that χ^2 belongs to the interval $(\chi^2, \chi^2+d(\chi^2))$ is equal to

$$p\{\chi^2\}d(\chi^2) = \frac{1}{2^{\frac{f}{2}}\Gamma\left(\frac{f}{2}\right)}(\chi^2)^{\frac{f}{2}-1}e^{-\frac{\chi^2}{2}}\,d(\chi^2) \qquad (0 \leq \chi^2 < \infty), \qquad (10.2.2)$$

where for $f > 2$

$$\Gamma\left(\frac{f}{2}\right) = \left(\frac{f}{2}-1\right)! = \begin{cases} \left(\frac{f}{2}-1\right)\left(\frac{f}{2}-2\right)\ldots 3\cdot 2\cdot 1 \text{ for } f \text{ even} \\ \left(\frac{f}{2}-1\right)\left(\frac{f}{2}-2\right)\ldots \frac{3}{2}\cdot\frac{1}{2}\cdot\sqrt{\pi} \text{ for } f \text{ odd}, \end{cases} \qquad (10.2.3)$$

and further $\Gamma(1) = 1$, and $\Gamma(\frac{1}{2}) = \sqrt{\pi}$, as proved in § 10.4. $\Gamma(x)$ denotes the Gamma-function. As f is a positive integer, we here need only the Gamma-function for the arguments given in (10.2.3).

Some of the properties of the distribution curve of χ^2 are as follows. For $f = 1$, we have

$$p\{\chi^2\} = \frac{1}{\sqrt{2\pi}}(\chi^2)^{-\frac{1}{2}}e^{-\frac{1}{2}\chi^2}. \qquad (10.2.4)$$

The distribution curve is steadily decreasing, the abscissa axis forming the asymptote. For $\chi^2 \to 0$ we obtain $p\{\chi^2\} \to \infty$, see Fig. 10.1.

For $f = 2$, we have

$$p\{\chi^2\} = \tfrac{1}{2}e^{-\frac{1}{2}\chi^2}. \qquad (10.2.5)$$

The distribution curve is a decreasing exponential curve with initial point $(0, \frac{1}{2})$ and the abscissa axis as an asymptote.

For $f = 3$, we have

$$p\{\chi^2\} = \frac{1}{\sqrt{2\pi}}(\chi^2)^{\frac{1}{2}}e^{-\frac{1}{2}\chi^2}. \qquad (10.2.6)$$

The distribution curve originates at the point (0, 0) and increases until $\chi^2 = 1$, whereafter it decreases with the abscissa axis as asymptote.

When the value of f increases ($f > 3$) the distribution curve takes a course similar to that for $f = 3$, see Fig. 10.1, which gives the distribution curves for $f = 4$, 10, and 20.

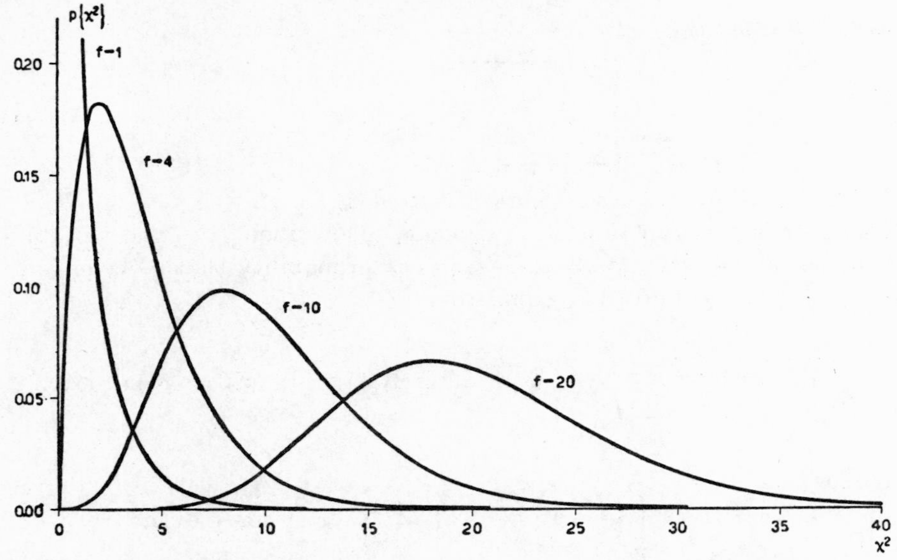

FIG. 10.1. Distribution curves of χ^2 for 1, 4, 10, and 20 degrees of freedom.

The abscissa of the maximum of the curve, *the mode*, is equal to $f-2$, as proved in § 10.4.

Since $\mathcal{V}\{u_i\} = \mathcal{M}\{u_i^2\} = 1$ it follows that

$$\mathcal{M}\{\chi^2\} = \sum_{i=1}^{f} \mathcal{M}\{u_i^2\} = f\,, \tag{10.2.7}$$

i. e., *the mean of χ^2 is equal to the number of degrees of freedom.*

Fig. 10.1 illustrates how the mode and the mean of the distribution increase as the number of degrees of freedom increases, and how at the same time the skewness of the curve becomes less and less pronounced.

10.3. The Cumulative Distribution Function and the Fractiles.

The cumulative distribution function $P\{\chi^2\}$ is obtained by integrating the distribution function

$$P\{\chi^2\} = \frac{1}{2^{\frac{f}{2}}\Gamma\left(\frac{f}{2}\right)} \int_0^{\chi^2} x^{\frac{f}{2}-1} e^{-\frac{x}{2}}\, dx\,. \tag{10.3.1}$$

For values of f from $f = 1$ to $f = 100$ the value of this integral may be calculated from K. PEARSON: *Tables of the Incomplete Gamma-function*, Biometrika Office, London, 1922. W. P. ELDERTON has given a table of $1-P\{\chi^2\}$ for $f = 2(1)29$ and $\chi^2 = 1(1)30(10)70$ in Biometrika, 1, 1902, 155–163, and K. PEARSON gives the same table in *Tables for Statisticians and Biometricians*, I, Biometrika Office, London, 1924.

The fractiles, χ^2_P, corresponding to suitably chosen values of P, may be calculated from the cumulative distribution function. The P-fractile is defined by the relation

$$\boxed{P\{\chi^2 < \chi^2_P\} = P\,,} \tag{10.3.2}$$

i. e., the probability of a χ^2-value less than or equal to the P-fractile is equal to P.

The fractiles are functions of both P and f. If we wish to emphasize the dependence upon f, the P-fractile is denoted $\chi^2_P(f)$ instead of χ^2_P.

Table V gives the fractiles corresponding to 23 values of P and to all values of f from 1 to 100. For example, the 95%-fractile for $f = 10$ is equal to 18.3, i. e.,

$$P\{\chi^2 < 18{\cdot}3\} = 95\% \text{ for } f = 10\,.$$

Bilateral limits for χ^2 may by obtained from the relation

$$P\{\chi^2_{P_1} < \chi^2 < \chi^2_{P_2}\} = P_2 - P_1\,. \tag{10.3.3}$$

For $P_1 = 2{\cdot}5\%$ and $P_2 = 97{\cdot}5\%$ and $f = 10$, we find in Table V

$$P\{3{\cdot}25 < \chi^2 < 20{\cdot}5\} = 95\%\,.$$

If we consider a population of samples each including 10 observations from a normally distributed population with parameters (0, 1), and we calculate the corresponding values of $\chi^2 = \sum_{i=1}^{10} u_i^2$, we obtain a population of values of χ^2, 95% of which belong in the interval (3·25, 20·5), while 2·5% are smaller than 3·25 and 2·5% larger than 20·5.

In connection with the distribution of s^2, we often require the value of χ^2_P/f. Some values of this quantity are given in Table VI.

In accordance with the central limit theorem, explained in § 8.2, we find that, for large values of f, χ^2 will be approximately normally distributed with parameters $(f, 2f)$, since

$$\mathcal{M}\{u_i^2\} = 1$$

and

$$\mathcal{V}\{u_i^2\} = \mathcal{M}\{(u_i^2-1)^2\} = \mathcal{M}\{u_i^4 - 2u_i^2 + 1\} = 3-2+1 = 2\,. \tag{10.3.4}$$

Hence χ^2/f will be approximately normally distributed with parameters $(1, 2/f)$, so that χ^2/f converges in probability to 1 for $f \to \infty$, cf. Table VI.

The χ^2-distribution, however, converges fairly slowly to the normal distribution; attempts have therefore been made to find transformations of χ^2 which are approximately normally distributed for small values of f.

The simplest transformation has been given by R. A. FISHER, who has demonstrated that, for $f > 30$, $\sqrt{2\chi^2}$ is approximately normally distributed with mean $\sqrt{2f-1}$ and variance 1, i. e.,

$$u \simeq \sqrt{2\chi^2}-\sqrt{2f-1} \quad (f > 30)\,. \tag{10.3.5}$$

If this equation is solved with respect to χ^2, we obtain

$$\boxed{\chi_P^2 \simeq \tfrac{1}{2}(\sqrt{2f-1}+u_P)^2 \text{ for } f > 30\,.} \tag{10.3.6}$$

Formula (10.3.6) shows that the difference between the two fractiles, $\chi_{P_2}^2-\chi_{P_1}^2$, increases almost in proportion with $\sqrt{f}$ for large values of f, as

$$\chi_{P_2}^2-\chi_{P_1}^2 \simeq \tfrac{1}{2}(u_{P_2}^2-u_{P_1}^2+2(u_{P_2}-u_{P_1})\sqrt{2f-1})\,. \tag{10.3.7}$$

If we choose values of P_1 and P_2 that are symmetrical with respect to 50%, e. g., $P_1 = 2{\cdot}5\%$ and $P_2 = 97{\cdot}5\%$, we have

$$\chi_{\cdot 975}^2-\chi_{\cdot 025}^2 \simeq 2\times 1{\cdot}96\times\sqrt{2f-1} \simeq 5{\cdot}5\sqrt{f}\,. \tag{10.3.8}$$

The approximation (10.3.6) is not very accurate for small or large values of P even for large f. A more accurate approximation is given by[1])

$$\boxed{\chi_P^2 \simeq f\left(1-\frac{2}{9f}+u_P\sqrt{\frac{2}{9f}}\right)^3.} \tag{10.3.9}$$

10.4. Derivation of the χ^2-Distribution.

The χ^2-distribution may be derived by induction. In the following proof the letter y is used instead of χ^2.

For $f = 1$ we have $y = u^2$, which leads to

$$\begin{aligned} P\{y < y_0\} &= P\{u^2 < y_0\} \\ &= P\{-\sqrt{y_0} < u < \sqrt{y_0}\} \\ &= \Phi(\sqrt{y_0})-\Phi(-\sqrt{y_0}) \\ &= 2\Phi(\sqrt{y_0})-1\,. \end{aligned} \tag{10.4.1}$$

Differentiation with respect to y_0 gives

[1]) WILSON, E. B., and HILFERTY, M. M.: *The Distribution of Chi-Square*, Proc. Nat. Acad. Sci., Washington, 17, 1931, 684—688.

$$
\begin{aligned}
p\{y_0\} &= 2\frac{d\Phi(\sqrt{y_0})}{d(\sqrt{y_0})}\frac{d(\sqrt{y_0})}{dy_0} \\
&= 2\varphi(\sqrt{y_0})\tfrac{1}{2}y_0^{-\frac{1}{2}} \\
&= \frac{1}{\sqrt{2\pi}}y_0^{-\frac{1}{2}}e^{-\frac{1}{2}y_0}, \qquad (10.4.2)
\end{aligned}
$$

which is identical with (10.2.4) if χ^2 is substituted for y_0.

For $f = 2$, we have $y = u_1^2+u_2^2$, where u_1 and u_2 are stochastically independent. If we put $y_1 = u_1^2$ and $y_2 = u_2^2$, we obtain $y = y_1+y_2$, where y_1 and y_2 both are distributed according to (10.4.2) and the problem is now to find the distribution of the sum of two variables.

According to the multiplication formula, the probability that y_1 and y_2 belong to the rectangle defined by the intervals (y_1, y_1+dy_1) and (y_2, y_2+dy_2) is equal to

$$
p\{y_1\}dy_1\ p\{y_2\}dy_2\,. \qquad (10.4.3)
$$

The probability that $y_1+y_2 < y$ may be found by integrating (10.4.3) over the corresponding domain, i. e.,

$$
P\{y_1+y_2 < y\} = P\{y\} = \int_0^y p\{y_1\}\int_0^{y-y_1} p\{y_2\}dy_2\,dy_1\,. \qquad (10.4.4)
$$

If this expression is differentiated with respect to y, we obtain the required formula for the distribution function of a sum of two stochastic variables

$$
p\{y\} = \int_0^y p\{y_1\}[p\{y_2\}]_{y_2=y-y_1}dy_1\,. \qquad (10.4.5)
$$

If we introduce

$$
p\{y_1\} = \frac{1}{\sqrt{2\pi}}y_1^{-\frac{1}{2}}e^{-\frac{1}{2}y_1}
$$

and

$$
[p\{y_2\}]_{y_2=y-y_1} = \frac{1}{\sqrt{2\pi}}(y-y_1)^{-\frac{1}{2}}e^{-\frac{1}{2}(y-y_1)},
$$

we get

$$
p\{y\} = \frac{1}{2\pi}e^{-\frac{1}{2}y}\int_0^y y_1^{-\frac{1}{2}}(y-y_1)^{-\frac{1}{2}}dy_1\,. \qquad (10.4.6)
$$

Writing $y_1 = yt$ this becomes

$$
\begin{aligned}
p\{y\} &= \frac{1}{2\pi}e^{-\frac{1}{2}y}\int_0^1 t^{-\frac{1}{2}}(1-t)^{-\frac{1}{2}}dt \\
&= c\cdot\frac{1}{2\pi}e^{-\frac{1}{2}y},
\end{aligned}
$$

where c denotes a constant, since the integral

$$\int_0^1 t^{-\frac{1}{2}}(1-t)^{-\frac{1}{2}}\,dt = c \tag{10.4.7}$$

is independent of y. From (5.2.4) we know that

$$\int_0^\infty p\{y\}\,dy = 1\,,$$

and the constant c may therefore be determined from the equation

$$\frac{c}{2\pi}\int_0^\infty e^{-\frac{1}{2}y}\,dy = \frac{c}{\pi} = 1\,,$$

the result being that $c = \pi$. The distribution function of y is thus

$$p\{y\} = \tfrac{1}{2}e^{-\frac{1}{2}y}\,, \tag{10.4.8}$$

which agrees with (10.2.5) for $y = \chi^2$.

As this distribution function is so simple, the general formula is derived by combining the distribution functions for f and 2 degrees of freedom, respectively.

Assuming that the distribution function for $y_1 = \sum_{i=1}^{f} u_i^2$ is equal to

$$p\{y_1\} = \frac{1}{2^{\frac{f}{2}}\,\Gamma\left(\frac{f}{2}\right)}\,y_1^{\frac{f}{2}-1}\,e^{-\frac{1}{2}y_1}\,, \tag{10.4.9}$$

see (10.2.2) for $y_1 = \chi^2$, we shall derive the distribution function for a sum of squares, y, which includes $f+2$ terms

$$y = \sum_1^{f+2} u_i^2 = y_1 + y_2\,,$$

where $y_2 = u_{f+1}^2 + u_{f+2}^2$, and all the u's are stochastically independent. Application of (10.4.5) leads to

$$p\{y\} = \frac{1}{2^{\frac{f}{2}+1}\,\Gamma\left(\frac{f}{2}\right)}\int_0^y y_1^{\frac{f}{2}-1}\,e^{-\frac{1}{2}y_1}\,e^{-\frac{1}{2}(y-y_1)}\,dy_1$$

$$= \frac{1}{2^{\frac{f}{2}+1}\,\Gamma\left(\frac{f}{2}\right)}\,e^{-\frac{1}{2}y}\int_0^y y_1^{\frac{f}{2}-1}\,dy_1$$

$$= \frac{1}{2^{\frac{f}{2}+1}\Gamma\left(\frac{f}{2}\right)} e^{-\frac{1}{2}y} \frac{y^{\frac{f}{2}}}{\frac{f}{2}}$$

$$= \frac{1}{2^{\frac{f+2}{2}}\Gamma\left(\frac{f+2}{2}\right)} y^{\frac{f+2}{2}-1} e^{-\frac{1}{2}y} . \tag{10.4.10}$$

Thus, $p\{y\}$ is obtained from $p\{y_1\}$ by replacing f by $f+2$. As we have already shown that (10.4.9) holds for $f = 1$ and for $f = 2$ it holds in general.

By differentiating $p\{y\}$ with respect to y we get

$$\frac{dp\{y\}}{dy} = \frac{1}{2^{\frac{f}{2}}\Gamma\left(\frac{f}{2}\right)} e^{-\frac{y}{2}} y^{\frac{f}{2}-2}\left(\frac{f}{2}-1-\frac{y}{2}\right) \quad \text{for} \quad f > 2 .$$

Putting this equal to 0, we get an equation for the determination of the *mode*:

$$\frac{f}{2}-1-\frac{y}{2} = 0$$

or

$$y = f-2 \quad \text{for} \quad f > 2 .$$

10.5. The Addition Theorem for the χ^2-Distribution.

It follows directly from the definition of χ^2 that

$$\chi^2 = u_1^2+u_2^2+\ldots+u_f^2$$

may be interpreted as a sum of χ^2-values. If the sum of the first f_1 terms is denoted χ_1^2,

$$\chi_1^2 = u_1^2+u_2^2+\ldots+u_{f_1}^2,$$

the sum of the next f_2 terms χ_2^2,

$$\chi_2^2 = u_{f_1+1}^2+u_{f_1+2}^2+\ldots+u_{f_1+f_2}^2,$$

and so forth, we have

$$\chi^2 = \chi_1^2+\chi_2^2+\ldots+\chi_k^2$$

and

$$f = f_1+f_2+\ldots+f_k ,$$

which means that the following *addition theorem holds for the χ^2-distribution*: *If $\chi_1^2, \chi_2^2, \ldots, \chi_k^2$ are stochastically independent and have χ^2-distributions with $f_1, f_2, \ldots, f_k$ degrees of freedom, respectively, then the sum*

$$\chi^2 = \chi_1^2+\chi_2^2+\ldots+\chi_k^2$$

also has a χ^2-distribution with $f = f_1+f_2+\ldots+f_k$ degrees of freedom.

10.6. The Partition Theorem for the χ^2-Distribution.

The distribution function of the estimate

$$\hat{s}^2 = \frac{1}{n} \sum_{i=1}^{n} (x_i - \xi)^2$$

may be derived from the χ^2-distribution by the transformation

$$\hat{s}^2 = \sigma^2 \frac{\chi^2}{f}, \quad f = n\,,$$

since for each observation x_i we have a value of $u_i = \dfrac{x_i - \xi}{\sigma}$, and the sum of squares $\chi^2 = \sum_{i=1}^{n} u_i^2$ has a χ^2-distribution with $f = n$ degrees of freedom.

In practice ξ is unknown and $\hat{s}^2$ is replaced by the estimate

$$s^2 = \frac{1}{n-1} \sum_{i=1}^{n} (x_i - \bar{x})^2\,,$$

which does not contain ξ.

Between the two sums of squares, $\sum_{i=1}^{n} (x_i - \xi)^2$ and $\sum_{i=1}^{n} (x_i - \bar{x})^2$, that appear in $\hat{s}^2$ and s^2, respectively, the following relationship exists

$$\sum_{i=1}^{n} (x_i - \xi)^2 = \sum_{i=1}^{n} \big((x_i - \bar{x}) + (\bar{x} - \xi)\big)^2$$

$$= \sum_{i=1}^{n} (x_i - \bar{x})^2 + n(\bar{x} - \xi)^2\,, \tag{10.6.1}$$

since $\sum_{i=1}^{n} (x_i - \bar{x}) = 0$. Division by σ^2 leads to

$$\sum_{i=1}^{n} \left(\frac{x_i - \xi}{\sigma}\right)^2 = \left(\frac{\bar{x} - \xi}{\sigma/\sqrt{n}}\right)^2 + \sum_{i=1}^{n} \left(\frac{x_i - \bar{x}}{\sigma}\right)^2. \tag{10.6.2}$$

If we introduce

$$u_i = \frac{x_i - \xi}{\sigma}$$

and define

$$\bar{u} = \frac{1}{n} \sum_{i=1}^{n} u_i = \frac{\bar{x} - \xi}{\sigma}\,,$$

then (10.6.2) becomes

$$\chi^2 = \sum_{i=1}^{n} u_i^2 = (\sqrt{n}\,\bar{u})^2 + \sum_{i=1}^{n} (u_i - \bar{u})^2\,, \tag{10.6.3}$$

and χ^2 is thus partitioned into a sum of two terms. The first term,

$$Q_1 = (\sqrt{n}\,\bar{u})^2 = \hat{u}_1^2, \tag{10.6.4}$$

has a χ^2-distribution with $f = 1$ degree of freedom, because $\bar{u}$ is normally distributed about 0 with standard deviation $\frac{1}{\sqrt{n}}$ and hence $\hat{u}_1 = \sqrt{n}\,\bar{u}$ is normally distributed with parameters (0, 1). According to the addition theorem of the χ^2-distribution it now seems reasonable to expect that the second term of (10.6.3),

$$Q_2 = \sum_{i=1}^{n} (u_i - \bar{u})^2, \tag{10.6.5}$$

has a χ^2-distribution with $f = n-1$ degrees of freedom, and that Q_2 is stochastically independent of Q_1. It will now be proved that this is really the case.

Q_2 may be written

$$Q_2 = \sum_{i=1}^{n} l_i^2, \tag{10.6.6}$$

where

$$l_i = u_i - \bar{u}, \quad i = 1, 2, \ldots, n. \tag{10.6.7}$$

The n variables, $l_1, l_2, \ldots, l_n$, are linear functions of the n standardized variables, $u_1, u_2, \ldots, u_n$, because

$$l_i = u_i - \bar{u} = -\frac{1}{n}u_1 - \ldots - \frac{1}{n}u_{i-1} + \left(1 - \frac{1}{n}\right)u_i - \frac{1}{n}u_{i+1} - \ldots - \frac{1}{n}u_n. \tag{10.6.8}$$

Further, we have a linear relationship between the n variables, $l_1, l_2, \ldots, l_n$,

$$l_1 + l_2 + \ldots + l_n = 0, \tag{10.6.9}$$

and may therefore call these variables *linearly dependent. The number of degrees of freedom for the n variables is defined as the number of variables minus the number of linear relations (constraints) between them.* In the case in question the number of degrees of freedom is thus equal to $n-1$. *The number of degrees of freedom for the corresponding sum of squares, $Q_2 = \sum_{i=1}^{n} l_i^2$, is defined as the number of degrees of freedom for the n variables included in the sum of squares.*

As there is no relation between the variables, $u_1, u_2, \ldots, u_n$, the number of degrees of freedom is here equal to n, and therefore the number of degrees of freedom for $\chi^2 = \sum_{i=1}^{n} u_i^2$ is equal to n, in agreement with the definition in § 10.2. Contrary to the l's, the n variables $u_1, u_2, \ldots, u_n$ are linearly independent.

As the l's are linear functions of the normally distributed u's, it follows from the addition theorem for the normal distribution that the l's are normally distributed with the following means and variances:

$$\mathcal{M}\{l_i\} = \mathcal{M}\{u_i\} - \mathcal{M}\{\bar{u}\} = 0-0 = 0\,, \tag{10.6.10}$$

and

$$\begin{aligned} \mathcal{V}\{l_i\} &= \frac{1}{n^2}\,\mathcal{V}\{u_1\} + \ldots + \frac{1}{n^2}\,\mathcal{V}\{u_{i-1}\} + \left(1-\frac{1}{n}\right)^2 \mathcal{V}\{u_i\} \\ &\qquad + \frac{1}{n^2}\,\mathcal{V}\{u_{i+1}\} + \ldots + \frac{1}{n^2}\,\mathcal{V}\{u_n\} \\ &= (n-1)\frac{1}{n^2} + \left(1-\frac{1}{n}\right)^2 = \frac{n-1}{n}. \end{aligned} \tag{10.6.11}$$

From (5.16.4) and (10.6.10) we further obtain the covariance

$$\mathcal{V}\{l_i, l_j\} = \mathcal{M}\{l_i l_j\} - \mathcal{M}\{l_i\}\mathcal{M}\{l_j\} = \mathcal{M}\{l_i l_j\}\,.$$

The product $l_i l_j = (u_i - \bar{u})\,(u_j - \bar{u})$, $(i \neq j)$, includes the quadratic terms

$$\frac{1}{n^2}u_1^2 + \ldots + \frac{1}{n^2}u_{i-1}^2 - \frac{1}{n}\left(1-\frac{1}{n}\right)u_i^2 + \frac{1}{n^2}u_{i+1}^2 + \ldots + \frac{1}{n^2}u_{j-1}^2 - \frac{1}{n}\left(1-\frac{1}{n}\right)u_j^2$$
$$+ \frac{1}{n^2}u_{j+1} + \ldots + \frac{1}{n^2}u_n^2\,.$$

Besides these terms, $l_i l_j$ also contains products of the form $k_{rs}u_r u_s$, $(r \neq s)$, where k_{rs} is a constant. As the u's are stochastically independent, (5.16.8) gives us

$$\mathcal{M}\{k_{rs}u_r u_s\} = k_{rs}\,\mathcal{M}\{u_r\}\,\mathcal{M}\{u_s\} = 0\,, \quad (r \neq s)\,,$$

i. e., these terms vanish when we are to calculate $\mathcal{M}\{l_i l_j\}$. As $\mathcal{M}\{u^2\} = 1$, the quadratic terms lead to

$$\mathcal{M}\{l_i l_j\} = (n-2)\frac{1}{n^2} - 2\frac{1}{n}\left(1-\frac{1}{n}\right) = -\frac{1}{n},$$

and the covariance is therefore equal to

$$\mathcal{V}\{l_i, l_j\} = -\frac{1}{n}. \tag{10.6.12}$$

Thus, the n variables $l_1, l_2, \ldots, l_n$ are both linearly and stochastically dependent.

It will now be proved *that the sum of squares, Q_2, of these n variables can be transformed into a sum of squares of $n-1$ linearly and stochastically independent variables, $\hat{u}_2, \hat{u}_3, \ldots, \hat{u}_n$, which are normally distributed with parameters* $(0, 1)$ *and are linearly and stochastically independent of $\hat{u}_1 = \sqrt{n}\,\bar{u}$.*

The result will first be proved for $n = 2$ and $n = 3$.

For $n = 2$

$$Q_1 = \hat{u}_1^2 = \left(\frac{u_1+u_2}{\sqrt{2}}\right)^2$$

and

$$Q_2 = l_1^2+l_2^2 = 2l_2^2 = (\sqrt{2}\,l_2)^2 ,$$

as $l_1+l_2 = 0$. If we substitute

$$l_2 = u_2-\bar{u} = \frac{u_2-u_1}{2},$$

we obtain

$$Q_2 = \left(\frac{u_1-u_2}{\sqrt{2}}\right)^2 = \hat{u}_2^2 .$$

According to the addition theorem for the normal distribution, we have that the two quantities

$$\hat{u}_1 = \frac{u_1+u_2}{\sqrt{2}} \tag{10.6.13}$$

and

$$\hat{u}_2 = \frac{u_1-u_2}{\sqrt{2}} \tag{10.6.14}$$

are normally distributed with zero means and unit variances. Here we have determined the marginal distributions in the two-dimensional distribution of $(\hat{u}_1, \hat{u}_2)$. Since the variables $(\hat{u}_1, \hat{u}_2)$ are linear functions of the normally distributed variables (u_1, u_2), the distribution of $(\hat{u}_1, \hat{u}_2)$ is a two-dimensional normal distribution, cf. § 19.8 and § 5.12. In addition to the means and variances of the marginal distributions, this distribution is characterized by the covariance $\mathcal{V}\{\hat{u}_1, \hat{u}_2\}$. According to (5.16.4) we have

$$\begin{aligned}\mathcal{V}\{\hat{u}_1, \hat{u}_2\} &= \mathcal{M}\{\hat{u}_1\hat{u}_2\}-\mathcal{M}\{\hat{u}_1\}\,\mathcal{M}\{\hat{u}_2\}\\ &= \mathcal{M}\{\tfrac{1}{2}(u_1+u_2)(u_1-u_2)\}-0\\ &= \tfrac{1}{2}\mathcal{M}\{u_1^2-u_2^2\} = \tfrac{1}{2}(1-1) = 0 .\end{aligned}$$

Since the covariance of the two-dimensional normal distribution is zero the two variables are stochastically independent, cf. § 5.16.

The original sum of squares $\chi^2 = u_1^2+u_2^2$ may therefore be rearranged into a sum of two new squares, Q_1 and Q_2, which are stochastically independent, and each of which has a χ^2-distribution with one degree of freedom and has certain required properties.

For $n = 3$, we have

$$Q_1 = \left(\frac{u_1+u_2+u_3}{\sqrt{3}}\right)^2 = \hat{u}_1^2 \tag{10.6.15}$$

and
$$Q_2 = \sum_{i=1}^{3} (u_i - \bar{u})^2 = l_1^2 + l_2^2 + l_3^2$$
as well as $l_1 + l_2 + l_3 = 0$. If we substitute $l_1 = -(l_2 + l_3)$ we obtain
$$Q_2 = 2(l_2^2 + l_2 l_3 + l_3^2) ,$$
and we must now prove that this expression can be rewritten as a sum of squares containing only two terms,
$$Q_2 = \hat{u}_2^2 + \hat{u}_3^2 ,$$
where $\hat{u}_2$ and $\hat{u}_3$ are linear functions of l_2 and l_3, and hence of the u's. The two new variables, $\hat{u}_2$ and $\hat{u}_3$, must also be stochastically independent and normally distributed with parameters (0, 1), and, further, be stochastically independent of $\hat{u}_1$.

Q_2 may be expressed as a sum of squares with two terms by rewriting the terms including, e. g., l_2 as a sum of squares with a correction that is independent of l_2, i. e.,
$$l_2^2 + l_2 l_3 = (l_2 + \tfrac{1}{2} l_3)^2 - \tfrac{1}{4} l_3^2 .$$
This leads to
$$\begin{aligned} Q_2 &= 2(l_2 + \tfrac{1}{2} l_3)^2 + \tfrac{3}{2} l_3^2 \\ &= (\sqrt{2}(l_2 + \tfrac{1}{2} l_3))^2 + (\sqrt{\tfrac{3}{2}}\, l_3)^2 \\ &= \hat{u}_2^2 + \hat{u}_3^2 . \end{aligned}$$
If we substitute $l_i = u_i - \bar{u}$, we obtain
$$\begin{aligned} \hat{u}_2 &= \sqrt{2}(l_2 + \tfrac{1}{2} l_3) = \frac{1}{\sqrt{2}}(2l_2 + l_3) \\ &= \frac{1}{\sqrt{2}}(2u_2 + u_3 - 3\bar{u}) \qquad (10.6.16) \\ &= -\frac{1}{\sqrt{2}}(u_1 - u_2) , \end{aligned}$$
as $3\bar{u} = u_1 + u_2 + u_3$; further
$$\begin{aligned} \hat{u}_3 &= \sqrt{\frac{3}{2}}\, l_3 = \frac{1}{\sqrt{6}} 3l_3 \\ &= \frac{1}{\sqrt{6}}(3u_3 - 3\bar{u}) \\ &= -\frac{1}{\sqrt{6}}(u_1 + u_2 - 2u_3) . \qquad (10.6.17) \end{aligned}$$

According to the addition theorem for the normal distribution $\hat{u}_2$ and $\hat{u}_3$ are normally distributed with zero means, as $\mathcal{M}\{u_i\} = 0$, and unit variances, as

$$\mathcal{V}\{\hat{u}_2\} = \tfrac{1}{2}(\mathcal{V}\{u_1\}+\mathcal{V}\{u_2\}) = 1$$

and

$$\mathcal{V}\{\hat{u}_3\} = \tfrac{1}{6}(\mathcal{V}\{u_1\}+\mathcal{V}\{u_2\}+4\mathcal{V}\{u_3\}) = 1\ ,$$

since $\mathcal{V}\{u_i\} = 1$.

Stochastic independence between the three variables $\hat{u}_1$, $\hat{u}_2$, and $\hat{u}_3$ is conditioned by the three covariances $\mathcal{V}\{\hat{u}_1, \hat{u}_2\}$, $\mathcal{V}\{\hat{u}_1, \hat{u}_3\}$, and $\mathcal{V}\{\hat{u}_2, \hat{u}_3\}$ all being equal to 0. The product

$$\hat{u}_1\hat{u}_2 = -\frac{1}{\sqrt{6}}(u_1^2-u_2^2+u_1u_3-u_2u_3)$$

leads to

$$\mathcal{V}\{\hat{u}_1, \hat{u}_2\} = \mathcal{M}\{\hat{u}_1\hat{u}_2\} = -\frac{1}{\sqrt{6}}(1-1+0-0) = 0\ ,$$

as $\mathcal{M}\{u_i^2\} = \mathcal{V}\{u_i\} = 1$, and $\mathcal{M}\{u_iu_j\} = \mathcal{M}\{u_i\}\mathcal{M}\{u_j\} = 0$, $(i \neq j)$, as u_i and u_j are stochastically independent with zero means. Correspondingly it may be demonstrated that $\mathcal{M}\{\hat{u}_1\hat{u}_3\} = 0$ and $\mathcal{M}\{\hat{u}_2\hat{u}_3\} = 0$.

Hereby it has been proved that $\hat{u}_1$, $\hat{u}_2$, and $\hat{u}_3$ are stochastically independent and normally distributed with parameters (0,1), Q_1 and Q_2 consequently being stochastically independent and having χ^2-distributions with 1 and 2 degrees of freedom, respectively.

The general theorem may be proved in a similar manner. The equation (10.6.3) may be written as

$$\chi^2 = \sum_{i=1}^{n} u_i^2 = Q_1+Q_2\ ,$$

where

$$Q_1 = \left(\frac{u_1+u_2+\ldots+u_n}{\sqrt{n}}\right)^2 = \hat{u}_1^2,$$

and

$$Q_2 = \sum_{i=1}^{n}(u_i-\bar{u})^2 = \sum_{i=1}^{n} l_i^2\ ,$$

and further

$$\sum_{i=1}^{n} l_i = 0\ .$$

The last equation is used to *eliminate* l_1 *from* Q_2, since

$$l_1 = -(l_2+l_3+\ldots+l_n) \qquad (10.6.18)$$

and

$$\begin{aligned} l_1^2 = l_2^2&+2(l_2l_3+l_2l_4+\ldots+l_2l_n)\\ &+l_3^2+2(l_3l_4+\ldots+l_3l_n)\\ &\qquad\vdots\\ &\qquad\qquad+\ l_n^2\ . \end{aligned} \qquad (10.6.19)$$

If this result is inserted in Q_2, we obtain Q_2 as a *quadratic form in the* $n-1$ *variables* $l_2, l_3, \ldots, l_n$,

$$\begin{aligned} Q_2 = 2(l_2^2 &+ l_2 l_3 + l_2 l_4 + \ldots + l_2 l_n \\ &+ l_3^2 + l_3 l_4 + \ldots + l_3 l_n \\ &+ l_4^2 + \ldots + l_4 l_n \\ &\quad \vdots \\ &+ l_n^2) . \end{aligned} \tag{10.6.20}$$

Q_2 may now be expressed as a sum of $n-1$ *squares*. First we collect all terms that include l_2 and rewrite them as a square, together with corrections that are independent of l_2

$$\begin{aligned} &l_2^2 + l_2 l_3 + l_2 l_4 + \ldots + l_2 l_n \\ &= (l_2 + \tfrac{1}{2}(l_3 + l_4 + \ldots + l_n))^2 \\ &\quad - \tfrac{1}{4} l_3^2 - \tfrac{1}{2} l_3 l_4 - \ldots - \tfrac{1}{2} l_3 l_n \\ &\quad - \tfrac{1}{4} l_4^2 - \ldots - \tfrac{1}{2} l_4 l_n \\ &\quad \vdots \\ &\quad - \tfrac{1}{4} l_n^2 . \end{aligned}$$

Next, all terms including l_3 that do not appear in the square just formed are collected and treated in the same manner

$$\begin{aligned} &\tfrac{3}{4} l_3^2 + \tfrac{1}{2}(l_3 l_4 + \ldots + l_3 l_n) \\ &= \tfrac{3}{4}(l_3^2 + \tfrac{2}{3}(l_3 l_4 + \ldots + l_3 l_n)) \\ &= \tfrac{3}{4}[(l_3 + \tfrac{1}{3}(l_4 + \ldots + l_n))^2 \\ &\quad - \tfrac{1}{9} l_4^2 - \ldots - \tfrac{2}{9} l_4 l_n \\ &\quad \vdots \\ &\quad - \tfrac{1}{9} l_n^2] . \end{aligned}$$

Continuing in this manner Q_2 is rewritten as

$$Q_2 = \sum_{i=2}^{n} \hat{u}_i^2 , \tag{10.6.21}$$

where the $\hat{u}^2$'s are the successive squares defined above, i. e.,

$$\left.\begin{aligned} \hat{u}_2 &= \sqrt{2}(l_2 + \tfrac{1}{2}(l_3 + l_4 + \ldots + l_n)) \\ \hat{u}_3 &= \sqrt{\tfrac{3}{2}}(l_3 + \tfrac{1}{3}(l_4 + \ldots + l_n)) \\ &\vdots \\ \hat{u}_i &= \sqrt{\frac{i}{i-1}}\left(l_i + \frac{1}{i}(l_{i+1} + \ldots + l_n)\right) \\ &\vdots \\ \hat{u}_n &= \sqrt{\frac{n}{n-1}}\, l_n . \end{aligned}\right\} \tag{10.6.22}$$

If $l_i = u_i - \bar{u}$ is introduced in $\hat{u}_i$, we obtain

$$\hat{u}_i = \frac{1}{\sqrt{i(i-1)}}(il_i + l_{i+1} + \ldots + l_n)$$

$$= \frac{1}{\sqrt{i(i-1)}}(iu_i + u_{i+1} + \ldots + u_n - n\bar{u})$$

$$= \frac{-1}{\sqrt{i(i-1)}}(u_1 + u_2 + \ldots + u_{i-1} - (i-1)u_i). \qquad (10.6.23)$$

Since the $\hat{u}$'s are linear functions of the normally distributed u's, the distribution of $(\hat{u}_2, \ldots, \hat{u}_n)$ is an $(n-1)$-dimensional *normal* distribution, cf. § 20.1, and as such is fully determined by the means, variances, and covariances of the variables. In order to prove that the $\hat{u}$'s are stochastically independent we need only to prove that the covariances are zero. $\mathcal{M}\{\hat{u}_i\}$, $\mathcal{V}\{\hat{u}_i\}$, and $\mathcal{V}\{\hat{u}_i, \hat{u}_j\}$ are obtained as follows:

$$\mathcal{M}\{\hat{u}_i\} = -\frac{1}{\sqrt{i(i-1)}}(\mathcal{M}\{u_1\} + \mathcal{M}\{u_2\} + \ldots + \mathcal{M}\{u_{i-1}\} - (i-1)\mathcal{M}\{u_i\}) = 0, \qquad (10.6.24)$$

since $\mathcal{M}\{u_i\} = 0$.

$$\mathcal{V}\{\hat{u}_i\} = \frac{1}{i(i-1)}(\mathcal{V}\{u_1\} + \mathcal{V}\{u_2\} + \ldots + \mathcal{V}\{u_{i-1}\} + (i-1)^2\mathcal{V}\{u_i\})$$

$$= \frac{1}{i(i-1)}(i-1+(i-1)^2) = 1, \qquad (10.6.25)$$

since $\mathcal{V}\{u_i\} = 1$ and the u's are independent.

The covariance, $\mathcal{V}\{\hat{u}_i, \hat{u}_j\}$, is found as $\mathcal{M}\{\hat{u}_i\hat{u}_j\}$ since $\mathcal{M}\{\hat{u}_i\} = 0$. The product

$$\hat{u}_i\hat{u}_j = \frac{1}{\sqrt{i(i-1)j(j-1)}}(u_1 + \ldots + u_{i-1} - (i-1)u_i)(u_1 + \ldots + u_{j-1} - (j-1)u_j)$$

shows that, for $i < j$, $\hat{u}_i\hat{u}_j$ includes the following quadratic terms

$$u_1^2 + u_2^2 + \ldots + u_{i-1}^2 - (i-1)u_i^2,$$

and that $\hat{u}_i\hat{u}_j$ further includes product terms of the form $k_{rs}u_ru_s$, $(r \neq s)$, where k_{rs} denotes a constant. Since $\mathcal{M}\{k_{rs}u_ru_s\} = 0$ for $r \neq s$, and

$$\mathcal{M}\{u_1^2 + \ldots + u_{i-1}^2 - (i-1)u_i^2\} = (i-1) - (i-1) = 0, \qquad (10.6.26)$$

because $\mathcal{M}\{u_i^2\} = 1$, we get $\mathcal{V}\{\hat{u}_i, \hat{u}_j\} = 0$. Thus the $n-1$ variables are stochastically independent.

Consequently the sum of squares $Q_2 = \sum_{i=1}^{n} l_i^2$ with $n-1$ degrees of freedom has been transformed into the sum of squares $Q_2 = \sum_{i=2}^{n} \hat{u}_i^2$ of $n-1$ stochastically independent variables, $\hat{u}_2, \hat{u}_3, \ldots, \hat{u}_n$, and hence is distributed as χ^2 with $n-1$ degrees of freedom.

We have already shown that $\hat{u}_1^2$ is distributed as χ^2 with 1 degree of freedom, and it is easy to prove that $\mathcal{M}\{\hat{u}_1\hat{u}_i\} = 0$ for $i = 2, \ldots, n$, so that $\hat{u}_1$ is stochastically independent of the remaining $\hat{u}_i$'s.

Thus, the two terms, Q_1 and Q_2, into which χ^2 has been partitioned, are stochastically independent and distributed as χ^2 with 1 and $n-1$ degrees of freedom, respectively.

The above result is a simple case of the *partition theorem* which enables us to divide a sum of squares into several stochastically independent sums of squares.

Let us first give some definitions and examine the implications of this theorem before proving it.

The number of degrees of freedom for a set of variables, $l_1, l_2, \ldots, l_m$, will be defined as follows: *Let the m variables* $l_1, l_2, \ldots, l_m$ *be linear functions of n stochastically independent variables,* $u_1, u_2, \ldots, u_n$, *which are assumed to be normally distributed with parameters* (0, 1). *If r linear relations (constraints) exist between the m variables* $l_1, l_2, \ldots, l_m$, *the number of degrees of freedom is* $f = m-r$. *The number of degrees of freedom for the sum of squares*

$$Q = \sum_{i=1}^{m} l_i^2$$

is defined as the number of degrees of freedom for the m variables $l_1, l_2, \ldots, l_m$.

The assumption is made that the r linear relations between the l's are linearly independent, i. e., none of the relations may be derived from the others, it then being possible to consider r of the m variables as linear functions of the remaining $m-r$ (free) variables. (Therefore, the sum of squares may also be written as a quadratic form in the $m-r$ variables).

The partition theorem gives us the conditions under which a sum of squares of the type $Q = \sum_{i=1}^{m} l_i^2$ has a χ^2-distribution with $f = m-r$ degrees of freedom.

The partition theorem for the χ^2*-distribution*: *Let the sum of squares of the n variables* $u_1, u_2, \ldots, u_n$ *be partitioned into a sum of k sums of squares,* $Q_1, Q_2, \ldots, Q_k$, *with* $f_1, f_2, \ldots, f_k$ *degrees of freedom, respectively,*

$$\chi^2 = \sum_{i=1}^{n} u_i^2 = Q_1 + Q_2 + \ldots + Q_k.$$

The necessary and sufficient condition that $Q_1, Q_2, \ldots, Q_k$ are stochastically independent and distributed as χ^2 with $f_1, f_2, \ldots, f_k$ degrees of freedom, respectively, is that

$$f_1+f_2+\ldots+f_k = n \,.$$

Before proving this, let us examine two simple examples. Returning to the case just discussed where

$$\chi^2 = \sum_{i=1}^{n} u_i^2 = (\sqrt{n}\,\bar{u})^2 + \sum_{i=1}^{n} (u_i-\bar{u})^2 ,$$

the number of degrees of freedom for the quadratic term Q_1 is equal to 1, since there is only one variable $\sqrt{n}\bar{u}$, and the number of degrees of freedom for Q_2 is equal to $n-1$, as there are n variables, $u_1-\bar{u}, \ldots, u_n-\bar{u}$, and only one relation between them, namely $\sum_{i=1}^{n}(u_i-\bar{u}) = 0$. As $f_1+f_2 = n$, these two sums of squares are—according to the partition theorem—stochastically independent and distributed as χ^2 with 1 and $n-1$ degrees of freedom, respectively.

The detailed examination of the transformation of Q_2 showed that Q_2 may be partitioned into a sum of $n-1$ squares, which are stochastically independent and distributed as χ^2 with 1 degree of freedom each. This may also be seen from the partition theorem, since

$$\chi^2 = \sum_{i=1}^{n} u_i^2 = Q_1+Q_2+\ldots+Q_n \tag{10.6.27}$$

where

$$Q_1 = (\sqrt{n}\,\bar{u})^2 = \frac{1}{n}(u_1+u_2+\ldots+u_n)^2 \tag{10.6.28}$$

and

$$Q_i = \hat{u}_i^2 = \frac{1}{i(i-1)}(u_1+u_2+\ldots+u_{i-1}-(i-1)u_i)^2 \,, \quad i = 2, 3, \ldots n \,. \tag{10.6.29}$$

As regards the n Q's the number of degrees of freedom for each is 1, the total number is n, and in accordance with the partition theorem the Q's are stochastically independent and distributed as χ^2 with 1 degree of freedom.

It will be noted that the system of linear functions $\hat{u}_1, \ldots, \hat{u}_n$ has two basic properties: (1) the sum of squares of the coefficients of the u's in each function is 1, from which it follows that $\mathcal{V}\{\hat{u}_i\} = 1$, cf. (10.6.25), and (2) the sum of products of corresponding coefficients of any two linear functions is zero, from which it follows that $\mathcal{V}\{\hat{u}_i, \hat{u}_j\} = 0$, cf. (10.6.26). A system of linear functions with these properties is called *orthogonal*.

The partition theorem tells us at once *that the mean and the variance of n stochastically independent observations from a normally distributed population are stochastically independent.*

This important result is seen from the fact that $Q_1 = (\sqrt{n}\,\bar{u})^2$ and $Q_2 = \sum_{i=1}^{n}(u_i-\bar{u})^2$ are stochastically independent, multiplication of these two sums of squares by σ^2 leading to

$$\sigma^2 Q_1 = n(\sigma\bar{u})^2 = n(\bar{x}-\xi)^2 \tag{10.6.30}$$

and

$$\sigma^2 Q_2 = \sum_{i=1}^{n}(x_i-\bar{x})^2 = (n-1)s^2\,, \tag{10.6.31}$$

cf. (10.6.1) and (10.6.3).

The partition theorem forms the theoretical basis of the chapters on analysis of variance and regression analysis. These chapters contain several examples of partitioning $\sigma^2\chi^2 = \sum_{i=1}^{n}(x_i-\xi)^2$ into a sum of stochastically independent sums of squares, each having a $\sigma^2\chi^2$-distribution.

The following is an example of a form of partitioning which does not lead to χ^2-distributions

$$\chi^2 = \sum_{i=1}^{n} u_i^2 = (\sqrt{n}\,\bar{u})^2+(u_1-\bar{u})^2+\sum_{i=2}^{n}(u_i-\bar{u})^2\,. \tag{10.6.32}$$

If we put

$$Q_1 = (\sqrt{n}\,\bar{u})^2+(u_1-\bar{u})^2 \tag{10.6.33}$$

and

$$Q_2 = \sum_{i=2}^{n}(u_i-\bar{u})^2\,, \tag{10.6.34}$$

the numbers of degrees of freedom are 2 and $n-1$, respectively, as there is no relation between the two variables $\sqrt{n}\,\bar{u}$ and $u_1-\bar{u}$ nor among the $n-1$ variables $u_2-\bar{u}, \ldots, u_n-\bar{u}$. As $f_1+f_2 = 2+(n-1) = n+1$, Q_1 and Q_2 are neither stochastically independent nor distributed as χ^2.

Proof of the partition theorem.

The addition theorem for the χ^2-distribution shows us at once that the condition laid down is necessary.

It will now be shown that the condition is sufficient.

We have assumed that each of the sums of squares $Q_1, \ldots, Q_k$ is of the type

$$Q = \sum_{i=1}^{m} l_i^2\,, \tag{10.6.35}$$

there being r linear relations between the m variables. By eliminating r of the l's, which may be written as linear functions of the other $f=m-r$ l's, the sum of squares is transformed into a quadratic form in f variables, cf. (10.6.20),

$$Q = \sum_{i=1}^{f} \sum_{j=1}^{f} a_{ij} l_i l_j \,, \tag{10.6.36}$$

a_{ij} denoting a constant. By linear transformation this quadratic form leads to another quadratic form in f variables, $\hat{u}_1, \ldots, \hat{u}_f$, which has no product terms

$$Q = \sum_{i=1}^{f} b_i \hat{u}_i^2 \,, \tag{10.6.37}$$

and in which b_i is equal to $+1$ or -1[1]). The f $\hat{u}$'s are linear functions of the f l's, and therefore also linear functions of the n u's. This holds for each of the k sums of squares, $Q_1, \ldots, Q_k$.

If

$$\sum_{j=1}^{k} f_j = n \,,$$

we obtain

$$\sum_{i=1}^{n} u_i^2 = \sum_{j=1}^{k} Q_j = \sum_{i=1}^{n} b_i \hat{u}_i^2 \,, \tag{10.6.38}$$

f_1 $\hat{u}$'s being introduced for the transformation of Q_1, f_2 $\hat{u}$'s for the transformation of Q_2, etc. As $\Sigma u_i^2 \geq 0$ all the b's must be equal to $+1$, as otherwise $\Sigma b_i \hat{u}_i^2$ might become negative. Thus, the sum of squares Q_j of m_j l's is transformed by a linear transformation into a sum of squares of $f_j = m_j - r_j$ $\hat{u}$'s. As the n $\hat{u}$'s are linear functions of the u's, and as

$$\sum_{i=1}^{n} u_i^2 = \sum_{i=1}^{n} \hat{u}_i^2 \,, \tag{10.6.39}$$

the $\hat{u}_i$'s, as functions of the u's, must form an *orthogonal* transformation, i. e.,

$$\hat{u}_i = c_{i1} u_1 + c_{i2} u_2 + \ldots + c_{in} u_n \,, \quad i = 1, 2, \ldots, n \,, \tag{10.6.40}$$

where

$$c_{i1}^2 + c_{i2}^2 + \ldots + c_{in}^2 = 1 \,, \quad i = 1, 2, \ldots, n \,, \tag{10.6.41}$$

and

$$c_{i1} c_{j1} + c_{i2} c_{j2} + \ldots + c_{in} c_{jn} = 0 \,, \quad i \neq j \,. \tag{10.6.42}$$

These conditions are satisfied by the transformation (10.6.23).

According to the addition theorem for the normal distribution, $\hat{u}_i$ is normally distributed with mean

$$\mathcal{M}\{\hat{u}_i\} = c_{i1} \mathcal{M}\{u_1\} + \ldots + c_{in} \mathcal{M}\{u_n\} = 0 \tag{10.6.43}$$

and variance

$$\begin{aligned} \mathcal{V}\{\hat{u}_i\} &= c_{i1}^2 \mathcal{V}\{u_1\} + \ldots + c_{in}^2 \mathcal{V}\{u_n\} \\ &= c_{i1}^2 + \ldots + c_{in}^2 = 1 \,. \end{aligned} \tag{10.6.44}$$

---

[1]) See section on transformation of quadratic forms in any standard textbook on matrices or algebra.

The covariance $\mathcal{V}\{\hat{u}_i, \hat{u}_j\}$ is determined as the mean of $\hat{u}_i\hat{u}_j$. This product

$$\hat{u}_i\hat{u}_j = (c_{i1}u_1+c_{i2}u_2+\ldots+c_{in}u_n)(c_{j1}u_1+c_{j2}u_2+\ldots+c_{jn}u_n) \quad (10.6.45)$$

contains the quadratic terms

$$c_{i1}c_{j1}u_1^2+c_{i2}c_{j2}u_2^2+\ldots+c_{in}c_{jn}u_n^2$$

and product terms of the form $c_{i\nu}u_\nu c_{j\mu}u_\mu$ for $\nu \neq \mu$. As

$$\mathcal{M}\{c_{i\nu}c_{j\mu}u_\nu u_\mu\} = c_{i\nu}c_{j\mu}\mathcal{M}\{u_\nu\}\mathcal{M}\{u_\mu\} = 0 ,$$

and the mean of the quadratic terms is

$$c_{i1}c_{j1}\mathcal{M}\{u_1^2\}+c_{i2}c_{j2}\mathcal{M}\{u_2^2\}+\ldots+c_{in}c_{jn}\mathcal{M}\{u_n^2\}$$
$$= c_{i1}c_{j1}+c_{i2}c_{j2}+\ldots+c_{in}c_{jn} = 0 , \quad (10.6.46)$$

according to (10.6.42), then $\mathcal{V}\{\hat{u}_i, \hat{u}_j\} = 0$. Since all the covariances in the n-dimensional normal distribution of $(\hat{u}_1, \hat{u}_2, \ldots, \hat{u}_n)$ are 0, the n variables are stochastically independent.

Each of the sums of squares, $Q_1, \ldots, Q_k$, may thus be transformed to a sum of squares of $f_1, f_2, \ldots, f_k$ stochastically independent variables, respectively, which are normally distributed with parameters (0, 1). Therefore, $Q_1, \ldots, Q_k$ are stochastically independent and each of them has a χ^2-distribution with $f_1 = m_1-r_1, \ldots, f_k = m_k-r_k$ degrees of freedom, respectively.

10.7. Notes and References.

The induction proof given in § 10.4 for the χ^2-distribution is due to F. R. Helmert: *Ueber die Wahrscheinlichkeit der Potenzsummen der Beobachtungsfehler und über einige damit im Zusammenhange stehende Fragen*, Zeitschrift für Mathematik und Physik, 21, 1876, 192–218.

The proof of the special case of the partition theorem discussed in § 10.6 is also due to Helmert, who in *Die Genauigkeit der Formel von Peters zur Berechnung des wahrscheinlichen Beobachtungsfehlers direchter Beobachtungen gleicher Genauigkeit*, Astronomische Nachrichten, Bd. 88, Nr. 2096–97, 1876, 113–132, has shown that by the transformation (10.6.23)

$$s^2 = \frac{1}{n-1}\sum_{i=1}^{n}(x_i-\bar{x})^2$$

may be expressed as a sum of squares of $n-1$ stochastically independent, normally distributed variables.

The notation χ^2, now used everywhere, is due to K. Pearson, who derived the χ^2-distribution in 1900, when solving another problem, cf. Chapter 23.

"STUDENT" in *The Probable Error of a Mean*, Biometrika, 6, 1908, 1–25, derived the distribution function of s^2 independently of HELMERT. "STUDENT's" proof was very incomplete, and a correct proof was given by R. A. FISHER in *Frequency Distribution of the Values of the Correlation Coefficient in Samples from an Indefinitely Large Population*, Biometrika, 10, 1915, 507–21. FISHER's proof is based on geometric reasoning.

In *Applications of "Student's" Distribution*, Metron, 5, No. 3, 1925, 90–104, R. A. FISHER emphasized the fundamental importance of these results, and formulated and proved the partition theorem.

The proof of the partition theorem given in § 10.6 is due to W. G. COCHRAN: *The Distribution of Quadratic Forms in a Normal System*, Proc. Camb. Phil. Soc., 30, 1934, 178–91. In order to render the exposition as simple as possible, the formulation of the partition theorem in § 10.6 is not quite as general as by COCHRAN, e. g., the formal definition of the number of degrees of freedom is slightly more simple. This definition is due to G. RASCH (not published).

A simple proof of the distribution function of s^2 and of the stochastic independence of the mean and the variance has been given by W. BEHRENS: *Ein Beitrag zur Fehlerberechnung bei wenigen Beobachtungen*, Landwirtschaftliche Jahrbücher 68, 1929, 807–37, and by J. F. STEFFENSEN: *Free Functions and the "Student"-Fisher Theorem*, Skandinavisk Aktuarietidskrift, 1936, 108–125. These proofs are based on transformations of variables in multi-dimensional distributions, using functional determinants.

R. A. FISHER and F. YATES: *Statistical Tables*, Oliver and Boyd, London, include a table of χ^2_P with 3 decimals for 14 values of P and for $f = 1\,(1)\,30$. Biometrika, 1942, gives χ^2_P with 5 decimals for 12 values of P and for $f = 1\,(1)\,30\,(10)\,100$.

CHAPTER 11.

THE DISTRIBUTION OF THE VARIANCE

11.1. The Distribution of the Variance.

If we have n stochastically independent observations, $x_1, x_2, \ldots, x_n$, from a normally distributed population with parameters (ξ, σ^2), an estimate $\hat{s}^2$ of σ^2 is computed—assuming that ξ is known—from the formula

$$\hat{s}^2 = \frac{1}{n} \sum_{i=1}^{n} (x_i - \xi)^2 \approx \sigma^2 . \tag{11.1.1}$$

If ξ is unknown, the estimate s^2 of σ^2 is computed from

$$s^2 = \frac{1}{n-1} \sum_{i=1}^{n} (x_i - \bar{x})^2 \approx \sigma^2 , \tag{11.1.2}$$

where

$$\bar{x} = \frac{1}{n} \sum_{i=1}^{n} x_i .$$

According to § 10.6 the two stochastic variables, $\hat{s}^2$ and s^2, may both be expressed in terms of χ^2,

$$\hat{s}^2 = \sigma^2 \frac{\chi^2}{f} , \text{ where } f = n , \tag{11.1.3}$$

and

$$s^2 = \sigma^2 \frac{\chi^2}{f} , \text{ where } f = n-1 , \tag{11.1.4}$$

which means that the distributions of both $\hat{s}^2$ and s^2 may be derived from the χ^2-distribution by the same transformation.

In the following the distribution of the variable

$$\boxed{s^2 = \sigma^2 \frac{\chi^2}{f}} \tag{11.1.5}$$

will be termed *the distribution of the variance or the s^2-distribution with parameters* (σ^2, f).

If (11.1.5) is solved with regard to χ^2, we have

$$\boxed{\chi^2 = \frac{fs^2}{\sigma^2}.} \tag{11.1.6}$$

The fractiles of the s^2-distribution, s_P^2, are obtained from the fractiles of the χ^2-distribution since

$$s_P^2 = \sigma^2 \frac{\chi_P^2}{f}. \tag{11.1.7}$$

For example, Table V shows that, for $f = 10$ and $P = 95\%$, $\chi^2_{\cdot 95} = 18{\cdot}3$, which gives us

$$s^2_{\cdot 95} = \sigma^2 \frac{18{\cdot}3}{10} = 1{\cdot}83\sigma^2 \text{ for } f = 10,$$

see also Table VI, in which χ_P^2/f has been tabulated for suitable values of P less than or equal to 5% and greater than or equal to 95%. We thus see that the probability that a sample variance with 10 degrees of freedom is less than 1·83 times the population variance is 95%.

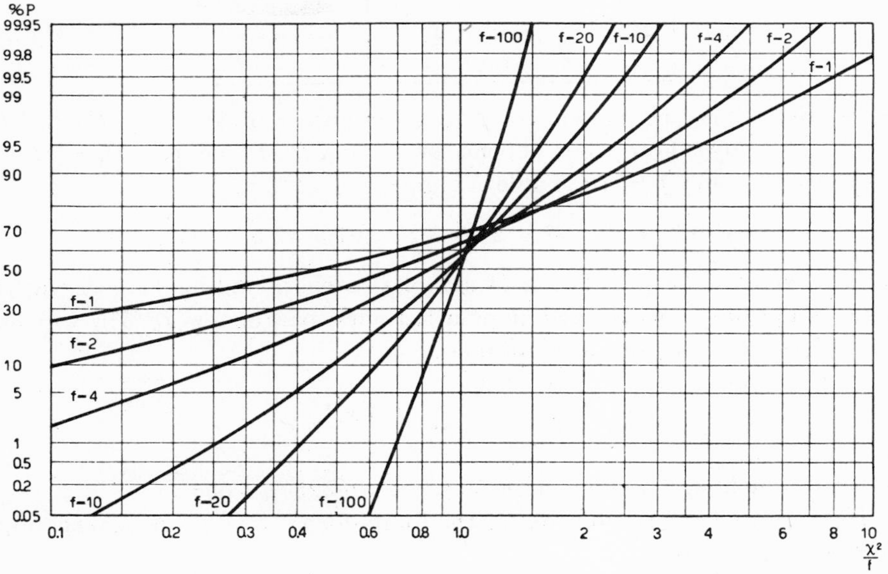

FIG. 11.1. The cumulative distribution function of $s^2/\sigma^2 = \chi^2/f$ plotted on logarithmic probability paper for $f = 1, 2, 4, 10, 20$, and 100.

Fig. 11.1 shows the cumulative distribution function of $s^2/\sigma^2 = \chi^2/f$ plotted on logarithmic probability paper for $f = 1, 2, 4, 10, 20$, and 100. The figure shows how the range of variation of s^2/σ^2 decreases as f increases.

It follows from (10.2.7) and (10.3.4) that

$$\boxed{\mathcal{M}\{s^2\} = \sigma^2} \tag{11.1.8}$$

and

$$\boxed{\mathcal{V}\{s^2\} = \frac{2\sigma^4}{f}} \tag{11.1.9}$$

for any value of f.

Since the χ^2-distribution for $f \to \infty$ converges to the normal distribution with parameters $(f, 2f)$, the s^2-distribution will also converge to the normal distribution with parameters $\left(\sigma^2, \dfrac{2\sigma^4}{f}\right)$.

11.2. The Fractiles.

Limits for s^2 may be derived from the fractiles, since

$$P\{s^2_{P_1} < s^2 < s^2_{P_2}\} = P_2 - P_1\,, \tag{11.2.1}$$

which, according to (11.1.7), may be written

$$\boxed{P\left\{\sigma^2 \frac{\chi^2_{P_1}}{f} < s^2 < \sigma^2 \frac{\chi^2_{P_2}}{f}\right\} = P_2 - P_1.} \tag{11.2.2}$$

These limits may be calculated from given values of σ^2 and f, P_1 and P_2. Table VI shows how they depend on f and P. If we put $P_1 = 2{\cdot}5\%$, $P_2 = 97{\cdot}5\%$, and $f = 10$, Table VI gives

$$P\{0{\cdot}325\ \sigma^2 < s^2 < 2{\cdot}05\ \sigma^2\} = 97{\cdot}5 - 2{\cdot}5 = 95\%\,.$$

This means that a population of samples, each including 11 observations from a normally distributed population with parameters (ξ, σ^2), will lead to a population of sample variances

$$s^2 = \frac{1}{10} \sum_{i=1}^{11} (x_i - \bar{x})^2,$$

95% of which will belong to the interval $(0{\cdot}325\sigma^2, 2{\cdot}05\sigma^2)$.

If we extract the square root of the limits for s^2, we obtain corresponding limits for s.

The *variability* of an industrial product may be controlled by regularly sampling n units of the product, measuring them, and calculating the variance corresponding to each sample,

$$s^2 = \frac{1}{n-1} \sum_{i=1}^{n} (x_i - \bar{x})^2\,.$$

These variances are plotted on a *control chart*, the number of the sample being used as abscissa and the variance as ordinate, see Fig. 11.2 where the 0·1%, 2·5%, 50%, 97·5% and 99·9% fractiles are used as control limits.

As was the case when dealing with the control limits for the mean, cf. § 9.3, the limits must be chosen to suit the type of problem dealt with. The number of sets of limits corresponding to a given probability P is infinite, as the equation

$$P\left\{\sigma^2 \frac{\chi^2_{P_1}}{f} < s^2 < \sigma^2 \frac{\chi^2_{P_2}}{f}\right\} = P_2 - P_1 = P \tag{11.2.3}$$

has an infinite number of solutions. If $P_1 = 0$, there is a unilateral upper limit, if $P_2 = 1$ a unilateral lower limit.

Very often we are mainly interested in preventing the population variance from increasing and we therefore use the upper fractile for s^2, $\sigma^2 \frac{\chi^2_{\cdot 999}}{f}$ say, as a control limit. If a sample variance exceeds this limit it is taken as an indication of an increase in the population variance and such an observation should at once be followed by an investigation of the production process.

Example 11.1. The object of the control chart may for instance be to control the effectiveness of a mixing process. The process is planned to yield a product in which the concentration of a certain component is normally distributed with variance σ^2_M. After a standard quantity has been mixed, a few samples, say 5, are taken at random, and the concentration in each sample is determined by duplicate analyses. The uncertainty of the analytical method is denoted by σ^2_A. If the variance of the 5 analytical determinations (means of the duplicate analyses) is calculated, we obtain an estimate of the total variance σ^2, which according to the addition theorem for the normal distribution is

$$\sigma^2 = \sigma^2_M + \frac{\sigma^2_A}{2}.$$

We then calculate the control limits

$$s^2_P = \sigma^2 \frac{\chi^2_P}{f} \text{ for } f = 4 .$$

For a mixing plant for "raw meal" in a cement plant the value of σ_M was reckoned to be 0·1% $CaCO_3$, and σ_A to be 0·15% $CaCO_3$, the total variance being $\sigma^2 = 0{\cdot}0100 + 0{\cdot}0113 = 0{\cdot}0213$, i.e., $\sigma = 0{\cdot}15\%$ $CaCO_3$. According to Tables V and VI the control limits for the sample variances for $f = 4$ are

$$s^2_{.001} = 0{\cdot}0213 \times 0{\cdot}0227 = 0{\cdot}0005, \quad s_{.001} = 0{\cdot}02\% \ CaCO_3 ,$$
$$s^2_{.025} = 0{\cdot}0213 \times 0{\cdot}121 = 0{\cdot}0026, \quad s_{.025} = 0{\cdot}05\% \ CaCO_3 ,$$
$$s^2_{.50} = 0{\cdot}0213 \times 0{\cdot}840 = 0{\cdot}0179, \quad s_{.50} = 0{\cdot}13\% \ CaCO_3 ,$$
$$s^2_{.975} = 0{\cdot}0213 \times 2{\cdot}79 = 0{\cdot}0594, \quad s_{.975} = 0{\cdot}24\% \ CaCO_3 ,$$
$$s^2_{.999} = 0{\cdot}0213 \times 4{\cdot}62 = 0{\cdot}0984, \quad s_{.999} = 0{\cdot}31\% \ CaCO_3 .$$

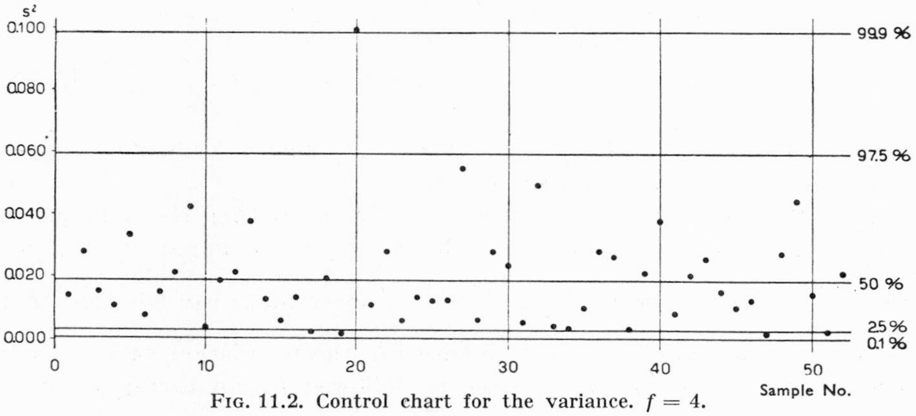

FIG. 11.2. Control chart for the variance. $f = 4$.

Fig. 11.2 shows the variances for 52 sets of samples plotted on a control chart. The variances found vary at random within the range of the control limits, and only a few single values are below the 2·5% or above the 97·5% limit. Theoretically, the number of values expected to lie outside these limits is $0{\cdot}025 \times 52 = 1{\cdot}3$, i. e., we must expect that among 52 empirical variances some few will lie outside the 2·5% and 97·5% limits. As regards the working of the plant, first of all the upper limit is of interest, too large a number of variances near this limit or above it indicating that the mixing is not satisfactory, which may be due to the quality of the raw materials or to the machinery employed. For the sake of completeness it must be pointed out that an increase in the spread of the variances may also be due to increased inaccuracies in the methods of analysis, which should therefore be controlled regularly.

11.3. The χ^2-Test of Significance.

The question whether a sample variance s^2 deviates significantly from a presupposed population variance σ_0^2 may be answered by means of a test of significance based on the χ^2-distribution, using the principles discussed in § 9.4.

The test hypothesis specifies the population variance as $\mathcal{M}\{s^2\} = \sigma_0^2$. Let us first treat the simple case where *the alternative hypothesis is* $\sigma^2 > \sigma_0^2$, whereas the possibility $\sigma^2 < \sigma_0^2$ is excluded a priori or is of no practical importance.

If the test hypothesis is correct, then s^2/σ_0^2 is distributed as χ^2/f. We want to determine a constant c so that according to whether $s^2/\sigma_0^2 > c$ or $s^2/\sigma_0^2 < c$ we reject or accept the test hypothesis. Choosing a level of significance α we determine c such that

$$P\left\{\frac{s^2}{\sigma_0^2} > c\ ;\ \mathcal{M}\{s^2\} = \sigma_0^2\right\} = \alpha\ , \tag{11.3.1}$$

i. e., if the test hypothesis is true, then the criterion $s^2/\sigma_0^2 > c$ will lead to the rejection of the hypothesis on only $100\alpha\%$, say 5%, of occasions. The solution of (11.3.1) is $c = \chi_{1-\alpha}^2/f$ since

$$P\left\{\frac{s^2}{\sigma_0^2} > \frac{\chi_{1-\alpha}^2}{f}\ ;\ \mathcal{M}\{s^2\} = \sigma_0^2\right\} = P\{\chi^2 > \chi_{1-\alpha}^2\} = \alpha\ .$$

Thus, *the critical region at the* α *level of significance* is defined by

$$\boxed{\frac{s^2}{\sigma_0^2} > \frac{\chi_{1-\alpha}^2}{f}}\ . \tag{11.3.2}$$

The *power of the test* with respect to the alternative hypothesis $\mathcal{M}\{s^2\} = \sigma^2$ is

$$\begin{aligned}\pi(\sigma^2) &= P\left\{\frac{s^2}{\sigma_0^2} > \frac{\chi_{1-\alpha}^2}{f}\ ;\ \mathcal{M}\{s^2\} = \sigma^2\right\}\\ &= P\left\{\frac{fs^2}{\sigma^2} > \frac{\sigma_0^2}{\sigma^2}\chi_{1-\alpha}^2;\ \mathcal{M}\{s^2\} = \sigma^2\right\}\\ &= P\left\{\chi^2 > \frac{1}{\lambda^2}\chi_{1-\alpha}^2\right\}\end{aligned} \tag{11.3.3}$$

where

$$\lambda^2 = \frac{\sigma^2}{\sigma_0^2}\ . \tag{11.3.4}$$

The power function is an increasing function of λ^2 or σ^2 since $\chi_{1-\alpha}^2/\lambda^2$ is a decreasing function of λ^2.

To determine the value of λ^2 for which the power function takes on a certain value $1-\beta$ we have to solve the equation

$$\pi(\sigma^2) = P\left\{\chi^2 > \frac{1}{\lambda^2}\chi_{1-\alpha}^2\right\} = 1-\beta\ . \tag{11.3.5}$$

It follows that

$$\frac{1}{\lambda^2}\chi_{1-\alpha}^2 = \chi_\beta^2$$

or

$$\lambda^2 = \frac{\chi^2_{1-\alpha}}{\chi^2_\beta}. \tag{11.3.6}$$

By means of this relation it is easy to determine a sufficient number of points to draw the power curve. For given f and α we only need to compute the ratio λ^2 between $\chi^2_{1-\alpha}$ and χ^2_β for selected values of β as found in Table V and then to plot the points $(\lambda^2, 1-\beta)$. Fig. 11.3 shows for different values of f the power curves at the 5% level of significance plotted on logarithmic probability paper. It will be seen that we need at most five points to be able to draw a power curve with sufficient accuracy for practical purposes, see Table 11.1 where such a set of points is given for different degrees of freedom and for α equal to 0·05.

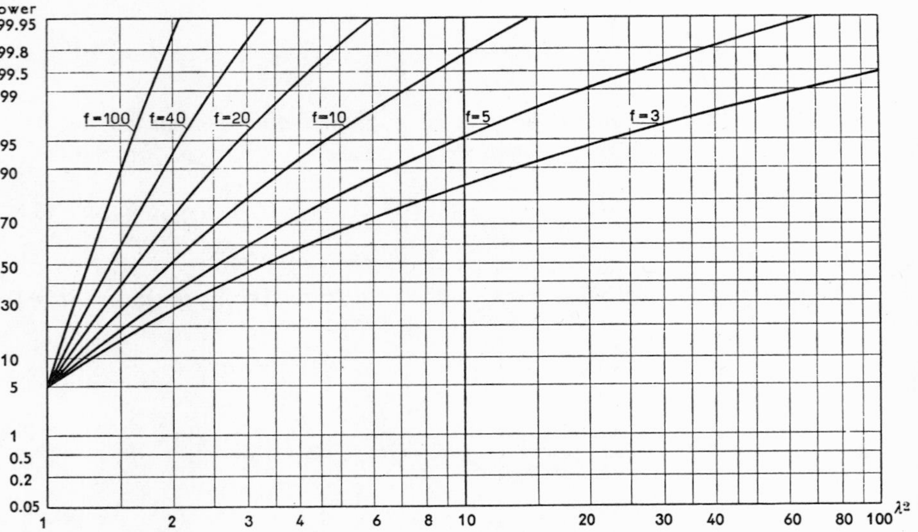

FIG. 11.3. Power curves for the one-tailed χ^2-test at the 5% level of significance. $\lambda^2 = \sigma^2/\sigma_0^2$.

It has been proved that the above test of the hypothesis $\mathcal{M}\{s^2\} = \sigma_0^2$ against the alternative $\sigma^2 > \sigma_0^2$ is a uniformly most powerful test.

Suppose for example that s^2 is based on 30 degrees of freedom. At the 5% level of significance the critical region is

$$\frac{s^2}{\sigma_0^2} > \frac{\chi^2_{\cdot 95}}{f} = 1{\cdot}46\,, \quad f = 30\,,$$

see Table VI. The power of the test for $\lambda^2 = 2$ is according to (11.3.3)

$$P\,\{\chi^2 > \tfrac{1}{2}\chi^2_{\cdot 95}\} = P\,\{\chi^2 > 21{\cdot}9\} = 85\% \quad \text{for} \quad f = 30\,,$$

i. e., the probability of accepting σ_0^2 when $\mathcal{M}\{s^2\} = 2\sigma_0^2$ is 15%. If we re-

TABLE 11.1.

Power function for the one-tailed χ^2-test at the 5% level of significance.

Reject the test hypothesis $\mathcal{M}\{s^2\} = \sigma_0^2$ when $\frac{s^2}{\sigma_0^2} > \frac{\chi^2_{.95}}{f}$.

$$\pi(\sigma^2) = P\left\{\chi^2 > \frac{1}{\lambda^2}\chi^2_{.95}\right\} \quad \text{for} \quad \lambda^2 = \frac{\sigma^2}{\sigma_0^2}.$$

Power		0·20	0·50	0·80	0·95	0·995
f	$\frac{\chi^2_{.95}}{f}$	Values of λ^2				
2	3·00	1·80	4·32	13·4	58·4	598
3	2·61	1·68	3·30	7·81	22·2	109
4	2·37	1·58	2·83	5·75	13·3	45·8
5	2·21	1·52	2·54	4·74	9·66	26·9
6	2·10	1·47	2·35	4·10	7·70	18·6
7	2·01	1·44	2·22	3·69	6·49	14·2
8	1·94	1·41	2·11	3·38	5·67	11·5
9	1·88	1·39	2·03	3·14	5·09	9·75
10	1·83	1·37	1·96	2·96	4·65	8·49
12	1·75	1·33	1·85	2·69	4·02	6·84
14	1·69	1·30	1·78	2·50	3·60	5·81
16	1·64	1·28	1·71	2·35	3·30	5·11
18	1·60	1·27	1·67	2·24	3·07	4·61
20	1·57	1·26	1·62	2·15	2·89	4·23
25	1·51	1·23	1·55	1·99	2·58	3·58
30	1·46	1·21	1·49	1·87	2·37	3·18
40	1·39	1·18	1·42	1·73	2·10	2·69
50	1·35	1·16	1·37	1·63	1·94	2·41
60	1·32	1·15	1·33	1·56	1·83	2·23
70	1·29	1·14	1·31	1·51	1·75	2·09
80	1·27	1·13	1·28	1·47	1·68	1·99
90	1·26	1·12	1·27	1·44	1·64	1·91
100	1·24	1·11	1·25	1·41	1·60	1·85

quire the test to discriminate better between σ_0^2 and $2\sigma_0^2$ we have to increase f. Choosing $\beta = 0{\cdot}01$, say, we find from (11.3.6) for $\lambda^2 = 2$ the equation

$$\frac{\chi^2_{.95}(f)}{\chi^2_{.01}(f)} = 2$$

from which we get $f = 70$ by means of Table V.

For large values of f we may use the approximation formula (10.3.6) which inserted into (11.3.6) leads to

$$f \simeq \tfrac{1}{2} + \tfrac{1}{2}\left(\frac{u_{1-\beta}\lambda + u_{1-\alpha}}{\lambda - 1}\right)^2. \tag{11.3.7}$$

In the above case we find $u_{1-\beta} = u_{.99} = 2{\cdot}33$ and $u_{1-\alpha} = u_{.95} = 1{\cdot}64$ so that

$$f \simeq \tfrac{1}{2} + \tfrac{1}{2}\left(\frac{2{\cdot}33\sqrt{2} + 1{\cdot}64}{\sqrt{2} - 1}\right)^2 = 71\,.$$

If *the alternative to the test hypothesis is* $\sigma^2 < \sigma_0^2$ we proceed in a similar way obtaining

$$\frac{s^2}{\sigma_0^2} < \frac{\chi_\alpha^2}{f} \tag{11.3.8}$$

as the critical region at the α level of significance and

$$\pi(\sigma^2) = P\left\{\chi^2 < \frac{1}{\lambda^2}\chi_\alpha^2\right\} \quad \text{for} \quad \lambda^2 = \frac{\sigma^2}{\sigma_0^2}. \tag{11.3.9}$$

Putting $\pi(\sigma^2) = 1-\beta$ we get

$$\lambda^2 = \frac{\chi_\alpha^2}{\chi_{1-\beta}^2} \tag{11.3.10}$$

analogous to (11.3.6).

When *the alternative to the test hypothesis is* $\sigma^2 \neq \sigma_0^2$ no uniformly most powerful test exists. The usual practice is to use *a critical region based on equal tail areas*, i. e., we reject the test hypothesis when s^2/σ_0^2 satisfies either of the inequalities

$$\frac{s^2}{\sigma_0^2} < \frac{\chi_{\frac{\alpha}{2}}^2}{f} \quad \text{or} \quad \frac{s^2}{\sigma_0^2} > \frac{\chi_{1-\frac{\alpha}{2}}^2}{f}. \tag{11.3.11}$$

The power function then becomes

$$\pi(\sigma^2) = P\left\{\frac{s^2}{\sigma_0^2} < \frac{\chi_{\frac{\alpha}{2}}^2}{f}\,;\; m\{s^2\} = \sigma^2\right\} + P\left\{\frac{s^2}{\sigma_0^2} > \frac{\chi_{1-\frac{\alpha}{2}}^2}{f}\,;\; m\{s^2\} = \sigma^2\right\}$$

$$= P\left\{\chi^2 < \frac{1}{\lambda^2}\chi_{\frac{\alpha}{2}}^2\right\} + P\left\{\chi^2 > \frac{1}{\lambda^2}\chi_{1-\frac{\alpha}{2}}^2\right\} \tag{11.3.12}$$

which for given f and α may be calculated as a function of λ^2 by means of Table V. Fig. 11.4 shows some power curves at the 5% level of significance. It is seen that the power is less than 5% for values of λ^2 in an interval to the left of $\lambda^2 = 1$, so that the probability of rejecting the test hypothesis when it is false for these values of λ^2 is less than 5%. NEYMAN and PEARSON[1]) have shown how to construct a uniformly most powerful

[1]) J. NEYMAN and E. S. PEARSON: *Contributions to the Theory of Testing Statistical Hypotheses*, Statistical Research Memoirs, 1, 1936, pp. 1–37, see especially p. 18.

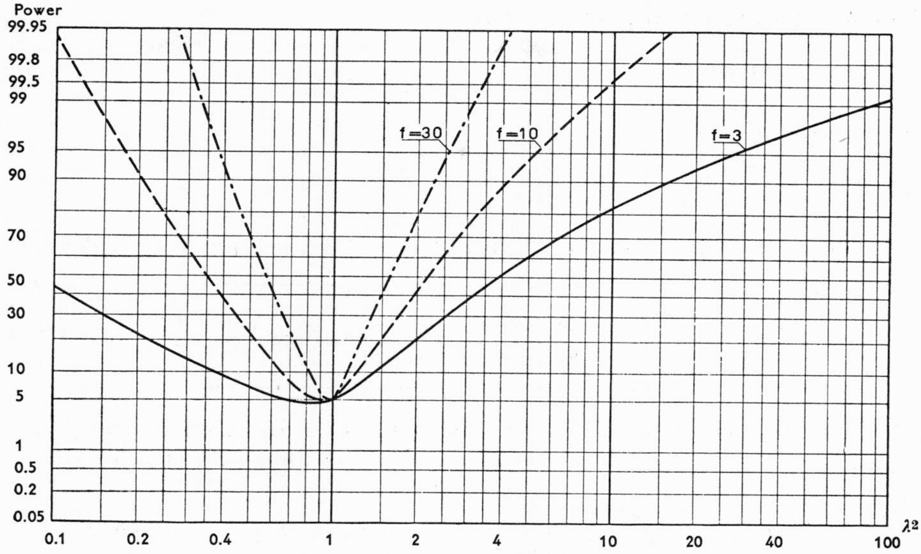

FIG. 11.4. Power curves for the two-tailed χ^2-test at the 5% level of significance. $\lambda^2 = \sigma^2/\sigma_0^2$.

unbiased test (a test is called unbiased if its power curve has a minimum at $\sigma^2 = \sigma_0^2$). This leads to a critical region defined as

$$\frac{s^2}{\sigma_0^2} < \frac{\chi^2_{\alpha_1}}{f} \quad \text{and} \quad \frac{s^2}{\sigma_0^2} > \frac{\chi^2_{1-\alpha_2}}{f}$$

where $\alpha_1+\alpha_2 = \alpha$ and $\alpha_1 > \alpha_2$, i. e., $\alpha_1 > \alpha/2$. For values of $f > 10$, however, the two tests do not differ greatly.

Example 11.2. After mixing a standard quantity, as described in Example 11.1, 30 samples have been collected and the variance computed, resulting in $s^2 = 0{\cdot}0349$, $f = 29$. Does this sample variance deviate significantly from $\sigma_0^2 = 0{\cdot}0213$, which is the specified standard?

As we are mainly interested in controlling the mixing from being worse than the standard, we want to test the hypothesis $\sigma^2 \leq \sigma_0^2$ against the alternative $\sigma^2 > \sigma_0^2$. We therefore use the one-tailed test (11.3.2) and compute

$$\frac{s^2}{\sigma_0^2} = \frac{0{\cdot}0349}{0{\cdot}0213} = 1{\cdot}64\ .$$

From Table VI it is seen that 1·64 lies between the 97·5% and the 99% fractiles. Thus, the population variance corresponding to $s^2 = 0{\cdot}0349$ is likely to be greater than σ_0^2, which suggests that the mixing process is not up to the standard specified.

11.4. Confidence limits for σ^2.

In the relation

$$P\{\chi^2_{P_1} < \chi^2 < \chi^2_{P_2}\} = P_2 - P_1$$

we introduce

$$\chi^2 = \frac{fs^2}{\sigma^2},$$

and solve the inequality with respect to σ^2. We hereby obtain *the confidence limits for* σ^2

$$P\left\{s^2 \frac{f}{\chi^2_{P_2}} < \sigma^2 < s^2 \frac{f}{\chi^2_{P_1}}\right\} = P_2 - P_1 . \qquad (11.4.1)$$

Table VI shows how these confidence limits depend on P and f. If we put $P_1 = 2{\cdot}5\%$, $P_2 = 97{\cdot}5\%$, and $f = 10$, we find in Table VI

$$P\left\{\frac{1}{2{\cdot}048} s^2 < \sigma^2 < \frac{1}{0{\cdot}3247} s^2\right\} = P\{0{\cdot}488\, s^2 < \sigma^2 < 3{\cdot}080\, s^2\}$$
$$= 97{\cdot}5 - 2{\cdot}5 = 95\%,$$

which indicates that, from a large number of samples, each of 11 observations, drawn from a normally distributed population with parameters (ξ, σ^2), we may compute a corresponding number of variances and intervals $(0{\cdot}488\, s^2, 3{\cdot}080\, s^2)$, and that about 95% of these intervals will contain σ^2, see the analogous discussion of the confidence interval for ξ in § 9.5.

Example 11.3. Confidence limits for σ^2 may be calculated from the estimate $s^2 = 0{\cdot}0349$, $f = 29$, given in Example 11.2.

Table VI gives us

$$\frac{\chi^2_{\cdot 025}}{29} = 0{\cdot}5533 \text{ and } \frac{\chi^2_{\cdot 975}}{29} = 1{\cdot}577,$$

whence the bilateral 95% confidence limits for σ^2 are, according to (11.4.1),

$$\frac{0{\cdot}0349}{1{\cdot}577} = 0{\cdot}0221 \text{ and } \frac{0{\cdot}0349}{0{\cdot}5533} = 0{\cdot}0631 .$$

If we want the unilateral 95% lower confidence limit we get

$$\frac{0{\cdot}0349}{1{\cdot}4675} = 0{\cdot}0239,$$

which is larger than $\sigma_0^2 = 0{\cdot}0213$ in agreement with the result of the one-tailed test in Example 11.2.

11.5. The Addition Theorem for the s^2-Distribution.

An addition theorem for the s^2-distribution may be derived from the addition theorem for the χ^2-distribution, cf. § 10.5.

From k samples, drawn from normally distributed populations with the *same variance* σ^2, but not necessarily with the same mean, we compute k sample variances $s_1^2, s_2^2, \ldots, s_k^2$ with $f_1, f_2, \ldots, f_k$ degrees of freedom, respectively. As

$$\chi_i^2 = \frac{f_i s_i^2}{\sigma^2} \qquad (i = 1, 2, \ldots, k) \tag{11.5.1}$$

has a χ^2-distribution with f_i degrees of freedom, the sum

$$\chi^2 = \sum_{i=1}^{k} \chi_i^2 = \frac{1}{\sigma^2} \sum_{i=1}^{k} f_i s_i^2 \tag{11.5.2}$$

has—according to the addition theorem for the χ^2-distribution—a χ^2-distribution with $f = \sum_{i=1}^{k} f_i$ degrees of freedom. Multiplying (11.5.2) by $\frac{\sigma^2}{f}$ we obtain

$$\sigma^2 \frac{\chi^2}{f} = \frac{\sum_{i=1}^{k} f_i s_i^2}{\sum_{i=1}^{k} f_i}, \tag{11.5.3}$$

and therefore the estimate

$$\boxed{s^2 = \frac{\sum_{i=1}^{k} f_i s_i^2}{\sum_{i=1}^{k} f_i}} \tag{11.5.4}$$

has an s^2-distribution with parameters (σ^2, f). *Consequently, a pooled estimate, s^2, of σ^2, is calculated from k sample variances with the same theoretical value σ^2, by determining the weighted average of the k variances, the numbers of degrees of freedom being used as weights.*

Since $f_i s_i^2$ is the sum of squares of deviations, SSD_i, s^2 may be computed as $\sum_{i=1}^{k} SSD_i$ divided by $\sum_{i=1}^{k} f_i$. The computations may be arranged as shown in the following table.

Sample No.	Observations	Population variance	Sum of squares of deviations	Number of degrees of freedom
1	$x_{11}, x_{12}, \ldots, x_{1n_1}$	σ^2	SSD_1	n_1-1
2	$x_{21}, x_{22}, \ldots, x_{2n_2}$	σ^2	SSD_2	n_2-1
$\vdots$	$\vdots$	$\vdots$	$\vdots$	$\vdots$
k	$x_{k1}, x_{k2}, \ldots, x_{kn_k}$	σ^2	SSD_k	n_k-1
		σ^2	$\sum_{i=1}^{k} SSD_i$	$\sum_{i=1}^{k}(n_i-1)$
$s^2 = \dfrac{\sum_{i=1}^{k} SSD_i}{\sum_{i=1}^{k}(n_i-1)} \approx \sigma^2.$				

In the ith sample, the sum of squares of deviations is

$$SSD_i = \sum_{\nu=1}^{n_i} (x_{i\nu}-\bar{x}_i)^2 , \tag{11.5.5}$$

$\bar{x}_i$ denoting the mean of the observations in the ith sample.

In this manner *it is possible from a large number of small samples with the same theoretical variance, σ^2, to construct an estimate, s^2, of σ^2, which is based on a large number of degrees of freedom.* If, for instance, we wish to estimate with considerable accuracy the error of a particular method of measurement, it is not necessary to have a *large* number of measurements of the *same* object in order to obtain an estimate s^2 of σ^2. According to (11.5.4) it is sufficient to have a series of comparatively small sets of measurements of different objects, if all sets of measurements have the same theoretical variance, i. e., if the error of measurement is independent of the object measured. This means that it is unnecessary to make a special series of measurements to determine the error; this may be important, as we often find that such a series is carried out more carefully than routine work, and the result is that the routine error of measurement is underestimated.

If we consider the particular case $n_1 = n_2 = \ldots = n_k = 2$, i. e., k duplicate determinations, we have

$$SSD_i = (x_{i1}-\bar{x}_i)^2+(x_{i2}-\bar{x}_i)^2 = \frac{(x_{i1}-x_{i2})^2}{2},$$

since

$$\bar{x}_i = \frac{x_{i1}+x_{i2}}{2}.$$

Introducing

$$d_i = x_{i1} - x_{i2} ,$$

we find

$$SSD_i = \tfrac{1}{2} d_i^2 , \tag{11.5.6}$$

and

$$s^2 = \frac{1}{2k} \sum_{i=1}^{k} d_i^2 \text{ and } f = k , \tag{11.5.7}$$

i. e., *from k duplicate determinations with the same theoretical variance σ^2 we may compute an estimate s^2 of σ^2 as the sum of squares of the k differences divided by $2k$. For this estimate the number of degrees of freedom is equal to k.*

Example 11.4. Table 11.2 gives the results of 30 duplicate titrations, and the differences between corresponding numbers. According to (11.5.7) we calculate

$$\sum_{i=1}^{30} d_i^2 = 1{\cdot}5410$$

and

$$s^2 = \frac{1}{60} \sum_{i=1}^{30} d_i^2 = 0{\cdot}02568 , \ f = 30 .$$

TABLE 11.2.

30 duplicate titrations of raw meal. The results denote the concentration of $CaCO_3$ in %.

Set No.	Titration		Difference	Set No.	Titration		Difference
	No. 1	No. 2			No. 1	No. 2	
1	75·68	75·58	0·10	16	76·35	76·23	0·12
2	75·98	76·30	−0·32	17	76·33	76·30	0·03
3	74·93	74·98	−0·05	18	76·45	76·33	0·12
4	76·23	76·45	−0·22	19	76·40	76·33	0·07
5	76·50	76·28	0·22	20	76·68	76·28	0·40
6	76·48	76·25	0·23	21	76·33	76·45	−0·12
7	76·43	76·48	−0·05	22	76·40	76·38	0·02
8	77·20	76·48	0·72	23	76·28	76·43	−0·15
9	76·45	76·60	−0·15	24	76·58	76·45	0·13
10	76·38	76·73	−0·35	25	76·65	76·60	0·05
11	76·25	76·50	−0·25	26	76·40	76·40	0·00
12	76·55	76·35	0·20	27	77·03	76·80	0·23
13	76·65	76·30	0·35	28	76·90	76·95	−0·05
14	76·55	76·40	0·15	29	74·83	74·88	−0·05
15	76·15	76·38	−0·23	30	75·28	75·25	0·03

According to (11.4.1) and Table VI the bilateral 95% confidence limits for σ^2 are

$$\frac{0{\cdot}02568}{1{\cdot}5660} = 0{\cdot}01640 \quad \text{and} \quad \frac{0{\cdot}02568}{0{\cdot}5597} = 0{\cdot}04588\,.$$

The corresponding limits for σ are 0·13% $CaCO_3$ and 0·21% $CaCO_3$.

Example 11.5. Table 7.10, p. 184, gives 4 distributions of the compressive strength of cement-mortar cubes after curing for 1, 3, 7, and 28 days, respectively. The following table gives the *SSD*, f and s^2 of the logarithms of the compressive strength.

Logarithm of compressive strength, cf. Table 7.10.

Curing for days	$10^6 \times SSD$	f	$10^6 \times s^2$
1	558	4	140
3	363	4	91
7	318	4	80
28	453	4	113
Total	1692	16	

According to the addition theorem for the s^2-distribution the pooled estimate of σ^2 is

$$s^2 = \frac{0{\cdot}001692}{16} = 0{\cdot}0001058\,, \quad f = 16\,,$$

which gives us $s = 0{\cdot}0103$.

The upper 95% confidence limit for σ^2 is

$$\frac{0{\cdot}0001058}{0{\cdot}4976} = 0{\cdot}0002126\,,$$

the corresponding limit for σ being 0·0146, see (11.4.1) and Table VI.

11.6. A Test for the Equality of k Theoretical Variances.

An assumption underlying the addition theorem for the s^2-distribution is that the k empirical variances $s_1^2, s_2^2, \ldots, s_k^2$ are all estimates of the same theoretical variance, σ^2. It is therefore necessary to test the hypothesis

$$\sigma_1^2 = \sigma_2^2 = \ldots = \sigma_k^2 = \sigma^2\,, \tag{11.6.1}$$

where $\sigma_i^2 = \mathcal{M}\{s_i^2\}$, before we can pool the k empirical variances to form the single estimate s^2 of σ^2.

If the test hypothesis is correct, the quantity

$$s^2 = \frac{\sum_{i=1}^{k} f_i s_i^2}{\sum_{i=1}^{k} f_i}$$

has an s^2-distribution with parameters (σ^2, f), f being equal to $\sum_{i=1}^{k} f_i$. As both s^2 and s_i^2 $(i = 1, 2, \ldots, k)$ are distributed as $\sigma^2\chi^2/f$ with the adequate number of degrees of freedom, the distribution of the quotient s_i^2/s^2 depends solely on f_i and f, since σ^2 vanishes in the quotient. The distribution of a function of $s_1^2/s^2, \ldots, s_k^2/s^2$ therefore depends only on $f_1, \ldots, f_k$ and k. M. S. BARTLETT has shown that

$$-\frac{1}{c}\sum_{i=1}^{k} f_i \ln \frac{s_i^2}{s^2}, \tag{11.6.2}$$

where

$$c = 1 + \frac{1}{3(k-1)}\left(\sum_{i=1}^{k}\frac{1}{f_i} - \frac{1}{f}\right), \tag{11.6.3}$$

has *approximately a χ^2-distribution with $k-1$ degrees of freedom*. The approximation is, however, a poor one for $f_i \leq 2$[1]).

For computations (11.6.2) is written

$$\boxed{\chi^2 \simeq \frac{2\cdot 3026}{c}\left(f \log s^2 - \sum_{i=1}^{k} f_i \log s_i^2\right)}, \tag{11.6.4}$$

in which we may put

$$\log s_i^2 = \log SSD_i - \log f_i\,. \tag{11.6.5}$$

For large values of $f_1, \ldots, f_k$, c is practically equal to 1.

If we introduce the geometric mean, s_g^2, of $s_1^2, \ldots, s_k^2$, from

$$\log s_g^2 = \frac{\sum_{i=1}^{k} f_i \log s_i^2}{\sum_{i=1}^{k} f_i}, \tag{11.6.6}$$

(11.6.4) may be written

---

[1]) M. S. BARTLETT: *Properties of Sufficiency and Statistical Tests*, Proc. Roy. Soc., A., 160, 1937, 268–82.

D. J. BISHOP and U. S. NAIR: *A Note on Certain Methods of Testing for the Homogeneity of a Set of Estimated Variances*, Suppl. Journ. Roy. Stat. Soc., 6, 1939, 89–99.

H. O. HARTLEY: *Testing the Homogeneity of a Set of Variances*, Biometrika, 31, 1940, 249–255.

$$\frac{2{\cdot}3026f}{c}\log\frac{s^2}{s_g^2}. \tag{11.6.7}$$

This shows that the test is based on a comparison of the weighted arithmetic and geometric means of $s_1^2, \ldots, s_k^2, f_i/f$ *being the weights.*

Considering the special case where $f_1 = f_2 = \ldots = f_k = f_0$, so that $f = \sum f_i = kf_0$, we obtain

$$\frac{2{\cdot}3026}{c}kf_0\left(\log s^2 - \frac{1}{k}\sum_{i=1}^{k}\log s_i^2\right), \tag{11.6.8}$$

where

$$c = 1 + \frac{k+1}{3kf_0}. \tag{11.6.9}$$

If the k observed variances and the corresponding value of s^2 are substituted in (11.6.4), we obtain a χ^2-value which may be compared with χ_P^2 for $k-1$ degrees of freedom. If the computed value of χ^2 exceeds $\chi_{\cdot 95}^2$, say, the test hypothesis is rejected. When the k values of s^2 deviate only slightly from one another, we get a low value of χ^2 and therefore the test hypothesis is rejected only if χ^2 is large. A small value of χ^2 may, however, indicate that the assumptions underlying the test are not fulfilled, e. g., because the k empirical variances are not stochastically independent, or because particularly large variances have been arbitrarily discarded, or because observations that deviated markedly from the mean of a certain sample have been omitted, thus artificially diminishing the variance.

The above test should be supplemented by a detailed *investigation of how the k s^2-values are distributed.* If σ^2 were known, it would be possible to calculate k χ^2-values

$$\chi_i^2 = \frac{f_i s_i^2}{\sigma^2} = \frac{SSD_i}{\sigma^2}, \tag{11.6.10}$$

the distribution of which might be compared with the theoretical χ^2-distribution given in Table V. The theoretical variance is, however, not known, and σ^2 is therefore replaced by s^2 so that the k χ^2-values are replaced by k "false" χ^2-values

$$\chi_i'^2 = \frac{f_i s_i^2}{s^2} = \frac{SSD_i}{s^2}. \tag{11.6.11}$$

If $f = \sum f_i$ is large compared with $f_1, f_2, \ldots, f_k$, the distribution of $\chi_i'^2$ is approximately equal to the distribution of χ^2. For each $\chi_i'^2$-value we find in Table V the two values of χ_P^2 between which lies the value of χ'^2 in question and register the corresponding "P-interval". If the test hypothesis is correct, it is to be expected that the number of χ'^2-values lying in suitably chosen P-intervals deviates only at random from the theoretical number, cf. Example 11.7. When the distribution found and the theoretical distribution are compared it is often possible to identify variances causing significant deviations from the test hypothesis.

If the numbers of degrees of freedom for all s_i^2-values are the same, $f_1 = f_2 = \ldots = f_k$, it is possible to make a graphical comparison of the distribution of the observed s_i^2-values with the corresponding χ^2-distribution.

According to (11.1.6) we have

$$\chi_P^2 = \frac{fs_P^2}{\sigma^2}, \tag{11.6.12}$$

so that a diagram where s_P^2 is plotted as abscissa and χ_P^2 as ordinate gives a straight line through the point (0, 0) with slope f/σ^2, cf. the corresponding relation (6.6.6) for the normal distribution, and § 5.6. To make use of this relationship we write the observed values of s_i^2 in order of magnitude, $s_{(1)}^2, s_{(2)}^2, \ldots, s_{(k)}^2$, and compute the corresponding cumulative frequencies, $H\{s^2\}$, cf. (3.5.1). As

$$H\{s^2\} \approx P\{s^2\} \tag{11.6.13}$$

the step-curve

$$\chi^2 = \chi_{H\{s^2\}}^2 \tag{11.6.14}$$

will form an estimate of the straight line

$$\chi^2 = \chi_{P\{s^2\}}^2 = \frac{fs^2}{\sigma^2}, \tag{11.6.15}$$

i. e.,

$$\chi_{H\{s^2\}}^2 \approx \frac{fs^2}{\sigma^2}, \tag{11.6.16}$$

cf. the corresponding mode of reasoning for the normal distribution in § 6.7.

If the observed variances constitute a random sample from an s^2-distributed population with parameters (σ^2, f), the points $\left(s_{(i)}^2, \chi_{\frac{i-1/2}{k}}^2\right)$ will deviate at random from a straight line through (0, 0) with slope f/σ^2. As was the case for the normal distribution, it is possible to construct a probability scale, so that the P-values may be plotted as ordinates instead of χ_P^2, but the scale depends on the number of degrees of freedom.

The values of $\chi_{\frac{i-1/2}{k}}^2$ may be determined by graphical interpolation, plotting the fractiles χ_P^2 from Table V on normal probability paper with logarithmic scale on the abscissa axis and reading the values of χ^2 corresponding to $P = \frac{i - 1/2}{k}$ off the curve passing through the points plotted.

Example 11.6. The hypothesis that $\sigma_1^2 = \ldots = \sigma_4^2$ is tested for the four variances in Example 11.5. According to (11.6.8) and (11.6.9) we compute

$$\sum_{i=1}^{4} \log s_i^2 = 8{\cdot}0614-4\times 6,\ \tfrac{1}{4}\sum_{i=1}^{4} \log s_i^2 = 2{\cdot}0154-6\,,$$

$$\log s^2 = 2{\cdot}0245-6\,,$$

$$c = 1+\frac{5}{3\times 4\times 4} = 1{\cdot}104\,,$$

and

$$\chi^2 \simeq \frac{2{\cdot}303}{1{\cdot}104}\times 4\times 4\times(2{\cdot}0245-2{\cdot}0154) = 0{\cdot}30\,, f = 3\,.$$

For 3 degrees of freedom we find in Table V $\chi^2_{\cdot 025} = 0{\cdot}22$ and $\chi^2_{\cdot 05} = 0{\cdot}35$; thus, our computed value of χ^2 corresponds to $2{\cdot}5\% < P < 5\%$ so that we do not reject the test hypothesis on the basis of the given data.

Example 11.7. Table 11.3 gives the values of $SSD_i = f_i s_i^2$, $f_i = 4$, for the 52 sets of samples mentioned in Example 11.1, taken after the mixing of raw meal. According to the addition theorem for the s^2-distribution, we have

$$s^2 = \frac{\sum_{i=1}^{52} SSD_i}{\sum_{i=1}^{52} f_i} = \frac{4{\cdot}0642}{52\times 4} = 0{\cdot}01954\,.$$

To see if this pooling is justified we use Bartlett's test. According to (11.6.8) and (11.6.9) we compute

$$\frac{1}{k}\sum_{i=1}^{k} \log s_i^2 = \frac{1}{k}\sum_{i=1}^{k} (\log SSD_i - \log f_i)$$

$$= \frac{1}{52}\sum_{i=1}^{52} (\log(10^4 SSD_i) - 4 - \log 4) = 0{\cdot}1350-2\,,$$

$$\log\, s^2 - \frac{1}{k}\sum_{i=1}^{k} \log s_i^2 = \bar{2}{\cdot}2909 - \bar{2}{\cdot}1350 = 0{\cdot}1559\,,$$

$$c = 1+\frac{53}{3\times 52\times 4} = 1{\cdot}085\,,$$

and

$$\chi^2 \simeq \frac{2{\cdot}303}{1{\cdot}085}\times 52\times 4\times 0{\cdot}1559 = 68{\cdot}8, f = 51\,.$$

For 51 degrees of freedom, Table V gives us $\chi^2_{\cdot 95} = 68{\cdot}7$. Thus, it is doubtful whether the pooling is justified or not.

A more detailed analysis of the distribution of the 52 values of s^2 is given in Table 11.3. According to (11.6.11) we compute

$$\chi_i'^2 = \frac{SSD_i}{s^2} = \frac{SSD_i \times 10^4}{195{\cdot}4}.$$

The χ^2_P-intervals corresponding to the values of χ'^2 are found in Table V. For example for $\chi'^2 = 2{\cdot}82$ we find

$$\chi^2_{\cdot 40} = 2{\cdot}75 < 2{\cdot}82 < 3{\cdot}36 = \chi^2_{\cdot 50}\,, \; f = 4\,.$$

These P-intervals have been entered in the fourth column of Table 11.3. The table shows that only one value is extremely large, namely χ'^2 for set No. 20, where $\chi'^2 > \chi^2_{\cdot 9995}$.

TABLE 11.3.

SSD values, in terms of per cent concentration of $CaCO_3$, for 52 sets of 5 samples each, taken from a mixing plant for raw meal.

Set No.	$10^4 \times SSD$	χ'^2	$P(\%)$	Set No.	$10^4 \times SSD$	χ'^2	$P(\%)$
1	551	2·82	40—50	27	2194	11·23	97·5—99
2	1107	5·67	70—80	28	255	1·31	10—20
3	605	3·10	40—50	29	1146	5·86	70—80
4	426	2·18	20—30	30	951	4·87	60—70
5	1325	6·78	80—90	31	189	0·97	5—10
6	305	1·56	10—20	32	1969	10·08	95—97·5
7	601	3·08	40—50	33	173	0·89	5—10
8	845	4·32	60—70	34	152	0·78	5—10
9	1707	8·74	90—95	35	409	2·09	20—30
10	115	0·59	2·5—5	36	1141	5·84	70—80
11	747	3·82	50—60	37	1077	5·51	70—80
12	839	4·29	60—70	38	135	0·69	2·5—5
13	1501	7·68	80—90	39	849	4·34	60—70
14	518	2·65	30—40	40	1524	7·80	90—95
15	235	1·20	10—20	41	341	1·75	20—30
16	525	2·69	30—40	42	831	4·25	60—70
17	89	0·46	1—2·5	43	1039	5·32	70—80
18	790	4·04	50—60	44	615	3·15	40—50
19	62	0·32	1—2·5	45	422	2·16	20—30
20	3981	20·37	99·95—100	46	509	2·60	30—40
21	445	2·28	30—40	47	81	0·41	1—2·5
22	1123	5·75	70—80	48	1098	5·62	70—80
23	239	1·22	10—20	49	1781	9·11	90—95
24	531	2·72	30—40	50	593	3·03	40—50
25	489	2·50	30—40	51	131	0·67	2·5—5
26	491	2·51	30—40	52	845	4·32	60—70

TABLE 11.4.

The distribution of the χ'^2-values from Table 11.3.

$P(\%)$	χ^2_P	Observed number	Theoretical number
0–1	0·000–0·297	0	0·52
1–5	0·297–0·711	6	2·08
5–10	0·711–1·06	3	2·60
10–30	1·06–2·19	8	10·40
30–50	2·19–3·36	12	10·40
50–70	3·36–4·88	8	10·40
70–90	4·88–7·78	9	10·40
90–95	7·78–9·49	3	2·60
95–99	9·49–13·3	2	2·08
99–100	13·3–∞	1	0·52
Total		52	52·00

Table 11.4 gives an enumeration of the χ'^2-values at suitable intervals. For comparison, the theoretical number of values corresponding to each interval has been computed. For instance,

$$P\{1{\cdot}06 < \chi^2 < 2{\cdot}19\} = 30-10 = 20\%$$

means that the expected number of χ^2-values lying between 1·06 and 2·19 is 20% of 52, i. e., 10·4.

A first glance at the observed and the theoretical distributions indicates a reasonable agreement. The most conspicuous deviation is for the interval 1–5%, the observed number here being 6, the theoretical 2·08. This relatively large accumulation of small values of χ'^2 is possibly due to too large a value of the divisor, $s^2 = 0{\cdot}01954$, owing to the possibility that the value from set No. 20 does not really belong to the distribution, which suggests that in this experiment the mixing process has been subject to some disturbing influence which has increased the variance.

The deviation in the interval 1–5% may be examined in greater detail by aid of the binomial distribution. If the distribution in Table 11·4 is considered in two groups only, 0–5% and 5–100%, the corresponding observed numbers are 6 and 46. According to the binominal distribution, cf. (2.1.2), the probability that x of the 52 χ^2-values belong to the interval $(\chi^2_{\cdot 00}, \chi^2_{\cdot 05})$ is

$$p\{x\} = \binom{52}{x} 0{\cdot}05^x \, 0{\cdot}95^{52-x} .$$

Thus, the probability of the observed number, or a larger number, is

$$\sum_{x=6}^{52} \binom{52}{x} 0{\cdot}05^x \, 0{\cdot}95^{52-x} ,$$

which is readily computed from

$$1 - \sum_{x=0}^{5} p\{x\} = 1 - 0{\cdot}955 = 0{\cdot}045 .$$

It is therefore doubtful whether the observed number deviates significantly from the theoretical as it is situated very near to the 95% significance limit.

If, all the same, we try the effect of rejecting set No. 20, we obtain the revised χ'^2-values given in Table 11.5. The corresponding value for s^2 is

$$s^2 = \frac{3{\cdot}6661}{204} = 0{\cdot}01797 .$$

Here, the observed distribution agrees rather better with the theoretical one than that in Table 11.4.

TABLE 11.5.

The distribution of the χ'^2-values for $s^2 = 0{\cdot}01797$.

$P(\%)$	χ^2_P	Observed number	Theoretical number
0 – 1	0·000 – 0·297	0	0·51
1 – 5	0·297 – 0·711	4	2·04
5 – 10	0·711 – 1·06	5	2·55
10 – 30	1·06 – 2·19	5	10·20
30 – 50	2·19 – 3·36	13	10·20
50 – 70	3·36 – 4·88	9	10·20
70 – 90	4·88 – 7·78	9	10·20
90 – 95	7·78 – 9·49	2	2·55
95 – 99	9·49 – 13·3	4	2·04
99 – 100	13·3 – ∞	0	0·51
Total		51	51·00

If χ^2 is computed according to BARTLETT's test—after rejection of set No. 20—we obtain

$$\chi^2 \simeq 58{\cdot}9$$

with 50 degrees of freedom. Table V gives $\chi^2_{\cdot 80} = 58{\cdot}2$ and $\chi^2_{\cdot 90} = 63{\cdot}2$. Thus, after rejecting set No. 20, the computed value of χ^2 is no longer significant. It is, however, doubtful whether set No. 20 should be rejected.

Example 11.8. In a plant where boxes of tablets are packed by hand the variations in weight of the full boxes are controlled at suitable time intervals by random choice of a sample of about 100 boxes, which are weighed individually. For each sample, the weights are normally distributed. The observed variances of 9 samples are given in Table 11.6.

TABLE 11.6.

Comparison of variances for the control of the variation of the weight of boxes of tablets packed by hand.

Sample No.	*SSD*	*f*	s^2	$\log s^2$	$1/f$	χ'^2	*P*(%)
1	8·59	99	0·0868	$\bar{2}$·9385	0·0101	69·5	1–2·5
2	9·67	99	0·0977	$\bar{2}$·9899	0·0101	78·2	5–10
3	14·32	99	0·1446	$\bar{1}$·1602	0·0101	115·9	80–90
4	10·90	99	0·1101	$\bar{1}$·0418	0·0101	88·2	20–30
5	26·13	131	0·1995	$\bar{1}$·3000	0·0076	211·4	99·999–100
6	10·57	131	0·0807	$\bar{2}$·9069	0·0076	85·5	0·1–0·5
7	18·33	139	0·1319	$\bar{1}$·1202	0·0072	148·3	70–80
8	6·14	86	0·0714	$\bar{2}$·8537	0·0116	49·7	0·05–0·1
9	16·43	97	0·1694	$\bar{1}$·2289	0·0103	132·9	99–99·5
	121·08	980			0·0847		

$s^2 = 121{\cdot}08/980 = 0{\cdot}1236. \quad \chi'^2 = SSD/0{\cdot}1236.$

According to the addition theorem for the s^2-distribution, we find $s^2 = 0{\cdot}1236$ with 980 degrees of freedom. According to (11.6.3) and (11.6.4) the following quantities are computed in order to test the identity of the nine theoretical variances:

$$f \log s^2 = 0{\cdot}16-890\,,$$

$$\sum_{i=1}^{9} f_i \log s_i^2 = 0{\cdot}34-914\,,$$

$$c = 1 + \frac{1}{3 \times 8} 0{\cdot}0837 = 1{\cdot}003\,,$$

and

$$\chi^2 \simeq \frac{2{\cdot}303}{1{\cdot}003} 23{\cdot}82 = 54{\cdot}7\,,\ f = 8\,.$$

Table V gives $\chi^2_{\cdot 9995} = 27{\cdot}9$ for $f = 8$, which means that the value found is highly significant; consequently the test hypothesis is rejected.

A more detailed analysis shows that in this case the significance is not due to a single very large variance, but that two variances are large and two small, viz. samples Nos. 5 and 9, and samples Nos. 6 and 8. If the *P*-value for sample No. 5 is computed according to (10.3.5), we have

$$u \simeq \sqrt{2\chi^2} - \sqrt{2f-1} = \sqrt{422{\cdot}8} - \sqrt{261} = 4{\cdot}40\,,$$

and according to Table II

$$P = \Phi(u) = 0{\cdot}9999946\,.$$

11.7. The Distribution of the Standard Deviation.

The fractiles for s are easily derived from the fractiles for χ^2 since

$$s = \sigma\sqrt{\frac{\chi^2}{f}} = \frac{\sigma\chi}{\sqrt{f}}\,, \tag{11.7.1}$$

see (11.1.7). The same transformation may also be used to derive the distribution function for s from the χ^2-distribution, cf. (5.5.8). For many applications, however, it will be sufficiently accurate to replace the exact sampling distribution of s by a normal distribution with the same mean and variance as the s-distribution, see §§ 11.8 and 11.9.

From (11.7.1) we find the mean and variance of s as

$$\mathcal{M}\{s\} = \frac{\sigma}{\sqrt{f}}\mathcal{M}\{\chi\} = \alpha_f\sigma \tag{11.7.2}$$

and

$$\mathcal{V}\{s\} = \frac{\sigma^2}{f}\mathcal{V}\{\chi\} = \beta_f^2\sigma^2\,. \tag{11.7.3}$$

According to (5.8.1) and (10.2.2) we have

$$\mathcal{M}\{\chi^r\} = \frac{1}{2^{\frac{f}{2}}\Gamma\left(\frac{f}{2}\right)}\int_0^\infty \chi^r (\chi^2)^{\frac{f}{2}-1} e^{-\frac{\chi^2}{2}} d(\chi^2)$$

$$= \frac{1}{2^{\frac{f}{2}}\Gamma\left(\frac{f}{2}\right)}\int_0^\infty (\chi^2)^{\frac{f+r}{2}-1} e^{-\frac{\chi^2}{2}} d(\chi^2)$$

$$= \frac{2^{\frac{f+r}{2}}\Gamma\left(\frac{f+r}{2}\right)}{2^{\frac{f}{2}}\Gamma\left(\frac{f}{2}\right)} = 2^{\frac{r}{2}}\frac{\Gamma\left(\frac{f+r}{2}\right)}{\Gamma\left(\frac{f}{2}\right)} \tag{11.7.4}$$

since

$$\frac{1}{2^{\frac{f+r}{2}}\Gamma\left(\frac{f+r}{2}\right)}\int_0^\infty (\chi^2)^{\frac{f+r}{2}-1} e^{-\frac{\chi^2}{2}} d(\chi^2) = 1 .$$

For $r = 1$ we get

$$m\{\chi\} = \sqrt{2}\,\frac{\Gamma\left(\frac{f+1}{2}\right)}{\Gamma\left(\frac{f}{2}\right)} \tag{11.7.5}$$

and for $r = 2$ we find the well-known result $m\{\chi^2\} = f$, so that

$$v\{\chi\} = f-2\left(\frac{\Gamma\left(\frac{f+1}{2}\right)}{\Gamma\left(\frac{f}{2}\right)}\right)^2, \tag{11.7.6}$$

which leads to

$$\alpha_f = \frac{m\{\chi\}}{\sqrt{f}} = \sqrt{\frac{2}{f}}\,\frac{\Gamma\left(\frac{f+1}{2}\right)}{\Gamma\left(\frac{f}{2}\right)} \simeq \sqrt{1-\frac{1}{2f}} \tag{11.7.7}$$

and

$$\beta_f^2 = \frac{v\{\chi\}}{f} = 1-\frac{2}{f}\left(\frac{\Gamma\left(\frac{f+1}{2}\right)}{\Gamma\left(\frac{f}{2}\right)}\right)^2 = 1-\alpha_f^2 \simeq \frac{1}{2f} \, . \tag{11.7.8}$$

Thus, it follows that s *is approximately normally distributed with mean*

$$\boxed{m\{s\} = \alpha_f\sigma \simeq \sqrt{1-\frac{1}{2f}}\;\sigma \simeq \sigma} \tag{11.7.9}$$

and variance

$$\boxed{v\{s\} = \beta_f^2\sigma^2 \simeq \frac{\sigma^2}{2f}} \tag{11.7.10}$$

for large values of f.

This result also follows directly from (10.3.5) when χ^2 is replaced by

fs^2/σ^2. As an approximation to the P-fractile we consequently get

$$s_P \simeq \sigma\left(\sqrt{1-\frac{1}{2f}}+\frac{u_P}{\sqrt{2f}}\right). \tag{11.7.11}$$

11.8. The Distribution of the Coefficient of Variation.

For a normally distributed population with parameters (ξ, σ^2) the coefficient of variation is defined as $\gamma = \sigma/\xi$. The sample coefficient of variation is $c = s/\bar{x}$.

The coefficient of variation may be a useful characteristic of the variability in cases where samples are drawn from populations with different means and variances but with the same coefficient of variation, as is for instance the case for the strength of concrete, see Example 7.7, p. 185. The relative variability may then be controlled by computing the sample coefficients of variation and plotting these on a control chart. For the determination of the control limits we need the distribution of c.

To find an approximation to the distribution of c we remark that c is a function of two variables, $\bar{x}$ and s, and that (1) $\bar{x}$ is normally distributed with parameters $(\xi, \sigma^2/n)$, (2) s is approximately normally distributed with parameters $(\sigma, \sigma^2/2f)$ for large values of f, and (3) $\bar{x}$ and s are stochastically independent. According to (5.17.6) and (9.9.11) we get

$$\boxed{\mathcal{M}\{c\} \simeq \frac{\sigma}{\xi} = \gamma} \tag{11.8.1}$$

and

$$\begin{aligned} \mathcal{C}^2\{c\} &\simeq \mathcal{C}^2\{\bar{x}\}+\mathcal{C}^2\{s\} \\ &\simeq \frac{\sigma^2}{n\xi^2}+\frac{1}{2f} \\ &= \frac{1}{2f}\left(1+2\gamma^2\frac{f}{n}\right) \\ &\simeq \frac{1}{2f}(1+2\gamma^2)\,, \end{aligned} \tag{11.8.2}$$

or

$$\boxed{\mathcal{V}\{c\} \simeq \frac{\gamma^2}{2f}(1+2\gamma^2)\,.} \tag{11.8.3}$$

The coefficient of variation may thus be considered as approximately normally distributed with mean and variance given by (11.8.1) *and* (11.8.3) *for large values of f and small values of* γ. The P-fractile is accordingly found as

$$c_P \simeq \gamma \left(1+\frac{u_P}{\sqrt{2f}}\sqrt{1+2\gamma^2}\right). \tag{11.8.4}$$

A better approximation to the distribution of c may be obtained by solving the equation

$$P\{c < c_P\} = P \tag{11.8.5}$$

or

$$P\{s-\bar{x}c_P < 0\} = P\,, \tag{11.8.6}$$

considering the variable

$$z = s-\bar{x}c_P \tag{11.8.7}$$

as approximately normally distributed with mean

$$\mathcal{M}\{z\} \simeq \sigma-\xi c_P \tag{11.8.8}$$

and variance

$$\mathcal{V}\{z\} \simeq \sigma^2 \left(\frac{1}{2f}+\frac{c_P^2}{n}\right). \tag{11.8.9}$$

The distribution of the variable z will usually deviate less from the normal distribution than the distribution of c because z is a *linear* function of a normally distributed and an approximately normally distributed variable whereas c is the quotient between the same two variables.

By means of the distribution of z we get

$$P\{s-\bar{x}c_P < 0\} \simeq \Phi\left(\frac{\xi c_P-\sigma}{\sigma\sqrt{\dfrac{1}{2f}+\dfrac{c_P^2}{n}}}\right)$$

$$= \Phi\left(\frac{c_P/\gamma-1}{\sqrt{\dfrac{1}{2f}+\dfrac{c_P^2}{n}}}\right) = P \tag{11.8.10}$$

which has the solution

$$\frac{c_P}{\gamma} \simeq 1+u_P\sqrt{\frac{1}{2f}+\frac{c_P^2}{n}}\,. \tag{11.8.11}$$

Solving this equation for c_P we find

$$c_P \simeq \gamma\, \frac{1+u_P\sqrt{\dfrac{1}{2f}\left(1-\dfrac{\gamma^2 u_P^2}{n}\right)+\dfrac{\gamma^2}{n}}}{1-\dfrac{\gamma^2 u_P^2}{n}}\,. \tag{11.8.12}$$

Example 11.9. On the basis of experience for a long time a laboratory has found an estimate, 2·4%, of the population coefficient of variation for the compressive strength of samples of concrete made and cured under standard conditions. Accepting this estimate as a standard, what upper limit for future sample coefficients of variation should be used for $n = 5$? For $\gamma = 0{\cdot}024$, $f = 4$ and $P = 0{\cdot}975$, say, we first find $u_{\cdot 975} = 1{\cdot}96$ and

$$1 - \frac{\gamma^2 u_P^2}{n} = 1 - \frac{0{\cdot}0^3576 \times 3{\cdot}84}{5} = 1{\cdot}000$$

so that (11.8.12) leads to

$$c_{\cdot 975} \simeq 0{\cdot}024(1 + 1{\cdot}96\sqrt{0{\cdot}125})$$
$$= 0{\cdot}024 \times 1{\cdot}69 = 0{\cdot}041 \,.$$

11.9. The Control of Proportion Defective by Means of a Linear Function of $\bar{x}$ and s.

The quality of an industrial product may often be judged by measuring a quality characteristic x of each item and studying the distribution of x. If the distribution of x for a population (a lot) of items is normal, then the quality variation as expressed by this single characteristic is fully described by means of the two parameters ξ and σ. (In practice where the lots are finite the assumption of normality can be only approximately fulfilled. It is assumed in the following that the sample contains at most 10% of the items of the lot so that the lot is large as compared to the sample). How to control that ξ and σ do not deviate essentially from a given standard by means of sampling inspection has been discussed in § 9.3 and § 11.2. Sometimes, however, we are mainly interested in controlling the proportion defective in the populations, whereas the values of ξ and σ are of less importance.

Let us suppose that *all items are classified as defective or nondefective on the basis of the measurable quality characteristic* x, for instance so that an item is classified as defective if $x > L$, where L is some limit specified on the basis of technical and commercial considerations. If the proportion defective θ of a population is known but ξ and σ are unknown, then there is an infinity of populations with different values of ξ and σ which may all lead to the given proportion defective. These populations are defined by the equation

$$P\{x > L\} = 1 - \Phi\left(\frac{L-\xi}{\sigma}\right) = \theta \qquad (11.9.1)$$

which has the solution

$$\frac{L-\xi}{\sigma} = u_{1-\theta} \qquad (11.9.2)$$

or
$$\xi+\sigma u_{1-\theta} = L\,. \tag{11.9.3}$$

A population with a large mean and a small standard deviation may thus give the same proportion defective as a population with a small mean and a large standard deviation.

Suppose now that a large number of lots with different ξ's and σ's are submitted for inspection and that we want to accept all lots with a proportion defective less than or equal to θ_0 and reject all lots with a proportion defective larger than θ_0. The criterion $\theta > \theta_0$ or

$$P\{x > L\} > \theta_0 \tag{11.9.4}$$

is the same as
$$\xi+\sigma u_{1-\theta_0} > L \tag{11.9.5}$$

where ξ and σ denote the parameters of a population with a proportion defective equal to θ, i. e., ξ and σ satisfy (11.9.3). Instead of classifying the lots directly on the basis of their proportions defective we may use the linear function $\xi+\sigma u_{1-\theta_0}$ of ξ and σ, rejecting all lots for which $\xi+\sigma u_{1-\theta_0}$ is larger than L and accepting all other lots.

Usually, however, it is not feasible or economical to inspect all the items of a lot to decide whether or not it should be accepted, but the decision must be based on a random sample of items from the lot. From the sample mean and standard deviation we may compute $\bar{x}+s u_{1-\theta_0}$ as an estimate of $\xi+\sigma u_{1-\theta_0}$. In repeated sampling from the same population this estimate will be approximately normally distributed about $\xi+\sigma u_{1-\theta_0}$ so that we cannot be sure that $\bar{x}+s u_{1-\theta_0}$ will be larger than L even if $\xi+\sigma u_{1-\theta_0} > L$. We therefore have to investigate what multiple, t_0 say, of the sample standard deviation we should choose to be reasonably sure that the criterion

$$\bar{x}+s t_0 > L \tag{11.9.6}$$

will lead to the right decision.

Before answering this question we shall find the distribution of the variable
$$z = \bar{x}+s t_0 \tag{11.9.7}$$

where t_0 denotes a constant. As in § 11.8 we make use of the approximation formula for the distribution of s so that z will be approximately normally distributed with parameters

$$\mathcal{M}\{z\} \simeq \xi+\sigma t_0 \tag{11.9.8}$$

and
$$\mathcal{V}\{z\} \simeq \sigma^2\left(\frac{1}{n}+\frac{t_0^2}{2f}\right) \tag{11.9.9}$$

for $n > 4$, say.

Let us suppose that the proportion defective of the population sampled is θ so that $\xi+\sigma u_{1-\theta} = L$. From each sample we compute $\bar{x}$ and s and further

$$t = \frac{L-\bar{x}}{s} \tag{11.9.10}$$

which vary at random about the "true" value $u_{1-\theta}$. The P-fractile for t is found from the relation

$$P\{t < t_P\} = P\left\{\frac{L-\bar{x}}{s} < t_P\right\} = P$$

which leads to

$$P\{\bar{x}+st_P > L\} \simeq \Phi\left(\frac{\xi+\sigma t_P-L}{\sigma\sqrt{\frac{1}{n}+\frac{t_P^2}{2f}}}\right) = P\,,$$

according to (11.9.8) and (11.9.9). Introducing $\xi+\sigma u_{1-\theta}$ instead of L we get

$$P\{t < t_P\} \simeq \Phi\left(\frac{t_P-u_{1-\theta}}{\sqrt{\frac{1}{n}+\frac{t_P^2}{2f}}}\right) = P \tag{11.9.11}$$

which has the solution

$$t_P-u_{1-\theta} \simeq u_P\sqrt{\frac{1}{n}+\frac{t_P^2}{2f}}\,. \tag{11.9.12}$$

Squaring both sides of this and solving with respect to t_P we get

$$t_P = t_P(\theta) \simeq \frac{u_{1-\theta}+u_P\sqrt{\frac{1}{n}\left(1-\frac{u_P^2}{2f}\right)+\frac{u_{1-\theta}^2}{2f}}}{1-\frac{u_P^2}{2f}}\,. \tag{11.9.13}$$

By means of this formula we may compute the fractiles for t and set up the control limits corresponding to any preassigned proportion defective θ and suitably chosen values of P. It will be noted that *bad quality leads to a small value of* t since $\theta > \theta_0$ involves $u_{1-\theta} < u_{1-\theta_0}$.

To test the hypothesis $\theta = \theta_0$ against the alternative hypothesis $\theta > \theta_0$ at the α level of significance we have to determine a critical region $\bar{x}+st_0 > L$ so that

$$P\{\bar{x}+st_0 > L;\ \theta_0\} = \alpha \tag{11.9.14}$$

or

$$P\left\{\frac{L-\bar{x}}{s} < t_0\,;\ \theta_0\right\} = P\{t < t_0;\ \theta_0\} = \alpha\,.$$

The solution of this equation is $t_0 = t_\alpha(\theta_0)$, see (11.9.13).

The power function is found as

$$\pi(\theta) = P\,\{\bar{x}+st_\alpha(\theta_0) > L;\ \theta\}$$

$$\simeq \Phi\left(\frac{\xi+\sigma t_\alpha(\theta_0)-L}{\sigma\sqrt{\dfrac{1}{n}+\dfrac{t_\alpha^2(\theta_0)}{2f}}}\right)$$

$$= \Phi\left(\frac{t_\alpha(\theta_0)-u_{1-\theta}}{\sqrt{\dfrac{1}{n}+\dfrac{t_\alpha^2(\theta_0)}{2f}}}\right) \tag{11.9.15}$$

since $L = \xi+\sigma u_{1-\theta}$. Putting $\pi(\theta) = 1-\beta$ we get

$$u_{1-\theta} = t_\alpha(\theta_0)+u_\beta\sqrt{\frac{1}{n}+\frac{t_\alpha^2(\theta_0)}{2f}}\ . \tag{11.9.16}$$

The power function may thus be tabulated by computing $u_{1-\theta}$ for suitably chosen values of $\beta = 1-\pi$ and finding θ from $u_{1-\theta}$ by means of Table II. It will be noticed that $\pi(\theta)$ is an increasing function of θ and further that the power curve will be a straight line when plotted on graph paper with normal probability scales on both axes. Thus we need to compute only two values of $\pi(\theta)$ when we use this special graph paper.

Usually there will not be any sharp distinction between good and bad lots. It may, however, be possible from practical considerations to set up two numbers θ_1 and θ_2, $\theta_1 < \theta_2$, to represent good and bad lots, respectively, so that if $\theta \leq \theta_1$ we want to be reasonably sure that the sampling procedure will lead to the acceptance of the lot, and if $\theta \geq \theta_2$ it will lead to the rejection of the lot. If the probability of rejecting a lot with a proportion defective of θ_1 is chosen as α (the producer's risk) and the probability of accepting a lot with a proportion defective of θ_2 is chosen as β (the consumer's risk), then we get the following two equations for the determination of the critical region and the sample size:

$$P\{\bar{x}+st_0 > L;\ \theta = \theta_1\} = \alpha \tag{11.9.17}$$

and

$$P\{\bar{x}+st_0 > L;\ \theta = \theta_2\} = 1-\beta\ . \tag{11.9.18}$$

By analogy with (11.9.12) we obtain

$$t_0-u_{1-\theta_1} = u_\alpha\sqrt{\frac{1}{n}+\frac{t_0^2}{2f}} \tag{11.9.19}$$

and

$$t_0-u_{1-\theta_2} = u_{1-\beta}\sqrt{\frac{1}{n}+\frac{t_0^2}{2f}}\ . \tag{11.9.20}$$

Dividing (11.9.19) by (11.9.20) and solving for t_0 we find

$$t_0 = \frac{u_{1-\alpha}u_{1-\theta_2}+u_{1-\beta}u_{1-\theta_1}}{u_{1-\alpha}+u_{1-\beta}}, \tag{11.9.21}$$

i. e., t_0 is the weighted mean of the two "true" values $u_{1-\theta_1}$ and $u_{1-\theta_2}$. If $\alpha = \beta$ we get $t_0 = \frac{1}{2}(u_{1-\theta_1}+u_{1-\theta_2})$.

Subtracting (11.9.19) from (11.9.20) we get

$$u_{1-\theta_1}-u_{1-\theta_2} = (u_{1-\beta}-u_{\alpha})\sqrt{\frac{1}{n}+\frac{t_0^2}{2f}}$$

$$\simeq (u_{1-\beta}-u_{\alpha})\sqrt{\frac{1}{n}}\sqrt{1+\frac{t_0^2}{2}},$$

where $f = n-1$ has been replaced by n. Solving for n we obtain

$$n \simeq \left(1+\frac{t_0^2}{2}\right)\left(\frac{u_{1-\alpha}+u_{1-\beta}}{u_{1-\theta_1}-u_{1-\theta_2}}\right)^2. \tag{11.9.22}$$

The power function of the test becomes

$$\pi(\theta) \simeq \Phi\left(\frac{t_0-u_{1-\theta}}{\sqrt{\frac{1}{n}+\frac{t_0^2}{2f}}}\right) \tag{11.9.23}$$

analogous to (11.9.15). It is seen that α is the producer's maximum risk, i. e., lots with proportions defective less than θ_1 will be rejected with a probability less than α, and, similarly, that β is the consumer's maximum risk.

If a defective item is defined as an item for which $x < L$ (instead of $x > L$) then an analogous sampling plan may be developed. The criterion for rejecting a lot now becomes $\bar{x}-st_0 < L$, where t_0 and n may be determined from (11.9.21) and (11.9.22).

The proportion defective in the lots may also be controlled by computing the proportion defective in the samples. Usually, it is much easier to decide whether an item is defective or not than to measure each item and it is therefore less expensive to determine the percent defective than to determine $\bar{x}+st_0$ for a sample of given size. Measurement of each item and computation of $\bar{x}+st_0$ will, however, give more precise information about the proportion defective in the lot than will the proportion defective in the sample and this fact has to be taken into account when choosing between the two types of sampling plans. This problem will be discussed further in connection with sampling plans based on percent defective, cf. § 21.12.

The above results are based on three papers:

(1) W. J. JENNETT and B. L. WELCH: *The Control of Proportion Defective as Judged by a Single Quality Characteristic Varying on a Continuous Scale*, Suppl. Journ. Roy. Stat. Soc., 6, 1939, 80–88.

(2) N. L. JOHNSON and B. L. WELCH: *Applications of the Non-Central t-Distribution*, Biometrika, 31, 1940, 362–389.

(3) W. ALLEN WALLIS: *Use of Variables in Acceptance Inspection for Percent Defective*, Chapter 1 (pp. 7–93) in *Techniques of Statistical Analysis* by the STATISTICAL RESEARCH GROUP, COLUMBIA UNIVERSITY, McGraw-Hill, New York, 1947.

The second paper contains tables from which the exact distribution of t may be derived. The third paper contains a detailed exposition of the theory and applications of the test $\bar{x}+st_0 > L$ together with tables of t_0 and n for given θ_1, θ_2, α, and β, and also a discussion of the corresponding two-sided test where nondefective items are defined by $L_1 < x < L_2$.

Example 11.10. Let us suppose that rivets with a diameter larger than 13·575 mm are classified as defective and that we want to control the proportion defective in a production of rivets with a standard proportion defective of $\theta_0 = 0{\cdot}06$.

Since $u_{\cdot 94} = 1{\cdot}555$ the lots of standard quality must satisfy the equation

$$\frac{13{\cdot}575-\xi}{\sigma} = 1{\cdot}555 \, .$$

The control is carried out by taking samples of 10 rivets each, computing

$$t = \frac{13{\cdot}575-\bar{x}}{s}$$

for each sample, and plotting the t-values on a control chart. The control limits may be found from (11.9.13). For $P = 97{\cdot}5\%$, for example, we get $u_{\cdot 975} = 1{\cdot}960$,

$$1-\frac{1{\cdot}960^2}{18} = 0{\cdot}7866, \text{ and } \frac{1{\cdot}555^2}{18} = 0{\cdot}1343,$$

so that

$$t_{\cdot 975} = \frac{1{\cdot}555+1{\cdot}960\sqrt{0{\cdot}1\times 0{\cdot}7866+0{\cdot}1343}}{0{\cdot}7866} = 3{\cdot}13 \, .$$

Similarly, we obtain $t_{\cdot 025} = 0{\cdot}83$, $t_{\cdot 999} = 6{\cdot}11$, and $t_{\cdot 001} = 0{\cdot}51$.

Reading the 200 observations in Table 3.1 as 20 samples of 10 observations each and computing the means and standard deviations for each sample we get the results given in Table 11.7. The values of t are plotted on the control chart in Fig. 11.5. Only one t-value falls outside the 95% limits which is exactly what should be expected for 20 samples.

TABLE 11.7.

Computation of $t = (13{\cdot}575 - \bar{x})/s$ for 20 samples of 10 rivets each.

Sample No.	$\bar{x}$	s	t	Number of defectives
1	13·404	0·0782	2·19	0
2	13·446	0·1398	0·92	1
3	13·426	0·0980	1·52	0
4	13·376	0·1008	1·97	1
5	13·381	0·1083	1·79	0
6	13·450	0·1182	1·06	1
7	13·434	0·0776	1·82	1
8	13·442	0·1370	0·97	2
9	13·470	0·1281	0·82	2
10	13·414	0·1123	1·43	1
11	13·450	0·1131	1·11	1
12	13·401	0·1041	1·67	1
13	13·380	0·0907	2·15	0
14	13·445	0·0814	1·60	1
15	13·384	0·1343	1·42	1
16	13·407	0·1322	1·27	1
17	13·417	0·0508	3·11	0
18	13·405	0·1438	1·18	2
19	13·433	0·1095	1·30	0
20	13·364	0·0850	2·48	0

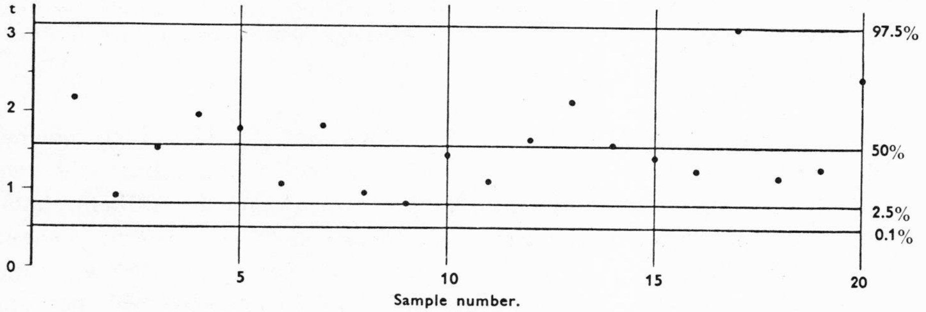

FIG. 11.5. Control chart for $t = (13{\cdot}575 - \bar{x})/s$.

For each sample the number of defective rivets is shown in Table 11.7. It will be noted that a large number of defective rivets corresponds to a small value of t and further that the proportion defective in a sample is a much coarser expression for the quality of the lot than is t.

The power of the test corresponding to the control chart is

$$\pi(\theta) = P\{t < t_{\cdot 025};\ \theta\} + P\{t > t_{\cdot 975};\ \theta\}$$

$$\simeq \Phi\left(\frac{0{\cdot}83 - u_{1-\theta}}{0{\cdot}372}\right) + \Phi\left(\frac{u_{1-\theta} - 3{\cdot}13}{0{\cdot}803}\right)$$

where

$$\sqrt{0 \cdot 1+\frac{0 \cdot 83^2}{18}}=0 \cdot 372 \quad \text{and} \quad \sqrt{0 \cdot 1+\frac{3 \cdot 13^2}{18}}=0 \cdot 803\,,$$

see (11.9.11).

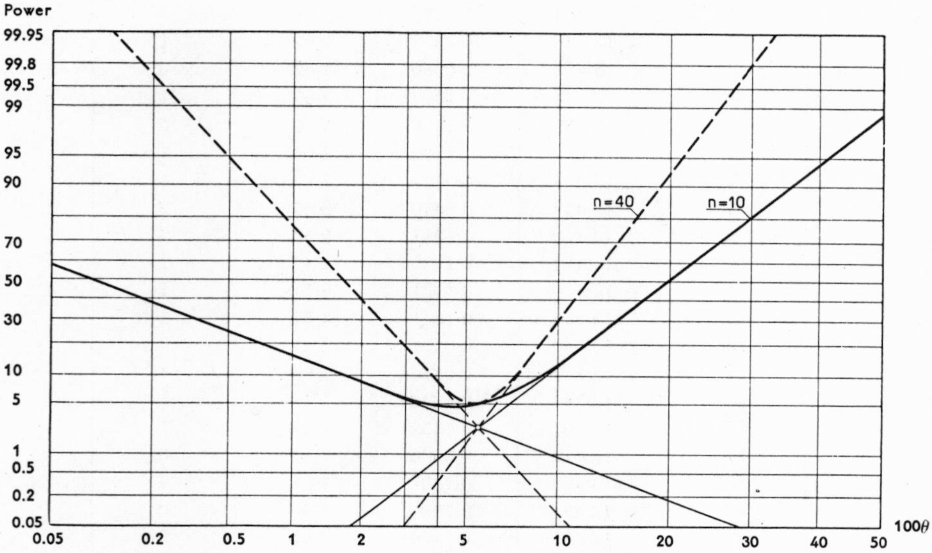

FIG. 11.6. Power curves for the test based on $t = (13 \cdot 575 - \bar{x})/s$ for $f = 10$ and $f = 40$ at the 5% level of significance. Proportion defective $= \theta$, $\theta_0 = 0 \cdot 06$. Normal probability scales on both axes.

The corresponding power curve is drawn on Fig. 11.6 with normal probability scales on both axes. The curve is constructed from the two straight lines representing the two terms of $\pi(\theta)$. We know beforehand that each term takes on the value 0·025 for $\theta = 0 \cdot 06$. Further, it is easily seen that the first term takes on the value 0·50 for $u_{1-\theta} = 0 \cdot 83$ and that the second term takes on this value for $u_{1-\theta} = 3 \cdot 13$. The corresponding values of θ are found to be $\Phi(-0 \cdot 83) = 0 \cdot 20$ and $\Phi(-3 \cdot 13) = 0 \cdot 0^3 87$, see Table II. Thus, the first line is determined from the points (0·06, 0·025) and (0·20, 0·50), and the second line from (0·06, 0·025) and ($0 \cdot 0^3 87$, 0·50). After drawing the lines, the two Φ-values (the ordinates in %) corresponding to a given abscissa are read off and the sum is plotted as ordinate to the given abscissa thus giving a point on the power curve.

It will be noted that for $\theta = 0 \cdot 10$ the power is 0·12, i. e., 12% of the sample points are expected to lie outside the 95% control limits if the submitted lots all have 10% defective items (11% of the t-values will fall below the lower limit and 1% above the upper limit). If the proportion defective in the submitted lots is 20% then we shall expect 50% of the points to fall below the lower limit.

If the discriminating power as described by the above power curve is not satisfactory we have to increase the sample size. Fig. 11.6 also shows the power curve corresponding to the 95% control limits for $n = 40$. The limits are $t_{\cdot 025} = 1{\cdot}15$ and $t_{\cdot 975} = 2{\cdot}12$. It is seen that the power function now takes on the value 50% for $\theta = \Phi(-2{\cdot}12) = 0{\cdot}017$ and for $\theta = \Phi(-1{\cdot}15) = 0{\cdot}125$.

Let us now construct a sampling plan with the following properties: For each sample we compute $\bar{x}+st_0$ and if this quantity is larger than 13·575 mm we reject the lot. Further, we require that at most 5% of submitted lots with a proportion defective less than 3% shall be rejected and that at most 10% of submitted lots with a proportion defective larger than 8% shall be accepted, i.e., we require that $\pi(0{\cdot}03)=0{\cdot}05$ and $\pi(0{\cdot}08)=0{\cdot}90$. From this specification we may determine t_0 and n according to (11.9.21) and (11.9.22).

We have

$$\theta_1 = 0{\cdot}03, \quad u_{\cdot 97} = 1{\cdot}881\,,$$

$$\theta_2 = 0{\cdot}08, \quad u_{\cdot 92} = 1{\cdot}405\,,$$

and

$$\alpha = 0{\cdot}05, \quad u_{\cdot 95} = 1{\cdot}645\,,$$

$$\beta = 0{\cdot}10, \quad u_{\cdot 90} = 1{\cdot}282\,.$$

According to (11.9.21) we find

$$t_0 = \frac{1{\cdot}645\times 1{\cdot}405+1{\cdot}282\times 1{\cdot}881}{1{\cdot}645+1{\cdot}282} = 1{\cdot}613\,.$$

Introducing this result into (11.9.22) we obtain

$$n \simeq \left(1+\frac{1{\cdot}613^2}{2}\right)\left(\frac{1{\cdot}645+1{\cdot}282}{1{\cdot}881-1{\cdot}405}\right)^2 = 87\,.$$

Thus we get the discriminating power required by taking a sample of 87 items from each lot, computing $\bar{x}+1{\cdot}613s$, and rejecting all lots for which this quantity exceeds 13·575 mm. The complete power curve may be obtained by drawing the straight line through the two given points (0·03, 0·05) and (0·08, 0·90) on graph paper with probability scales on both axes.

11.10. Tolerance Limits.

In the control of the quality of an industrial product a pair of fixed limits, L_1 and L_2, are often specified together with the requirement that only a small proportion of the produced items must fall outside these limits. L_1 and L_2 are called *tolerance limits.*

Assuming that the production process is stable, L_1 and L_2 in statistical terminology correspond to certain fractiles in the distribution of the quality characteristic in question. Consider a normally distributed population with parameters (ξ, σ^2). Corresponding to a given number $1-2\varepsilon$ we find the *symmetrical tolerance limits* as $\xi \pm \sigma u_{1-\varepsilon}$ since

$$\int_{\xi-\sigma u_{1-\varepsilon}}^{\xi+\sigma u_{1-\varepsilon}} p\{x\}dx = 1-2\varepsilon\,, \tag{11.10.1}$$

where $p\{x\}$ represents the normal distribution function. For example, $1-2\varepsilon = 0{\cdot}95$ gives the tolerance limits $\xi \pm 1{\cdot}96\sigma$, i. e., 95% of the population lie between these limits.

Usually we do not know ξ and σ but only the estimates $\bar{x}$ and s. If we replace ξ and σ by $\bar{x}$ and s we get $\bar{x} \pm 1{\cdot}96s$. In repeated sampling from the same population these limits will vary about the population tolerance limits and for some samples the limits will include less than 95% of the population and for some other samples more than 95%. To be reasonably sure that at least 95% of the population lie between the sample tolerance limits we must find a number $l > 1{\cdot}96$ such that there is a high probability that $\bar{x} \pm sl$ will include at least 95% of the population. The area under the normal distribution curve corresponding to the above limits is

$$A = A(\bar{x}, s, l) = \int_{\bar{x}-sl}^{\bar{x}+sl} p\{x\}dx\,. \tag{11.10.2}$$

For any given l this area will vary from sample to sample because $\bar{x}$ and s vary. From the distribution of A we can find the probability that at least $1-2\varepsilon$ of the population is included between the sample tolerance limits. For a given value of this probability P we may determine l from the equation

$$P\{A > 1-2\varepsilon\} = P\,. \tag{11.10.3}$$

Let us first consider the simpler problem of finding a one-sided tolerance limit $\bar{x}+st_0$ so that the probability is P that at least the proportion $1-\theta$ of the population lies below $\bar{x}+st_0$. In this case the population tolerance limit is $\xi+\sigma u_{1-\theta}$ since

$$\int_{-\infty}^{\xi+\sigma u_{1-\theta}} p\{x\}dx = 1-\theta\,.$$

If $\bar{x}+st_0 > \xi+\sigma u_{1-\theta}$, then

$$A(\bar{x}, s, t_0) = \int_{-\infty}^{\bar{x}+st_0} p\{x\}dx > 1-\theta\,. \tag{11.10.4}$$

It therefore follows that

$$P\{A > 1-\theta\} = P\{\bar{x}+st_0 > \xi+\sigma u_{1-\theta}\}$$

$$\simeq \Phi\left(\frac{t_0-u_{1-\theta}}{\sqrt{\dfrac{1}{n}+\dfrac{t_0^2}{2f}}}\right) \tag{11.10.5}$$

according to (11.9.11). Setting $P\{\mathrm{A} > 1-\theta\} = P$ we obtain $t_0 = t_P(\theta)$, see (11.9.13). Thus we may state with *a confidence coefficient of P* that *the one-sided tolerance limit corresponding to the proportion* $1-\theta$ *of the population is* $\bar{x}+st_P(\theta)$. The meaning of this statement is as follows: Consider a large number of samples of n observations each from the same population. For each sample $\bar{x}+st_P(\theta)$ and the corresponding $A(\bar{x}, s, t_P)$ are computed. In the long run the proportion P of the number of samples will give A's larger than $1-\theta$.

In practice we have usually only one sample and if we want to be practically certain that the tolerance limit computed from this sample shall exceed the population tolerance limit we must choose a high value for P.

Consider now the problem of finding two-sided tolerance limits when ξ is known but σ unknown. The sample tolerance limits then are $\xi \pm sl$. The condition that $\xi \pm sl$ will include at least $1-2\varepsilon$ of the population is that $sl > \sigma u_{1-\varepsilon}$ or

$$\frac{s}{\sigma} > \frac{u_{1-\varepsilon}}{l} \tag{11.10.6}$$

so that

$$P\{A > 1-2\varepsilon\} = P\left\{\frac{s}{\sigma} > \frac{u_{1-\varepsilon}}{l}\right\}. \tag{11.10.7}$$

Since $s/\sigma = \sqrt{\chi^2/f}$ the equation

$$P\left\{\frac{s}{\sigma} > \frac{u_{1-\varepsilon}}{l}\right\} = P \tag{11.10.8}$$

has the solution

$$\sqrt{\frac{\chi^2_{1-P}}{f}} = \frac{u_{1-\varepsilon}}{l}, \tag{11.10.9}$$

which leads to

$$l = u_{1-\varepsilon}\sqrt{\frac{f}{\chi^2_{1-P}}}. \tag{11.10.10}$$

Thus the sample tolerance limits are

$$\xi \pm su_{1-\varepsilon}\sqrt{\frac{f}{\chi^2_{1-P}}}, \tag{11.10.11}$$

i. e., there is a probability P that at least $1-2\varepsilon$ of the population is included between these limits in repeated sampling.

In the general case where both ξ and σ are unknown it may be shown that the corresponding value of l is given approximately by

$$\boxed{l \simeq u_{1-\varepsilon} \sqrt{\frac{f}{\chi^2_{1-P}}} \left(1+\frac{1}{2n}\right),} \tag{11.10.12}$$

i. e., the result for ξ known multiplied by $1+\dfrac{1}{2n}$.

On the basis of a given sample it is thus possible to predict within what limits future observations from the same population will lie.

The above results may also be used to calculate limits for future sample means, the tolerance limits for a sample mean based on m observations being $\bar{x}\pm l\dfrac{s}{\sqrt{m}}$, where $\bar{x}$, s, and l are found from the given sample of n observations.

Table 11.8 shows some values of l according to (11.10.12). It will be noted that l is considerably larger than $u_{1-\varepsilon}$ for small values of n. If, for example, we have a sample of 21 observations and want to be 99% confident that the tolerance limits will include at least 95% of the population, we find $l = 3{\cdot}12$ so that the tolerance limits become $\bar{x}\pm 3{\cdot}12s$. For $n = 200$ we get $l = 2{\cdot}22$.

The results of this paragraph are based on the following papers:

(1) W. A. Shewhart: *Statistical Method from the Viewpoint of Quality Control*, Chapter 2 (pp. 50–79), The Graduate School, The Department of Agriculture, Washington, 1939.

(2) A. Wald and J. Wolfowitz: *Tolerance Limits for a Normal Distribution*, Ann. Math. Stat., 17, 1946, 208–215.

(3) A. H. Bowker: *Computation of Factors for Tolerance Limits on a Normal Distribution When the Sample is Large*, Ann. Math. Stat., 17, 1946, 238–240.

(4) A. H. Bowker: *Tolerance Limits for Normal Distributions*, Chapter 2 in *Techniques of Statistical Analysis* by the Statistical Research Group, Columbia University, McGraw-Hill, New York, 1947.

In (1) a thorough discussion of the concept of tolerance limits is given. (4) contains an extensive table of values of l for $P = 0{\cdot}75$, $0{\cdot}90$, $0{\cdot}95$, $0{\cdot}99$ and $1-2\varepsilon = 0{\cdot}75$, $0{\cdot}90$, $0{\cdot}95$, $0{\cdot}99$, $0{\cdot}999$.

Example 11.11. The sample of 500 diameters given in Table 4.4 may be used to set up a pair of 95% tolerance limits, say. Choosing a confidence

TABLE 11.8.

$$\text{Table of } l = u_{1-\varepsilon}\sqrt{\frac{f}{\chi^2_{1-P}}}\left(1+\frac{1}{2n}\right).$$

P denotes the probability that at least $1-2\varepsilon$ of the normal distribution will be included between the tolerance limits $\bar{x}\pm sl$, where $\bar{x}$ and s are the mean and the standard deviation of the n observations and $f = n-1$.

Degrees of freedom f	Confidence coefficient $P = 0{\cdot}90$				Confidence coefficient $P = 0{\cdot}95$				Confidence coefficient $P = 0{\cdot}99$			
	Proportion of distribution $1-2\varepsilon$				Proportion of distribution $1-2\varepsilon$				Proportion of distribution $1-2\varepsilon$			
	0·90	0·95	0·99	0·999	0·90	0·95	0·99	0·999	0·90	0·95	0·99	0·999
4	3·51	4·18	5·49	7·02	4·29	5·11	6·72	8·58	6·64	7·92	10·40	13·29
5	3·14	3·74	4·92	6·28	3·72	4·44	5·83	7·45	5·35	6·38	8·38	10·71
6	2·91	3·47	4·55	5·82	3·38	4·02	5·29	6·75	4·62	5·51	7·24	9·25
7	2·75	3·27	4·30	5·50	3·14	3·74	4·92	6·28	4·15	4·95	6·51	8·31
8	2·63	3·13	4·12	5·26	2·97	3·54	4·65	5·94	3·83	4·56	5·98	7·66
9	2·54	3·02	3·97	5·08	2·84	3·39	4·45	5·69	3·59	4·27	5·62	7·17
10	2·47	2·94	3·86	4·93	2·74	3·26	4·29	5·48	3·40	4·05	5·32	6·80
12	2·36	2·81	3·69	4·72	2·59	3·08	4·05	5·18	3·13	3·73	4·90	6·26
14	2·28	2·72	3·57	4·56	2·49	2·96	3·89	4·96	2·95	3·52	4·61	5·89
16	2·22	2·65	3·48	4·44	2·40	2·86	3·76	4·80	2·81	3·35	4·40	5·62
18	2·17	2·59	3·40	4·35	2·34	2·79	3·66	4·68	2·70	3·22	4·23	5·41
20	2·14	2·54	3·34	4·27	2·29	2·72	3·58	4·57	2·62	3·12	4·10	5·24
25	2·07	2·46	3·23	4·13	2·19	2·61	3·43	4·39	2·47	2·94	3·87	4·94
30	2·02	2·40	3·16	4·04	2·13	2·54	3·33	4·26	2·37	2·82	3·71	4·74
40	1·95	2·33	3·06	3·91	2·05	2·44	3·20	4·09	2·24	2·67	3·50	4·47
50	1·91	2·28	3·00	3·83	1·99	2·37	3·12	3·99	2·16	2·57	3·38	4·31
60	1·89	2·25	2·95	3·77	1·96	2·33	3·06	3·91	2·10	2·50	3·29	4·20
70	1·86	2·22	2·92	3·73	1·93	2·30	3·02	3·85	2·06	2·45	3·22	4·11
80	1·85	2·20	2·89	3·69	1·91	2·27	2·98	3·81	2·02	2·41	3·17	4·05
90	1·83	2·18	2·87	3·67	1·89	2·25	2·96	3·78	2·00	2·38	3·13	3·99
100	1·82	2·17	2·85	3·64	1·87	2·23	2·93	3·75	1·98	2·35	3·09	3·95
200	1·76	2·10	2·76	3·53	1·80	2·14	2·82	3·60	1·87	2·22	2·92	3·73
300	1·74	2·07	2·73	3·48	1·77	2·11	2·77	3·54	1·82	2·17	2·85	3·64
400	1·73	2·06	2·70	3·45	1·75	2·08	2·74	3·50	1·79	2·14	2·81	3·59
500	1·72	2·05	2·69	3·43	1·74	2·07	2·72	3·48	1·78	2·12	2·78	3·55
600	1·71	2·04	2·68	3·42	1·73	2·06	2·71	3·46	1·76	2·10	2·76	3·53
800	1·70	2·03	2·66	3·40	1·72	2·05	2·69	3·43	1·75	2·08	2·74	3·50
1000	1·70	2·02	2·65	3·39	1·71	2·04	2·68	3·42	1·74	2·07	2·72	3·47
∞	1·64	1·96	2·58	3·29	1·64	1·96	2·58	3·29	1·64	1·96	2·58	3·29

coefficient of 0·99 we find from Table 11.8 $l = 2{\cdot}12$ which leads to the limits $13{\cdot}426 \pm 0{\cdot}115 \times 2{\cdot}12 = (13{\cdot}182,\ 13{\cdot}670)$ mm. If the production process remains stable we may thus expect that at least 95% of the future produced rivets will have diameters between 13·182 and 13·670 mm.

Example 11.12. In testing the quality of a batch of certain materials, for example the strength of cement, it is customary to pick out at random a small number of samples from the batch and determine the strength of the test-cubes made from these samples. From the strengths of the test cubes we may find a tolerance limit so that, for example, we are 90% confident that at most 1% of all test-cubes which could be made from this batch will have strengths less than the limit found.

In the setting up of specifications of the strength it seems reasonable therefore to require that a batch will be classified as acceptable only if the sample tolerance limit exceeds a preassigned standard strength.

According to (11.9.13) the condition for accepting a batch will be that $\bar{x}-st_P(\theta) > L$, where L is the preassigned strength below which at most the proportion θ of the population is allowed to fall. In this manner both the sample mean and the standard deviation as well as the number of test-cubes will influence the tolerance limits.

Choosing for example $\theta = 0{\cdot}01$, $P = 0{\cdot}90$, and $n = 6$ we find $t = 4{\cdot}05$ from (11.9.13), so that the criterion for accepting the batch becomes $\bar{x}-4{\cdot}05s > L$. For $n = 10$ we require $\bar{x}-3{\cdot}44s > L$.

11.11. The Distribution of the Estimate of σ in a Truncated Normal Distribution.

For a truncated or a censored normal distribution, the distribution of the estimate s, see (6.9.9) and (6.9.13), of the parameter σ is very complicated, but for large values of n the estimate will, according to the properties of maximum likelihood estimates, be approximately normally distributed about σ with variance

$$\mathcal{V}\{s\} = \frac{\sigma^2}{n}\mu_{22}(\zeta), \qquad (11.11.1)$$

so that

$$\boxed{u \simeq \frac{s-\sigma}{\sigma\sqrt{\dfrac{\mu_{22}(\zeta)}{n}}} = \frac{\dfrac{s}{\sigma}-1}{\sqrt{\dfrac{\mu_{22}(\zeta)}{n}}}.} \qquad (11.11.2)$$

The function $\mu_{22}(\zeta)$ has been tabulated in Tables IX and X for the truncated and the censored distributions, respectively.

For $\zeta \to -\infty$, when the degree of truncation converges to zero, we have $\mu_{22}(\zeta) \to 0{\cdot}5$, in agreement with the fact that, when s is calculated from a non-truncated distribution, it is normally distributed about σ when $n \to \infty$ with variance $\dfrac{\sigma^2}{2f} \simeq \dfrac{\sigma^2}{n}\, 0{\cdot}5$.

From (11.11.2) we obtain

$$P\left\{u_{P_1} < \frac{\frac{s}{\sigma}-1}{\sqrt{\frac{\mu_{22}(\zeta)}{n}}} < u_{P_2}\right\} = P_2-P_1, \tag{11.11.3}$$

from which we may calculate fractiles for s and confidence limits for σ. Solving the equation with respect to $\frac{s}{\sigma}$, we find

$$P\left\{1+u_{P_1}\sqrt{\frac{\mu_{22}(\zeta)}{n}} < \frac{s}{\sigma} < 1+u_{P_2}\sqrt{\frac{\mu_{22}(\zeta)}{n}}\right\} = P_2-P_1. \tag{11.11.4}$$

Thus, the confidence limits for σ are

$$P\left\{\frac{s}{1+u_{P_2}\sqrt{\frac{\mu_{22}(\zeta)}{n}}} < \sigma < \frac{s}{1+u_{P_1}\sqrt{\frac{\mu_{22}(\zeta)}{n}}}\right\} = P_2-P_1. \tag{11.11.5}$$

In contrast to the limits in (11.4.1), the above limits cannot be calculated exactly as they depend on the unknown ζ. In practical work, ζ is replaced by the estimate z, see (6.9.8) and (6.9.12).

The estimates of ξ and σ in a truncated normal distribution are not stochastically independent. For $n \to \infty$ the distribution function for these estimates converges to the two-dimensional normal distribution with the covariance $\mu_{12}(\zeta)$ and the correlation coefficient $\varrho(\zeta)$, as given in Tables IX and X.

Example 11.13. According to (11.11.1) the results in Table 6.6, p. 148, lead to

$$\mathcal{V}\{s\} \simeq \frac{\sigma^2}{310}\mu_{22}(-0{\cdot}310) = \frac{\sigma^2}{310}\,2{\cdot}947 = 0{\cdot}009506\,\sigma^2,$$

cf. Table IX. As s is equal to $0{\cdot}111$ and $\sqrt{\frac{\mu_{22}(z)}{n}} = 0{\cdot}0975$ the 95% bilateral confidence limits for σ will be

$$\left(\frac{0{\cdot}111}{1+1{\cdot}960\times 0{\cdot}0975} < \sigma < \frac{0{\cdot}111}{1-1{\cdot}960\times 0{\cdot}0975}\right)$$

$$= (0{\cdot}840\times 0{\cdot}111 < \sigma < 1{\cdot}236\times 0{\cdot}111)$$

or $(0{\cdot}093 < \sigma < 0{\cdot}137)$.

Correspondingly, we find from Table 6.7, p. 151, that

$$\mathcal{V}\{s\} \simeq \frac{\sigma^2}{500}\mu_{22}(-0{\cdot}306) = \frac{\sigma^2}{500}0{\cdot}955 = 0{\cdot}001910\,\sigma^2\,,$$

cf. Table X. As s is equal to 0·111 and $\sqrt{\frac{\mu_{22}(z)}{n}} = 0{\cdot}0437$, the bilateral 95% limits for σ are here

$$\left(\frac{0{\cdot}111}{1+1{\cdot}960\times 0{\cdot}0437} < \sigma < \frac{0{\cdot}111}{1-1{\cdot}960\times 0{\cdot}0437}\right)$$

$$= (0{\cdot}921\times 0{\cdot}111 < \sigma < 1{\cdot}094\times 0{\cdot}111)$$

or $(0{\cdot}102 < \sigma < 0{\cdot}121)$.

For the non-truncated distribution in Table 4.4, p. 77, we have

$$\mathcal{V}\{s\} \simeq \frac{\sigma^2}{998} = 0{\cdot}001002\,\sigma^2\,.$$

Here, as $s = 0{\cdot}115$ and $\sqrt{0{\cdot}001002} = 0{\cdot}0317$, we obtain from (11.7.11) as bilateral 95% limits for σ

$$\left(\frac{0{\cdot}115}{1+1{\cdot}960\times 0{\cdot}0317} < \sigma < \frac{0{\cdot}115}{1-1{\cdot}960\times 0{\cdot}0317}\right)$$

$$= (0{\cdot}942\times 0{\cdot}115 < \sigma < 1{\cdot}066\times 0{\cdot}115)$$

or $(0{\cdot}108 < \sigma < 0{\cdot}123)$. The same result may be obtained from Table VI, since

$$\sqrt{\frac{f}{\chi^2_{\cdot 975}}} = \sqrt{\frac{1}{1{\cdot}1278}} = 0{\cdot}942 \quad \text{for } f = 499$$

and

$$\sqrt{\frac{f}{\chi^2_{\cdot 025}}} = \sqrt{\frac{1}{0{\cdot}8798}} = 1{\cdot}066 \quad \text{for } f = 499\,,$$

the limits of σ being $0{\cdot}942\times 0{\cdot}115 = 0{\cdot}108$ and $1{\cdot}066\times 0{\cdot}115 = 0{\cdot}123$.

The influence of the truncation upon the limits for σ may be seen by comparing the three sets of factors for s:

0·840 and 1·236 for the truncated distribution,
0·921 and 1·094 for the censored distribution, and
0·942 and 1·066 for the non-truncated distribution.

CHAPTER 12.

THE DISTRIBUTION OF THE RANGE

12.1. The Distribution Function of the Range.

In § 11.2 we saw how the variability of, for example, a quality characteristic of an industrial product may be controlled by plotting successive sample variances on a control chart, after the control limits have been calculated from the population variance and the number of degrees of freedom for the sample variances. In the routine control of the variability of a manufactured product by sampling inspection, it is important that the arithmetic involved should be a minimum. This can be arranged by estimating the variability by means of *the range, which is defined as the difference between the largest and the smallest observation in a sample.*

Let $x_1, x_2, \ldots, x_n$ denote n stochastically independent observations from a population with the distribution function $p\{x\}$, and let $x_{(1)}, x_{(2)}, \ldots, x_{(n)}$ denote the same observations in ascending order of magnitude. *The range* w is defined as

$$\boxed{w = x_{(n)} - x_{(1)}\,.} \qquad (12.1.1)$$

If we have a population of samples, each including n observations, and if the range of each sample is calculated, we obtain a *population of ranges*, for which the cumulative distribution function $P_n\{w\}$ may be derived as follows: The probability that an observation will lie in the interval $(x, x+w)$ is $P\{x+w\}-P\{x\}$, and the probability that an observation will lie in the "infinitely small" interval $(x, x+dx)$ is $dP\{x\} = p\{x\}dx$, as shown in (5.2.6) and (5.2.8). If the number of observations is n, the probability that $x_1, x_2, \ldots, x_{n-1}$ will lie in the interval $(x, x+w)$ and x_n in the interval $(x, x+dx)$ is, according to the multiplication formula,

$$(P\{x+w\}-P\{x\})^{n-1}dP\{x\}\,.$$

Then, according to the addition formula, the probability that one of the n observations lies in the interval $(x, x+dx)$ and $n-1$ in the interval $(x, x+w)$ is equal to

$$n(P\{x+w\}-P\{x\})^{n-1}dP\{x\},$$

which gives the probability that the range is less than w, presuming that the smallest observation lies in the interval $(x, x+dx)$. Hence, according to the addition formula, the probability that the range is less than w for any value of the smallest observation becomes

$$P_n\{w\} = n\int_{-\infty}^{\infty}(P\{x+w\}-P\{x\})^{n-1}dP\{x\}. \qquad (12.1.2)$$

Differentiating $P_n\{w\}$ with regard to w, we obtain the distribution function

$$p_n\{w\} = n(n-1)\int_{-\infty}^{\infty}(P\{x+w\}-P\{x\})^{n-2}p\{x\}p\{x+w\}\,dx. \qquad (12.1.3)$$

This result could be obtained more directly, since (12.1.3) merely expresses the fact that the probability of obtaining one observation "equal to x", one "equal to $x+w$" and $n-2$ observations between x and $x+w$ is

$$n(n-1)p\{x\}\,dx\,p\{x+w\}\,dw\,(P\{x+w\}-P\{x\})^{n-2},$$

and $p_n\{w\}dw$ follows by integrating this expression with respect to x.

If $x_1, x_2, \ldots, x_n$ originate from a normally distributed population with parameters (ξ, σ^2), standardization leads to

$$\frac{w}{\sigma} = \frac{(x_{(n)}-\xi)-(x_{(1)}-\xi)}{\sigma} = u_{(n)}-u_{(1)},$$

or, if W denotes the range of the observations from a standardized normally distributed population,

$$\boxed{W = \frac{w}{\sigma} = u_{(n)}-u_{(1)}.} \qquad (12.1.4)$$

The distribution function and cumulative distribution function of W, $p_n\{W\}$ and $P_n\{W\}$, are obtained from (12.1.2) and (12.1.3) for $p\{x\}=\varphi(x)$. The cumulative distribution function, $P_n\{W\}$, has been tabulated in *Biometrika*, 32, 1942, 301–10, for $n = 2(1)20$.

Table VIII contains the fractiles $W_P = W_P(n)$ for $n = 2(1)20$ and suitable values of P.

According to (12.1.4), the P-fractiles of w are

$$w_P = \sigma W_P, \qquad (12.1.5)$$

so that

$$\boxed{P\{w < \sigma W_P\} = P.} \qquad (12.1.6)$$

Table VIII, for example, shows that

$$P\{w < 4{\cdot}79\ \sigma\} = 97{\cdot}5\% \text{ for } n = 10\,.$$

Fig. 12.1 gives the cumulative distribution function $P_n\{W\}$ plotted on probability paper for $n = 2, 5, 10$, and 20.

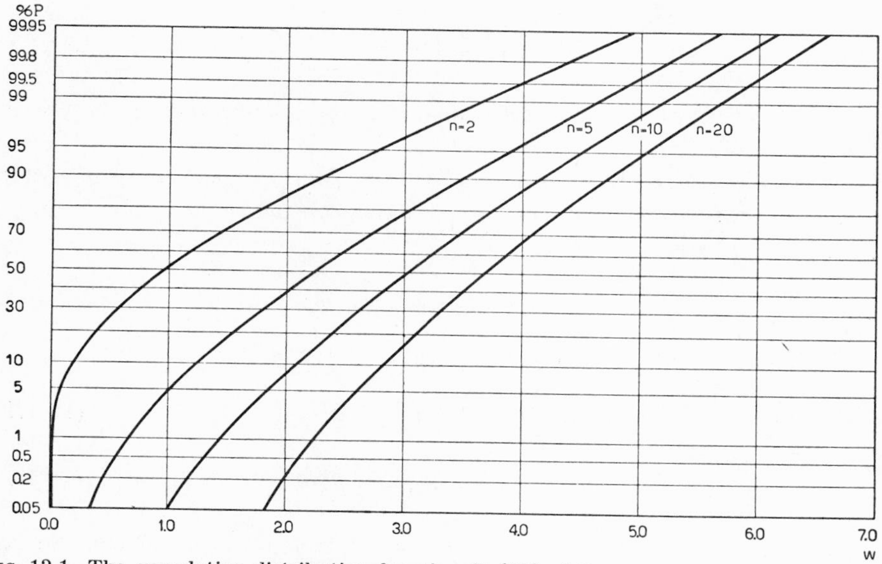

FIG. 12.1. The cumulative distribution function $P_n\{W\}$ of the range plotted on probability paper for $n = 2, 5, 10$, and 20.

According to (5.7.1) the mean of W is

$$m\{W\} = \int_0^\infty W p_n\{W\} dW = \alpha_n\,, \tag{12.1.7}$$

and the variance is, according to (5.9.2),

$$v\{W\} = \int_0^\infty (W-\alpha_n)^2 p_n\{W\} dW = \beta_n^2\,. \tag{12.1.8}$$

The functions α_n and β_n have been given in Table VIII. According to (12.1.4) we have

$$\boxed{m\{w\} = \alpha_n \sigma} \tag{12.1.9}$$

and

$$\boxed{\sigma\{w\} = \beta_n \sigma\,,} \tag{12.1.10}$$

where $\sigma = \sigma\{x\}$ denotes the standard deviation for the distribution of the observations. Considering characteristic values of α_n, we find that $\alpha_{10}=3{\cdot}08$, $\alpha_{30} = 4{\cdot}09$, $\alpha_{100} = 5{\cdot}02$, and $\alpha_{500} = 6{\cdot}07$, which means that, in samples of 10, 30, 100 and 500 observations, the mean of the range is about 3, 4, 5, and 6 times the standard deviation in the population.

Dividing (12.1.9) by α_n, we obtain

$$m\left\{\frac{w}{\alpha_n}\right\} = \sigma , \tag{12.1.11}$$

which means that *an unbiased estimate of* σ *may be obtained from an observed value of* w *as* w/α_n.

If we have k *samples, each of* n *observations, from* k *populations, all having the same standard deviation, and we calculate the mean* $\bar{w}$ *of the* k *ranges,* $w_1, w_2, \ldots, w_k$, i. e.,

$$\bar{w} = \frac{1}{k}(w_1+w_2+\ldots+w_k) , \tag{12.1.12}$$

it follows from (5.17.3) and (5.17.5) that

$$m\{\bar{w}\} = m\{w\} = \alpha_n\sigma \tag{12.1.13}$$

and

$$v\{\bar{w}\} = \frac{1}{k}v\{w\} = \frac{1}{k}\beta_n^2\sigma^2 . \tag{12.1.14}$$

According to the central limit theorem, if k is large, $\bar{w}$ will be approximately normally distributed with the above mean and variance. *For values of* n *between* 5 *and* 10 *the distribution of* $\bar{w}$ *may be replaced by the normal distribution for quite small values of* k. This implies that the quantity

$$\boxed{u \simeq \frac{\bar{w}-\alpha_n\sigma}{\beta_n\sigma/\sqrt{k}} = \frac{\dfrac{\bar{w}}{\alpha_n}-\sigma}{\gamma_n\sigma/\sqrt{k}},} \tag{12.1.15}$$

where $\gamma_n = \dfrac{\beta_n}{\alpha_n}$, is approximately normally distributed with parameters (0, 1). Thus, *the unbiased estimate* $\dfrac{\bar{w}}{\alpha_n}$ *of* σ *is approximately normally distributed about* σ *with standard deviation* $\gamma_n\sigma/\sqrt{k}$. γ_n has been tabulated in Table VIII.

12.2. The Control Chart for the Range.

From the fractiles we immediately obtain

$$P\{\sigma W_{P_1} < w < \sigma W_{P_2}\} = P_2 - P_1 . \tag{12.2.1}$$

For instance, for $n = 10$ we get from Table VIII

$$P\{1{\cdot}67\ \sigma < w < 4{\cdot}79\ \sigma\} = 97{\cdot}5 - 2{\cdot}5 = 95\%\ .$$

If σ is known, the control limits for w may be calculated in the same manner as the limits for s^2 and s, respectively, cf. (11.2.2).

The variability of the quality of an industrial product may be controlled by regularly taking samples of n observations, determining the corresponding ranges, and plotting these on a control chart, as in Fig. 12.2.

The control chart may be interpreted in the usual way. If a point lies above the 99·9% limit, say, this fact suggests that the variability of the quality characteristic has increased and *such a point therefore should lead to an investigation of the production process* in order to detect the cause of the suspected change of the theoretical variance. The consequences of applying such a rule follow from the principles laid down in § 9.4 where the basis for tests of significance was discussed, since the control chart is merely a graphical representation of a series of successive tests of significance, based on samples of the same size.

If the preassigned value of the population standard deviation is σ_0 and the alternative values are $\sigma > \sigma_0$, we find as *the critical region at the α level of significance*

$$w > w_{1-\alpha} = \sigma_0 W_{1-\alpha}\ , \tag{12.2.2}$$

since

$$P\{w > \sigma_0 W_{1-\alpha};\ \sigma = \sigma_0\} = P\{W > W_{1-\alpha}\} = \alpha\ . \tag{12.2.3}$$

The power of the test with respect to the alternative hypothesis $\sigma = \lambda\sigma_0$, $\lambda > 1$, is found as

$$\begin{aligned}\pi(\sigma) &= P\{w > \sigma_0 W_{1-\alpha};\ \sigma = \lambda\sigma_0\}\\ &= P\{\sigma W > \sigma_0 W_{1-\alpha}\}\\ &= P\left\{W > \frac{1}{\lambda}\,W_{1-\alpha}\right\}.\end{aligned} \tag{12.2.4}$$

Putting $\pi(\sigma) = 1-\beta$ we obtain

$$\lambda = \frac{W_{1-\alpha}}{W_\beta} \tag{12.2.5}$$

analogous to (11.3.6). By means of this relation it is easy to determine a sufficient number of points to draw the power curve. For given n and α we only need to compute the ratio λ between $W_{1-\alpha}$ and W_β for selected values of β as found in Table VIII and then to plot the points $(\lambda, 1-\beta)$.

Table 12.1 shows for $n = 5$ and for $n = 10$ some characteristic values of the power function at the 5% level of significance. For comparison the corresponding values for the one-sided test based on s^2 for $f = 4$ and $f = 9$ are given. It will be seen that the s^2-test is uniformly more powerful than the w-test; in fact the test based on s^2 is the uniformly most powerful test for the hypothesis $\sigma = \sigma_0$ against the one-sided alternative $\sigma > \sigma_0$. For small values of n, however, the differences between the two power functions are small, and for $n = 2$ the two tests are identical, since they both are based on the difference $x_1 - x_2$. The larger the value of n, the less efficient is the w-test.

TABLE 12.1.

Power functions of the w- and s^2-tests for $n = 5$ and $n = 10$ at the 5% level of significance.

π	β	$n = 5$ $\lambda = \frac{3{\cdot}86}{W_\beta}$	$n = 10$ $\lambda = \frac{4{\cdot}47}{W_\beta}$	$n = 5$ $\lambda = \sqrt{\frac{9{\cdot}49}{\chi^2_\beta}}$	$n = 10$ $\lambda = \sqrt{\frac{16{\cdot}9}{\chi^2_\beta}}$
0·05	0·95	1·00	1·00	1·00	1·00
0·20	0·80	1·27	1·20	1·26	1·18
0·50	0·50	1·71	1·48	1·68	1·42
0·80	0·20	2·46	1·87	2·40	1·77
0·95	0·05	3·75	2·40	3·65	2·25
0·99	0·01	5·85	3·04	5·65	2·84
0·999	0·001	10·4	4·14	10·2	3·83

To study the sensitivity of the control chart to changes in the population standard deviation let us suppose that σ changes from σ_0 to $1{\cdot}5\sigma_0$ and that we watch the frequency of points above the unilateral 95% limit. For $n = 10$ we obtain from Table VIII $W_{.95} = 4{\cdot}47$. Putting $\lambda = 1{\cdot}5$ in (12.2.4) we find

$$\pi(1{\cdot}5\sigma_0) = P\left\{W > \frac{4{\cdot}47}{1{\cdot}5}\right\} = P\{W > 2{\cdot}98\} = 0{\cdot}52$$

by interpolation in Table VIII. We thus see that, if the theoretical standard deviation increases with 50%, the expected frequency of points above the 95% limit on the control chart will increase from 5% to 52%.

In a similar manner we find for the control chart for s^2 that

$$\pi(2{\cdot}25\sigma_0^2) = P\{\chi^2 > 7{\cdot}51\} = 0{\cdot}58$$

from Table V for $f = 9$, see (11.3.3).

This shows that *the control chart for w is less sensitive to changes in σ than is the control chart for s^2*, since the w-chart will less frequently lead to detection of changes in σ than will the s^2-chart. As the difference between the corresponding power functions increases rapidly with n, w should be used only for values of n below 20, since, for larger values of n, w will not extract sufficient information from the sample.

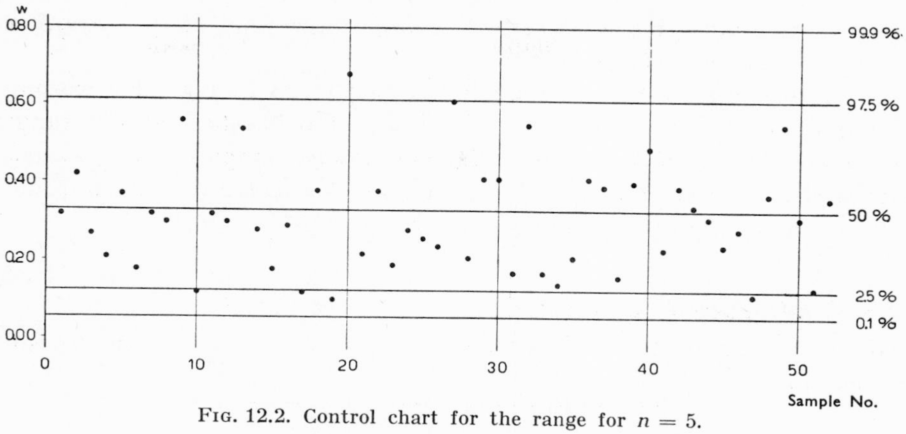

FIG. 12.2. Control chart for the range for $n = 5$.

Example 12.1. Fig. 12.2 gives the control chart for the range corresponding to the control chart for the variance in Fig. 11.2, p. 280. According to Table VIII the control limits calculated from $\sigma = 0{\cdot}146\%$ $CaCO_3$ are

$$w_{.001} = 0{\cdot}146 \times 0{\cdot}37 = 0{\cdot}054\% \ CaCO_3 ,$$
$$w_{.025} = 0{\cdot}146 \times 0{\cdot}85 = 0{\cdot}124\% \ CaCO_3 ,$$
$$w_{.50} = 0{\cdot}146 \times 2{\cdot}26 = 0{\cdot}330\% \ CaCO_3 ,$$
$$w_{.975} = 0{\cdot}146 \times 4{\cdot}20 = 0{\cdot}613\% \ CaCO_3 ,$$
$$w_{.999} = 0{\cdot}146 \times 5{\cdot}48 = 0{\cdot}800\% \ CaCO_3 .$$

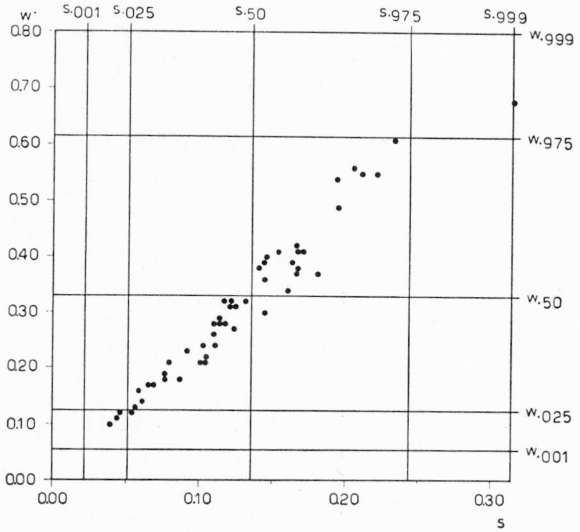

FIG. 12.3. The correlation between s and w.

The two diagrams give practically the same impression of the variations in the homogeneity of the raw meal.

If corresponding values of s and w are plotted as in Fig. 12.3, we may examine the correlation between the standard deviation and the range. The figure shows that s and w correspond closely; samples giving s-values outside the bilateral 95% control limits for s will as a rule also give w-values outside the 95% control limits for w.

12.3. Confidence Limits for σ.

If the inequality (12.2.1) is solved with respect to σ we obtain the confidence limits for σ

$$\boxed{P\left\{\frac{w}{W_{P_2}} < \sigma < \frac{w}{W_{P_1}}\right\} = P_2 - P_1\,.} \tag{12.3.1}$$

As an example, for $n = 10$ Table VIII gives

$$P\left\{\frac{w}{4\cdot79} < \sigma < \frac{w}{1\cdot67}\right\} = P\{0\cdot209\,w < \sigma < 0\cdot599\,w\}$$
$$= 97\cdot5 - 2\cdot5 = 95\%.$$

If we have *k samples each with n observations* the confidence limits for σ may be computed from the estimate $\dfrac{\bar{w}}{\alpha_n}$ by employing (12.1.15), which gives

$$P\left\{u_{P_1} < \frac{\dfrac{\bar{w}}{\alpha_n} - \sigma}{\gamma_n \sigma/\sqrt{k}} < u_{P_2}\right\} = P_2 - P_1\,. \tag{12.3.2}$$

Solving this inequality with respect to σ, we find

$$\boxed{P\left\{\frac{\bar{w}}{\alpha_n}\,\frac{1}{1+u_{P_2}\dfrac{\gamma_n}{\sqrt{k}}} < \sigma < \frac{\bar{w}}{\alpha_n}\,\frac{1}{1+u_{P_1}\dfrac{\gamma_n}{\sqrt{k}}}\right\} = P_2 - P_1\,.} \tag{12.3.3}$$

For $n = 5$, $k = 20$, and $P_1 = 100 - P_2 = 0\cdot5\%$, say, we have

$$1+u_{P_2}\frac{\gamma_n}{\sqrt{k}} = 1+2\cdot58\,\frac{0\cdot371}{\sqrt{20}} = 1\cdot214$$

and

$$1+u_{P_1}\frac{\gamma_n}{\sqrt{k}} = 1-2\cdot58\,\frac{0\cdot371}{\sqrt{20}} = 0\cdot786\,,$$

and therefore

$$P\left\{\frac{\bar{w}}{\alpha_n}\frac{1}{1\cdot 214} < \sigma < \frac{\bar{w}}{\alpha_n}\frac{1}{0\cdot 786}\right\} = P\left\{0\cdot 82\,\frac{\bar{w}}{\alpha_n} < \sigma < 1\cdot 27\,\frac{\bar{w}}{\alpha_n}\right\}$$
$$= 99\cdot 5 - 0\cdot 5 = 99\% \,.$$

The confidence limits for σ as computed from (12.3.3) may be compared with the confidence limits as given by (11.4.1), using similar observational data. According to the addition theorem for the s^2-distribution, an estimate s^2 of σ^2 may be computed from the k samples with $k(n-1)$ degrees of freedom. From (11.4.1) we then obtain

$$P\left\{s\sqrt{\frac{f}{\chi^2_{P_2}}} < \sigma < s\sqrt{\frac{f}{\chi^2_{P_1}}}\right\} = P_2 - P_1, \quad f = k(n-1)\,. \qquad (12.3.4)$$

On comparing (12.3.3) with (12.3.4) we find, with the help of Tables VI and VIII, that

$$\frac{1}{1+u_{P_2}\dfrac{\gamma_n}{\sqrt{k}}} < \sqrt{\frac{f}{\chi^2_{P_2}}} < \sqrt{\frac{f}{\chi^2_{P_1}}} < \frac{1}{1+u_{P_1}\dfrac{\gamma_n}{\sqrt{k}}} \quad \text{for } f = k(n-1)\,, \qquad (12.3.5)$$

i. e., the confidence limits for σ when computed from s are narrower than when computed from $\bar{w}/\alpha_n$.

For comparison with the above example for $n = 5$, $k = 20$, and $P_1 = 100 - P_2 = 0\cdot 5\%$ we find from (12.3.4) and Table VI that

$$P\left\{s\sqrt{\frac{1}{1\cdot 4540}} < \sigma < s\sqrt{\frac{1}{0\cdot 6396}}\right\} = P\{0\cdot 83\,s < \sigma < 1\cdot 25\,s\}$$
$$= 99\cdot 5 - 0\cdot 5 = 99\% \,.$$

If we have one sample with N observations, an estimate of σ based on the use of ranges may be computed by two different methods:

1. From the range $x_{(N)} - x_{(1)}$ we compute

$$\frac{x_{(N)} - x_{(1)}}{\alpha_N} \approx \sigma\,.$$

2. The observations may be divided at random into k groups, each including n observations, $N = kn$, the corresponding k ranges and their mean $\bar{w}$ are determined, and the estimate of σ is computed as $\dfrac{\bar{w}}{\alpha_n}$. It has been shown[1]) that it is advisable to divide the N observations into groups of 5 to 10, the best size being 8.

[1]) F. E. GRUBBS and C. L. WEAVER: *The Best Unbiased Estimate of Population Standard Deviation Based on Group Ranges*, Journ. Amer. Stat. Ass., 42, 1947, 224–241.

It can be proved that a better estimate of σ is obtained by using the second method.

We thus see *that the range may be used instead of s in control charts for $n < 20$, and also for computing an estimate of σ and confidence limits for σ, if based on the mean of the ranges of several small samples.* For further statistical analysis, however, s is to be preferred to w.

Example 12.2. From the four distributions of the logarithms of the compressive strengths in Table 7.10, p. 184, the following ranges are obtained: 0·030, 0·024, 0·023 and 0·027. The mean range is

$$\bar{w} = \frac{0{\cdot}104}{4} = 0{\cdot}026$$

and the estimate of σ is

$$\frac{\bar{w}}{\alpha_5} = \frac{0{\cdot}026}{2{\cdot}33} = 0{\cdot}0112 \, .$$

According to (12.3.3) the unilateral 95% limit of σ is

$$0{\cdot}0112 \frac{1}{1 - 1{\cdot}64 \frac{0{\cdot}371}{\sqrt{4}}} = 0{\cdot}0112 \times 1{\cdot}44 = 0{\cdot}0161 \, .$$

From Example 11.5, p. 290, we see that, when calculated from the standard deviation, the corresponding results were $s = 0{\cdot}0103$ and $0{\cdot}0103 \times 1{\cdot}42 = 0{\cdot}0146$ as the 95% limit for σ.

Example 12.3. The ranges of the 52 samples mentioned in Example 12.1 are given in Table 12.2. The mean of the 52 ranges is

$$\bar{w} = \frac{16{\cdot}18}{52} = 0{\cdot}311 \, ,$$

and therefore the estimate of σ is

$$\frac{\bar{w}}{\alpha_5} = \frac{0{\cdot}311}{2{\cdot}33} = 0{\cdot}133\% \ CaCO_3 \, .$$

The corresponding s-value is 0·140% $CaCO_3$, see Example 11.7, p. 294.

The above use of the mean range may be justified by examining the distribution of the 52 ranges. The fractiles W_P for $n = 5$ are found in Table VIII for suitable values of P, see Table 12.3. These values are multiplied by the estimate, 0·133, of σ, which gives the estimate of the fractiles in the w-distribution. Enumeration of the w-values in Table 12.2 which lie between the limits given in Table 12.3 leads to the distribution shown in the 4th column of Table 12.3. The agreement between the observed and the theoretical distribution is fair, compare Example 11.7, p. 294. The largest value of w, 0·68, lies in the interval 99·5–99·9%.

TABLE 12.2.
The values of the ranges, in terms of % concentration of $CaCO_3$, for 52 sets of 5 samples each, taken from a mixing plant for raw meal.

0·32	0·28	0·61	0·49
0·42	0·18	0·21	0·23
0·27	0·29	0·41	0·39
0·21	0·12	0·41	0·34
0·37	0·38	0·17	0·31
0·18	0·10	0·55	0·24
0·32	0·68	0·17	0·28
0·30	0·22	0·14	0·11
0·56	0·38	0·21	0·37
0·12	0·19	0·41	0·55
0·32	0·28	0·39	0·31
0·30	0·26	0·16	0·13
0·54	0·24	0·40	0·36

TABLE 12.3.
Grouped distribution of 52 ranges.

$P(\%)$	w_P	$0{\cdot}133\,w_P$	Observed number	Theoretical number
0 – 1	0·00 – 0·66	0·000 – 0·088	0	0·52
1 – 5	0·66 – 1·03	0·088 – 0·137	5	2·08
5 – 10	1·03 – 1·26	0·137 – 0·168	2	2·60
10 – 30	1·26 – 1·82	0·168 – 0·242	12	10·40
30 – 50	1·82 – 2·26	0·242 – 0·301	8	10·40
50 – 70	2·26 – 2·73	0·301 – 0·363	7	10·40
70 – 90	2·73 – 3·48	0·363 – 0·463	11	10·40
90 – 95	3·48 – 3·86	0·463 – 0·513	1	2·60
95 – 99	3·86 – 4·60	0·513 – 0·612	5	2·08
99 – 100	4·60 – ∞	0·612 – ∞	1	0·52
		Total	52	52·00

12.4. The Distribution of the Largest Observation.

In Example 5.3, p. 94, the distribution function and the cumulative distribution function are given for the largest observation $x_{(n)}$ among n stochastically independent observations with the same distribution function $p\{x\}$, see (5.2.14) and (5.2.15).

From the cumulative distribution function

$$P\{x_{(n)} < x\} = (P\{x\})^n \tag{12.4.1}$$

the fractiles $x_{(n)P}$ are calculated by solving the equation

$$P\{x_{(n)} < x_{(n)P}\} = P \tag{12.4.2}$$

or

$$(P\{x\})^n = P \text{ for } x = x_{(n)P} . \tag{12.4.3}$$

This leads to

$$P\{x\} = P^{1/n} = P_1 \text{ for } x = x_{(n)P} . \tag{12.4.4}$$

The equation

$$P\{x\} = P_1 ,$$

however, has the solution $x = x_{P_1}$, and therefore

$$\boxed{x_{(n)P} = x_{P_1} \quad \text{for} \quad P_1 = P^{1/n} .} \tag{12.4.5}$$

For instance, for $P = 95\%$, we obtain the values of P_1 given in Table 12.4; from this table we see, for example, that the 95% fractile for the largest observation of 51 observations is equal to the 99·9% fractile in the distribution of the observations.

TABLE 12.4.

n	$P_1 = 0{\cdot}95^{1/n}$
1	0·95
2	0·975
5	0·990
10	0·995
51	0·999
102	0·9995

For n observations from a normally distributed population with parameters (ξ, σ^2), we have

$$\boxed{x_{(n)P} = \xi + \sigma u_{(n)P} = \xi + \sigma u_{P_1} \quad \text{for} \quad P_1 = P^{1/n} .} \tag{12.4.6}$$

From Table 12.4 and Table III we obtain, for instance,

$$u_{(51)\cdot 95} = u_{\cdot 999} = 3{\cdot}09 ,$$

i. e., the probability that the largest of 51 observations is less than $\xi + 3{\cdot}09\sigma$ is 95%.

Because of the symmetry of the normal distribution, the fractiles for the smallest of n observations from a standardized normally distributed population are equal in magnitude but opposite in sign to the fractiles for the largest observation.

When (12.4.1)–(12.4.5) were derived, it was not assumed that the variable was normally distributed. Therefore, the result obtained also holds for the χ^2- and W-distributions. When we wish to decide whether a single large value of an empirical variance or range really belongs to an observed distribution of variances or ranges, determined from the same number of observations, as in Example 11.7, p. 294, and Example 12.3, p. 328, the distribution of the largest variance or range may yield an approximate evaluation of the probability of these largest values.

In Table 11.3, p. 295, for instance, the largest χ^2-value is 20·37. From Table V we see that $\chi^2_{.9995} = 20{\cdot}0$ for $f = 4$ and therefore the probability that the largest of 52 χ^2-values is less than 20·0 is $0{\cdot}9995^{52} = 0{\cdot}974$. This means that the probability that the largest χ^2-value among the 52 values is larger than $\chi^2_{.9995}$ is 2·6%, i. e., the value found has a small probability. This implies that for set No. 20 the theoretical variance is probably larger than for the other sets.

12.5. The Distribution of the Largest Deviation from the Population Mean.

Let $x_1, x_2, \ldots, x_n$ denote n stochastically independent observations from a population with distribution function $p\{x\}$ and mean $\mathcal{M}\{x\} = \xi$. *The largest deviation from the population mean* ξ *is defined as the largest of the quantities* $|x_1-\xi|, |x_2-\xi|, \ldots, |x_n-\xi|$. It may also be defined as the larger of the two quantities $|x_{(1)}-\xi|$ and $|x_{(n)}-\xi|$.

The necessary and sufficient condition that the largest deviation is less than d is that all deviations are less than d. Then, according to the multiplication formula of the probability calculus, the probability $P\{d\}$ is equal to

$$P\{d\} = P\{|x_1-\xi| < d\}\ldots P\{|x_n-\xi| < d\}$$
$$= (P\{|x-\xi| < d\})^n . \qquad (12.5.1)$$

If the variable is normally distributed with parameters (ξ, σ^2) we have

$$P\{d\} = \left(1-2\Phi\left(-\frac{d}{\sigma}\right)\right)^n . \qquad (12.5.2)$$

Introducing $D = d/\sigma$, we get

$$P\{D\} = \left(1-2\Phi(-D)\right)^n . \qquad (12.5.3)$$

The fractiles, D_P, are determined from the equation

$$\left(1-2\Phi(-D)\right)^n = P , \qquad (12.5.4)$$

which leads to

$$\boxed{D_P = u_{P_1} \text{ for } P_1 = \tfrac{1}{2}(1+P^{1/n}).} \qquad (12.5.5)$$

For $n = 10$ and $P = 0{\cdot}95$, we obtain $P_1 = \frac{1}{2}\left(1+0{\cdot}95^{\frac{1}{10}}\right) = \frac{1}{2}\times 1{\cdot}99488 =$

0·99744, whence $D_{.95} = u_{.99744} = 2{\cdot}80$, cf. Table III, that is for $n = 10$ the 95% fractile for the largest deviation is $2{\cdot}80\sigma$,

$$P\{d < 2{\cdot}80\,\sigma\} = 0{\cdot}95 \;.$$

The largest deviation may be applied in control charts and for the calculation of confidence limits for σ in a manner similar to that described for the range. As, however, the population mean is usually unknown, the practical value of the largest deviation is far less than that of the range.

12.6. The Studentized Range.

W. S. GOSSET, writing under the pseudonym of "STUDENT", studied the distribution of the variable $\sqrt{n}(\bar{x}-\xi)/s$, which was obtained from the standardized variable $\sqrt{n}(\bar{x}-\xi)/\sigma$ by substitution of s for σ, see Chapter 15. GOSSET also suggested study of the "studentized" range defined as $q=w/s$, where w and s are stochastically independent, i. e., w and s are computed from two independent random samples from the same normally distributed population. Under this assumption w will be distributed as σW and s as $\sigma\sqrt{\chi^2/f}$ so that

$$q = \frac{w}{s} = \frac{W}{\sqrt{\chi^2/f}} = \frac{u_{(n)}-u_{(1)}}{\sqrt{\chi^2/f}} \,. \tag{12.6.1}$$

Hence the distribution of q depends only on n, the number of observations in the sample from which w is found, and f, the number of degrees of freedom for s.

Tables for calculating the distribution of q and tables of the 1%, 5%, 95%, and 99% fractiles of q for $n = 2(1)20$ are given by E. S. PEARSON and H. O. HARTLEY in *Biometrika*, 33, 1943, 89–99.

An approximation to the 95% and 99% fractiles may be found as

$$q_P \simeq W_P\left(1+\frac{3}{f}\right) \quad \text{for} \quad P = {\cdot}95 \text{ and } {\cdot}99 \,, \tag{12.6.2}$$

where W_P may be found in Table VIII.

12.7. Largest Observation Minus Sample Mean.

The distribution of the variable

$$\boxed{k = \frac{x_{(n)}-\bar{x}}{\sigma} ,} \tag{12.7.1}$$

the standardized deviation between the largest observation and the sample mean, has been studied by A. T. MCKAY, *Biometrika*, 27, 1935, 466–71, who derived the distribution function and gave the following approximation formula for the fractiles for $P > \cdot 90$

$$k_P \simeq u_{P_1} \sqrt{\frac{n-1}{n}} \quad \text{for} \quad P_1 = 1 - \frac{1-P}{n}. \tag{12.7.2}$$

Further, F. E. GRUBBS, *Ann. Math. Stat.*, 21, 1950, 27–58, tabulated the cumulative distribution function for $n = 2(1)25$ and the exact 90%, 95%, 99%, and 99·5% fractiles, which showed that McKAY's formula gave a very good approximation for all values of n.

The corresponding studentized variable

$$m = \frac{x_{(n)} - \bar{x}}{s_f}, \tag{12.7.3}$$

where s_f is computed from a second independent sample from the same population, has been studied by K. R. NAIR, *Biometrika*, 35, 1948, 118–44, who tabulated the distributions of k and m for $n = 3(1)9$.

An approximate formula is

$$m_P \simeq k_P \left(1 + \frac{3}{f}\right) \quad \text{for} \quad P = \cdot 95 \quad \text{and} \quad \cdot 99\,. \tag{12.7.4}$$

GRUBBS also studied the distribution of

$$g = \frac{x_{(n)} - \bar{x}}{s}, \tag{12.7.5}$$

where $x_{(n)}$, $\bar{x}$, and s are computed from the same sample, and tabulated the fractiles for $n = 2(1)25$ and $P = 90$, 95, 97·5, and 99%.

Analogous formulas hold for the smallest observation minus the sample mean.

12.8. Criteria for Rejection of Outlying Observations.

In practice the question is often raised whether or not certain observations in a sample may be rejected as "outliers". A typical example of an outlying observation may be found in a series of measurements where a blunder leads to a single observation deviating considerably from the other observations. If the experimenter knows that a blunder has occurred the measurement must be rejected irrespective of its magnitude. In some cases, however, only a suspicion exists and then a test of significance may support or weaken this suspicion.

In general the problem is whether the observations are randomly drawn from a single population or the larger part of the observations are drawn from the same population and the remaining few observations from another or several other populations.

The only safe procedure for rejecting observations is based on a careful examination of the conditions underlying the observations, rejecting observations carried out under conditions essentially different from the standard conditions. By this procedure observations are rejected without taking their magnitude into account.

In many cases, however, it may be difficult to decide whether or not the conditions have changed essentially, and therefore a test of significance may give useful supplementary guidance.

If a whole series of samples is given, outlying observations may be localized by studying the distribution of the sample variances or sample ranges as shown in § 11.6 and § 12.3.

If only a single sample is given, the problem is more delicate. As alternative hypothesis to the test hypothesis that the observations are randomly drawn from the same normally distributed population we will consider first the case of a single outlier (A) and then the case of one outlier in each end of the distribution (B).

A. Tests of significance for a single outlier.

Limits of significance for $x_{(n)}$ may be found as

$$x_{(n)P} = \xi + \sigma u_{P_1} \quad \text{for} \quad P_1 = P^{1/n} \tag{12.8.1}$$

if ξ and σ are known, see (12.4.6).

If ξ is unknown but σ is known we have for $P > \cdot 90$

$$x_{(n)P} \simeq \bar{x} + \sigma u_{P_1} \sqrt{\frac{n-1}{n}} \quad \text{for} \quad P_1 = 1 - \frac{1-P}{n}, \tag{12.8.2}$$

according to (12.7.2).

Further, if both ξ and σ are unknown, but an estimate, s_f, of σ from past experience is known, we obtain according to (12.7.4)

$$x_{(n)P} \simeq \bar{x} + s_f u_{P_1} \left(1 + \frac{3}{f}\right) \sqrt{\frac{n-1}{n}} \quad \text{for} \quad P = \cdot 95 \quad \text{and} \quad \cdot 99\,. \tag{12.8.3}$$

If the test must be based solely on the sample values, a criterion

$$x_{(n)P} = \bar{x} + s g_P \tag{12.8.4}$$

may be obtained from GRUBBS' table[1]).

J. O. IRWIN, *Biometrika*, 17, 1925, 238–50, has proposed a criterion based on the difference between $x_{(n)}$ and $x_{(n-1)}$, viz.,

---

[1]) F. E. GRUBBS: *Sample Criteria for Testing Outlying Observations*, Ann. Math. Stat., 21, 1950, 27–58.

$$\lambda = \frac{x_{(n)} - x_{(n-1)}}{\sigma}, \tag{12.8.5}$$

and has tabulated the cumulative distribution function of λ. From this table the following table of fractiles has been obtained.

TABLE 12.5.

Fractiles of λ.

n	$\lambda_{\cdot 95}$	$\lambda_{\cdot 99}$
2	2·8	3·7
3	2·2	2·9
10	1·5	2·0
20	1·3	1·8
30	1·2	1·7
50	1·1	1·6
100	1·0	1·5
400	0·9	1·3
1000	0·8	1·2

B. Tests of significance for one outlier at each end of the distribution.

If σ^2 is known, s^2 or w may be computed from the sample and

$$s_P^2 = \sigma^2 \chi_P^2 / f \tag{12.8.6}$$

or

$$w_P = \sigma W_P \tag{12.8.7}$$

may be used as limits of significance.

If σ is unknown, but an estimate, s_f, based on f degrees of freedom is known, we may use

$$s_P^2 = s_f^2 v_P^2 \tag{12.8.8}$$

where the distribution of v^2 will be discussed in Chapter 14, or

$$w_P \simeq s_f W_P \left(1 + \frac{3}{f}\right) \quad \text{for} \quad P = \cdot 95 \quad \text{and} \quad \cdot 99 \tag{12.8.9}$$

according to (12.6.2).

These criteria may also be used in case A but are usually less sensitive than the criteria given there.

Further, a test based on the largest deviation from the sample mean is needed but the distribution of this variable has not been tabulated.

If the sample contains more than one outlying observation repeated applications of the above criteria may lead to successive rejections of the outliers. GRUBBS has given a criterion for the simultaneous rejection of several outliers.

Example 12.4. The following 12 observations represent the distribution of a quality characteristic of 12 randomly chosen items from a mass production.

$$3, 4, 5, 6, 7, 7, 8, 9, 9, 10, 11, 17, \quad \bar{x} = 8{\cdot}0, \quad s^2 = 13{\cdot}8, \quad s = 3{\cdot}71 .$$

The problem is whether or not the largest observation should be rejected. This will be investigated by means of several tests according to the information available on the population parameters.

1. Suppose that items produced under standard conditions will be normally distributed with mean $\xi = 7{\cdot}0$ and standard deviation $\sigma = 2{\cdot}5$. The 99% fractile of $x_{(n)}$ may then be found as $7{\cdot}0+2{\cdot}5\times 3{\cdot}16 = 14{\cdot}9$, since $P_1 = 0{\cdot}99^{1/12} = 0{\cdot}9992$ and $u_{P_1} = 3{\cdot}16$. Hence, the largest observation is significant at the 99% level.

2. If only $\sigma = 2{\cdot}5$ is known we may use the sample mean instead of ξ. We then obtain as limit of significance $8{\cdot}0+2{\cdot}5\times 3{\cdot}16\sqrt{\dfrac{11}{12}} = 15{\cdot}6$, since $P_1 = 1-\dfrac{0{\cdot}01}{12} = 0{\cdot}9992$ and $u_{P_1} = 3{\cdot}16$. Also by this criterion the largest observation will be rejected at the 99% level of significance.

3. Suppose that a previously drawn random sample of 21 observations has given the estimate $s = 3{\cdot}0$ of σ, based on 20 degrees of freedom. The 99% limit of significance then becomes $8{\cdot}0+3{\cdot}0\times 3{\cdot}16\times 1{\cdot}15\sqrt{\dfrac{11}{12}} = 18{\cdot}4$, so that the largest observation is not significant.

4. Irwin's criterion leads to a significant result since

$$\lambda = \frac{17-11}{2{\cdot}5} = 2{\cdot}4 ,$$

see Table 12.5. If we do not know σ, however, and use the sample standard deviation, $s = 3{\cdot}7$, instead, we find $\lambda' = (17-11)/3{\cdot}7 = 1{\cdot}6$, which is significant at the 95% level but not at the 99% level.

5. From the χ^2-test we obtain $s^2/\sigma^2 = 13{\cdot}8/6{\cdot}25 = 2{\cdot}21$ which is just below the 99% fractile of χ^2/f for $f = 11$.

6. From the W-distribution we find $w/\sigma = (17-3)/2{\cdot}5 = 5{\cdot}6$ which is just above the 99·5% limit of significance.

Rejecting the largest observation the sample mean and standard deviation become 7·2 and 2·52, respectively.

12.9. References.

The following are important publications dealing with the range:

L. H. C. Tippett: *On the Extreme Individuals and the Range of Samples*

Taken from a Normal Population, Biometrika, 17, 1925, 364–87. This paper includes a table of $\mathcal{M}\{W\} = a_n$ from $n = 1 - 1000$.

STUDENT: *Errors of Routine Analysis*, Biometrika, 19, 1927, 151–164. The author discusses the criterion $\frac{w}{\sigma} > W_P$ as a basis for rejecting outlying observations.

E. S. PEARSON: *The Percentage Limits for the Distribution of Range in Samples from a Normal Population*, ($n \leq 100$), Biometrika, 24, 1932, 404–17.

O. L. DAVIES and E. S. PEARSON: *Methods of Estimating From Samples the Population Standard Deviation*, Suppl. Journ. Roy. Stat. Soc., 1, 1934, 76–93. A comparison of confidence limits for σ calculated from s and from w.

E. S. PEARSON and J. HAINES: *The Use of Range in Place of Standard Deviation in Small Samples*, Suppl. Journ. Roy. Stat. Soc., 2, 1935, 83–98. A discussion of control diagrams for s and w and of the correlation between s and w.

E. S. PEARSON: *The Probability Integral of the Range in Samples of n Observations From a Normal Population*, Biometrika, 32, 1942, 301–08. Table of $P_n\{W\}$ for $n = 2(1)20$ and of W_P for $n = 2(1)12$.

E. S. PEARSON and H. O. HARTLEY: *Tables of the Probability Integral of the Studentized Range*, Biometrika, 33, 1943, 89–99. A table of the cumulative distribution function of $q = w/s$ for $n = 2(1)20$ and of q_P for $P =$ 1, 5, 95, and 99%.

CHAPTER 13.

STATISTICAL CONTROL

13.1. Introduction.

In the following paragraphs will be discussed the concept of statistical control and criteria for detecting lack of control, especially for continuous variables. These methods have been developed and applied first of all in the control of quality of manufactured products, but their applicability is much wider, and all tests of significance presuppose a state of statistical control.

The fundamental principles in the theory and practice of statistical quality control are due to W. A. SHEWHART.

13.2. The State of Statistical Control. Randomness.

As stated in § 5.4 a sample of n observations, $x_1, x_2, \ldots, x_n$, from a population with distribution function $p\{x\}$ is called *a random sample from that population* if

$$P\{x_1, x_2, \ldots, x_n\} = P\{x_1\}P\{x_2\}\ldots P\{x_n\} . \qquad (13.2.1)$$

It follows that the $n!$ different possible orders of n given sample values are all equally likely when the sampling is random, since the value of $P\{x_1, x_2, \ldots, x_n\}$ is independent of permutations of the x's when (13.2.1) is satisfied. A general definition of randomness of a sequence of n observations from the same population in terms of the magnitude and order of these observations therefore seems impossible.

If, however, an alternative hypothesis can be specified, it is possible to decide upon what sequences shall be taken as indicating that the n observations are not randomly drawn from the same population. The alternative hypothesis may, for instance, state that the population mean has increased during the drawing of the sample. The $n!$ different sequences may then be classified in two groups, the one group (the critical region) containing sequences showing an upward trend, i. e., sequences where the n successive observations occur nearly in ascending order of magnitude, and the other group containing all other sequences. If the sequence observed is a member of the first group, the test hypothesis will be rejected.

In general, the alternative hypothesis will lead to a critical region defined by a certain *relationship between order and magnitude* of the observations in the sequence.

The definition (13.2.1) specifies a *mathematical operation* which forms the basis for the whole statistical theory of random sampling distributions. As all other definitions in mathematics it is a purely formal definition and (as usual) it is impossible a *priori* to decide whether or not a mathematical model based on this concept will be suitable for the description and prediction of physical phenomena. *Experience, however, has shown that certain physical operations exist which by repetition lead to results that may be predicted from probability theory, using the concept* (13.2.1) *of randomness.* Such operations are called random operations, and a series of results produced by repetition of a random operation is called a random sample.

As examples of random operations we may mention the classical *games of chance* which were the first objects of the probability calculus. It is a long established fact within this domain that the results produced by repetition of such games may be predicted by means of the "Laws of Chance".

Within *other fields*, however, it is not advisable to assume a priori that the repetition of a given operation will lead to results showing only random variations, even if the repetitions are performed under what is customarily called "the same essential conditions".

As a simple *standard example of a random operation* we consider the following experiment: N physically similar chips, on each of which is written a number, are placed in a bowl. After thorough mixing, a chip is drawn, the number recorded, and the chip replaced in the bowl. Repeating this operation n times, we get a random sample of n numbers, and proceeding in this way we get an infinite series of random samples, each of n observations. SHEWHART[1]) has reported the results of 4000 drawings from each of three bowls where the chips have been marked with three different sets of numbers. Further, SHEWHART has given the values of the means and standard deviations and the ratio of these numbers for each set of 4 observations taken in the order in which they have been drawn from the bowl. The results of a similar experiment consisting of 15000 drawings from a bowl with 10 chips marked with the numbers 0, 1, . . . , 9 have been reported in Table XIX, the table of random sampling numbers.

A probability model of this standard experiment may easily be made, assigning to each chip the same probability, $1/N$, of being drawn. The probabilities of the various possible samples of n observations may then

[1]) W. A. SHEWHART: *Economic Control of Quality of Manufactured Product*, Appendix II, Van Nostrand, New York, 1931.

be computed by means of the probability calculus, using the assumption (13.2.1). In this way we may derive "probability limits" for the sample means and variances or other statistics calculated from the samples, as shown in Chapters 8–12. Experience has shown that the variations of such sample statistics may be predicted from probability theory when the sample values are the results of drawings from a bowl in the manner described above.

TABLE 13.1.
200 sums of four random numbers.

20	19	18	19	22	29	26	12	15	16
17	13	18	20	11	26	16	26	17	18
26	20	30	24	17	9	18	8	14	23
12	19	20	17	23	26	19	23	25	21
11	14	15	17	19	23	12	16	12	12
11	20	21	20	28	15	24	17	10	15
16	17	21	9	15	22	18	13	17	14
28	23	23	19	17	20	12	22	18	19
8	20	9	16	21	23	9	19	18	16
24	24	28	21	14	5	26	26	19	18
23	12	14	26	18	21	18	24	21	15
26	15	23	24	11	17	27	28	22	19
13	21	24	12	21	16	23	12	23	16
11	10	21	19	14	7	23	18	19	15
21	24	14	16	10	17	17	12	18	20
14	18	13	12	16	24	17	24	18	6
24	11	23	14	17	17	12	27	21	21
21	23	16	12	21	22	17	14	17	27
23	23	29	12	17	16	13	22	30	27
18	20	24	21	19	11	18	13	19	20

As an example, consider the 200 sums of four random numbers given in Table 13.1. In Example 8.2 the agreement between the frequency distribution of these 200 sums and the corresponding probability distribution has been demonstrated. In the frequency distribution, however, the observations are ranked according to magnitude, whereas the *order* of the observations has been disregarded. Table 13.1 gives an example of random order, and for further illustration the first 100 sums are plotted on a control chart in Fig. 13.1. (The sums are recorded in the order found in Table XIX as explained in Ex. 8.2).

Probability theory predicts that in the long run 99·9% of the number of sums will fall within the interval $2 \leq z \leq 34$ and that 95·8% will fall within the interval $7 \leq z \leq 29$. These limits are shown on Fig. 13.1, and further results of this kind may be derived from the distribution of z.

The control chart shows a typical example of random order, where *no simple relationship seems to exist between the sample number* (*the order of drawing*) *and the magnitude of the result.*

A simple characteristic of random order may be obtained by considering *runs* of points above or below the median and runs up and down. If the

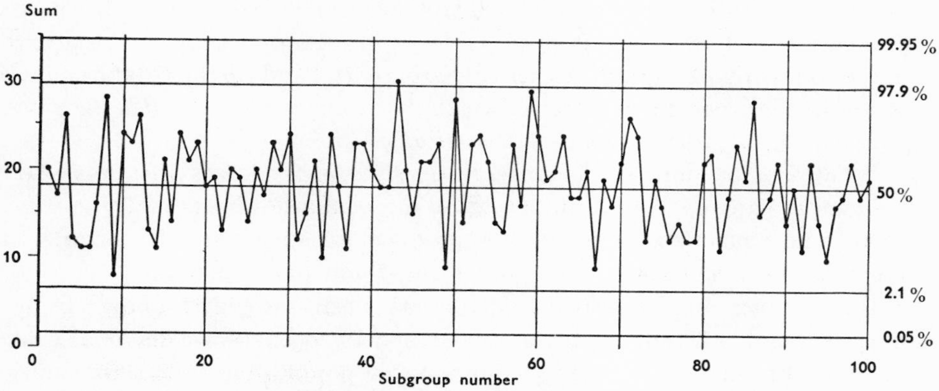

FIG. 13.1. Control chart for 100 sums of four random numbers.

observations are randomly drawn from the same population we do not expect very long runs of either type, and the occurrence of such runs will therefore usually be taken as indicating nonrandomness. Tests of significance based on runs will be discussed in §§ 13.3 and 13.4.

Since the numbers written on the chips may be chosen at will it follows that the *form* of the distribution function has nothing to do with the concept of a random operation.

A phenomenon is usually said to be under control when it is possible to predict its future course. Accordingly a phenomenon will be called *statistically controlled* when prediction is possible in the probability sense, i. e., when *the probability of the phenomenon falling within specified limits may be predicted. The results of a random operation will therefore be statistically controlled.*

A series of observations produced by repetition of a given operation will usually not be under statistical control, even if the conditions are assumed to be essentially the same. It is always necessary to test the series for randomness to find out whether or not the conditions really are essentially the same.

In a production process many causes or factors which produce variations in the quality of the product may not be completely under control. As examples of such causes may be mentioned tool wear, different sources of raw materials, differences in machines and operators, fluctuations in temperature and humidity, etc. These causes may produce trends in the quality of the product or sudden fluctuations in the quality level, the result being that the quality variation can not be predicted.

SHEWHART has introduced the terminology that the system of (unknown) causes corresponding to a state of statistical control is called *a constant system of chance causes*, whereas the unknown causes of variability which disturb the state of statistical control are called *assignable* causes. Experience has shown that it is possible by means of certain criteria given in the following paragraphs to *detect lack of control* and further that *assignable causes* in these cases *may be found and brought under control so that the process gradually approaches a state of statistical control.*

What can be done by means of statistics is to detect the presence of assignable causes, to find out at what moment the effects of these causes occur, and what type of effect they produce. This may help the engineer to identify the assignable causes and bring them under control.

Randomness or statistical control has been discussed above from a narrow point of view in theoretical respects, because we have discussed only the problem of sampling from a stable population. If the distribution function depends on a parameter, time say, in a known way, (13.2.1) may be replaced by

$$P\{x_1, x_2, \ldots, x_n;\ t_1, t_2, \ldots, t_n\} = P\{x_1;\ t_1\}P\{x_2;\ t_2\}\ldots P\{x_n;\ t_n\} \quad (13.2.2)$$

where $P\{x;\, t\}$ denotes the cumulative distribution function of x for given t. For example, the quality of a product may depend in a known way on tool wear. Problems of this kind are not essentially different from the above and may be treated by means of regression analysis as developed in Chapters 18 and 20.

13.3. Runs above and below the Median.

Consider a random arrangement of n elements consisting of n_1 a's and n_2 b's, $n = n_1+n_2$, such as the following arrangement of 11 a's and 14 b's:

$$b\,a\,b\,a\,a\,a\,a\,b\,a\,b\,b\,b\,a\,a\,b\,a\,b\,b\,b\,b\,b\,a\,b\,b\,a\,. \quad (13.3.1)$$

A sequence of elements of the same kind is called a run, and *the length of a run* is given by the number of elements defining the run. The sequence above begins with a run of one b, then follows a run of one a, a run of one b, a run of 4 a's, a run of one b, etc. The sequence contains a total number of runs of 14, namely 7 runs of a's and 7 runs of b's. The difference between the numbers of runs of the two kinds of elements must be either one or zero, depending on whether the whole sequence begins and ends with elements of the same kind or of different kinds.

Further, we introduce the notation:

r_{1i} = the number of runs of a's of length i,

r_{2i} = the number of runs of b's of length i,

r_i = $r_{1i}+r_{2i}$ = the number of runs of both kinds of elements of length i,

$R_{1k} = \sum_{i=k}^{n_1} r_{1i}$ = the number of runs of a's of length k or more,

$R_{2k} = \sum_{i=k}^{n_2} r_{2i}$ = the number of runs of b's of length k or more,

R_k = $R_{1k}+R_{2k}$ = the number of runs of both kinds of elements of length k or more,

R_1 = R = the total number of runs of both kinds of elements.

For the sequence (13.3.1) we find the results given in Table 13.2. The 14 runs are distributed according to length, 9 runs being of length 1, 2 runs of length 2, and 1 run of each of the lengths 3, 4 and 5.

TABLE 13.2.

Distribution of runs of a's and b's according to length of run.

i	r_{1i}	r_{2i}	r_i	R_{1i}	R_{2i}	R_i
1	5	4	9	7	7	14
2	1	1	2	2	3	5
3	0	1	1	1	2	3
4	1	0	1	1	1	2
5	0	1	1	0	1	1

Assuming that all possible arrangements occur with the same probability we may find the distribution function of the r_{ij}'s by means of combinatorial theory. We will only give the most important results without proofs; references to the adequate literature may be found in § 13.10.

The mean number of runs of a's of length i in random arrangements of n elements is given by

$$\mathcal{M}\{r_{1i}\} = \frac{n_1^{(i)}(n_2+1)^{(2)}}{n^{(i+1)}}, \quad i = 1, 2, \ldots, n_1\,, \tag{13.3.2}$$

from which it follows that

$$\mathcal{M}\{R_{1k}\} = \frac{n_1^{(k)}(n_2+1)}{n^{(k)}}, \quad k = 1, 2, \ldots, n_1\,, \tag{13.3.3}$$

where

$$n^{(i)} = n(n-1)\ldots(n-i+1)\,. \tag{13.3.4}$$

Introducing

$$\frac{n_1}{n} = \theta_1 \quad \text{and} \quad \frac{n_2}{n} = \theta_2, \quad \theta_1+\theta_2 = 1\,, \tag{13.3.5}$$

we find for large values of n

$$\mathcal{M}\{r_{1i}\} \simeq n \cdot \theta_1^i \cdot \theta_2^2 \tag{13.3.6}$$

and

$$\mathcal{M}\{R_{1k}\} \simeq n \cdot \theta_1^k \cdot \theta_2\,. \tag{13.3.7}$$

From these results the means of the other quantities defined above may be found.

In what follows we will only treat the case where $n_1 = n_2 = m$, i. e., $n = 2m$ and $\theta_1 = \theta_2 = \frac{1}{2}$. The formulas above lead to

$$\mathcal{M}\{r_{1i}\} = \mathcal{M}\{r_{2i}\} = \frac{(m+1)^{(i+1)}m}{n^{(i+1)}} \simeq \frac{n}{2^{i+2}} \tag{13.3.8}$$

and

$$\mathcal{M}\{R_{1k}\} = \mathcal{M}\{R_{2k}\} = \frac{(m+1)^{(k+1)}}{n^{(k)}} \simeq \frac{n}{2^{k+1}}\,. \tag{13.3.9}$$

Successive values of $\mathcal{M}\{R_{1k}\}$ may easily be computed from the formula

$$\mathcal{M}\{R_{1,k+1}\} = \mathcal{M}\{R_{1k}\}\frac{m-k}{n-k}\,, \mathcal{M}\{R_{11}\} = \frac{m+1}{2}, \tag{13.3.10}$$

whereafter

$$\mathcal{M}\{r_{1i}\} = \mathcal{M}\{R_{1i}\} - \mathcal{M}\{R_{1,i+1}\}\,. \tag{13.3.11}$$

For small values of n the distribution function of R_{1i} may be tabulated by writing down all possible arrangements of a's and b's, as shown in Table 13.3 for $n = 6$. The number of different arrangements is $\binom{n}{m} = \binom{6}{3} = 20$, but from the 10 arrangements beginning with an a the remaining 10 may be written down interchanging a and b. The grouped distribution in Table 13.4 has been derived from the ungrouped distribution in Table 13.3. The distribution of the total number of runs of both kinds of elements is a symmetrical unimodal distribution. The same holds for the distributions of the total number of runs of each kind of elements, whereas the distributions of the number of runs of length 2 or more are skew.

The expected number of runs of both kinds of elements of length k or more is

$$\mathcal{M}\{R_k\} = 2\mathcal{M}\{R_{1k}\} = \frac{2(m+1)^{(k+1)}}{n^{(k)}} \simeq \frac{n}{2^k}, \tag{13.3.12}$$

from which we get the expected total number of runs as

TABLE 13.3.
Ungrouped distribution of number of runs for $n = 2m = 6$.

No.	Arrangement	R	R_{11}	R_{12}	R_{13}	R_{21}	R_{22}	R_{23}
1	*a a a b b b*	2	1	1	1	1	1	1
2	*a a b b b a*	3	2	1	0	1	1	1
3	*a b b b a a*	3	2	1	0	1	1	1
4	*a a b a b b*	4	2	1	0	2	1	0
5	*a a b b a b*	4	2	1	0	2	1	0
6	*a b a a b b*	4	2	1	0	2	1	0
7	*a b b a a b*	4	2	1	0	2	1	0
8	*a b a b b a*	5	3	0	0	2	1	0
9	*a b b a b a*	5	3	0	0	2	1	0
10	*a b a b a b*	6	3	0	0	3	0	0
11	*b a b a b a*	6	3	0	0	3	0	0
12	*b a a b a b*	5	2	1	0	3	0	0
13	*b a b a a b*	5	2	1	0	3	0	0
14	*b a a b b a*	4	2	1	0	2	1	0
15	*b a b b a a*	4	2	1	0	2	1	0
16	*b b a a b a*	4	2	1	0	2	1	0
17	*b b a b a a*	4	2	1	0	2	1	0
18	*b a a a b b*	3	1	1	1	2	1	0
19	*b b a a a b*	3	1	1	1	2	1	0
20	*b b b a a a*	2	1	1	1	1	1	1
	Total:	80	40	16	4	40	16	4
	Mean:	4	2	0·8	0·2	2	0·8	0·2

TABLE 13.4.
Distribution of number of runs for $n = 2m = 6$.

Number of runs (R)	$20p\{R\}$	$20p\{R_{11}\}$	$20p\{R_{12}\}$	$20p\{R_{13}\}$
0	—	—	4	16
1	—	4	16	4
2	2	12	—	—
3	4	4	—	—
4	8	—	—	—
5	4	—	—	—
6	2	—	—	—
Total:	20	20	20	20
Mean:	4	2	0·8	0·2

$$\boxed{m\{R\} = \frac{n+2}{2}.} \tag{13.3.13}$$

It follows from (13.3.12) that the sample size for which the expected number of runs of both kinds of elements of length k or more equals one is approximately $n \simeq 2^k$ for large k. In random arrangements of 64 a's and 64 b's, say, runs of length 7 or more therefore must be expected to occur on the average once per sample.

The distribution of R, the total number of runs, is approximately normal for $n > 20$ with mean given by (13.3.13) and variance

$$\boxed{v\{R\} = \frac{n}{4}\left(1 - \frac{1}{n-1}\right) \simeq \frac{n-1}{4}.} \tag{13.3.14}$$

Taking into account the discontinuity of R we find

$$P\{R \leq R_P\} \simeq \Phi\left(\frac{R_P + \frac{1}{2} - \frac{n+2}{2}}{\frac{1}{2}\sqrt{n-1}}\right) = P \tag{13.3.15}$$

which leads to

$$R_P \simeq \tfrac{1}{2}(n+1+u_P\sqrt{n-1}). \tag{13.3.16}$$

For $n = 20$ and $P = 0{\cdot}025$, say, we find

$$R_{.025} \simeq \tfrac{1}{2}(21 - 1{\cdot}96\sqrt{19}) = 6{\cdot}2,$$

so that, taking $R_P = 6$, P will be slightly less than 2·5%. The probability of getting 6 or less runs in random arrangements of 10 a's and 10 b's is consequently less than 2·5%.

Similarly, the distribution of R_{11}, the total number of runs of a's, is approximately normal for $n > 20$ with mean

$$m\{R_{11}\} = \frac{n+2}{4} \tag{13.3.17}$$

and variance

$$v\{R_{11}\} = \frac{n+2}{16}\left(1 - \frac{1}{n-1}\right) \simeq \frac{n+1}{16}, \tag{13.3.18}$$

which leads to

$$R_{11,P} \simeq \tfrac{1}{4}(n+u_P\sqrt{n+1}). \tag{13.3.19}$$

Since

$$R = R_{11} + R_{21} = 2R_{11} + \varepsilon, \tag{13.3.20}$$

where $\varepsilon = R_{21} - R_{11}$ is a random variable with mean zero, taking on only the values -1, 0 and $+1$, the above results might be expected from (13.3.13)–(13.3.16).

The distribution of R_{1k} for $k \geq 2$ is skew, but converges toward the normal distribution for $n \to \infty$ with mean given by (13.3.9) and variance

$$\mathcal{V}\{R_{1k}\} \simeq \frac{n}{2^{k+1}}\left(1-\frac{1}{2^k}\left(1+\frac{k^2}{2}\right)\right) \quad \text{for } n \text{ large}. \qquad (13.3.21)$$

How large n has to be before the normal approximation may be used has not been investigated, but it seems reasonable to require that $n > 2^{k+4}$, i. e., $\mathcal{M}\{R_{1k}\} > 9$. This, however, is so large a sample size, even for moderately large values of k, that the approximation formula is of no practical importance.

For $k \geq 8$ and moderately large values of n, however, we have approximately

$$\mathcal{M}\{R_{1k}\} \simeq \mathcal{V}\{R_{1k}\} \simeq \frac{n}{2^{k+1}}, \qquad k \geq 8. \qquad (13.3.22)$$

This indicates that the POISSON-distribution may be used as an approximation to $p\{R_{1k}\}$ (the POISSON-distribution will be dealt with in detail in Chapter 22), so that the probability of getting ν runs of a's of length k or more is

$$P\{R_{1k} = \nu\} \simeq \frac{\xi^\nu}{\nu!} \cdot e^{-\xi}, \quad \text{where } \xi = \mathcal{M}\{R_{1k}\}. \qquad (13.3.23)$$

Hence the probability of getting at least one run of a's of length k or more is

$$P\{R_{1k} \geq 1\} = 1-P\{R_{1k} = 0\} \simeq 1-e^{-\mathcal{M}\{R_{1k}\}}. \qquad (13.3.24)$$

From this formula we may determine a value of k, so that $P\{R_{1k} \geq 1\}$ takes on a preassigned small value. For example,

$$P\{R_{1k} \geq 1\} \simeq 1-e^{-\frac{n}{2^{k+1}}} = 0{\cdot}05 \qquad (13.3.25)$$

leads to

$$n \simeq 0{\cdot}0513 \times 2^{k+1} \qquad (13.3.26)$$

or

$$k_{\cdot 05} \simeq 3{\cdot}3\ (\log n+1). \qquad (13.3.27)$$

For $n = 30$, say, we obtain

$$k_{\cdot 05} \simeq 3{\cdot}3\,(\log 30+1) = 8{\cdot}2,$$

i. e., the probability is only 5% of getting at least one run of 8 or more a's in random arrangements of 15 a's and 15 b's.

Formula (13.3.27) gives values of $k_{\cdot 05}$ larger than the correct values, because $\mathcal{V}\{R_{1k}\}$ is less than $\mathcal{M}\{R_{1k}\}$ even for $k \geq 8$, see (13.3.21). On comparison with the exact values, however, it will be found that the approximation formula leads to the correct results for $n = 20$, 30, 40 and 50, if the values found are *rounded down* instead of to the nearest integer. It is believed that rounding should follow the usual rules for $k_{\cdot 05} > 11$.

It is not advisable to use the POISSON-approximation for values of P considerably less than 0·05.

For large values of k the correlation between R_{2k} and R_{1k} will be small, so that $m\{R_k\} \simeq v\{R_k\} \simeq n/2^k$. The solution of the equation $P\{R_k \geq 1\} =$ 0·05 will therefore be $k_{\cdot 05}+1$.

P. S. OLMSTEAD[1]) has tabulated the exact probabilities $P\{R_{1k} \geq 1\}$ as a function of k for $n = 10$, 20, and 40. By graphical interpolation and extrapolation in this table we may for each k determine the largest (even) integer for which $P\{R_{1k} \geq 1\} \leq 0{\cdot}05$. Similarly, we may determine the corresponding n's when runs of either a's or b's or both, or runs of both a's and b's, are considered, i. e., the largest (even) integers for which

$$P\{R_{1k} \geq 1 \text{ or } R_{2k} \geq 1 \text{ or both}\} = P\{R_k \geq 1\} \leq 0{\cdot}05$$

and

$$P\{R_{1k} \geq 1 \text{ and } R_{2k} \geq 1\} \leq 0{\cdot}05 \,.$$

The results are given in Table 13.5.

TABLE 13.5.

The largest sample size for which the probability is less than or equal to 5% of getting at least one run of length k or more of one kind, either kind, and each kind of elements (Approximate values).

Length of run	Either kind	One kind	Each kind
5	10	10	16
6	14	18	32
7	22	28	64
8	34	48	120
9	54	80	230
10	86	130	
11	140	230	
12	230	420	

It follows from this table that runs of 8 or more a's, say, may be expected to occur on the average with a frequency less than 5% in random arrangements of m a's and m b's, $n = 2m$, for $n \leq 48$.

Comparing these results with (13.3.26) it will be found that (13.3.26) gives too small values of n. For example, for $k = 8$ we find

$$n \simeq 0{\cdot}0513 \times 2^9 = 26$$

whereas the correct value is about 48.

[1]) Published by F. MOSTELLER: *Note on an Application of Runs to Quality Control Charts*, Ann. Math. Stat., 12, 1941, 228–232.

The theory above may be applied to samples from a population with a continuous distribution function by *classifying sample values larger than the sample median as a's and values smaller than the median as b's*. If the number of observations is odd, $n = 2m+1$, the median is found as $x_{(m+1)}$, the $(m+1)$st of the observations ranked according to magnitude, and if $n = 2m$ the median is any number between $x_{(m)}$ and $x_{(m+1)}$. For $n = 2m+1$ we ignore $x_{(m+1)}$, so that in both cases we have m numbers above (a's) and m numbers below (b's) the sample median.

Under the assumption of statistical control, we therefore know the distribution of the total number of runs and of the number of very long runs.

Suppose the hypothesis alternative to the hypothesis of statistical control is that the level of the distribution has shifted during the drawing of the sample. If, for example, a shift to a higher level takes place, then one or more long runs may be expected to occur, and the total number of runs decreases. Therefore, *the hypothesis of statistical control may be tested against the alternative hypothesis of shifts in the level of the distribution by means of the total number of runs, a small number of runs being significant, and further by means of the length of runs, the occurrence of very long runs being significant.* It must be remembered, however, that these tests are not independent.

In some cases *the population median* is known or a hypothetical value is tested. Sample values are then classified as below or above the population median. It will be noted that in this case the number of a's will be a random variable, having a binomial distribution with $\theta = 0{\cdot}5$. In spite of this difference the distribution of the number of runs will be very nearly the same as that found above, even for moderately large values of n. We give the two formulas analogous to (13.3.9) and (13.3.21):

$$\mathcal{M}\{R_{1k}\} = \mathcal{M}\{R_{2k}\} = \frac{n-k+2}{2^{k+1}} \simeq \frac{n}{2^{k+1}} \tag{13.3.28}$$

and

$$\mathcal{V}\{R_{1k}\} = \mathcal{V}\{R_{2k}\} \simeq \frac{n}{2^{k+1}}\left(1-\frac{2k+1}{2^{k+1}}\right). \tag{13.3.29}$$

Example 13.1. Consider the sequence of 200 diameters of rivet heads in the order of observation as shown in Table 3.1, reading down the successive columns, see also Fig. 8.4. It follows from Table 3.2 that the median diameter is 13·41 mm, 97 diameters being less than 13·41, 4 equal to 13·41, and 99 larger than 13·41. To obtain an equal number of observations on each side of the median two of the observations equal to the median are counted as below the median. The two observations are chosen among the four in such a way that the number of runs below the median does not increase, i. e., the two observations are chosen in the way least favourable to the hypothesis of statistical control. For the 99 observations on each side of the median we obtain the distribution of runs given in Table 13.6.

TABLE 13.6.
Distribution of runs below and above the median according to length of run.

Length of run	Number of runs			Expected number of runs		
i	Below r_{1i}	Above r_{2i}	Both r_i	Below (Above) $\mathcal{m}\{r_{1i}\} = \mathcal{m}\{r_{2i}\}$	Both $\mathcal{m}\{r_i\}$	Cumulative $\mathcal{m}\{R_{1k}\}$
1	21	26	47	25·1	50·2	50·0
2	16	11	27	12·6	25·2	24·9
3	6	5	11	6·2	12·4	12·3
4	7	5	12	3·1	6·2	6·1
5	0	2	2	1·5	3·0	3·0
≥ 6	0	1	1	1·5	3·0	1·5
Total:	50	50	100	50·0	100·0	—

The one run of length ≥ 6 is of length 6.

The expected number of runs may be found from (13.3.10) for $m = 99$ and $n = 198$. First we find $\mathcal{m}\{R_{11}\} = 50$, and then

$$\mathcal{m}\{R_{1,k+1}\} = \mathcal{m}\{R_{1k}\}\frac{99-k}{198-k},$$

so that $\mathcal{m}\{R_{12}\} = 50\times 98/197 = 24{\cdot}9$, $\mathcal{m}\{R_{13}\} = 24{\cdot}9\times 97/196 = 12{\cdot}3$, etc., as shown in the last column of Table 13.6. From these results $\mathcal{m}\{r_{1i}\}$ are obtained by computing the differences between successive numbers.

By accident the total number of runs equals the expected total number of runs, so there is no need for carrying out the test based on this criterion.

The run of greatest length is a run above the median of length 6, but this is not an exceptional length since (13.3.27) for $n = 198$ gives $k_{\cdot 05} = 11$, see also Table 13.5.

The most conspicuous deviation between the observed and expected numbers of runs is found for $i = 4$, where 12 runs are observed but only 6 expected. If there is any disagreement between the data and the hypothesis of statistical control, it seems as if the data show more regularity than expected.

Example 13.2. In an experiment with a mixing apparatus two portions of the same substance were placed in the mixer, the lower layer having a high concentration of a certain component and the upper layer a low concentration. After mixing, the substance was tapped off through a hole in the bottom of the mixer and 58 samples were taken at regular intervals. Table 13.7 shows the concentration for these samples.

The problem is to test the homogeneity of the substance, i. e., to test whether or not the samples represent a state of statistical control. The alternative hypothesis will be that the concentration is high for the first samples and low for the last since a poor mixing will not be able to destroy the original two layers completely.

TABLE 13.7.

Concentration of a certain component in 58 successive samples.

38	*33*	*29*	*16*	*44*	*21*	*16*	*17*	*19*	1	*22*	*28*	*22*	14	7
13	*21*	15	*34*	*23*	15	*19*	*32*	*24*	14	13	*22*	8	*30*	11
15	*24*	*26*	14	11	*25*	*17*	10	*19*	5	6	*16*	7	10	1
5	2	8	14	14	15	*16*	13	11	9	11	*19*	*21*		

Since 29 observations are less than or equal to 15, the median is between 15 and 16. Runs above the median are italicized in Table 13.7, where successive observations are placed along the lines. The observations are shown on Fig. 13.2 together with the median.

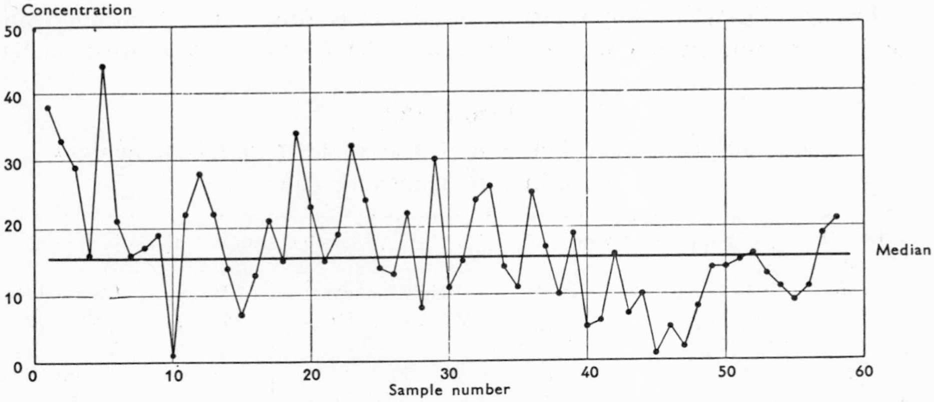

FIG. 13.2. Concentration of a certain component in 58 successive samples.

The distribution of the runs according to length is given in Table 13.8. The two runs of length ≥ 5 are both of length 9.

The total number of runs is five less than the expected number, $R - \mathcal{M}\{R\} = 25 - 30 = -5$. From (13.3.14), it follows that $\mathcal{V}\{R\} \simeq 14{\cdot}25$ so that $\sigma\{R\} = 3{\cdot}8$. Hence, the deviation is not significant.

TABLE 13.8.
Distribution of runs according to length.

Length of run	Number of runs			Expected number of runs		
i	Below r_{1i}	Above r_{2i}	Both r_i	Below (Above) $m\{r_{1i}\}$	Both $m\{r_i\}$	Cumulative $m\{R_{1k}\}$
1	5	6	11	7·6	15·2	15·0
2	4	4	8	3·8	7·6	7·4
3	1	2	3	1·9	3·8	3·6
4	1	0	1	0·9	1·8	1·7
≥ 5	1	1	2	0·8	1·6	0·8
Total:	12	13	25	15·0	30·0	—

From (13.3.27) we obtain

$$k_{\cdot 05} \simeq 3{\cdot}3\ (\log 58+1) = 9{\cdot}1$$

so that the occurrence of one run of length 9 on each side of the median may be considered highly significant. Table 13.5 shows that the 5% significance limit for k is $k = 7$ for $34 \leq n \leq 64$, when the critical region consists of long runs on each side of the median.

Thus the mixing has not been quite satisfactory.

Example 13.3. The hypothetical distribution function of the 200 sums of four random numbers given in Table 13.1 is symmetrical with median

TABLE 13.9.
Distribution of runs below and above the population median according to length of run.

Length of run	Number of runs			Expected number of runs		
i	Below r_{1i}	Above r_{2i}	Both r_i	Below (Above) $m\{r_{1i}\} = m\{r_{2i}\}$	Both $m\{r_i\}$	Cumulative $m\{R_{1k}\}$
1	28	31	59	25·3	50·5	50·3
2	11	6	17	12·6	25·2	25·0
3	6	8	14	6·2	12·4	12·4
4	4	1	5	3·1	6·2	6·2
5	2	4	6	1·6	3·2	3·1
6	0	0	0	0·7	1·4	1·5
7	0	1	1	0·4	0·8	0·8
≥ 8	0	1	1	0·4	0·8	0·4
Total:	51	52	103	50·3	100·5	—

equal to 18. The hypothesis of statistical control may therefore be tested by means of the distribution of runs above and below the population median. Observations equal to the median may be classified as above or below by tossing a coin. From Table 13.1 we obtain the distribution shown in Table 13.9.

The one run of length ≥ 8 is of length 8.

The deviation between the total number of runs and the expected number is $103-100{\cdot}5 = 2{\cdot}5$, whereas the standard deviation is $\sigma\{R\} \simeq \sqrt{200/4} = 7{\cdot}1$, so that the deviation is not significant.

The run of greatest length is a run above the median of length 8, which is not exceptional since (13.3.27) for $n = 200$ gives $k_{.05} = 11$.

Thus the run tests do not lead to rejection of the hypothesis of statistical control.

13.4. Runs up and down.

Consider a sequence of n different observations, $x_1, x_2, \ldots, x_n$, and the *sequence of signs* (+ or −) of the $(n-1)$ differences $x_{i+1}-x_i$. A sequence of successive + signs is called a *run up* and a sequence of successive − signs a *run down*. The length of a run is given by the number of equal signs defining the run. The total number of runs is denoted by R, the number of runs of length i by r_i, and the number of runs of length k or more by $R_k = \sum_{i=k}^{n-1} r_i$.

For example, the sequence of 15 elements:

$$39,\ 42,\ 38,\ 53,\ 51,\ 30,\ 40,\ 28,\ 43,\ 46,\ 52,\ 55,\ 29,\ 24,\ 34\ , \qquad (13.4.1)$$

leads to the following sequence of signs of differences between successive elements:

$$+\ -\ +\ -\ -\ +\ -\ +\ +\ +\ +\ -\ -\ + \qquad (13.4.2)$$

The sequence is thus characterized by 4 runs up of length 1, one run up of length 4, 2 runs down of length 1, and 2 runs down of length 2, giving a total of 9 runs.

Assuming that all $n!$ possible arrangements of the n numbers occur with the same probability, we have the following mean number of runs:

$$\mathcal{m}\{r_i\} = \frac{2}{(i+3)!}[n(i^2+3i+1)-(i^3+3i^2-i-4)]\,, \quad i \leq n-2\,, \qquad (13.4.3)$$

and

$$\mathcal{m}\{R_k\} = \frac{2}{(k+2)!}[n(k+1)-(k^2+k-1)]\,, \quad k \leq n-1\,, \qquad (13.4.4)$$

see Table 13.10.

TABLE 13.10.

The expected number of runs up and down of length k or more in random arrangements of n different numbers.

k	$\mathcal{m}\{R_k\}$
1	$\frac{1}{3}(2n-1)$
2	$\frac{1}{12}(3n-5)$
3	$\frac{1}{60}(4n-11)$
4	$\frac{1}{360}(5n-19)$
5	$\frac{1}{2520}(6n-29)$
6	$\frac{1}{20160}(7n-41)$
7	$\frac{1}{181440}(8n-55)$

From Table 13.10 may be found the number of observations giving $\mathcal{m}\{R_k\} = 1$ for given k. For $k = 5$, say, we find $n = 2549/6 \simeq 425$, i. e., in random arrangements of 425 observations we expect only one run up or down of length 5 or more per arrangement.

For the total number of runs we have

$$\boxed{\mathcal{m}\{R\} = \tfrac{1}{3}(2n-1)} \tag{13.4.5}$$

and

$$\boxed{\mathcal{v}\{R\} = \tfrac{1}{90}(16n-29)\ .} \tag{13.4.6}$$

For $n > 20$, R may be regarded as normally distributed with good approximation.

A test of significance based on long runs may be found by means of the POISSON approximation since

$$\mathcal{v}\{R_k\} \simeq \mathcal{m}\{R_k\} \quad \text{for} \quad k \geq 5\ . \tag{13.4.7}$$

This leads to

$$P\{R_k \geq 1\} \simeq 1-e^{-\mathcal{m}\{R_k\}}, \qquad k \geq 5\ . \tag{13.4.8}$$

Inserting (13.4.4) and solving with respect to n we obtain

$$n \simeq k - \frac{1}{k+1} - \frac{k+2}{2} k! \log_e (1-P\{R_k \geq 1\})\ . \tag{13.4.9}$$

Thus, for given P and k, we may determine the value of n required to give a probability of P that a random arrangement will have at least one run of length k or more.

For $P = 0{\cdot}05$ and $k = 5$, say, we find

$$n \simeq 5 - \tfrac{1}{6} + \tfrac{7}{2} \times 120 \times 0{\cdot}05130 = 26{\cdot}4\,,$$

i. e.,

$$P\{R_5 \geq 1\} \leq 0{\cdot}05 \quad \text{for} \quad n \leq 26\,. \tag{13.4.10}$$

Similarly,

$$P\{R_6 \geq 1\} \leq 0{\cdot}05 \quad \text{for} \quad n \leq 153\,, \tag{13.4.11}$$

and

$$P\{R_7 \geq 1\} \leq 0{\cdot}05 \quad \text{for} \quad n \leq 1170\,. \tag{13.4.12}$$

Runs of length 7 or more will therefore be expected to occur very seldom in samples of size less than 1170.

Suppose the hypothesis alternative to the hypothesis of statistical control is that *a gradual change in the level of the distribution* has taken place during the drawing of the sample. Such a change will produce trends or cycles in the observations, so that one or more long runs may be expected to occur and the total number of runs will be small. The hypothesis of statistical control may therefore be tested by means of the total number of runs, a small number of runs being significant, and further by means of the length of runs, very long runs being significant.

Example 13.4. From Table 3.1 we find the distribution of runs up and down shown in Table 13.11. In the 5 cases where two successive diameters are equal we disregard the one observation.

TABLE 13.11.
Distribution of runs up and down.

Length of run	Number of runs			Expected number	
i	Up	Down	Both	$\mathcal{m}\{r_i\}$	$\mathcal{m}\{R_k\}$
1	40	30	70	81·4	129·7
2	11	22	33	35·5	48·3
3	8	6	14	10·1	12·8
4	2	1	3	2·2	2·7
≥ 5	0	1	1	0·5	0·5
Total:	61	60	121	129·7	—

The one run of length ≥ 5 is of length 5.

From Table 13.10 we compute $\mathcal{m}\{R_k\}$ for $n = 195$ and obtain $\mathcal{m}\{r_i\}$ by subtraction. The deviation between the observed and expected number of runs is $121 - 129{\cdot}7 = -8{\cdot}7$. The corresponding variance is $\mathcal{V}\{R\} = 34{\cdot}3$, which gives $\sigma\{R\} = 5{\cdot}9$, so that the deviation is not significant.

It follows directly from (13.4.10) that a run of length 5 is not significant in a random sample of 195 observations.

Example 13.5. From a continuous production of a certain material samples are taken every half hour and the quality is recorded as shown in Table 13.12 and Fig. 13.3 for a period of 30 hours.

TABLE 13.12.
Quality of 60 successive samples.

68	*96*	72	*78*	73	*74*	*76*	72	64	*68*	*70*	*71*	*76*	*79*	*84*
81	72	71	*76*	74	*81*	74	73	72	*73*	*74*	73	71	*76*	*84*
62	*73*	*77*	68	*79*	62	*63*	*83*	77	50	*77*	72	71	*72*	*73*
70	*71*	*83*	77	*81*	*82*	*95*	85	83	52	*82*	76	74	*76*	72

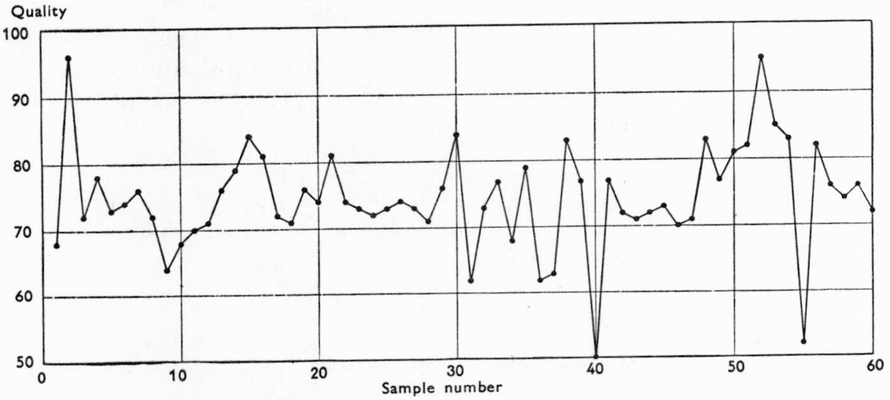

FIG. 13.3. Quality of 60 successive samples.

It is known a priori that the raw materials show periodic fluctuations with respect to a quality characteristic which is supposed to influence the quality of the final product. The problem therefore is to test the hypoth-

TABLE 13.13.
Distribution of runs up and down.

Length of run	Number of runs			Expected number	
i	Up	Down	Both	$m\{r_i\}$	$m\{R_k\}$
1	8	9	17	25·1	39·7
2	7	5	12	10·8	14·6
3	1	3	4	3·0	3·8
≥ 4	1	0	1	0·8	0·8
Mean:	17	17	34	39·7	—

esis of statistical control against the alternative hypothesis of cyclical changes in quality.

The distribution of runs up and down is given in Table 13.13. (Runs up are italicized in Table 13.12).

The one run of length ≥ 4 is of length 6.

The distribution of the total number of runs is approximately normal with mean 39·67 and standard deviation 3·22, according to (13.4.5) and (13.4.6) for $n = 60$. Hence

$$P\{R \leq 34\} \simeq \Phi\left(\frac{34{\cdot}5-39{\cdot}67}{3{\cdot}22}\right) = \Phi(-1{\cdot}61) = 0{\cdot}054\,,$$

i. e., the observed number of runs is barely significant.

It follows from (13.4.11), however, that a run of length 6 may be considered significant in a sample of 60 observations. Thus, the production process does not seem to be in a state of statistical control, but probably reflects the variations in the raw materials.

13.5. The Mean Square Successive Difference.

The mean square successive difference is defined as

$$\frac{1}{n-1}\sum_{i=1}^{n-1}(x_{i+1}-x_i)^2\,. \tag{13.5.1}$$

If $x_1, x_2, \ldots, x_n$ represent a random sample from a normally distributed population with parameters (ξ, σ^2), then the successive differences

$$d_i = x_{i+1}-x_i\,, \qquad i = 1, 2, \ldots, n-1\,, \tag{13.5.2}$$

will be normally distributed with parameters $(0, 2\sigma^2)$ so that

$$\mathcal{M}\{d_i^2\} = 2\sigma^2\,. \tag{13.5.3}$$

Defining

$$\boxed{q^2 = \frac{1}{2(n-1)}\sum_{i=1}^{n-1}(x_{i+1}-x_i)^2} \tag{13.5.4}$$

it follows that

$$\mathcal{M}\{q^2\} = \sigma^2\,, \tag{13.5.5}$$

i. e., q^2 is an unbiased estimate of σ^2.

Further, it has been shown that

$$\mathcal{V}\{q^2\} = \frac{3n-4}{(n-1)^2}\,\sigma^4\,. \tag{13.5.6}$$

Since

$$s^2 = \frac{1}{n-1} \sum_{i=1}^{n} (x_i - \bar{x})^2 \tag{13.5.7}$$

is an unbiased and efficient estimate of σ^2, and

$$\mathcal{V}\{s^2\} = \frac{2\sigma^4}{n-1}, \tag{13.5.8}$$

the estimate q^2 will have an efficiency of

$$\frac{\mathcal{V}\{s^2\}}{\mathcal{V}\{q^2\}} = \frac{2(n-1)}{3n-4} = \frac{2}{3}\left(1+\frac{1}{3n-4}\right). \tag{13.5.9}$$

In some cases, however, where the mean of the population, from which the observations are successively drawn, is gradually changing but the variance is constant, the estimate s^2 of σ^2 will tend to be too large because s^2 includes the whole variation of the population mean. If the change in the population mean between two successive observations is small as compared with σ, the effect of this change on the estimate q^2 will be relatively small since q^2 includes only the differences between successive values. The estimate s^2, therefore, is much more sensitive to gradual changes in the population mean, such as trends or cyclical movements, than is the estimate q^2.

If the hypothesis alternative to the hypothesis of statistical control is that *a gradual change in the population mean has taken place* we may use the ratio

$$\boxed{r = \frac{q^2}{s^2}} \tag{13.5.10}$$

as a test, small values of r being significant.

It has been proved that

$$\boxed{\mathcal{M}\{r\} = 1\,,} \tag{13.5.11}$$

and

$$\boxed{\mathcal{V}\{r\} = \frac{n-2}{(n-1)(n+1)} = \frac{1}{n+1}\left(1-\frac{1}{n-1}\right)} \tag{13.5.12}$$

and further that r is approximately normally distributed for $n > 20$, so that

$$r_P \simeq 1 + \frac{u_P}{\sqrt{n+1}}.$$

For $n < 20$ some fractiles in the distribution of r are given in Table 13.14[1]).

TABLE 13.14.
Fractiles in the distribution of $r = q^2/s^2$.

n	Probability in per cent 0·1	1·0	5·0
4	0·295	0·313	0·390
5	0·208	0·269	0·410
6	0·182	0·281	0·445
7	0·185	0·307	0·468
8	0·202	0·331	0·491
9	0·221	0·354	0·512
10	0·241	0·376	0·531
11	0·260	0·396	0·548
12	0·278	0·414	0·564
13	0·295	0·431	0·578
14	0·311	0·447	0·591
15	0·327	0·461	0·603
16	0·341	0·475	0·614
17	0·355	0·487	0·624
18	0·368	0·499	0·633
19	0·381	0·510	0·642
20	0·393	0·520	0·650

Example 13.6. During a production period of 45 days in a cement plant samples were taken each day and the mean compressive strength of test cubes was determined. The results are shown in Table 13.15 and Fig. 13.4.

If the process is not under control cyclical fluctuations must be expected. The hypothesis of statistical control may be tested by means of the mean square successive difference test.

TABLE 13.15.
Compressive strength—400 in kg/cm² for each day during a production period of 45 days.

40	33	75	18	62	33	38	69	65	100	124	91	79	42	63
23	47	52	98	97	73	85	88	40	42	51	23	75	52	126
90	111	92	109	72	28	56	17	52	68	75	102	107	77	45

[1]) This table has been computed from the values given in a table by B. I. HART: *Significance Levels for the Ratio of the Mean Square Successive Difference to the Variance*, Ann. Math. Stat., 13, 1942, 445–447, by multiplication with $(n-1)/2n$. HART's table includes values for $n = 4(1)\,60$.

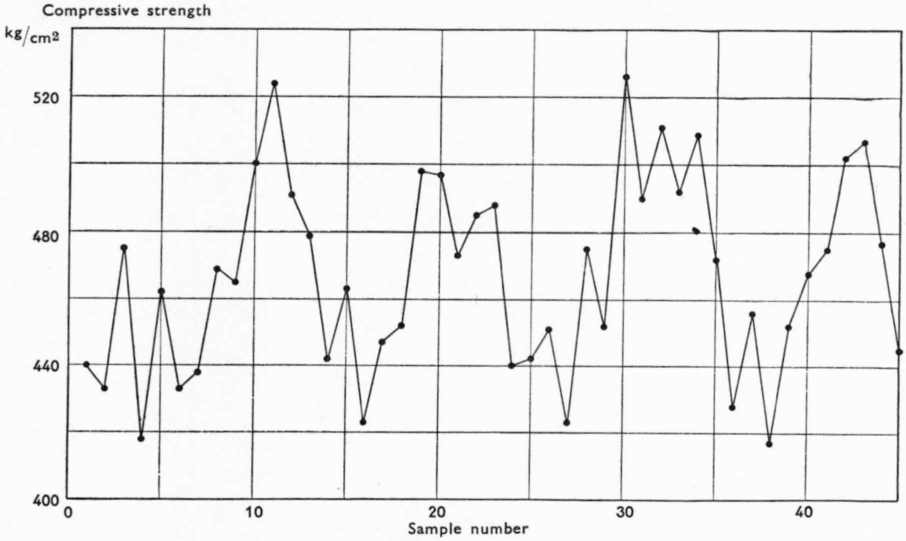

FIG. 13.4. Compressive strength of test cubes for each day during a production period of 45 days.

From the 45 observations we find

$$\sum_{i=1}^{45}(x_i-\bar{x})^2 = 37336 \qquad \text{and} \qquad \sum_{i=1}^{44}(x_{i+1}-x_i)^2 = 42819$$

leading to $s^2 = 848{\cdot}5$ and $q^2 = 486{\cdot}6$, respectively.

Thus $r = 486{\cdot}6/848{\cdot}5 = 0{\cdot}5735$ and $u = (0{\cdot}5735-1)\sqrt{46} = -2{\cdot}89$, so that the observed value of r deviates significantly from the expected value and the test hypothesis is rejected. Hence, the quality of the cement is not under statistical control and assignable causes influencing the quality figures may be found in the manufacturing process itself or in the conditions under which the laboratory control of the strength is carried out.

13.6. Detecting Lack of Control in Respect to a Specified Distribution.

Consider an operation in a state of statistical control and let us assume that the corresponding distribution function is normal with known mean and variance. Such a situation may be approximated to in practice, where a production process has been analyzed and assignable causes of variation removed by means of the methods described in § 13.7, and the process then for a long time has been in a state of statistical control. From a large number of observations in the period of statistical control the parameters may be estimated with a high degree of precision and these estimates may then be used as hypothetical values of the parameters for the future. The problem now is to find *a simple method of testing whether the process remains in a state of statistical control with the given parameters.*

It has been found that *the control charts for the mean and the range* (or the variance) are simple and useful tools for controlling the process. The minimum sample size necessary to be reasonably sure to detect a change in parameter of given magnitude may be determined from the power function. The result obtained from the power function, however, may be modified from practical considerations regarding costs, testing facilities, speed, etc. For the control chart to be of highest value the points should be plotted on the chart as soon as possible after the sample has been obtained so that action to correct the process can be taken immediately after the trouble has occurred. Also it has been found that assignable causes of variation often will occur in an erratic fashion, disturbing the process for a while, then disappearing and returning again. For these reasons the period for collecting and testing the sample has to be short. Therefore, *the most common sample size in the quality control of a manufactured product* seems to be 4 to 10.

The *American Standards Association* uses "three-sigma" limits as control limits, i. e., the limits for $\bar{x}$ are $\xi \pm 3\sigma/\sqrt{n}$ and the limits for w are $\alpha_n\sigma \pm 3\beta_n\sigma$ even if the distribution of w is skew.

The *British Standards Institution* uses the 0·001 and 0·999 fractiles as outer control limits and the 0·025 and 0·975 fractiles as inner control limits or warning limits.

This difference is not of practical importance for the mean since the 0·001 and 0·999 fractiles equal $\xi \pm 3{\cdot}09\sigma/\sqrt{n}$. For the range (or the variance) the disagreement may be larger, but in both cases the essential point is that the probability of getting a point outside the control limits is very small if the process is under control.

In cases where the distribution function is non-normal, the control limits for the mean and the variance will usually be computed from the same formulas as for the normal distribution. The reason is that application of the exact formulas will be too difficult in routine control, and further that even if the probability of getting a point outside the usual control limits is not known exactly we do know that this probability will be very small as long as the process remains under control. Even if the distribution deviates considerably from the normal, the distribution of the mean will be fairly normal for quite small samples, so that the usual formulas are valid with good approximation. For the variance and especially the range larger deviations from the sampling distribution, derived under the assumption of normality, may be expected.

If the distribution deviates from the normal, the mean will usually not be the statistic which is most sensitive to changes in the level of the distribution. For a logarithmic normal distribution, say, the mean of the logarithms ought to be used to test changes in the level of the distribution but the computations involved will usually be too laborious for applications in routine control.

In cases where samples are not of the same size the control limits will vary. For the sample mean, say, the distance between the control limits will be inversely proportional to $\sqrt{n}$.

Example 13.7. From previous investigations of a production of rivets it has been found that the process is in a state of statistical control with regard to diameter of rivet heads and that the diameters are normally distributed with mean 13·42 mm and standard deviation 0·12 mm. Accepting these values as standards for the future production we may test the hypothesis of statistical control by means of control charts for the mean and range of samples of 5 diameters of rivets, randomly chosen from the production at regular intervals.

Considering the 200 successive observations given in Table 3.1 as 40 samples of 5 diameters each, we find the means and ranges as shown in Tables 13.16 and 13.17.

TABLE 13.16.

Mean diameters—13·00 in millimetres for 40 samples of 5 rivets each.

·446	·362	·414	·478	·376	·476	·424	·328	·378	·384
·478	·422	·442	·426	·358	·526	·478	·462	·430	·398
·426	·474	·380	·422	·346	·414	·480	·410	·366	·402
·400	·414	·426	·408	·432	·378	·410	·456	·344	·384

TABLE 13.17.

Ranges in millimetres for 40 samples of 5 rivets each.

·15	·15	·31	·32	·31	·10	·25	·19	·29	·23
·30	·33	·26	·14	·22	·33	·20	·40	·14	·43
·29	·33	·17	·36	·14	·22	·18	·19	·34	·38
·27	·42	·18	·09	·23	·50	·36	·12	·28	·11

From the given mean and standard deviation we compute the inner control limits for the mean as $13{\cdot}42 \pm 1{\cdot}96 \times 0{\cdot}12/\sqrt{5} = (13{\cdot}315,\ 13{\cdot}525)$ mm, and the outer control limits as $13{\cdot}42 \pm 3{\cdot}09 \times 0{\cdot}12/\sqrt{5} = (13{\cdot}254,\ 13{\cdot}586)$ mm.

Similarly, we find for the range the median as $2{\cdot}26 \times 0{\cdot}12 = 0{\cdot}27$ mm, the inner control limits as $(0{\cdot}85 \times 0{\cdot}12,\ 4{\cdot}20 \times 0{\cdot}12) = (0{\cdot}10,\ 0{\cdot}50)$ mm and the outer limits as $(0{\cdot}37 \times 0{\cdot}12,\ 5{\cdot}48 \times 0{\cdot}12) = (0{\cdot}04,\ 0{\cdot}66)$ mm.

These limits and the 40 means and ranges have been plotted on a control chart as shown in Fig. 13·5. The control chart does not indicate any change in the process.

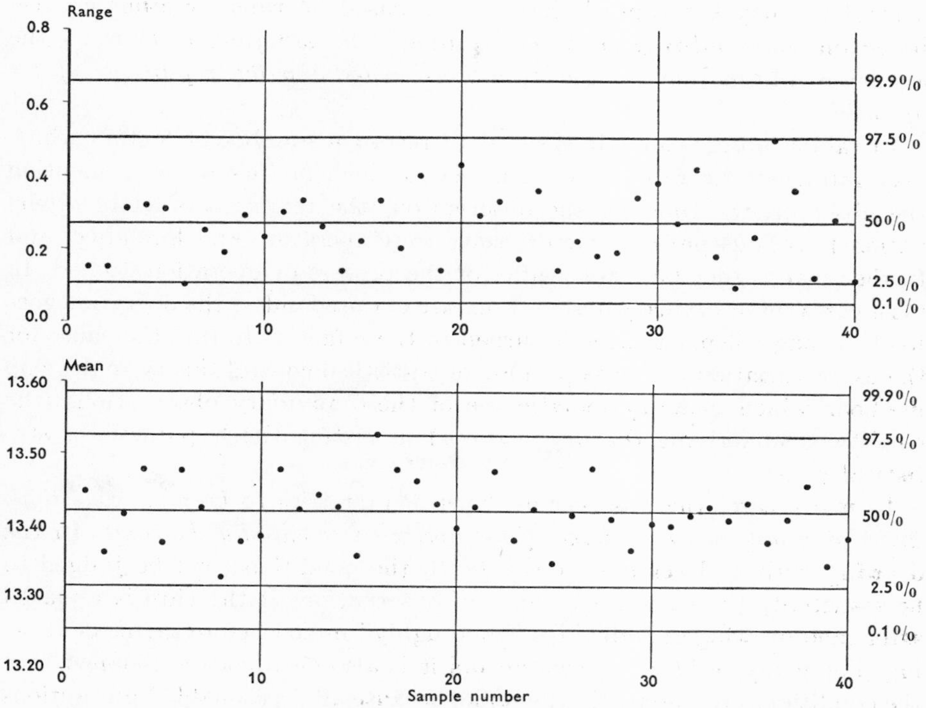

FIG. 13.5. Control charts for means and ranges of 40 samples of five rivets each.

13.7. Detecting Lack of Control When the Distribution is Unknown.

In contrast to the problem in § 13.6 where we have a process in a state of statistical control with a specified distribution function and want to test whether the process will remain stable, we here have the problem of testing whether or not a process is in a state of statistical control. Since the distribution function is unknown we cannot a priori determine the control limits but we must analyze a sequence of observations to see whether it shows any signs of lack of control. If the data indicate lack of control assignable causes of trouble may usually be found and eliminated. By means of statistical methods, it is possible to point out where or when the assignable causes occur and also to indicate the effect of these causes, thus giving the experimenter or engineer some advice in his work of locating and eliminating the assignable causes.

In general, data obtained from repetitions of any given operation will show signs of lack of control if the operation has not been analyzed previously and assignable causes removed.

To test the hypothesis that the operation is in a state of statistical control we may use control charts, tests based on runs, or other criteria based on some *ordering of the observations*. The ordering depends on the available information on the *conditions underlying the repetitions of the operation*.

In taking observations it is usual to record a number of factors which may influence the results. As examples of such factors we may mention the observer, the time for the observation (the temporal order of observations), various physical factors such as temperature and humidity, and further many other factors specific for the process or operation studied. In most cases these auxiliary observations are not used unless the operation goes badly wrong; then attention is turned to these factors to find the cause for the extra variation. It is the purpose of statistical control theory to develop methods which make systematic use of these auxiliary observations (the conditions underlying the observations) in testing the hypothesis of statistical control.

In many cases the observer or experimenter tries to *keep the conditions underlying any two repetitions of the operation essentially the same*. In the drawing with replacement from a bowl, the conditions will be judged to be essentially the same for a series of observations if the chip is replaced after each drawing and all chips "thoroughly" mixed before the next drawing. For many series of measurements it is also customarily assumed that the conditions are essentially the same, because all "reasonable" precautions have been taken to keep conditions alike. Nevertheless it is always advisable to test the hypothesis of statistical control. In such cases the conditions do not suggest any specific ordering of the observations other than *the temporal order* since the conditions are judged to be essentially the same.

In other cases the observer or experimenter may suspect the repetitions of not being carried out under the same essential conditions so that *the resulting observations could be ordered according to criteria based on the observed conditions*.

SHEWHART[1]) mentions a case where relay springs were cut from a long strip and a priori reasons existed for expecting both trends and erratic effects in the thickness along the strip. Therefore, the measurements of the thickness on the springs were arranged in the order of their original position in the strip.

The most common procedure is to *divide the observations into rational subgroups* on the basis of the conditions such that *the conditions are essen-*

[1]) W. A. SHEWHART: *Contribution of Statistics to the Science of Engineering*, Bell Telephone System, B–1319, 1941.

tially the same within each subgroup. In the quality control of a manufactured product rational subgroups are usually based on *order of production* after grouping according to more obvious causes of variation, such as machines, operators, and batches of raw material, has been carried out.

In cases where the experimenter knows the process well, he may often be able to formulate a hypothesis regarding what causes or conditions are likely to disturb the process, so that the observations may be grouped together to test this hypothesis. If no such hypothesis exists, the grouping is made on the basis of *observation order, because conditions usually will be less different for successive observations than for observations far apart.*

Consider a sequence of observations divided into rational subgroups of n observations each and let us assume that statistical control exists *within* each subgroup, but that conditions have changed *between* subgroups, i. e., from one subgroup to another. The data may be presented as shown in Table 13.18. If we further assume that the variability within subgroups

TABLE 13.18.

Subgroup No.	Observations	Sample mean	Sample range	Population mean	Population variance
1	$x_{11}, x_{12}, \ldots, x_{1n}$	$\bar{x}_1$	w_1	ξ_1	σ^2
2	$x_{21}, x_{22}, \ldots, x_{2n}$	$\bar{x}_2$	w_2	ξ_2	σ^2
.	. . .	.	.	.	.
.	. . .	.	.	.	.
.	. . .	.	.	.	.
k	$x_{k1}, x_{k2}, \ldots, x_{kn}$	$\bar{x}_k$	w_k	ξ_k	σ^2

is normal and the same for all groups, i. e., $\sigma_1^2 = \sigma_2^2 = \ldots = \sigma_k^2 = \sigma^2$, then lack of control will exist only if the population means are different.

From the variability within subgroups we may estimate σ, either from the mean range, see (12.1.12), or from the pooled *SSD*'s, see (11.5.4). If the process is in a state of statistical control, the sample means will be normally distributed about the common population mean ξ with variance σ^2/n, i. e., the variation between subgroups may be predicted from the variation within subgroups. Hence, *the hypothesis of statistical control may be tested by plotting the k sample means on a control chart with the central line placed at $\bar{x}$, the grand mean, and the control limits at distances $\pm 3{\cdot}09s/\sqrt{n}$ from the central line, s denoting the within-group estimate of σ.* If one or more points are found outside the control limits it will usually be possible to find and eliminate assignable causes of variation and in this way successively approach a state of statistical control.

Under the conditions stated the control chart will be more sensitive to changes in the population means, the larger the size of the subgroups, because the standard error of the mean is inversely proportional to $\sqrt{n}$.

In practice, however, the contrast between the within-group and the between-group variation will not be as great as supposed above, because the division of the data into rational subgroups usually does not lead to subgroups where statistical control exists within each group. If *the population mean changes from one observation to the next*, then the range of variation for any subgroup will depend on both the random variation (for given population mean) and the variation of the population means within the subgroup. Hence *a large size of subgroups will lead to a large variation within subgroups making the control chart with the usual limits insensitive to changes in the population mean.* For this reason small subgroups must be employed. From a theoretical point of view subgroups of size one would be the best, but since the within-group variance has to be estimated from the observations, subgroups of at least two observations must be employed. On the basis of wide experience it has been found that *subgroups of size* 4 *or* 5 *usually are to be preferred*. (In the case of gradual changes in the population mean subgroups of size one may be employed using the more complicated technique of regression analysis, see Chapter 18).

A *shift in level of the distribution* may thus be detected by means of a *control chart for means of rational subgroups of observations* where the control limits are determined from the within-group estimate of variability. This criterion may be supplemented by *criteria based on runs*. Sudden and irregular fluctuations in level may usually be detected by means of runs above and below the median either of the observations or of the subgroup means. A gradual change in level, a trend, may be detected by means of both runs above and below the median and runs up and down. Periodic fluctuations in level, cyclical movements, may be detected by means of runs up and down and the test based on the mean square successive difference.

Shifts in variability may be detected by means of *control charts for the range or standard deviation* of subgroups where the control limits are determined from the pooled estimate of within-group variability. The run tests may also be applied to these control charts to detect sudden fluctuations, gradual changes, etc., in variability.

Repeated applications of these criteria to sequences of observations of the same operation will usually lead to the identification of assignable causes, which may be eliminated, so that the operation gradually approaches a state of statistical control. To be reasonably sure that a state of statistical control has been obtained it is customary to require that *at least* 25 *rational subgroups of* 4 *observations each do not show any signs of lack of control.*

The methods described above have found widespread applications in industry not only in the routine control of quality of manufactured products but also in research work.

All tests of significance, however, are based on the assumption that the given set of observations is randomly drawn from the same population or from populations with parameters depending on known conditions. Hence, *the validity of any test of significance presupposes a state of statistical control.*

Example 13.8. The cyclical movement of the compressive strength in a cement plant, see Ex. 13.6, aroused the suspicion that the test conditions in the laboratory, although conforming to those laid down in standard specifications, were not under statistical control. The testing technique was then investigated in the following way. In order to eliminate variations in the quality of the cement a large quantity of cement was thoroughly mixed and divided into 64 equal samples which were carefully stored in sealed tins. Each day two samples were randomly chosen and the compressive strengths of test cubes made from these samples were determined. The testing was performed under conditions which were judged to be essentially the same.

Table 13.19 shows the mean compressive strength for 32 successive days, the results for 4 days being grouped together.

TABLE 13.19.
The compressive strengths in kg/cm² for test cubes of cement.

Subgroup number	Compressive strength				Mean	Range
1	90	88	89	76	85·75	14
2	90	90	87	88	88·75	3
3	95	87	89	81	88·00	14
4	74	78	77	80	77·25	6
5	81	82	85	89	84·25	8
6	88	83	87	90	87·00	7
7	91	89	86	92	89·50	6
8	91	77	88	78	83·50	14
Mean:					85·50	9

From the mean range, $\bar{w} = 9$, we obtain an estimate of the within-group standard deviation as $\dfrac{\bar{w}}{\alpha_4} = \dfrac{9}{2{\cdot}059} = 4{\cdot}37 \approx \sigma$.

The control limits for the mean then become $85{\cdot}50 \pm 3{\cdot}09 \times 4{\cdot}37/\sqrt{4} =$ (78·73, 92·27). It will be seen that the mean for subgroup number 4 falls below the lower control limit, thus indicating lack of control.

Investigating the original sequence of 32 observations for runs, a run of length 8 below the median is found, beginning with observation number 12. This supports the evidence from the control chart of means, since a run of length 8 is significant at the 5% level for a sample of less than 34 observations, see Table 13.5.

Example 13.9. The observations from Example 13.2 may also be analyzed by the control chart method, as shown in Table 13.20, where subgroups of 4 successive observations from Table 13.7 have been formed.

TABLE 13.20.
Concentration of a certain component in 56 successive samples.

Subgroup number	Concentration				Mean	Range
1	38	33	29	16	29·00	22
2	44	21	16	17	24·50	28
3	19	1	22	28	17·50	27
4	22	14	7	13	14·00	15
5	21	15	34	23	23·25	19
6	15	19	32	24	22·50	17
7	14	13	22	8	14·25	14
8	30	11	15	24	20·00	19
9	26	14	11	25	19·00	15
10	17	10	19	5	12·75	14
11	6	16	7	10	9·75	10
12	1	5	2	8	4·00	7
13	14	14	15	16	14·75	2
14	13	11	9	11	11·00	4
Mean:					16·88	15·21

$$\frac{\bar{w}}{a_4} = \frac{15{\cdot}21}{2{\cdot}059} = 7{\cdot}39 \approx \sigma .$$

$$\frac{\sigma}{\sqrt{4}} \approx \frac{7{\cdot}39}{2} = 3{\cdot}69, \qquad 3{\cdot}09 \times 3{\cdot}69 = 11{\cdot}4$$

Lower control limit for $\bar{x}$: $16{\cdot}9 - 11{\cdot}4 = 5{\cdot}5$
Upper control limit for $\bar{x}$: $16{\cdot}9 + 11{\cdot}4 = 28{\cdot}3$
Lower control limit for w: $7{\cdot}39 \times 0{\cdot}20 = 1{\cdot}5$
Upper control limit for w: $7{\cdot}39 \times 5{\cdot}31 = 39{\cdot}2$

The computation of control limits for $\bar{x}$ and w will be found in Table 13.20. Fig. 13.6 shows the corresponding control charts.

The two means outside the control limits correspond to the two long runs above and below the median respectively which were observed in Example 13.2. The control chart indicates a downward trend.

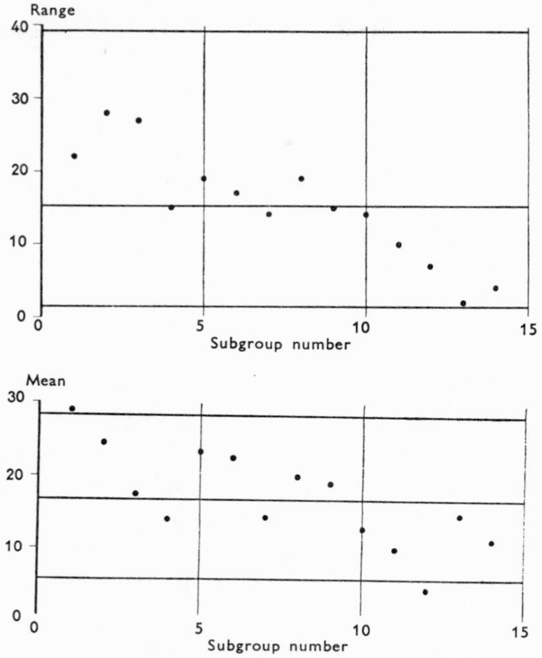

FIG. 13.6. Control charts for means and ranges of 14 subgroups, each of four observations.

Further, the control chart for the range indicates a change in variability. The run down of length 5, beginning with subgroup number 8, is significant at the 5% level according to (13.4.10).

Hence, the control chart analysis supports the conclusion reached in Example 13.2, that the mixing has not been satisfactory.

13.8. The State of Maximum Control.

The above definition of statistical control is based on the concept of a random sample drawn from a stable population, whereas the form of the distribution of that population is of no importance. Hence the criteria for detecting lack of control are based on the *order* of the observations.

When a state of statistical control has been obtained an entirely new problem arises. First the distribution function is studied by means of the methods given in Chapters 6 and 7, and after a reliable estimate of the distribution function has been found the problem is whether the state of statistical control obtained is a state of *maximum control*, i. e., will it be possible by changing the process to obtain a new state of statistical control with a variability essentially smaller than the old one?

In solving this problem the *form* of the distribution function may often be helpful, indicating that one or a group of causes in the constant system of chance causes produces a predominating effect leading to a bimodal or a very flat distribution curve.

An example of a cause system of this kind has been given in Example 2.3, C, where $(n-1)$ causes each produces the effects $+\varepsilon$ or $-\varepsilon$ and one cause produces the effects $+(1+\alpha)\varepsilon$ or $-(1+\alpha)\varepsilon$. As shown in Example 8.1 this one cause is responsible for the fraction

$$\frac{(1+\alpha)^2}{(1+\alpha)^2+n-1}$$

of the total variance. If α is large as compared with n the distribution will be bimodal and usually the cause may be identified and the process modified so that a new state of statistical control with a smaller variance is obtained.

Other examples of this kind have been given in § 6.10 under the heading: Heterogeneous Distributions. A bimodal distribution of the quality of an industrial product may, for example, be due to differences in the setting of two machines. Usually such a cause will easily be discovered and removed.

In other cases, however, it may be difficult to find out what causes are responsible for the larger part of the total variation. In the following chapters methods will be developed for *partitioning the total variance into component parts corresponding to different groups of causes.*

Conducting suitable experiments, it will usually be possible by means of this analysis of variance to discover the causes giving the dominating contributions to the total variance and afterwards modify the process in a suitable way.

If the constant system of chance causes consists of *a large number of causes of which no single cause produces a predominating effect* it will generally not be economical to go further in the elimination of causes and we then have reached a state of maximum control. Let us consider Example 2.3, A, where the standard deviation of the distribution is $\sigma_n = \varepsilon\sqrt{n}$. For $n = 25$ we get $\sigma_{25} = 5\varepsilon$. If it was possible to exclude one of the causes we would find $\sigma_{24} = \varepsilon\sqrt{24} = 4{\cdot}90\varepsilon$, i. e., the standard deviation has decreased by 2%. To demonstrate a change of this size by an experiment would require a very large number of observations and the elimination of this single cause would hardly be worth while.

13.9. Specification Limits and Statistical Control.

In mass production limits are often specified for the quality of the produced items. For example, upper and lower limits may be specified for the dimension of an item, a lower limit may be specified for the tensile strength, and so forth. The reason for these specification limits is to get the product screened, rejecting all nonconforming items, so that the quality of each single item of the final product will satisfy the specification.

If the process is in a state of statistical control the variability of the quality characteristic may be described by means of a stable distribution function. In the special case where the distribution is normal we may find tolerance limits for the quality characteristic from formulas (11.9.13) and (11.10.12). Comparing the tolerance and the specification limits it may be decided whether or not the process is satisfactory.

For example, if we have a sample of 1000 observations from a process in a state of statistical control (this may be obtained by pooling several smaller samples) we may be 99% confident that at least 99·9% of the items produced under the same conditions will lie within the tolerance limits $\bar{x} \pm 3{\cdot}5s$. If these limits are within the specification limits, the product may be accepted without inspection. It may happen that the upper tolerance limit, $\bar{x}+3{\cdot}5s$ say, exceeds the upper specification limit, and at the same time the tolerance range $7s$ is less than the specification range. The remedy then is to change the process so that the mean quality approximately equals the mean of the specification limits.

If the tolerance range is considerably larger than the specification range, a large percentage of the product will be nonconforming and the product must be screened through inspection, unless the process can be essentially changed so that a smaller variability results.

In the case where only a single limit is specified similar results may be obtained by comparing the one-sided tolerance limit and the specification limit.

13.10. Notes and References.

The application of statistical methods to the analysis of industrial data and especially to the control of quality of mass produced items has been growing rapidly since 1920. In the twenties two main lines of approach were developed: the graphical analysis of heterogeneous distributions and the control chart method.

The analysis of heterogeneous distributions as a clue to assignable causes of variation has been discussed primarily by K. Daeves in *Praktische Grosszahl-Forschung*, VDI-Verlag, Berlin, 1933, and in K. Daeves und A. Beckel: *Auswertung durch Grosszahl-Forschung*, Die Chemische Fabrik, 14, 1941, 131–43.

The more important control chart method and other fundamental principles in the theory and practice of statistical control are due to W. A.

SHEWHART. The publications of W. A. SHEWHART: *Economic Control of Quality of Manufactured Product*, Van Nostrand, New York, 1931, 501 pp., *Statistical Method from the View-point of Quality Control*, The Department of Agriculture, Washington, 1939, 155 pp., and *Contribution of Statistics to the Science of Engineering*, Bell Telephone System, B-1319, 1941, 28 pp., are still the most important on the subject, and the present chapter is to a large extent based on SHEWHART's results. Under the influence of SHEWHART's work a large number of books on statistical quality control have been published. Some of these books are mentioned below:

E. S. PEARSON: *The Application of Statistical Methods to Industrial Standardisation and Quality Control*, British Standard 600, 1935, 161 pp.

B. P. DUDDING and W. J. JENNETT: *Quality Control Charts*, British Standard 600 R, 1942, 89 pp.

E. L. GRANT: *Statistical Quality Control*, McGraw-Hill, New York, 1946, 563 pp.

W. B. RICE: *Control Charts in Factory Management*, J. Wiley, New York, 1947, 149 pp.

P. PEACH: *An Introduction to Industrial Statistics and Quality Control*, Edwards Broughton Co., Raleigh, 1947, 236 pp.

Further, AMERICAN STANDARDS ASSOCIATION has published three short manuals on quality control: *American War Standards* Z. 1. 1, Z. 1. 2 (1941) and Z. 1. 3 (1942).

The exposition of the theory of runs is based on the following papers where proofs and also more general results may be found.

A. WALD and J. WOLFOWITZ: *On a Test Whether Two Samples are From the Same Population*, Ann. Math. Stat., 11, 1940, 147–162. (The test is based on the total number of runs above and below a certain value).

F. S. SWED and C. EISENHART: *Tables for Testing Randomness of Grouping in a Sequence of Alternatives*, Ann. Math. Stat., 14, 1943, 66–87. (Tables of fractiles in the distribution of the total number of runs above and below a certain value).

A. M. MOOD: *The Distribution Theory of Runs*, Ann. Math. Stat. 11, 1940, 367–392. (A general exposition of the distribution of runs of any length above and below any value).

F. MOSTELLER: *Note on an Application of Runs to Quality Control Charts*, Ann. Math. Stat., 12, 1941, 228–232. (The distribution of the number of runs of length k or more above and below the median).

H. LEVENE and J. WOLFOWITZ: *The Covariance Matrix of Runs up and down*, Ann. Math. Stat., 15, 1944, 58–69.

J. WOLFOWITZ: *Asymptotic Distribution of Runs up and down*, Ann. Math. Stat., 15, 1944, 163–172.

P. S. OLMSTEAD: *Distribution of Sample Arrangements for Runs up and down*, Ann. Math. Stat., 17, 1946, 24–33. (The distribution of the number of runs of length k or more. Tables.)

The test based on the mean square successive difference has been developed in the following papers:

J. VON NEUMANN, R. H. KENT, H. R. BELLINSON and B. I. HART: *The Mean Square Successive Difference*, Ann. Math. Stat., 12, 1941, 153–162. (Derivation of the distribution of the m. sq. successive difference).

L. C. YOUNG: *On Randomness in Ordered Sequences*, Ann. Math. Stat., 12, 1941, 293–300. (Derivation of the moments of $1-r$ under the assumption of randomness).

J. VON NEUMANN: *Distribution of the Ratio of the Mean Square Successive Difference to the Variance*, Ann. Math. Stat., 12, 1941, 367–395.

B. I. HART: *Significance Levels for the Ratio of the Mean Square Successive Difference to the Variance*, Ann. Math. Stat., 13, 1942, 445–447. (Table of fractiles for $n = 4\ (1)\ 60$).

CHAPTER 14.

THE DISTRIBUTION OF THE VARIANCE RATIO

14.1. The v^2-Distribution.

In § 11.6 it has been shown how the hypothesis $\sigma_1^2 = \sigma_2^2 = \ldots = \sigma_k^2$ may be tested by BARTLETT's test, which, however, is only an approximation. R. A. FISHER has given an exact test, the v^2-test, for the hypothesis $\sigma_1^2 = \sigma_2^2$[1].

Let s_1^2 and s_2^2 denote two stochastically independent variables, both having s^2-distributions with parameters (σ^2, f_1) and (σ^2, f_2), respectively, so that

$$s_1^2 = \sigma^2 \frac{\chi_1^2}{f_1} \qquad (14.1.1)$$

and

$$s_2^2 = \sigma^2 \frac{\chi_2^2}{f_2}. \qquad (14.1.2)$$

From these variables we form the variance ratio v^2

$$\boxed{v^2 = \frac{s_1^2}{s_2^2} = \frac{\chi_1^2/f_1}{\chi_2^2/f_2}} \qquad (0 \leq v^2 < \infty)\,. \qquad (14.1.3)$$

As this quotient does not contain the theoretical variance, the distribution of the quotient depends solely on the numbers of degrees of freedom f_1 and f_2.

The distribution function of v^2, which will be derived in § 14.4, is

$$p\{v^2\} = \frac{\Gamma\left(\frac{f_1+f_2}{2}\right)}{\Gamma\left(\frac{f_1}{2}\right)\Gamma\left(\frac{f_2}{2}\right)} f_1^{\frac{f_1}{2}} f_2^{\frac{f_2}{2}} \frac{(v^2)^{\frac{f_1}{2}-1}}{(f_2+f_1v^2)^{\frac{f_1+f_2}{2}}} \qquad (0 \leq v^2 < \infty). \qquad (14.1.4)$$

---

[1]) R. A. FISHER: *On a Distribution Yielding the Error Functions of Several Well-Known Statistics,* Proceedings of the International Mathematical Congress, Toronto, 1924, 805–813.

From this we find that *the mean of* v^2 *is*

$$m\{v^2\} = \frac{f_2}{f_2 - 2}, \quad f_2 > 2, \tag{14.1.5}$$

which for large values of f_2 is approximately equal to 1.

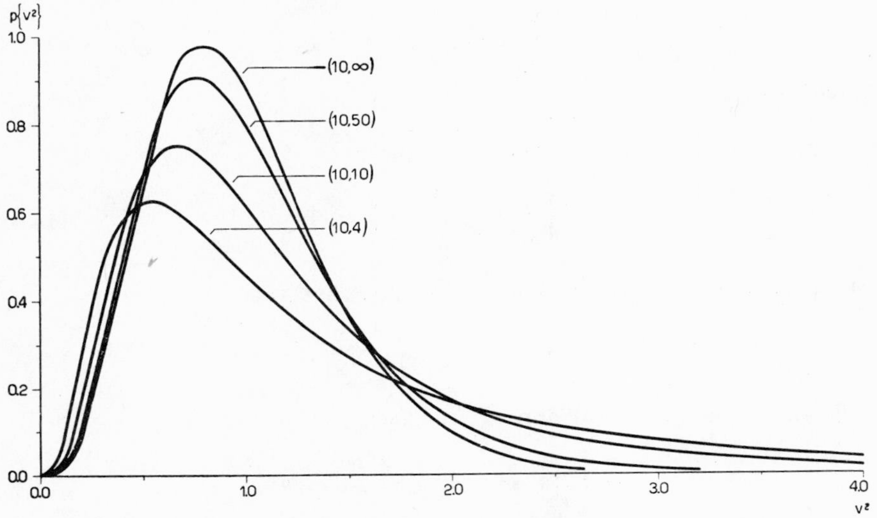

Fig. 14.1. Distribution curves of v^2 for $(f_1, f_2) = (10, 4), (10, 10), (10, 50)$, and $(10, \infty)$.

Fig. 14.1 shows the distribution curves of v^2 for $(f_1, f_2) = (10, 4), (10, 10), (10, 50)$, and $(10, \infty)$.

The fractiles of the v^2*-distribution*

$$v_P^2 = v_P^2(f_1, f_2),$$

besides depending on P, depend on f_1 and f_2, f_1 *denoting the number of degrees of freedom of the numerator of* v^2 *and* f_2 *the number of degrees of freedom of the denominator.*

TABLE 14.1.

$P(\%)$	$v_P^2(10, 4)$
50	1·11
70	1·80
90	3·92
95	5·96
97·5	8·84
99·0	14·5
99·5	21·0
99·9	48·0
99·95	68·3

In Table VII $v_P^2(f_1, f_2)$ has been given as a function of f_1 and f_2 for various values of P. Table 14.1 shows the fractiles given in Table VII for $(f_1, f_2) = (10, 4)$.

Table VII only gives v_P^2 values for $P \geq 50\%$. For $P \leq 50\%$, v_P^2 may be calculated as follows: The relation

$$P\left\{\frac{s_1^2}{s_2^2} < v_P^2(f_1, f_2)\right\} = P \tag{14.1.6}$$

entails that

$$P\left\{\frac{s_2^2}{s_1^2} > \frac{1}{v_P^2(f_1, f_2)}\right\} = P$$

or

$$P\left\{\frac{s_2^2}{s_1^2} < \frac{1}{v_P^2(f_1, f_2)}\right\} = 1-P. \tag{14.1.7}$$

If we consider the distribution of the variance ratio

$$v^2 = v^2(f_2, f_1) = \frac{s_2^2}{s_1^2},$$

we obtain, from the definition of fractiles,

$$P\left\{\frac{s_2^2}{s_1^2} < v_{1-P}^2(f_2, f_1)\right\} = 1-P. \tag{14.1.8}$$

On comparing (14.1.7) and (14.1.8) we find

$$v_{1-P}^2(f_2, f_1) = \frac{1}{v_P^2(f_1, f_2)}$$

or

$$\boxed{v_P^2(f_1, f_2) = \frac{1}{v_{1-P}^2(f_2, f_1)}.} \tag{14.1.9}$$

Hence *the P-fractiles may be calculated as the reciprocals of the* $(1-P)$-*fractiles when the number of degrees of freedom of the numerator and the denominator are interchanged.*

For example we get

$$v_{.30}^2(10, 4) = \frac{1}{v_{.70}^2(4, 10)} = \frac{1}{1{\cdot}41} = 0{\cdot}71.$$

In this manner fractiles may be calculated for $P < 0{\cdot}5$, as shown in Table 14.2, which supplements Table 14.1.

Together, Tables 14.1 and 14.2 give 17 fractiles for v^2 (10, 4). If the 17 corresponding values of v^2 and P are plotted on probability paper or logarithmic probability paper and a curve drawn through them, we get a good representation of the cumulative distribution function of v^2, as shown in Fig. 14.2.

Table 14.1 shows that

$$P\left\{\frac{s_1^2}{s_2^2} < 5{\cdot}96\right\} = 95\%\,, \quad (f_1, f_2) = (10, 4)\,.$$

Hence, the probability that the ratio between two empirical variances with 10 and 4 degrees of freedom, respectively, is smaller than 5·96 is equal to 95%. In spite of the fact that the two sample variances are estimates of the same population variance, we see that the variance ratio often takes on fairly large values.

TABLE 14.2.

$P(\%)$	$v_{1-P}^2(4, 10)$	$v_P^2(10, 4)$
50	0·899	1·11
30	1·41	0·71
10	2·61	0·38
5	3·48	0·29
2·5	4·47	0·22
1·0	5·99	0·17
0·5	7·34	0·14
0·1	11·3	0·088
0·05	13·4	0·075

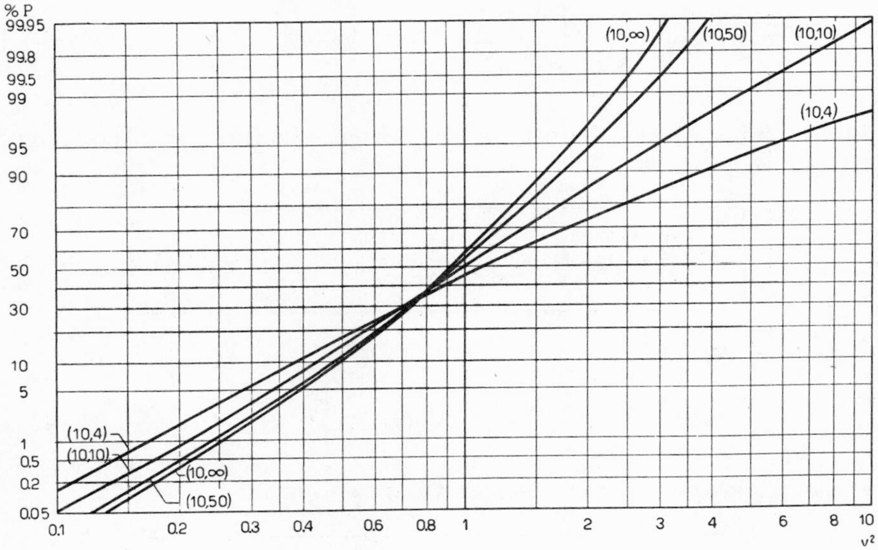

FIG. 14.2. The cumulative distribution function of v^2, plotted on logarithmic probability paper for $(f_1, f_2) = (10, 4), (10, 10), (10, 50)$, and $(10, \infty)$.

It follows from the distribution of s^2 that the deviations of the two variances from the theoretical value will decrease with increasing number of degrees of freedom, and for $(f_1, f_2) \to (\infty, \infty)$, s_1^2 and s_2^2 will converge in probability towards σ^2. Accordingly, Table VII shows that $v_{.95}^2$, for example, decreases as f_1 and f_2 increase and consequently $v_{.05}^2$ increases as f_1 and f_2 increase, so that the distance between the two fractiles decreases as the numbers of degrees of freedom increase. For $(f_1, f_2) \to (\infty, \infty)$, $v_P^2 \to 1$.

Until now we have assumed that the two sample variances are estimates of the same population variance. Now let s_1^2 and s_2^2 denote estimates of σ_1^2 and σ_2^2, respectively, so that

$$s_1^2 = \sigma_1^2 \frac{\chi_1^2}{f_1} \tag{14.1.10}$$

and

$$s_2^2 = \sigma_2^2 \frac{\chi_2^2}{f_2}. \tag{14.1.11}$$

The variance ratio then becomes

$$\boxed{\frac{s_1^2}{s_2^2} = \frac{\sigma_1^2}{\sigma_2^2} \frac{\chi_1^2/f_1}{\chi_2^2/f_2} = \frac{\sigma_1^2}{\sigma_2^2} v^2(f_1, f_2).} \tag{14.1.12}$$

Hence *the fractiles for s_1^2/s_2^2 are derived from v^2 by multiplication by σ_1^2/σ_2^2.* From this follows that

$$\boxed{P\left\{\frac{s_1^2}{s_2^2} < \frac{\sigma_1^2}{\sigma_2^2} v_P^2(f_1, f_2)\right\} = P.} \tag{14.1.13}$$

Limits for s_1^2/s_2^2 are derived directly from the fractiles. According to (14.1.13) we have

$$\boxed{P\left\{\frac{\sigma_1^2}{\sigma_2^2} v_{P_1}^2(f_1, f_2) < \frac{s_1^2}{s_2^2} < \frac{\sigma_1^2}{\sigma_2^2} v_{P_2}^2(f_1, f_2)\right\} = P_2 - P_1.} \tag{14.1.14}$$

If $\sigma_1^2 = \sigma_2^2$, we get

$$P\left\{v_{P_1}^2(f_1, f_2) < \frac{s_1^2}{s_2^2} < v_{P_2}^2(f_1, f_2)\right\} = P_2 - P_1. \tag{14.1.15}$$

Solving the inequality with respect to s_1^2, we obtain

$$\boxed{P\{s_2^2 v_{P_1}^2(f_1, f_2) < s_1^2 < s_2^2 v_{P_2}^2(f_1, f_2)\} = P_2 - P_1.} \tag{14.1.16}$$

This formula implies that at the moment where we have observed s_2^2 but have not yet observed s_1^2 we are entitled to use (14.1.16) to predict limits between which we are confident that s_1^2 will lie, *under the condition that we have prior reasons for thinking that* $\sigma_1^2 = \sigma_2^2$. We are, however, not entitled to forecast a succession of values of s_1^2 by this method, using the same s_2^2 throughout, for we are essentially employing the variable $v^2 = s_1^2/s_2^2$, the theory for which demands that both numerator and denominator vary at random from one value of v^2 to the next.

For $f_2 \to \infty$ and $f_1 = f$ the result given by (14.1.16) is identical with (11.2.2), see also (14.5.1).

14.2. The v^2-Test of Significance.

A hypothesis concerning the ratio

$$\lambda^2 = \frac{\sigma_1^2}{\sigma_2^2}. \tag{14.2.1}$$

may be tested from the corresponding sample ratio s_1^2/s_2^2 by means of the v^2-distribution. The test is analogous to the χ^2-test given in § 11.3.

Usually we are interested in testing the hypothesis $\lambda^2 = 1$, i. e., $\sigma_1^2 = \sigma_2^2$, against the one-sided alternative $\sigma_1^2 > \sigma_2^2$ or the double-sided alternative $\sigma_1^2 \neq \sigma_2^2$.

Let us first treat the simple case where *the alternative hypothesis is* $\sigma_1^2 > \sigma_2^2$. *If the test hypothesis is true*, then s_1^2/s_2^2 is distributed as $v^2(f_1, f_2)$ so that *the critical region at the* α *level of significance* becomes

$$\boxed{\frac{s_1^2}{s_2^2} > v_{1-\alpha}^2(f_1, f_2)} \tag{14.2.2}$$

since the probability of this inequality being fulfilled is α for $\sigma_1^2 = \sigma_2^2$.

The *power of the test* with respect to the alternative hypothesis $\lambda^2 > 1$ is

$$\begin{aligned}\pi(\lambda^2) &= P\left\{\frac{s_1^2}{s_2^2} > v_{1-\alpha}^2(f_1, f_2);\ \lambda^2\right\}\\ &= P\left\{\frac{s_1^2/\sigma_1^2}{s_2^2/\sigma_2^2} > \frac{1}{\lambda^2}v_{1-\alpha}^2(f_1, f_2);\ \lambda^2\right\}\\ &= P\left\{v^2(f_1, f_2) > \frac{1}{\lambda^2}v_{1-\alpha}^2(f_1, f_2)\right\}.\end{aligned} \tag{14.2.3}$$

The power function is an increasing function of λ^2 since $v_{1-\alpha}^2/\lambda^2$ is a decreasing function of λ^2.

To determine the value of λ^2 for which the power function takes on a certain value $1-\beta$ we have to solve the equation

$$\pi(\lambda^2) = P\left\{ v^2(f_1, f_2) > \frac{1}{\lambda^2} v^2_{1-\alpha}(f_1, f_2) \right\} = 1-\beta \tag{14.2.4}$$

which leads to

$$\lambda^2 = \frac{v^2_{1-\alpha}(f_1, f_2)}{v^2_{\beta}(f_1, f_2)} = v^2_{1-\alpha}(f_1, f_2) v^2_{1-\beta}(f_2, f_1) , \tag{14.2.5}$$

analogous to (11.3.6). By means of this relation and the fractiles given in Table VII it is easy to determine a sufficient number of points to draw the power curve.

An extensive table of λ^2 corresponding to the 5% and 1% levels of significance has been given by CHURCHILL EISENHART in Chapter 8 of *Selected Techniques of Statistical Analysis*, STATISTICAL RESEARCH GROUP, Columbia University, McGraw-Hill, New York, 1947.

Suppose, for example, that $f_1 = 30$, $f_2 = 10$, and $\alpha = 5\%$. The critical region then becomes $s_1^2/s_2^2 > v^2_{\cdot 95}(30, 10) = 2{\cdot}70$, i. e., only if the observed s_1^2 based on 30 degrees of freedom exceeds 2·70 times the observed s_2^2 based on 10 degrees of freedom will the test hypothesis be rejected. The power of the test for $\lambda^2 = 2$, say, is

$$P\{v^2(30, 10) > 1{\cdot}35\} = \text{ca. } 30\% ,$$

i. e., the probability of accepting $\sigma_1^2 = \sigma_2^2$ when $\sigma_1^2 = 2\sigma_2^2$ is ca. 70%. If we require the test to discriminate better between the two hypotheses we have to increase f_1 and f_2.

Suppose that we have to choose between the two hypotheses $\sigma_1^2 = \sigma_2^2$ and $\sigma_1^2 = 2\sigma_2^2$. How many observations must be made from each population, assuming $f_1 = f_2$, if we want to be reasonably sure to choose the right hypothesis? Putting $\alpha=\beta=5\%$ and $\lambda^2=2$ we find from (14.2.5) $v^2_{\cdot 95}(f, f) = \sqrt{2}$ which from Table VII$_4$ gives $f_1 = f_2 = \text{ca. } 90$.

When *the alternative to the test hypothesis* is $\sigma_1^2 \neq \sigma_2^2$, without it a priori being possible to decide whether $\sigma_1^2 > \sigma_2^2$ or $\sigma_1^2 < \sigma_2^2$, we must use a double-sided criterion. The usual practice is to use *a critical region based on equal tail areas*, i. e., we reject the test hypothesis when either of the inequalities

$$\frac{s_1^2}{s_2^2} < v^2_{\frac{\alpha}{2}}(f_1, f_2) \quad \text{or} \quad \frac{s_1^2}{s_2^2} > v^2_{1-\frac{\alpha}{2}}(f_1, f_2) \tag{14.2.6}$$

is satisfied. We might just as well base the test on the reciprocal ratio s_2^2/s_1^2 which, however, leads to the same result as above on account of (14.1.9). Hence, the test may be based on one of the ratios only, for instance the larger variance ratio, which is then compared with the upper limit of significance, i. e., the test takes the form

$$\frac{\text{Larger variance}}{\text{Smaller variance}} > v^2_{1-\frac{\alpha}{2}} . \qquad (14.2.7)$$

Although the test presents itself in a unilateral form, it is really a bilateral test at the α level of significance. (This simple result depends essentially on the use of equal tail areas as critical region).

The power function becomes

$$\pi(\lambda^2) = P\left\{v^2 < \frac{1}{\lambda^2} v^2_{\frac{\alpha}{2}}\right\} + P\left\{v^2 > \frac{1}{\lambda^2} v^2_{1-\frac{\alpha}{2}}\right\} . \qquad (14.2.8)$$

Example 14.1. To compare the titration error of two technicians a series of duplicate titrations of the $CaCO_3$ contents of raw-meal was made by each technician. For technician No. 1 we find the variance according to (11.5.7) as

$$s_1^2 = 0{\cdot}0388 \qquad (s_1 = 0{\cdot}197\% \ CaCO_3)$$

with 32 degrees of freedom, and correspondingly for technician No. 2

$$s_2^2 = 0{\cdot}0177 \qquad (s_2 = 0{\cdot}133\% \ CaCO_3)$$

with $f_2 = 26$.

In order to test the hypothesis $\sigma_1^2 = \sigma_2^2$ we calculate

$$v^2 = \frac{0{\cdot}0388}{0{\cdot}0177} = 2{\cdot}19 .$$

Table VII shows that $v^2_{.975}(32, 26) = 2{\cdot}14$ and $v^2_{.99}(32, 26) = 2{\cdot}48$. Thus, the observed value of v^2 is just significant at the 5% level.

14.3. Confidence Limits for σ_1^2/σ_2^2.

Confidence limits for σ_1^2/σ_2^2 may be obtained by solving the inequality in (14.1.14) with respect to σ_1^2/σ_2^2:

$$P\left\{\frac{s_1^2}{s_2^2}\frac{1}{v^2_{P_2}(f_1, f_2)} < \frac{\sigma_1^2}{\sigma_2^2} < \frac{s_1^2}{s_2^2}\frac{1}{v^2_{P_1}(f_1, f_2)}\right\} = P_2 - P_1 . \qquad (14.3.1)$$

14.4. Derivation of the v^2-Distribution.

The cumulative distribution function of $z = y_1/y_2$, where y_1 and y_2 are two positive, stochastically independent variables, is obtained by integrating the probability element

$$p\{y_1, y_2\}\,dy_1\,dy_2 = p\{y_1\}\,dy_1\,p\{y_2\}\,dy_2$$

over the domain defined by $\dfrac{y_1}{y_2} \leq z$, i. e.,

$$P\left\{\frac{y_1}{y_2} \leq z\right\} = P\{z\} = \int_{y_2=0}^{\infty}\int_{y_1=0}^{zy_2} p\{y_1\}\,p\{y_2\}\,dy_1\,dy_2\,. \tag{14.4.1}$$

Differentiation with respect to z leads to

$$p\{z\} = \int_0^{\infty}[p\{y_1\}]_{y_1=zy_2}\,y_2\,p\{y_2\}\,dy_2\,. \tag{14.4.2}$$

In the present case we put $y_1 = \dfrac{\chi_1^2}{f_1}$ and $y_2 = \dfrac{\chi_2^2}{f_2}$ which makes z equal to v^2.

From

$$p\{\chi^2\} = \frac{1}{2^{\frac{f}{2}}\,\Gamma\left(\frac{f}{2}\right)}(\chi^2)^{\frac{f}{2}-1}\,e^{-\frac{\chi^2}{2}} \tag{14.4.3}$$

we obtain the distribution of $y = \chi^2/f$, using (5.5.5), as

$$p\{y\} = \frac{f^{\frac{f}{2}}}{2^{\frac{f}{2}}\,\Gamma\left(\frac{f}{2}\right)}\,y^{\frac{f}{2}-1}\,e^{-\frac{fy}{2}}. \tag{14.4.4}$$

If we introduce the corresponding distributions of y_1 and y_2 in (14.4.2) we obtain

$$p\{z\} = c\int_0^{\infty}(zy_2)^{\frac{f_1}{2}-1}\,e^{-\frac{f_1 zy_2}{2}}\,y_2^{\frac{f_2}{2}}\,e^{-\frac{f_2 y_2}{2}}\,dy_2$$

$$= cz^{\frac{f_1}{2}-1}\int_0^{\infty}y_2^{\frac{f}{2}-1}\,e^{-\frac{1}{2}y_2(f_1z+f_2)}\,dy_2\,, \tag{14.4.5}$$

where

$$c = \frac{f_1^{\frac{f_1}{2}}\,f_2^{\frac{f_2}{2}}}{2^{\frac{f}{2}}\,\Gamma\left(\frac{f_1}{2}\right)\Gamma\left(\frac{f_2}{2}\right)} \tag{14.4.6}$$

and

$$f = f_1 + f_2\,. \tag{14.4.7}$$

Introducing

$$y = y_2(f_1z+f_2)$$

as a new variable, we find

$$p\{z\} = c\,\frac{z^{\frac{f_1}{2}-1}}{(f_1z+f_2)^{\frac{f}{2}}}\int_0^{\infty}y^{\frac{f}{2}-1}\,e^{-\frac{y}{2}}\,dy\,. \tag{14.4.8}$$

According to (10.3.1) we also have

$$\frac{1}{2^{\frac{f}{2}}\,\Gamma\left(\frac{f}{2}\right)}\int_0^\infty y^{\frac{f}{2}-1}e^{-\frac{y}{2}}\,dy = 1\,, \tag{14.4.9}$$

which together with (14.4.8) gives the final result

$$p\{z\} = \frac{\Gamma\left(\frac{f}{2}\right)}{\Gamma\left(\frac{f_1}{2}\right)\Gamma\left(\frac{f_2}{2}\right)}\,f_1^{\frac{f_1}{2}}\,f_2^{\frac{f_2}{2}}\,\frac{z^{\frac{f_1}{2}-1}}{(f_2+f_1z)^{\frac{f}{2}}}\,, \tag{14.4.10}$$

cf. (14.1.4) for $z = v^2$.

The v^2-distribution may be expressed by the well-known B-function in the following way.

Introducing the variable y by the transformation

$$z = \frac{f_2}{f_1}\,\frac{y}{1-y} \qquad (0 \le y \le 1)\,, \tag{14.4.11}$$

i. e.,

$$y = \frac{f_1 z}{f_2+f_1z} \quad \text{and} \quad 1-y = \frac{f_2}{f_2+f_1z}\,, \tag{14.4.12}$$

we obtain from (14.4.10), using (5.5.5),

$$p\{y\} = \frac{\Gamma\left(\frac{f_1+f_2}{2}\right)}{\Gamma\left(\frac{f_1}{2}\right)\Gamma\left(\frac{f_2}{2}\right)}\,y^{\frac{f_1}{2}-1}\,(1-y)^{\frac{f_2}{2}-1} \qquad (0 \le y \le 1)\,, \tag{14.4.13}$$

which is called the Beta-distribution, since the function

$$\mathrm{B}(s,\,t) = \int_0^1 y^{s-1}(1-y)^{t-1}\,dy = \frac{\Gamma(s)\,\Gamma(t)}{\Gamma(s+t)} \tag{14.4.14}$$

is called the B-function. The integral

$$\mathrm{B}_y(s,\,t) = \int_0^y y^{s-1}(1-y)^{t-1}\,dy \qquad (0 \le y \le 1) \tag{14.4.15}$$

is called the incomplete B-function. The ratio between the complete and the incomplete B-function

$$I_y(s,\,t) = \frac{\mathrm{B}_y(s,\,t)}{\mathrm{B}(s,\,t)} \tag{14.4.16}$$

has been tabulated in K. PEARSON: *Tables of the Incomplete Beta-Function*, Biometrika Office, London, 1934, for $s \le 50$ and $t \le 50$.

From (14.4.13)–(14.4.16) it follows that

$$P\{y\} = I_y\left(\frac{f_1}{2}, \frac{f_2}{2}\right), \tag{14.4.17}$$

so that the fractiles of the y-distribution—and therefore also of the v^2-distribution—may be computed from the above-mentioned table.

From (14.4.10) and (14.4.11) we obtain

$$m\{v^2\} = \int_0^\infty zp\{z\}\,dz$$

$$= \frac{\Gamma\left(\frac{f_1+f_2}{2}\right)}{\Gamma\left(\frac{f_1}{2}\right)\Gamma\left(\frac{f_2}{2}\right)} \frac{f_2}{f_1} \int_0^1 \frac{y}{1-y} y^{\frac{f_1}{2}-1} (1-y)^{\frac{f_2}{2}-1} dy$$

$$= \frac{\Gamma\left(\frac{f_1+f_2}{2}\right)}{\Gamma\left(\frac{f_1}{2}\right)\Gamma\left(\frac{f_2}{2}\right)} \frac{f_2}{f_1} \frac{\Gamma\left(\frac{f_1}{2}+1\right)\Gamma\left(\frac{f_2}{2}-1\right)}{\Gamma\left(\frac{f_1+f_2}{2}\right)} = \frac{f_2}{f_2-2}. \tag{14.4.18}$$

14.5. Special Cases of the v^2-Distribution.

The definition of v^2

$$v^2(f_1, f_2) = \frac{\chi_1^2/f_1}{\chi_2^2/f_2}$$

shows that v^2 converges[1]) towards χ_1^2/f_1 for $f_2 \to \infty$, since χ_2^2/f_2 in this case converges towards 1, as shown in § 10.3. Hence

$$v^2(f_1, \infty) = \frac{\chi_1^2}{f_1}$$

or

$$\boxed{v_P^2(f_1, \infty) = \frac{\chi_P^2(f_1)}{f_1},} \tag{14.5.1}$$

i. e., the χ^2-distribution is a special case of the v^2-distribution. If we compare Tables VI and VII, we see how the values agree with (14.5.1). For example for $f_1 = 8$ and $P = 95\%$ we find

$$v_{\cdot 95}^2(8, \infty) = 1{\cdot}94 = \frac{\chi_{\cdot 95}^2(8)}{8}.$$

[1]) Convergence in § 14.5 is understood as "convergence in probability". Proofs of the theorems, where either f_1 or $f_2 \to \infty$, require finding the limit of the v^2-distribution.

Correspondingly, for $f_1 \to \infty$

$$v^2(\infty, f_2) = \frac{f_2}{\chi_2^2},$$

so that

$$P\{v^2 < v_P^2\} = P\left\{\frac{f_2}{\chi_2^2} < v_P^2\right\}$$

$$= P\left\{\frac{\chi_2^2}{f_2} > \frac{1}{v_P^2}\right\} = P.$$

Hence

$$P\left\{\frac{\chi_2^2}{f_2} < \frac{1}{v_P^2}\right\} = 1-P,$$

which means that $\frac{1}{v_P^2}$ is equal to the $(1-P)$-fractile for χ_2^2/f_2, i. e.,

$$\frac{1}{v_P^2} = \frac{\chi_{1-P}^2(f_2)}{f_2}$$

or

$$\boxed{v_P^2(\infty, f_2) = \frac{f_2}{\chi_{1-P}^2(f_2)}.} \tag{14.5.2}$$

This result may also be obtained by applying (14.1.9) to (14.5.1). For example, comparison of Tables VI and VII leads to

$$v_{\cdot 95}^2(\infty, 6) = 3{\cdot}67 = \frac{1}{\cdot 2725} = \frac{6}{\chi_{\cdot 05}^2(6)}.$$

For $f_1 = 1$, χ_1^2/f_1 is equal to u^2, cf. (10.2.1), and therefore

$$v^2(1, f_2) = \frac{u^2}{\chi_2^2/f_2}. \tag{14.5.3}$$

If we introduce

$$t = t(f_2) = \frac{u}{\sqrt{\chi_2^2/f_2}} \qquad (-\infty < t < \infty) \tag{14.5.4}$$

(14.5.3) may be written

$$v^2(1, f_2) = t^2(f_2)\,. \tag{14.5.5}$$

This t-distribution, which—like the χ^2-distribution—may be regarded as a special case of the v^2-distribution, will be dealt with in Chapter 15.

If, for example, we introduce

$$u = \frac{\bar{x}-\xi}{\sigma/\sqrt{n}} \tag{14.5.6}$$

and

$$\frac{\chi_2^2}{f_2} = \frac{s^2}{\sigma^2} \tag{14.5.7}$$

in (14.5.4) we obtain

$$t = \frac{\bar{x}-\xi}{s/\sqrt{n}}, \tag{14.5.8}$$

i. e., t is equal to the deviation of the sample mean from the population mean divided by the estimate of the standard error of the mean.

The relation

$$P\{v^2(1, f_2) < v_P^2(1, f_2)\} = P$$

is equal to

$$P\{-\sqrt{v_P^2} < t(f_2) < \sqrt{v_P^2}\} = P.$$

As the t-distribution is symmetrical about $t = 0$, as shown in § 15.1, we obtain

$$P\{t(f_2) < -\sqrt{v_P^2}\} = \tfrac{1}{2}(1-P),$$

and therefore

$$P\{t(f_2) < \sqrt{v_P^2}\} = \tfrac{1}{2}(1-P)+P = \tfrac{1}{2}(1+P).$$

This leads to

$$t_{\frac{1}{2}(1+P)}(f_2) = \sqrt{v_P^2},$$

or

$$\boxed{v_P^2(1, f_2) = t^2_{\frac{1}{2}(1+P)}(f_2)\,.} \tag{14.5.9}$$

The fractiles of the t-distribution are given in Table IV. From this table and Table VII, we see, for example, that

$$v^2_{.95}(1,\ 12) = 4{\cdot}75 = 2{\cdot}179^2 = t^2_{.975}(12)\,.$$

For $f_2 = 1$ we obtain in a corresponding manner, or from (14.1.9),

$$\boxed{v_P^2(f_1,\ 1) = \frac{1}{t^2_{\frac{1}{2}P}(f_1)}.} \tag{14.5.10}$$

Finally, for $f_1 = 1$ and $f_2 \to \infty$ we find

$$v^2(1, \infty) = u^2\,,$$

so that

$$\boxed{v_P^2(1,\ \infty) = u^2_{\frac{1}{2}(1+P)}} \tag{14.5.11}$$

and

$$\boxed{v_P^2(\infty,\ 1) = \frac{1}{u^2_{\frac{1}{2}P}}.} \tag{14.5.12}$$

These results are arranged in the following table which shows the relationship between the v^2-fractiles and the u-, t-, and χ^2-fractiles.

Table of $v_P^2(f_1, f_2)$.

$f_2 \backslash f_1$	1	…	f_1	…	∞
1	$t^2_{\frac{1}{2}(1+P)}(1) = \dfrac{1}{t^2_{\frac{1}{2}P}(1)}$	…	$\dfrac{1}{t^2_{\frac{1}{2}P}(f_1)}$	…	$\dfrac{1}{u^2_{\frac{1}{2}P}}$
⋮	⋮		⋮		⋮
f_2	$t^2_{\frac{1}{2}(1+P)}(f_2)$	…	$v_P^2(f_1, f_2)$	…	$\dfrac{f_2}{\chi^2_{1-P}(f_2)}$
⋮	⋮		⋮		⋮
∞	$u^2_{\frac{1}{2}(1+P)}$	…	$\dfrac{\chi_P^2(f_1)}{f_1}$	…	1

CHAPTER 15.

THE t-DISTRIBUTION

15.1. The t-Distribution.

In Chapter 9 we have seen that the mean, $\bar{x}$, of n stochastically independent observations from a normally distributed population with parameters (ξ, σ^2) is normally distributed with parameters $(\xi, \sigma^2/n)$, i. e.,

$$u = \frac{\bar{x}-\xi}{\sigma/\sqrt{n}} \tag{15.1.1}$$

is normally distributed with parameters (0, 1). According to (9.5.3) we obtain as confidence limits for ξ

$$\bar{x}-u_{P_2}\frac{\sigma}{\sqrt{n}} < \xi < \bar{x}-u_{P_1}\frac{\sigma}{\sqrt{n}}, \tag{15.1.2}$$

which inequality depends on σ. As σ is usually unknown, these limits cannot be calculated. It is therefore necessary to derive confidence limits for ξ in which σ is replaced by an estimate, s, calculated from the observations.

If we introduce the estimate s instead of σ in (15.1.1), the variable u is replaced by

$$\boxed{t = \frac{\bar{x}-\xi}{s/\sqrt{n}}.} \tag{15.1.3}$$

The distribution of this quantity, *the t-distribution*, will now be dealt with in more detail. Writing

$$t = \frac{\bar{x}-\xi}{\dfrac{s}{\sqrt{n}}} = \frac{\bar{x}-\xi}{\dfrac{\sigma}{\sqrt{n}}\dfrac{s}{\sigma}}$$

$$= \frac{u}{s/\sigma} = \frac{u}{\sqrt{\chi^2/f}}, \tag{15.1.4}$$

we see that the t-distribution is independent of both ξ and σ, since the distributions of both u and χ^2/f are independent of these parameters. Thus, *the t-distribution depends only on the number of degrees of freedom, f, for s^2.*

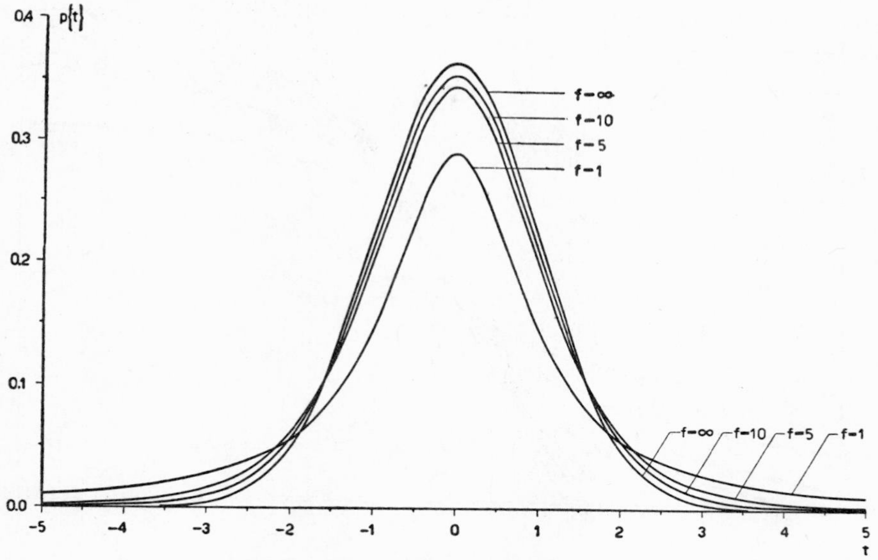

FIG. 15.1. Distribution curves of t for $f = 1, 5, 10$, and ∞.

According to (14.5.5) we have

$$v^2(1, f) = t^2(f) ,$$

from which we see that the distribution function of t may be derived from the distribution of v^2 by means of the transformation

$$t(f) = \pm\sqrt{v^2(1, f)} ,$$

which leads to

$$p\{t\} = \frac{1}{\sqrt{\pi f}} \frac{\Gamma\left(\frac{f+1}{2}\right)}{\Gamma\left(\frac{f}{2}\right)} \left(1+\frac{t^2}{f}\right)^{-\frac{f+1}{2}} \quad (-\infty < t < \infty) . \qquad (15.1.5)$$

Like the u-distribution, the t-distribution is symmetrical about zero. For $f \rightarrow \infty$ the t-distribution tends to the u-distribution, as s converges in probability to σ. Fig. 15.1 shows four distribution curves, corresponding to different values of f.

The fractiles of the t-distribution

$$t_P = t_P(f)$$

are given in Table IV, which shows how t_P converges towards u_P for $f \rightarrow \infty$. For $f > 30$, t_P is practically equal to u_P, except for extreme values of P.

The deviations of the cumulative distribution function from the standardized, normal distribution are illustrated in Fig. 15.2, in which the

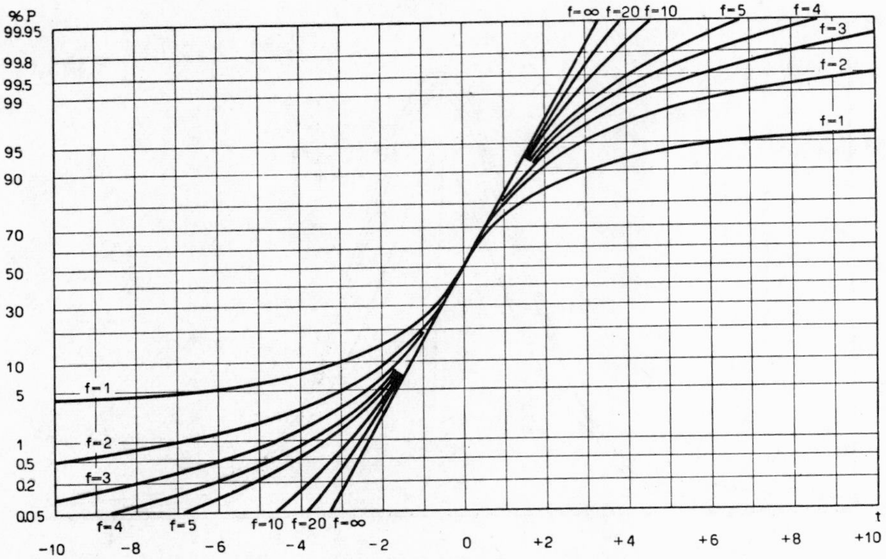

FIG. 15.2. Cumulative distribution functions of t for $f = 1, 2, 3, 4, 5, 10, 20$, and ∞ plotted on probability paper.

cumulative distribution functions for different values of f have been plotted on probability paper.

The symmetrical form of the distribution shows that

$$t_P = -t_{1-P} \,. \tag{15.1.6}$$

For example, Table IV gives $t_{\cdot 95} = -t_{\cdot 05} = 1{\cdot}86$ for $f = 8$, which means that

$$P\{t < 1{\cdot}86\} = 95\% \quad \text{and} \quad P\{t < -1{\cdot}86\} = 5\%\,,$$

and hence

$$P\{|t| < 1{\cdot}86\} = 90\% \quad \text{for} \quad f = 8\,.$$

If we introduce (15.1.3) we have

$$P\left\{-1{\cdot}86 < \frac{\bar{x} - \xi}{s/\sqrt{n}} < 1{\cdot}86\right\} = 90\% \text{ for } f = 8\,.$$

In analogy with (9.3.1) we have the following general expression

$$\boxed{P\left\{\frac{\bar{x} - \xi}{s/\sqrt{n}} < t_P\right\} = P\,.} \tag{15.1.7}$$

From (15.1.4) it follows that the definition of t given by (15.1.3) may be regarded as a special case of the definition

$$\boxed{t = \frac{u}{\sqrt{\chi^2/f}} = \frac{u}{s/\sigma},} \tag{15.1.8}$$

and accordingly we see that any variable which can be written in the form (15.1.8), in which u *and* s *are stochastically independent*, has a t-distribution with f degrees of freedom.

If we consider the special case

$$u = \frac{x-\xi}{\sigma},$$

we obtain

$$\boxed{t = \frac{x-\xi}{s},} \tag{15.1.9}$$

which shows that, if σ is replaced by s, a variable which has a u-distribution will be replaced by one having a t-distribution; this holds good if s^2 is stochastically independent of x and has an s^2-distribution with parameters (σ^2, f).

15.2. The t-Test of Significance.

In § 9.4 it was demonstrated how the hypothesis $\xi = \xi_0$ might be tested from the sample mean $\bar{x}$ and the population standard deviation σ. As most often σ is unknown, an estimate s of σ must be used for the test, and hence the u-test is replaced by a t-test.

If the test hypothesis $\mathcal{M}\{x\} = \xi_0$ is true, the variable

$$t = \frac{\bar{x}-\xi_0}{s/\sqrt{n}} \tag{15.2.1}$$

has a t-distribution with f degrees of freedom, f denoting the number of degrees of freedom for s.

If the alternative hypothesis is unilateral, $\xi > \xi_0$ say, the critical region at the α level of significance becomes

$$\frac{\bar{x}-\xi_0}{s/\sqrt{n}} > t_{1-\alpha}\,. \tag{15.2.2}$$

In the bilateral case, $\xi \neq \xi_0$, the critical region is

$$\frac{\bar{x}-\xi_0}{s/\sqrt{n}} < t_{\frac{\alpha}{2}} \quad \text{and} \quad \frac{\bar{x}-\xi_0}{s/\sqrt{n}} > t_{1-\frac{\alpha}{2}}\,. \tag{15.2.3}$$

In the first case the power of the test becomes

$$\pi(\xi) = P\left\{\frac{\bar{x}-\xi_0}{s/\sqrt{n}} > t_{1-\alpha}\,;\ \mathcal{m}\{x\} = \xi\right\}. \tag{15.2.4}$$

Writing

$$\frac{\bar{x}-\xi_0}{s/\sqrt{n}} = \frac{\bar{x}-\xi}{s/\sqrt{n}} + \frac{\xi-\xi_0}{\sigma/\sqrt{n}}\frac{\sigma}{s}$$

$$= t+\lambda\sqrt{f/\chi^2}, \quad \text{where} \quad \lambda = \frac{\xi-\xi_0}{\sigma/\sqrt{n}}, \tag{15.2.5}$$

it will be seen that the power function depends on a combination of the t- and χ-distributions which has been called the non-central t-distribution. Tables for calculating the fractiles of this distribution have been given in N. L. JOHNSON and B. L. WELCH: *Applications of the Non-Central t-Distribution*, Biometrika, 31, 1940, 362–389.

To find an approximation based on the normal distribution (15.2.4) is written as

$$\pi(\xi) = P\left\{\bar{x}-t_{1-\alpha}\frac{s}{\sqrt{n}} > \xi_0;\ \mathcal{m}\{x\} = \xi\right\}. \tag{15.2.6}$$

The auxiliary variable

$$z = \bar{x}-t_{1-\alpha}\frac{s}{\sqrt{n}} \tag{15.2.7}$$

will be approximately normally distributed for quite small values of f with mean and variance given by

$$\mathcal{m}\{z\} \simeq \xi-t_{1-\alpha}\frac{\sigma}{\sqrt{n}} \tag{15.2.8}$$

and

$$\mathcal{v}\{z\} = \frac{\sigma^2}{n}\left(1+\frac{t_{1-\alpha}^2}{2f}\right), \tag{15.2.9}$$

see § 11.7 and similar results in § 11.8 and § 11.9. Hence, the power function becomes

$$\pi(\xi) \simeq \Phi\left(\frac{t_\alpha+\lambda}{\sqrt{1+\dfrac{t_{1-\alpha}^2}{2f}}}\right), \quad \lambda = \frac{\xi-\xi_0}{\sigma/\sqrt{n}}, \tag{15.2.10}$$

which is analogous to (9.4.13).

Similarly, for the bilateral test we obtain

$$\pi(\xi) \simeq \Phi\left(\frac{t_{\frac{\alpha}{2}}-\lambda}{\sqrt{1+\dfrac{t_{\frac{\alpha}{2}}^2}{2f}}}\right) + \Phi\left(\frac{t_{\frac{\alpha}{2}}+\lambda}{\sqrt{1+\dfrac{t_{\frac{\alpha}{2}}^2}{2f}}}\right). \tag{15.2.11}$$

Example 15.1. In the manufacture of electricity meters of the rotating disc type the meters are adjusted to make them work synchronously with a standard meter. After adjustment, a sample of 10 meters is tested by determining their constants by precision watt-meters and stop-watches. Table 15.1 gives the results of such a test, the constant of the standard meter being 1·0000. The question is now: Can the deviations from the standard be regarded as random or do the results indicate that the constants of the adjusted meters deviate systematically from the constant of the standard meter? This question may be answered by testing the hypothesis that the ten measurements are a random sample taken from a normally distributed population with mean $\xi_0 = 1{\cdot}0000$.

Table 15.1 shows how the mean and variance of the ten measurements have been calculated according to the scheme set up in Table 4.3, p. 74; further the square of the standard error of the mean, s^2/n, the deviation $\bar{x}-\xi_0$, and the deviation divided by the standard error have been computed. As the t-value, $t = -2{\cdot}35$, lies between $t_{\cdot 025}$ and $t_{\cdot 01}$, the test hypothesis must be regarded as dubious, and it seems reasonable to investigate further the question of a possibly systematic error by taking a larger sample.

TABLE 15.1.

Control measurements from 10 adjusted meters. The constant of the standard meter is 1·0000. x denotes the meter constants multiplied by 1000.

x	$x-995$		$x-995$
983	−12	S	−10
1002	7	SS	594
998	3	S^2/n	10
996	1	SSD	584
1002	7	f	9
983	−12	s^2	64·89
994	−1	S/n	−1·0
991	−4	$\bar{x}$	994·0
1005	10	s^2/n	6·489
986	−9	$\sqrt{s^2/n}$	2·55

$$t = \frac{994{\cdot}0-1000{\cdot}0}{2{\cdot}55} = -2{\cdot}35.$$

$$t_{\cdot 025} = -2{\cdot}26, \quad t_{\cdot 01} = -2{\cdot}82.$$

$$P\{|t| > 2{\cdot}35\} = \text{ca. } 5\%.$$

15.3. Confidence Limits for ξ.

The relation

$$P\left\{t_{P_1} < \frac{\bar{x}-\xi}{s/\sqrt{n}} < t_{P_2}\right\} = P_2 - P_1$$

gives us confidence limits for ξ by solving the inequality with respect to ξ:

$$\boxed{P\left\{\bar{x}-t_{P_2}\frac{s}{\sqrt{n}} < \xi < \bar{x}-t_{P_1}\frac{s}{\sqrt{n}}\right\} = P_2 - P_1 .} \tag{15.3.1}$$

If these limits are compared with the analogous limits in (9.5.2) we see that the introduction of the estimate s instead of the theoretical value σ leads to t_P, instead of u_P, t_P being numerically larger than u_P. For $f > 30$, however, the difference between t_P and u_P is small.

The confidence limits for ξ given by (15.3.1) depend solely on the empirical values $\bar{x}$ and s, and on n, P_1 and P_2.

15.4. Comparison of Two Sets of Stochastically Independent Observations.

Let $x_{11}, x_{12}, \ldots, x_{1n_1}$, and $x_{21}, x_{22}, \ldots, x_{2n_2}$, be two sets of observations, randomly chosen from two normally distributed populations with parameters (ξ_1, σ_1^2) and (ξ_2, σ_2^2). (If the observations themselves are not normally distributed, they must be transformed, as in Chapter 7, and the computations carried out with the transformed, normally distributed observations). Estimates of the parameters are computed from each set of observations as

$$\bar{x}_1 = \frac{1}{n_1}\sum_{\nu=1}^{n_1} x_{1\nu} = \frac{S_1}{n_1} \approx \xi_1 , \tag{15.4.1}$$

$$s_1^2 = \frac{1}{n_1-1}\sum_{\nu=1}^{n_1}(x_{1\nu}-\bar{x}_1)^2 = \frac{SSD_1}{f_1} \approx \sigma_1^2,\ f_1 = n_1-1 , \tag{15.4.2}$$

$$\bar{x}_2 = \frac{1}{n_2}\sum_{\nu=1}^{n_2} x_{2\nu} = \frac{S_2}{n_2} \approx \xi_2, \tag{15.4.3}$$

and

$$s_2^2 = \frac{1}{n_2-1}\sum_{\nu=1}^{n_2}(x_{2\nu}-\bar{x}_2)^2 = \frac{SSD_2}{f_2} \approx \sigma_2^2,\ f_2 = n_2-1 . \tag{15.4.4}$$

According to § 9.2 the two means, $\bar{x}_1$ and $\bar{x}_2$, are normally distributed with parameters $(\xi_1, \sigma_1^2/n_1)$ and $(\xi_2, \sigma_2^2/n_2)$ and the difference between the two means, $d = \bar{x}_1-\bar{x}_2$, is normally distributed about $\delta = \xi_1-\xi_2$ with the variance

$$\mathcal{V}\{d\} = \frac{\sigma_1^2}{n_1} + \frac{\sigma_2^2}{n_2}.$$

The various combinations of the four population parameters that can occur may be classified as follows:

	Variances	Means	
A	$\sigma_1^2 = \sigma_2^2$	$\xi_1 = \xi_2$	A_1
		$\xi_1 \neq \xi_2$	A_2
B	$\sigma_1^2 \neq \sigma_2^2$	$\xi_1 = \xi_2$	B_1
		$\xi_1 \neq \xi_2$	B_2

We distinguish between cases A and B by means of the v^2-test, the hypothesis $\sigma_1^2 = \sigma_2^2$ being tested by calculation of the variance ratio

$$v^2(f_1, f_2) = \frac{s_1^2}{s_2^2}, \tag{15.4.5}$$

cf. § 14.2.

Case A_1. If s_1^2 does not differ significantly from s_2^2, we set up *the test hypothesis* that $\xi_1 = \xi_2 = \xi$ and $\sigma_1^2 = \sigma_2^2 = \sigma^2$. If this hypothesis is correct, $d = \bar{x}_1 - \bar{x}_2$ is normally distributed about $\delta = \xi_1 - \xi_2 = 0$ with variance

$$\mathcal{V}\{d\} = \sigma^2\left(\frac{1}{n_1} + \frac{1}{n_2}\right),$$

and therefore

$$u = \frac{d}{\sigma\{d\}} = \frac{\bar{x}_1 - \bar{x}_2}{\sigma\sqrt{\dfrac{1}{n_1} + \dfrac{1}{n_2}}} \tag{15.4.6}$$

is normally distributed with parameters (0, 1).

According to the addition theorem for the s^2-distribution we compute the following estimate of σ^2:

$$s^2 = \frac{f_1 s_1^2 + f_2 s_2^2}{f_1 + f_2} = \frac{SSD_1 + SSD_2}{f_1 + f_2} \simeq \sigma^2,\ f = f_1 + f_2 = n_1 + n_2 - 2. \tag{15.4.7}$$

If in (15.4.6) we introduce s instead of σ, we obtain

$$\boxed{t = \frac{\bar{x}_1 - \bar{x}_2}{s\sqrt{\dfrac{1}{n_1} + \dfrac{1}{n_2}}},\ f = n_1 + n_2 - 2,} \tag{15.4.8}$$

which according to (15.1.8) has a t-distribution with n_1+n_2-2 degrees of freedom.

A significant value of t is usually interpreted as indicating that $\xi_1 \neq \xi_2$ and $\sigma_1^2 = \sigma_2^2$, but in principle the significant value may be caused by the fact that $\xi_1 = \xi_2$ and $\sigma_1^2 \neq \sigma_2^2$, or $\xi_1 \neq \xi_2$ and $\sigma^2 \neq \sigma_1^2$, as the v^2-test does not exclude the possibility $\sigma_1^2 \neq \sigma_2^2$, even though the value of v^2 was not significant. In experimental work where each set of observations represents the result of a certain treatment, the aim of the t-test may be to decide whether two different treatments have caused the two sets of observations to differ, while it is of secondary importance to learn whether the difference lies in the means or the variances or possibly in both.

If the test hypothesis is not rejected an estimate of the parameters (ξ, σ^2) based on both sets of observations may be required. As the n_1+n_2 observations may be regarded as samples from the same population, the mean $\bar{x}$ and the variance s_0^2 may be calculated from the formulas

$$\bar{x} = \frac{S_1+S_2}{n_1+n_2} = \frac{n_1\bar{x}_1+n_2\bar{x}_2}{n_1+n_2} \approx \xi \tag{15.4.9}$$

and

$$s_0^2 = \frac{\sum_{\nu=1}^{n_1} (x_{1\nu}-\bar{x})^2 + \sum_{\nu=1}^{n_2} (x_{2\nu}-\bar{x})^2}{n_1+n_2-1} \approx \sigma^2, \; f = n_1+n_2-1\,. \tag{15.4.10}$$

The numerator of s_0^2 may be calculated from the formula

$$\begin{aligned}\sum_{\nu=1}^{n_1} (x_{1\nu}-\bar{x})^2 + \sum_{\nu=1}^{n_2} (x_{2\nu}-\bar{x})^2 &= SSD_1+SSD_2+n_1(\bar{x}_1-\bar{x})^2+n_2(\bar{x}_2-\bar{x})^2 \\ &= SSD_1+SSD_2+\frac{S_1^2}{n_1}+\frac{S_2^2}{n_2}-\frac{(S_1+S_2)^2}{n_1+n_2}\,.\end{aligned} \tag{15.4.11}$$

As $\bar{x}$ is normally distributed about ξ with variance $\sigma^2/(n_1+n_2)$, we see that

$$t = \frac{\bar{x}-\xi}{s_0/\sqrt{n_1+n_2}}, \; f = n_1+n_2-1\,, \tag{15.4.12}$$

has a t-distribution with n_1+n_2-1 degrees of freedom. From this we may calculate confidence limits for ξ. For large values of f, s^2 according to (15.4.7) is often used instead of s_0^2, as the limits for ξ and σ^2 will show practically no difference whether calculated from s_0^2 or s^2.

The conditions underlying the t-test are (1) that the observations within each set are randomly drawn from the same population, see Chapter 13, and (2) that the two populations are normally distributed, see Chapters

6–7. The first condition is essential for the validity of the t-test[1]), whereas moderate deviations from normality usually are of no great importance[2]).

Case A_2. If, according to (15.4.8), t is significant, the hypothesis $\xi_1 = \xi_2$ and $\sigma_1^2 = \sigma_2^2$ is rejected. Sometimes the observations are further analyzed on basis of the hypothesis $\xi_1 \neq \xi_2$ and $\sigma_1^2 = \sigma_2^2 = \sigma^2$. Assuming this hypothesis to be correct, $d = \bar{x}_1 - \bar{x}_2$ is normally distributed about $\delta = \xi_1 - \xi_2$ with variance $\sigma^2 \left(\frac{1}{n_1} + \frac{1}{n_2}\right)$, and therefore the quantity

$$t = \frac{d - \delta}{s \sqrt{\frac{1}{n_1} + \frac{1}{n_2}}}, \quad f = n_1 + n_2 - 2, \tag{15.4.13}$$

where s is calculated from (15.4.7), has a t-distribution with $n_1 + n_2 - 2$ degrees of freedom. From this formula confidence limits for $\delta = \xi_1 - \xi_2$ may be calculated.

Case B. If $\sigma_1^2 \neq \sigma_2^2$, the t-test cannot be applied directly. The difference $d = \bar{x}_1 - \bar{x}_2$ is normally distributed about $\delta = \xi_1 - \xi_2$ with variance

$$\mathcal{V}\{d\} = \frac{\sigma_1^2}{n_1} + \frac{\sigma_2^2}{n_2},$$

and therefore

$$u = \frac{\bar{x}_1 - \bar{x}_2 - \delta}{\sqrt{\frac{\sigma_1^2}{n_1} + \frac{\sigma_2^2}{n_2}}} \tag{15.4.14}$$

is normally distributed with parameters (0, 1). If σ_1^2 and σ_2^2 are replaced by s_1^2 and s_2^2, u is not transformed to t, since the denominator is not distributed as $\sqrt{\chi^2/f}$.

The distribution of the quantity

$$t = \frac{\bar{x}_1 - \bar{x}_2 - \delta}{\sqrt{\frac{s_1^2}{n_1} + \frac{s_2^2}{n_2}}} \tag{15.4.15}$$

has been studied by B. L. WELCH[3]) and the 5% and 1% unilateral limits of significance as functions of f_1, f_2, and

---

1) J. E. WALSH: *Concerning the Effect of Intraclass Correlation on Certain Significance Tests*, Ann. Math. Stat., 18, 1947, 88–96.

2) A. K. GAYEN: *The Distribution of* STUDENT'S *t in Random Samples of Any Size Drawn from Non-Normal Universes*, Biometrika, 36, 1949, 353–369.

3) B. L. WELCH: *The Generalization of Student's Problem When Several Different Population Variances are Involved*, Biometrika, 34, 1947, 28–35.

$$c = \frac{\dfrac{s_1^2}{n_1}}{\dfrac{s_1^2}{n_1}+\dfrac{s_2^2}{n_2}} \tag{15.4.16}$$

have been tabulated by A. A. ASPIN.[1]) Further, WELCH[2]) has shown that a good approximation to these limits may be obtained from the t-distribution, as *the variable given by* (15.4.15) *is distributed approximately as t with f degrees of freedom, where*

$$\frac{1}{f} = \frac{c^2}{f_1} + \frac{(1-c)^2}{f_2}, \tag{15.4.17}$$

i. e., f lies between the smaller of the two numbers f_1 and f_2 and the sum of f_1 and f_2. Hence, if both f_1 and f_2 are larger than 30, (15.4.15) may be regarded as normally distributed with good approximation.

Example 15.2. The two distributions of reaction times in Table 7.5, p. 173, correspond to 30 and 60 seconds contact time, respectively. In order to examine whether the contact time has influenced the distribution of the reaction times, we first test the hypothesis $\sigma_1^2 = \sigma_2^2$.

The variance ratio is

$$v^2(15,14) = \frac{0{\cdot}2080}{0{\cdot}1930} = 1{\cdot}08 \simeq v^2_{.50}(15,14).$$

As the two s^2-values do not differ significantly, we compute an estimate, s^2, of σ^2 according to the addition theorem for the s^2-distribution

$$s^2 = \frac{3{\cdot}1204+2{\cdot}7024}{15+14} = \frac{5{\cdot}8228}{29} = 0{\cdot}2008\,.$$

The difference, d, between the two means is

$$d = 1{\cdot}219-1{\cdot}179 = 0{\cdot}040\,,$$

and the estimate of the variance of this difference is

$$s^2\left(\frac{1}{n_1}+\frac{1}{n_2}\right) = 0{\cdot}2008\left(\frac{1}{16}+\frac{1}{15}\right) = 0{\cdot}02594\,,$$

and therefore

$$t = \frac{0{\cdot}040}{\sqrt{0{\cdot}02594}} = \frac{0{\cdot}040}{0{\cdot}161} = 0{\cdot}25 \simeq t_{.60}\,,\ f = 29\,.$$

---

1) ALICE A. ASPIN: *Tables for Use in Comparisons Whose Accuracy Involves Two Variances, Separately Estimated*, Biometrika, 36, 1949, 290–293.

2) B. L. WELCH: *Further Note on Mrs. Aspin's Tables and on Certain Approximations to the Tabled Function*, Biometrika, 36, 1949, 293–296.

As we cannot prove any difference between the two distributions of the reaction times, we calculate the total mean

$$\bar{x} = \frac{19{\cdot}506+17{\cdot}689}{16+15} = \frac{37{\cdot}195}{31} = 1{\cdot}200$$

and the variance

$$s_0^2 = \frac{5{\cdot}8228+0{\cdot}0123}{29+1} = \frac{5{\cdot}8351}{30} = 0{\cdot}1945\,.$$

The estimate of the standard error of the mean then becomes

$$\sqrt{\frac{0{\cdot}1945}{31}} = 0{\cdot}0792\,.$$

Example 15.3. In order to test whether a special treatment of concrete has an effect on the strength of the concrete, a small experiment was made. From the given batch of raw materials 6 samples were taken, the samples being made as similar as possible. Then the 6 samples were divided *at random* into two groups, each of three samples, and from each sample a test cube was made, samples from group II receiving a special treatment.

TABLE 15.2.

Compressive strength of test cubes of concrete, kg/cm².

	Concrete I	Concrete II
	290	309
	311	318
	284	318
$\bar{x}$	295	315
SSD	402	54
f	2	2
s^2	201	27

$$v^2 = \frac{201}{27} = 7{\cdot}44 < v^2_{.90}(2,2) = 9{\cdot}00.$$

$$s^2 = \frac{402+54}{2+2} = 114, \quad f = 4.$$

$$d = 315-295 = 20$$

$$s^2\left(\frac{1}{n_1}+\frac{1}{n_2}\right) = 114\times\frac{2}{3} = 76, \quad \sqrt{76} = 8{\cdot}7.$$

$$t = \frac{20}{8{\cdot}7} = 2{\cdot}3, \quad f = 4.$$

$$t_{.95} = 2{\cdot}1 \text{ and } t_{.975} = 2{\cdot}8 \text{ for } f = 4.$$

$$P\{|t| > 2{\cdot}3\} = 8\,\%.$$

After 28 days' curing the compressive strengths of the 6 test cubes were found. The strengths and the computations used for carrying out the t-test for comparison of the two sets of observations are given in Table 15.2. As there is no a priori reason to consider concrete II as stronger than concrete I, we apply a bilateral test. The t-value found lies between the bilateral 90% and 95% significance limits and can therefore not be regarded as significant.

TABLE 15.3.

Tensile strength in kg of 2·52 mm wire.

	Cable 1	Cable 2
	348	338
	345	350
	350	335
	344	333
	342	345
	346	331
	349	354
	345	339
	341	339
	347	343
	344	347
	339	340
	341	336
	348	342
	352	345
	345	338
	346	335
	349	339
$\bar{x}$	345·61	340·50
SSD	202·28	610·50
f	17	17
s^2	11·90	35·91

$$v^2 = \frac{35{\cdot}91}{11{\cdot}90} = 3{\cdot}02 > v^2_{.975}$$

$$\frac{s_1^2}{n_1} + \frac{s_2^2}{n_2} = \frac{11{\cdot}90}{18} + \frac{35{\cdot}91}{18} = 2{\cdot}656$$

$$d = 345{\cdot}61 - 340{\cdot}50 = 5{\cdot}11$$

$$t \simeq \frac{5{\cdot}11}{\sqrt{2{\cdot}656}} = 3{\cdot}13\,.$$

$$c = \frac{0{\cdot}6611}{2{\cdot}656} = 0{\cdot}249\,,\ \frac{1}{f} = \frac{0{\cdot}249^2}{17} + \frac{0{\cdot}751^2}{17} = 0{\cdot}0362,\ f = 27{\cdot}6\,.$$

Example 15.4. For a high-voltage net, cables of great and equal tensile strength are required. Each cable is composed of wires, which are manufactured in one length. In order to measure the tensile strength a sample is taken from each wire and tested. Table 15.3 shows the distribution of the tensile strength of 2×18 samples taken from two cables.

The variance ratio shows that the difference between the two s^2-values is significant, so that we cannot apply the ordinary t-test. Applying the approximate t-test given by (15.4.15) we find $t \simeq 3 \cdot 13$ with 27 degrees of freedom, which value exceeds the bilateral 99% limit of significance. Hence, the wires of cable No. 2 are of lesser tensile strength and are less uniform than those of cable No. 1.

15.5. Comparison of the Effects of Two Treatments.

In experimental work we often meet the problem of determining whether two treatments will produce different effects under essentially the same conditions. The word treatment is used in a wide sense; for instance the two treatments may denote two different temperatures for a chemical process, two different speeds of a machine, two different operators performing the same operation, two different methods of measurement, etc. Let us denote the treatments by A and B.

Two cases may arise:

1. *The units available for treatment are essentially similar* in the sense that treatment A applied to any unit will produce the same result apart from *random*, normally distributed deviations from the mean effect. The same holds for treatment B. Hence, before applying the treatments, the units may be divided into two groups which are considered as two random samples from the same population.

2. The units available for treatment are not essentially similar as defined above, but we expect, or have prior knowledge, that *differences in some properties of the units will cause them to react in different ways to treatment* A, and similarly to treatment B. On the basis of these properties, we therefore divide the units into groups within which the units are essentially similar. In the present case we consider only groups of two units each, i. e., *each pair of units is essentially similar with respect to the properties which we expect may influence the effects of the treatments*. If treatment A is applied to the two units of a pair it will produce the same effect apart from random, normally distributed deviations from the mean effect, and similarly for treatment B, i. e., the units may be divided into n pairs, and the pairs are considered as random samples of two drawn from n different populations.

For instance, when comparing the quality of a product manufactured by two different processes, we may carry out an experiment by using raw material as uniform as possible for the whole experiment, or we may use raw material from different batches, taking two samples from each batch, and assign the two processes at random to the two samples. Similarly, when comparing two methods of measurement we may measure the same object several times by each method or we may measure different objects, representative of the different "conditions" under which the two methods may be used in practice.

To test whether the two treatments produce different effects we may design the experiment in two different ways according to which situation exists.

1. *Among the units to be treated two samples of n are chosen at random* and are subjected to treatments A and B, respectively. Defining the true means of the effects of A and B as ξ_1 and ξ_2, respectively, we test the hypothesis $\xi_1 = \xi_2$ by the technique developed in § 15.4, i. e., we compute

$$t = \frac{\bar{x}_1 - \bar{x}_2}{s\sqrt{\frac{2}{n}}} = \frac{(\bar{x}_1 - \bar{x}_2)\sqrt{n}}{s\sqrt{2}}, \; f = 2(n-1) \,, \tag{15.5.1}$$

where $\bar{x}_1$, $\bar{x}_2$, s_1^2, and s_2^2 according to the notation in § 15.4 denote the means and variances of the two samples and s^2 the pooled variance, which may be written as

$$s^2 = \tfrac{1}{2}(s_1^2 + s_2^2) \tag{15.5.2}$$

because $n_1 = n_2 = n$, see (15.4.7) and (15.4.8). To apply this method we must assume that $\sigma_1^2 = \sigma_2^2$.

If $\sigma_1^2 \neq \sigma_2^2$ we may use the approximate t-test (15.4.15), which leads to

$$t \simeq \frac{(\bar{x}_1 - \bar{x}_2)\sqrt{n}}{\sqrt{s_1^2 + s_2^2}}, \; f = \frac{n-1}{c^2 + (1-c)^2}, \tag{15.5.3}$$

where $c = s_1^2/(s_1^2 + s_2^2)$, the only differences from (15.5.1) being the approximation involved and a decrease in the number of degrees of freedom, since f lies between $(n-1)$ and $2(n-1)$.

2. *The units to be treated are grouped into n pairs* as described above, and in each pair the treatments are assigned to the units by drawing lots. Let $(x_{11}, x_{12}), \ldots, (x_{n1}, x_{n2})$ be the corresponding observations, where x_{i1} refers to treatment A and x_{i2} to treatment B.

According to the test hypothesis corresponding observations are assumed to have the same value apart from random variations, from which it follows that the differences $d_i = x_{i1} - x_{i2}$ all have the true value zero. Further, the differences are assumed to be normally distributed with variance σ_d^2, so that the mean difference

$$\bar{d} = \frac{1}{n} \sum_{i=1}^{n} d_i \tag{15.5.4}$$

is normally distributed with parameters $(0, \sigma_d^2/n)$, and

$$u = \frac{\bar{d}}{\sigma_d/\sqrt{n}} \tag{15.5.5}$$

is normally distributed with parameters (0, 1). As an estimate of σ_d^2 we compute the variance

$$s_d^2 = \frac{1}{n-1}\sum_{i=1}^{n}(d_i-\bar{d})^2 . \tag{15.5.6}$$

Hence, it follows from (15.1.8) that

$$\boxed{t = \frac{\bar{d}}{s_d/\sqrt{n}},\ f = n-1,} \tag{15.5.7}$$

has a t-distribution with $n-1$ degrees of freedom. If the computed value of t is significant we reject the hypothesis that the two treatments have the same mean effects.

From the first component of the n pairs of observations we may compute the variance

$$s_1^2 = \frac{1}{n-1}\sum_{i=1}^{n}(x_{i1}-\bar{x}_1)^2 , \tag{15.5.8}$$

and similarly from the second component

$$s_2^2 = \frac{1}{n-1}\sum_{i=1}^{n}(x_{i2}-\bar{x}_2)^2 . \tag{15.5.9}$$

Considering s_d^2 we see that

$$d_i-\bar{d} = (x_{i1}-\bar{x}_1)-(x_{i2}-\bar{x}_2)$$

so that

$$s_d^2 = s_1^2+s_2^2-2s_{12} , \tag{15.5.10}$$

where

$$s_{12} = \frac{1}{n-1}\sum_{i=1}^{n}(x_{i1}-\bar{x}_1)(x_{i2}-\bar{x}_2) , \tag{15.5.11}$$

i. e., the covariance of (x_{i1}, x_{i2}).

Thus, the variance of the differences depends on the correlation between pairs of the treatment effects, and for this reason we selected our pairs in such a way that we expect to get a high *positive correlation* with the intention of minimizing the variance.

If there is the possibility of choosing between the two cases, when designing the experiment, we usually choose case 2 for two reasons:

(a) The t-test in case 2 is often more powerful, i. e., it leads more often to the rejection of the test hypothesis when it is false, than the t-test in case 1, because the positive correlation between the two treatment effects, resulting from the pairing of similar units, leads to a smaller variance which more than balances the disadvantage that the t-value in case 2 has only half the number of degrees of freedom that it has in case 1.

(b) By choosing the pairs so that the properties of the units within each pair are similar and the properties of the units differ widely from one pair to another, we get the possibility of investigating the difference between the two treatment effects under the different conditions that are likely to arise in practice.

Variations *between the pairs* will not influence the variance of the mean difference, because this variance depends only on the variations of the differences *between units within the pairs*.

The above is a simple example of how it pays to plan an experiment carefully in advance. Some further aspects of the theory of the design of experiments are treated in Chapters 16 and 17.

In case 2, the assumptions underlying the test should be tested *by plotting*

TABLE 15.4.
Starch content of potatoes.

Sample No.	Method I % Starch	Method II % Starch	Difference % Starch
1	21·7	21·5	0·2
2	18·7	18·7	0·0
3	18·3	18·3	0·0
4	17·5	17·4	0·1
5	18·5	18·3	0·2
6	15·6	15·4	0·2
7	17·0	16·7	0·3
8	16·6	16·9	−0·3
9	14·0	13·9	0·1
10	17·2	17·0	0·2
11	21·7	21·4	0·3
12	18·6	18·6	0·0
13	17·9	18·0	−0·1
14	17·7	17·6	0·1
15	18·3	18·5	−0·2
16	15·6	15·5	0·1
		$\bar{d}$	0·075
		s_d^2	0·02867
		f	15
		$s_d/\sqrt{n}$	0·042
		t	0·075/0·042 = 1·8
		$P\{\lvert t\rvert > 1{\cdot}8\}$	ca. 10%

the n pairs of numbers in a diagram, using x_{i1} *as abscissa and* x_{i2} *as ordinate.* If the points of the diagram are randomly scattered with constant variation about a line with slope 1, it seems reasonable to assume that the differences, $d_i = x_{i1} - x_{i2}$, are distributed with a true mean δ and standard deviation σ, and then to test the hypothesis $\delta = 0$ as indicated above. If, however, the points are scattered about a line through the point (0, 0) and with a slope different from 1 and the variation around the line increases with increasing distance from the origin, the above theory does not apply directly, but after a logarithmic transformation of the observations it may be tried whether ($\log x_{i1}$, $\log x_{i2}$) fulfill the assumptions, and if so, the hypothesis $\mathcal{M}\{\log x_{i1}\} = \mathcal{M}\{\log x_{i2}\}$ may be tested.

Example 15.5. Two methods for determining the starch content of potatoes were to be compared. Sixteen potatoes with widely differing starch contents were taken, and the two methods of measurement were applied to each potato[1]). Table 15.4 gives the analytical results and the corresponding t-test for the differences.

Thus, the test does not indicate any significant difference between the true values of the analytical results obtained by the two methods.

Example 15.6. To test the effect of using a special seed drill[2]), ten plots of land were sown with an ordinary seed drill and ten with the special one, and the subsequent yields of grain compared. The twenty plots of equal size were chosen in pairs, two neighbouring plots forming a pair. The decision as to which of two neighbouring plots was to be treated with the special machine was made by tossing a coin. The results are given in Table 15.5.

As $t = 3{\cdot}2$ and $P\{|t| > 3{\cdot}2\}$ is about 1%, see Table IV, the difference is regarded as significant, i. e., we conclude that the special machine increases the yield.

The design of the experiment and the corresponding statistical handling of the resulting data have eliminated a major part of the variations in soil fertility, comparisons being made only between neighbouring plots, the yields of which according to Table 15.5 are strongly and positively correlated.

If, however, the experiment had been designed in such a manner that the 10 plots to be sown with one machine or the other had been chosen at

---

[1]) C. von Scheele, G. Svensson and J. Rasmusson: *Om Bestämning av Potatisens Stärkelse och Torrsubstanshalt med Tillhjälp av dess specifika Vikt*, Nordisk Jordbrugsforskning, 1935, p. 22.

[2]) J. Wishart: *Statistics in Agricultural Research*, Suppl. Journ. Roy. Stat. Soc., 1, 1934, 26–51.

TABLE 15.5.
Weight of yield per plot.

Pair No.	Special machine	Ordinary machine	Difference in yield
1	8·0	5·6	2·4
2	8·4	7·4	1·0
3	8·0	7·3	0·7
4	6·4	6·4	0·0
5	8·6	7·5	1·1
6	7·7	6·1	1·6
7	7·7	6·6	1·1
8	5·6	6·0	−0·4
9	5·6	5·5	0·1
10	6·2	5·5	0·7
		$\bar{d}$	0·83
		s_d^2	0·667
		f	9
		$s_d/\sqrt{n}$	0·258
		t	$0{\cdot}83/0{\cdot}258 = 3{\cdot}2$
		$P\{\lvert t\rvert > 3{\cdot}2\}$	ca. 1%

random and in such a manner that their mutual situations had been independent, the difference would probably not have been significant. If the results of such an experiment are analyzed according to the method in § 15.4 the t derived from (15.4.8) will include the variances from the two sets of observations, and these variances will become comparatively large as they depend on the variations in the fertility of the soil as well as on the experimental error.

If we imagine for a moment that the plots had not been paired, then we would use the method of § 15.4. Computing the pooled estimate of the variance, we obtain

$$s^2 = \frac{SSD_1+SSD_2}{f_1+f_2} = \frac{11{\cdot}936+5{\cdot}569}{9+9} = 0{\cdot}9725\ .$$

Further

$$s\sqrt{\frac{1}{n_1}+\frac{1}{n_2}} = \sqrt{0{\cdot}9725\left(\frac{1}{10}+\frac{1}{10}\right)} = 0{\cdot}441\ ,$$

and

$$t = \frac{7{\cdot}22-6{\cdot}39}{0{\cdot}441} = \frac{0{\cdot}83}{0{\cdot}441} = 1{\cdot}9,\ f = 18.$$

As $P\{|t| > 1{\cdot}9\} > 5\%$, the difference is regarded as insignificant.

15.6. Combination of Tests of Significance.

With more complicated investigations the data are often divided into groups, it being assumed that all the observations in a group are drawn from the same population, as in Example 7.2, p. 171 (grouping according to contact time), Example 7.6, p. 184 (grouping according to grinding time), etc. When a hypothesis has been tested for each group we sometimes want to combine all the evidence we have thus obtained.

If, for instance, the data include 100 groups, and a u-test is applied to each group, the 100 u-values must—if the hypothesis is true—be normally distributed with parameters (0, 1), so that about 5 groups may be expected to give results outside the usual 95% limits of significance for individual groups Hence, it is not correct to reject the test hypothesis for the individual groups, which taken separately appear to yield a significant result, unless the number of such groups is considerably larger than 5.

On the other hand we will find cases in which the test hypothesis must be rejected even though none of the individual u-values are significant. For instance, this may be the case if all deviations have the same sign or present some other systematic feature.

If we have many u-values, we may obtain an estimate of their distribution through a fractile diagram, cf. § 6.7. It is very seldom, however, that the number of groups is large enough to allow a proper evaluation to be made from such a diagram and it is often difficult to decide how much importance to attach to extreme values. It will therefore usually be necessary to confirm the graphical examination by a numerical one.

Let the observations be divided into k groups, and let us imagine a u-value has been calculated for each group, e. g., for the ith group

$$u_i = \frac{\bar{x}_i - \xi_i}{\sigma_i / \sqrt{n_i}}, \tag{15.6.1}$$

as discussed in § 9.4.

If the test hypothesis is true, the mean

$$\bar{u} = \frac{1}{k} \sum_{i=1}^{k} u_i \tag{15.6.2}$$

is normally distributed with parameters $(0, 1/k)$, and therefore

$$u = \frac{\bar{u} - 0}{1/\sqrt{k}} = \bar{u}\sqrt{k} = \frac{\sum_{i=1}^{k} u_i}{\sqrt{k}} \tag{15.6.3}$$

is normally distributed with parameters (0, 1). Further, the variance

$$s_u^2 = \frac{1}{k-1} \sum_{i=1}^{k} (u_i - \bar{u})^2 \tag{15.6.4}$$

has an s^2-distribution with parameters $(1, k-1)$, i. e., the quantity

$$\chi^2 = \sum_{i=1}^{k} (u_i - \bar{u})^2 , \quad f = k-1 , \tag{15.6.5}$$

has a χ^2-distribution with $k-1$ degrees of freedom.

If u in (15.6.3) is significant or χ^2 in (15.6.5) is significant, the test hypothesis is rejected.

If the test used within each of the k-groups is a χ^2-test, the k χ^2-values may be combined according to the addition theorem for the χ^2-distribution. For the t- and v^2-distributions we have no addition theorem similar to those that may be applied to the u- and χ^2-distributions. For values of $f > 30$, however, the t-test may be handled as though it were a u-test.

R. A. FISHER[1]) has given a general method for combining the probabilities of several mutually independent tests.

Let $y = f(x_1, x_2, \ldots, x_n)$ denote the function of the observations applied in the test. For example in the bilateral u-test we have

$$y = \left| \frac{\bar{x} - \xi}{\sigma / \sqrt{n}} \right| . \tag{15.6.6}$$

If we introduce

$$z = -2 \ln (1 - P\{y\}) , \qquad (0 < z < \infty) , \tag{15.6.7}$$

as a new variable, we get

$$\frac{dz}{dy} = 2 \frac{p\{y\}}{1 - P\{y\}} .$$

Hence, from (5.5.5),

$$p\{z\} = \left[p\{y\} \frac{1 - P\{y\}}{2p\{y\}} \right]_{z = -2 \ln (1 - P\{y\})}$$

$$= \frac{1}{2} e^{-\frac{z}{2}} , \tag{15.6.8}$$

i. e., z has a χ^2-distribution with two degrees of freedom, cf. (10.2.5).

This result is utilized to combine the probabilities $P\{y_1\}, P\{y_2\}, \ldots, P\{y_k\}, y_i$ denoting the value of y observed in the ith group. If we calculate $z_1, z_2, \ldots, z_k$ according to (15.6.7) and apply the addition theorem for the χ^2-distribution we obtain the quantity

$$\boxed{\chi^2 = \sum_{i=1}^{k} z_i = -4{\cdot}605 \sum_{i=1}^{k} \log_{10}(1 - P\{y_i\}) , \ f = 2k ,} \tag{15.6.9}$$

---

[1]) R. A. FISHER: *Statistical Methods for Research Workers*, Oliver and Boyd, Edinburgh, 9th Edition, 1944, § 21.1.

E. S. PEARSON: *The Probability Integral Transformation for Testing Goodness of Fit and Combining Independent Tests of Significance*, Biometrika, 30, 1938, 134–148.

which has a χ^2-distribution with $2k$ degrees of freedom. As a significant deviation from the test hypothesis leads to large values of $P\{y_i\}$, the criterion of significance in the above combination is $\chi^2 > \chi^2_P$.

Tests may be combined in several other ways than those described here, and each combination will lead to a different result as each is based on a different function of the observations. According to the general theory of testing statistical hypotheses, the choice among the different possible combinations should be made on account of the alternative hypothesis. It may, however, be difficult to solve this problem, and in such cases FISHER's "omnibus"-test may be used, even if it is not the most powerful.

Example 15.7. Imagine that the u-test applied to four groups of data gives the following values:

$$\begin{aligned} u_1 &= 1{\cdot}2\,, & P_1 &= 2\times 0{\cdot}115 &&= 0{\cdot}230\,,\\ u_2 &= 1{\cdot}0\,, & P_2 &= 2\times 0{\cdot}159 &&= 0{\cdot}318\,,\\ u_3 &= 0{\cdot}9\,, & P_3 &= 2\times 0{\cdot}184 &&= 0{\cdot}368\,,\\ u_4 &= 1{\cdot}7\,, & P_4 &= 2\times 0{\cdot}0446 &&= 0{\cdot}0892\,. \end{aligned}$$

When bilateral testing is used, none of the single values found is significant, the bilateral 95% limit being 1·96. If we combine the four values according to (15.6.3) we obtain

$$u = \frac{1{\cdot}2+1{\cdot}0+0{\cdot}9+1{\cdot}7}{\sqrt{4}} = \frac{4{\cdot}8}{2} = 2{\cdot}4\,.$$

As $P\{|u| > 2{\cdot}4\} = 1{\cdot}6\%$, the mean of the four observed u-values deviates significantly from the theoretical value 0.

If we use the combination indicated in (15.6.9) we obtain

$$\chi^2 = -4{\cdot}605 \sum \log_{10} P_i = 4{\cdot}605\times 2{\cdot}619 = 12{\cdot}1,\ f = 8\,.$$

As $\chi^2 = 12{\cdot}1$ lies between $\chi^2_{\cdot 80}$ and $\chi^2_{\cdot 90}$ this value cannot be considered significant.

The difference between the answers given by these two different ways of combining the tests lies in the fact that (15.6.9) in contrast to (15.6.3) does not take the sign of the u-values into account. If u_1 and u_2 had been negative, $u_1 = -1{\cdot}2$ and $u_2 = -1{\cdot}0$, P_1 and P_2 would have been the same, and (15.6.9) would have led to the same result. In this case, however, (15.6.3) gives

$$u = \frac{-1{\cdot}2-1{\cdot}0+0{\cdot}9+1{\cdot}7}{\sqrt{4}} = \frac{0{\cdot}4}{2} = 0{\cdot}2\,,$$

a value which is not significant.

If it is possible to decide a priori, i. e., independently of the observations, that the test hypothesis is only to be rejected if u is significantly *larger* than 0, (15.6.3) and (15.6.9) lead to almost the same results. As here "$(1-P\{y_i\})$" is equal to half the above probabilities, we obtain from (15.6.9)

$$\chi^2 = 4{\cdot}605 \times 3{\cdot}824 = 17{\cdot}6 \simeq \chi^2_{\cdot 975}\,,\; f = 8\,,$$

to which corresponds $u = 2{\cdot}4$ and $P\{u > 2{\cdot}4\} = 0{\cdot}8\%$.

15.7. Comparison of Two Truncated Distributions.

Let $(\bar{x}_1, s_1)$ and $(\bar{x}_2, s_2)$ denote the estimates of the corresponding parameters in two truncated normal distributions, cf. § 6.9.

The hypothesis $\xi_1 = \xi_2$ and $\sigma_1 = \sigma_2$ is tested by calculating the difference

$$d = \bar{x}_1 - \bar{x}_2\,, \tag{15.7.1}$$

which for large values of n_1 and n_2 (the number of observations) is normally distributed with mean value $\xi_1 - \xi_2 = 0$ and variance

$$\mathcal{V}\{d\} = \mathcal{V}\{\bar{x}_1\} + \mathcal{V}\{\bar{x}_2\} = \sigma^2\left(\frac{1}{n_1} + \frac{1}{n_2}\right)\mu_{11}(\zeta)\,, \tag{15.7.2}$$

cf. § 9.8. From this it follows that

$$u \simeq \frac{\bar{x}_1 - \bar{x}_2}{\sigma\sqrt{\left(\frac{1}{n_1} + \frac{1}{n_2}\right)\mu_{11}(\zeta)}}\,. \tag{15.7.3}$$

Instead of σ we introduce the estimate

$$s = \frac{n_1 s_1 + n_2 s_2}{n_1 + n_2}\,, \tag{15.7.4}$$

which is stochastically independent of d, and instead of ζ the estimate z, calculated from (6.9.8) or (6.9.12), combining the two sets of data. If

$$u \simeq \frac{\bar{x}_1 - \bar{x}_2}{s\sqrt{\left(\frac{1}{n_1} + \frac{1}{n_2}\right)\mu_{11}(z)}} \tag{15.7.5}$$

is significant, the test hypothesis is rejected. The variable u is only approximately normally distributed with parameters (0, 1). The larger the degree of truncation, the larger n_1 and n_2 must be before it is safe to apply (15.7.5).

15.8. References.

The derivation of the t-distribution and its application to the comparison of observations taken in pairs is due to W. S. Gosset, writing under the pseudonym "STUDENT" in *The Probable Error of a Mean*, Biometrika, 6, 1908, 1–25. The t-distribution is therefore often termed "STUDENT'S" distribution.

As "STUDENT'S" proof was incomplete, R. A. FISHER later derived the distribution in *Applications of* "STUDENT'S" *Distribution*, Metron, 5, 1925, 90–104, in which paper the general character of the distribution and its extensive field of applicability have been emphasized.

The principles underlying a test of significance for the case B of § 15.4 is still a matter of discussion. R. A. FISHER applies (15.4.15) but with limits of significance differing slightly from those found by WELCH. A discussion of the two points of view will be found in the following papers:

R. A. FISHER: *The Fiducial Argument in Statistical Inference*, Annals of Eugenics, 6, 1935, 391–398.

B. L. WELCH: *The Significance of the Difference Between Two Means When the Population Variances Are Unequal*, Biometrika, 29, 1937, 350–362.

R. A. FISHER: *The Comparison of Samples with Possibly Unequal Variances*, Annals of Eugenics, 9, 1939, 174–180.

J. NEYMAN: *Fiducial Argument and the Theory of Confidence Intervals*, Biometrika, 32, 1941, 128–150.

R. A. FISHER: *The Asymptotic Approach to* BEHRENS' *Integral with Further Tables for the d Test of Significance*, Annals of Eugenics, 11, 1941, 141–172.

H. SCHEFFÉ: *On Solutions of the* BEHRENS–FISHER *Problem Based on the t-Distribution*, Ann. of Math. Stat., 14, 1943, 35–44.

B. L. WELCH: *The Generalization of* "STUDENT'S" *Problem When Several Different Population Variances Are Involved*, Biometrika, 34, 1947, 28–35.

G. A. BARNARD: *On the* FISHER–BEHRENS *Test*, Biometrika, 37, 1950, 203–207.

CHAPTER 16.

ANALYSIS OF VARIANCE

16.1. Analysis of Variance and the *t*-Test.

A comparison of *several* sets of observations may be carried out by means of *the analysis of variance*. This analysis is based on the partition theorem for the χ^2-distribution and on the v^2-test.

A simple example, illustrating some of the fundamental ideas we shall meet with, is the following, which deals with a *single* set of observations.

Let $x_1, x_2, \ldots, x_n$ denote a set of stochastically independent observations from a normally distributed population with parameters (ξ, σ^2). From (10.6.1) we have

$$\sum_{i=1}^{n} (x_i-\xi)^2 = \sum_{i=1}^{n} (x_i-\bar{x})^2+n(\bar{x}-\xi)^2 . \qquad (16.1.1)$$

According to the partition theorem for the χ^2-distribution the two terms on the right are stochastically independent and distributed as $\sigma^2\chi^2$ with $n-1$ and 1 degrees of freedom, respectively. We can therefore calculate two stochastically independent estimates, s_1^2 and $s_{2(\xi)}^2$, of σ^2 where

$$s_1^2 = \frac{1}{n-1}\sum_{i=1}^{n} (x_i-\bar{x})^2 , \; f = n-1 , \qquad (16.1.2)$$

and

$$s_{2(\xi)}^2 = n(\bar{x}-\xi)^2 , \; f = 1 . \qquad (16.1.3)$$

Hence, the ratio

$$v^2(1, n-1) = \frac{s_{2(\xi)}^2}{s_1^2} \qquad (16.1.4)$$

has a v^2-distribution with $(1, n-1)$ degrees of freedom.

The estimate s_1^2 is based on *the variation of the observations about the sample mean* and is thus *independent of the population mean* ξ, while the estimate $s_{2(\xi)}^2$ is based on *the deviation of the sample mean from the theoretical value* ξ. If a hypothetical value ξ_0 is inserted in (16.1.3) we get

$$\begin{aligned} s_{2(\xi_0)}^2 &= n(\bar{x}-\xi_0)^2 = n\big((\bar{x}-\xi)+(\xi-\xi_0)\big)^2 \\ &= n(\bar{x}-\xi)^2+n(\xi-\xi_0)^2+2n(\bar{x}-\xi)(\xi-\xi_0) . \end{aligned} \qquad (16.1.5)$$

Since

$$\mathcal{M}\{\bar{x}-\xi\} = 0$$

we find

$$\mathcal{M}\{n(\bar{x}-\xi_0)^2\} = \mathcal{M}\{n(\bar{x}-\xi)^2\}+n(\xi-\xi_0)^2 \, . \tag{16.1.6}$$

Further

$$\mathcal{M}\{n(\bar{x}-\xi)^2\} = \mathcal{M}\{s^2_{2(\xi)}\} = \sigma^2 \, ,$$

so that

$$\mathcal{M}\{s^2_{2(\xi_0)}\} = \sigma^2+n(\xi-\xi_0)^2 \, , \tag{16.1.7}$$

i. e., $s^2_{2(\xi_0)}$ has an expected value which is *larger* than σ^2. Thus, if $\xi_0 \neq \xi$, $s^2_{2(\xi_0)}$ will tend to be significantly larger than s_1^2, and therefore a significant value of $v^2 = s^2_{2(\xi_0)}/s_1^2$ indicates that the hypothetical value ξ_0 cannot be accepted as population mean.

If we take the root of v^2 in (16.1.4) for $\xi = \xi_0$ we obtain

$$\sqrt{v^2(1, n-1)} = \frac{s_{2(\xi_0)}}{s_1} = \frac{\sqrt{n}(\bar{x}-\xi_0)}{s_1} \, , \tag{16.1.8}$$

which is the variable employed in the t-test of § 15.2.

In the case where *two sets of observations are compared*, the analysis of variance is identical with the t-test as dealt with in § 15.4, case A. An alternative proof of this test, using the methods of analysis of variance, will now be given, as an introduction to the more general analysis employed later.

Let

$$x_{11}, x_{12}, \ldots, x_{1n_1} \, ,$$

and

$$x_{21}, x_{22}, \ldots, x_{2n_2} \, ,$$

denote two sets of stochastically independent observations from two normally distributed populations with the same variance σ^2 and with means ξ_1 and ξ_2. As in (16.1.1) the sums of squares for the two sets of observations are partitioned as follows:

$$\sum_{\nu=1}^{n_1} (x_{1\nu}-\xi_1)^2 = \sum_{\nu=1}^{n_1} (x_{1\nu}-\bar{x}_1)^2+n_1(\bar{x}_1-\xi_1)^2 \tag{16.1.9}$$

and

$$\sum_{\nu=1}^{n_2} (x_{2\nu}-\xi_2)^2 = \sum_{\nu=1}^{n_2} (x_{2\nu}-\bar{x}_2)^2+n_2(\bar{x}_2-\xi_2)^2 \, , \tag{16.1.10}$$

where

$$\bar{x}_1 = \frac{1}{n_1}\sum_{\nu=1}^{n_1} x_{1\nu} \quad \text{and} \quad \bar{x}_2 = \frac{1}{n_2}\sum_{\nu=1}^{n_2} x_{2\nu} \, . \tag{16.1.11}$$

Partitioning in this manner leads to four stochastically independent estimates, cf. (16.1.2) and (16.1.3), of the theoretical variance σ^2, two of which do not depend on the population means ξ_1 and ξ_2. Using the addition theorem for the s^2-distribution, these two estimates give the pooled estimate

$$s_1^2 = \frac{\sum_{\nu=1}^{n_1} (x_{1\nu} - \bar{x}_1)^2 + \sum_{\nu=1}^{n_2} (x_{2\nu} - \bar{x}_2)^2}{n_1 + n_2 - 2}, \quad f = n_1 + n_2 - 2\,. \tag{16.1.12}$$

This estimate of σ^2 is based on the variation of the observations about the two sample means, i. e., s_1^2 is based on the variation *within* the sets and is therefore independent of all hypotheses regarding the magnitude of ξ_1 and ξ_2.

The two remaining estimates of σ^2, $n_1(\bar{x}_1 - \xi_1)^2$ and $n_2(\bar{x}_2 - \xi_2)^2$, have s^2-distributions each with one degree of freedom, so that the estimate

$$\tfrac{1}{2}\left(n_1(\bar{x}_1 - \xi_1)^2 + n_2(\bar{x}_2 - \xi_2)^2\right) \tag{16.1.13}$$

has an s^2-distribution with two degrees of freedom. A pair of hypothetical values, ξ_{10} and ξ_{20}, of ξ_1 and ξ_2 may now be tested by computing

$$v^2(2, n_1 + n_2 - 2) = \frac{\tfrac{1}{2}\left(n_1(\bar{x}_1 - \xi_{10})^2 + n_2(\bar{x}_2 - \xi_{20})^2\right)}{s_1^2}. \tag{16.1.14}$$

The practical value of this test is, however, small as usually it is not possible to specify the theoretical values.

When the aim of the analysis is to *test the hypothesis that the two sets of observations come from the same normal population*, as in § 15.4, case A, a test may be obtained by partitioning (16.1.13) as follows.

The mean of all the observations

$$\bar{x} = \frac{\sum_{\nu=1}^{n_1} x_{1\nu} + \sum_{\nu=1}^{n_2} x_{2\nu}}{n_1 + n_2} = \frac{n_1\bar{x}_1 + n_2\bar{x}_2}{n_1 + n_2} \tag{16.1.15}$$

is—according to the addition theorem for the normal distribution—normally distributed with mean

$$\mathcal{M}\{\bar{x}\} = \bar{\xi} = \frac{n_1\xi_1 + n_2\xi_2}{n_1 + n_2} \tag{16.1.16}$$

and variance

$$\mathcal{V}\{\bar{x}\} = \frac{\sigma^2}{n_1 + n_2}, \tag{16.1.17}$$

and therefore

$$u = \frac{\bar{x} - \bar{\xi}}{\sigma/\sqrt{n_1 + n_2}} \tag{16.1.18}$$

is normally distributed with parameters (0, 1).

Introducing $\bar{x}$ and $\bar{\xi}$ into the differences $\bar{x}_1 - \xi_1$ and $\bar{x}_2 - \xi_2$, we get

$$\bar{x}_1-\xi_1 = (\bar{x}_1-\bar{x})-(\xi_1-\bar{\xi})+(\bar{x}-\bar{\xi})$$

$$= (\bar{x}_1-\bar{x}-\eta_1)+(\bar{x}-\bar{\xi})$$

and

$$\bar{x}_2-\xi_2 = (\bar{x}_2-\bar{x}-\eta_2)+(\bar{x}-\bar{\xi})\ ,$$

where

$$\eta_1 = \xi_1-\bar{\xi}$$

and

$$\eta_2 = \xi_2-\bar{\xi}\ .$$

These quantities satisfy the following relations

$$n_1(\bar{x}_1-\bar{x})+n_2(\bar{x}_2-\bar{x}) = 0 \qquad (16.1.19)$$

and

$$n_1(\xi_1-\bar{\xi})+n_2(\xi_2-\bar{\xi}) = n_1\eta_1+n_2\eta_2 = 0\ . \qquad (16.1.20)$$

Whence

$$n_1(\bar{x}_1-\xi_1)^2+n_2(\bar{x}_2-\xi_2)^2$$

$$= [n_1(\bar{x}_1-\bar{x}-\eta_1)^2+n_2(\bar{x}_2-\bar{x}-\eta_2)^2]+(n_1+n_2)(\bar{x}-\bar{\xi})^2, \qquad (16.1.21)$$

since the product sum

$$(\bar{x}-\bar{\xi})\big(n_1(\bar{x}_1-\bar{x}-\eta_1)+n_2(\bar{x}_2-\bar{x}-\eta_2)\big) = 0\ .$$

According to the partition theorem for the χ^2-distribution the two right-hand terms of (16.1.21) are stochastically independent and distributed as $\sigma^2\chi^2$ with one degree of freedom each, the two variances

$$s^2_{2(\eta)} = n_1(\bar{x}_1-\bar{x}-\eta_1)^2+n_2(\bar{x}_2-\bar{x}-\eta_2)^2,\ f = 1\ , \qquad (16.1.22)$$

and

$$s^2_{3(\bar{\xi})} = (n_1+n_2)(\bar{x}-\bar{\xi})^2,\ f = 1\ , \qquad (16.1.23)$$

therefore having s^2-distributions with parameters (σ^2, 1).

A hypothesis regarding the difference between ξ_1 and ξ_2 or between the η's may be tested by comparing $s^2_{2(\eta)}$ and s^2_1, while a hypothesis regarding $\bar{\xi}$ may be tested by comparing $s^2_{3(\bar{\xi})}$ and s^2_1.

If the two sets of observations belong to the same population, i. e., if the hypothesis $\xi_1 = \xi_2$ or $\eta_1 = \eta_2 = 0$ is true, it follows from (16.1.22) that

$$s^2_2 = n_1(\bar{x}_1-\bar{x})^2+n_2(\bar{x}_2-\bar{x})^2\ . \qquad (16.1.24)$$

This estimate of σ^2 is based on the variation *between* the sets of observations—in contrast to the estimate s^2_1, which is based on the variation *within* the sets. While s^2_1 is independent of ξ_1 and ξ_2, s^2_2 has an s^2-distribution with mean σ^2 only if $\xi_1 = \xi_2$, and therefore the hypothesis may be tested by computing

$$v^2(1,\ n_1+n_2-2) = \frac{s^2_2}{s^2_1}, \qquad (16.1.25)$$

which has a v^2-distribution if the hypothesis is correct.

If the hypothesis $\xi_1 = \xi_2$ *is false*, the mean of s_2^2 is equal to σ^2 plus a quantity which depends on the difference between the means of the two populations. From (16.1.22) we get

$$n_1(\bar{x}_1-\bar{x}-\eta_1)^2+n_2(\bar{x}_2-\bar{x}-\eta_2)^2$$
$$= n_1(\bar{x}_1-\bar{x})^2+n_2(\bar{x}_2-\bar{x})^2+n_1\eta_1^2+n_2\eta_2^2-2n_1(\bar{x}_1-\bar{x})\eta_1-2n_2(\bar{x}_2-\bar{x})\eta_2 \,.$$

Since
$$\mathcal{M}\{n_1(\bar{x}_1-\bar{x}-\eta_1)^2+n_2(\bar{x}_2-\bar{x}-\eta_2)^2\} = \mathcal{M}\{s_{2(\eta)}^2\} = \sigma^2 \,,$$
$$\mathcal{M}\{\bar{x}_1-\bar{x}\} = \eta_1 \,,$$
and
$$\mathcal{M}\{\bar{x}_2-\bar{x}\} = \eta_2 \,,$$
we get
$$\mathcal{M}\{n_1(\bar{x}_1-\bar{x})^2+n_2(\bar{x}_2-\bar{x})^2\}-n_1\eta_1^2-n_2\eta_2^2 = \sigma^2 \,,$$
or
$$\mathcal{M}\{s_2^2\} = \sigma^2+n_1\eta_1^2+n_2\eta_2^2 \,. \tag{16.1.26}$$
The quantity
$$n_1\eta_1^2+n_2\eta_2^2 = \frac{n_1 n_2}{n_1+n_2}(\xi_1-\xi_2)^2 \tag{16.1.27}$$

depends on the difference between the means of the two populations.

Thus s_2^2 has an s^2-distribution with mean value σ^2 only if $\xi_1 = \xi_2$. If the test hypothesis is incorrect, the mean of s_2^2 is larger than σ^2. The hypothesis is therefore rejected if s_2^2 *is significantly larger than* s_1^2, and we test this by means of the v^2-test, cf. (16.1.25).

The relation to the t-test in § 15.4 may be seen by introducing (16.1.15) in (16.1.24) which leads to

$$s_2^2 = \frac{n_1 n_2}{n_1+n_2}(\bar{x}_1-\bar{x}_2)^2 \,. \tag{16.1.28}$$

Inserting this result in (16.1.25), we obtain

$$v^2 = \frac{s_2^2}{s_1^2} = \frac{(\bar{x}_1-\bar{x}_2)^2}{s_1^2\left(\dfrac{1}{n_1}+\dfrac{1}{n_2}\right)}, \tag{16.1.29}$$

which is the square of t in (15.4.8).

If the hypothesis $\xi_1 = \xi_2$ is not rejected, we may according to the addition theorem for the s^2-distribution calculate a pooled estimate, s_0^2, of σ^2 as

$$s_0^2 = \frac{(n_1+n_2-2)s_1^2+s_2^2}{n_1+n_2-1} = \frac{\sum_{\nu=1}^{n_1}(x_{1\nu}-\bar{x})^2+\sum_{\nu=1}^{n_2}(x_{2\nu}-\bar{x})^2}{n_1+n_2-1}, \tag{16.1.30}$$

cf. (15.4.10).

In this case we obtain from the estimate $s^2_{3(\bar{\xi})}$

$$v^2(1,\, n_1+n_2-1) = \frac{s^2_{3(\bar{\xi})}}{s_0^2} = \frac{(\bar{x}-\bar{\xi})^2}{s_0^2/(n_1+n_2)}, \tag{16.1.31}$$

which is equal to the square of t in (15.4.12).

The above analysis of variance is usefully summarized in Table 16.1, the form of which is due to R. A. FISHER, who originally developed this technique.

TABLE 16.1.
Analysis of variance for two sets of observations.

Variation	Sum of squares (*SSD*)	Degrees of freedom	Variance	Test
Between sets	$n_1(\bar{x}_1-\bar{x})^2+n_2(\bar{x}_2-\bar{x})^2$	1	s_2^2	$v^2 = \dfrac{s_2^2}{s_1^2}$
Within sets	$\sum_{\nu=1}^{n_1}(x_{1\nu}-\bar{x}_1)^2+\sum_{\nu=1}^{n_2}(x_{2\nu}-\bar{x}_2)^2$	n_1+n_2-2	s_1^2	
Total	$\sum_{\nu=1}^{n_1}(x_{1\nu}-\bar{x})^2+\sum_{\nu=1}^{n_2}(x_{2\nu}-\bar{x})^2$	n_1+n_2-1		

16.2. Two Types of Problem Dealt with by Means of Analysis of Variance.

Let a given set of data consist of k sets of stochastically independent observations, each set containing n observations. Two simple examples are:

1. From a given product kn items are taken at random. The k groups of n items each are stored under different conditions, and after storage the water content of each item is determined.

2. From the output of a process for manufacturing sheets of building material, k sheets are taken at random. From each sheet n pieces are cut out at random and the permeability determined. By permeability is meant the time taken for water to soak through the material.

In the first example, we consider the water contents of the n items of a set as a random sample drawn from a normally distributed population with a mean and variance which characterize the effect of the particular storage conditions upon the given product. Taking another random sample of n items would after storage under the same conditions lead to a similar set of observations, which might be regarded as another sample drawn from the same population, and so forth. The other sets of observations corresponding to the other storage conditions are interpreted in an analogous way, i. e., corresponding to each of the k storage conditions we imagine a normally distributed population from which the k sets of water contents are assumed to be randomly drawn. These populations are characterized by the k means and variances, which are estimated by the k sample means and variances.

A repetition of the whole experiment implies the setting up of the same k storage conditions, as the purpose of the experiment is to analyze the effect of just these conditions.

The equality of the k theoretical variances can be tested by means of BARTLETT's test, see § 11.6. It is assumed in the following that these variances are identical. Thus, the basis for this type of analysis of variance is the specification that we have k normally distributed populations with means $\xi_1, \xi_2, \ldots, \xi_k$ and the same variance σ^2. This specification contains $k+1$ unknown parameters, namely k means and one variance, and the purpose of the analysis of variance is to estimate these parameters, to find the sampling distributions of the estimates and to test certain hypotheses regarding the values of the parameters.

In the above example we would first of all test the hypothesis that the k population means are all equal, i. e., that the differences in storage conditions do not affect the water content.

The main feature of this type of analysis of variance is that each observation is regarded as a sum of two components, namely the unknown mean of the population from which the observation is drawn and which is common to all observations within a set, and the deviation from this mean. The first component is considered as *an unknown constant* and is called the *systematic* component, whereas the second component is considered as a *random* variable. The k population means characterize the *systematic (fixed) differences* between the populations and express the effects of the k "causes" or "treatments" corresponding to the given classification.

Consider now the second example mentioned above. From each sheet we take n pieces at random and determine their permeability. These measurements deviate at random owing to the impossibility of manufacturing an absolutely uniform sheet. The n measurements are assumed to be a random sample drawn from the hypothetical infinite population of normally distributed measurements of permeability of the same sheet. For the ith sheet the mean of the corresponding population, ξ_i say, is a measure of the over-all permeability of that sheet, and the variance, σ^2, is a measure of the non-uniformity of the sheet. (The variance is assumed to be the same for all sheets.)

Owing to variations in the manufacturing process the over-all permeability of the sheets may vary from sheet to sheet. Hence, the k sheets are regarded as a random sample drawn from the hypothetical infinite population of sheets which might be manufactured by the given process. Assuming that the distribution of the over-all permeability of sheets in this population is normal, we regard the ξ_i's as a random sample from this population which may be characterized by its mean ξ and its variance ω^2.

A repetition of the experiment does not imply that the test-pieces are taken from the first set of k sheets, but a new sample of k sheets may be used, as the purpose of the experiment is to characterize the manufacturing process as a whole and not just to compare k particular sheets.

Thus, the basis for this type of analysis of variance is the specification that each measure of permeability has two *random* components: (1) the deviation of the measurement from the ξ_i in question, and (2) the deviation of ξ_i from ξ. This specification contains three unknown parameters, namely one mean and two variances, and the purpose of the analysis of variance is to estimate these parameters, to find the sampling distributions of the estimates and to test certain hypotheses regarding the values of the parameters.

In the above example we would first of all estimate the three parameters and then compare the estimates of the two variances to find out which source of variation is the more important.

The main feature of this type of analysis of variance is that each observation is regarded as a sum of three components, namely *an unknown constant*, the mean, which is common to all observations, and two *random* components, which characterize the variability arising from different "causes", one group of causes producing the variation within the sets of observations and another group the variation between the sets of observations.

In the three following sections the analysis of variance corresponding to the two specifications mentioned above will be discussed. After that more complex cases of analysis of variance will be dealt with. In all cases it is important to *distinguish between systematic and random components*.

16.3. Partitioning the Total Sum of Squares.

Let us consider k sets of stochastically independent observations from normally distributed populations, the variance of all k populations being σ^2 and the means $\xi_1, \xi_2, \ldots, \xi_k$, respectively, see Table 16.2, which will be explained fully later. It is further assumed that the ith sample contains n_i observations, which is a rather more general case than the one mentioned in § 16.2.

Hence the observations may be written as

$$\boxed{x_{i\nu} = \xi_i + z_{i\nu},} \qquad (16.3.1)$$

where the constants ξ_i, $i = 1, \ldots, k$, represent the population means, and the variables $z_{i\nu}$, $i = 1, \ldots, k$ and $\nu = 1, \ldots, n_i$, are independent and normally distributed with parameters $(0, \sigma^2)$.

Corresponding to each set of observations we have a sum of squares of deviations from the population mean. In analogy with (16.1.1) these sums of squares are partitioned, which for the ith set leads to

$$\sum_{\nu=1}^{n_i} (x_{i\nu}-\xi_i)^2 = \sum_{\nu=1}^{n_i} (x_{i\nu}-\bar{x}_i)^2 + n_i(\bar{x}_i-\xi_i)^2 . \tag{16.3.2}$$

From the partition theorem we know that the two terms on the right-hand side are stochastically independent and distributed as $\sigma^2\chi^2$ with n_i-1 and 1 degrees of freedom, respectively. By summing all the sums of squares, we obtain

$$\sum_{i=1}^{k}\sum_{\nu=1}^{n_i} (x_{i\nu}-\xi_i)^2 = \sum_{i=1}^{k}\sum_{\nu=1}^{n_i} (x_{i\nu}-\bar{x}_i)^2 + \sum_{i=1}^{k} n_i(\bar{x}_i-\xi_i)^2 . \tag{16.3.3}$$

From the addition theorem for the χ^2-distribution we know that the two sums of squares on the right-hand side are distributed as $\sigma^2\chi^2$ with $\sum_{i=1}^{k} n_i-k$ and k degrees of freedom, respectively.

The first of these sums of squares gives us an estimate s_1^2 of σ^2:

$$\boxed{s_1^2 = \frac{\sum_{i=1}^{k}\sum_{\nu=1}^{n_i} (x_{i\nu}-\bar{x}_i)^2}{\sum_{i=1}^{k} n_i-k} , \quad f = \sum_{i=1}^{k} n_i-k ,} \tag{16.3.4}$$

based on the variation *within* the sets of observations, and therefore independent of the population means $\xi_1, \xi_2, \ldots, \xi_k$, cf. Table 16.2 and § 11.5.

The second sum of squares in (16.3.3) gives us an estimate

$$\frac{1}{k}\sum_{i=1}^{k} n_i(\bar{x}_i-\xi_i)^2 \tag{16.3.5}$$

of σ^2, which has an s^2-distribution with k degrees of freedom and which depends on the population means $\xi_1, \xi_2, \ldots, \xi_k$. A set of hypothetical means may be tested by substituting such values in (16.3.5) and comparing this estimate of σ^2 with the estimate s_1^2.

When the aim of the analysis is to test the hypothesis that the k population means are equal we proceed as follows.

The mean of all the observations

TABLE 16.2.

Set No.	Observations	Sample mean	Population mean	Deviation	Deviation	Sum of squares (SSD)	Degrees of freedom (f)	Sample variance	Population variance
1	$x_{11}, x_{12}, \ldots, x_{1n_1}$	$\bar{x}_1$	ξ_1	$\bar{x}_1 - \bar{x}$	$\eta_1 = \xi_1 - \bar{\xi}$	$\sum_{\nu=1}^{n_1} (x_{1\nu} - \bar{x}_1)^2$	$n_1 - 1$	s_{11}^2	σ^2
2	$x_{21}, x_{22}, \ldots, x_{2n_2}$	$\bar{x}_2$	ξ_2	$\bar{x}_2 - \bar{x}$	$\eta_2 = \xi_2 - \bar{\xi}$	$\sum_{\nu=1}^{n_2} (x_{2\nu} - \bar{x}_2)^2$	$n_2 - 1$	s_{21}^2	σ^2
$\vdots$	$\vdots \quad \vdots \quad \vdots$	$\vdots$	$\vdots$	$\vdots$	$\vdots$	$\vdots$	$\vdots$	$\vdots$	$\vdots$
k	$x_{k1}, x_{k2}, \ldots, x_{kn_k}$	$\bar{x}_k$	ξ_k	$\bar{x}_k - \bar{x}$	$\eta_k = \xi_k - \bar{\xi}$	$\sum_{\nu=1}^{n_k} (x_{k\nu} - \bar{x}_k)^2$	$n_k - 1$	s_{k1}^2	σ^2
Total		$\sum_{i=1}^{k} n_i \bar{x}_i$	$\sum_{i=1}^{k} n_i \xi_i$	0	0	$\sum_{i=1}^{k} \sum_{\nu=1}^{n_i} (x_{i\nu} - \bar{x}_i)^2$	$\sum_{i=1}^{k} n_i - k$		
Mean		$\bar{x}$	$\bar{\xi}$	0	0			s_1^2	σ^2

$$\bar{x} = \frac{\sum_{i=1}^{k}\sum_{\nu=1}^{n_i} x_{i\nu}}{\sum_{i=1}^{k} n_i} = \frac{\sum_{i=1}^{k} n_i\bar{x}_i}{\sum_{i=1}^{k} n_i}, \tag{16.3.6}$$

is, according to the addition theorem for the normal distribution, normally distributed with mean

$$m\{\bar{x}\} = \bar{\xi} = \frac{\sum_{i=1}^{k} n_i\xi_i}{\sum_{i=1}^{k} n_i} \tag{16.3.7}$$

and variance

$$v\{\bar{x}\} = \sigma^2 \Big/ \sum_{i=1}^{k} n_i, \tag{16.3.8}$$

so that

$$u = \frac{\bar{x}-\bar{\xi}}{\sigma/\sqrt{\sum n_i}} \tag{16.3.9}$$

is normally distributed with parameters (0, 1).

The deviations of the k sample means from the total mean, $\bar{x}_1-\bar{x},\ldots,\bar{x}_k-\bar{x}$, form estimates of the corresponding theoretical deviations $\eta_1 = \xi_1-\bar{\xi}, \ldots, \eta_k = \xi_k-\bar{\xi}$. According to (16.3.6) and (16.3.7) these two sets of deviations satisfy the relations

$$\sum_{i=1}^{k} n_i(\bar{x}_i-\bar{x}) = 0 \tag{16.3.10}$$

and

$$\sum_{i=1}^{k} n_i\eta_i = \sum_{i=1}^{k} n_i(\xi_i-\bar{\xi}) = 0. \tag{16.3.11}$$

The total sample mean, $\bar{x}$, and the corresponding population mean, $\bar{\xi}$, are now introduced into the sum of squares (16.3.5), the deviations, $\bar{x}_i-\xi_i$, being written

$$\begin{aligned}\bar{x}_i-\xi_i &= (\bar{x}_i-\bar{x})-(\xi_i-\bar{\xi})+(\bar{x}-\bar{\xi})\\ &= (\bar{x}_i-\bar{x}-\eta_i)+(\bar{x}-\bar{\xi}).\end{aligned} \tag{16.3.12}$$

Squaring and summing leads to

$$\sum_{i=1}^{k} n_i(\bar{x}_i-\xi_i)^2 = \sum_{i=1}^{k} n_i(\bar{x}_i-\bar{x}-\eta_i)^2 + \left(\sum_{i=1}^{k} n_i\right)(\bar{x}-\bar{\xi})^2, \tag{16.3.13}$$

since

$$(\bar{x}-\bar{\xi})\sum_{i=1}^{k} n_i(\bar{x}_i-\bar{x}-\eta_i) = 0$$

according to (16.3.10) and (16.3.11).

From the partition theorem we know that the two terms on the right-hand side of (16.3.13) are stochastically independent and distributed as $\sigma^2\chi^2$ with $k-1$ and 1 degrees of freedom, respectively, and consequently the two variances

$$s^2_{2(\eta)} = \frac{\sum_{i=1}^{k} n_i(\bar{x}_i - \bar{x} - \eta_i)^2}{k-1}, \quad f = k-1, \tag{16.3.14}$$

and

$$s^2_{3(\bar{\xi})} = \left(\sum_{i=1}^{k} n_i\right)(\bar{x} - \bar{\xi})^2, \quad f = 1, \tag{16.3.15}$$

have s^2-distributions with mean value σ^2.

A hypothesis regarding the η's may thus be tested by comparing $s^2_{2(\eta)}$ and s^2_1, while a hypothesis regarding $\bar{\xi}$ may be tested by comparing $s^2_{3(\bar{\xi})}$ and s^2_1.

The above analysis is summarized in Table 16.3.

TABLE 16.3.

Sum of squares	f	s^2	$\mathcal{M}\{s^2\}$	Test
$\left(\sum_{i=1}^{k} n_i\right)(\bar{x} - \bar{\xi})^2$	1	$s^2_{3(\bar{\xi})}$	σ^2	$v^2 = \frac{s^2_{3(\bar{\xi})}}{s^2_1}$
$\sum_{i=1}^{k} n_i(\bar{x}_i - \bar{x} - \eta_i)^2$	$k-1$	$s^2_{2(\eta)}$	σ^2	$v^2 = \frac{s^2_{2(\eta)}}{s^2_1}$
$\sum_{i=1}^{k}\sum_{\nu=1}^{n_i} (x_{i\nu} - \bar{x}_i)^2$	$\sum_{i=1}^{k} n_i - k$	s^2_1	σ^2	
$\sum_{i=1}^{k}\sum_{\nu=1}^{n_i} (x_{i\nu} - \bar{\xi} - \eta_i)^2$	$\sum_{i=1}^{k} n_i$			

16.4. A Test for the Equality of the Means of k Normal Populations with the Same Variance.

If the k sets of observations have been taken from the same population, *i. e., if the hypothesis* $\xi_1 = \xi_2 = \ldots = \xi_k$ *is true*, it follows from (16.3.14) that

$$s^2_2 = \frac{\sum_{i=1}^{k} n_i(\bar{x}_i - \bar{x})^2}{k-1}, \quad f = k-1, \tag{16.4.1}$$

is an estimate of σ^2, as $\eta_1 = \eta_2 = \ldots = \eta_k = 0$.

The estimate s_2^2 is based on the variation *between* the sets of observations, in contrast to the estimate s_1^2, which is based on the variation *within* the sets. The variance s_2^2 estimates the variance of the population from *the variations of the sample means about the total mean* (it is based on the relation that $\mathcal{V}\{\bar{x}_i\} = \sigma^2/n_i$), whereas s_1^2 uses *the variations of the individual observations about their sample means.* If all the observations are drawn from the same population these two variances must be of about the same magnitude. While s_1^2 is independent of $\xi_1, \xi_2, \ldots, \xi_k$, s_2^2 only has an s^2-distribution with mean σ^2, if $\xi_1 = \xi_2 = \ldots = \xi_k$. Therefore the hypothesis can be tested by computing

$$v^2\left(k-1, \sum_{i=1}^{k} n_i - k\right) = \frac{s_2^2}{s_1^2}, \tag{16.4.2}$$

which has a v^2-distribution if the hypothesis is correct.

If the hypothesis $\xi_1 = \xi_2 = \ldots = \xi_k$ *is false*, then $\mathcal{M}\{s_2^2\}$ is equal to σ^2 plus a quantity which depends on the differences between the population means. From the numerator of (16.3.14) we obtain

$$\sum_{i=1}^{k} n_i(\bar{x}_i - \bar{x} - \eta_i)^2 = \sum_{i=1}^{k} n_i(\bar{x}_i - \bar{x})^2 + \sum_{i=1}^{k} n_i\eta_i^2 - 2\sum_{i=1}^{k} n_i(\bar{x}_i - \bar{x})\eta_i,$$

and further

$$\mathcal{M}\left\{\sum_{i=1}^{k} n_i(\bar{x}_i - \bar{x} - \eta_i)^2\right\} = (k-1)\sigma^2 = \mathcal{M}\left\{\sum_{i=1}^{k} n_i(\bar{x}_i - \bar{x})^2\right\} - \sum_{i=1}^{k} n_i\eta_i^2,$$

since

$$\mathcal{M}\{\bar{x}_i - \bar{x}\} = \xi_i - \bar{\xi} = \eta_i\,.$$

Hence

$$\mathcal{M}\left\{\sum_{i=1}^{k} n_i(\bar{x}_i - \bar{x})^2\right\} = (k-1)\sigma^2 + \sum_{i=1}^{k} n_i\eta_i^2, \tag{16.4.3}$$

or, dividing by $k-1$,

$$\boxed{\mathcal{M}\{s_2^2\} = \sigma^2 + \frac{1}{k-1}\sum_{i=1}^{k} n_i\eta_i^2 = \sigma^2 + \frac{1}{k-1}\sum_{i=1}^{k} n_i(\xi_i - \bar{\xi})^2\,.} \tag{16.4.4}$$

Thus, only if $\xi_1 = \xi_2 = \ldots = \xi_k$, the variance s_2^2 has an s^2-distribution with parameters $(\sigma^2, k-1)$. *If the test hypothesis is not true,* $\mathcal{M}\{s_2^2\}$ *is larger than* σ^2 *and therefore the hypothesis is rejected if* s_2^2 *is significantly larger than* s_1^2.

The last term on the right-hand side of (16.4.4) describes the variation between the population means by a single number. For further analysis we introduce the mean number of observations per sample

$$\bar{n} = \frac{1}{k}\sum_{i=1}^{k} n_i \tag{16.4.5}$$

and the "variance"

$$\varkappa^2 = \frac{1}{\bar{n}(k-1)} \sum_{i=1}^{k} n_i(\xi_i - \bar{\xi})^2 \tag{16.4.6}$$

so that

$$\mathcal{M}\{s_2^2\} = \sigma^2 + \bar{n}\varkappa^2 . \tag{16.4.7}$$

The power function of the above variance ratio test is given by

$$\pi(\lambda^2) = P\left\{\frac{s_2^2}{s_1^2} > v_{1-\alpha}^2(f_2, f_1);\ \lambda^2\right\} \tag{16.4.8}$$

where

$$\lambda^2 = \frac{\mathcal{M}\{s_2^2\}}{\mathcal{M}\{s_1^2\}} = 1 + \frac{\bar{n}\varkappa^2}{\sigma^2}, \tag{16.4.9}$$

cf. (14.2.3). If $\lambda^2 > 1$, s_2^2 will not be distributed as $\mathcal{M}\{s_2^2\}\chi^2/(k-1)$, since the k sample means do not have the same population mean, and therefore s_2^2/s_1^2 will not be distributed as v^2. It has been proved[1]), however, that s_2^2/s_1^2 in this case is distributed approximately as $\lambda^2 v^2$ with (f_2', f_1) degrees of freedom, where

$$f_2' = f_2 \frac{\lambda^4}{2\lambda^2 - 1}. \tag{16.4.10}$$

It follows, see (14.2.3), that

$$\pi(\lambda^2) \simeq P\left\{v^2(f_2', f_1) > \frac{1}{\lambda^2} v_{1-\alpha}^2(f_2, f_1)\right\}. \tag{16.4.11}$$

For $\pi = 1-\beta$ we obtain

$$\lambda^2 \simeq v_{1-\alpha}^2(f_2, f_1) v_{1-\beta}^2(f_1, f_2') . \tag{16.4.12}$$

For $n_i = n$ we find

$$\lambda^2 = 1 + \frac{n\varkappa^2}{\sigma^2} = v_{1-\alpha}^2\big(k-1, k(n-1)\big) v_{1-\beta}^2\big(k(n-1), f_2'\big) \tag{16.4.13}$$

where

$$\varkappa^2 = \frac{1}{k-1} \sum_{i=1}^{k} (\xi_i - \bar{\xi})^2 . \tag{16.4.14}$$

In this case the power function depends only on n, k, σ^2, and $\varkappa^2$, so that one of these quantities may be determined from the three others, given α and β.

The analysis of variance is summarized in Table 16.4.

The relation

$$\sum_{i=1}^{k} \sum_{\nu=1}^{n_i} (x_{i\nu} - \bar{x})^2 = \sum_{i=1}^{k} \sum_{\nu=1}^{n_i} (x_{i\nu} - \bar{x}_i)^2 + \sum_{i=1}^{k} n_i(\bar{x}_i - \bar{x})^2 \tag{16.4.15}$$

is proved by squaring and summing

$$x_{i\nu} - \bar{x} = (x_{i\nu} - \bar{x}_i) + (\bar{x}_i - \bar{x}) . \tag{16.4.16}$$

---

[1]) P. B. Patnaik: *The Non-Central χ^2- and F-Distributions and Their Applications*, Biometrika, 36, 1949, 202–232.

TABLE 16.4.
Analysis of variance for k sets of observations.

Variation	*SSD*	*f*	s^2	$m\{s^2\}$	Test
Between sets	$\sum_{i=1}^{k} n_i(\bar{x}_i-\bar{x})^2$	$k-1$	s_2^2	$\sigma^2+\frac{\Sigma n_i(\xi_i-\bar{\xi})^2}{k-1}$	$v^2=\frac{s_2^2}{s_1^2}$
Within sets	$\sum_{i=1}^{k}\sum_{\nu=1}^{n_i}(x_{i\nu}-\bar{x}_i)^2$	$\sum_{i=1}^{k} n_i-k$	s_1^2	σ^2	
Total	$\sum_{i=1}^{k}\sum_{\nu=1}^{n_i}(x_{i\nu}-\bar{x})^2$	$\sum_{i=1}^{k} n_i-1$			

For $k = 2$ we obtain the result stated in Table 16.1.

The computations are carried out according to the scheme given in Table 16.5.

S, SS, S^2/n, SSD, and s^2 are computed for each set of observations ac-

TABLE 16.5.
Computations for an analysis of variance.

Set No.	Observations	n	S	SS	S^2/n	SSD	f	s^2
1	$x_{11}, x_{12}, \ldots, x_{1n_1}$	n_1	S_1	SS_1	S_1^2/n_1	SSD_1	n_1-1	s_{11}^2
2	$x_{21}, x_{22}, \ldots, x_{2n_2}$	n_2	S_2	SS_2	S_2^2/n_2	SSD_2	n_2-1	s_{21}^2
⋮	⋮ ⋮ ⋮	⋮	⋮	⋮	⋮	⋮	⋮	⋮
k	$x_{k1}, x_{k2}, \ldots, x_{kn_k}$	n_k	S_k	SS_k	S_k^2/n_k	SSD_k	n_k-1	s_{k1}^2
Total		Σn_i	ΣS_i	ΣSS_i	$\Sigma S_i^2/n_i$	ΣSSD_i	Σn_i-k	

$$\sum_{i=1}^{k} n_i(\bar{x}_i-\bar{x})^2 = \sum_{i=1}^{k} S_i^2/n_i - \left(\sum_{i=1}^{k} S_i\right)^2 \Big/ \sum_{i=1}^{k} n_i, \quad f = k-1$$

$$\sum_{i=1}^{k}\sum_{\nu=1}^{n_i}(x_{i\nu}-\bar{x})^2 = \sum_{i=1}^{k} SS_i - \left(\sum_{i=1}^{k} S_i\right)^2 \Big/ \sum_{i=1}^{k} n_i, \quad f = \sum_{i=1}^{k} n_i-1$$

$$\sum_{i=1}^{k}\sum_{\nu=1}^{n_i}(x_{i\nu}-\bar{x}_i)^2 = \sum_{i=1}^{k}\sum_{\nu=1}^{n_i}(x_{i\nu}-\bar{x})^2 - \sum_{i=1}^{k} n_i(\bar{x}_i-\bar{x})^2 = \sum_{i=1}^{k} SSD_i, \quad f = \sum_{i=1}^{k} n_i-k$$

cording to the rules given in §§ 4.3–4.4. The sums of squares required are then obtained by summation as shown in Table 16.5.

If v^2 is not significant, a pooled estimate s_0^2 of σ^2 may be computed, viz.,

$$s_0^2 = \frac{\sum_{i=1}^{k}\sum_{\nu=1}^{n_i}(x_{i\nu}-\overline{x})^2}{\sum_{i=1}^{k} n_i - 1}, \quad f = \sum_{i=1}^{k} n_i - 1 . \tag{16.4.17}$$

If s_0 is substituted for σ in (16.3.9) we have, according to (15.1.8),

$$t = \frac{\overline{x}-\xi}{s_0/\sqrt{\Sigma n_i}}, \quad f = \sum_{i=1}^{k} n_i - 1 , \tag{16.4.18}$$

from which confidence limits for ξ may be determined.

In this analysis we have developed a test for the hypothesis that the k population means are equal, under the assumption that all the populations have the same variance. If we have any doubt as to the validity of this assumption we must test the equality of the population variances by applying BARTLETT'S test, see § 11.6, to the k sample variances $s_{11}^2, s_{21}^2, \ldots, s_{k1}^2$, given in Table 16.2.

In spite of taking these precautions the fact remains that a significant value of v^2 may in principle be due not only to different population means but also to different population variances or a combination of these cases.

If the population variances are not equal, an attempt may be made to *transform the observations* in such a manner that the population variances of the transformed variables are equal, cf. p. 176; then the analysis of variance is performed on the transformed observations.

The analysis of variance should be supplemented by *a detailed analysis of the deviations of each single mean from the total mean*. The variance ratio in (16.4.2) may be written as

$$v^2 = \frac{s_2^2}{s_1^2} = \frac{1}{k-1}\sum_{i=1}^{k}\left(\frac{\overline{x}_i-\overline{x}}{s_1/\sqrt{n_i}}\right)^2 . \tag{16.4.19}$$

The k quantities in this sum of squares

$$\frac{\overline{x}_i-\overline{x}}{s_1/\sqrt{n_i}}, \quad i = 1, 2, \ldots, k ,$$

should be computed, and the variation in their signs and size examined, cf. Chapter 13. If the test hypothesis is correct, the variables

$$u_i = \frac{\overline{x}_i-\xi}{\sigma/\sqrt{n_i}}, \quad i = 1, 2, \ldots, k ,$$

are independently and normally distributed with parameters (0, 1), and for large values of k it is therefore to be expected that the above k variables are approximately normally distributed with parameters (0, 1).

If the test hypothesis $\xi_1 = \xi_2 = \ldots = \xi_k$ is rejected, confidence limits for the separate population means may be established from the t-distribution, since the variable

$$t = \frac{\bar{x}_i - \xi_i}{s_1/\sqrt{n_i}}, \quad f = \sum_{i=1}^{k} n_i - k, \tag{16.4.20}$$

has a t-distribution with $\sum_{i=1}^{k} n_i - k$ degrees of freedom.

Similarly, when comparing two sets of observations, we find that

$$t = \frac{\bar{x}_i - \bar{x}_j - (\xi_i - \xi_j)}{s_1\sqrt{\dfrac{1}{n_i} + \dfrac{1}{n_j}}}, \quad f = \sum_{i=1}^{k} n_i - k, \tag{16.4.21}$$

has a t-distribution with $\sum_{i=1}^{k} n_i - k$ degrees of freedom, cf. § 15.4.

If there are a priori reasons for classifying the k populations into two groups, A and B say, we may wish to *test the hypothesis that the population means are identical within each group.*

Let n_A, n_B, x_A, and $\bar{x}_B$ denote the number of observations and their means for the two groups, so that

$$n = n_A + n_B \tag{16.4.22}$$

and

$$n\bar{x} = n_A\bar{x}_A + n_B\bar{x}_B\,. \tag{16.4.23}$$

Similarly, $\bar{\xi}_A$ and $\bar{\xi}_B$ denote the weighted population means, i. e.,

$$n\bar{\xi} = n_A\bar{\xi}_A + n_B\bar{\xi}_B\,, \tag{16.4.24}$$

and k_A and k_B denote the number of populations within each group so that

$$k = k_A + k_B\,. \tag{16.4.25}$$

The variations of the sample means about the group means give

$$\sum_A n_i(\bar{x}_i - \bar{x}_A)^2, \quad f = k_A - 1, \tag{16.4.26}$$

and

$$\sum_B n_i(\bar{x}_i - \bar{x}_B)^2, \quad f = k_B - 1, \tag{16.4.27}$$

from which the corresponding variances may be derived. Dividing these variances with s_1^2 we get two variables, each distributed as v^2, which may be used to test the hypothesis.

As a control of the independence of the variances involved in analyses of this kind, a complete analysis of variance should always be carried out. Introducing the new mean $\bar{x}_A$ in the k_A deviations $\bar{x}_i-\bar{x}$ we find

$$\sum_A n_i(\bar{x}_i-\bar{x})^2 = \sum_A n_i(\bar{x}_i-\bar{x}_A)^2+n_A(\bar{x}_A-\bar{x})^2\,, \qquad (16.4.28)$$

and similarly for B. Summing these relations we obtain

$$\sum_{i=1}^{k} n_i(\bar{x}_i-\bar{x})^2 = \sum_A n_i(\bar{x}_i-\bar{x}_A)^2+\sum_B n_i(\bar{x}_i-\bar{x}_B)^2+\frac{n_A n_B}{n}(\bar{x}_A-\bar{x}_B)^2. \qquad (16.4.29)$$

Thus, the SSD for the k means has been partitioned into three SSD's, the two first representing the variations of the means within the groups, and the third representing the difference between the two groups. Counting the degrees of freedom for the three SSD's we find k_A-1, k_B-1, and 1, giving $k-1$ as a total, so that the three terms are independently distributed as $\sigma^2\chi^2$, see § 10.6.

Each SSD may be computed from a formula analogous to

$$\Sigma\, n_i(\bar{x}_i-\bar{x})^2 = \Sigma\frac{S_i^2}{n_i} - \frac{(\Sigma S_i)^2}{\Sigma n_i}. \qquad (16.4.30)$$

The complete analysis of variance is shown in Table 16.6.

TABLE 16.6.

Analysis of variance. Partition of the variation between sets in Table 16.4.

Variation	SSD	f	s^2	$\mathcal{M}\{s^2\}$
Between group A and B	$\frac{n_A n_B}{n}(\bar{x}_A-\bar{x}_B)^2$	1	s_4^2	$\sigma^2+\frac{n_A n_B}{n}(\bar{\xi}_A-\bar{\xi}_B)^2$
Between sets within A	$\sum_A n_i(\bar{x}_i-\bar{x}_A)^2$	k_A-1	s_3^2	$\sigma^2+\frac{\sum_A n_i(\xi_i-\bar{\xi}_A)^2}{k_A-1}$
Between sets within B	$\sum_B n_i(\bar{x}_i-\bar{x}_B)^2$	k_B-1	s_2^2	$\sigma^2+\frac{\sum_B n_i(\xi_i-\bar{\xi}_B)^2}{k_B-1}$
Within sets	$\sum_{i=1}^{k}\sum_{\nu=1}^{n_i}(x_{i\nu}-\bar{x}_i)^2$	$\sum_{i=1}^{k} n_i-k$	s_1^2	σ^2
Total	$\sum_{i=1}^{k}\sum_{\nu=1}^{n_i}(x_{i\nu}-\bar{x})^2$	$\sum_{i=1}^{k} n_i-1$		

If s_2^2 and s_3^2 are not significantly larger than s_1^2, the two groups of populations may be characterized by their means $\bar{\xi}_A$ and $\bar{\xi}_B$. Finally, the hypothesis $\bar{\xi}_A = \bar{\xi}_B$ may be tested from $v^2 = s_4^2/s_1^2$.

The above partitioning of the *SSD* for the means

$$\sum_{i=1}^{k} n_i(\bar{x}_i - \bar{x})^2 \tag{16.4.31}$$

is only one *example* of how such a sum of squares may be partitioned. In principle any *SSD* with f degrees of freedom may be partitioned into f stochastically independent squares each with one degree of freedom, as shown in § 10.6. The *SSD* in (16.4.31) may for example be partitioned in analogy with (10.6.21), i. e., $\bar{x}_2$ is compared with $\bar{x}_1$, $\bar{x}_3$ with the weighted mean of $\bar{x}_1$ and $\bar{x}_2$, $\bar{x}_4$ with the weighted mean of $\bar{x}_1$, $\bar{x}_2$, and $\bar{x}_3$, etc. In each case *the partitioning is determined by what hypotheses are to be tested.*

The general rule for partitioning the *SSD* for the means into components, each having one degree of freedom, is that the square root of *such components must be mutually orthogonal linear functions of the observations.*

Consider first a linear function

$$q = \sum_{i=1}^{k} c_i n_i \bar{x}_i = \sum_{i=1}^{k} c_i S_i \,. \tag{16.4.32}$$

It follows from the addition theorem for the normal distribution that q is normally distributed with mean

$$\mathcal{M}\{q\} = \xi \sum_{i=1}^{k} c_i n_i \tag{16.4.33}$$

and variance

$$\mathcal{V}\{q\} = \sigma^2 \sum_{i=1}^{k} c_i^2 n_i \,. \tag{16.4.34}$$

For q^2 to be distributed as $\sigma^2\chi^2$ with one degree of freedom we must choose the coefficients c_i so that

$$\boxed{\sum_{i=1}^{k} c_i n_i = 0} \tag{16.4.35}$$

and

$$\boxed{\sum_{i=1}^{k} c_i^2 n_i = 1 \,.} \tag{16.4.36}$$

For two linear functions

$$q_1 = \sum_{i=1}^{k} c_{i1} n_i \bar{x}_i \tag{16.4.37}$$

and

$$q_2 = \sum_{i=1}^{k} c_{i2} n_i \bar{x}_i \tag{16.4.38}$$

to be *independently* distributed as $\sigma^2\chi^2$ with one degree of freedom each, the coefficients must fulfill the conditions (16.4.35) and (16.4.36) and the further condition

$$\boxed{\sum_{i=1}^{k} c_{i1}c_{i2}n_i = 0} \tag{16.4.39}$$

resulting from the requirement of independence, i. e.,

$$\mathcal{V}\{q_1, q_2\} = \mathcal{M}\{q_1q_2\} = 0\,, \tag{16.4.40}$$

see § 10.6. In this manner we may build up successively a set of $k-1$ mutually orthogonal functions, thus partitioning the *SSD* completely.

As examples we will show some partitions for $k = 3$ and $k = 4$, assuming that all $n_i = n$. The three conditions then become $\Sigma c_i = 0$, $\Sigma c_i^2 = 1$, and $\Sigma c_{i1}c_{i2} = 0$, disregarding the common factor n.

For $k = 3$ we have already in § 10.6 found the functions

	$\bar{x}_1$	$\bar{x}_2$	$\bar{x}_3$	(Divisor)^2
q_1	+1	−1	0	2
q_2	+1	+1	−2	6

i. e.,

$$q_1 = (\bar{x}_1-\bar{x}_2)/\sqrt{2} \tag{16.4.41}$$

and

$$q_2 = (\bar{x}_1+\bar{x}_2-2\bar{x}_3)/\sqrt{6} \tag{16.4.42}$$

so that

$$\sum_{i=1}^{3} (\bar{x}_i-\bar{x})^2 = q_1^2+q_2^2\,. \tag{16.4.43}$$

The (Divisor)^2 is found as the sum of squares of the coefficients in the table.

Consider the following three sets of orthogonal functions for $k = 4$.

		$\bar{x}_1$	$\bar{x}_2$	$\bar{x}_3$	$\bar{x}_4$	(Divisor)^2
Set 1	q_1	+1	−1	0	0	2
Set 1	q_2	+1	+1	−2	0	6
Set 1	q_3	+1	+1	+1	−3	12
Set 2	q_1	+1	−1	0	0	2
Set 2	q_2	0	0	+1	−1	2
Set 2	q_3	+1	+1	−1	−1	4
Set 3	q_1	+1	+1	−1	−1	4
Set 3	q_2	+1	−1	+1	−1	4
Set 3	q_3	+1	−1	−1	+1	4

The first set has been found by adding a third function to the two already found for $k = 3$, cf. § 10.6.

The second set is constructed to test the significance of the differences $\bar{x}_1 - \bar{x}_2$ and $\bar{x}_3 - \bar{x}_4$, and if no significant difference exists to test the difference $\frac{1}{2}(\bar{x}_1 + \bar{x}_2) - \frac{1}{2}(\bar{x}_3 + \bar{x}_4)$.

In the third set the observed means are taken together in pairs and the significance of differences between such pairs is tested.

Further examples will be given in § 17.10.

Example 16.1. From a given product 14 samples as similar as possible were taken and at random referred to 5 groups (of predetermined size), which were stored under different conditions. After being stored the water content of each sample was determined, see Table 16.7.

Assuming that the population variances are equal, we want to test the hypothesis that the population means are equal, i. e., that the differences in storage conditions do not affect the water content.

TABLE 16.7.
Water content in per cent after storage.

Method of storing	Water content in per cent					n	S	SS	S^2/n	$\bar{x}$
1	7·3	8·3	7·6	8·4	8·3	5	39·9	319·39	318·40	8·0
2	5·4	7·4	7·1			3	19·9	134·33	132·00	6·6
3	8·1	6·4				2	14·5	106·57	105·13	7·3
4	7·9	9·5	10·0			3	27·4	252·66	250·25	9·1
5	7·1					1	7·1	50·41	50·41	7·1
Total						14	108·8	863·36	856·19	
$(\Sigma S_i)^2/\Sigma n_i$							$108{\cdot}8^2/14 =$	845·53	845·53	
$\Sigma\Sigma(x_{i\nu} - \bar{x})^2$ and $\Sigma n_i(\bar{x}_i - \bar{x})^2$								17·83	10·66	

Based on the computations in Table 16.7 the following Table 16.8 is arranged in analogy with Table 16.4.

Table VII$_4$ shows that $v^2_{.95} = 3{\cdot}63$ for $f_1 = 4$ and $f_2 = 9$. Thus, the variation between the sets is not significantly greater than the variation within the sets, from which we draw the conclusion that, as regards water content after storage, no one method of storage can be preferred to any of the others on the basis of the present experiment, which, however, contains very few observations, so that its discriminating power is small.

TABLE 16.8.
Analysis of variance for water content.

Variation	*SSD*	f	s^2	Test
Between storage conditions	10·66	4	2·665	$v^2 = 3{\cdot}35$
Within storage conditions	7·17	9	0·7967	$P\{v^2 > 3{\cdot}35\} = 6\,\%$
Total	17·83	13		

Let us compute the power function for $\varkappa^2/\sigma^2 = 2$. First, we find $\bar{n} = 14/5 = 2{\cdot}8$, $\lambda^2 = 1+2{\cdot}8\times 2 = 6{\cdot}6$, and

$$f_2' = 4\times 6{\cdot}6^2/12{\cdot}2 = 14{\cdot}3\,.$$

Hence, at the 5% level of significance

$$\pi \simeq P\left\{v^2(14,\,9) > \frac{3{\cdot}63}{6{\cdot}6}\right\} = P\{v^2(9,\,14) < 1{\cdot}82\} = 84\%\,.$$

We may also determine the number of observations per sample necessary to obtain a power of 90% for $\varkappa^2/\sigma^2 = 1$, say. From (16.4.13) we find

$$\lambda^2 = 1+n = v^2_{\cdot 95}(4,\,5(n-1))v^2_{\cdot 90}(5(n-1),\,f_2')$$

where $f_2' = 4\lambda^4/(2\lambda^2-1)$. For $n = 4$ and $n = 5$ we get

n	$1+n$	f_2'	$v^2_{\cdot 95}(4,\,5(n-1))$	$v^2_{\cdot 90}(5(n-1),\,f_2')$	$v^2_{\cdot 95}v^2_{\cdot 90}$
4	5	11·1	3·06	2·16	6·61
5	6	13·1	2·87	2·01	5·77

so that five observations for each storage condition will give a power larger than 90%, since $1+n > v^2_{\cdot 95}v^2_{\cdot 90}$ for $n = 5$.

The differences between the population means are usually not fully determined from the equation $\varkappa^2/\sigma^2 = 1$. As a special case, where the equation has only one solution, we may assume that the five population means are equidistant so that they take on the values ξ, $\xi\pm\delta$, and $\xi\pm 2\delta$, giving $\varkappa^2 = 2{\cdot}5\delta^2$. This leads to the solution $\delta = \sigma\sqrt{0{\cdot}4}$. If $\sigma^2 = 0{\cdot}9$, say ($s_1^2 = 0{\cdot}7969$, see Table 16.8), we get $\delta = 0{\cdot}6$. Hence, if the population means corresponding to the five storage conditions are 6·8, 7·4, 8·0, 8·6, and 9·2% water, respectively, and the experimental error is $\sigma = \sqrt{0{\cdot}9} = 0{\cdot}95\%$ water, then an experiment with five observations for each storage condition will lead to rejection of the test hypothesis (no difference between the population means) with a probability of more than 90%.

Example 16.2. For a high voltage network, uniform cables of great tensile strength are required. Each cable is composed of wires which are manufactured in one length. In order to examine the tensile strength a sample is taken from each wire and tested. Table 16.9 shows the tensile strength of 9 cables each with 12 wires. As the tensile strength of each wire is about 340 kgs, 340 has been subtracted from all the observations in order to simplify the computations shown at the bottom of the table.

TABLE 16.9.
Tensile strength, in kgs, of the wires in 9 cables with 12 wires each.
Computing origin: 340 kgs.

	Cable No.									
	1	2	3	4	5	6	7	8	9	
	5	−11	0	−12	7	1	−1	−1	2	
	−13	−13	−10	4	1	0	0	0	6	
	−5	−8	−15	2	5	−5	2	7	7	
	−2	8	−12	10	0	−4	1	5	8	
	−10	−3	−2	−5	10	−1	−4	10	15	
	−6	−12	−8	−8	6	0	2	8	11	
	−5	−12	−5	−12	5	2	7	1	−7	
	0	−10	0	0	2	5	5	2	7	
	−3	5	−4	−5	0	1	1	−3	10	
	2	−6	−1	−3	−1	−2	0	6	7	
	−7	−12	−5	−3	−10	6	−4	0	8	
	−5	−10	−11	0	−2	7	2	5	1	Total
S	−49	−84	−73	−32	23	10	11	40	75	−79
SS	471	1120	725	540	345	162	121	314	811	4609
S^2/n	200·1	588·0	444·1	85·3	44·1	8·3	10·1	133·3	468·8	1982·1
SSD	270·9	532·0	280·9	454·7	300·9	153·7	110·9	180·7	342·2	2626·9
$\log SSD$	2·4328	2·7259	2·4485	2·6577	2·4784	2·1867	2·0449	2·2569	2·5342	21·7660

$$\Sigma n_i(\bar{x}_i-\bar{x})^2 = 1982{\cdot}1-79^2/108 = 1924{\cdot}3$$

$$\Sigma\Sigma(x_{i\nu}-\bar{x})^2 = 4609-79^2/108 = 4551{\cdot}2$$

First, we test the equality of the theoretical variances of the 9 distributions using BARTLETT's test, see § 11.6. We compute

$$\frac{1}{k}\Sigma \log s_i^2 = \frac{1}{k}\Sigma \log SSD_i - \log 11 = \frac{1}{9}\times 21{\cdot}7660-1{\cdot}0415 = 1{\cdot}3769,$$

$$\log s^2 = \log 2626{\cdot}9 - \log 99 = 3{\cdot}4195 - 1{\cdot}9956 = 1{\cdot}4239 ,$$

$$c = 1 + \frac{10}{3 \times 9 \times 11} = 1{\cdot}0337 ,$$

and

$$\chi^2 \simeq \frac{2{\cdot}3026}{1{\cdot}0337} \times 99 \times (1{\cdot}4239 - 1{\cdot}3769) = 10{\cdot}4 , \quad f = 8 .$$

Since $\chi^2 = 10{\cdot}4$ is less than $\chi^2_{\cdot 80} = 11{\cdot}0$, we do not reject the hypothesis. The variation within the sets is therefore estimated by the pooled variance

$$s_1^2 = \frac{2626{\cdot}9}{99} = 26{\cdot}53 .$$

Table 16.10 gives the analysis of variance.

TABLE 16.10.
Analysis of variance for tensile strength of electric wires.

Variation	*SSD*	*f*	s^2	Test
Between cables	1924·3	8	240·5	$v^2 = 9{\cdot}07$
Within cables	2626·9	99	26·53	
Total	4551·2	107		

As the variance ratio

$$v^2 = \frac{240{\cdot}5}{26{\cdot}53} = 9{\cdot}07$$

far exceeds $v^2_{\cdot 99} = 2{\cdot}69$ for $f_1 = 8$ and $f_2 = 99$, it is unreasonable to assume that the strength of the cables is the same, apart from random variations. This can also be seen directly from Table 16.9 where we find that for cables Nos. 1–4 the tensile strength of individual wires is in most cases less than 340 kgs while for cables Nos. 5–9 most of the wires have a tensile strength of more than 340 kgs.

A more detailed examination of the manufacturing process revealed that the cables had been manufactured from raw materials taken from two different lots, cables Nos. 1–4 having been made from lot A and cables Nos. 5–9 from lot B.

It would therefore be reasonable to divide the variation between cables into 3 portions: (1) the variation between the cables from lot A, (2) the variation between the cables from lot B, and (3) the variation between lot A and lot B, see Table 16.6.

This analysis of variance is shown in Tables 16.11 and 16.12. Most of the computations have already been carried out in Table 16.9. The three *SSD*'s are computed from formula (16.4.30).

TABLE 16.11.
Computation of sums of squares for the analysis of variance.

	Between cables from lot A		Between cables from lot B		Between lot A and lot B	
	S_i	S_i^2/n_i	S_i	S_i^2/n_i	S_i	S_i^2/n_i
	−49	200·1	23	44·1	−238	1180·1
	−84	588·0	10	8·3	159	421·4
	−73	444·1	11	10·1		
	−32	85·3	40	133·3		
			75	468·8		
S	−238		159		−79	
$\Sigma S_i^2/n_i$		1317·5		664·6		1601·5
$S^2/\Sigma n_i$	$238^2/4\times 12 =$	1180·1	$159^2/5\times 12 =$	421·4	$79^2/108 =$	57·8
SSD		137·4		243·2		1543·7
f		3		4		1

TABLE 16.12.
Partitioning the variation between cables in Table 16.10.

Variation	SSD	f	s^2
Between cables from lot A (cables Nos. 1–4)	137·4	3	45·8
Between cables from lot B (cables Nos. 5–9)	243·2	4	60·8
Between lot A and lot B	1543·7	1	1543·7
Total	1924·3	8	

If we compare these three variances with the variance 26·5, which gives the variation within cables, we get the three ratios

$$v^2 = \frac{45\cdot 8}{26\cdot 5} = 1\cdot 73\,, \quad f_1 = 3 \text{ and } f_2 = 99\,,$$

$$v^2 = \frac{60\cdot 8}{26\cdot 5} = 2\cdot 29\,, \quad f_1 = 4 \text{ and } f_2 = 99\,,$$

and

$$v^2 = \frac{1543\cdot 7}{26\cdot 5} = 58\cdot 3\,, \quad f_1 = 1 \text{ and } f_2 = 99\,.$$

The first two are both smaller than the corresponding 95% fractiles, 2·70 and 2·46, see Table VII_4, and we therefore conclude that there is no essential difference between strength of cables manufactured from the same lot. Since the third variance ratio is far greater than the 99·95% fractile we conclude that cables manufactured from the two different lots of raw materials do have different mean strength.

Thus, our analysis suggests that cables Nos. 1–4 may be characterized by a mean tensile strength of $340-\frac{238}{48} = 335\cdot 0$ kgs, while cables Nos. 5–9 may be characterized by a mean tensile strength of $340+\frac{159}{60} = 342\cdot 7$ kgs. The standard errors of these two means are $\sqrt{26\cdot 5/48} = 0\cdot 74$ kg and $\sqrt{26\cdot 5/60} = 0\cdot 65$ kg, respectively.

16.5. Analysis of the Variance into Two Components Representing Different Sources of Random Variation.

As mentioned in the second example of § 16.2 cases occur where each observation contains two random components. Another example is found in Example 9.2, p. 217, where the total variation is made up of the variation of the quality of the product and the uncertainty of the measurements. If each item is measured several times the observations appear as in Table 16.2, the k sets corresponding to k items, the ith item being measured n_i times. The variation within the sets is due to errors of measurement, and the variation between sets is due to errors of measurement plus the variation in quality from item to item, which is supposed to be random and normally distributed.

The observation $x_{i\nu}$ may be written

$$x_{i\nu} = \xi_i + z_{i\nu} = \xi + y_i + z_{i\nu}\,, \tag{16.5.1}$$

where $z_{i\nu}$ represents the variation within the sets and is assumed to be normally distributed with parameters $(0, \sigma^2)$, and y_i represents the variation between the true means of the sets and is assumed to be normally distributed with parameters $(0, \omega^2)$, and further that every z is stochastically independent of every y. In what follows we first assume that each of the k sets contains the same number, n, of observations.

Thus, we imagine that the kn observations are generated by the following *iterated sampling process*: (1) k elements are chosen at random from a normally distributed population with parameters (ξ, ω^2) giving the values $\xi_1 = \xi + y_1, \ldots, \xi_k = \xi + y_k$. (2) Each of these k values represents the mean of a normally distributed population with variance σ^2, and from each of the k populations n elements are chosen at random giving the values $x_{i\nu} = \xi_i + z_{i\nu}$ for $i = 1, \ldots, k$ and $\nu = 1, \ldots, n$. This model contains three parameters, ξ, ω^2, and σ^2.

The sum of squares

$$\sum_{i=1}^{k} \sum_{\nu=1}^{n} (x_{i\nu} - \xi_i)^2 = \sum_{i=1}^{k} \sum_{\nu=1}^{n} z_{i\nu}^2 \tag{16.5.2}$$

has a $\sigma^2\chi^2$-distribution with kn degrees of freedom. By analogy with (16.3.3) we obtain

$$\sum_{i=1}^{k} \sum_{\nu=1}^{n} (x_{i\nu} - \xi_i)^2 = \sum_{i=1}^{k} \sum_{\nu=1}^{n} (x_{i\nu} - \bar{x}_i)^2 + n \sum_{i=1}^{k} (\bar{x}_i - \xi_i)^2, \tag{16.5.3}$$

in which, on the right-hand side, only the second term depends upon the ξ_i's. Hence

$$s_1^2 = \frac{\sum_{i=1}^{k} \sum_{\nu=1}^{n} (x_{i\nu} - \bar{x}_i)^2}{k(n-1)} \tag{16.5.4}$$

has an s^2-distribution with parameters $(\sigma^2, k(n-1))$ for any set of ξ_i's. Further, $\bar{z}_i = \bar{x}_i - \xi_i$ is stochastically independent of s_1^2 and normally distributed with parameters $(0, \sigma^2/n)$. So far the results are identical with those of § 16.3.

Consider now the sampling distribution of $\bar{x}_i$. A repetition of the whole sampling process will lead to a new $\bar{x}_i$ corresponding to the new y_i and $\bar{z}_i$ found, and so on. Writing

$$\bar{x}_i = \xi + y_i + \bar{z}_i \tag{16.5.5}$$

it follows from the addition theorem for the normal distribution that $\bar{x}_i$ *is normally distributed with mean*

$$\mathcal{M}\{\bar{x}_i\} = \xi \tag{16.5.6}$$

and variance

$$\boxed{\mathcal{V}\{\bar{x}_i\} = \mathcal{V}\{y_i\} + \mathcal{V}\{\bar{z}_i\} = \omega^2 + \frac{\sigma^2}{n}.} \tag{16.5.7}$$

As estimates of these parameters we compute from the k sample means the total mean

$$\bar{x} = \frac{1}{k} \sum_{i=1}^{k} \bar{x}_i = \frac{1}{kn} \sum_{i=1}^{k} \sum_{\nu=1}^{n} x_{i\nu} \approx \xi \tag{16.5.8}$$

and the variance

$$\frac{1}{k-1} \sum_{i=1}^{k} (\bar{x}_i - \bar{x})^2 \approx \omega^2 + \frac{\sigma^2}{n}. \tag{16.5.9}$$

It follows that the variance

$$s_2^2 = \frac{n \sum_{i=1}^{k} (\bar{x}_i - \bar{x})^2}{k-1} \tag{16.5.10}$$

has an s^2-distribution with parameters $(n\omega^2+\sigma^2, k-1)$, and that $\bar{x}$ is independently and normally distributed with parameters $(\xi, (n\omega^2+\sigma^2)/kn)$.

Formally the whole analysis of variance may be carried out as shown in Table 16.4 with the only modification that $\mathcal{M}\{s_2^2\} = \sigma^2+n\omega^2$. *The hypothesis* $\omega^2 = 0$ *may be tested from the variance ratio* s_2^2/s_1^2. The power function (14.2.3) may be applied directly putting

$$\lambda^2 = \frac{\mathcal{M}\{s_2^2\}}{\mathcal{M}\{s_1^2\}} = 1+\frac{n\omega^2}{\sigma^2}. \tag{16.5.11}$$

For $\pi = 1-\beta$ we obtain from (14.2.5)

$$\lambda^2 = 1+\frac{n\omega^2}{\sigma^2} = v_{1-\alpha}^2\big(k-1,\, k(n-1)\big)\, v_{1-\beta}^2\big(k(n-1),\, k-1\big)\,. \tag{16.5.12}$$

As estimates of the three parameters we obtain

$$\bar{x} = \frac{1}{kn} \sum_{i=1}^{k} \sum_{\nu=1}^{n} x_{i\nu} \approx \xi\,, \tag{16.5.13}$$

$$s_1^2 = \frac{1}{k(n-1)} \sum_{i=1}^{k} \sum_{\nu=1}^{n} (x_{i\nu}-\bar{x}_i)^2 \approx \sigma^2\,, \tag{16.5.14}$$

and

$$\frac{1}{n}\,(s_2^2 - s_1^2) \approx \omega^2\,. \tag{16.5.15}$$

Confidence limits for ξ may be calculated from the t-distribution, since

$$u = \frac{\bar{x}-\xi}{\sqrt{\dfrac{1}{kn}\,(n\omega^2+\sigma^2)}} \tag{16.5.16}$$

is normally distributed with parameters (0, 1), and therefore

$$t = \frac{\bar{x}-\xi}{s_2/\sqrt{kn}}\,, \quad f = k-1\,, \tag{16.5.17}$$

has a t-distribution with $k-1$ degrees of freedom.

Confidence limits for σ^2 are calculated from the χ^2-distribution, since $\chi^2 = fs_1^2/\sigma^2$, $f = k(n-1)$.

According to (14.1.12) the ratio s_2^2/s_1^2 is distributed as

$$\frac{n\omega^2+\sigma^2}{\sigma^2}v^2 = \left(n\frac{\omega^2}{\sigma^2}+1\right)v^2 \tag{16.5.18}$$

with $(k-1, k(n-1))$ degrees of freedom, so that confidence limits for ω^2/σ^2 may be calculated from the relation

$$P\left\{\left(n\frac{\omega^2}{\sigma^2}+1\right)v_{P_1}^2 < \frac{s_2^2}{s_1^2} < \left(n\frac{\omega^2}{\sigma^2}+1\right)v_{P_2}^2\right\} = P_2-P_1,$$

which leads to

$$P\left\{\frac{s_2^2}{s_1^2}\frac{1}{v_{P_2}^2}-1 < \frac{n\omega^2}{\sigma^2} < \frac{s_2^2}{s_1^2}\frac{1}{v_{P_1}^2}-1\right\} = P_2-P_1. \tag{16.5.19}$$

The distribution function of the estimate $\frac{1}{n}(s_2^2-s_1^2)$ of ω^2 is more complicated and so far has not been tabulated. For values of k that are greater than about 30, s_2^2 and s_1^2 may be considered as approximately normally distributed with variances

$$\mathcal{V}\{s_2^2\} = \frac{2(n\omega^2+\sigma^2)^2}{k-1} \tag{16.5.20}$$

and

$$\mathcal{V}\{s_1^2\} = \frac{2\sigma^4}{k(n-1)}, \tag{16.5.21}$$

cf. (11.1.9), and therefore

$$\mathcal{V}\left\{\frac{1}{n}(s_2^2-s_1^2)\right\} = \frac{2(n\omega^2+\sigma^2)^2}{n^2(k-1)} + \frac{2\sigma^4}{n^2k(n-1)}. \tag{16.5.22}$$

If we employ the estimate s_2^2 instead of $n\omega^2+\sigma^2$ and s_1^2 instead of σ^2, we obtain

$$\frac{2}{n^2}\left(\frac{s_2^4}{k-1}+\frac{s_1^4}{k(n-1)}\right) \approx \mathcal{V}\left\{\frac{1}{n}(s_2^2-s_1^2)\right\}. \tag{16.5.23}$$

By using this estimate we can compute approximate confidence limits for ω^2, based on the normal distribution, for $k > 30$.

In the case where the k samples contain different numbers of observations the arithmetic of the analysis of variance is as shown in Table 16.4. However,

$$\mathcal{M}\{s_2^2\} = \sigma^2+\bar{n}\omega^2, \tag{16.5.24}$$

where

$$\bar{n} = \frac{1}{k-1}\left(\sum_{i=1}^{k} n_i - \frac{\sum_{i=1}^{k} n_i^2}{\sum_{i=1}^{k} n_i}\right), \tag{16.5.25}$$

and the distribution of s_2^2 is only approximately an s^2-distribution when $\omega^2 \neq 0$.

The total mean, $\bar{x}$, is normally distributed with mean ξ and

$$\mathcal{V}\{\bar{x}\} = \frac{\sum_{i=1}^{k} n_i^2}{\left(\sum_{i=1}^{k} n_i\right)^2}\omega^2 + \frac{\sigma^2}{\sum_{i=1}^{k} n_i}. \qquad (16.5.26)$$

Example 16.3. When emptying a mixer for raw meal, 40 samples were collected at regular intervals. The $CaCO_3$ content of the raw meal was determined by duplicate titrations, with the results shown in Table 16.13, which gives the 80 numbers for the $CaCO_3$ percentages minus 76%, which has been subtracted in order to simplify the computations.

Here, the variation within the sets indicates the uncertainty of the

TABLE 16.13.

The $CaCO_3$ contents in %, minus 76·00%, of 40 samples of raw meal taken when emptying a mixer.

Sample No.	Titration No. 1	Titration No. 2	Difference	Sum	Sample No.	Titration No. 1	Titration No. 2	Difference	Sum
1	0·35	0·35	0·00	0·70	21	0·30	0·40	−0·10	0·70
2	0·20	0·25	−0·05	0·45	22	0·35	0·35	0·00	0·70
3	0·40	0·20	0·20	0·60	23	0·45	0·55	−0·10	1·00
4	0·50	0·45	0·05	0·95	24	0·50	0·40	0·10	0·90
5	0·50	0·25	0·25	0·75	25	0·75	0·95	−0·20	1·70
6	0·35	0·50	−0·15	0·85	26	0·70	0·35	0·35	1·05
7	0·80	0·75	0·05	1·55	27	0·60	0·75	−0·15	1·35
8	0·55	0·60	−0·05	1·15	28	0·60	0·55	0·05	1·15
9	0·35	0·35	0·00	0·70	29	0·20	0·40	−0·20	0·60
10	0·60	0·55	0·05	1·15	30	0·35	0·35	0·00	0·70
11	0·70	0·85	−0·15	1·55	31	0·60	0·90	−0·30	1·50
12	0·15	0·05	0·10	0·20	32	0·40	0·45	−0·05	0·85
13	0·30	0·50	−0·20	0·80	33	0·20	0·25	−0·05	0·45
14	0·15	0·25	−0·10	0·40	34	0·55	0·50	0·05	1·05
15	0·25	0·15	0·10	0·40	35	0·20	0·20	0·00	0·40
16	0·85	0·75	0·10	1·60	36	0·50	0·50	0·00	1·00
17	0·40	0·30	0·10	0·70	37	0·35	0·50	−0·15	0·85
18	0·40	0·35	0·05	0·75	38	0·60	0·55	0·05	1·15
19	0·30	0·30	0·00	0·60	39	0·40	0·50	−0·10	0·90
20	0·60	0·80	−0·20	1·40	40	0·50	0·35	0·15	0·85
							S	−0·50	36·10
							SS	0·7050	37·9050
							S^2/n		32·5803
							SSD		5·3247

titrations, and according to (11.5.7) this uncertainty is determined by computing the sum of squares of the differences between corresponding titration results, divided by $2k = 80$. This leads to the variance

$$s_1^2 = \frac{0{\cdot}7050}{80} = 0{\cdot}008813\ , \quad f = 40\ ,$$

and the standard deviation

$$s_1 = 0{\cdot}094\%\ \mathrm{CaCO_3}\ .$$

The variation between sets is obtained by calculating the variance of the sums of corresponding titration results and dividing this variance by $n=2$.

Arranged in the usual manner we have the following analysis:

TABLE 16.14.
Analysis of variance for results of a mixing process.

Variation	*SSD*	*f*	s^2	Test
Between sets	2·6624	39	0·06827	$v^2 = 7{\cdot}75$
Within sets	0·3525	40	0·008813	
Total	3·0149	79		

As the variance ratio $v^2 = 7{\cdot}75$ is larger than $v^2_{.99} = 2{\cdot}12$, $f_1 = 39$ and $f_2 = 40$, the variation of the $CaCO_3$ percentages from sample to sample is far larger than can be explained by the uncertainty of the titrations, i. e., the raw meal is not completely homogeneous.

To investigate the variation between samples further, the mean titration for each sample has been plotted on Fig. 16.1 with the order of the sample as abscissa. The middle horizontal line indicates the total mean, 76·45%. Two lines have been drawn at a distance $\pm 1{\cdot}96\ s_1/\sqrt{2} = \pm 0{\cdot}13\%$ from the middle line, and they thus delimit a domain within which we would expect to find about 95% of the points if the variation was due only to the uncertainty of titration. A far higher percentage of the observed points lies outside this domain, which confirms the result obtained above, namely that the variation between the samples considerably exceeds the uncertainty of the titrations.

Fig. 16.1 suggests that the $CaCO_3$ percentages are distributed at random, and investigations of the distributions of runs above and below the median and runs up and down, according to the criteria given in Chapter 13, do not lead to rejection of the hypothesis of randomness. Hence, the mixing process may be characterized by the standard deviation ω of the distribu-

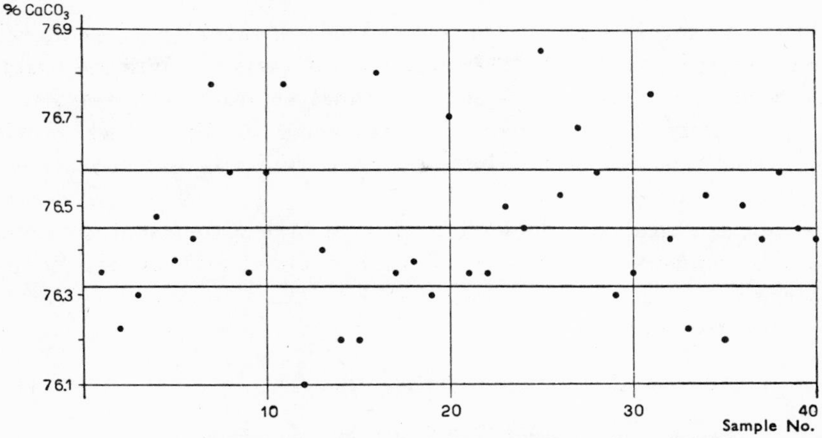

FIG. 16.1. Diagram for the means of the determinations of $CaCO_3$.

tion of the true $CaCO_3$ percentages of the samples. A large standard deviation means poor mixing, and a small standard deviation good mixing; in particular a standard deviation of value zero would indicate that the raw meal had been perfectly mixed. A fractile diagram for the 40 means shows that their distribution may be considered normal, so that the assumptions underlying the analysis of variance seem to be fulfilled.

The value of a control chart must be emphasized, especially in cases where the samples are taken at successive intervals of time or space. The chart gives a description of the sequence of events, while the analysis of variance completely disregards the order of the means.

Consider Fig. 16.2, in which the data of Fig. 16.1 have been arranged

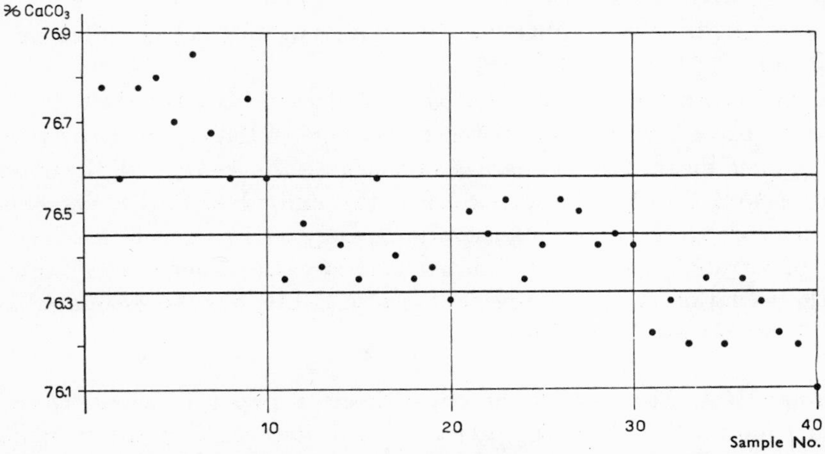

FIG. 16.2. Diagram for the means of the determinations of $CaCO_3$. The values are the same as in Fig. 16.1, but have been plotted in another order.

in a different (and imaginary) sequence. This rearrangement has no effect whatever on the arithmetic of the analysis of variance, but the chart, if it were correct, would show us that this analysis must be discarded: the $CaCO_3$ percentages would clearly be decreasing as the mixer is being emptied, and this fact would have to be included in our mathematical model, see Chapter 18.

However, as already emphasized, in this case the sequence of percentages seems to be random, and we may therefore proceed with the simpler type of analysis.

From the two variances

$$s_1^2 = 0{\cdot}008813 \approx \sigma^2, \quad f = 40,$$

$$s_2^2 = 0{\cdot}06827 \approx 2\omega^2+\sigma^2, \quad f = 39,$$

we compute the estimate of ω^2 as

$$\frac{1}{n}(s_2^2-s_1^2) = \frac{1}{2}(0{\cdot}06827-0{\cdot}008813) = 0{\cdot}02973 \approx \omega^2 .$$

Thus, the mixing process is characterized by the standard deviation $\sqrt{0{\cdot}02973} = 0{\cdot}17\%$ $CaCO_3$.

According to (16.5.23) we find

$$\frac{2}{4}\left(\frac{0{\cdot}06827^2}{39}+\frac{0{\cdot}008813^2}{40}\right) = 0{\cdot}00006073 \approx \mathcal{V}\left\{\frac{1}{n}(s_2^2-s_1^2)\right\},$$

so that we may approximate to the 95% confidence limits for ω^2 by computing $0{\cdot}02973 \pm 1{\cdot}96\sqrt{0{\cdot}0^46073} = (0{\cdot}01446, 0{\cdot}04500)$, which gives us the limits $(0{\cdot}12, 0{\cdot}21)\%$ $CaCO_3$ for ω.

The following conclusions may now be drawn from the data in Table 16.13: As the mixer is emptied the percentage of $CaCO_3$ varies at random about a fixed mean. The estimate of this mean is the mean of all the observations, 76·45% $CaCO_3$. The variation of the individual $CaCO_3$ percentages is partly due to the errors of titration, partly to incomplete mixing. The error of titration has the estimated standard deviation 0·094% $CaCO_3$, and the variation in the true percentages of $CaCO_3$ has the estimated standard deviation 0·17% $CaCO_3$.

Example 16.4. The quality of paper from a paper mill was tested at regular intervals by taking samples and determining their breaking strength. We consider the variation of breaking strength from one set to another to be random and wish to estimate the three parameters

TABLE 16.15.
The breaking strength of paper.

Set No.	Breaking strength in kgs	S	SS	S^2/n	SSD	$\log SSD$	$\bar{x}$	$\frac{(\bar{x}-8{\cdot}364)}{0{\cdot}240}$
1	9·1 8·5 8·6 7·2 7·6	41·0	338·62	336·20	2·42	0·384	8·20	−0·68
2	8·1 8·2 7·6 7·7 8·0	39·6	313·90	313·63	0·27	$\bar{1}$·431	7·92	−1·85
3	8·5 9·0 8·1 8·5 8·0	42·1	355·11	354·48	0·63	$\bar{1}$·799	8·42	0·23
4	7·4 7·2 8·3 8·5 8·5	39·9	319·99	318·40	1·59	0·201	7·98	−1·60
5	9·0 7·9 8·2 7·6 8·7	41·4	344·10	342·79	1·31	0·117	8·28	−0·35
6	8·6 8·8 9·2 8·8 9·8	45·2	409·52	408·61	0·91	$\bar{1}$·959	9·04	2·82
7	8·5 8·1 8·8 8·6 8·5	42·5	361·51	361·25	0·26	$\bar{1}$·415	8·50	0·57
8	7·8 8·8 7·6 8·9 8·7	41·8	350·94	349·45	1·49	0·173	8·36	−0·02
9	8·7 8·2 8·7 9·2 8·7	43·5	378·95	378·45	0·50	$\bar{1}$·699	8·70	1·40
10	8·5 8·9 7·7 8·1 9·0	42·2	357·36	356·17	1·19	0·076	8·44	0·32
11	7·3 7·9 7·9 8·5 9·2	40·8	335·00	332·93	2·07	0·316	8·16	−0·85
Total		460·0	3865·00	3852·36	12·64	$\bar{1}$·570		
$(\Sigma S_i)^2/\Sigma n_i$		$460{\cdot}0^2/55 =$	3847·27	3847·27				
$\Sigma\Sigma(x_{iv}-\bar{x})^2$ and $\Sigma n_i(\bar{x}_i-\bar{x})^2$			17·73	5·09				

of the corresponding mathematical model. Table 16.15 gives the results of testing 11 sets of samples, 5 determinations of breaking strength having been made in each set.

We first apply BARTLETT's test, to test the hypothesis that the theoretical variance is the same for all 11 sets. Applying (11.6.8) we compute

$$\frac{1}{k}\Sigma \log s_i^2 = \frac{1}{11}\Sigma \log SSD - \log 4 = \frac{10{\cdot}570}{11} - 1 - 0{\cdot}602 = -0{\cdot}641\ ,$$

$$\log s^2 = \log 12{\cdot}64 - \log 44 = 1{\cdot}102 - 1{\cdot}643 = -0{\cdot}541\ ,$$

$$c = 1 + \frac{12}{3\times 11\times 4} = 1{\cdot}091\ ,$$

and

$$\chi^2 \simeq \frac{2{\cdot}303}{1{\cdot}091}\, 44\times(-0{\cdot}541+0{\cdot}641) = 9{\cdot}29\,,\ f = 10\,.$$

Table V shows that, for $f = 10$, $\chi^2_{\cdot 40} = 8{\cdot}30$ and $\chi^2_{\cdot 50} = 9{\cdot}34$, so that the hypothesis cannot be rejected.

From the computations in Table 16.15 the following scheme is set up for the analysis of variance:

TABLE 16.16.
Analysis of variance for the breaking strength of paper.

Variation	*SSD*	*f*	s^2	Test
Between sets	5·09	10	0·509	$v^2 = 1{\cdot}77$
Within sets	12·64	44	0·287	$P\{v^2 > 1{\cdot}77\} = 10\,\%$
Total	17·73	54		

As the total number of observations is $11\times 5 = 55$, the number of degrees of freedom for the total sum of squares is $55-1 = 54$, $11-1 = 10$ belonging to the variation between the 11 sets and $11\times 4 = 44$ to the variation within the sets.

Table VII$_4$ shows that $v^2_{.95} = 2{\cdot}05$ for $f_1 = 10$ and $f_2 = 44$. Thus, the variation between the sets is not significantly larger than the variation within the sets, i. e., we cannot reject the hypothesis $\omega = 0$.

The total mean

$$\bar{x} = \frac{460{\cdot}0}{55} = 8{\cdot}364 \text{ kgs}$$

is an estimate of ξ, and as an estimate of σ^2 we have $s_1^2 = 0{\cdot}287$ giving the standard deviation $s_1 = 0{\cdot}536$ kg.

As an estimate of ω^2 we get

$$\tfrac{1}{5}(0{\cdot}509-0{\cdot}287) = 0{\cdot}044$$

which as shown above is not significantly different from zero.

The last column of Table 16.15 shows the computation of the quantities

$$\frac{\bar{x}_i-\bar{x}}{s_1/\sqrt{n_i}} = \frac{\bar{x}_i-8{\cdot}364}{\sqrt{0{\cdot}287/5}} = \frac{\bar{x}_i-8{\cdot}364}{0{\cdot}240}.$$

(These quantities may also be computed from

$$\frac{\bar{x}_i-\bar{x}}{s_1/\sqrt{n_i}} = \frac{S_i-n_i\bar{x}}{s_1\sqrt{n_i}} = \frac{S_i-41{\cdot}82}{1{\cdot}20}\Bigg).$$

The sum of squares of these quantities divided by $k-1 = 10$ is equal to $v^2 = 1{\cdot}77$. We see that 5 values are positive and 6 negative, and that the variation in signs and in size seems to be random. This impression will be confirmed by plotting the 11 means on a control chart with the middle line at $\bar{x} = 8{\cdot}364$, inner control limits at $8{\cdot}364 \pm 1{\cdot}96\times 0{\cdot}240 = (7{\cdot}89,\ 8{\cdot}83)$, and outer limits at $8{\cdot}364 \pm 3{\cdot}09 \times 0{\cdot}240 = (7{\cdot}62,\ 9{\cdot}11)$, cf. § 13.7.

Suppose that we want to determine n, the number of observations per sample, so that we can be reasonably sure to detect variations of the true means larger than 10% of the total mean. Since 10% of 8·36 is 0·84 we put $2\omega = 0\cdot84$, which gives $\omega = 0\cdot42$ and $\omega^2 = 0\cdot18$. Using the estimate $s_1^2 = 0\cdot29$ instead of σ^2 the above requirement may be formulated as follows: n must be determined so that the power is about 90% for $k = 11$ and $\omega^2/\sigma^2 = 0\cdot18/0\cdot29 = 0\cdot62$. From (16.5.12) we obtain for $n = 5$, $\lambda^2 = 1+5\times0\cdot62 = 4\cdot10$ and $v_{.95}^2 v_{.90}^2 = 4\cdot37$. For $n = 6$ the corresponding numbers are 4·72 and 4·26. Hence $n = 5$ will give the discriminating power required.

16.6. Multi-Stage Grouping.

In the previous sections we have pointed out that there are two types of problems in the analysis of variance. Further we have given the two standard procedures for handling data according to these principles and illustrated these procedures by examples.

Most problems met with in practice are far more complicated, yet their solutions are based on the same principles. In the rest of this chapter we discuss certain generalized procedures which have been found useful in practice. *It is pointed out, however, that the treatment is not exhaustive and cannot be so. We must always be prepared to modify these procedures in order to meet the requirements of individual problems.*

We have just dealt with the case where the data present themselves as k sets of observations and studied the variation *within* and *between* sets. Often a further analysis of the variation within each set is necessary. For instance, when considering k machines that are working day and night we may want to split up the total variation into the variation within shifts, the variation between shifts, and the variation between machines. Consequently, for each machine the observations are grouped according to shifts. This is an example of two-stage grouping.

Two-stage grouping where all the groups contain the same number of observations is shown in Table 16.17. The notation employed in this table is analogous to that of Table 16.2. The k groups within the first stage are called *groups of the first order* and the subgroups within these groups are called *groups of the second order*, and so forth.

The fundamental assumptions made in the case of *systematic components* are that all observations are stochastically independent and that in the ijth group they are normally distributed with mean ξ_{ij} and variance σ_{ij}^2, i. e., the observations may be written

$$x_{ijv} = \xi_{ij} + z_{ijv}\,, \tag{16.6.1}$$

TABLE 16.17.
Two-stage grouping.

First stage	Second stage	Observations	Sample mean	Population mean	SSD	Sample variance
1	1.1	$x_{111}, x_{112}, \ldots, x_{11n}$	$\bar{x}_{11}$	$\xi_1+\eta_{11}$	$\Sigma(x_{11\nu}-\bar{x}_{11})^2$	s^2_{111}
	1.2	$x_{121}, x_{122}, \ldots, x_{12n}$	$\bar{x}_{12}$	$\xi_1+\eta_{12}$	$\Sigma(x_{12\nu}-\bar{x}_{12})^2$	s^2_{121}
	⋮	⋮ ⋮ ⋮	⋮	⋮	⋮	⋮
	1.m	$x_{1m1}, x_{1m2}, \ldots, x_{1mn}$	$\bar{x}_{1m}$	$\xi_1+\eta_{1m}$	$\Sigma(x_{1m\nu}-\bar{x}_{1m})^2$	s^2_{1m1}
⋮	⋮					
k	k.1	$x_{k11}, x_{k12}, \ldots, x_{k1n}$	$\bar{x}_{k1}$	$\xi_k+\eta_{k1}$	$\Sigma(x_{k1\nu}-\bar{x}_{k1})^2$	s^2_{k11}
	k.2	$x_{k21}, x_{k22}, \ldots, x_{k2n}$	$\bar{x}_{k2}$	$\xi_k+\eta_{k2}$	$\Sigma(x_{k2\nu}-\bar{x}_{k2})^2$	s^2_{k21}
	⋮	⋮ ⋮ ⋮	⋮	⋮	⋮	⋮
	$k.m$	$x_{km1}, x_{km2}, \ldots, x_{kmn}$	$\bar{x}_{km}$	$\xi_k+\eta_{km}$	$\Sigma(x_{km\nu}-\bar{x}_{km})^2$	s^2_{km1}

where $z_{ij\nu}$ is normally distributed with parameters $(0, \sigma^2_{ij})$. Within each first-order group the m population means are written as

$$\xi_{ij} = \xi_i+\eta_{ij} \quad \text{for} \quad j = 1, 2, \ldots, m\,, \tag{16.6.2}$$

where

$$\xi_i = \frac{1}{m}\sum_{j=1}^{m} \xi_{ij}\,, \tag{16.6.3}$$

and hence

$$\sum_{j=1}^{m} \eta_{ij} = 0\,, \tag{16.6.4}$$

cf. Table 16.2.

Assuming that the population variances are equal within each first-order group, the second-order groups within each first-order group may be compared by the methods given in § 16.4. For example, for group No. 1 we obtain

$$s^2_{11} = \frac{1}{m(n-1)}\sum_{j=1}^{m}\sum_{\nu=1}^{n}(x_{1j\nu}-\bar{x}_{1j})^2 = \frac{1}{m}\sum_{j=1}^{m} s^2_{1j1}\,, \tag{16.6.5}$$

$$s^2_{21} = \frac{n}{m-1}\sum_{j=1}^{m}(\bar{x}_{1j}-\bar{x}_1)^2 \tag{16.6.6}$$

$$v^2_1 = \frac{s^2_{21}}{s^2_{11}}\,, \tag{16.6.7}$$

(The m values of s^2 used for computing s^2_{11} are used for testing the hypothesis that $\sigma^2_{11} = \sigma^2_{12} = \ldots = \sigma^2_{1m}$). In this manner we obtain k stochastically independent variance ratios, each distributed as v^2 with $(m-1, m(n-1))$ degrees of freedom if there is no difference between the populations within the first-order groups. The evidence from these k tests of significance may be combined according to the principles given in § 15.6.

If, however, all the km population variances are equal, $\sigma_{ij}^2 = \sigma^2$ say, the above k tests may be combined by pooling the variances as shown in Table 16.18, where the variation between the first-order groups has also been included.

TABLE 16.18.

Analysis of variance of observations grouped in two stages.

Variation	SSD	f	s^2
Between first-order groups	$nm\sum_{i=1}^{k}(\bar{x}_i-\bar{x})^2$	$k-1$	s_3^2
Between second-order groups (within first-order groups)	$n\sum_{i=1}^{k}\sum_{j=1}^{m}(\bar{x}_{ij}-\bar{x}_i)^2$	$k(m-1)$	s_2^2
Within second-order groups	$\sum_{i=1}^{k}\sum_{j=1}^{m}\sum_{\nu=1}^{n}(x_{ij\nu}-\bar{x}_{ij})^2$	$km(n-1)$	s_1^2
Total	$\sum_{i=1}^{k}\sum_{j=1}^{m}\sum_{\nu=1}^{n}(x_{ij\nu}-\bar{x})^2$	$kmn-1$	

The variance s_1^2, which characterizes the variation within the second-order groups, is an average of the km variances $s_{111}^2, \ldots, s_{km1}^2$. This variance is independent of any hypothesis concerning the population means.

The variance s_2^2, which characterizes the variation between the second-order groups within the first-order groups, is an average of the k variances which are computed for each of the first-order groups in analogy with (16.6.6) for group No. 1. If the m population means for the second-order groups within each first-order group are equal, s_2^2 will have an s^2-distribution with parameters $(\sigma^2, k(m-1))$, and therefore the ratio $v^2 = s_2^2/s_1^2$ will have a v^2-distribution with $(k(m-1), km(n-1))$ degrees of freedom. This value of v^2 represents the above-mentioned combination of the k v^2-values.

If the value of $v^2 = s_2^2/s_1^2$ *is not significant*, all the observations within a first-order group may be considered as drawn from the same population. In this case the two s^2-values s_1^2 and s_2^2 may be combined to an estimate of σ^2, which gives us

$$s^2 = \frac{f_1 s_1^2 + f_2 s_2^2}{f_1 + f_2} = \frac{1}{k(mn-1)}\left(\sum\sum\sum(x_{ij\nu} - \bar{x}_{ij})^2 + n\sum\sum(\bar{x}_{ij} - \bar{x}_i)^2\right)$$

$$= \frac{1}{k(mn-1)} \sum_{i=1}^{k} \sum_{j=1}^{m} \sum_{\nu=1}^{n} (x_{ij\nu} - \bar{x}_i)^2 . \qquad (16.6.8)$$

However, the v^2-test does not *prove* that the means are identical and it may be safer not to pool the two variances.

Finally the equality of the population means of the first-order groups is tested by computing the ratio $v^2 = s_3^2/s_1^2$, which has a v^2-distribution with $(k-1, km(n-1))$ degrees of freedom if the hypothesis is true.

If, however, the value of $v^2 = s_2^2/s_1^2$ *is significant*, the means of the second-order populations must be considered as differing.

Each second-order sample mean is an estimate of the corresponding ξ_{ij}, and confidence limits may be found in the usual way. Further, in analogy with (16.4.4) we have

$$\mathcal{M}\{s_2^2\} = \sigma^2 + \frac{n}{k(m-1)} \sum_{i=1}^{k} \sum_{j=1}^{m} \eta_{ij}^2 \qquad (16.6.9)$$

and

$$\mathcal{M}\{s_3^2\} = \sigma^2 + \frac{mn}{k-1} \sum_{i=1}^{k} (\xi_i - \bar{\xi})^2 . \qquad (16.6.10)$$

In the case where we regard *the means of the second-order populations as stochastic variables*, the observations may be written

$$x_{ij\nu} = \xi_i + y_{ij} + z_{ij\nu} , \qquad (16.6.11)$$

where $z_{ij\nu}$, which indicates the variation within the second-order groups, is assumed to be normally distributed with parameters $(0, \sigma^2)$, and y_{ij}, which indicates the variation between the second-order groups within the first-order groups, is assumed to be normally distributed with parameters $(0, \omega^2)$. From this it follows that the k variances

$$s_{2i}^2 = \frac{n \sum_{j=1}^{m} (\bar{x}_{ij} - \bar{x}_i)^2}{m-1} , \quad i = 1, 2, \ldots, k , \qquad (16.6.12)$$

have the same theoretical value, $\sigma^2 + n\omega^2$. (This may be tested by means of BARTLETT'S test, cf. § 11.6).

In analogy with § 16.5, we see that it follows from the above assumptions that the variance s_1^2 has an s^2-distribution with parameters $(\sigma^2, km(n-1))$, and the variance s_2^2 has an s^2-distribution with parameters $(\sigma^2 + n\omega^2, k(m-1))$. Further, if the hypothesis $\xi_1 = \xi_2 = \ldots = \xi_k$ is correct, the variance s_3^2 has an s^2-distribution with parameters $(\sigma^2 + n\omega^2, k-1)$, and this hypothesis may therefore be tested by computing the variance ratio

$$v^2 = \frac{s_3^2}{s_2^2} \, . \tag{16.6.13}$$

In this case the variation between the first-order groups is caused by the random variation both within and between the second-order groups, and the hypothesis is therefore tested by comparing s_3^2 *and* s_2^2*, and not as in the case of systematic components* s_3^2 *and* s_1^2*.*

Example 16.5. When studying the manufacture of sheets of building material, three sheets were taken at random from the production every day. Eight pieces were cut out at random from each sheet, and the permeability was determined for each piece. Examination of a large series of measurements showed that the logarithm of the permeability may be considered normally distributed. Table 16.19 gives the measurements for three sheets together with the computations of the sums and *SSD*'s for the logarithms of the permeabilities.

TABLE 16.19.
Permeabilities (in seconds) and their logarithms for samples from three sheets from the first day.

Day (i)	1					
Sheets (j)	1		2		3	
Samples (ν)	Sec.	Log	Sec.	Log	Sec.	Log
1	40	1·602	40	1·602	20	1·301
2	65	1·813	45	1·653	15	1·176
3	175	2·243	50	1·699	15	1·176
4	35	1·544	35	1·544	15	1·176
5	45	1·653	40	1·602	25	1·398
6	50	1·699	35	1·544	40	1·602
7	105	2·021	60	1·778	25	1·398
8	95	1·978	25	1·398	20	1·301
S		14·553		12·820		10·528
SS		26·8843		20·6354		14·0093
S^2/n		26·4737		20·5441		13·8548
SSD		0·4106		0·0913		0·1545

Table 16.20 gives similar results of the computations for each sheet during four days. The data given in the first three rows of this table are the results of the computations given in Table 16.19. Thus, the observations are divided into four first-order groups (days), which are again divided into three second-order groups (sheets), each including eight measurements.

TABLE 16.20.
Computation of the variation within sheets.

Day (i)	Sheets (j)	S_{ij}	SS_{ij}	S^2_{ij}/n	SSD_{ij}	log SSD_{ij}
	1	14·553	26·8843	26·4737	0·4106	$\bar{1}$·613
1	2	12·820	20·6354	20·5441	0·0913	$\bar{2}$·960
	3	10·528	14·0093	13·8548	0·1545	$\bar{1}$·189
	1	10·577	14·1771	13·9841	0·1930	$\bar{1}$·286
2	2	13·452	22·7843	22·6195	0·1648	$\bar{1}$·217
	3	14·323	25·7152	25·6435	0·0717	$\bar{2}$·856
	1	11·771	17·4166	17·3196	0·0970	$\bar{2}$·987
3	2	10·658	14·3405	14·1991	0·1414	$\bar{1}$·151
	3	10·778	14·5533	14·5207	0·0326	$\bar{2}$·513
	1	11·084	15·4487	15·3569	0·0918	$\bar{2}$·963
4	2	12·297	19·0005	18·9020	0·0985	$\bar{2}$·993
	3	10·972	15·2169	15·0481	0·1688	$\bar{1}$·227
Total		143·813	220·1821	218·4661	1·7160	$\overline{12}$·955
$(\Sigma\Sigma S_{ij})^2/\Sigma n_i$		$143{\cdot}813^2/96 = 215{\cdot}4394$				
Difference		$\sum_{i=1}^{k}\sum_{j=1}^{m}\sum_{\nu=1}^{n}(x_{ij\nu}-\bar{x})^2 = 4{\cdot}7427$				

The hypothesis that the theoretical variance is the same for all 12 sheets is tested with BARTLETT's test by computing

$$\frac{1}{12}\sum\sum \log s^2_{ij} = \frac{1}{12}\sum\sum \log SSD_{ij} - \log 7 = \frac{0{\cdot}955}{12} - 1 - 0{\cdot}8451 = -1{\cdot}7655,$$

$$\log s^2 = \log 1{\cdot}7160 - \log 84 = -1{\cdot}6898\,,$$

$$c = 1 + \frac{13}{3\times 12\times 7} = 1{\cdot}0516\,,$$

and

$$\chi^2 \simeq \frac{2{\cdot}3026}{1{\cdot}0516}\times 84\times(-1{\cdot}6898+1{\cdot}7655) = 13{\cdot}9\,, \quad f = 11\,.$$

Since $\chi^2_{\cdot 80} = 14{\cdot}6$ we do not reject this hypothesis. As an estimate of the theoretical variance within a sheet we obtain the total SSD within sheets, 1·7160, divided by the corresponding number of degrees of freedom, $12\times 7 = 84$, i. e.,

$$s_1^2 = \frac{1{\cdot}7160}{84} = 0{\cdot}02043\,.$$

At the same time this table gives us the total *SSD*, 4·7427, corresponding to $12 \times 8 - 1 = 95$ degrees of freedom.

The difference between the number of degrees of freedom for each of these sums of squares, $95 - 84 = 11$, corresponds to $4 \times 2 = 8$ degrees of freedom for the variation between sheets within days plus $4 - 1 = 3$ degrees of freedom for the variation between days. The corresponding *SSD*'s have been computed in Table 16.21.

TABLE 16.21.
Computation of the variation between sheets within days.

Day	$S_i = \sum_{j=1}^{3} S_{ij}$	$\sum_{j=1}^{3} S_{ij}^2/n$	S_i^2/nm	SSD_i
1	37·901	60·8726	59·8536	1·0190
2	38·352	62·2471	61·2865	0·9606
3	33·207	46·0394	45·9460	0·0934
4	34·353	49·3070	49·1720	0·1350
Total	143·813	218·4661	216·2581	2·2080
$(\Sigma S_i)^2/kmn$	$143{\cdot}813^2/96 =$		215·4394	
Difference	$nm \sum_{i=1}^{k} (\bar{x}_i - \bar{x})^2 = 0{\cdot}8187$			

For each day we find

$$SSD_i = n \sum_{j=1}^{m} (\bar{x}_{ij} - \bar{x}_i)^2 = \sum_{j=1}^{m} \frac{S_{ij}^2}{n} - \frac{S_i^2}{nm}$$

where

$$\sum_{j=1}^{m} S_{ij}^2/n \quad \text{and} \quad S_i = \sum_{j=1}^{m} S_{ij}$$

is computed by summation in Table 16.20. By adding the four *SSD*'s we obtain the sum of squares of the deviations between sheets within days.

Finally, the *SSD* for the variation between days is computed from the following formula, cf. Table 16.21:

$$nm \sum_{i=1}^{k} (\bar{x}_i - \bar{x})^2 = \sum_{i=1}^{k} S_i^2/nm - \left(\sum_{i=1}^{k} S_i\right)^2 /nmk \, .$$

The analysis has been arranged in the following table, which is analogous to Table 16.18.

TABLE 16.22.
Analysis of variance for the logarithm of permeability.

Variation	*SSD*	f	s^2
Between days	0·8187	3	0·2729
Between sheets within days	2·2080	8	0·2760
Within sheets	1·7160	84	0·02043
Total	4·7427	95	

Comparison of the variation within sheets and between sheets leads to

$$v^2 = \frac{0{\cdot}2760}{0{\cdot}02043} = 13{\cdot}5,$$

which by far exceeds the 99% fractile of the v^2-distribution, cf. Table VII$_6$.

If the variation from sheet to sheet may be considered random—which can be decided only by examining the distribution of the means from a large number of sheets, cf. Chapter 13—the variance $s_2^2 = 0{\cdot}2760$ forms an estimate of the theoretical variance $n\omega^2+\sigma^2 = 8\omega^2+\sigma^2$, ω^2 denoting the variance in the distribution of the population means (the variation of the means from sheet to sheet), and σ^2 denoting the variance in the distribution of the observations from the same sheet.

As $s_3^2 = 0{\cdot}2729$ *is of the same size as* $s_2^2 = 0{\cdot}2760$ *there is no reason to assume that the variation from one day to another is due to special factors which are not present within each day.* During the production period examined here, the variation between days may well be explained by the variation between sheets.

As estimates of the two variances, σ^2 and ω^2, we obtain

$$s_1^2 = 0{\cdot}02043 \approx \sigma^2$$

and

$$\frac{1}{n}(s_2^2-s_1^2) = \frac{1}{8}(0{\cdot}2760-0{\cdot}0204) = 0{\cdot}03195 \approx \omega^2,$$

respectively.

The variance of the mean of 8 determinations of the permeability of samples from the *same* sheet is

$$\frac{s_1^2}{8} = \frac{0{\cdot}02043}{8} = 0{\cdot}002554 \approx \frac{\sigma^2}{8},$$

i. e., by repeatedly determining the permeability of the *same* sheet 8 determinations lead to a mean, the distribution of which is characterized by the standard deviation $\sqrt{0{\cdot}002554} = 0{\cdot}0505$.

The variation between the means from *different* sheets is characterized by the variance

$$\frac{s_2^2}{8} = \frac{0{\cdot}2760}{8} = 0{\cdot}03450 \approx \omega^2 + \frac{\sigma^2}{8}$$

and the corresponding standard deviation $\sqrt{0{\cdot}03450} = 0{\cdot}186$.

The quality of the sheets as far as permeability is concerned may be controlled by plotting the mean for each sheet on a control chart with limits computed from the standard deviation $\sqrt{\omega^2 + \frac{\sigma^2}{8}}$.

If we compare the estimates of ω^2 and $\sigma^2/8$, we find that $\sigma^2/8$ is very small compared with ω^2. This means that the control of the permeability of the sheets has been wrongly designed, the number of measurements made on each sheet being too many in consideration of the more important variation from sheet to sheet.

The variation between the means based on n measurements per sheet is characterized by the variance $\omega^2 + \frac{\sigma^2}{n}$. As an estimate of this variance we have

$$0{\cdot}03195 + \frac{0{\cdot}02043}{n} \approx \omega^2 + \frac{\sigma^2}{n}.$$

This is a decreasing function of n, which for $n \to \infty$ converges to ω^2, as shown in the following table:

Number of measurements per sheet n	Estimate of $\omega^2 + \frac{\sigma^2}{n}$ $0{\cdot}03195 + \frac{0{\cdot}02043}{n}$	Estimate of $\sqrt{\omega^2 + \frac{\sigma^2}{n}}$
1	0·05238	0·229
2	0·04217	0·205
4	0·03706	0·193
8	0·03450	0·186
16	0·03323	0·182
∞	0·03195	0·179

The standard error of the mean is practically constant for values of n between 4 and ∞, and decreases but little from $n = 1$ to $n = 4$. This means that if only one or two measurements are made per sheet the control of the permeability of the sheets is practically as good as when 8 measurements are made.

A more efficient control of the production process using the same number of measurements, viz. 24 per day, may be obtained by taking 12 sheets and making two measurements from each. In the first case the variance of the mean permeability of the measurements within a day is

$$\frac{1}{3}\left(\omega^2+\frac{\sigma^2}{8}\right)\approx\frac{0{\cdot}03450}{3}=0{\cdot}0115\ .$$

In the second case it is

$$\frac{1}{12}\left(\omega^2+\frac{\sigma^2}{2}\right)\approx\frac{0{\cdot}04217}{12}=0{\cdot}0035\ ,$$

which is less than $^1/_3$ of the variance in the first case.

If in all $n_1 \times n_2$ measurements are made, n_2 measurements being carried out on each of n_1 sheets, the variance of the mean of these measurements becomes

$$\mathcal{V}\{\bar{x}\}=\frac{1}{n_1}\left(\omega^2+\frac{\sigma^2}{n_2}\right).$$

The final choice of the number of sheets to be taken per day and the number of measurements to be made per sheet should take the costs into account, see § 17.4.

16.7. Two-Way Classification.

A. *Formulation of the problem.*

As explained in the previous section a two-stage classification is a classification where the one criterion is subordinate to the other. Another important case of classification according to two criteria is the two-way classification. In this case the data may be arranged in a rectangular table with a certain number, n, of observations corresponding to each combination of the two criteria, as in Table 16.23, which shows *one* observation for each of the km combinations of the two criteria.

The k rows may, for example, correspond to k machines of the same type and the m columns to m production periods, for instance corresponding

TABLE 16.23.

Two-way classification with one observation per cell.

Row No.	Column No. 1	2	...	m
1	x_{11}	x_{12}	...	x_{1m}
2	x_{21}	x_{22}	...	x_{2m}
⋮	⋮	⋮		⋮
k	x_{k1}	x_{k2}	...	x_{km}

to m batches of raw materials, m shifts, or simply a division of the production period into days or weeks. The observations x_{ij} now denote a certain property of the finished product, e. g., strength, purity, or the like, corresponding to machine No. i and batch of raw material No. j. If several samples are taken from the same machine and the same batch we obtain an estimate of the random variation. By aid of an analysis of variance the variation between machines (between rows) and the variation between batches of raw material (between columns) are compared with the random variation in order to search for possible variations in the properties of the product due to differences between the machines or between the batches.

B. *Partitioning the total sum of squares.*

Let a set of data be given corresponding to Table 16.23, except that now we have n observations corresponding to each of the km combinations of the two criteria, as in Table 16.24, in which the νth observation of each of these km sets has been indicated.

TABLE 16.24.

The νth observation in each set, $\nu = 1, 2, \ldots, n$, in a two-way classification of the observations.

Row No.	Column No. 1	2	...	m
1	$x_{11\nu}$	$x_{12\nu}$	...	$x_{1m\nu}$
2	$x_{21\nu}$	$x_{22\nu}$	...	$x_{2m\nu}$
⋮	⋮	⋮		⋮
k	$x_{k1\nu}$	$x_{k2\nu}$	...	$x_{km\nu}$

TABLE 16.25.

The mean of the observations within each set in a two-way classification.

Row No.	Column No. 1	2	...	m	Row means
1	$\bar{x}_{11}$	$\bar{x}_{12}$	...	$\bar{x}_{1m}$	$\bar{x}_{1\cdot}$
2	$\bar{x}_{21}$	$\bar{x}_{22}$	...	$\bar{x}_{2m}$	$\bar{x}_{2\cdot}$
⋮	⋮	⋮		⋮	⋮
k	$\bar{x}_{k1}$	$\bar{x}_{k2}$	...	$\bar{x}_{km}$	$\bar{x}_{k\cdot}$
Column means	$\bar{x}_{\cdot 1}$	$\bar{x}_{\cdot 2}$	...	$\bar{x}_{\cdot m}$	$\bar{x}_{\cdot\cdot}$

For each set of observations, the mean and the *SSD* are computed, cf. Tables 16.25 and 16.26.

TABLE 16.26.
The sum of squares of the deviations from the mean within each set in a two-way classification.

Row No.	Column No. 1	. . .	m
1	$\sum_{\nu=1}^{n}(x_{11\nu}-\bar{x}_{11})^2$	. . .	$\sum_{\nu=1}^{n}(x_{1m\nu}-\bar{x}_{1m})^2$
⋮	⋮		⋮
k	$\sum_{\nu=1}^{n}(x_{k1\nu}-\bar{x}_{k1})^2$	. . .	$\sum_{\nu=1}^{n}(x_{km\nu}-\bar{x}_{km})^2$

The following analysis is based on the assumption that the km sets of observations are random samples from km normally distributed populations, each having its own mean, but the variance being the same for all populations, i. e.,

$$x_{ij\nu} = \xi_{ij}+z_{ij\nu}\ , \qquad (16.7.1)$$

where $z_{ij\nu}$ is normally distributed with parameters $(0, \sigma^2)$.

The hypothesis that the variance is the same for all km populations may be tested by the km variances corresponding to the *SSD*'s in Table 16.26, each of which has $n-1$ degrees of freedom. Accepting the hypothesis, then, as an estimate of σ^2, we obtain

$$s_1^2 = \frac{1}{km(n-1)}\sum_{i=1}^{k}\sum_{j=1}^{m}\sum_{\nu=1}^{n}(x_{ij\nu}-\bar{x}_{ij})^2\ , \qquad (16.7.2)$$

which has an s^2-distribution with $km(n-1)$ degrees of freedom.

The total sum of squares of deviations

$$\sum_{i=1}^{k}\sum_{j=1}^{m}\sum_{\nu=1}^{n}(x_{ij\nu}-\xi_{ij})^2 \qquad (16.7.3)$$

which is distributed as $\sigma^2\chi^2$ with kmn degrees of freedom is thus partitioned into two sums of squares, the deviations $x_{ij\nu}-\xi_{ij}$ being written

$$x_{ij\nu}-\xi_{ij} = (x_{ij\nu}-\bar{x}_{ij})+(\bar{x}_{ij}-\xi_{ij})\ . \qquad (16.7.4)$$

This leads to

$$\sum_{i=1}^{k}\sum_{j=1}^{m}\sum_{\nu=1}^{n}(x_{ij\nu}-\xi_{ij})^2 = \sum_{i=1}^{k}\sum_{j=1}^{m}\sum_{\nu=1}^{n}(x_{ij\nu}-\bar{x}_{ij})^2+n\sum_{i=1}^{k}\sum_{j=1}^{m}(\bar{x}_{ij}-\xi_{ij})^2\ , \qquad (16.7.5)$$

the first term of which is independent of the population means and gives *the estimate* s_1^2 of σ^2 *based on the variation within sets*, while the second term shows that the means

$$\bar{x}_{ij} = \frac{1}{n}\sum_{\nu=1}^{n} x_{ij\nu} = \xi_{ij}+\bar{z}_{ij}, \quad i = 1,\ldots, k \quad \text{and} \quad j = 1,\ldots, m\,, \tag{16.7.6}$$

are stochastically independent of s_1^2 and normally distributed with parameters $(\xi_{ij}, \sigma^2/n)$, so that $\bar{z}_{ij}$ is normally distributed with parameters $(0, \sigma^2/n)$.

We now proceed to set up a simple hypothesis concerning the structure of the ξ_{ij}'s. First consider Table 16.27, where all the population means and the corresponding row means and column means are given.

TABLE 16.27.

Population means in a two-way classification.

Row No.	Column No. 1	2	...	m	Row means	Deviations
1	ξ_{11}	ξ_{12}	...	ξ_{1m}	$\xi_{1\cdot}$	$\eta_1 = \xi_{1\cdot}-\xi_{\cdot\cdot}$
2	ξ_{21}	ξ_{22}	...	ξ_{2m}	$\xi_{2\cdot}$	$\eta_2 = \xi_{2\cdot}-\xi_{\cdot\cdot}$
$\vdots$	$\vdots$	$\vdots$		$\vdots$	$\vdots$	$\vdots$
k	ξ_{k1}	ξ_{k2}	...	ξ_{km}	$\xi_{k\cdot}$	$\eta_k = \xi_{k\cdot}-\xi_{\cdot\cdot}$
Column means	$\xi_{\cdot 1}$	$\xi_{\cdot 2}$	...	$\xi_{\cdot m}$	$\xi_{\cdot\cdot}$	0
Deviations	$\zeta_1 = \xi_{\cdot 1}-\xi_{\cdot\cdot}$	$\zeta_2 = \xi_{\cdot 2}-\xi_{\cdot\cdot}$	...	$\zeta_m = \xi_{\cdot m}-\xi_{\cdot\cdot}$	0	

The row means are defined as

$$\xi_{i\cdot} = \frac{1}{m}\sum_{j=1}^{m} \xi_{ij}\,, \quad i = 1, 2, \ldots, k\,, \tag{16.7.7}$$

and the column means are defined similarly.

The mean ξ_{ij} may then be written

$$\begin{aligned}\xi_{ij} &= \xi_{\cdot\cdot}+(\xi_{i\cdot}-\xi_{\cdot\cdot})+(\xi_{\cdot j}-\xi_{\cdot\cdot})+(\xi_{ij}-\xi_{i\cdot}-\xi_{\cdot j}+\xi_{\cdot\cdot}) \\ &= \xi_{\cdot\cdot}+\eta_i+\zeta_j+\vartheta_{ij}\,,\end{aligned} \tag{16.7.8}$$

where

$$\xi_{\cdot\cdot} = \frac{1}{km}\sum_{i=1}^{k}\sum_{j=1}^{m} \xi_{ij} \tag{16.7.9}$$

denotes the total mean of the km population means,

$$\eta_i = \xi_{i\cdot}-\xi_{\cdot\cdot} \tag{16.7.10}$$

is the deviation of the ith row mean from the total mean,

$$\zeta_j = \xi_{\cdot j} - \xi_{\cdot\cdot} \tag{16.7.11}$$

the deviation of the jth column mean from the total mean, and

$$\vartheta_{ij} = \xi_{ij} - \xi_{i\cdot} - \xi_{\cdot j} + \xi_{\cdot\cdot} = \xi_{ij} - (\xi_{\cdot\cdot} + \eta_i + \zeta_j) \tag{16.7.12}$$

is *the remainder term* which shows to what extent ξ_{ij} differs from the simple expression $\xi_{\cdot\cdot} + \eta_i + \zeta_j$.

From the definitions of η, ζ, and ϑ, we obtain the following linear relationships between these quantities

$$\sum_{i=1}^{k} \eta_i = 0 \,, \tag{16.7.13}$$

$$\sum_{j=1}^{m} \zeta_j = 0 \,, \tag{16.7.14}$$

$$\sum_{j=1}^{m} \vartheta_{ij} = 0 \,, \quad \text{for} \quad i = 1, 2, \ldots, k \,, \tag{16.7.15}$$

and

$$\sum_{i=1}^{k} \vartheta_{ij} = 0 \,, \quad \text{for} \quad j = 1, 2, \ldots, m \,. \tag{16.7.16}$$

The two sets of relations for the ϑ's are not linearly independent, since the relationship

$$\sum_{i=1}^{k} \sum_{j=1}^{m} \vartheta_{ij} = 0 \tag{16.7.17}$$

is got by summing either (16.7.15) or (16.7.16), so that the number of independent relations for the ϑ's is only $k+m-1$. Hence, *instead of the original km parameters* ξ_{ij} *we now have introduced km new parameters which are linear functions of the* ξ_{ij}'s, namely the parameter $\xi_{\cdot\cdot}$, $k-1$ mutually independent η's, $m-1$ mutually independent ζ's, and $km-(k+m-1) = (k-1)(m-1)$ mutually independent ϑ's.

The reasons for doing this are as follows. By writing

$$\xi_{ij} = \xi_{\cdot\cdot} + \eta_i + \zeta_j + \vartheta_{ij} \tag{16.7.18}$$

we express the parameter for the ijth cell as a constant $\xi_{\cdot\cdot}$ plus η_i which is the same for all cells in row i, plus ζ_j which is the same for all cells in column j, plus the individual deviation ϑ_{ij} which is particular to the ijth cell. If the ϑ's are negligible or zero, we then have the cell mean expressed as the sum of two parts, one due to the "row effect", the other due to the "column effect", together with a constant, see Table 16.28. This is a simple and often a reasonable model of the data.

TABLE 16.28.
Population means for a two-way classification when $\vartheta_{ij} = 0$.

Row No.	Column No. 1	2	...	m	Row means
1	$\xi_{..}+\eta_1+\zeta_1$	$\xi_{..}+\eta_1+\zeta_2$	...	$\xi_{..}+\eta_1+\zeta_m$	$\xi_{..}+\eta_1$
2	$\xi_{..}+\eta_2+\zeta_1$	$\xi_{..}+\eta_2+\zeta_2$	...	$\xi_{..}+\eta_2+\zeta_m$	$\xi_{..}+\eta_2$
$\vdots$	$\vdots$	$\vdots$		$\vdots$	$\vdots$
k	$\xi_{..}+\eta_k+\zeta_1$	$\xi_{..}+\eta_k+\zeta_2$	...	$\xi_{..}+\eta_k+\zeta_m$	$\xi_{..}+\eta_k$
Column means	$\xi_{..}+\zeta_1$	$\xi_{..}+\zeta_2$	...	$\xi_{..}+\zeta_m$	$\xi_{..}$

In the following sections we will develop a test for the hypothesis $\vartheta_{ij}=0$. However, a preliminary analysis of the data will often show whether this hypothesis is obviously untenable. Such a preliminary analysis may, for example, be carried out by plotting the row means $\bar{x}_{i.}$ as abscissas and the corresponding means of each column $\bar{x}_{ij}$ as ordinates. If the hypothesis is true, the points should be randomly distributed about m straight lines with slope one, see Table 16.28 where the true values are given as

$$(\xi_{..}+\eta_i, \xi_{..}+\eta_i+\zeta_j) \quad \text{for} \quad i = 1, 2, \ldots, k .$$

If the hypothesis $\vartheta_{ij} = 0$ is true, the number of parameters necessary to represent the population means is reduced from km to $k+m-1$, which obviously leads to *a considerable simplification of the description and interpretation of the observations.*

If the ϑ's are not negligible we face the problem of devising a new hypothesis that is plausible. This may be based on general knowledge of the background for the observations, but very often such knowledge is lacking or is too vague to lead to a specific model for the data. The model is then chosen to agree with the behaviour of our observations, usually after a careful graphical analysis.

If the ϑ's differ from zero, the effects of the factors corresponding to the two criteria are not independent, but a certain "interaction" between them exists. Some combinations of the two factors lead to a favorable result, others to an unfavorable result as compared with the results calculated from the row and column means. Hence, ϑ_{ij} will be termed the *interaction* associated with the ijth cell.

Whatever we may finally decide about the ϑ's we can, at this stage, obtain the following estimates of the four components into which ξ_{ij} has been partitioned:

$$\bar{x}_{..} = \frac{1}{km}\sum_{i=1}^{k}\sum_{j=1}^{m}\bar{x}_{ij} \approx \xi_{..}\,, \tag{16.7.19}$$

$$\bar{x}_{i\cdot}-\bar{x}_{..} \approx \xi_{i\cdot}-\xi_{..} = \eta_i\,, \tag{16.7.20}$$

$$\bar{x}_{\cdot j}-\bar{x}_{..} \approx \xi_{\cdot j}-\xi_{..} = \zeta_j\,, \tag{16.7.21}$$

and

$$\bar{x}_{ij}-\bar{x}_{i\cdot}-\bar{x}_{\cdot j}+\bar{x}_{..} \approx \xi_{ij}-\xi_{i\cdot}-\xi_{\cdot j}+\xi_{..} = \vartheta_{ij}\,, \tag{16.7.22}$$

cf. Tables 16.25 and 16.27.

We may now partition the deviation $\bar{x}_{ij}-\xi_{ij}$ as follows:

$$\bar{x}_{ij}-\xi_{ij} = (\bar{x}_{ij}-\bar{x}_{i\cdot}-\bar{x}_{\cdot j}+\bar{x}_{..}-\vartheta_{ij}) + (\bar{x}_{i\cdot}-\bar{x}_{..}-\eta_i)+(\bar{x}_{\cdot j}-\bar{x}_{..}-\zeta_j)+(\bar{x}_{..}-\xi_{..})\,,$$

from which by squaring and summing we obtain

$$n\sum_{i=1}^{k}\sum_{j=1}^{m}(\bar{x}_{ij}-\xi_{ij})^2 = n\sum_{i=1}^{k}\sum_{j=1}^{m}(\bar{x}_{ij}-\bar{x}_{i\cdot}-\bar{x}_{\cdot j}+\bar{x}_{..}-\vartheta_{ij})^2$$

$$+nm\sum_{i=1}^{k}(\bar{x}_{i\cdot}-\bar{x}_{..}-\eta_i)^2+nk\sum_{j=1}^{m}(\bar{x}_{\cdot j}-\bar{x}_{..}-\zeta_j)^2+nmk(\bar{x}_{..}-\xi_{..})^2\,. \tag{16.7.23}$$

The product terms have vanished, on account of relations (16.7.13) to (16.7.16) and the following corresponding relations for the estimates:

$$\sum_{i=1}^{k}(\bar{x}_{i\cdot}-\bar{x}_{..}) = 0\,, \tag{16.7.24}$$

$$\sum_{j=1}^{m}(\bar{x}_{\cdot j}-\bar{x}_{..}) = 0\,, \tag{16.7.25}$$

$$\sum_{j=1}^{m}(\bar{x}_{ij}-\bar{x}_{i\cdot}-\bar{x}_{\cdot j}+\bar{x}_{..}) = 0 \quad \text{for} \quad i = 1, 2, \ldots, k\,, \tag{16.7.26}$$

and

$$\sum_{i=1}^{k}(\bar{x}_{ij}-\bar{x}_{i\cdot}-\bar{x}_{\cdot j}+\bar{x}_{..}) = 0 \quad \text{for} \quad j = 1, 2, \ldots, m\,. \tag{16.7.27}$$

The number of degrees of freedom for the four sums of squares on the right side of the sign of equality are, according to the above relations, $(k-1)(m-1)$, $k-1$, $m-1$, and 1, which add up to the km degrees of freedom for the sum of squares on the left side. According to the partition theorem for the χ^2-distribution we therefore have that the four sums of squares are stochastically independent and distributed as $\sigma^2\chi^2$ with the respective numbers of degrees of freedom. The corresponding values of s^2:

$$s^2_{2(\vartheta)} = \frac{n\sum_{i=1}^{k}\sum_{j=1}^{m}(\bar{x}_{ij}-\bar{x}_{i\cdot}-\bar{x}_{\cdot j}+\bar{x}_{..}-\vartheta_{ij})^2}{(k-1)(m-1)}\,, \tag{16.7.28}$$

$$s^2_{3(\eta)} = \frac{nm \sum_{i=1}^{k} (\bar{x}_{i\cdot} - \bar{x}_{\cdot\cdot} - \eta_i)^2}{k-1}, \tag{16.7.29}$$

$$s^2_{4(\zeta)} = \frac{nk \sum_{j=1}^{m} (\bar{x}_{\cdot j} - \bar{x}_{\cdot\cdot} - \zeta_j)^2}{m-1} \tag{16.7.30}$$

and

$$s^2_{5(\xi..)} = nmk(\bar{x}_{\cdot\cdot} - \xi_{\cdot\cdot})^2 \tag{16.7.31}$$

are stochastically independent and all have the mean value σ^2.

Any hypothesis concerning specified values of the ϑ's, the η's, the ζ's, and $\xi_{\cdot\cdot}$ may be tested by comparing these variances with s_1^2.

The whole analysis may be summarized as shown in Table 16.29.

TABLE 16.29.

Sum of squares	Degrees of freedom	Variance
$nmk(\bar{x}_{\cdot\cdot} - \xi_{\cdot\cdot})^2$	1	$s^2_{5(\xi..)}$
$nk \sum_{j=1}^{m} (\bar{x}_{\cdot j} - \bar{x}_{\cdot\cdot} - \zeta_j)^2$	$m-1$	$s^2_{4(\zeta)}$
$nm \sum_{i=1}^{k} (\bar{x}_{i\cdot} - \bar{x}_{\cdot\cdot} - \eta_i)^2$	$k-1$	$s^2_{3(\eta)}$
$n \sum_{i=1}^{k} \sum_{j=1}^{m} (\bar{x}_{ij} - \bar{x}_{i\cdot} - \bar{x}_{\cdot j} + \bar{x}_{\cdot\cdot} - \vartheta_{ij})^2$	$(k-1)(m-1)$	$s^2_{2(\vartheta)}$
$\sum_{i=1}^{k} \sum_{j=1}^{m} \sum_{\nu=1}^{n} (x_{ij\nu} - \bar{x}_{ij})^2$	$km(n-1)$	s_1^2
$\sum_{i=1}^{k} \sum_{j=1}^{m} \sum_{\nu=1}^{n} (x_{ij\nu} - \xi_{\cdot\cdot} - \eta_i - \zeta_j - \vartheta_{ij})^2$	kmn	

C. *Tests of significance in the case of systematic components.*

1. *Testing the hypothesis* $\vartheta_{ij} = 0$ for $i = 1, 2, \ldots, k$ and $j = 1, 2, \ldots, m$.

If the hypothesis that $\vartheta_{ij} = 0$ *for* $i = 1, 2, \ldots, k$ *and* $j = 1, 2, \ldots, m$ *is true*, the variance

$$s_2^2 = \frac{n \sum_{i=1}^{k} \sum_{j=1}^{m} (\bar{x}_{ij} - \bar{x}_{i\cdot} - \bar{x}_{\cdot j} + \bar{x}_{\cdot\cdot})^2}{(k-1)(m-1)} \tag{16.7.32}$$

has an s^2-distribution with parameters $(\sigma^2, (k-1)(m-1))$, so that

$$v^2 = \frac{s_2^2}{s_1^2} \tag{16.7.33}$$

has a v^2-distribution with $((k-1)(m-1), km(n-1))$ degrees of freedom.

If the hypothesis $\vartheta_{ij} = 0$ *is false*, the mean value of s_2^2 is equal to σ^2 plus a quantity which depends on the interaction. From the numerator in (16.7.28) we obtain

$$n \sum_{i=1}^{k} \sum_{j=1}^{m} (\bar{x}_{ij} - \bar{x}_{i\cdot} - \bar{x}_{\cdot j} + \bar{x}_{\cdot\cdot} - \vartheta_{ij})^2 = n \sum_{i=1}^{k} \sum_{j=1}^{m} (\bar{x}_{ij} - \bar{x}_{i\cdot} - \bar{x}_{\cdot j} + \bar{x}_{\cdot\cdot})^2$$
$$-2n \sum_{i=1}^{k} \sum_{j=1}^{m} (\bar{x}_{ij} - \bar{x}_{i\cdot} - \bar{x}_{\cdot j} + \bar{x}_{\cdot\cdot})\vartheta_{ij} + n \sum_{i=1}^{k} \sum_{j=1}^{m} \vartheta_{ij}^2 ,$$

which leads to

$$\mathcal{M}\left\{n \sum_{i=1}^{k} \sum_{j=1}^{m} (\bar{x}_{ij} - \bar{x}_{i\cdot} - \bar{x}_{\cdot j} + \bar{x}_{\cdot\cdot})^2\right\} = (k-1)(m-1)\sigma^2 + n \sum_{i=1}^{k} \sum_{j=1}^{m} \vartheta_{ij}^2,$$

and

$$\mathcal{M}\{s_2^2\} = \sigma^2 + \frac{n}{(k-1)(m-1)} \sum_{i=1}^{k} \sum_{j=1}^{m} \vartheta_{ij}^2 . \tag{16.7.34}$$

Thus we see that the mean of s_2^2 is σ^2 only when all the ϑ's are zero, otherwise the mean of s_2^2 is larger than σ^2. Therefore, *if* s_2^2 *is significantly larger than* s_1^2, the hypothesis is rejected. The power of the test may be found from a formula analogous to (16.4.11).

If the interaction is significant we cannot use the standard methods of analysis of variance further but a new model must be formulated. (If the interaction may be considered as a stochastic variable we proceed as shown under D). In the case where the two criteria for classification are quantitative we may perhaps use the methods of regression analysis, see § 20.6. Denoting the two independent variables (classification criteria) by (a_1, a_2) the problem is to specify ξ_{ij} as a function of (a_{1i}, a_{2j}), as far as possible in such a manner that the unknown parameters enter linearly into the known function. For example, we may use the specification

$$\xi_{ij} = \xi_{\cdot\cdot} + \eta_i + \zeta_j + \vartheta g(a_{1i}, a_{2j})$$

where the interaction term

$$\vartheta_{ij} = \vartheta g(a_{1i}, a_{2j})$$

is proportional to the *known* function $g(a_{1i}, a_{2j})$ so that we only have to estimate the parameter ϑ.

In some cases a transformation of the observations will make the interaction zero. If $\xi_{ij} = c_0 \alpha_i^{c_1} \beta_j^{c_2}$ we find $\log \xi_{ij} = \log c_0 + c_1 \log \alpha_i + c_2 \log \beta_j$, i.e., the interaction between the logarithms is zero.

If the interaction is insignificant we may test the row and column effects as shown under 2 and 3 below.

2. *Test of the hypothesis* $\eta_1 = \eta_2 = \ldots = \eta_k = 0$.
This hypothesis is tested by computing

$$s_3^2 = \frac{nm \sum_{i=1}^{k} (\bar{x}_{i\cdot} - \bar{x}_{\cdot\cdot})^2}{k-1} \tag{16.7.35}$$

and

$$v^2 = \frac{s_3^2}{s_1^2}. \tag{16.7.36}$$

If s_3^2 is significantly larger than s_1^2, the hypothesis is rejected.

3. *Test of the hypothesis* $\zeta_1 = \zeta_2 = \ldots = \zeta_m = 0$.
This hypothesis is tested by computing

$$s_4^2 = \frac{nk \sum_{j=1}^{m} (\bar{x}_{\cdot j} - \bar{x}_{\cdot\cdot})^2}{m-1} \tag{16.7.37}$$

and

$$v^2 = \frac{s_4^2}{s_1^2}. \tag{16.7.38}$$

If s_4^2 is significantly larger than s_1^2, the hypothesis is rejected.

Table 16.30 summarizes this analysis.

TABLE 16.30.
Analysis of variance in a two-way classification.

Variation	*SSD*	f	s^2	$\mathcal{M}\{s^2\}$
Between columns	$nk \sum_{j=1}^{m} (\bar{x}_{\cdot j} - \bar{x}_{\cdot\cdot})^2$	$m-1$	s_4^2	$\sigma^2 + \frac{nk}{f_4} \Sigma \zeta_j^2$
Between rows	$nm \sum_{i=1}^{k} (\bar{x}_{i\cdot} - \bar{x}_{\cdot\cdot})^2$	$k-1$	s_3^2	$\sigma^2 + \frac{nm}{f_3} \Sigma \eta_i^2$
Inter-action	$n \sum_{i=1}^{k} \sum_{j=1}^{m} (\bar{x}_{ij} - \bar{x}_{i\cdot} - \bar{x}_{\cdot j} + \bar{x}_{\cdot\cdot})^2$	$(k-1)(m-1)$	s_2^2	$\sigma^2 + \frac{n}{f_2} \Sigma\Sigma \vartheta_{ij}^2$
Within sets	$\sum_{i=1}^{k} \sum_{j=1}^{m} \sum_{\nu=1}^{n} (x_{ij\nu} - \bar{x}_{ij})^2$	$km(n-1)$	s_1^2	σ^2
Total	$\sum_{i=1}^{k} \sum_{j=1}^{m} \sum_{\nu=1}^{n} (x_{ij\nu} - \bar{x}_{\cdot\cdot})^2$	$kmn-1$		

The total SSD and the SSD's corresponding to s_1^2, s_3^2, and s_4^2 are computed according to the rules given in § 4.4. The sum of squares for the interaction is then computed as the difference between the total SSD and the three other SSD's.

Introducing the notation

$$S_{ij} = \sum_{\nu=1}^{n} x_{ij\nu}, \qquad SS_{ij} = \sum_{\nu=1}^{n} x_{ij\nu}^2,$$

and

$$S_{i\cdot} = \sum_{j=1}^{m} S_{ij}, \qquad S_{\cdot j} = \sum_{i=1}^{k} S_{ij}, \qquad S_{\cdot\cdot} = \sum_{i=1}^{k} \sum_{j=1}^{m} S_{ij},$$

we find

$$SSD_{ij} = SS_{ij} - \frac{S_{ij}^2}{n}, \tag{16.7.39}$$

$$nk \sum_{j=1}^{m} (\bar{x}_{\cdot j} - \bar{x}_{\cdot\cdot})^2 = \sum_{j=1}^{m} \frac{S_{\cdot j}^2}{nk} - \frac{S_{\cdot\cdot}^2}{nkm}, \tag{16.7.40}$$

and

$$nm \sum_{i=1}^{k} (\bar{x}_{i\cdot} - \bar{x}_{\cdot\cdot})^2 = \sum_{i=1}^{k} \frac{S_{i\cdot}^2}{nm} - \frac{S_{\cdot\cdot}^2}{nmk}. \tag{16.7.41}$$

If the hypotheses are rejected, confidence limits for the population means may be established with the aid of the t-test, the quantities

$$t = \frac{\bar{x}_{ij} - \xi_{ij}}{s_1/\sqrt{n}}, \tag{16.7.42}$$

$$t = \frac{\bar{x}_{i\cdot} - \xi_{i\cdot}}{s_1/\sqrt{nm}}, \tag{16.7.43}$$

and

$$t = \frac{\bar{x}_{\cdot j} - \xi_{\cdot j}}{s_1/\sqrt{nk}} \tag{16.7.44}$$

having t-distributions with $km(n-1)$ degrees of freedom.

When comparing, say, two row means, we similarly find that

$$t = \frac{\bar{x}_{i_1\cdot} - \bar{x}_{i_2\cdot} - (\xi_{i_1\cdot} - \xi_{i_2\cdot})}{s_1\sqrt{\dfrac{2}{nm}}} \tag{16.7.45}$$

has a t-distribution with $km(n-1)$ degrees of freedom, cf. § 15.5.

From (16.7.34) we obtain

$$\mathcal{m}\left\{\frac{1}{n}(s_2^2 - s_1^2)\right\} = \frac{1}{(k-1)(m-1)} \sum_{i=1}^{k} \sum_{j=1}^{m} \vartheta_{ij}^2. \tag{16.7.46}$$

Similarly we have
$$\mathcal{M}\left\{\frac{1}{nm}(s_3^2-s_1^2)\right\}=\frac{1}{k-1}\sum_{i=1}^{k}\eta_i^2 \tag{16.7.47}$$
and
$$\mathcal{M}\left\{\frac{1}{nk}(s_4^2-s_1^2)\right\}=\frac{1}{m-1}\sum_{j=1}^{m}\zeta_j^2. \tag{16.7.48}$$

As stated in § 16.4, p. 428, each of the sums of squares may be further partitioned.

If there is *only one observation in each set*, i. e., $n = 1$, as in Table 16.23, the sum of squares corresponding to the variation within sets does not exist.

If, however, we know a priori that the theoretical interaction is zero ($\vartheta_{ij}=0$), which may sometimes be inferred from our knowledge of the two factors causing the variation, then the variance corresponding to the interaction supplies an estimate of σ^2 for comparison with the variances corresponding to the rows and the columns. In this case the analysis of variance will be as shown in Table 16.31, cf. Table 16.30 for $n = 1$.

TABLE 16.31.

Analysis of variance for a two-way classification with only one observation per set.

Variation	*SSD*	f	s^2	$\mathcal{M}\{s^2\}$
Between columns	$k\sum_{j=1}^{m}(\bar{x}_{\cdot j}-\bar{x}_{\cdot\cdot})^2$	$m-1$	s_4^2	$\sigma^2+\frac{k}{f_4}\Sigma\zeta_j^2$
Between rows	$m\sum_{i=1}^{k}(\bar{x}_{i\cdot}-\bar{x}_{\cdot\cdot})^2$	$k-1$	s_3^2	$\sigma^2+\frac{m}{f_3}\Sigma\eta_i^2$
Inter-action	$\sum_{i=1}^{k}\sum_{j=1}^{m}(x_{ij}-\bar{x}_{i\cdot}-\bar{x}_{\cdot j}+\bar{x}_{\cdot\cdot})^2$	$(k-1)(m-1)$	s_2^2	σ^2
Total	$\sum_{i=1}^{k}\sum_{j=1}^{m}(x_{ij}-\bar{x}_{\cdot\cdot})^2$	$km-1$		

For $m = 2$ this analysis includes the "STUDENT" test of § 15.5. If, as in § 15.5, we choose the notation $d_i = x_{i1}-x_{i2}$, and $\bar{d} = \bar{x}_{\cdot 1}-\bar{x}_{\cdot 2}$, we find

$$s_2^2=\frac{1}{2}\frac{\sum_{i=1}^{k}(d_i-\bar{d})^2}{k-1}=\frac{1}{2}s_d^2,$$

$$s_4^2=\frac{k}{2}\bar{d}^2,$$

and

$$v^2 = \frac{s_4^2}{s_2^2} = \frac{k\bar{d}^2}{s_d^2} = \left(\frac{\bar{d}}{s_d/\sqrt{k}}\right)^2 = t^2 \, , \tag{16.7.49}$$

cf. (15.5.7).

In the analysis of variance the estimates of the parameters may be derived by *the method of least squares* in the same manner as the estimate $\bar{x}$ was derived as an estimate of ξ. If, e. g., we have km observations, x_{ij}, $(i = 1, 2, \ldots, k, j = 1, 2, \ldots, m)$, and the mean values ξ_{ij} are represented by the relation

$$\xi_{ij} = \xi_{..} + \eta_i + \zeta_j \, ,$$

where $\Sigma\eta_i = 0$ and $\Sigma\zeta_i = 0$, the estimates may be derived as the values of $\xi_{..}$, η_i, and ζ_j which minimize the sum of squares

$$\sum_{i=1}^{k} \sum_{j=1}^{m} (x_{ij} - \xi_{..} - \eta_i - \zeta_j)^2$$

when the conditional equations $\Sigma\eta_i = 0$ and $\Sigma\zeta_j = 0$ are taken into consideration. This minimization leads to the same estimates $\bar{x}_{..} \approx \xi_{..}$, $\bar{x}_{i.} - \bar{x}_{..} \approx \eta_i$, and $\bar{x}_{.j} - \bar{x}_{..} \approx \zeta_j$ as we employ here.

D. *Tests of significance in the case of random components.*

If, according to the analysis of variance developed in part C, it is necessary to reject the hypothesis that $\vartheta_{ij} = 0$, we must now decide whether the interaction is of a systematic or of a stochastic nature. If the interaction is assumed to be random the observational result $x_{ij\nu}$ may be written

$$x_{ij\nu} = \xi_{ij} + z_{ij\nu} = \xi_{..} + \eta_i + \zeta_j + y_{ij} + z_{ij\nu} \, , \tag{16.7.50}$$

where $z_{ij\nu}$—as in part C—is assumed to be normally distributed with parameters $(0, \sigma^2)$, and further y_{ij} is normally distributed with parameters $(0, \omega^2)$, and every y is independent of every z. The populations from which the observations originate are characterized by the parameters $\xi_{..}, \eta_1, \ldots, \eta_k$, and $\zeta_1, \ldots, \zeta_m$, where $\sum_{i=1}^{k} \eta_i = 0$ and $\sum_{j=1}^{m} \zeta_j = 0$, together with the parameters (the variances) ω^2 and σ^2.

The mean $\bar{x}_{ij}$ may be written

$$\bar{x}_{ij} = \xi_{ij} + \bar{z}_{ij} = \xi_{..} + \eta_i + \zeta_j + y_{ij} + \bar{z}_{ij} \, , \tag{16.7.51}$$

where

$$\bar{x}_{ij} - \xi_{ij} = \bar{z}_{ij} = \frac{1}{n} \sum_{\nu=1}^{n} z_{ij\nu}$$

is normally distributed with parameters $(0, \sigma^2/n)$. If the sampling distribution of $\bar{x}_{ij}$ is considered, we see that, according to the addition theorem for the normal distribution, $\bar{x}_{ij}$ is normally distributed with mean

$$\mathcal{M}\{\bar{x}_{ij}\} = \xi_{..}+\eta_i+\zeta_j \tag{16.7.52}$$

and variance

$$\mathcal{V}\{\bar{x}_{ij}\} = \mathcal{V}\{y_{ij}\}+\mathcal{V}\{\bar{z}_{ij}\} = \omega^2+\frac{\sigma^2}{n}. \tag{16.7.53}$$

It follows that the sum of squares

$$n\sum_{i=1}^{k}\sum_{j=1}^{m}(\bar{x}_{ij}-\xi_{..}-\eta_i-\zeta_j)^2 \tag{16.7.54}$$

is distributed as $(n\omega^2+\sigma^2)\chi^2$ with km degrees of freedom. If we introduce the estimates (16.7.19)–(16.7.21) of $\xi_{..}$, η_i, and ζ_j, we obtain

$$n\sum_{i=1}^{k}\sum_{j=1}^{m}(\bar{x}_{ij}-\xi_{..}-\eta_i-\zeta_j)^2 = n\sum_{i=1}^{k}\sum_{j=1}^{m}(\bar{x}_{ij}-\bar{x}_{i\cdot}-\bar{x}_{\cdot j}+\bar{x}_{..})^2$$

$$+nm\sum_{i=1}^{k}(\bar{x}_{i\cdot}-\bar{x}_{..}-\eta_i)^2+nk\sum_{j=1}^{m}(\bar{x}_{\cdot j}-\bar{x}_{..}-\zeta_j)^2+nmk(\bar{x}_{..}-\xi_{..})^2. \tag{16.7.55}$$

According to the partition theorem for the χ^2-distribution the four terms on the right are stochastically independent and distributed as $(n\omega^2+\sigma^2)\chi^2$ with $(k-1)(m-1)$, $k-1$, $m-1$, and 1 degrees of freedom, respectively. From this it follows that

$$s_2^2 = \frac{n\sum_{i=1}^{k}\sum_{j=1}^{m}(\bar{x}_{ij}-\bar{x}_{i\cdot}-\bar{x}_{\cdot j}+\bar{x}_{..})^2}{(k-1)(m-1)}, \tag{16.7.56}$$

$$s_{3(\eta)}^2 = \frac{nm\sum_{i=1}^{k}(\bar{x}_{i\cdot}-\bar{x}_{..}-\eta_i)^2}{k-1}, \tag{16.7.57}$$

$$s_{4(\zeta)}^2 = \frac{nk\sum_{j=1}^{m}(\bar{x}_{ij}-\bar{x}_{..}-\zeta_j)^2}{m-1}, \tag{16.7.58}$$

and

$$s_{5(\xi_{..})}^2 = nmk(\bar{x}_{..}-\xi_{..})^2 \tag{16.7.59}$$

have s^2-distributions with mean $n\omega^2+\sigma^2$.

The hypothesis $\eta_1 = \eta_2 = \ldots = \eta_k = 0$ may therefore be tested by computing

$$s_3^2 = \frac{nm\sum_{i=1}^{k}(\bar{x}_{i\cdot}-\bar{x}_{..})^2}{k-1} \tag{16.7.60}$$

and the variance ratio

$$v^2 = \frac{s_3^2}{s_2^2}. \tag{16.7.61}$$

If s_3^2 is significantly larger than s_2^2, the hypothesis is rejected. The power of the test may be obtained from (14.2.3). Thus, if the interaction is a stochastic variable, the variance s_3^2, which indicates the variation between rows, must be compared with s_2^2, in contrast with the comparison in the test in paragraph C, where we compared s_3^2 with s_1^2.

Correspondingly, we test the hypothesis $\zeta_1 = \zeta_2 = \ldots = \zeta_m = 0$ by computing the variance ratio $v^2 = s_4^2/s_2^2$, where s_4^2 is derived from $s_{4(\zeta)}^2$ for $\zeta_1 = \zeta_2 = \ldots = \zeta_m = 0$.

If the variation between rows and columns is not significant, the observations may be looked upon as originating from km populations with means that vary at random. These populations may in this case be characterized by the mean $\xi_{..}$, the variance ω^2 in the distribution of the population means, and the variance σ^2 in the distribution of the observations within a set.

We have the following estimates of these three parameters:

$$\bar{x}_{..} \approx \xi_{..}, \quad \mathcal{V}\{\bar{x}_{..}\} = \frac{1}{km}\left(\omega^2 + \frac{\sigma^2}{n}\right), \tag{16.7.62}$$

$$s_1^2 \approx \sigma^2, \tag{16.7.63}$$

and

$$\frac{1}{n}(s_2^2 - s_1^2) \approx \omega^2. \tag{16.7.64}$$

Confidence limits for these parameters may be obtained by analogy with the formulas (16.5.17)–(16.5.23).

If the variation between rows and columns is significant, we obtain the following estimates of the parameters:

$$\bar{x}_{..} \approx \xi_{..}, \quad \mathcal{V}\{\bar{x}_{..}\} = \frac{1}{km}\left(\omega^2 + \frac{\sigma^2}{n}\right), \tag{16.7.65}$$

$$\bar{x}_{i.} \approx \xi_{i.}, \quad \mathcal{V}\{\bar{x}_{i.}\} = \frac{1}{m}\left(\omega^2 + \frac{\sigma^2}{n}\right), \tag{16.7.66}$$

$$\bar{x}_{.j} \approx \xi_{.j}, \quad \mathcal{V}\{\bar{x}_{.j}\} = \frac{1}{k}\left(\omega^2 + \frac{\sigma^2}{n}\right), \tag{16.7.67}$$

$$s_1^2 \approx \sigma^2, \tag{16.7.68}$$

and

$$\frac{1}{n}(s_2^2 - s_1^2) \approx \omega^2. \tag{16.7.69}$$

If the η's and the ζ's can be considered as stochastic variables, the analysis may be continued as in § 16.5.

Example 16.6. During the manufacture of sheets of building material the permeability was determined for three sheets from each of three machines on each day, cf. Example 16.5. Table 16.32 gives the logarithms of the permeability in seconds for sheets selected from the three machines during a production period of nine days, i. e., $9 \times 3 \times 3 = 81$ observations. (Each of these results is the average of eight measurements, see Example 16.5). The three machines received their raw materials from a common store.

We are thus dealing with 81 observations which are grouped according to two criteria, days and machines, giving 9×3 classes with three observations in each class. Hence the data are in a form corresponding to Table 16.24.

Table 16.32 shows the computation of the total *SSD*, 4·3509, with $9 \times 3 \times 3 - 1 = 80$ degrees of freedom, together with the *SSD* for each set of three observations, and their sum, $\Sigma\Sigma SSD_{ij} = 2{\cdot}0150$. The number of degrees of freedom for the sum is $9 \times 3 \times 2 = 54$, each of the 27 sets of observations yielding two degrees of freedom. The corresponding variance within sets is $s_1^2 = 2{\cdot}0150/54 = 0{\cdot}03731$.

The difference, $4{\cdot}3509 - 2{\cdot}0150 = 2{\cdot}3359$, with $80 - 54 = 26$ degrees of freedom gives the variation between the 27 means. This is partitioned into three *SSD*'s corresponding to the variation between days, between machines, and to the interaction, respectively.

Table 16.33 includes the 27 sums S_{ij} from Table 16.32, arranged as the means in Table 16.25. It will be seen that machine No. 2 makes sheets with permeability smaller than that of both machine No. 1 and No. 3, the differences $S_{i1} - S_{i2}$ and $S_{i3} - S_{i2}$ nearly all being positive. We may obtain a preliminary rough evaluation of the size of the latter quantities compared with the random variation by computing the variance

$$\mathcal{V}\{S_{i1} - S_{i2}\} = \mathcal{V}\{S_{i1}\} + \mathcal{V}\{S_{i2}\} = 6\sigma^2 .$$

As an estimate of this quantity, we obtain

$$6s_1^2 = 0{\cdot}2239 \approx \mathcal{V}\{S_{i1} - S_{i2}\} ,$$

the corresponding standard deviation being $\sqrt{0{\cdot}2239} = 0{\cdot}473$. As several of the differences are 3–4 times this standard deviation there is no doubt that the permeability is smaller for sheets from machine 2 than for sheets from machines 1 and 3.

In order to analyze the interaction further, we compute the "expected" values, $3(\bar{x}_{i\cdot} + \bar{x}_{\cdot j} - \bar{x}_{\cdot\cdot})$, under the assumption of no interaction, see Table 16.34. Table 16.35 gives the differences between the observed and the "expected" values. Remembering that $\mathcal{V}\{3\bar{x}_{ij}\} = 3\sigma^2 \approx 0{\cdot}1119$, which gives a standard deviation of 0·33, it seems as if the deviations vary at random in accordance with the hypothesis of no interaction. This result might also

TABLE 16.32.

The logarithms of the permeabilities in seconds for 81 sheets taken during 9 days from 3 machines. Computation of the variation within sets.

Day	Machine	Log of permeability						
i	j	x_{ij1}	x_{ij2}	x_{ij3}	S_{ij}	SS_{ij}	S_{ij}^2/n	SSD_{ij}
	1	1·404	1·346	1·618	4·368	6·4009	6·3598	0·0411
1	2	1·306	1·628	1·410	4·344	6·3441	6·2901	0·0540
	3	1·932	1·674	1·399	5·005	8·4921	8·3500	0·1421
	1	1·447	1·569	1·820	4·836	7·8680	7·7956	0·0724
2	2	1·241	1·185	1·516	3·942	5·2426	5·1798	0·0628
	3	1·426	1·768	1·859	5·053	8·6152	8·5109	0·1043
	1	1·914	1·477	1·894	5·285	9·4322	9·3104	0·1218
3	2	1·506	1·575	1·649	4·730	7·4679	7·4576	0·0103
	3	1·382	1·690	1·361	4·433	6·6183	6·5505	0·0678
	1	1·887	1·485	1·392	4·764	7·7037	7·5652	0·1385
4	2	1·673	1·372	1·114	4·159	5·9223	5·7658	0·1565
	3	1·721	1·528	1·371	4·620	7·1763	7·1148	0·0615
	1	1·772	1·728	1·545	5·045	8·5130	8·4840	0·0290
5	2	1·227	1·397	1·531	4·155	5·8011	5·7547	0·0464
	3	1·320	1·489	1·336	4·145	5·7444	5·7270	0·0174
	1	1·665	1·539	1·680	4·884	7·9631	7·9512	0·0119
6	2	1·404	1·452	1·627	4·483	6·7266	6·6991	0·0275
	3	1·633	1·612	1·359	4·604	7·1121	7·0656	0·0465
	1	1·918	1·931	2·129	5·978	11·9401	11·9122	0·0279
7	2	1·229	1·508	1·436	4·173	5·8466	5·8046	0·0420
	3	1·328	1·802	1·385	4·515	6·9290	6·7951	0·1339
	1	1·845	1·790	2·042	5·677	10·7779	10·7428	0·0351
8	2	1·583	1·627	1·282	4·492	6·7965	6·7260	0·0705
	3	1·689	2·248	1·795	5·732	11·1283	10·9519	0·1764
	1	1·540	1·428	1·704	4·672	7·3144	7·2759	0·0385
9	2	1·636	1·067	1·384	4·087	5·7304	5·5679	0·1625
	3	1·703	1·370	1·839	4·912	8·1590	8·0426	0·1164
Total					127·093	203·7661	201·7511	2·0150
					$127{\cdot}093^2/81 =$	199·4152	199·4152	
Difference						4·3509	2·3359	

TABLE 16.33.
Table of $S_{ij} = 3\bar{x}_{ij}$, see Table 16.32.

Day	Machine (j)					
(i)	1	2	3	$S_{i\cdot} = 9\bar{x}_{i\cdot}$	$3\bar{x}_{i\cdot}$	$3(\bar{x}_{i\cdot} - \bar{x}_{\cdot\cdot})$
1	4·368	4·344	5·005	13·717	4·5723	−0·1348
2	4·836	3·942	5·053	13·831	4·6103	−0·0968
3	5·285	4·730	4·433	14·448	4·8160	0·1089
4	4·764	4·159	4·620	13·543	4·5143	−0·1928
5	5·045	4·155	4·145	13·345	4·4483	−0·2588
6	4·884	4·483	4·604	13·971	4·6570	−0·0501
7	5·978	4·173	4·515	14·666	4·8887	0·1816
8	5·677	4·492	5·732	15·901	5·3003	0·5932
9	4·672	4·087	4·912	13·671	4·5570	−0·1501
$S_{\cdot j} = 27\bar{x}_{\cdot j}$	45·509	38·565	43·019	127·093		
$3\bar{x}_{\cdot j}$	5·0566	4·2850	4·7799		$3\bar{x}_{\cdot\cdot} = 4{\cdot}7071$	0·0003
$3(\bar{x}_{\cdot j} - \bar{x}_{\cdot\cdot})$	0·3495	−0·4221	0·0728		0·0002	

TABLE 16.34.
Table of $3(\bar{x}_{i\cdot} + \bar{x}_{\cdot j} - \bar{x}_{\cdot\cdot})$.

Day	Machine (j)			
(i)	1	2	3	$3\bar{x}_{i\cdot}$
1	4·922	4·150	4·645	4·572
2	4·960	4·188	4·683	4·610
3	5·166	4·394	4·889	4·816
4	4·864	4·092	4·587	4·514
5	4·798	4·026	4·521	4·448
6	5·007	4·235	4·730	4·657
7	5·238	4·467	4·962	4·889
8	5·650	4·878	5·373	5·300
9	4·907	4·135	4·630	4·557
$3\bar{x}_{\cdot j}$	5·057	4·285	4·780	4·707

be expected a priori since there does not seem to be any reason to assume that the three machines should react differently to the influences (e. g., variations in the qualities of the raw materials) which cause variations between days.

Following this examination of the data, we proceed with the analysis of variance by computing the *SSD*'s for rows and columns according to (16.7.40) and (16.7.41). The correction term becomes

$$S_{..}^2/81 = 127{\cdot}093^2/81 = 199{\cdot}4152 \, .$$

Then

$$\Sigma S_{i.}^2/9 = 1799{\cdot}7177/9 = 199{\cdot}9686$$

and

$$\Sigma S_{.j}^2/27 = 5408{\cdot}9627/27 = 200{\cdot}3320 \, ,$$

which lead to the two *SSD*'s as 0·5534 and 0·9168 by subtraction of the correction term.

The resulting analysis of variance is shown in Table 16.36.

TABLE 16.35.
Deviations between observed and "expected" values.
Table of $3(\bar{x}_{ij} - \bar{x}_{i.} - \bar{x}_{.j} + \bar{x}_{..})$.

Day	Machine (j)			Total
(i)	1	2	3	
1	−0·554	0·194	0·360	0·000
2	−0·124	−0·246	0·370	0·000
3	0·119	0·336	−0·456	−0·001
4	−0·100	0·067	0·033	0·000
5	0·247	0·129	−0·376	0·000
6	−0·123	0·248	−0·126	−0·001
7	0·740	−0·294	−0·447	−0·001
8	0·027	−0·386	0·359	0·000
9	−0·235	−0·048	0·282	−0·001
Total	−0·003	0·000	−0·001	−0·004

TABLE 16.36.
Analysis of variance for the logarithm of the permeability.

Variation	*SSD*	f	s^2
Between machines	0·9168	2	0·4584
Between days	0·5534	8	0·06918
Interaction	0·8657	16	0·05411
Within machines and days	2·0150	54	0·03731
Total	4·3509	80	

The sum of squares of the deviations corresponding to the interaction is computed as the difference between the total *SSD* and the three other *SSD*'s. It may also be computed directly from the definition, see Table 16.30, using the deviations in Table 16.35, which gives 0·8662 (the difference between this figure and that of Table 16.36 is due to rounding errors in Table 16.35).

First, the hypothesis that the interaction is zero is tested. This is done by computing the ratio

$$v^2 = \frac{0{\cdot}05411}{0{\cdot}03731} = 1{\cdot}45 .$$

As the 95% fractile for v^2 with $(f_1, f_2) = (16, 54)$ is equal to 1·83, we do not reject this hypothesis.

Then the hypothesis that the theoretical variation between days is zero is tested by computing the ratio

$$v^2 = \frac{0{\cdot}06918}{0{\cdot}03731} = 1{\cdot}85 .$$

As the 95% fractile for v^2 with $(f_1, f_2) = (8,54)$ is 2·11 we do not reject this hypothesis. (Instead of the denominator 0·03731 we might employ the variance which results from combining the sums of squares corresponding to the variation within machines and days and the interaction, since we have shown above that these variances do not differ significantly. This leads to the denominator $(0{\cdot}8657+2{\cdot}0150)/70 = 0{\cdot}04115$ with 70 degrees of freedom. The variance ratio is now $0{\cdot}06918/0{\cdot}04115 = 1{\cdot}68$, and the 95% fractile is 2·07, i. e., the variation between days is not significant).

Finally the variation between machines is compared with the random variation, which leads to

$$v^2 = \frac{0{\cdot}4584}{0{\cdot}03731} = 12{\cdot}3 ,$$

and this quantity exceeds the 99·95% fractile for $(f_1, f_2) = (2,54)$. We thus see that the permeability of the sheets from the three machines is not the same, as already indicated above. The following are the means of the logarithms, cf. Table 16.33.

Machine	Mean of log	Permeability in seconds, antilog
1	1·686	48·5
2	1·428	26·8
3	1·593	39·2

As an estimate of the standard error of these means we have

$$\sqrt{\frac{s_1^2}{27}} = \sqrt{\frac{0{\cdot}03731}{27}} = 0{\cdot}0372 \approx \sigma\{\bar{x}_{\cdot j}\} .$$

The above analysis has thus led to the following conclusions:

1. The sheets from machine 2 are considerably less waterproof than the sheets from machines 1 and 3.

2. During the observed production period the variation from day to day is random.

3. The random variation is characterized by the variance $s_1^2 = 0{\cdot}03731$. A control chart for future variations of the means of the three determinations made for each machine during a day may be set up using the variance s_1^2 and the total mean for each of the three machines for computing the control limits.

Example 16.7. In a factory, where the finished product is packed in sacks of 50 kgs each, the weighing and packaging is carried out by 6 machines which are fed from the same store. At certain intervals a control sack is taken from each machine. Table 16.37 gives such a series of control weights.

TABLE 16.37.

16 control weights of 50-kg packages from 6 automatic packaging and weighing machines. Computing origin: 50 kgs. Computing unit: 0·1 kg.

Sample No.	Machine No. 1	2	3	4	5	6	$S_{i\cdot}$	$\bar{x}_{i\cdot}$	$\bar{x}_{i\cdot} - \bar{x}_{\cdot\cdot}$
1	6	5	7	2	5	4	29	4·83	3·12
2	1	4	5	2	1	4	17	2·83	1·12
3	1	6	8	4	7	8	34	5·67	3·96
4	2	4	10	2	0	5	23	3·83	2·12
5	1	5	7	1	3	3	20	3·33	1·62
6	4	3	8	3	−1	4	21	3·50	1·79
7	2	4	10	4	2	0	22	3·67	1·96
8	0	4	8	4	1	3	20	3·33	1·62
9	0	5	4	3	3	6	21	3·50	1·79
10	0	2	0	−4	−7	2	−7	−1·17	−2·88
11	1	1	3	−1	1	−2	3	0·50	−1·21
12	−2	1	0	−1	−3	1	−4	−0·67	−2·38
13	−2	1	2	0	−3	1	−1	−0·17	−1·88
14	−1	3	−1	1	−2	2	2	0·33	−1·38
15	1	3	3	3	0	1	11	1·83	0·12
16	−8	−7	−7	−8	−7	−10	−47	−7·83	−9·54
$S_{\cdot j}$	6	44	67	15	0	32	164		
$\bar{x}_{\cdot j}$	0·38	2·75	4·19	0·94	0·00	2·00		$\bar{x}_{\cdot\cdot} = 1{\cdot}71$	
$\bar{x}_{\cdot j} - \bar{x}_{\cdot\cdot}$	−1·33	1·04	2·48	−0·77	−1·71	0·29			

The variations in weights may be caused by the following three groups of factors:

1. The weighing uncertainty of the machines.

2. Variations in the materials, i. e., their temperature, humidity, fineness, specific gravity, etc., may vary.

3. Variations in the adjustment of the machines (systematic errors in weighing).

The uncertainty of weighing will appear as random variations in all the weights, while the variations in the material will influence the results in a row, and differences in adjustment of the machines will influence the figures in a column. From Table 16.37, it will be seen that, e. g., the weights of samples No. 16 are considerably below those of samples No. 1. This appears in all the machines and is due to the fact that the material has changed. It is also clear that the sacks from machine No. 3 on the average weigh 0·4 kg more than the sacks from machine No. 1, and when the individual differences between the weights in the two columns are studied, 14 of the 16 differences between the weights of the sacks from machine No. 3 and machine No. 1 are positive.

It is therefore reasonable to assume that the theoretical interaction is zero, i. e., the true weight is the sum of two components, characterizing the material at the moment the sample was taken and the adjustment of the machine. The empirical interaction may thus be looked upon as representing the uncertainty of weighing.

Table 16.38 shows the analysis of variance, cf. Table 16.31.

TABLE 16.38.
Analysis of variance for control weighings.

Variation	*SSD*	f	s^2
Between samples	952	15	63·5
Between machines	202	5	40·4
Interaction	262	75	3·49
Total	1416	95	

The computations are as follows:

$$S_{..} = 164, \quad S_{..}^2/96 = 280 ,$$

$$\Sigma\Sigma(x_{ij}-\bar{x}_{..})^2 = \Sigma\Sigma x_{ij}^2-280 = 1416 ,$$

$$m\Sigma(\bar{x}_{i.}-\bar{x}_{..})^2 = \Sigma S_{i.}^2/6-280 = 952 ,$$

and

$$k\Sigma(\bar{x}_{.j}-\bar{x}_{..})^2 = \Sigma S_{.j}^2/16-280 = 202 .$$

As an estimate of the standard deviation corresponding to the uncertainty of weighing we obtain

$$0{\cdot}1\sqrt{3{\cdot}49} = 0{\cdot}19 \text{ kg} .$$

Both the variation between samples and that between machines is significant, the ratios

$$v^2 = \frac{63{\cdot}5}{3{\cdot}49} = 18{\cdot}2 > 3{\cdot}2 = v_{.9995}^2\ (15,75)$$

and

$$v^2 = \frac{40 \cdot 4}{3 \cdot 49} = 11 \cdot 6 > 5 \cdot 0 = v^2_{.9995}\ (5,75)$$

being larger than the 99·95% fractile for v^2.

When the materials vary as much as in the above case, it is necessary to adjust the machines fairly often. Adjustment of the machines may, e. g., take place according to the following rules: A normal weight is established, which denotes the true mean weight of the sacks when conditions are ideal. As the uncertainty in weighing is characterized by the standard deviation 0·19 kg, a random variation in the weights of the sacks of up to 0·4 kg (95% limit) is to be expected; therefore, in order to take a single deviation as an indication that the machine needs adjusting, the deviation

TABLE 16.39. Analysis of variance f

Variation	*SSD*
Between A-groups	$nqm \sum_{i=1}^{k} (\bar{x}_{i..} - \bar{x}_{...})^2$
Between B-groups	$nkq \sum_{j=1}^{m} (\bar{x}_{.j.} - \bar{x}_{...})^2$
Between C-groups	$nkm \sum_{r=1}^{q} (\bar{x}_{..r} - \bar{x}_{...})^2$
Interaction *AB*	$nq \sum_{i=1}^{k} \sum_{j=1}^{m} (\bar{x}_{ij.} - \bar{x}_{i..} - \bar{x}_{.j.} + \bar{x}_{...})^2$
Interaction *AC*	$nm \sum_{i=1}^{k} \sum_{r=1}^{q} (\bar{x}_{i.r} - \bar{x}_{i..} - \bar{x}_{..r} + \bar{x}_{...})^2$
Interaction *BC*	$nk \sum_{j=1}^{m} \sum_{r=1}^{q} (\bar{x}_{.jr} - \bar{x}_{.j.} - \bar{x}_{..r} + \bar{x}_{...})^2$
Interaction *ABC*	$n \sum_{i=1}^{k} \sum_{j=1}^{m} \sum_{r=1}^{q} (\bar{x}_{ijr} - \bar{x}_{ij.} - \bar{x}_{i.r} - \bar{x}_{.jr} + \bar{x}_{i..} + \bar{x}_{.j.} + \bar{x}_{..r} - \bar{x}_{...})^2$
Within sets	$\sum_{i=1}^{k} \sum_{j=1}^{m} \sum_{r=1}^{q} \sum_{\nu=1}^{n} (x_{ijr\nu} - \bar{x}_{ijr})^2$
Total	$\sum_{i=1}^{k} \sum_{j=1}^{m} \sum_{r=1}^{q} \sum_{\nu=1}^{n} (x_{ijr\nu} - \bar{x}_{...})^2$

must exceed 0·4 kg. The standard deviation of the sum of 6 weights is $0{\cdot}1\sqrt{6\times 3{\cdot}49} = 0{\cdot}46$ kg, so that we can allow the sum of a set of 6 control weights to present random variations of as much as about 0·9 kg (95% limit) from the figure which gives 6 times the normal weight of a sack. If in a set of control weights we find that all the deviations from the normal weight are less than 0·4 kg, and the sum of the deviations is less than 0·9, there is no reason to readjust the machines, but if this is not the case, readjustment must take place. If each of the weights deviates less than 0·4 kg from the normal weight, while the sum of the deviations exceeds 0·9 kg, it is necessary to make further series of controls in order to find out which machine is to be adjusted; this we find by calculating the average or the sum of the deviations for each machine.

ssifications according to three criteria.

f	s^2	$m\{s^2\}$
$k-1$	s_8^2	$\sigma^2+(n\sigma^2_{ABC}+nq\sigma^2_{AB}+nm\sigma^2_{AC})+nqm\sigma^2_A$
$m-1$	s_7^2	$\sigma^2+(n\sigma^2_{ABC}+nq\sigma^2_{AB}+nk\sigma^2_{BC})+nkq\sigma^2_B$
$q-1$	s_6^2	$\sigma^2+(n\sigma^2_{ABC}+nm\sigma^2_{AC}+nk\sigma^2_{BC})+nkm\sigma^2_C$
$(k-1)(m-1)$	s_5^2	$\sigma^2+(n\sigma^2_{ABC})+nq\sigma^2_{AB}$
$(k-1)(q-1)$	s_4^2	$\sigma^2+(n\sigma^2_{ABC})+nm\sigma^2_{AC}$
$(m-1)(q-1)$	s_3^2	$\sigma^2+(n\sigma^2_{ABC})+nk\sigma^2_{BC}$
$(k-1)(m-1)(q-1)$	s_2^2	$\sigma^2+n\sigma^2_{ABC}$
$kmq(n-1)$	s_1^2	σ^2
$kmqn-1$		

16.8. Three-Way Classification.

When we are dealing with observations that have been classified according to three criteria, the analysis of variance is analogous to the theory developed in § 16.7 for the two-way classification. The observations x_{ijrv} are partitioned as follows:

$$x_{ijrv} = \xi_{ijr} + z_{ijrv} = \xi_{\cdots} + \eta_{i\cdot\cdot} + \eta_{\cdot j\cdot} + \eta_{\cdot\cdot r} + \zeta_{ij\cdot} + \zeta_{i\cdot r} + \zeta_{\cdot jr} + \vartheta_{ijr} + z_{ijrv}, \quad (16.8.1)$$

where z_{ijrv} is normally distributed with parameters $(0, \sigma^2)$, and $\sum_i \eta_{i\cdot\cdot} = 0$, $\sum_j \eta_{\cdot j\cdot} = 0$, $\sum_r \eta_{\cdot\cdot r} = 0$, $\sum_i \zeta_{ij\cdot} = 0$, $\sum_j \zeta_{ij\cdot} = 0$, $\sum_i \zeta_{i\cdot r} = 0$, $\sum_r \zeta_{i\cdot r} = 0$, $\sum_j \zeta_{\cdot jr} = 0$, $\sum_r \zeta_{\cdot jr} = 0$, $\sum_i \vartheta_{ijr} = 0$, $\sum_j \vartheta_{ijr} = 0$, and $\sum_r \vartheta_{ijr} = 0$, in analogy with (16.7.8) and (16.7.13)–(16.7.16).

Table 16.39 gives the corresponding analysis of variance.

The table of $\mathcal{M}\{s^2\}$ should be interpreted as follows:

1. *All components of x_{ijrv} are random.*

The variance σ^2_{ABC} represents the variance of the random component corresponding to ϑ_{ijr} in (16.8.1), σ^2_{BC} represents the variance of the random component corresponding to $\zeta_{\cdot jr}$, and so forth. (σ^2_{ABC} is analogous to the variance denoted by ω^2 in § 16.7, part D).

2. *All components of x_{ijrv}, apart from z_{ijrv}, are systematic.*

The variance σ^2_{ABC} is defined as

$$\sigma^2_{ABC} = \frac{1}{f_2} \Sigma\Sigma\Sigma\vartheta^2_{ijr} ,$$

see the analogous expression in (16.7.34). Similarly,

$$\sigma^2_{BC} = \frac{1}{f_3} \Sigma\Sigma\zeta^2_{\cdot jr} ,$$

etc.

In this case *all terms in parentheses should be disregarded*, see the analogous Table 16.30.

3. *Some components of x_{ijrv} are random and some are systematic.*

The interpretation of the σ^2's for each kind of component is as under 1 and 2. If, for instance, ϑ_{ijr} is a random component, σ^2_{ABC} will be included in *all* the variances, i. e., terms in parentheses corresponding to random components should be retained while terms corresponding to systematic components should be discarded.

For a given specification the variance ratios appropriate for testing the different components may be found by comparisons of the $\mathcal{M}\{s^2\}$'s in Table 16.39.

Example 16.8. When the compressive strengths of test cubes of cement are to be determined, the moulds in which the samples are made and the press in which the samples are crushed influence the strengths obtained. In order to obtain an idea of what part the moulds and the press play, the strength of test cubes made from the same batch of cement was determined with old and new moulds and press-sheets at a time when they were to be replaced by new ones, the four possible combinations of old and new moulds and press-sheets being applied. Corresponding to each combination 4 sets of 6 samples were made and cured for 3, 7, and 28 days, and 28 days under special conditions before testing their strength. At all stages of the experiment the test cubes were processed in random order to ensure that possible systematic variations due to lack of experimental control might have the same chance of affecting all treatments. An analysis of the 16 sets of observations shows that the logarithms of the strengths may be considered normally distributed, the variance being the same for all sets, see also Example 7.7, p. 185. Table 16.40 gives the sums of the logarithms of the strengths of each set of 6 observations, and Table 16.41 the corresponding sums of squares of the deviations.

BARTLETT's test is used to test the hypothesis that the theoretical variances of the 16 sets are the same, cf. § 11.6. We find $\chi^2 = 12{\cdot}7$ with 15 degrees of freedom, which means that we cannot reject the hypothesis. Table 16.41 gives the computation of the pooled estimate $s_1^2 = 0{\cdot}0001731$ of the population variance σ^2. The estimate of the variance of the sum of the 6 observations thus becomes $6 \times 0{\cdot}0001731 = 0{\cdot}001039$, which implies a standard deviation of $0{\cdot}032$.

Table 16.40 shows that in every case a new press-sheet leads to a greater

TABLE 16.40.

The sums of the logarithms of the compressive strengths in kg/cm² for 96 test cubes of cement, grouped in 16 sets of 6 strengths each.

Day	Press-sheet	Mould old	Mould new
3	old	15·959	16·028
	new	16·151	16·221
7	old	16·443	16·451
	new	16·636	16·607
28_1	old	16·780	16·788
	new	16·859	16·948
28_2	old	17·086	17·090
	new	17·309	17·318

strength than an old one. As the differences between the sums for new and for old press-sheets are of the order of magnitude 0·200, these differences are undoubtedly significant, the standard error of such a difference being $0{\cdot}032\sqrt{2} = 0{\cdot}045$.

TABLE 16.41.

Sums of squares of deviations (SSD) of logarithms of compressive strength in kg/cm^2 for 96 test cubes of cement, grouped in 16 sets of 6 strengths each. The table gives $10^4 \times SSD$.

Day	Press-sheet	Mould old	Mould new
3	old	7·99	8·89
	new	14·43	10·05
7	old	5·30	15·53
	new	4·11	3·75
28_1	old	2·45	14·36
	new	9·99	4·91
28_2	old	3·85	19·09
	new	4·11	9·67

Total: $\Sigma 10^4 \times SSD = 138{\cdot}48$

Degrees of freedom: $16 \times 5 = 80$

$s_1^2 = 10^{-4} \times 138{\cdot}48/80 = 0{\cdot}0001731$

In seven out of the eight cases, the sums corresponding to new moulds are larger than those corresponding to old ones, but as the differences are comparatively small further analysis is necessary in order to determine whether the differences are significant.

Tables 16.42 and 16.43 give the computations of the SSD's corresponding to classification according to one and two criteria, respectively. All the SSD's may be computed from formulas similarly to

$$SSD = n \sum_{i=1}^{k} (\bar{x}_i - \bar{x})^2 = \sum_{i=1}^{k} \frac{S_i^2}{n} - \frac{S^2}{kn},$$

where S_i represents the sum of n observations and S denotes the sum of all $kn = 96$ observations. Using this formula directly the correction term S^2/kn will be the same for all SSD's. It may, however, as in the present case be advantageous to use different origins for computing the SSD's, in

TABLE 16.42.

Sums of logarithms of strength in kg/cm², classified according to one criterion, cf. Table 16.40.

Day	Sum	Press-sheet	Sum	Mould	Sum
3	64·359	old	132·625	old	133·223
7	66·137	new	134·049	new	133·451
28_1	67·375				
28_2	68·803				
Total	266·674		266·674		266·674
Computing					
origin	64·000		132·000		133·000
k	4		2		2
S	10·674		2·674		0·674
SS	39·155084		4·589026		0·253130
S^2/k	28·483569		3·575138		0·227138
Difference	10·671515		1·013888		0·025992
SSD	0·444646		0·021123		0·000542

TABLE 16.43.

Sums of logarithms of strength in kg/cm², classified according to two criteria, cf. Table 16.40.

Day	Press-sheet old	new	Total	Day	Mould old	new	Total	Press-sheet	Mould old	new	Total
3	31.987	32·372	64·359	3	32·110	32·249	64·359	old	66·268	66·357	132·625
7	32·894	33·243	66·137	7	33·079	33·058	66·137	new	66·955	67·094	134·049
28_1	33·568	33·807	67·375	28_1	33·639	33·736	67·375				
28_2	34·176	34·627	68·803	28_2	34·395	34·408	68·803				
Total	132·625	134·049	266·674	Total	133·223	133·451	266·674	Total	133·223	133·451	266·674
Computing											
origin	31·000				32·000				66·000		
k	8				8				4		
S	18·674				10·674				2·674		
SS	49·190816				19·592212				2·308134		
S^2/k	43·589785				14·241785				1·787569		
Difference	5·601031				5·350427				0·520565		
SSD	0·466753				0·445869				0·021690		
	−0·444646				−0·444646				−0·021123		
	−0·021123				−0·000542				−0·000542		
Interaction	0·000984				0·000681				0·000025		

which case the correction term takes on different values. The computations are then carried out according to the formula

$$SSD = \frac{1}{n}\left(\sum_{i=1}^{k} S_i^2 - \frac{S^2}{k}\right) = \frac{k}{96}\left(\sum_{i=1}^{k} S_i^2 - \frac{S^2}{k}\right),$$

where a convenient origin has been used for the computation of $SS = \Sigma S_i^2$ and $S = \Sigma S_i$.

The interactions of the first order are obtained by subtracting the *SSD*'s for the two one-way classifications (Table 16.42) from the *SSD* for the two-way classification, see Table 16.43. Similarly, the interaction of the second order is obtained by subtracting the *SSD*'s for the three one-way classifications and the interactions for the three two-way classifications from the *SSD* for the three-way classification, see Table 16.44.

TABLE 16.44.

Computation of the interaction of the second order.
The S_i's are to be found in Table 16.40. Computing origin: 16·00.

k	16
S	10·674
SS	9·930732
S^2/k	7·120892
Difference	2·809840
SSD	0·468307
Days	−0·444646
Press-sheets	−0·021123
Moulds	−0·000542
Days × Press-sheets	−0·000984
Days × Moulds	−0·000681
Press-sheets × Moulds	−0·000025
Interaction (Difference)	0·000306

The results of the computations have been summarized in Table 16.45.

It is obvious that the three factors must be considered as producing systematic effects. Hence, to test the effects of the factors the corresponding variances must be compared with the variance within sets, as shown in the last column of Table 16.45. First we test the significance of the interactions. According to Table VII, $v^2_{.95}(1,80) = 3{\cdot}96$ and $v^2_{.95}(3,80) = 2{\cdot}72$, so that none of the interactions are significant. This means that *changing from old to new press-sheets produces the same effect in the logarithms of the compressive strength irrespective of the days or moulds considered, and further that corresponding results hold for moulds and days.*

TABLE 16.45.
Analysis of variance for the logarithms of the compressive strength in kg/cm² classified according to days (time of curing), press-sheets, and moulds.

Variation	SSD	f	s^2	v^2
Between days	0·444646	3	0·148215	857
Between press-sheets	0·021123	1	0·021123	122
Between moulds	0·000542	1	0·000542	3·13
Interactions:				
Days × Press-sheets	0·000984	3	0·000328	1·90
Days × Moulds	0·000681	3	0·000227	1·31
Press-sheets × Moulds	0·000025	1	0·000025	0·14
Days × Press-sheets × Moulds	0·000306	3	0·000102	0·59
Within sets	0·013848	80	0·000173	
Total	0·482155	95		

The 4 sums of squares which correspond to the interactions may be combined with the sum of squares within sets, but as the latter already is based on 80 degrees of freedom, the resultant gain in reliability is hardly worth bothering about.

The variation between moulds is not significant, as the variance ratio, 3·13, is smaller than $v^2_{.95}$. On the other hand, the variation between press-sheets is definitely significant, and so is the variation between days. This means that the strengths of test cubes made in old and in new moulds do not differ significantly, while the new press-sheets give higher strengths than the old sheets. Finally, the significant variation between days confirms the well-known fact that the strength increases with the length of the curing time.

Thus, this investigation has led to the following results: (1) The uncertainty of the determination of the compressive strength of a test cube of cement (the errors of measurement) is characterized by the standard deviation $\sqrt{0{\cdot}0^31731} = 0{\cdot}0132$ of the logarithm of the compressive strength. This corresponds to a coefficient of variation of 3·0%. (2) When new press-sheets are used for determining the strength, the results are higher than with old sheets. According to Table 16.42 we obtain an estimate of the difference as $1/48(134{\cdot}049 - 132{\cdot}625) = 0{\cdot}0297$, the variance being $0{\cdot}0001731(1/48+1/48) = 0{\cdot}000007213$, and the standard error 0·0027. Thus, the strength found with the new press-sheets is on the average 7·1% larger than with the old sheets (Antilog $0{\cdot}0297 = 1{\cdot}071$). The t-distribution furnishes us with the limits for the true value of the increase. In the t-distribution with 80 degrees of freedom the bilateral 95% limits are 1·99, which figure multiplied by the standard error 0·0027 gives 0·0054, the confidence limits of the increase thus being $0{\cdot}0297 \pm 0{\cdot}0054 = (0{\cdot}0243, 0{\cdot}0351)$. In percentages the limits of the true increase in compressive strength are found to be 5·7% and 8·4%.

16.9. Miscellaneous Remarks.

Use of the v^2-test for several variance ratios.

The analysis of variance will usually lead to a comparison of several variances with a single residual variance. Using the corresponding variance ratios it must be remembered that several tests are carried out simultaneously and further that these tests are not independent, because the same variance is employed as denominator several times. If we had as many independent residual variances as numerators, the tests would be independent and their evidence might be combined by the methods given in § 15.6. The larger the number of variances to be compared, the greater the range of variation to be tolerated.

D. J. Finney has shown that approximations to the significance levels for *the largest variance ratio* among a number of variance ratios with common denominator may be obtained by assuming the variance ratios to be independent, when the degrees of freedom for the denominator exceed 10, see *The Joint Distribution of Variance Ratios Based on a Common Error Mean Square*, Ann. of Eugenics, 11, 1941, 136–140. If all the numerators have the same number of degrees of freedom we may use (12.4.5), which leads to

$$v^2_{(k)P} = v^2_{P_1} \qquad \text{for} \qquad P_1 = P^{1/k} .$$

A test for additivity.

If several observations are given for each class in a two-way classification we may test the hypothesis of no interaction or additivity by comparison of the interaction variance with the variance within classes. J. W. Tukey has given a test for additivity in the case of a single observation per class, see *One Degree of Freedom for Non-additivity*, Biometrics, 5, 1949, 232–242.

Ordering of means.

When a significant variance ratio has been found the problem of ordering the corresponding means into distinguishable groups arises. Some tests for solving this problem have been proposed by J. W. Tukey: *Comparing Individual Means in the Analysis of Variance*, Biometrics, 5, 1949, 99–114.

16.10. Notes and References.

The analysis of variance was developed by R. A. Fisher in connection with the planning and the statistical analysis of field experiments about

1920. The two books of R. A. FISHER: *Statistical Methods* and *Design of Experiments*, Oliver and Boyd, Edinburgh, contain numerous examples, mainly from the fields of biology and agriculture. (The application of the analysis of variance in the design of experiments will be discussed in Chapter 17).

A mathematical exposition of the analysis of variance has been given by W. G. COCHRAN: *The Distribution of Quadratic Forms in a Normal System with Applications to the Analysis of Covariance*, Proc. Camb. Phil. Soc., 30, 1934, 178–91, and J. O. IRWIN: *Mathematical Theorems Involved in the Analysis of Variance*, Journ. Roy. Stat. Soc., 94, 1931, 284–300, and *On the Independence of the Constituent Items in the Analysis of Variance*, Journ. Roy. Stat. Soc., Suppl. 1, 1934, 236–52. These papers form the basis of the present chapter.

A number of examples of the application of the analysis of variance to technical problems may be found in *Suppl. Journ. Roy. Stat. Soc.*, see, e. g., B. L. WELCH: *The Specification of Rules for Rejecting too Variable a Product, with Particular Reference to an Electric Lamp Problem*, 3, 1936, 29–48, C. E. GOULD and W. M. HAMPTON: *Statistical Methods Applied to the Manufacture of Spectacle Glasses*, 3, 1936, 137–177, A. W. BAYES: *Some Considerations on the Variability of Cotton Cloth Strength*, 4, 1937, 61–93, H. E. DANIELS: *Some Problems of Statistical Interest in Wool Research*, 5, 1938, 89–128, T. N. HOBLYN: *A Study of the Variation in Keeping Quality of Apples in Store*, 5, 1938, 129–170, and H. E. DANIELS: *The Estimation of Components of Variance*, 6, 1939, 186–197.

A very large number of examples with a careful discussion of computations and interpretation, but without proofs, are to be found in G. W. SNEDECOR: *Statistical Methods*, The Iowa State College Press, Fourth Edition, 1946.

CHAPTER 17.

DESIGNS OF SAMPLING INVESTIGATIONS AND EXPERIMENTS

17.1. Introduction.

The present chapter contains some basic principles for the planning of sampling investigations and experiments from a statistical point of view. The main part of the exposition is limited to designs involving only one observed variable so that the analysis of variance technique applies. Statistical techniques for analyzing the interdependence between two or more variables will be discussed in Chapters 18–20.

In general, the observations or suitably transformed values are assumed to be independent and normally distributed, so that the previously developed theory applies. If this assumption is not fulfilled, as in the case of sampling from finite populations, it is assumed that the number of observations is so large that the sample mean may be considered as practically normally distributed. Further we assume that we are interested primarily in the population mean, and that for convenience we want to use the sample mean as estimate of the population mean, even if the sample mean perhaps is not the most efficient estimate.

17.2. Random Sampling from a Finite Population.

Suppose that a population consists of N elements with the associated numbers $X_1, \ldots, X_N$, the mean being $\xi = \Sigma X_i/N$. Drawing a random sample of n elements without replacements means that every set of n elements among the $\binom{N}{n}$ different sets has the same probability of being drawn. The n elements may be drawn simultaneously or successively. In the case of successive drawings, the $N-i$ remaining elements of the population after drawing number i should all have the same probability, $1/(N-i)$, of being chosen in the next drawing. Denoting the n sample values by $x_1, \ldots, x_n$ we have

$$p\{x_1, x_2, \ldots, x_n\} = 1/N(N-1)\ldots(N-n+1)\ , \qquad (17.2.1)$$

where the x's take on the values $X_1, \ldots, X_N$, each X occurring only once in a sample. Since this probability depends on N and n, but is independent of the *values* of the elements, the $n!$ permutations of the sample values all have the same probability. Hence the probability of obtaining a given set of values, irrespective of their order, will be $n!$ times the above probability or $1/\binom{N}{n}$.

It follows that the *marginal* distributions of $x_1, x_2, \ldots$, or x_n, respectively, are identical, they all being equal to

$$p\{x = X_i\} = 1/N, \quad i = 1, \ldots, N. \tag{17.2.2}$$

Whence

$$\mathcal{M}\{x\} = \xi = \sum_{i=1}^{N} X_i/N \tag{17.2.3}$$

and

$$\mathcal{V}\{x\} = \sigma^2 = \sum_{i=1}^{N} (X_i - \xi)^2/N, \tag{17.2.4}$$

cf. (8.5.14)–(8.5.16).

Further every pair of sample values has the same two-dimensional (marginal) distribution, viz.

$$p\{(x_1, x_2) = (X_i, X_j)\} = \begin{cases} 0 & \text{for} \quad i = j \\ 1/N(N-1) & \text{for} \quad i \neq j \end{cases} \tag{17.2.5}$$

from which we may find the covariance as

$$\begin{aligned} \mathcal{V}\{x_1, x_2\} &= \mathcal{M}\{(x_1 - \xi)(x_2 - \xi)\} \\ &= \sum_{i=1}^{N} \sum_{\substack{j=1 \\ j \neq i}}^{N} (X_i - \xi)(X_j - \xi)/N(N-1). \end{aligned} \tag{17.2.6}$$

Since

$$\sum_{j=1}^{N} (X_j - \xi) = 0$$

we have

$$\sum_{\substack{j=1 \\ j \neq i}}^{N} (X_j - \xi) = -(X_i - \xi)$$

leading to

$$\mathcal{V}\{x_1, x_2\} = -\sum_{i=1}^{N} (X_i - \xi)^2/N(N-1) = -\sigma^2/N-1 \tag{17.2.7}$$

or $\varrho\{x_1, x_2\} = -1/(N-1)$. Thus every pair of sample values is (slightly) negatively correlated.

We may now determine the mean and variance of the sample mean, as

$$\mathcal{M}\{\bar{x}\} = \frac{1}{n} \sum_{i=1}^{n} \mathcal{M}\{x_i\} = \xi \tag{17.2.8}$$

and

$$\mathcal{V}\{\bar{x}\} = \frac{1}{n^2}\left(\sum_{i=1}^{n} \mathcal{V}\{x_i\} + 2\sum_{i=1}^{n-1}\sum_{j=i+1}^{n} \mathcal{V}\{x_i, x_j\}\right)$$

according to (5.17.3) and (5.17.4). Inserting $\mathcal{V}\{x_i\} = \sigma^2$ and $\mathcal{V}\{x_i, x_j\} = -\sigma^2/(N-1)$ we obtain

$$\boxed{\mathcal{V}\{\bar{x}\} = \frac{\sigma^2}{n}\left(1 - \frac{n-1}{N-1}\right) \simeq \frac{\sigma^2}{n}(1-a),} \tag{17.2.9}$$

where $a = n/N$ denotes *the sampling fraction*. The factor

$$1 - \frac{n-1}{N-1} \simeq 1 - a$$

is often called *the finite population correction* since, apart from this factor, the variance of the mean is given by the usual formula σ^2/n for sampling from an infinite population. For $a < 0{\cdot}1$ the correction will be negligible for many practical purposes. *In the following we will assume that the finite population correction with sufficient accuracy may be replaced by* $1-a$.

Hence *by random sampling from a finite population the sample mean will be an unbiased estimate of the population mean with variance* $\frac{\sigma^2}{n}(1-a)$.

Further it may be proved that

$$\mathcal{M}\{s^2\} = \frac{N}{N-1}\sigma^2, \tag{17.2.10}$$

wherefore $\frac{s^2}{n}(1-a)$ will be an unbiased estimate of $\mathcal{V}\{\bar{x}\}$.

In the special case where the population consists of only two kinds of elements, $X_1 = \ldots = X_M = 1$ and $X_{M+1} = \ldots = X_N = 0$ say, we find from the above formulas

$$\xi = M/N = \theta, \tag{17.2.11}$$

i. e., the frequency of elements of the one kind in the population, and

$$\sigma^2 = \frac{1}{N}\left(M(1-\theta)^2 + (N-M)\theta^2\right) = \theta(1-\theta). \tag{17.2.12}$$

Similarly, the sample mean becomes *the frequency of elements of the one kind in the sample*, h say, and from (17.2.8)–(17.2.9) we find $\mathcal{M}\{h\} = \theta$ and

$$\mathcal{V}\{h\} = \frac{\theta(1-\theta)}{n}\left(1 - \frac{n-1}{N-1}\right) \simeq \frac{\theta(1-\theta)}{n}(1-a), \tag{17.2.13}$$

cf. (8.5.22) and (8.5.23).

To obtain a random sample from a finite population a table of *random sampling numbers* may be used. Suppose that a random sample of size $n = 18$ is to be drawn from a population of size $N = 1783$. From the first four columns of Table XIX we obtain a series of four-figure random numbers, rejecting all numbers larger than 1783. If the same number occurs more than once it should be rejected at the second and following occurrences. In this manner we obtain the following numbers: 1577, 1326, 1055, 506, 1415, 1324, 452, etc., rejecting approximately 82% of the numbers in the table. A more expeditious procedure would be to use each set of numbers 2000–3999, 4000–5999, 6000–7999, and 8000–9999 in the same way as 0000–1999, subtracting the first number in each set from all numbers of the set. The same result will be obtained by dividing all four-figure numbers by 2000 using only the remainders between 1 and 1783. Applying this method the first four columns of Table XIX give: 1577, 540, 769, 1326, 1055, 506, 550, etc., rejecting only about 10% of the numbers in the table.

The sample size necessary to obtain a predetermined standard error of the mean depends on σ, which is generally unknown. We may often, however, obtain an idea of the magnitude of σ from previous investigations. If we know the range of variation for x, $\alpha < x < \beta$, we may also often obtain an upper bound for σ^2 from the rectangular distribution over the interval (α, β) which has the variance

$$\mathcal{V}\{x\} = \frac{(\beta - \alpha)^2}{12}. \tag{17.2.14}$$

For unimodal distributions this upper bound will usually be much on the safe side.

For given σ^2 and the $\mathcal{V}\{\bar{x}\}$ required we find from (17.2.9)

$$\frac{1}{n} = \frac{\mathcal{V}\{\bar{x}\}}{\sigma^2} + \frac{1}{N}\left(1 - \frac{\mathcal{V}\{\bar{x}\}}{\sigma^2}\right) \tag{17.2.15}$$

for the determination of n.

17.3. Stratified Random Sampling.

Suppose that a given population consists of or may be divided into k *subpopulations or strata* in the proportions $w_1 : w_2 : \ldots : w_k$, $\Sigma w_i = 1$. Denoting the distribution function for the ith subpopulation by $p_i\{x\}$ and the corresponding mean and variance by ξ_i and σ_i^2, the heterogeneous distribution function will be

$$p\{x\} = \sum_{i=1}^{k} w_i p_i\{x\} \tag{17.3.1}$$

with mean

$$\mathcal{m}\{x\} = \xi = \sum_{i=1}^{k} w_i \xi_i \tag{17.3.2}$$

and variance

$$\mathcal{v}\{x\} = \sigma^2 = \sum_{i=1}^{k} w_i \sigma_i^2 + \sum_{i=1}^{k} w_i (\xi_i - \xi)^2 \,. \tag{17.3.3}$$

Hence the total mean equals the weighted average of the strata means, and the total variance equals the weighted average of the within-strata variances plus a term depending on the differences among the strata means (the variation between strata means).

For a finite population consisting of N elements the stratification has been illustrated in Table 17.1. The weights are here $w_i = N_i/N$, and as usual

$$\xi_i = \sum_{\nu=1}^{N_i} X_{i\nu}/N_i \tag{17.3.4}$$

and

$$\sigma_i^2 = \sum_{\nu=1}^{N_i} (X_{i\nu} - \xi_i)^2/N_i \,. \tag{17.3.5}$$

TABLE 17.1.
Stratified random sampling.

Stratum number	Population values	Population			Sample		
		size	mean	variance	size	mean	variance
1	$X_{11}, \ldots, X_{1N_1}$	N_1	ξ_1	σ_1^2	n_1	$\bar{x}_1$	s_1^2
2	$X_{21}, \ldots, X_{2N_2}$	N_2	ξ_2	σ_2^2	n_2	$\bar{x}_2$	s_2^2
⋮	⋮ ⋮	⋮	⋮	⋮	⋮	⋮	⋮
k	$X_{k1}, \ldots, X_{kN_k}$	N_k	ξ_k	σ_k^2	n_k	$\bar{x}_k$	s_k^2

To estimate ξ we may draw a random sample of n elements from the heterogeneous populations, obtaining a sample mean with variance $\sigma^2(1-\alpha)/n$, where $\alpha = 0$ if the population is infinite.

However, *if the weights* $w_1, \ldots, w_k$ *are known, and if the sampling may be carried out within each stratum*, we may eliminate the variation between strata. Drawing a random sample of size n_i from stratum number i for $i = 1, \ldots, k$ we obtain from the given weights and the k sample means the weighted mean

$$\bar{x} = \sum_{i=1}^{k} w_i \bar{x}_i \tag{17.3.6}$$

as an unbiased estimate of ξ since $\mathcal{M}\{\bar{x}_i\} = \xi_i$. The variance of this estimate becomes

$$\mathcal{V}\{\bar{x}\} = \sum_{i=1}^{k} w_i^2 \mathcal{V}\{\bar{x}_i\} = \sum_{i=1}^{k} w_i^2 \frac{\sigma_i^2}{n_i}(1-a_i) \tag{17.3.7}$$

since the k sample means are independent. The sampling fraction for stratum number i has been denoted by $a_i = n_i/N_i$.

This method of sampling where *a random sample is drawn from each stratum* instead of drawing a single random sample from the total population has been called *stratified random sampling*. The variance of the weighted sample mean depends on *the variances within strata* whereas the variation between strata has been eliminated. If the strata may be formed in various ways the formation giving the most homogeneous strata (the smallest variances within strata) should be preferred.

Consider now for *a given total sample size*, $n = \Sigma n_i$, two special types of stratified sampling, viz. proportional sampling and optimum allocation of the sample.

In *proportional sampling* the sample size for stratum number i is proportional to the weight of the stratum, i. e., $n_i = w_i n$. For a finite population this means that *the sampling fractions are the same for all strata* or $n_i = aN_i$, where $a = n/N$ denotes the sampling fraction for the total population. Inserting $n_i = w_i n$ into (17.3.6) and (17.3.7) we obtain

$$\bar{x} = \Sigma n_i \bar{x}_i / \Sigma n_i \tag{17.3.8}$$

and

$$\mathcal{V}\{\bar{x}\} = \frac{\Sigma w_i \sigma_i^2}{n}(1-a) \, . \tag{17.3.9}$$

To determine *the optimum allocation of the sample*, i. e., the values of $n_1, \ldots, n_k$, which give the minimum value of $\mathcal{V}\{\bar{x}\}$ for given $n = \Sigma n_i$, we have to find the minimum of

$$f(n_1, \ldots, n_{k-1}) = \mathcal{V}\{\bar{x}\} = \sum_{i=1}^{k} w_i^2 \sigma_i^2 \left(\frac{1}{n_i} - \frac{1}{N_i}\right),$$

where

$$n_k = n - n_1 - \ldots - n_{k-1} \, .$$

Differentiating with respect to n_i and putting the derivative equal to zero we obtain

$$\frac{\partial f}{\partial n_i} = -\frac{w_i^2 \sigma_i^2}{n_i^2} + \frac{w_k^2 \sigma_k^2}{n_k^2} = 0, \quad i = 1, 2, \ldots, k-1 \, ,$$

which has the solution

$$n_i = \frac{w_i \sigma_i}{\Sigma w_i \sigma_i} n, \quad i = 1, 2, \ldots, k \, . \tag{17.3.10}$$

Inserting this result in (17.3.7) we find the minimum value of the variance as

$$\mathcal{V}\{\bar{x}\} = \frac{(\Sigma w_i \sigma_i)^2}{n} - \frac{\Sigma w_i \sigma_i^2}{N} = \frac{(\Sigma w_i \sigma_i)^2}{n}\left(1 - a\frac{\Sigma w_i \sigma_i^2}{(\Sigma w_i \sigma_i)^2}\right). \tag{17.3.11}$$

Hence *optimum allocation* of the sample requires that the *number of observations from every stratum should be proportional to both the weight and the standard deviation of the stratum.* This result is due to J. NEYMAN[1]).

Comparing the variances for unrestricted random sampling, $\mathcal{V}_1$, proportional sampling, $\mathcal{V}_2$, and stratified sampling with optimum allocation, $\mathcal{V}_3$, respectively, we obtain

$$\mathcal{V}_1 = \mathcal{V}_2 + \frac{\Sigma w_i (\xi_i - \xi)^2}{n}(1-a) \tag{17.3.12}$$

and

$$\mathcal{V}_2 = \mathcal{V}_3 + \frac{\Sigma w_i (\sigma_i - \bar{\sigma})^2}{n}, \tag{17.3.13}$$

since

$$\Sigma w_i (\sigma_i - \bar{\sigma})^2 = \Sigma w_i \sigma_i^2 - (\Sigma w_i \sigma_i)^2, \tag{17.3.14}$$

where $\bar{\sigma} = \Sigma w_i \sigma_i$.

This result shows that the gain in precision due to proportional sampling as compared with unrestricted random sampling depends on the variation between the strata means. The further gain in precision due to optimum allocation as compared with proportional sampling depends on the variation between the strata standard deviations. If all strata have the same variance optimum allocation will be identical with proportional sampling.

The above result regarding optimum allocation may easily be modified taking costs into account. If, apart from the overhead cost c_0, the cost is c_i per observation from stratum number i, the total cost becomes

$$c = c_0 + \Sigma c_i n_i\,. \tag{17.3.15}$$

Minimizing $\mathcal{V}\{\bar{x}\}$ for given total cost leads to

$$\boxed{n_i = \frac{w_i \sigma_i / \sqrt{c_i}}{\Sigma w_i \sigma_i / \sqrt{c_i}}\, n\,.} \tag{17.3.16}$$

To use this formula approximate values of σ_i and c_i must be known or guessed at before the sampling is carried out.

The general conclusion of the above discussion is that in sampling investigations for determining an estimate of the mean of a given population it is usually advisable to *divide the given population into strata, which are as*

[1]) J. NEYMAN: *On the Two Different Aspects of the Representative Method: the Method of Stratified Sampling and the Method of Purposive Selection*, Journ. Roy. Stat. Soc., 97, 1934, 558–625.

homogeneous as possible, and then allocate the sample to the strata, drawing *a random sample from each stratum*. The variance of the weighted sample mean depends on the variance within strata, whereas the variation between strata means has been eliminated. Hence the variance will often be considerably smaller than the variance of the mean of an unrestricted random sample from the total population. Optimum allocation of the sample requires that the number of observations taken at random from a given stratum should be proportional to the weight and the standard deviation of the stratum and inversely proportional to the square root of the cost per observation for the stratum.

Suppose, for example, that the total output of a plant consists of the output of k machines, some being of an older model than others and therefore giving a smaller output of a poorer quality. However, the output will be mixed before leaving the plant. To estimate the mean quality of the total output, the sampling may be done from the output of each machine. The weight w_i for a given machine represents the ratio between the output of that machine and the total output. Since the standard deviations in such a case usually will not differ very much, proportional sampling may be applied with good results.

E. S. PEARSON[1]) has illustrated the difference between unrestricted random sampling and proportional sampling for estimating the mean quality of bricks produced in a kiln. Owing to the conditions of the firing the quality of the bricks depends on their position in the kiln. After dividing the kiln into nine regions, giving widely differing mean qualities of the bricks, a random sample of bricks was drawn from each region. Since the variance within regions was only $^1/_3$ of the total variance, unrestricted random sampling required a sample size three times the sample size of proportional sampling to obtain the same precision of the estimated mean quality.

The general experience that elements close together in time or space will often be more similar than elements far apart may often be successfully used in the formation of strata.

In sampling investigations of industrial products stratified sampling is often useful[2]). For example, when a lot is being loaded a random sample of items may be taken from every truckload, the truckloads being the strata.

A very popular method of sampling is *the systematic sampling with a random start*. Suppose that the N elements of a given population form an ordered sequence. The sequence may then be divided into n strata, each containing N/n successive elements, and one element may be drawn at

[1]) E. S. PEARSON: *Sampling Problems in Industry*, Suppl. Journ. Roy. Stat. Soc., 1, 1934, 107–36.

[2]) W. E. DEMING: *On the Sampling of Physical Materials*, Revue de l'Institut Intern. de Stat., 50, 1950.

random from every stratum. However, the random shifts in position from stratum to stratum may be inconvenient and therefore only the element from the first stratum is chosen at random whereafter elements are chosen with the constant sampling interval N/n. If the elements of the sequence are ordered at random with respect to the characteristic studied the mean of a systematic sample will have the same precision as the mean of a random sample. If, however, the elements vary at random within the strata and essential differences between strata exist then the systematic sample will give a higher precision than the random sample. In general, the variance of the mean of a systematic sample depends on the correlation between successive elements, which is generally unknown. Before using systematic sampling it is therefore advisable to study what patterns of variation the elements sampled usually present.

17.4. Two-Stage Sampling.

Two-stage sampling is very common in industry where the items to be sampled often appear in bales, sacks, boxes, or similar containers which may be used as primary sampling units. The first stage consists of *selecting a random sample of n_1 primary units* from the population of primary units. In the second stage a *random sample of n_2 secondary units is selected from each primary unit.* In Example 16.5 the sheets represent the primary units and the measurements on each sheet the secondary units.

Hence the observations may be written as

$$x_{i\nu} = \xi + y_i + z_{i\nu} , \tag{17.4.1}$$

where y_i and $z_{i\nu}$ denote the contributions of the primary and secondary units, respectively, to the total variation.

For a finite population consisting of N_1 primary units, each containing N_2 secondary units, taking on the values $X_{i\nu}$, $i=1, \ldots, N_1$ and $\nu=1, \ldots, N_2$, we find

$$\sigma^2 = \frac{1}{N_1 N_2} \sum_{i=1}^{N_1} \sum_{\nu=1}^{N_2} (X_{i\nu} - \xi)^2 = \frac{1}{N_1} \sum_{i=1}^{N_1} (\xi_i - \xi)^2 + \frac{1}{N_1 N_2} \sum_{i=1}^{N_1} \sum_{\nu=1}^{N_2} (X_{i\nu} - \xi_i)^2$$

$$= \sigma_1^2 + \sigma_2^2 , \tag{17.4.2}$$

where $\xi_i = \sum_{\nu} X_{i\nu}/N_2$ and $\xi = \sum_i \sum_{\nu} X_{i\nu}/N_1 N_2$. The two variances σ_1^2 and σ_2^2 represent the variation between the true means of primary units and the average of the variances within primary units, respectively.

Drawing a random sample of n_1 primary units we have for every y in (17.4.1) that $p\{y = \xi_i\} = 1/N_1$ and according to (17.2.9)

$$\mathcal{V}\{\bar{y}\} = \frac{\sigma_1^2}{n_1}(1-\alpha_1)\,, \quad \alpha_1 = \frac{n_1}{N_1}. \tag{17.4.3}$$

For a given primary unit, number i say, a random sample of n_2 secondary units is drawn so that for every $z_{i\nu}$ we have $p\{z = X_{i\nu}-\xi_i\} = 1/N_2$ for given i and $\nu = 1, \ldots, N_2$. According to (17.2.9) we have for given i

$$\mathcal{V}\{\bar{z}_i\} = \frac{\sigma_{2i}^2}{n_2}(1-\alpha_2)\,, \quad \alpha_2 = \frac{n_2}{N_2}, \tag{17.4.4}$$

where

$$\sigma_{2i}^2 = \frac{1}{N_2}\sum_{\nu=1}^{N_2}(X_{i\nu}-\xi_i)^2\,. \tag{17.4.5}$$

Combining these results we find for the sample mean

$$\bar{x} = \frac{1}{n_1 n_2}\Sigma\Sigma x_{i\nu} = \xi+\frac{1}{n_1}\Sigma y_i+\frac{1}{n_1 n_2}\Sigma\Sigma z_{i\nu} = \xi+\bar{y}+\frac{1}{n_1}\Sigma\bar{z}_i \tag{17.4.6}$$

the following variance

$$\boxed{\mathcal{V}\{\bar{x}\} = \frac{\sigma_1^2}{n_1}(1-\alpha_1)+\frac{\sigma_2^2}{n_1 n_2}(1-\alpha_2)\,,} \tag{17.4.7}$$

where σ_2^2 as defined above denotes the mean of all σ_{2i}^2's.

For a given total number of observations, $n = n_1 n_2$, this variance attains its minimum value for $n_1 = n$ and $n_2 = 1$, i. e., the largest possible number of primary units should be sampled and only one secondary unit taken from each primary unit.

If the cost is c_1 per primary unit and c_2 per secondary unit the total cost becomes

$$c = c_1 n_1+c_2 n_1 n_2 = n_1(c_1+c_2 n_2)\,. \tag{17.4.8}$$

Minimizing $\mathcal{V}\{\bar{x}\}$ for given total cost leads to

$$\boxed{n_2 = \frac{\sigma_2}{\sigma_1}\sqrt{\frac{c_1}{c_2}}\Bigg/\sqrt{1-\frac{\sigma_2^2}{N_2\sigma_1^2}}.} \tag{17.4.9}$$

For infinite populations we have $n_2 = \dfrac{\sigma_2}{\sigma_1}\sqrt{\dfrac{c_1}{c_2}}$[1]. It will be noted that n_2 does not depend on the total cost but only on the ratios c_1/c_2 and σ_2/σ_1. The optimum value of n_1 may be found from (17.4.8) for given c and n_2.

---

[1]) See W. A. Shewhart: *Economic Control of Quality of Manufactured Product*, D. Van Nostrand, New York, 1931, p. 388 ff., and L. H. C. Tippett: *The Methods of Statistics*, Williams and Norgate, London, 1931, p. 207 ff.

In Example 16.5 we have $c_1/c_2 \simeq 10$. For $\sigma_1^2 \approx 0{\cdot}032$ and $\sigma_2^2 \approx 0{\cdot}020$ we obtain $n_2 = 2{\cdot}52$, the optimal number of determinations per sheet thus being two or three. Further examples may be found in L. TANNER and W. E. DEMING: *Some Problems in the Sampling of Bulk Materials*, Proceedings of the American Society for Testing Materials, 49, 1949.

17.5. The Ratio Estimate.

Consider a finite population where each element is associated with a pair of numbers, $(X_1, Y_1), \ldots, (X_N, Y_N)$, and a random sample of n elements $(x_1, y_1), \ldots, (x_n, y_n)$.

As usual we introduce the two means (ξ, η), the two variances (σ_x^2, σ_y^2), the covariance

$$\sigma_{xy} = \frac{1}{N} \sum_{i=1}^{N} (X_i - \xi)(Y_i - \eta) \tag{17.5.1}$$

and further the ratio

$$\beta = \frac{\eta}{\xi} = \frac{\sum_{i=1}^{N} Y_i}{\sum_{i=1}^{N} X_i}. \tag{17.5.2}$$

If ξ is known and an estimate b of β may be obtained from the sample we may use $b\xi$ as an estimate of η.

Many different estimates of β may be obtained, using different weights for the individual ratios y_i/x_i, see § 18.6. Here we will consider only the estimate

$$\boxed{b = \frac{\bar{y}}{\bar{x}} = \frac{\sum_{i=1}^{n} y_i}{\sum_{i=1}^{n} x_i}.} \tag{17.5.3}$$

Introducing the auxiliary variable

$$D_i = Y_i - \beta X_i\,, \quad i = 1, \ldots, N\,, \tag{17.5.4}$$

i. e., the Y-deviations from the straight line $Y = \beta X$, we find for a random sample of n elements

$$\bar{d} = \bar{y} - \beta\bar{x}, \quad \mathcal{M}\{\bar{d}\} = 0\,, \tag{17.5.5}$$

and

$$\mathcal{V}\{\bar{d}\} = \frac{\sigma_d^2}{n}(1-\alpha)\,, \tag{17.5.6}$$

where

$$\sigma_d^2 = \frac{1}{N}\sum_{i=1}^{N} D_i^2 = \frac{1}{N}\sum_{i=1}^{N}(Y_i - \beta X_i)^2$$

$$= \frac{1}{N}\sum_{i=1}^{N}(Y_i - \eta - \beta(X_i - \xi))^2$$

$$\doteq \sigma_y^2 + \beta^2\sigma_x^2 - 2\beta\sigma_{xy} . \tag{17.5.7}$$

Since the knowledge of the extra variable X permits us to reduce the variation of the Y-variable from σ_y^2 to σ_d^2 it may be expected that $\mathcal{V}\{b\xi\} \simeq \mathcal{V}\{\bar{d}\}$. This supposition may be proved by means of the results given in § 9.9. First it follows that $\mathcal{M}\{b\} \simeq \beta$, if the coefficients of variation for $\bar{x}$ and $\bar{y}$ are small. Next we obtain from (9.9.11)

$$\mathcal{C}^2\{b\} \simeq \mathcal{C}^2\{\bar{x}\} + \mathcal{C}^2\{\bar{y}\} - 2\varrho\,\mathcal{C}\{\bar{x}\}\mathcal{C}\{\bar{y}\} , \tag{17.5.8}$$

where $\varrho = \varrho\{\bar{x}, \bar{y}\}$. It follows easily from the definition of ϱ that $\varrho\{\bar{x}, \bar{y}\} = \sigma_{xy}/\sigma_x\sigma_y$. Introducing

$$\mathcal{C}^2\{\bar{x}\} = \frac{\mathcal{V}\{\bar{x}\}}{\xi^2} = \frac{1}{\xi^2}\frac{\sigma_x^2}{n}(1-a)$$

and the analogous expression for $\mathcal{C}^2\{\bar{y}\}$ we find

$$\mathcal{V}\{b\} \simeq \frac{1}{\xi^2}(\beta^2\sigma_x^2 + \sigma_y^2 - 2\beta\sigma_{xy})\frac{1-a}{n} = \frac{1}{\xi^2}\frac{\sigma_d^2}{n}(1-a) , \tag{17.5.9}$$

and the final result

$$\boxed{\mathcal{V}\{b\xi\} \simeq \frac{\sigma_d^2}{n}(1-a) .} \tag{17.5.10}$$

Hence the ratio estimate $b\xi$ will have a smaller variance than the estimate y obtained from a random sample of Y's if $\sigma_d^2 < \sigma_y^2$, which is the same as

$$\varrho_{xy} > \frac{1}{2}\frac{\mathcal{C}\{x\}}{\mathcal{C}\{y\}} . \tag{17.5.11}$$

If the two coefficients of variation are equal, the correlation coefficient must be larger than $\frac{1}{2}$.

In general the precision of the ratio estimate increases with increasing values of the coefficient of correlation.

The ratio estimate has been used successfully for the redetermination of quality or weight of industrial products which have been stored for a while or transported from one country to another[1]). If the original quality, X,

[1]) W. E. Deming: *Some Theory of Sampling*, p. 183 ff., J. Wiley and Sons, New York, 1950.

of each item is known, the ratio estimate may be applied directly after new quality determinations, Y, for the items of a random sample have been carried out.

17.6. Experiments and Statistical Control. Randomization.

The first step in the planning of an experiment is to choose the experimental factors and the levels for these factors. We will use the term experimental *treatments* to denote the combinations of experimental factors which are to be compared. If, for example, an experiment contains three experimental factors each at two levels, there will be eight treatments.

The purpose of an experiment is to *compare the effects of the experimental treatments*, but these effects will always be obscured by *experimental errors*. Therefore, statistical methods are necessary to evaluate an estimate of the error variance, which may be used for testing the significance of differences among the treatment effects. In addition statistics may help to reduce the experimental error by an adequate design and analysis of the experiment.

The classical theory of errors assumed that the errors were independent and normally distributed, see § 8.2. This theory was developed mainly as a model for the distribution of *errors of measurements* occurring in astronomy and geodesy where the sources of variability were to be found in the non-uniformity of the experimental technique. Using the terminology of § 13.2 this variation is produced by a constant system of chance causes.

In industrial experimentation another source of variation will usually be found in *the heterogeneity of the experimental materials* or in variations of *factors* (*conditions*) *not being under control* (*assignable causes*). In what follows we will, for the sake of argument, denote this group of causes as "heterogeneity of the experimental material". In contrast to the first-mentioned source of variability this heterogeneity will often produce a *systematic* and relatively large variation of the experimental results. If, for example, experimental materials from different batches are used or time trends in a production process influence the experimental results, it may be difficult to decide whether an observed difference is due to the treatments or to the unknown systematic variation. Therefore, *if the variations caused by the non-experimental factors are not in a state of statistical control, they must be forced to be so*, if valid conclusions are to be drawn by means of statistical methods. The remedy is to *allot the treatments to the experimental units at random*. This idea of *randomized* experiments is due to R. A. Fisher.

In what follows different designs of randomized experiments will be discussed by means of an example which has been constructed to illustrate the basic principles. Imagine that five treatments, A, B, C, D, and E, are to be compared and that 35 experimental units are available for the

TABLE 17.2.
An example of four experimental designs.

Experimental unit no.	Variation of experimental units	Random variations	Sum	Completely randomized	Randomized blocks	Latin square	Systematic arrangement
1	9	−1·4	7·6	*D*	*B*	*C*	*D*
2	12	2·8	14·8	*D*	*E*	*E*	*E*
3	11	3·6	14·6	*B*	*A*	*A*	*A*
4	12	−1·7	10·3	*E*	*D*	*D*	*B*
5	16	0·4	16·4	*C*	*C*	*B*	*C*
6	17	2·9	19·9	*D*	*E*	*B*	*D*
7	14	0·4	14·4	*B*	*C*	*D*	*E*
8	14	1·4	15·4	*A*	*A*	*E*	*A*
9	17	−2·0	15·0	*C*	*D*	*C*	*B*
10	16	1·6	17·6	*B*	*B*	*A*	*C*
11	13	0·0	13·0	*C*	*B*	*A*	*D*
12	12	−2·3	9·7	*A*	*A*	*C*	*E*
13	12	−3·1	8·9	*C*	*D*	*D*	*A*
14	17	−0·1	16·9	*A*	*C*	*B*	*B*
15	20	−0·4	19·6	*D*	*E*	*E*	*C*
16	20	−1·0	19·0	*A*	*B*	*E*	*D*
17	20	−3·2	16·8	*E*	*C*	*B*	*E*
18	22	1·6	23·6	*D*	*A*	*C*	*A*
19	23	1·4	24·4	*B*	*E*	*A*	*B*
20	24	2·7	26·7	*B*	*D*	*D*	*C*
21	23	1·8	24·8	*E*	*C*	*D*	*D*
22	26	1·5	27·5	*B*	*E*	*A*	*E*
23	26	−1·7	24·3	*A*	*B*	*B*	*A*
24	27	0·5	27·5	*A*	*D*	*E*	*B*
25	26	0·6	26·6	*D*	*A*	*C*	*C*
26	25	0·3	25·3	*C*	*D*		*D*
27	26	−0·4	25·6	*C*	*B*		*E*
28	29	3·5	32·5	*B*	*C*		*A*
29	29	0·4	29·4	*C*	*E*		*B*
30	28	2·5	30·5	*E*	*A*		*C*
31	30	−1·2	28·8	*E*	*D*		*D*
32	31	0·3	31·3	*A*	*A*		*E*
33	30	−2·7	27·3	*E*	*B*		*A*
34	30	−3·8	26·2	*D*	*E*		*B*
35	28	4·5	32·5	*E*	*C*		
Mean	21·00	0·28	21·28				
Variance	46·18	4·404	52·49				

experiment, so that each treatment may be applied seven times. If, for instance, each treatment affects the whole production of a plant and has to be applied a day, say, the experimental unit will be the plant, the operators, and the raw materials used for a day's production and the whole experiment will cover a period of 35 days.

Let us imagine that the experimental units under "normal conditions" would have given the results shown in column 4 of Table 17.2. Each number is composed of two numbers, the first showing the "true" systematic variation resulting from differences among the experimental units, the second the normally distributed errors of measurements. Hence the experimental material displays a large systematic variation which may invalidate the conclusions to be drawn from the experiment if special precautions are not taken. The minimum error variance obtainable in this experiment will be $s^2 = 4{\cdot}40$ which gives the variance of the 35 errors of measurements in Table 17.2.

Further, suppose that treatment A changes the normal quality of the product by $+3$, irrespective of what experimental unit treatment A will be applied to. For example, applying treatment A to the first unit the observed result will be $9+3-1{\cdot}4 = 10{\cdot}6$, 9 being the result produced under normal conditions, 3 being added as the effect of treatment A, and $-1{\cdot}4$ being the error of measurement. Similarly, for unit no. 35, the result of applying treatment A will be $28+3+4{\cdot}5 = 35{\cdot}5$. Thus *the effects of the treatments are supposed to be constant and additive, independent of the quality of the experimental units.* For the five treatments we assume that the effects are $+3{\cdot}0$, $+1{\cdot}5$, $0{\cdot}0$, $-1{\cdot}5$, and $-3{\cdot}0$, respectively.

In practice the variation of the experimental material will generally be unknown. Designing an experiment without taking possible systematic variations into account may therefore be dangerous, if previous experimentation does not ensure that systematic variations are of no importance in this special field. It is easy to see that if we would apply treatment A to the first seven units, treatment B to the next seven, and so forth, the results would be quite misleading, giving the mean effects: 17·0, 15·3, 22·1, 25·5, 26·4, the total mean being 21·3. Because treatments A and B have been applied to the first units, where the quality level is low, it seems as if the effects of A and B are negative as compared with the effects of D and E. The systematic variation of the experimental material and the effects of the treatments have been tied together so that an estimate of the treatment effects alone cannot be obtained.

To split up this connection and equalize the conditions under which the treatments are applied, the treatments may be allotted to the experimental units at random, thus obtaining a *completely randomized experiment.*

A random arrangement of the seven A's, seven B's, etc., may be obtained by drawing 35 chips successively after thorough mixing in a bowl, seven chips having been marked with A's, seven with B's, and so forth. Similar results may be obtained from a table of random sampling numbers.

Table 17.2 shows the resulting random arrangement. Adding the fictitious treatment effects to the numbers in Table 17.2 we obtain the results given in Table 17.3, which illustrates the analysis of a completely randomized experiment.

TABLE 17.3.

Analysis of the results of a completely randomized experiment.

	Treatment				
	A	B	C	D	E
	18·4	16·1	16·4	6·1	7·3
	12·7	15·9	15·0	13·3	13·8
	19·9	19·1	13·0	18·4	21·8
	22·0	25·9	8·9	18·1	27·5
	27·3	28·2	25·3	22·1	25·8
	30·5	29·0	25·6	25·1	24·3
	34·3	34·0	29·4	24·7	29·5
S_i	165·1	168·2	133·6	127·8	150·0
$\bar{x}_i$	23·59	24·03	19·09	18·26	21·43
$\bar{x}_i - \bar{x}$	2·31	2·75	−2·19	−3·02	0·15

$S =$	744·7	$\bar{x} =$	21·28
$SS =$	17684·13	$\Sigma S_i^2/7 =$	16033·00
$S^2/35 =$	15845·09		15845·09
Difference =	1839·04		187·91

Analysis of variance

Variation	SSD	f	s^2
Between treatments	187·91	4	46·98
Residual	1651·13	30	55·04
Total	1839·04	34	54·09

The effect of the randomization is to transform the unknown systematic variations of the experimental material to independent and random variations in relation to the treatments. Every set of seven experimental units has the same probability of receiving a specified treatment.

Hence the assumption of normality which underlies the analysis of variance will not be fulfilled, but it has been proved that randomization leads to a distribution of the variance ratio which may be approximated to by the v^2-distribution.

The error variance, the variance within treatments, contains the contributions from both sources of variation: the technical errors and the unknown (systematic) variation. In the present example the contribution from the second source of variation dominates, the variance of the numbers in columns 2 and 3 of Table 17.2 being 46·18 and 4·40, respectively. The total error variance, estimated from Table 17.3, becomes 55·04. Because of the considerable variation of the experimental material the existence of the treatment effects cannot be demonstrated, the estimated standard error of a treatment mean being $\sqrt{55{\cdot}04/7} = 2{\cdot}80$.

A decrease in the standard error may possibly be obtained by repeating each treatment more than seven times. This, however, requires more experimental material, which may not be obtainable, but even if a larger experiment could be carried out the mean error would decrease only if the new experimental material was not more heterogeneous than that already used.

Another way to obtain a smaller standard error is to diminish the numerator, the error variance, instead of augmenting the denominator, the number of replications. In many cases the error variance may be successfully diminished by a procedure analogous to stratified sampling in § 17.3, i. e., the experimental units are grouped together in strata or blocks of similar units and the randomization is performed within the blocks, as shown in the following paragraph.

The general plan for a completely randomized experiment is as follows. First the number of times each treatment is to be applied, the number of replications, must be determined, and then the treatments are allotted to the experimental units at random. If the experiment consists of several stages, randomization must be carried out independently for every stage. The results may be analyzed as shown in Table 16.4.

Completely randomized experiments are generally used when the experimental material does not exhibit large heterogeneity, such as in many laboratory experiments under standard conditions. In cases where large heterogeneity may be expected more complicated designs are generally used with the purpose of eliminating a part of the heterogeneity from the error variance.

Examples of completely randomized experiments may be found in Examples 16.1 and 16.8.

17.7. Randomized Blocks.

The plan for an experiment with k treatments arranged in n randomized blocks is as follows: The kn experimental units are divided into n groups or blocks, each containing k units as similar as possible. Within each block the k treatments are assigned to the k units at random.

The grouping of the experimental units into blocks of similar units serves the same purpose as in stratified sampling. Only the variation between units within blocks contributes to the error variance, whereas the variation between blocks is eliminated.

Every treatment will be replicated n times, each block representing a replication of all treatments. By applying the same treatment to two plots within each block, the number of replications for that treatment may be doubled.

The statistical technique for analyzing a randomized block experiment has been given in Table 16.31, where "between columns" and "between rows" now becomes "between blocks" and "between treatments", respectively. The interaction variance may be used as error variance if the treatment effects are independent of the levels of the block means. The error variance

TABLE 17.4.
Analysis of the results of a randomized block experiment.

Block no.	Treatment *A*	*B*	*C*	*D*	*E*	$S_{.j}$	$\bar{x}_{.j}$
1	17·6	9·1	16·4	8·8	11·8	63·7	12·74
2	18·4	19·1	14·4	13·5	16·9	82·3	16·46
3	12·7	14·5	16·9	7·4	16·6	68·1	13·62
4	26·6	20·5	16·8	25·2	21·4	110·5	22·10
5	29·6	25·8	24·8	26·0	24·5	130·7	26·14
6	33·5	27·1	32·5	23·8	26·4	143·3	28·66
7	34·3	28·8	32·5	27·3	23·2	146·1	29·22
$S_{i.}$	172·7	144·9	154·3	132·0	140·8	744·7	
$\bar{x}_{i.}$	24·67	20·70	22·04	18·86	20·11		21·28
$\bar{x}_{i.}-\bar{x}$	3·39	−0·58	0·76	−2·42	−1·17		

	$SS = 17702{\cdot}43$	$\Sigma S_{i.}^2/7 = 15982{\cdot}63$	$\Sigma S_{.j}^2/5 = 17328{\cdot}29$
	$S^2/35 = 15845{\cdot}09$	15845·09	15845·09
Difference =	1857·34	137·54	1483·20

Analysis of variance

Variation	*SSD*	*f*	s^2	v^2
Between blocks	1483·20	6	247·20	
Between treatments	137·54	4	34·39	3·49
Residual	236·60	24	9·858	
Total	1857·34	34		

of the corresponding completely randomized experiment would have included both the interaction and the "between blocks" variances.

Column 6 of Table 17.2 shows the arrangement of 5 treatments in 7 randomized blocks. Each block of five successive units contains the five letters A, B, C, D, and E in a random order. Adding the fictitious treatment effects we find the results shown in Table 17.4.

As compared with the completely randomized experiment the error variance has decreased from 55·04 to 9·86, the result being that the existence of the treatment effects may now be demonstrated. The estimated standard error of the treatment means is $\sqrt{9{\cdot}858/7} = 1{\cdot}19$.

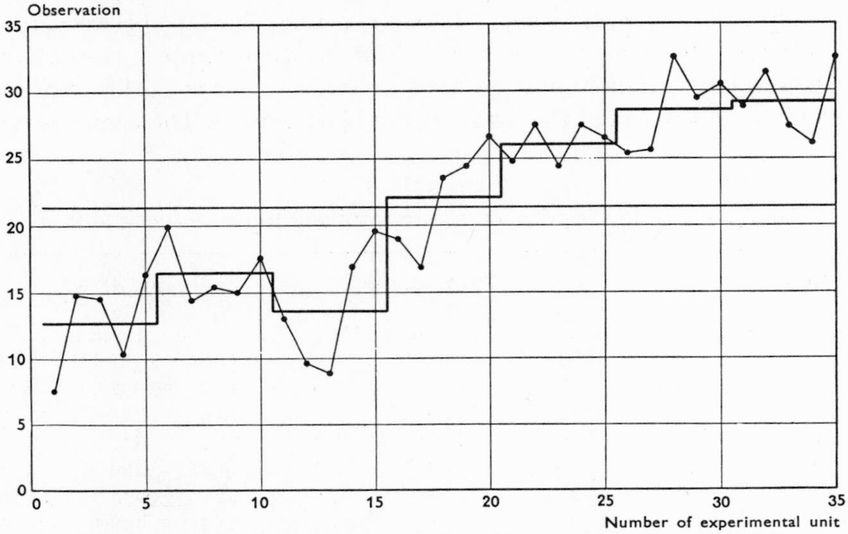

FIG. 17.1. The variations of the experimental units plus the errors of measurements as compared with the block means.

The elimination of the differences between blocks has been illustrated in Fig. 17.1 where the step-curve represents the block means. The error variance contains the (randomized) deviations from the step-curve, whereas the error variance of the completely randomized experiment contains the deviations from the total mean, represented by the horizontal line.

Examples of experiments using randomized blocks may be found in Example 15.6, where a block consists of two neighbouring plots, and in Example 15.5, samples from the same potato forming a block.

In general, samples from the same batch will be more similar than samples from different batches, wherefore experimental units from the same batch usually are grouped together in a block. Similarly, experiments carried out as close as possible in time or space and by the same observer will usually be more alike than experiments spread out over a larger interval and carried out by different observers. Hence a short time interval, a day say, and an observer may be the block so that time trends between days and differences between observers may be eliminated.

17.8. Latin Squares.

In many cases the single classification of experimental units in blocks will eliminate the larger part of the variation due to heterogeneity of the experimental material. This elimination may, however, be carried further by classification according to two or more criteria. Experiments with two-way classification of the experimental units have been called *Latin squares.*

The plan for a Latin square with n treatments is as follows: n^2 experimental units are classified according to two criteria, forming a square with n rows and n columns. The n replications of the n treatments are distributed at random over the square, satisfying the condition that *each treatment must occur only once in each row and once in each column.*

A Latin square with five treatments may be as follows:

C	*E*	*A*	*D*	*B*
B	*D*	*E*	*C*	*A*
A	*C*	*D*	*B*	*E*
E	*B*	*C*	*A*	*D*
D	*A*	*B*	*E*	*C*

Fisher and Yates have given a complete enumeration of Latin squares of sizes 4×4 to 6×6 and examples of squares up to 12×12, see their *Statistical Tables.* The correct procedure is to choose one square at random out of all possible squares of given size for each experiment. A simpler and in most cases satisfactory procedure, which, however, excludes some squares from being chosen, is to write down a Latin square and permute all rows and columns at random. The square above has been obtained in this way by permutations of rows and columns as indicated for the square below.

	5	1	4	2	3
2	*A*	*B*	*C*	*D*	*E*
3	*E*	*A*	*B*	*C*	*D*
4	*D*	*E*	*A*	*B*	*C*
5	*C*	*D*	*E*	*A*	*B*
1	*B*	*C*	*D*	*E*	*A*

Applying this arrangement to the first 25 units of Table 17.2 we find the results given in Table 17.5. The two criteria for classifying the experimental units are here *block number* and *order within blocks.* If the systematic variation between units within blocks had been the same for every block it would have been eliminated completely by this double classification. A linear trend, for example, would have been eliminated. In the present case the increasing trend has in the main been eliminated by means of the block arrangement, wherefore the further classification presumably not will lead to a smaller error variance, because no characteristic trend persists within blocks.

TABLE 17.5.
Analysis of the results of a Latin square experiment.

Block no.	Order within blocks 1	2	3	4	5	Sum	Mean
1	*C* 7·6	*E* 11·8	*A* 17·6	*D* 8·8	*B* 17·9	63·7	12·74
2	*B* 21·4	*D* 12·9	*E* 12·4	*C* 15·0	*A* 20·6	82·3	16·46
3	*A* 16·0	*C* 9·7	*D* 7·4	*B* 18·4	*E* 16·6	68·1	13·62
4	*E* 16·0	*B* 18·3	*C* 23·6	*A* 27·4	*D* 25·2	110·5	22·10
5	*D* 23·3	*A* 30·5	*B* 25·8	*E* 24·5	*C* 26·6	130·7	26·14
Sum	84·3	83·2	86·8	94·1	106·9	455·3	
Mean	16·86	16·64	17·36	18·82	21·38		18·21

	Treatments *A*	*B*	*C*	*D*	*E*
Sum	112·1	101·8	82·5	77·6	81·3
Mean	22·42	20·36	16·50	15·52	16·26
Deviation	4·21	2·15	−1·71	−2·69	−1·95

	Total *SS*	$\Sigma S^2/5$ for Blocks	Order	Treatments
	9331·31	8952·27	8369·08	8473·47
$S^2/25$	8291·92	8291·92	8291·92	8291·92
Difference	1039·39	660·35	77·16	181·55

Analysis of variance

Variation	*SSD*	*f*	s^2	v^2
Between blocks	660·35	4	165·09	
Between order	77·16	4	19·29	1·92
Between treatments	181·55	4	45·39	4·53
Residual	120·33	12	10·03	
Total	1039·39	24		

Table 17.5 shows the results of the analysis of variance. The partitioning of the total *SSD* is as shown in Table 17.6, where $\bar{x}_t$ denotes the treatment means. The *SSD* for treatments and the residual has been obtained by partitioning the interaction-*SSD* of Table 16.31.

TABLE 17.6.
Analysis of variance for a Latin square experiment.

Variation	SSD	f	s^2
Between columns	$n\Sigma(\bar{x}_{.j}-\bar{x}_{..})^2$	$n-1$	s_4^2
Between rows	$n\Sigma(\bar{x}_{i.}-\bar{x}_{..})^2$	$n-1$	s_3^2
Between treatments	$n\Sigma(\bar{x}_t-\bar{x}_{..})^2$	$n-1$	s_2^2
Residual	$\Sigma\Sigma(x_{ij}-\bar{x}_{i.}-\bar{x}_{.j}-\bar{x}_t+2\bar{x}_{..})^2$	$(n-1)(n-2)$	s_1^2
Total	$\Sigma\Sigma(x_{ij}-\bar{x}_{..})^2$	n^2-1	

Table 17.5 shows that the variation between means corresponding to order within blocks is not significant. Hence nothing has been gained by the second classification. As in Table 17.4 the treatment effects are clearly significant. The estimated standard error of the treatment means is $\sqrt{10{\cdot}03/5} = 1{\cdot}42$.

The elimination of the systematic variation due to blocks and order within blocks has been illustrated in Fig. 17.2 where the differences between the two curves represent the residual variation. The curve representing the block and order effects has been constructed by adding the order means to each block mean and subtracting the total mean. The estimated (insignificant) trend within blocks may be found as the order means minus the total mean, i. e., $-1{\cdot}35$, $-1{\cdot}57$, $-0{\cdot}85$, $+0{\cdot}61$, $+3{\cdot}17$.

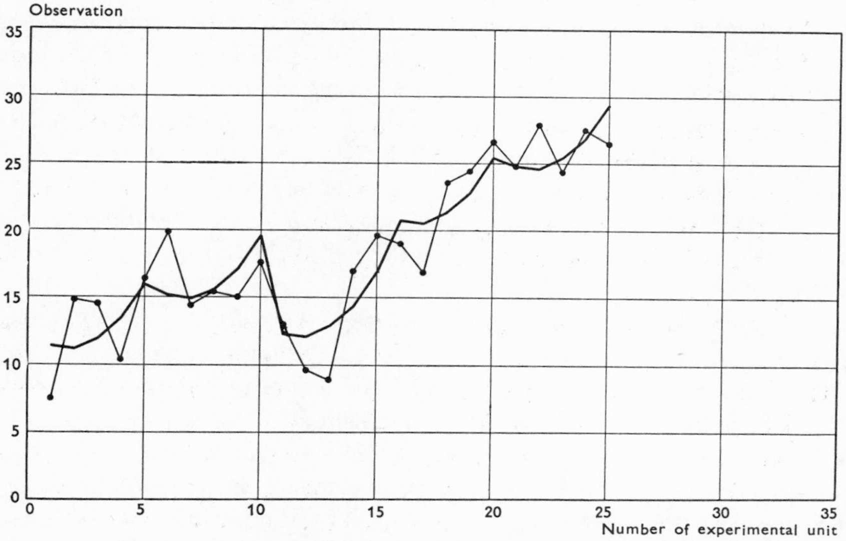

FIG. 17.2. The variations of the experimental units plus the errors of measurements as compared with the block and order within block effects.

17.9. Systematic Arrangements.

The randomized block experiment, the Latin square, and similar designs of experiments are based on *a combination of systematic and random arrangements of the treatments*, in such a manner that part of the variation due to disturbing factors not under control may be eliminated through the systematic structure of the experiment *and* the corresponding analysis of variance, while the remaining variation is included in the experimental error through randomization.

In general, randomization is necessary to ensure that the conditions for applying the statistical tests of significance will be fulfilled. However, two cases exist where randomization is not necessary, so that the treatments may be arranged in any convenient way: (1) The variation caused by the non-experimental factors has been studied in previous experiments and found to be in a *state of statistical control*. (2) The systematic variation produced by the non-experimental factors may be eliminated by means of statistical methods, *the residual variation being random.*

In both cases previous experience is necessary to indicate whether randomization may be omitted or not. The possibility of eliminating the systematic variation effectively depends on whether the variation is smooth or irregular fluctuations occur.

As an example, consider the systematic arrangement $ABCDE \mid ABCDE \mid$, etc. The last column of Table 17.2 shows this arrangement for six blocks, together with two auxiliary units at each end of the arrangement, see also Table 17.7, where the treatment effects have been added.

To eliminate the systematic variation we first compute moving averages of length 5, i. e., all possible averages of 5 successive observations, and then we subtract the moving averages from the corresponding observations, see Table 17.7. The moving averages represent the general trend of the observations since every average includes the mean of all five treatment effects, i. e., the moving averages are independent of the differences among the treatment effects. Further the moving averages will display a smaller random variation than the original observations because they contain the means of five successive random errors.

The problem is how the moving average transforms the systematic variation. *If the moving average contains the systematic component unaltered*, then the elimination may be performed by subtracting the moving average from the original observations. This condition will be satisfied exactly if the systematic variation is linear (and also in some more complicated cases) and approximately if the systematic variation is smooth and slow. Experience has shown that this method of eliminating the systematic variation works satisfactorily in many agricultural experiments, where the experimental units (plots of land) have been arranged in a row over the

TABLE 17.7.
The elimination of the systematic variation by means of moving averages.

Experimental unit no.	Observations	Moving sums	Moving averages	Deviations d_{vi}	Residuals $d_{vi}-\bar{d}_{.i}$
1	6·1				
2	11·8				
3	17·6	63·7	12·74	4·86	1·93
4	11·8	76·0	15·20	−3·40	−4·82
5	16·4	75·6	15·12	1·28	−0·41
6	18·4	76·4	15·28	3·12	4·45
7	11·4	81·1	16·22	−4·82	−0·54
8	18·4	82·3	16·46	1·94	−0·99
9	16·5	75·4	15·08	1·42	0·00
10	17·6	70·7	14·14	3·46	1·77
11	11·5	64·2	12·84	−1·34	−0·01
12	6·7	66·1	13·22	−6·52	−2·24
13	11·9	68·1	13·62	−1·72	−4·65
14	18·4	74·1	14·82	3·58	2·16
15	19·6	81·2	16·24	3·36	1·67
16	17·5	95·9	19·18	−1·68	−0·35
17	13·8	103·4	20·68	−6·88	−2·60
18	26·6	110·5	22·10	4·50	1·57
19	25·9	116·3	23·26	2·64	1·22
20	26·7	127·0	25·40	1·30	−0·39
21	23·3	127·7	25·54	−2·24	−0·91
22	24·5	130·8	26·16	−1·66	2·62
23	27·3	130·7	26·14	1·16	−1·77
24	29·0	131·2	26·24	2·76	1·34
25	26·6	129·3	25·86	0·74	−0·95
26	23·8	137·5	27·50	−3·70	−2·37
27	22·6	139·4	27·88	−5·28	−1·00
28	35·5	143·3	28·66	6·84	3·91
29	30·9	146·8	29·36	1·54	0·12
30	30·5	152·5	30·50	0·00	−1·69
31	27·3	147·3	29·46	−2·16	−0·83
32	28·3	144·1	28·82	−0·52	3·76
33	30·3				
34	27·7				

experimental field. Similar results may be expected in industrial experimentation where the systematic variation is due to slow and regular (time) trends, whereas discontinuities arising from the use of different batches of raw materials may be eliminated by using a randomized block design.

After eliminating the systematic component the differences are grouped according to treatments and the *mean differences* computed as estimates of the treatment effects, see Table 17.8.

The sum of squares of deviations may be partitioned as follows:

$$\sum_{\nu=1}^{n}\sum_{i=1}^{k} d_{\nu i}^2 = \sum_{\nu=1}^{n}\sum_{i=1}^{k}(d_{\nu i}-\bar{d}_{.i})^2+n\sum_{i=1}^{k}\bar{d}_{.i}^2$$

where $d_{\nu i}$ denotes the difference between the observation for treatment i in block ν and the corresponding moving average. Hence the error variance becomes

$$s_1^2 = \frac{1}{(n-1)(k-1)}\sum_{\nu=1}^{n}\sum_{i=1}^{k}(d_{\nu i}-\bar{d}_{.i})^2$$

and the variation between treatment effects becomes

$$s_2^2 = \frac{n}{k-1}\sum_{i=1}^{k}\bar{d}_{.i}^2 .$$

The variance ratio s_2^2/s_1^2 will be distributed approximately as v^2 with $(k-1, (n-1)(k-1))$ degrees of freedom.

TABLE 17.8.

Deviations between the observed values and the moving averages. Estimation of the treatment effects and the error variance.

Block no.	Treatment A	B	C	D	E
1	4·86	−3·40	1·28	3·12	−4·82
2	1·94	1·42	3·46	−1·34	−6·52
3	−1·72	3·58	3·36	−1·68	−6·88
4	4·50	2·64	1·30	−2·24	−1·66
5	1·16	2·76	0·74	−3·70	−5·28
6	6·84	1·54	0·00	−2·16	−0·52
S	17·58	8·54	10·14	−8·00	−25·68
Mean, $\bar{d}_{.i}$	2·93	1·42	1·69	−1·33	−4·28

Total $SS = \Sigma\Sigma d_{\nu i}^2 = 350{\cdot}9188$.

$\Sigma S^2/6 = n\Sigma\bar{d}_{.i}^2 = 201{\cdot}3783$, $f = 4$, $s_2^2 = 50{\cdot}34$

Difference $= 149{\cdot}5405$, $f = 20$, $s_1^2 = 7{\cdot}477$

$v^2 = 6{\cdot}73$

Table 17.8 shows the differences from Table 17.7 arranged according to treatments, and the computations of the treatment means and the error variance, which becomes smaller than those previous found. The differences between treatment effects are clearly significant. The estimated standard error of the treatment means is $\sqrt{7{\cdot}477/6} = 1{\cdot}12$.

The error variance, $s_1^2 = 7{\cdot}477$, is larger than the variance of the random component of Table 17.2 which was found to be 4·40. Hence the systematic variation has not been completely eliminated. The reason is that the moving average is smoother than the systematic variation in question, wherefore some small irregular fluctuations of the latter will be included in the error variance. Fig. 17.3 shows the deviations between the original observations (without the treatment effects added) and the moving average.

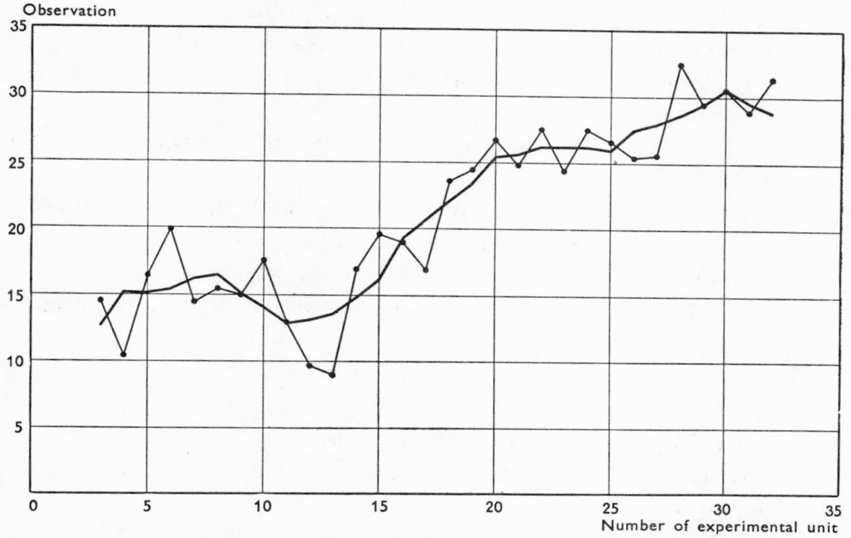

Fig. 17.3. The variations of the experimental units plus the errors of measurements as compared with the moving averages.

The "residuals" have been recorded in Table 17.7. In actual experimentation the series of residuals must always be tested for randomness, for example by means of run tests even if the conditions for applying these tests are not strictly fulfilled on account of the restraints resulting from using the moving average.

17.10. Factorial Experiments.

According to the classical method of designing experiments, only one experimental factor is varied at a time. R. A. Fisher has, however, at the same time as he developed the analysis of variance and the above-mentioned randomized designs, stressed the importance of *investigating all possible combinations of the experimental factors* in the same experiment. An experiment which has been designed in this manner may yield more information, and information that rests on a sounder basis, than an experiment of the same size but designed on the classical principles, because the possible interactions between the experimental factors are explored.

Let us imagine an experiment which aims at examining the influence of the variation of two factors, A and B, in a certain production process. The factors may, e. g., be temperature, viscosity, reaction time, speed or settings of machines, the quality or quantity of raw materials used, etc., and the effect (the observations) may, e. g., be read from the yield of the process or the quality of the product. Let the "normal" values of the two experimental factors be A_1 and B_1, and let the aim of the experiment be to examine the effect of a change in A from A_1 to A_2 and a change in B from B_1 to B_2. According to the classical design of experiments the procedure would probably have been as follows:

1. m "control" experiments are made, i. e., m experiments in which the combination of factors is A_1B_1.

2. The effect of varying A is elucidated by making m experiments in which A has changed from A_1 to A_2, while B retains the value B_1, i. e., m experiments in which the combination of factors is A_2B_1.

3. The effect of B is elucidated by m experiments in which the combination of factors is A_1B_2.

Let the means of the above three sets of observations be $\overline{X}_{11}, \overline{X}_{21}$, and $\overline{X}_{12}$, respectively, and let the variance within the sets be the same, σ^2 say, the variance of the three means then being σ^2/m.

The effects of changing factors A and B, respectively, may then be estimated from the differences

$$D_A = \overline{X}_{21} - \overline{X}_{11} \qquad (17.10.1)$$

and

$$D_B = \overline{X}_{12} - \overline{X}_{11}, \qquad (17.10.2)$$

both having the variance

$$\mathcal{V}\{D_A\} = \mathcal{V}\{D_B\} = \frac{2\sigma^2}{m}. \qquad (17.10.3)$$

According to FISHER's principles experiments are made with *all combinations of the factors*, i. e., besides the above-mentioned combinations A_1B_1, A_2B_1, and A_1B_2, also the combination A_2B_2. Experiments of this type are called *factorial experiments* or multiple factor experiments in contrast to the classical single factor experiment. Performing n replications for each of the four treatment combinations we find the means $\bar{x}_{11}, \bar{x}_{21}, \bar{x}_{12}$, and $\bar{x}_{22}$. The effect of varying factor A may now be calculated partly for $B = B_1$, giving

$$d_{A|B_1} = \bar{x}_{21} - \bar{x}_{11}, \qquad (17.10.4)$$

being analogous with (17.10.1), and partly for $B = B_2$, which leads to

$$d_{A|B_2} = \bar{x}_{22} - \bar{x}_{12}. \qquad (17.10.5)$$

This latter quantity cannot be calculated according to the classical design, as it implies a change in both factors.

The main effect of changing factor A is found as the mean of the two conditional effects, i. e.,

$$d_A = \tfrac{1}{2}(d_{A|B_1}+d_{A|B_2}) = \tfrac{1}{2}((\bar{x}_{21}-\bar{x}_{11})+(\bar{x}_{22}-\bar{x}_{12})) = \bar{x}_{2\cdot}-\bar{x}_{1\cdot}\,. \qquad (17.10.6)$$

Correspondingly the main B-effect is calculated from the formula

$$d_B = \tfrac{1}{2}(d_{B|A_1}+d_{B|A_2}) = \tfrac{1}{2}((\bar{x}_{12}-\bar{x}_{11})+(\bar{x}_{22}-\bar{x}_{21})) = \bar{x}_{\cdot 2}-\bar{x}_{\cdot 1}\,. \qquad (17.10.7)$$

We thus see that *in the factorial experiment all the experimental results are utilized for the estimation of each main effect*, while in the classical experiment only $\frac{2}{3}$ of the total number of observations are used for calculating each of these effects.

Since the variance of each of the four means is σ^2/n, the variances of d_A and d_B are identical and equal to

$$\mathcal{V}\{d_A\} = \mathcal{V}\{d_B\} = \frac{\sigma^2}{n}. \qquad (17.10.8)$$

Further, the factorial experiment yields an estimate of the *interaction of the factors*, which is excluded by the classical design. If the two conditional effects d_{A/B_1} and d_{A/B_2} differ significantly, the A-effect must presumably be dependent on B, i. e., there exists an interaction between the effects of the two factors. This interaction is indicated by the difference

$$d_{AB} = \tfrac{1}{2}(d_{A|B_2}-d_{A|B_1}) = \tfrac{1}{2}(d_{B|A_2}-d_{B|A_1}) = \tfrac{1}{2}(\bar{x}_{22}-\bar{x}_{12}-\bar{x}_{21}+\bar{x}_{11}), \qquad (17.10.9)$$

where the conventional factor $\frac{1}{2}$ is used to obtain an expression analogous to (17.10.6) and (17.10.7). If, however, this quantity does not differ significantly from zero, we may accept the hypothesis that the effect of a change in A will be independent of the value of B, cf. p. 461.

In the case of no interaction d_A and D_A are both estimates of the same quantity with variances σ^2/n and $2\sigma^2/m$, respectively. If the total number of experiments made according to the two designs is the same, i. e., $3m=4n$, we find that in the factorial experiment the effects of the two factors have been determined with a variance which is only $\frac{2}{3}$ of the variance of the classical experiment. Or, in other words, according to FISHER's experimental design the total number of experiments needs to be only $\frac{2}{3}$ of the number necessary according to the classical design in order to obtain the same precision in the estimation of the main effects.

If interaction exists it may be estimated from the results of a factorial experiment, while the classical experimental design excludes the possible interactions from being investigated.

The above analysis of the factorial experiment may be arranged in a table corresponding to that for the analysis of variance as shown in Table 17.9, which is analogous with Table 16.25.

TABLE 17.9.
Means of observations in a 2×2 factorial experiment.

Factor	B_1	B_2	Mean
A_1	$\bar{x}_{11}$	$\bar{x}_{12}$	$\bar{x}_{1\cdot}$
A_2	$\bar{x}_{21}$	$\bar{x}_{22}$	$\bar{x}_{2\cdot}$
Mean	$\bar{x}_{\cdot 1}$	$\bar{x}_{\cdot 2}$	$\bar{x}_{\cdot\cdot}$

For each of the four treatment combinations we have n replications from which an *SSD* may be computed, cf. Table 16.26. Pooling the four *SSD*'s and the corresponding degrees of freedom we obtain

$$s_1^2 = \frac{1}{4(n-1)} \sum_{i=1}^{2} \sum_{j=1}^{2} \sum_{\nu=1}^{n} (x_{ij\nu} - \bar{x}_{ij})^2 , \; f = 4(n-1) . \qquad (17.10.10)$$

The whole analysis is summarized in the following analysis of variance table, which may be derived from Table 16.30 for $k = m = 2$ by simple reductions of the sums of squares.

TABLE 17.10.
Analysis of variance for a 2×2 factorial experiment.

Variation	*SSD*	f	s^2
Between A_1 and A_2	$nd_A^2 = n(\bar{x}_{1\cdot} - \bar{x}_{2\cdot})^2$	1	s_4^2
Between B_1 and B_2	$nd_B^2 = n(\bar{x}_{\cdot 1} - \bar{x}_{\cdot 2})^2$	1	s_3^2
Interaction	$nd_{AB}^2 = \frac{1}{4} n(\bar{x}_{22} - \bar{x}_{12} - \bar{x}_{21} + \bar{x}_{11})^2$	1	s_2^2
Within sets	$\sum_{i=1}^{2} \sum_{j=1}^{2} \sum_{\nu=1}^{n} (x_{ij\nu} - \bar{x}_{ij})^2$	$4(n-1)$	s_1^2
Total	$\sum_{i=1}^{2} \sum_{j=1}^{2} \sum_{\nu=1}^{n} (x_{ij\nu} - \bar{x}_{\cdot\cdot})^2$	$4n-1$	

The partitioning of the *SSD* for the four treatment means into the three terms in Table 17.10, each representing a single degree of freedom, may be elucidated by the following table of coefficients, giving the three orthogonal functions employed, cf. p. 431.

TABLE 17.11.
Coefficients of treatment means in the partitioning of the *SSD* for the treatment means. Divisor = 2.

Effects	Treatment combination			
	A_2B_2	A_2B_1	A_1B_2	A_1B_1
	$\bar{x}_{22}$	$\bar{x}_{21}$	$\bar{x}_{12}$	$\bar{x}_{11}$
d_A	+1	+1	−1	−1
d_B	+1	−1	+1	−1
d_{AB}	+1	−1	−1	+1

The coefficients for the A-effect are +1 for all treatment combinations including A_2 and −1 for all combinations including A_1, and similarly for the B-effect. The coefficients for the interaction may be found formally by multiplying the two corresponding rows of coefficients for A and B. All coefficients have to be divided by 2.

It will be seen that the observations from a factorial experiment with two factors which take on k and m values, respectively, may be analyzed according to the rules given in § 16.7.

The advantage of the factorial experiment over the classical one increases as the number of factors to be examined increases. The results obtained in factorial experiments with three factors or more should be analyzed according to the rules stated in § 16.8. Here we shall merely outline the procedure for an experiment with three factors, A, B, and C, which take on two values each, the 8 possible combinations being $A_1B_1C_1$, $A_1B_1C_2$, $A_1B_2C_1$, $A_1B_2C_2$, $A_2B_1C_1$, $A_2B_1C_2$, $A_2B_2C_1$, and $A_2B_2C_2$. Making n replications for each of the eight treatment combinations we find the means as shown in Table 17.12.

TABLE 17.12.
Means of observations in a $2\times2\times2$ factorial experiment.

	A_1		A_2	
	B_1	B_2	B_1	B_2
C_1	$\bar{x}_{111}$	$\bar{x}_{121}$	$\bar{x}_{211}$	$\bar{x}_{221}$
C_2	$\bar{x}_{112}$	$\bar{x}_{122}$	$\bar{x}_{212}$	$\bar{x}_{222}$

A_1 is present in four of the eight combinations, together with all possible combinations of the other two factors, and similarly for A_2. The main effect of A is therefore estimated from the difference

$$d_A = \bar{x}_{2\cdot\cdot} - \bar{x}_{1\cdot\cdot} \,. \qquad (17.10.11)$$

Because of the symmetrical arrangement of the experiment the effect of B and C presents itself in the same manner in the two means $\bar{x}_{2\cdot\cdot}$ and $\bar{x}_{1\cdot\cdot}$, wherefore the effect of these two factors is eliminated in the difference. Correspondingly, the main effects of B and C may be calculated as

$$d_B = \bar{x}_{\cdot 2\cdot} - \bar{x}_{\cdot 1\cdot}$$

and

$$d_C = \bar{x}_{\cdot\cdot 2} - \bar{x}_{\cdot\cdot 1} .$$

The difference d_A may also be derived as the mean of the four conditional A-effects, i. e.,

$$d_A = \tfrac{1}{4}(d_{A|B_1C_1} + d_{A|B_1C_2} + d_{A|B_2C_1} + d_{A|B_2C_2}) , \qquad (17.10.12)$$

where

$$d_{A|B_1C_1} = \bar{x}_{211} - \bar{x}_{111} ,$$

$$d_{A|B_1C_2} = \bar{x}_{212} - \bar{x}_{112} ,$$

$$d_{A|B_2C_1} = \bar{x}_{221} - \bar{x}_{121} ,$$

and

$$d_{A|B_2C_2} = \bar{x}_{222} - \bar{x}_{122} ,$$

giving

$$d_A = \tfrac{1}{4}(\bar{x}_{222} - \bar{x}_{122} + \bar{x}_{221} - \bar{x}_{121} + \bar{x}_{212} - \bar{x}_{112} + \bar{x}_{211} - \bar{x}_{111}) = \bar{x}_{2\cdot\cdot} - \bar{x}_{1\cdot\cdot} . \qquad (17.10.13)$$

The conditional A-effects elucidate the interactions of A and B and of A and C. Calculation of the means of the first two and the last two differences leads to

$$d_{A|B_1\bar{C}} = \bar{x}_{21\cdot} - \bar{x}_{11\cdot}$$

and

$$d_{A|B_2\bar{C}} = \bar{x}_{22\cdot} - \bar{x}_{12\cdot} .$$

These two differences indicate the A-effect for $B = B_1$ and $B = B_2$, respectively, the factor C entering into these two differences in the same manner. The interaction between A and B is therefore tested by comparing $d_{A|B_1\bar{C}}$ and $d_{A|B_2\bar{C}}$, i. e., by calculating the difference

$$d_{AB} = \tfrac{1}{2}(\bar{x}_{22\cdot} - \bar{x}_{12\cdot} - \bar{x}_{21\cdot} + \bar{x}_{11\cdot}) , \qquad (17.10.14)$$

and making a t-test for this difference, the theoretical value being zero. Correspondingly the interaction AC is given by

$$d_{AC} = \tfrac{1}{2}(\bar{x}_{2\cdot 2} - \bar{x}_{1\cdot 2} - \bar{x}_{2\cdot 1} + \bar{x}_{1\cdot 1}) .$$

The interaction d_{AB} may also be found as the mean of the interactions of A and B, conditioned by $C = C_1$ and $C = C_2$, respectively, i. e.,

$$d_{AB} = \tfrac{1}{2}(d_{AB|C_1} + d_{AB|C_2}) , \qquad (17.10.15)$$

where

$$d_{AB|C_1} = \tfrac{1}{2}(\bar{x}_{221} - \bar{x}_{121} - \bar{x}_{211} + \bar{x}_{111}) \qquad (17.10.16)$$

and

$$d_{AB|C_2} = \tfrac{1}{2}(\bar{x}_{222} - \bar{x}_{122} - \bar{x}_{212} + \bar{x}_{112}) , \qquad (17.10.17)$$

leading to

$$d_{AB} = \tfrac{1}{4}(\bar{x}_{222}-\bar{x}_{122}-\bar{x}_{212}+\bar{x}_{112}+\bar{x}_{221}-\bar{x}_{121}-\bar{x}_{211}+\bar{x}_{111})$$

$$= \tfrac{1}{2}(\bar{x}_{22\cdot}-\bar{x}_{12\cdot}-\bar{x}_{21\cdot}+\bar{x}_{11\cdot}) \ . \qquad (17.10.18)$$

The interaction between A, B, and C may therefore be tested by comparing the conditional AB-interactions, i. e., by calculating the difference

$$d_{ABC}=\tfrac{1}{2}(d_{AB|C_2}-d_{AB|C_1}) = \tfrac{1}{4}(\bar{x}_{222}-\bar{x}_{122}-\bar{x}_{212}+\bar{x}_{112}-\bar{x}_{221}+\bar{x}_{121}+\bar{x}_{211}-\bar{x}_{111}) \ , \qquad (17.10.19)$$

and making a t-test for this difference.

The above expressions for main effects and interactions may be summarized as shown in the following table, which gives the coefficients of the eight treatment means.

TABLE 17.13.
Coefficients of treatment means in the partitioning of the *SSD* for the treatment means. Divisor = 4.

Effects	Treatment combination							
	$A_2B_2C_2$ $\bar{x}_{222}$	$A_2B_2C_1$ $\bar{x}_{221}$	$A_2B_1C_2$ $\bar{x}_{212}$	$A_2B_1C_1$ $\bar{x}_{211}$	$A_1B_2C_2$ $\bar{x}_{122}$	$A_1B_2C_1$ $\bar{x}_{121}$	$A_1B_1C_2$ $\bar{x}_{112}$	$A_1B_1C_1$ $\bar{x}_{111}$
d_A	+1	+1	+1	+1	−1	−1	−1	−1
d_B	+1	+1	−1	−1	+1	+1	−1	−1
d_C	+1	−1	+1	−1	+1	−1	+1	−1
d_{AB}	+1	+1	−1	−1	−1	−1	+1	+1
d_{AC}	+1	−1	+1	−1	−1	+1	−1	+1
d_{BC}	+1	−1	−1	+1	+1	−1	−1	+1
d_{ABC}	+1	−1	−1	+1	−1	+1	+1	−1

The coefficients for the main effects are +1 for all combinations including the higher level of the factor and −1 for the lower level. All other coefficients may be found by multiplying the corresponding coefficients for the two or three main effects. The divisor equals 4 in all cases.

The above analysis may be summarized as in Table 17.14, which results from Table 16.39 for $k = m = q = 2$.

The v^2-test is in the present case identical with the t-test since the variances $s_2^2, s_3^2, \ldots, s_8^2$ all have one degree of freedom.

It will be seen that the results of a factorial experiment with three factors at k, m, and q levels, respectively, may be analyzed according to the principles laid down in § 16.8.

TABLE 17.14.
Analysis of variance for a $2\times2\times2$ factorial experiment.

Variation	SSD	f	s^2
Between A_1 and A_2	$2nd_A^2$	1	s_8^2
Between B_1 and B_2	$2nd_B^2$	1	s_7^2
Between C_1 and C_2	$2nd_C^2$	1	s_6^2
Interaction AB	$2nd_{AB}^2$	1	s_5^2
Interaction AC	$2nd_{AC}^2$	1	s_4^2
Interaction BC	$2nd_{BC}^2$	1	s_3^2
Interaction ABC	$2nd_{ABC}^2$	1	s_2^2
Within sets	$\sum_i \sum_j \sum_r \sum_v (x_{ijrv}-\bar{x}_{ijr})^2$	$8(n-1)$	s_1^2
Total	$\sum_i \sum_j \sum_r \sum_v (x_{ijrv}-\bar{x}\ldots)^2$	$8n-1$	

A $2\times2\times4$ factorial experiment has been analyzed in Example 16.8. The factors of primary interest are the press-sheets and the moulds. The third factor, the curing time, has been included in the experiment to investigate whether or not the strength reacts in the same manner to changes in the press-sheets and moulds for all curing times employed in usual testing practice.

FISHER has pointed out *the importance of including such subsidiary factors in the experiment* to give the conclusions a broader background.

Factorial experiments should be designed in accordance with the principles discussed in the previous paragraphs. The n replications of a $2\times2\times2$ factorial experiment, say, may be arranged in n randomized blocks, each block containing the eight treatment combinations. If, however, the number of treatment combinations is large, the experimental units within blocks may display a considerable variation and cause a large error variance. FISHER has therefore developed a device known as *confounding*, whereby each replication covers two or more blocks, the error variance still being dependent only on variations within blocks. The consequence of this design is that some comparisons, interactions of high order say, will be linked together with block differences, it being impossible to estimate these components separately.

17.11. Notes and References.

The present chapter has only outlined some basic principles regarding the design and analysis of sampling investigations and experiments. For a more detailed discussion reference must be made to the more specialized works.

Two comprehensive works exist on the design and analysis of sampling investigations: W. E. DEMING: *Some Theory of Sampling*, John Wiley, New York, 1950, 602 pp., and F. YATES: *Sampling Methods for Censuses and Surveys*, Charles Griffin, London, 1949, 318 pp., the former being more mathematical and treating more applications to industrial problems than the latter.

The fundamental work on the design of experiments is R. A. FISHER: *The Design of Experiments*, Oliver and Boyd, Edinburgh (1935), 4th edt. 1947. Further we may mention F. YATES: *The Design and Analysis of Factorial Experiments*, Imperial Bureau of Soil Science, J. WISHART: *Field Trials: Their Lay-out and Statistical Analysis*, Imperial Bureau of Plant Breeding and Genetics, 1940, K. A. BROWNLEE: *Industrial Experimentation*, His Majesty's Stationery Office, London, 1946, which contains a detailed exposition without proofs of the application of the analysis of variance to factorial experiments in industry, and W. G. COCHRAN and G. M. COX: *Experimental Designs*, John Wiley, New York, 1950, which is mainly a handbook with detailed descriptions and examples of a wealth of experimental plans and the circumstances under which each plan may be employed.

CHAPTER 18.

LINEAR REGRESSION ANALYSIS WITH ONE INDEPENDENT VARIABLE

18.1. Introduction.

An objective analysis of the association between two (or more) variables must to a great extent be based on statistical methods. By statistical analysis we obtain a mathematical description of the relationship between the variables and a discussion of the uncertainty of the relationship found.

It is, however, necessary to distinguish between several ways of attacking the problem, depending on the characteristics of the observations and on the aim of the analysis. The following two chapters give an exposition of two statistical methods—regression analysis and correlation analysis—for the handling of two-dimensional observations. For certain problems other methods are necessary, but as the development of the theory involved is far from complete, only one of them, the confluence analysis, will be briefly described here. In Chapter 20 regression and correlation analysis are generalized to the case of more than two variables.

The literature on the problems dealt with in Chapters 18—20 is referred to collectively in § 20.9.

A short outline of the problems which form the basis of regression, correlation, and confluence analysis, respectively, is given in the following.

1. *Regression analysis.*

Consider the distribution of one of the two variables, y say, *for given values of the other variable*, x, i. e., the distribution functions $p\{y|x_1\}, \ldots, p\{y|x_k\}$. The variable y is supposed to be a *stochastic* variable, whose distribution function depends on the variable x. In particular the population mean of y will be a function of x, e.g., $\eta = \mathcal{M}\{y|x\} = f(x) = f(x;\, \alpha, \beta, \gamma, \ldots)$. In regression analysis the functional form of η is assumed to be known and estimates of the unknown parameters $\alpha, \beta, \gamma, \ldots$ are to be determined from the data together with an estimate of the variance $\mathcal{V}\{y|x\}$.

No assumptions are made regarding the distribution of the x's. Regression analysis is carried out in the same manner whether x is a stochastic or a non-stochastic variable, and whether x is able in principle to take all values in a certain interval or is only able to take discrete values.

Regression analysis is, for example, used to determine the association between two variables *in experiments where one of the variables is a non-stochastic variable, the values of which are predetermined when the experiments are planned.*

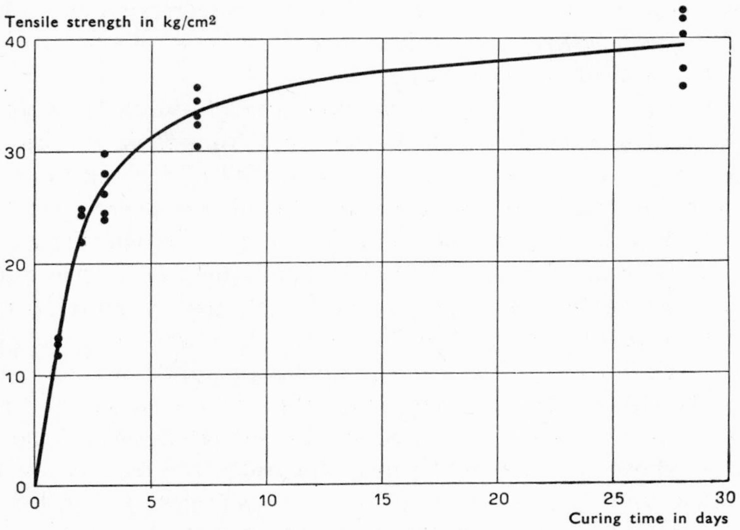

FIG. 18.1. The dependence of the tensile strength of test pieces of cement upon the curing time.

An example of such an experiment is the determination of the tensile strength of test pieces of cement, when we want to investigate how the strength depends on the curing time.

The curing time is the non-stochastic variable, the values of which are predetermined when the experiment is planned. Corresponding to a given curing time the tensile strength is a stochastic variable, and the mean values of the tensile strengths of the samples form an increasing function of the curing time, as shown in Fig. 18.1, in which the results of such an experiment have been plotted.

Regression analysis may also be applied to observations which do not originate from experiments but which result from a random choice of elements from a population, where each element is associated with a pair of numbers, cf. the following paragraph.

2. *Correlation analysis.*

Here *both variables are stochastic*, cf. Table 4.7, p. 82, which gives the association between the percentage of starch and the specific gravity of

560 *samples of potatoes, chosen at random*. The association between the variables may be described by *a two-dimensional distribution function.* In principle the association is characterized by *calculating estimates of the parameters of this two-dimensional distribution*, but this characteristic is often supplemented by *a regression analysis*, in which the mean value of y is determined as a function of x, either because variations in the size of x are the *cause* of variations in y, or because the regression curve furnishes us with a *suitable description* of the association between the two variables for certain practical purposes.

When buying potatoes for the manufacture of starch it is sensible to estimate the starch contents of the potatoes from their specific gravity, since the determination of the specific gravity is quicker and less expensive than determination of the starch content. In this case, therefore, the mean percentage of starch is expressed as a function of the specific gravity, cf. Fig. 4.4, p. 84, while for the purpose in hand there is no point in trying to express the average specific gravity as a function of the starch percentage.

Correlation analysis may often be applied to observations made in industrial plants during *normal working conditions*, when random variations in the properties of the raw materials or in certain factors pertaining to the production process cause variations in the properties of the finished product, as shown in Fig. 18.2, which illustrates the association between the vacuum in the glass-oven and the percentage of rejected glass[1]). This kind of analysis must, however, be made with the greatest caution;

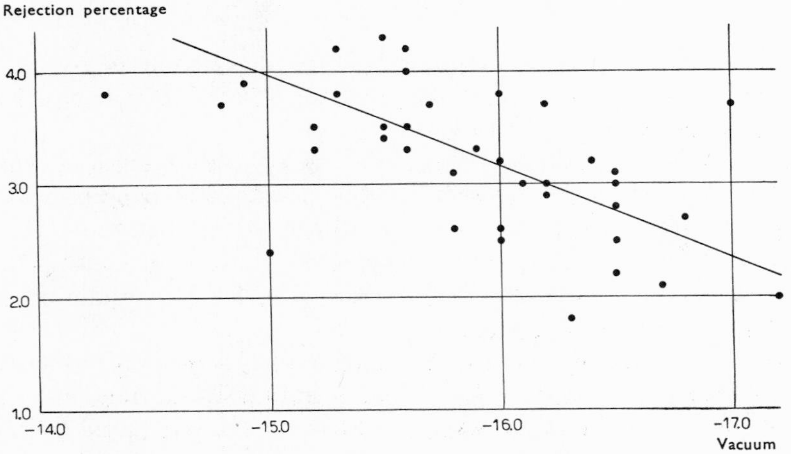

FIG. 18.2. Association between vacuum in the glass-oven and rejection percentage in glass manufacture.

[1]) PLAUT, H. C.: *Betriebliche Grosszahl-Forschung in der Glasindustrie*, Berichte der Fachausschüsse der Deutschen Glastechnischen Gesellschaft, Nr. 19, 1931.

other factors than the two observed ones may vary, and since in principle a statistical analysis can establish only a *stochastic* dependence, the question as to whether the dependence is also *causal* must be further investigated from a professional point of view, cf. § 1.6.

3. *Confluence analysis.*

Here each of the two variables is partitioned into a "structural component" and a stochastic component,

$$x = \xi + x'$$

$$y = \eta + y' \, ,$$

where $\mathcal{M}\{x\} = \xi$ and $\mathcal{M}\{y\} = \eta$, so that $\mathcal{M}\{x'\} = \mathcal{M}\{y'\} = 0$. The association between the two variables arises on account of an *association between the two structural components*, (ξ, η), i. e., the structural components satisfy a relation of the form $f(\xi, \eta) = f(\xi, \eta; \alpha, \beta, \gamma, \ldots) = 0$. In confluence analysis an estimate is determined of the parameters $\alpha, \beta, \gamma, \ldots$ in the structural equation, as well as of the parameters in the distribution of the stochastic components (x', y').

Regression analysis presents itself as a special case of confluence analysis, the variance of x' being equal to zero.

The confluence analysis may be applied to the determination of certain "laws" in technical work in cases where two properties are both subject to errors of measurement, and a description of their association is required. In this case the structural components denote the theoretical values of the two properties, while the stochastic components represent the errors of measurement.

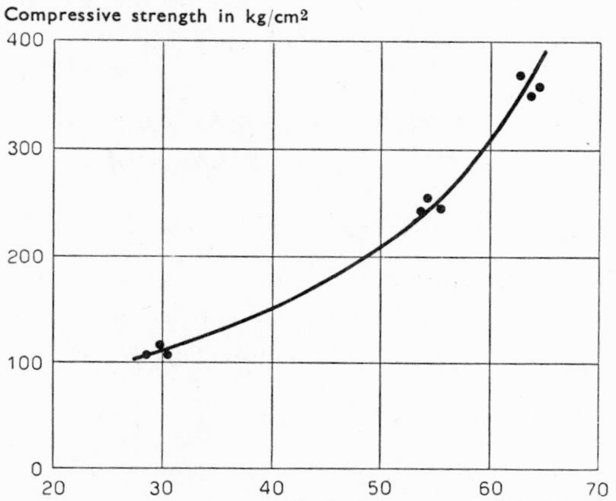

FIG. 18.3. Relation between the bending-stress strengths and the compressive strengths of test pieces of cement.

Fig. 18.3 shows the relation between the bending-stress and the compressive strengths of test pieces of cement. 3×3 samples are taken from a batch of cement, and tested after 3, 7, and 28 days' curing. The bending-stress strength is first tested for each sample. Hereby each test piece is broken into two pieces, and the compressive strength is determined for each piece separately, and lastly the mean is calculated for each pair. Thus, for each curing period we have three bending-stress strengths and three pairs of compressive strengths which give us estimates of the theoretical values. Through a confluence analysis an estimate is obtained of the relation between the bending-stress strengths and the compressive strengths.

It is characteristic of the above example that the structural components of the two variables both are functions of a third variable, namely the curing time. In the present case this variable has also been observed, and therefore it may be included in the analysis if we so wish. In other cases the factors leading to the confluence between the observed variables have not been observed or it is not possible to express them quantitatively.

Confluence analysis has been applied particularly in economics when the relation between two or more variables that are observed at successive times is to be examined, e. g., the relation between the demand for butter, the price of butter and the price of margarine. In most industries automatic registration of variations in the factors, which are meant to be of importance as regards the working of a plant, takes place at regular intervals. Probably statistical analysis of data of this kind will often call for the application of confluence analysis.

18.2. Hypotheses Underlying Regression Analysis.

In regression analysis one variable, x, is denoted the *independent* variable, and the other variable, y, the *dependent* variable.

In what follows it is assumed that

1. Corresponding to each value of x, *y is normally distributed.*
2. *The mean value of y, $\mathcal{M}\{y|x\}$, is a function of x,*

$$\eta = f(x) = f(x;\ \alpha, \beta, \gamma, \ldots)\ , \tag{18.2.1}$$

which includes certain unknown constants or *parameters*, $\alpha, \beta, \gamma, \ldots$. *The type of the function is known*, and the function *is linear as regards the parameters*, i. e., the function for example takes the form

$$f(x) = \alpha+\beta x+\gamma x^2 \tag{18.2.2}$$

or

$$f(x) = \alpha+\beta x+\gamma \log x\ . \tag{18.2.3}$$

The graphical representation of the function $\eta = f(x)$ is called *the theoretical regression curve.*

3. *The variance of y,* $\mathcal{V}\{y|x\}$, *is constant or proportional to a known function of* x.

In regression analysis we determine estimates of the parameters α, β, γ, ..., and the parameter in $\mathcal{V}\{y|x\}$.

Let us assume that the data are given in the form shown in Table 18.1, the observations being stochastically independent.

TABLE 18.1.

Independent variable	Dependent (stochastic) variable	Sample means	Population means
x_1	$y_{11}, y_{12}, \ldots, y_{1n_1}$	$\bar{y}_1$	$\eta_1 = f(x_1)$
x_2	$y_{21}, y_{22}, \ldots, y_{2n_2}$	$\bar{y}_2$	$\eta_2 = f(x_2)$
$\vdots$	$\vdots \quad \vdots \quad \vdots$	$\vdots$	$\vdots$
x_k	$y_{k1}, y_{k2}, \ldots, y_{kn_k}$	$\bar{y}_k$	$\eta_k = f(x_k)$

An estimate of the course of the function $\eta = f(x)$ may be obtained by plotting the k pairs of numbers $(x_i, \bar{y}_i)$, $i = 1, 2, \ldots, k$, in a coordinate system, see Fig. 4.4, p. 84. (In the special case when n_i is equal to one, i. e., only one value of the dependent variable corresponds to each value of the independent variable, we have $\bar{y}_i = y_{i1} = y_i$). Based on this diagram and our *theoretical* (professional) knowledge regarding the processes that lead to the relationship between the variables, we set up *a hypothesis regarding the type of the function* $\eta = f(x)$. Even though our theoretical knowledge is often very limited, e. g., we only know about the asymptotes of the function, it is very important to make use of such knowledge, as in this manner we are often able to exclude certain types of functions a priori.

It can be proved that the "best" estimates of the parameters in the equation of the regression curve are obtained by application of the method of least squares, i. e., by determining the values of $\alpha, \beta, \gamma, \ldots$ which minimize the sum of squares

$$\sum_{i=1}^{k} \sum_{\nu=1}^{n_i} (y_{i\nu} - f(x_i;\ \alpha, \beta, \gamma, \ldots))^2 , \qquad (18.2.4)$$

cf. § 8.8. By this method the determination of the empirical regression curve is such that the sum of squares of the deviations between the observed values of y and the corresponding values on the curve takes the least possible value. *The estimates of the parameters* obtained in this manner are "best" in the sense that they are *normally distributed with the required parameters as mean values and with the least possible variances.*

In the following we will deal with the theory of regression when the function $f(x)$ is a linear function of x. The cases where $f(x)$ takes the form (18.2.2), (18.2.3), or an analogous form will be dealt with in Chapter 20, in which regression analysis with several independent variables is developed. Since

$$f(x) = \alpha+\beta x+\gamma \log x$$

may be written

$$f(x) = g(x_1, x_2) = \alpha+\beta x_1+\gamma x_2 , \tag{18.2.5}$$

where $x_1 = x$ and $x_2 = \log x$, it follows that the theory of Chapter 20 applies.

Regression analysis may be considered as an extension of the analysis of variance. In the analysis of variance the hypothesis $\eta_1 = \eta_2 = \ldots = \eta_k$ is tested by comparing the variance within sets and between sets. If it is possible to give a quantitative expression for the criteria used for grouping, so that the number of the set may be substituted by an (independent) variable, a regression curve may be determined, and the variation between sets may be partitioned into the random variation about the regression curve and the variation between the values on the regression curve. The random variation about the regression curve may be combined with the variation within sets, and by comparing this variation with the variation between the values on the regression curve we may test the hypothesis $\eta_1 = \eta_2 = \ldots = \eta_k$, see Table 18.3.

18.3. Derivation of the Estimates of the Parameters.

The assumptions for the treatment of the *linear regression analysis with only one independent variable* as developed in the following paragraphs are:

1. y is normally distributed for every value of x.
2. The mean value of y corresponding to a given value of x is a *linear* function of x. This function may be written

$$\boxed{\mathcal{M}\{y|x\} = \eta = \alpha+\beta(x-\bar{x}),} \tag{18.3.1}$$

where x denotes the weighted mean of the x-values, i. e.,

$$\bar{x} = \frac{\sum_{i=1}^{k} n_i x_i}{\sum_{i=1}^{k} n_i}. \tag{18.3.2}$$

(The reason why η is written as in (18.3.1) instead of as $\eta = \alpha+\beta x$ will become apparent later).

3. The variance of y corresponding to a given value of x is constant

$$\boxed{\mathcal{V}\{y|x\} = \sigma^2 ,} \tag{18.3.3}$$

or proportional to a given function of x, i. e.,

$$\boxed{\mathcal{V}\{y|x\} = \sigma^2 h^2(x) .}$$

The latter case is dealt with in § 18.6.

4. The observations are stochastically independent.

We will now calculate estimates of the three parameters α, β, and σ^2 from the observations in Table 18.1.

Corresponding to each value of the independent variable we calculate the mean and the variance of the values of the dependent variable, i. e., corresponding to the independent variable $x = x_i$ we calculate the mean

$$\bar{y}_i = \frac{1}{n_i} \sum_{\nu=1}^{n_i} y_{i\nu} , \quad i = 1, 2, \ldots, k , \tag{18.3.4}$$

as an estimate of

$$\eta_i = \alpha + \beta(x_i - \bar{x}) , \quad i = 1, 2, \ldots, k , \tag{18.3.5}$$

and the variance

$$s_{1i}^2 = \frac{1}{n_i - 1} \sum_{\nu=1}^{n_i} (y_{i\nu} - \bar{y}_i)^2 , \quad f_i = n_i - 1 , \quad i = 1, 2, \ldots, k , \tag{18.3.6}$$

as an estimate of σ^2.

The hypothesis of a constant variance, σ^2, is tested by comparing the variances $s_{11}^2, s_{12}^2, \ldots, s_{1k}^2$, cf. § 11.6. If the hypothesis is not rejected, the k variances are combined to the estimate

$$\boxed{s_1^2 = \frac{\sum_{i=1}^{k} f_i s_{1i}^2}{\sum_{i=1}^{k} f_i} = \frac{\sum_{i=1}^{k} \sum_{\nu=1}^{n_i} (y_{i\nu} - \bar{y}_i)^2}{\sum_{i=1}^{k} n_i - k} .} \tag{18.3.7}$$

As in the analysis of variance, cf. (16.3.3), the total sum of squares of the deviations is partitioned by introducing the means in the following manner:

$$\sum_{i=1}^{k} \sum_{\nu=1}^{n_i} (y_{i\nu} - \eta_i)^2 = \sum_{i=1}^{k} \sum_{\nu=1}^{n_i} (y_{i\nu} - \bar{y}_i)^2 + \sum_{i=1}^{k} n_i (\bar{y}_i - \eta_i)^2 . \tag{18.3.8}$$

As $\Sigma\Sigma(y_{i\nu}-\eta_i)^2$ is distributed as $\sigma^2\chi^2$ with Σn_i degrees of freedom, and as the numbers of degrees of freedom for the two sums of squares on the right side are Σn_i-k and k, respectively, it follows from the partition theorem for the χ^2-distribution that the two sums of squares are stochastically independent, and distributed as $\sigma^2\chi^2$ with the respective numbers of degrees of freedom.

From the sum of squares $\Sigma\Sigma(y_{i\nu}-\bar{y}_i)^2$, which represents the *variation within sets*, we calculate the above estimate s_1^2 of σ^2. This estimate has an s^2-distribution with parameters $(\sigma^2, \Sigma n_i-k)$, and the estimate is *independent of the theoretical regression curve.*

From the sum of squares $\Sigma n_i(\bar{y}_i-\eta_i)^2$, which represents *the variation about the theoretical values,* we calculate the variance

$$s_{2(\eta)}^2 = \frac{1}{k}\sum_{i=1}^{k} n_i(\bar{y}_i-\eta_i)^2 , \tag{18.3.9}$$

which has an s^2-distribution with parameters (σ^2, k). By comparing $s_{2(\eta)}^2$ and s_1^2 a set of hypothetical values of $\eta_1, \eta_2, \ldots, \eta_k$, e. g., the values $\eta_i = \alpha+\beta(x_i-\bar{x})$, $i = 1, 2, \ldots, k$, may be tested. As a rule, however, it is not possible to specify the values of the η's, but we have only a hypothesis regarding the *form of the regression curve,* e. g., that it is linear. A test for this hypothesis will be developed later, see (18.3.28).

Introducing the notation

$$\boxed{Y = a+b(x-\bar{x})} \tag{18.3.10}$$

for the estimate of the theoretical regression line, and substituting the estimate $Y_i = a+b(x_i-\bar{x})$ for the theoretical value $\eta_i = \alpha+\beta(x_i-\bar{x})$ in the above sum of squares, we obtain the sum of squares

$$\sum_{i=1}^{k} n_i(\bar{y}_i-Y_i)^2 = \sum_{i=1}^{k} n_i(\bar{y}_i-a-b(x_i-\bar{x}))^2 \tag{18.3.11}$$

which represents the *variation about the empirical regression line,* the deviation $\bar{y}_i-Y_i$ denoting the vertical distance between the point $(x_i, \bar{y}_i)$ and the corresponding point (x_i, Y_i) on the empirical regression line. Fig. 18.4 gives an example of the relative positions of the points referred to.

The value of the sum of squares (18.3.11) depends on the values chosen for the estimates a and b. If these values are chosen according to the method of least squares, we must determine a and b in such a manner that the sum of squares takes the least possible value. Differentiating (18.3.11) with regard to a and b, respectively, we obtain

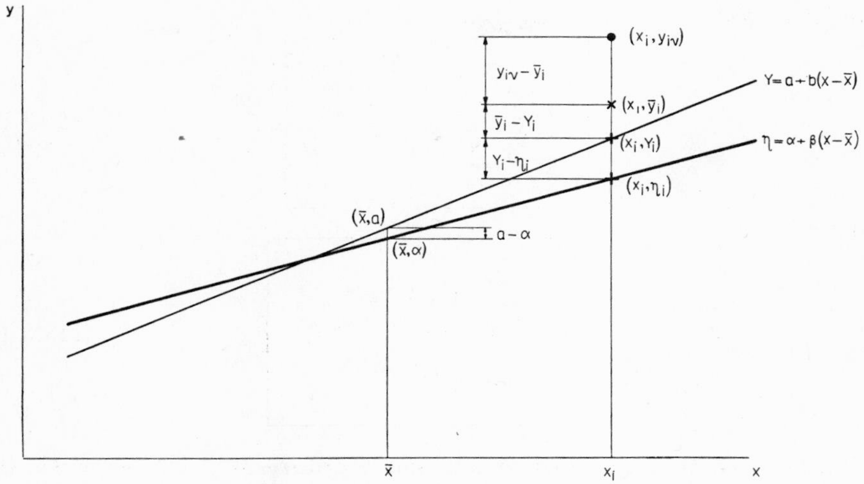

FIG. 18.4. Example of the relative positions of the theoretical and the empirical regression lines and the observations.

$$-2\sum_{i=1}^{k} n_i(\bar{y}_i - a - b(x_i - \bar{x}))$$

and

$$-2\sum_{i=1}^{k} n_i(\bar{y}_i - a - b(x_i - \bar{x}))(x_i - \bar{x}) .$$

Equating these two derivatives to zero, we obtain the following equations for the determination of a and b:

$$\sum_{i=1}^{k} n_i(\bar{y}_i - a - b(x_i - \bar{x})) = 0 \tag{18.3.12}$$

and

$$\sum_{i=1}^{k} n_i(\bar{y}_i - a - b(x_i - \bar{x}))(x_i - \bar{x}) = 0 . \tag{18.3.13}$$

Collecting the terms including a and b on the left side of the sign of equality, we have

$$a\sum_{i=1}^{k} n_i + b\sum_{i=1}^{k} n_i(x_i - \bar{x}) = \sum_{i=1}^{k} n_i\bar{y}_i \tag{18.3.14}$$

and

$$a\sum_{i=1}^{k} n_i(x_i - \bar{x}) + b\sum_{i=1}^{k} n_i(x_i - \bar{x})^2 = \sum_{i=1}^{k} n_i(x_i - \bar{x})\bar{y}_i . \tag{18.3.15}$$

Since from (18.3.2)

$$\sum_{i=1}^{k} n_i(x_i - \bar{x}) = 0$$

it follows that

$$\boxed{a = \frac{\sum_{i=1}^{k} n_i \bar{y}_i}{\sum_{i=1}^{k} n_i} = \bar{y} \approx \alpha} \tag{18.3.16}$$

and

$$\boxed{b = \frac{\sum_{i=1}^{k} n_i (x_i - \bar{x}) \bar{y}_i}{\sum_{i=1}^{k} n_i (x_i - \bar{x})^2} \approx \beta\,.} \tag{18.3.17}$$

Thus we see that the estimates a and b are *linear functions of the y's*, satisfying the linear relations given in (18.3.12) and (18.3.13).

The expression for the slope b may be transformed as follows:

$$b = \frac{\sum_{i=1}^{k} n_i (x_i - \bar{x})(\bar{y}_i - \bar{y})}{\sum_{i=1}^{k} n_i (x_i - \bar{x})^2} = \frac{\sum_{i=1}^{k} n_i (x_i - \bar{x})^2 \dfrac{\bar{y}_i - \bar{y}}{x_i - \bar{x}}}{\sum_{i=1}^{k} n_i (x_i - \bar{x})^2}. \tag{18.3.18}$$

If we introduce the notation

$$b_i = \frac{\bar{y}_i - \bar{y}}{x_i - \bar{x}}, \tag{18.3.19}$$

which denotes the slope of the straight line from the point $(\bar{x}, \bar{y})$ to the point $(x_i, \bar{y}_i)$, and also

$$w_i = n_i (x_i - \bar{x})^2\,, \tag{18.3.20}$$

b may be written

$$b = \frac{\sum_{i=1}^{k} w_i b_i}{\sum_{i=1}^{k} w_i}, \tag{18.3.21}$$

which means that the slope b may be interpreted as the weighted mean of the slopes b_i, $i = 1, 2, \ldots, k$, with weights w_i, which are proportional to *the number of observations*, n_i, and *the square of the distance between the abscissa* x_i *and the mean abscissa* $\bar{x}$. It follows that points with abscissas deviating considerably from the mean abscissa exert a comparatively large influence upon the slope of the empirical regression line.

As an estimate of the theoretical regression line $\eta = \alpha+\beta(x-\bar{x})$ we have the empirical regression line

$$\boxed{Y = \bar{y}+b(x-\bar{x})\ .} \tag{18.3.22}$$

According to (18.3.21) the equation for the empirical regression line may be interpreted as the weighted mean of the equations for the k lines which connect the points $(x_i, \bar{y}_i)$, $i = 1, 2, \ldots, k$, with the "mean point" $(\bar{x}, \bar{y})$, see Fig. 18.5, the equations of these lines being

$$y = \bar{y}+b_i(x-\bar{x})\ , \quad i = 1, 2, \ldots, k\ . \tag{18.3.23}$$

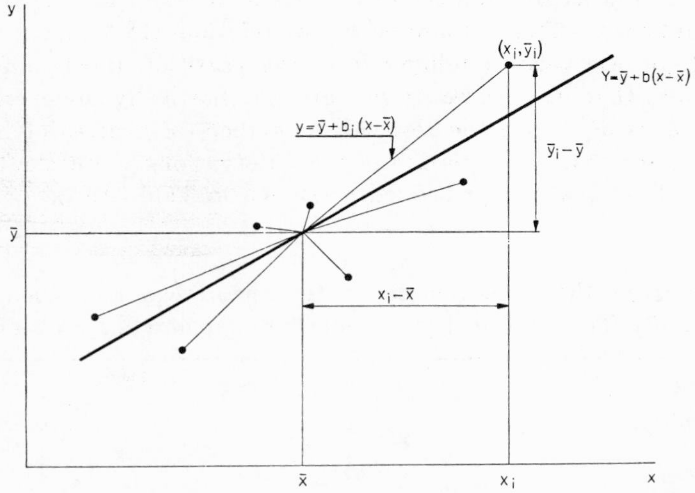

FIG. 18.5. The regression line and the lines which join the points $(x_i, \bar{y}_i)$ with the mean $(\bar{x}, \bar{y})$.

As the estimates of a and b are linear functions of the y's, the *distribution functions of a and b* may be derived by applying the addition theorem for the normal distribution, cf. § 9.1. It is, however, simpler to apply the partition theorem for the χ^2-distribution, whereby we can further prove that a and b are stochastically independent.

The deviations of the means from the theoretical values, $\bar{y}_i-\eta_i$, are partitioned by introducing the estimate Y_i

$$\begin{aligned}\bar{y}_i-\eta_i &= (\bar{y}_i-Y_i)+(Y_i-\eta_i)\\ &= (\bar{y}_i-Y_i)+(a-\alpha)+(b-\beta)(x_i-\bar{x})\ ,\end{aligned} \tag{18.3.24}$$

see Fig. 18.4. Squaring and summing leads to

$$\sum_{=1}^{k} n_i(\bar{y}_i-\eta_i)^2 = \sum_{i=1}^{k} n_i(\bar{y}_i-Y_i)^2+(a-\alpha)^2\sum_{i=1}^{k} n_i+(b-\beta)^2\sum_{i=1}^{k} n_i(x_i-\bar{x})^2 \quad (18.3.25)$$

where the product terms vanish because of (18.3.12) and (18.3.13), which may be written

$$\sum_{i=1}^{k} n_i(\bar{y}_i-Y_i) = 0 \quad (18.3.26)$$

and

$$\sum_{i=1}^{k} n_i(\bar{y}_i-Y_i)(x_i-\bar{x}) = 0\,. \quad (18.3.27)$$

As the sum of squares $\Sigma n_i(\bar{y}_i-\eta_i)^2$ is distributed as $\sigma^2\chi^2$ with k degrees of freedom, and as the number of degrees of freedom for the three terms on the right are $k-2$ (on account of the two relations (18.3.26) and (18.3.27)), 1, and 1, respectively, it follows from the partition theorem for the χ^2-distribution that the three terms are stochastically independent and distributed as $\sigma^2\chi^2$ with the respective numbers of degrees of freedom.

Thus we see that the variation of the observations about the theoretical regression line may be partitioned as shown in Table 18.2.

TABLE 18.2.

Partitioning of the variation about the theoretical regression line $\eta = \alpha+\beta(x-\bar{x})$. The empirical regression line is denoted $Y = a+b(x-\bar{x})$.

Variation	Sum of squares	Degrees of freedom	Variance
Deviation of a from α	$(a-\alpha)^2\sum_{i=1}^{k} n_i$	1	s_4^2
Deviation of b from β	$(b-\beta)^2\sum_{i=1}^{k} n_i(x_i-\bar{x})^2$	1	s_3^2
About the empirical regression line	$\sum_{i=1}^{k} n_i(\bar{y}_i-Y_i)^2$	$k-2$	s_2^2
Within sets	$\sum_{i=1}^{k}\sum_{\nu=1}^{n_i}(y_{i\nu}-\bar{y}_i)^2$	$\sum_{i=1}^{k} n_i-k$	s_1^2
About the theoretical regression line	$\sum_{i=1}^{k}\sum_{\nu=1}^{n_i}(y_{i\nu}-\eta_i)^2$	$\sum_{i=1}^{k} n_i$	

The variance

$$\boxed{s_2^2 = \frac{1}{k-2}\sum_{i=1}^{k} n_i(\bar{y}_i-Y_i)^2} \quad (18.3.28)$$

is based on the variation about the empirical regression line. If the hypothesis that the theoretical regression curve is linear is true, then s_2^2 will have an s^2-distribution with parameters $(\sigma^2, k-2)$. If, however, the hypothesis is false, we find in analogy with (16.4.4)

$$\mathcal{M}\{s_2^2\} = \sigma^2 + \frac{1}{k-2}\sum_{i=1}^{k} n_i(\eta_i - \bar{\eta} - \beta(x_i - \bar{x}))^2$$

where

$$\bar{\eta} = \Sigma n_i \eta_i / \Sigma n_i \quad \text{and} \quad \beta = \Sigma n_i (x_i - \bar{x})\eta_i / \Sigma n_i (x_i - \bar{x})^2,$$

the last term on the right representing the variation of the true values η_i about a straight line.

The hypothesis regarding the form of the regression curve is therefore tested by means of the variance ratio $v^2 = s_2^2/s_1^2$. If s_2^2 is significantly larger than s_1^2 the hypothesis of linearity must be rejected.

If the variance ratio is not significant, a pooled estimate s^2 of σ^2 may be calculated as

$$s^2 = \frac{\sum_{i=1}^{k}\sum_{\nu=1}^{n_i}(y_{i\nu} - \bar{y}_i)^2 + \sum_{i=1}^{k} n_i(\bar{y}_i - Y_i)^2}{\left(\sum_{i=1}^{k} n_i - k\right) + (k-2)} = \frac{\sum_{i=1}^{k}\sum_{\nu=1}^{n_i}(y_{i\nu} - Y_i)^2}{\sum_{i=1}^{k} n_i - 2}. \qquad (18.3.29)$$

The third and fourth terms of the partitioning of the variation about the theoretical regression line shows that

1. a is normally distributed about α with variance

$$\mathcal{V}\{a\} = \frac{\sigma^2}{\sum_{i=1}^{k} n_i}. \qquad (18.3.30)$$

2. b is normally distributed about β with variance

$$\mathcal{V}\{b\} = \frac{\sigma^2}{\sum_{i=1}^{k} n_i(x_i - \bar{x})^2}. \qquad (18.3.31)$$

3. a and b are *stochastically independent* and independent of s^2. For this reason we prefer the formula $Y = a + b(x - \bar{x})$ to $Y = (a - b\bar{x}) + bx = a_0 + bx$, as a_0 and b are not stochastically independent.

We thus see that the variance of a is inversely proportional to the total number of observations, while the variance of b is inversely proportional to the sum of squares of the x-deviations.

When *designing an experiment* where we wish to estimate a regression line the values of the independent variable should therefore be chosen as *far apart as possible* in order to determine the regression line with the least possible uncertainty. This holds good only if it is known a priori that the regression curve is linear. If the form of the regression curve is not known, it is generally preferable to distribute the x-values equidistantly over the total range of variation in order to obtain estimates of so many η-values that it is possible to set up and test a rational hypothesis regarding the form of the regression curve.

The estimate of the theoretical regression line $Y = a+b(x-\bar{x})$ is a linear function of a and b and in consequence Y itself is normally distributed about $\eta = \alpha+\beta(x-\bar{x})$ with variance

$$\boxed{\begin{aligned} \mathcal{V}\{Y\} &= \mathcal{V}\{a\}+\mathcal{V}\{b\}(x-\bar{x})^2 \\ &= \sigma^2\left(\frac{1}{\sum_{i=1}^{k} n_i}+\frac{(x-\bar{x})^2}{\sum_{i=1}^{k} n_i(x_i-\bar{x})^2}\right). \end{aligned}} \tag{18.3.32}$$

We thus see that the variance of Y is equal to σ^2 multiplied by a polynomial of the second degree in x, which takes its minimum value when $x = \bar{x}$. This means that the uncertainty of the estimate Y increases with the distance between the independent variable and $\bar{x}$.

As already stated, regression analysis may be interpreted as an extension of the analysis of variance. In the analysis of variance the deviation $y_{i\nu}-\bar{y}$ is partitioned into a sum of two deviations

$$y_{i\nu}-\bar{y} = (y_{i\nu}-\bar{y}_i)+(\bar{y}_i-\bar{y})\,,$$

by introducing the means of the k sets of observations, cf. (16.4.15) and (16.4.16) together with Table 16.4, which give the corresponding partitioning of the sum of squares.

In regression analysis the deviation $y_{i\nu}-\bar{y}$ is partitioned into a sum of three deviations

$$y_{i\nu}-\bar{y} = (y_{i\nu}-\bar{y}_i)+(\bar{y}_i-Y_i)+(Y_i-\bar{y})\,,$$

by introducing, besides the means of the k sets of observations, the corresponding k values on the regression line, see Fig. 18.4. Squaring and summing lead to the results given in Table 18.3, which is analogous to Table 16.4.

The variation between the values on the regression line may also be written as

$$\sum_{i=1}^{k} n_i(Y_i - \bar{y})^2 = b^2 \sum_{i=1}^{k} n_i(x_i - \bar{x})^2 .$$

The hypothesis regarding the linearity of the regression curve is tested by comparing s_1^2 and s_2^2. If v^2 is not significant s_1^2 and s_2^2 may be combined according to the addition theorem for the s^2-distribution to form the estimate s^2. Then the hypothesis $\beta = 0$ is tested by comparing s^2 and s_3^2, see (18.4.2). When comparing Table 18.2 and Table 18.3 we see that Table 18.3 may

TABLE 18.3.
Partitioning of the variation about the mean.

Variation	SSD	f	s^2	Tests
Between the values on the regression line	$\sum_{i=1}^{k} n_i(Y_i - \bar{y})^2$	1	s_3^2	$v^2 = \frac{s_3^2}{s^2}$
About the regression line	$\sum_{i=1}^{k} n_i(\bar{y}_i - Y_i)^2$	$k-2$	s_2^2	$v^2 = \frac{s_2^2}{s_1^2}$
Within sets	$\sum_{i=1}^{k}\sum_{\nu=1}^{n_i} (y_{i\nu} - \bar{y}_i)^2$	$\sum_{i=1}^{k} n_i - k$	s_1^2	
Total	$\sum_{i=1}^{k}\sum_{\nu=1}^{n_i} (y_{i\nu} - \bar{y})^2$	$\sum_{i=1}^{k} n_i - 1$		

be derived from Table 18.2 for $\beta = 0$, the following relation being valid:

$$\sum_{i=1}^{k}\sum_{\nu=1}^{n_i} (y_{i\nu} - \eta_i)^2 - (a - \alpha)^2 \sum_{i=1}^{k} n_i = \sum_{i=1}^{k}\sum_{\nu=1}^{n_i} (y_{i\nu} - \bar{y})^2 .$$

It often happens that *there is only one value of the dependent variable corresponding to each value of the independent variable*, the observations taking the form $(x_1, y_1), (x_2, y_2), \ldots, (x_k, y_k)$, which means that $n_i = 1$ and $y_{i1} = y_i$ for $i = 1, 2, \ldots, k$.

The empirical regression line $Y = a + b(x - \bar{x})$ is then determined by calculating the three quantities

$$\bar{x} = \frac{1}{k} \sum_{i=1}^{k} x_i , \tag{18.3.33}$$

$$a = \bar{y} = \frac{1}{k} \sum_{i=1}^{k} y_i , \tag{18.3.34}$$

and

$$b = \frac{\sum_{i=1}^{k} (x_i - \bar{x}) y_i}{\sum_{i=1}^{k} (x_i - \bar{x})^2}, \qquad (18.3.35)$$

cf. (18.3.2), (18.3.16), and (18.3.17) for $n_i = 1$.

The estimate a is normally distributed about α with variance σ^2/k, the estimate b is normally distributed about β with variance $\sigma^2/\Sigma(x_i - \bar{x})^2$, and the estimate Y is normally distributed about η with variance

$$\mathcal{V}\{Y\} = \sigma^2 \left(\frac{1}{k} + \frac{(x - \bar{x})^2}{\sum_{i=1}^{k} (x_i - \bar{x})^2} \right), \qquad (18.3.36)$$

see (18.3.30), (18.3.31), and (18.3.32).

As an estimate of the variance σ^2 we have

$$s_2^2 = \frac{1}{k-2} \sum_{i=1}^{k} (y_i - Y_i)^2 , \qquad (18.3.37)$$

cf. (18.3.29), since s_1^2, the variation within sets, does not exist. We therefore test the hypothesis regarding linearity by plotting the points (x_1, y_1), (x_2, y_2), . . ., (x_k, y_k) and the regression line $Y = a + b(x - \bar{x})$ in a coordinate system. A supplementary investigation should be carried out by studying the quantities $(y_i - Y_i)/s_2$, as for large values of k they are approximately normally distributed with parameters (0, 1) if the hypothesis regarding the linear regression with constant variance is correct.

If the hypothesis regarding the regression equation is $\mathcal{M}\{y|x\} = \beta x$ instead of (18.3.1), which means that the regression line passes through the origin, an analysis analogous to the one above may be made. As an estimate of the slope of the line we obtain

$$b = \frac{\Sigma n_i x_i \bar{y}_i}{\Sigma n_i x_i^2} \qquad (18.3.38)$$

with variance

$$\mathcal{V}\{b\} = \frac{\sigma^2}{\Sigma n_i x_i^2}. \qquad (18.3.39)$$

In Table 18.4 *a scheme for the computations pertaining to regression analysis* is given. The notation used is that given in Chapter 4. Moreover the symbol

$$SSD_{y|x} = \sum_{i=1}^{k} \sum_{\nu=1}^{n_i} (y_{i\nu} - Y_i)^2 , \quad f = N - 2 , \qquad (18.3.40)$$

and

$$SSD_{\bar{y}|x} = \sum_{i=1}^{k} n_i(\bar{y}_i - Y_i)^2 , \quad f = k-2 , \tag{18.3.41}$$

have been employed.

TABLE 18.4.
Computing scheme for regression analysis.

Observations		n	S	SS	S^2/n	SSD	f	s^2	S/n
x_1	$y_{11}, y_{12}, \ldots, y_{1n_1}$	n_1	S_1	SS_1	S_1^2/n_1	SSD_1	n_1-1	s_{11}^2	S_1/n_1
x_2	$y_{21}, y_{22}, \ldots, y_{2n_2}$	n_2	S_2	SS_2	S_2^2/n_2	SSD_2	n_2-1	s_{21}^2	S_2/n_2
$\vdots$	$\vdots \quad \vdots \quad \vdots$	$\vdots$	$\vdots$	$\vdots$	$\vdots$	$\vdots$	$\vdots$	$\vdots$	$\vdots$
x_k	$y_{k1}, y_{k2}, \ldots, y_{kn_k}$	n_k	S_k	SS_k	S_k^2/n_k	SSD_k	n_k-1	s_{k1}^2	S_k/n_k
Total		Σn_i	ΣS_i	ΣSS_i	$\Sigma S_i^2/n_i$	ΣSSD_i	$\Sigma n_i - k$		

$N = \Sigma n_i$	$S_x = \Sigma n_i x_i$	$S_{\bar{y}} = \Sigma n_i \bar{y}_i = \Sigma S_i$	
	$SS_x = \Sigma n_i x_i^2$	$SS_{\bar{y}} = \Sigma n_i \bar{y}_i^2 = \Sigma S_i^2/n_i$	$SP_{x\bar{y}} = \Sigma n_i x_i \bar{y}_i = \Sigma x_i S_i$
	S_x^2/N	$S_{\bar{y}}^2/N$	$S_x S_{\bar{y}}/N$
	$SSD_x = SS_x - S_x^2/N$	$SSD_{\bar{y}} = SS_{\bar{y}} - S_{\bar{y}}^2/N$	$SPD_{x\bar{y}} = SP_{x\bar{y}} - S_x S_{\bar{y}}/N$
	$= \Sigma n_i(x_i - \bar{x})^2$	$= \Sigma n_i(\bar{y}_i - \bar{y})^2$	$= \Sigma n_i(x_i - \bar{x})(\bar{y}_i - \bar{y})$
	$b = SPD_{x\bar{y}}/SSD_x$		
		$SPD_{x\bar{y}}^2/SSD_x = bSPD_{x\bar{y}}$	

$$SSD_{\bar{y}|x} = SSD_{\bar{y}} - \frac{SPD_{x\bar{y}}^2}{SSD_x} = \Sigma n_i(\bar{y}_i - Y_i)^2, \quad f = k-2 .$$

$$s_2^2 = \frac{SSD_{\bar{y}|x}}{k-2}, \quad s_1^2 = \frac{\Sigma SSD_i}{N-k}, \quad v^2 = \frac{s_2^2}{s_1^2} .$$

$$SSD_{y|x} = \Sigma SSD_i + SSD_{\bar{y}|x}, \quad f = N-2, \quad s^2 = \frac{SSD_{y|x}}{N-2} .$$

$$Y = a + b(x - \bar{x}), \quad a = \bar{y} = \frac{S_{\bar{y}}}{N}, \quad \bar{x} = \frac{S_x}{N} .$$

$$s_a^2 = \frac{s^2}{N}, \quad s_b^2 = \frac{s^2}{SSD_x} .$$

$$s_Y^2 = s^2\left(\frac{1}{N} + \frac{(x-\bar{x})^2}{SSD_x}\right) .$$

The sum of squares $SSD_{\bar{y}|x}$ may be computed as

$$SSD_{\bar{y}|x} = SSD_{\bar{y}} - bSPD_{x\bar{y}} = SSD_{\bar{y}} - \frac{SPD_{x\bar{y}}^2}{SSD_x} , \tag{18.3.42}$$

which leads to

$$SSD_{y|x} = SSD_y - \frac{SPD^2_{x\bar{y}}}{SSD_x}, \tag{18.3.43}$$

see Tables 18.3 and 18.4.

18.4. Tests of Significance. Confidence Limits.

From the distributions of the estimates a, b, Y, and s^2 we obtain the following tests of significance.

For the estimate a we have that the quantity

$$\boxed{t = \frac{a-\alpha}{s_a}, \quad s_a = \frac{s}{\sqrt{\Sigma n_i}}, \quad f = \Sigma n_i - 2,} \tag{18.4.1}$$

has a t-distribution with $\Sigma n_i - 2$ degrees of freedom. We thus see that a hypothetical value of α may be tested by means of the t-test, and that we may calculate confidence limits for α.

For the estimate b we have that the quantity

$$\boxed{t = \frac{b-\beta}{s_b}, \quad s_b = \frac{s}{\sqrt{\Sigma n_i(x_i-\bar{x})^2}}, \quad f = \Sigma n_i - 2,} \tag{18.4.2}$$

has a t-distribution with $\Sigma n_i - 2$ degrees of freedom. A hypothetical value of β may be tested by means of the t-test, and confidence limits may be calculated for β. As a special case the hypothesis that there is no relationship between x and the mean value of y may be tested by putting $\beta = 0$ in the formula (18.4.2).

For the estimate Y we have that the quantity

$$\boxed{t = \frac{Y-\eta}{s_Y}, \quad s_Y = s\sqrt{\frac{1}{\Sigma n_i} + \frac{(x-\bar{x})^2}{\Sigma n_i(x_i-\bar{x})^2}}, \quad f = \Sigma n_i - 2,} \tag{18.4.3}$$

has a t-distribution with $\Sigma n_i - 2$ degrees of freedom. Thus, we may determine confidence limits for η with the aid of the t-distribution. In particular it may be examined whether the value of the regression line for $x = 0$, i. e., $Y = a - b\bar{x}$, differs significantly from 0. Introducing t from (18.4.3) in the relation

$$P\{t_{P_1} < t < t_{P_2}\} = P_2 - P_1,$$

we obtain the following confidence limits for η:

$$P\left\{ Y - t_{P_2} s\sqrt{\frac{1}{\Sigma n_i} + \frac{(x-\bar{x})^2}{SSD_x}} < \eta < Y - t_{P_1} s\sqrt{\frac{1}{\Sigma n_i} + \frac{(x-\bar{x})^2}{SSD_x}} \right\} = P_2 - P_1, \tag{18.4.4}$$

where $Y = a+b(x-\bar{x})$ and $\eta = \alpha+\beta(x-\bar{x})$. The limits are closest for $x = \bar{x}$, and the distance between them increases as x diverges from $\bar{x}$.

If the *theoretical* regression line is given, e. g., 95% limits for individual values of y may be determined by drawing two lines parallel to the regression line at the vertical distance of $\pm 1{\cdot}96\sigma$.

From the empirical regression line and the estimated variance we may obtain *tolerance limits* using the coefficients found in § 11.10. In this way the empirical regression line may be used for predictions of future individual observations or means of y corresponding to given values of x.

Example 18.1. The observations given in Table 18.5 and plotted on Fig. 18.1 originate from an investigation of how the tensile strength of test pieces of cement depend on the curing time. As under the given conditions of curing the tensile strength cannot exceed a certain limit, which in principle is reached when the curing time becomes infinitely large, the

TABLE 18.5.
Dependence of tensile strength on curing time.

Curing time in days				
1	2	3	7	28
Tensile strength in kg/cm²				
13·0	21·9	29·8	32·4	41·8
13·3	24·5	28·0	30·4	42·6
11·8	24·7	24·1	34·5	40·3
		24·2	33·1	35·7
		26·2	35·7	37·3

regression curve must possess an upper, horizontal asymptote. Furthermore, for curing time 0 the function must take the value 0. If the strength is represented by the following function:

$$\text{Strength} = \gamma e^{-\frac{\beta}{t}},$$

t denoting the curing time, these conditions are fulfilled. This function is transformed to

$$\log \text{strength} = \log \gamma - \frac{\beta}{t} \log e ,$$

i. e., the logarithm of the tensile strength is a linear function of the reciprocal value of the curing time.

In Fig. 18.6. we test whether this type of function can be used to describe the observations; the reciprocal values of the curing times have been plotted as abscissas and the logarithms of the tensile strengths as ordinates. It seems that the points are scattered at random about a straight line, and we can therefore accept this type of function. A numerical test for linearity will be given later.

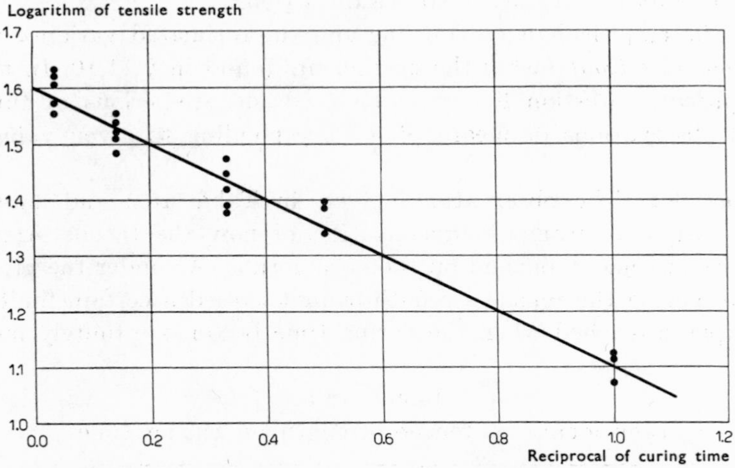

FIG. 18.6. Relationship between the reciprocal of curing time and the logarithm of tensile strength.

If we denote the logarithm of the tensile strength by y and the reciprocal of the curing time by x, we have to determine a regression line with the equation $Y = a+b(x-\bar{x})$ and an estimate of the corresponding variance. Table 18.6 gives the logarithms of the tensile strengths and shows how the mean and variance are computed for each set of observations. By means of BARTLETT's test, see § 11.6, it is found that the five variances do not differ significantly, so that the assumption that the variance is constant (independent of x) may be adopted. Combination of the five variances leads to the estimate $s_1^2 = 0{\cdot}001045$ of σ^2, cf. (18.3.7).

From this estimate we compute the standard errors of the means: $\sqrt{s_1^2/3} = 0{\cdot}0187$ and $\sqrt{s_1^2/5} = 0{\cdot}0145$. As the number of degrees of freedom for s_1^2 is 16, confidence limits for the population means may be computed from the above standard errors and the bilateral 95% limits for t for 16 degrees of freedom. Table IV shows that t is equal to 2·12, which multiplied by the above standard errors gives 0·040 and 0·031, respectively, leading to the 95% confidence limits: $(1{\cdot}103 \pm 0{\cdot}040) = (1{\cdot}063,\ 1{\cdot}143)$, $(1{\cdot}374 \pm 0{\cdot}040) = (1{\cdot}334,\ 1{\cdot}414)$, $(1{\cdot}421 \pm 0{\cdot}031) = (1{\cdot}390,\ 1{\cdot}452)$, $(1{\cdot}521 \pm 0{\cdot}031) = (1{\cdot}490,\ 1{\cdot}552)$, and $(1{\cdot}596 \pm 0{\cdot}031) = (1{\cdot}565,\ 1{\cdot}627)$. Fig. 18.7 shows the means and the confidence limits together with the empirical regression line.

Like Fig. 18.6, this diagram indicates that the type of function chosen may be used for describing the data under discussion.

TABLE 18.6.
Computation of the mean and the variance of the logarithms of the tensile strengths.

	Curing time in days				
	1	2	3	7	28
	Logarithms of tensile strengths				
	1·114	1·340	1·474	1·511	1·621
	1·124	1·389	1·447	1·483	1·629
	1·072	1·393	1·382	1·538	1·605
			1·384	1·520	1·553
			1·418	1·553	1·572
n	3	3	5	5	5
S	3·310	4·122	7·105	7·605	7·980
SS	3·653556	5·665370	10·102589	11·570063	12·740300
S^2/n	3·652033	5·663628	10·096205	11·567205	12·736080
SSD	0·001523	0·001742	0·006384	0·002858	0·004220
f	2	2	4	4	4
s^2	0·000762	0·000871	0·001596	0·000715	0·001055
S/n	1·103	1·374	1·421	1·521	1·596

$$s_1^2 = \frac{\Sigma SSD}{\Sigma f} = \frac{0{\cdot}016727}{16} = 0{\cdot}001045.$$

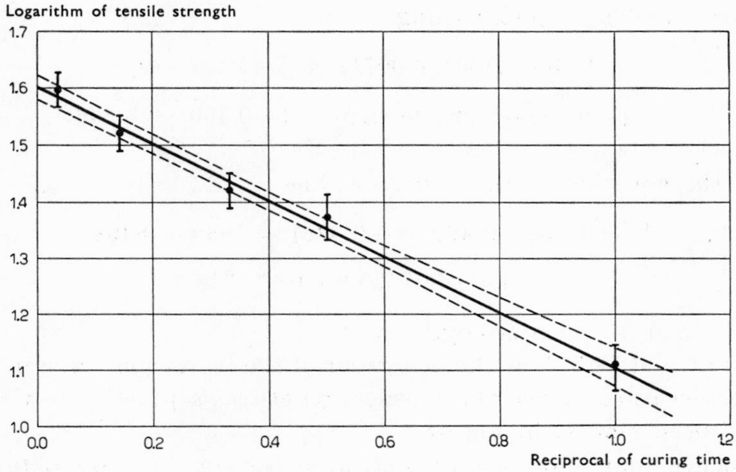

FIG. 18.7. Relationship between the reciprocal of the curing time and the logarithm of the tensile strength. The points indicate the sample means, and the intervals the 95 % confidence limits of the population means. The straight line is the empirical regression line, and the dashed lines denote the 95 % confidence limits for the theoretical regression line.

Table 18.7 gives the regression analysis, carried out according to the scheme given in Table 18.4; some of the computations have, however, already been given in Table 18.6. The quantities SS_x and $SP_{x\bar{y}}$ may be computed as $SS_x = \Sigma(n_i x_i)x_i$ and $SP_{x\bar{y}} = \Sigma(n_i \bar{y}_i)x_i$ or as $SS_x = \Sigma(n_i x_i)^2/n_i$ and $SP_{x\bar{y}} = \Sigma(n_i \bar{y}_i)(n_i x_i)/n_i$ analogous to $SS_{\bar{y}}$, which is computed as $\Sigma(n_i \bar{y}_i)^2/n_i = \Sigma S_i^2/n_i$, see Table 18.6. The quantity $n(\bar{y}-x)$ has been computed as a check, the SSD corresponding to this quantity, $SSD_{\bar{y}-x}$, being equal to $SSD_{\bar{y}} + SSD_x - 2SPD_{x\bar{y}}$, see also the corresponding method of control in Table 4.9, p. 88. (The computations may be somewhat simplified by using $84x$ as the independent variable instead of x, as then the values of the independent variable become integers).

The linearity of the regression curve is tested by computing $v^2 = s_2^2/s_1^2 = 1{\cdot}17$. As v^2 is not significant, the type of function used is regarded as satisfactory.

The three last columns of Table 18.7 give the comparisons between the observed means and the corresponding values on the regression line. The difference between corresponding values has been divided by the standard error of $\bar{y}$. The table shows that none of the differences are conspicuously large, which agrees with the result

$$v^2 = \frac{s_2^2}{s_1^2} = \frac{1}{k-2}\sum_{i=1}^{k} n_i \left(\frac{\bar{y}_i - Y_i}{s_1}\right)^2 = 1{\cdot}17.$$

As the number of degrees of freedom for s^2 is 19, the bilateral 95% limits in the t-distribution are $\pm 2{\cdot}09$, see Table IV, the 95% confidence limits for α and β therefore being

$$(1{\cdot}434 \pm 2{\cdot}09 \times 0{\cdot}0071) = (1{\cdot}419,\ 1{\cdot}449)$$

and

$$(-0{\cdot}498 \pm 2{\cdot}09 \times 0{\cdot}0229) = (-0{\cdot}450,\ -0{\cdot}546)\,,$$

respectively.

From the equation of the regression line

$$Y = 1{\cdot}434 - 0{\cdot}498\ (x - 0{\cdot}336) = 1{\cdot}601 - 0{\cdot}498x$$

we obtain

$$\text{strength} = 39{\cdot}9 \times 10^{-0{\cdot}498/t}$$

as $x = 1/t$ and $Y = \log$ strength.

Figs. 18.6 and 18.7 show the behaviour of the regression curve in relation to the observations, while the tensile strength as a function of the curing time has been plotted in Fig. 18.1.

Confidence limits for the theoretical regression line are obtained by means of the estimate of $\mathcal{V}\{Y\}$, see Table 18.7. For suitable values of the curing time the corresponding values of Y have been computed in Table 18.8 together with the standard errors and the confidence limits. The confidence limits have been illustrated in Figs. 18.7 and 18.8.

TABLE 18.7. *Regression Analysis.*

Relationship between the reciprocal of the curing time (x) and the logarithm of the tensile strength (y). (S_y and SS_y have been taken from Table 18.6).

x	n	nx	$n\bar{y}$	$n(\bar{y}-x)$	$\bar{y}$	Y	$(\bar{y}-Y)\sqrt{n}/s_1$
1·000	3	3·000	3·310	0·310	1·103	1·103	0·0
0·500	3	1·500	4·122	2·622	1·374	1·352	1·2
0·333	5	1·665	7·105	5·440	1·421	1·435	−1·0
0·143	5	0·715	7·605	6·890	1·521	1·530	−0·6
0·036	5	0·180	7·980	7·800	1·596	1·583	0·9
S	21	7·060	30·122	23·062			
SS		4·413170	43·715151	29·904801	9·111760 = SP		
$S^2/21$		2·373505	43·206423	25·326469	10·126730 = $S_x S_{\bar{y}}/21$		
SSD		2·039665	0·508728	4·578332	−1·014970 = SPD		
b	$SPD/SSD_x = -0{\cdot}498$						
SPD^2/SSD_x			0·505065				
$SSD_{\bar{y}\mid x}$			0·003663, $f = 3$				
s_2^2			0·001221				

$$v^2 = \frac{s_2^2}{s_1^2} = \frac{0{\cdot}001221}{0{\cdot}001045} = 1{\cdot}17$$

$$s^2 = \frac{0{\cdot}016727+0{\cdot}003663}{16+3} = \frac{0{\cdot}020390}{19} = 0{\cdot}001073$$

$S/21$: $\bar{x} = 0{\cdot}336$ $\quad \bar{y} = 1{\cdot}434 \quad$ 1·098

$$Y = 1{\cdot}434-0{\cdot}498(x-0{\cdot}336) = 1{\cdot}601-0{\cdot}498\,x$$

$$s_a^2 = \frac{s^2}{21} = 0{\cdot}00005110, \quad s_a = 0{\cdot}0071.$$

$$s_b^2 = \frac{s^2}{SSD_x} = 0{\cdot}0005261, \quad s_b = 0{\cdot}0229.$$

$$s_Y^2 = 0{\cdot}00005110+0{\cdot}0005261\,(x-0{\cdot}336)^2.$$

From a purely statistical point of view the regression curve yields only a description of the average tensile strength as a function of the curing time *within the range of the observations*, i. e., a curing time of from 1 to 28 days. When computing an estimate of the tensile strength corresponding to a curing time which lies outside this range, the choice of the type of the function becomes of vital importance, for *even though the function chosen does not deviate significantly from the true function within the range of the observations, there may be important deviations between the two functions when a larger interval is employed. Every extrapolation implies an element of*

TABLE 18.8.

Computation of the 95% confidence limits for the theoretical regression line.

t	$x = 1/t$	Y	s_Y	$Y-2{\cdot}09s_Y$	$Y+2{\cdot}09s_Y$	Antilog		
						$Y-2{\cdot}09s_Y$	Y	$Y+2{\cdot}09s_Y$
0	∞	$-\infty$	∞	$-\infty$	$-\infty$	0·0	0·0	0·0
0·5	2·000	0·605	0·0388	0·524	0·686	3·3	4·0	4·9
1	1·000	1·103	0·0168	1·068	1·138	11·7	12·7	13·7
2	0·500	1·352	0·0081	1·335	1·369	21·6	22·5	23·4
3	0·333	1·435	0·0072	1·420	1·450	26·3	27·2	28·2
5	0·200	1·502	0·0078	1·486	1·518	30·6	31·8	33·0
7	0·143	1·530	0·0084	1·512	1·548	32·5	33·9	35·3
10	0·100	1·552	0·0090	1·533	1·571	34·1	35·6	37·2
15	0·067	1·568	0·0094	1·548	1·588	35·3	37·0	38·7
20	0·050	1·576	0·0097	1·556	1·596	36·0	37·7	39·5
28	0·036	1·583	0·0099	1·562	1·604	36·5	38·3	40·2
∞	0·000	1·601	0·0105	1·579	1·623	37·9	39·9	42·0

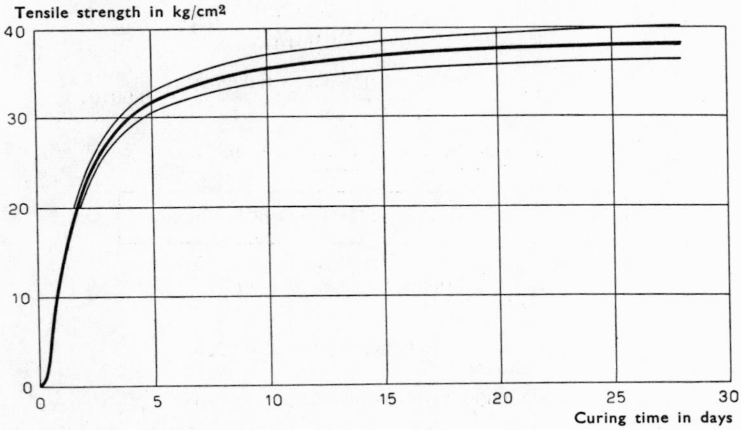

FIG. 18.8. The empirical regression curve for the tensile strength as a function of the curing time together with the 95% confidence limits for the theoretical regression curve.

uncertainty which cannot be statistically evaluated. It is therefore very important to utilize all professional knowledge regarding the relationship between the variables when choosing the type of function.

In the present case we have for $t \to \infty$, i. e., $x = 1/t \to 0$, that $Y \to 1{\cdot}601$, i. e., the strength 39·9 kg/cm². From Table 18.8 we see that the 95% confidence limits of this asymptotic strength are (37·9, 42·0) kg/cm². The value of these results depends on how close the chosen function for 28 days' curing or more lies to the theoretically correct function.

Example 18.2. An investigation concerning the preservation of ascorbic acid in vegetables during drying and storing gave the figures shown in Table 18.9 for the dry matter of fresh spinach and percentage preserved ascorbic acid after drying at 90° C.[1]) The observations have been plotted in Fig. 18.9.

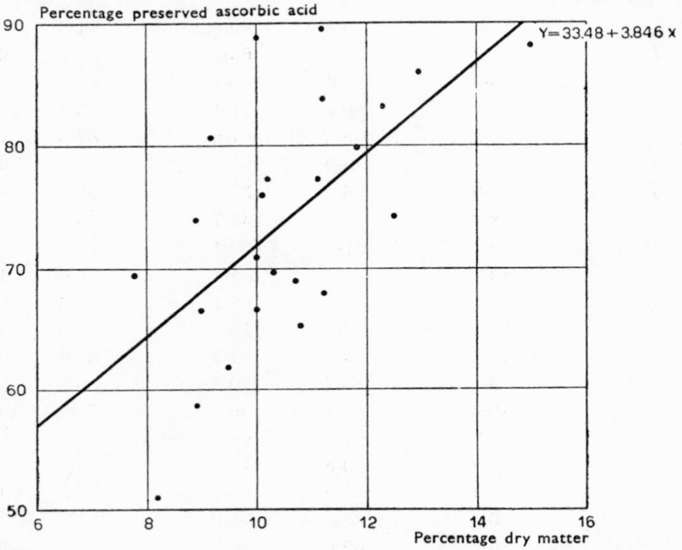

FIG. 18.9. Percentage dry matter in fresh spinach and percentage preserved ascorbic acid after drying at 90° C.

According to the diagram there seems to be some relationship between the content of dry matter and the preservation of ascorbic acid. The problem has been numerically examined in Table 18.9, in which a computation of the regression line is given.

The column $y-x$ in Table 18.9 has been included as a check. The two last columns in the table give the values on the regression line, Y, together with the deviations divided by the standard deviation. If the function type is correct the figures in the last column should be approximately normally distributed with parameters (0, 1). The table shows that this claim is reasonably satisfied. Furthermore, the variation of the signs of the deviations should be random when the deviations are arranged according to increasing values of x. Fig. 18.9, in which the regression line is drawn, shows that the variation of the observations about the computed values seems to be random, there being no tendency to a preponderance of either negative or positive deviations within any range of variation of the independent variable.

---

[1]) E. ETTRUP PETERSEN: *Studier over C-Vitaminets Nedbrydning ved Tørring og Lagring af Vegetabilier,* Thesis, Copenhagen, 1946, Table 22, p. 74.

TABLE 18.9. *Regression Analysis.*

Relation between the percentage dry matter (x) in fresh spinach and the percentage preserved ascorbic acid (y) after drying at 90° C.

	x	y	$y-x$	Y	$(y-Y)/s$
	10·0	70·9	60·9	71·9	−0·1
	8·9	74·0	65·1	67·7	0·8
	8·9	58·6	49·7	67.7	−1·1
	9·2	80·6	71·4	68·9	1·5
	7·8	69·4	61·6	63·5	0·7
	10·1	76·0	65·9	72·3	0·5
	9·0	66·4	57·4	68·1	−0·2
	8·2	50·9	42·7	65·0	−1·8
	9·5	61·9	52·4	70·0	−1·0
	10·8	65·2	54·4	75·0	−1·2
	11·1	77·2	66·1	76·2	0·1
	11·2	89·6	78·4	76·6	1·6
	12·5	74·2	61·7	81·6	−0·9
	12·3	83·1	70·8	80·8	0·3
	10·0	66·7	56·7	71·9	−0·6
	10·2	77·2	67·0	72·7	0·6
	11·2	83·8	72·6	76·6	0·9
	11·2	67·9	56·7	76·6	−1·1
	10·0	88·9	78·9	71·9	2·1
	10·7	69·0	58·3	74·6	−0·7
	10·3	69·8	59·5	73·1	−0·4
	12·9	86·0	73·1	83·1	0·4
	11·8	79·9	68·1	78·9	0·1
	14·9	88·2	73·3	90·8	−0·3
n	24				
S	252·7	1775·4	1522·7	1775·5	0·2
SS	2720·39	133644·24	98518·71	18922·96 = SP	
S^2/n	2660·72	131335·22	96608·97	18693·48 = S_xS_y/n	
SSD	59·67	2309·02	1909·74	229·48 = SPD	
b			SPD/SSD_x =	<u>3·846</u>	
SPD^2/SSD_x		882·54			
$SSD_{y\|x}$		1426·48			
f		22			
s^2		64·84, s = 8·05%.			
S/n	<u>10·53</u>	<u>73·98</u>	63·45		

$$Y = 73\cdot98+3\cdot846(x-10\cdot53) = 33\cdot48+3\cdot846\,x$$

$$s_a^2 = \frac{s^2}{n} = 2\cdot702, \quad s_a = 1\cdot644.$$

$$s_b^2 = \frac{s^2}{SSD_x} = 1\cdot087, \quad s_b = 1\cdot043.$$

$$s_Y^2 = 2\cdot702+1\cdot087(x-10\cdot53)^2.$$

Table 18.9 shows that the slope of the regression line is $b = 3{\cdot}846$, and that the estimate of the standard error of the slope is 1·043. By means of the t-distribution a test is made whether the slope deviates significantly from 0. If the computed values and $\beta = 0$ are introduced in equation (18.4.2), we obtain

$$t = \frac{3{\cdot}846}{1{\cdot}043} = 3{\cdot}69, \quad f = 22 .$$

Table IV gives that $P\{|t| > 3{\cdot}69\} \simeq 0{\cdot}001$, i. e., our value of b differs significantly from zero.

18.5. Solving the Regression Equation with Respect to the Independent Variable.

If the observations are a random sample from a two-dimensional, normally distributed population, both a regression of y upon x and a regression of x upon y can be computed, cf. § 19.7. These two regression lines differ, as they give the mean value of y as a function of x

$$\mathcal{M}\{y|x\} \approx Y = \bar{y} + b_{y|x}(x - \bar{x}), \quad b_{y|x} = \frac{SPD_{xy}}{SSD_x}, \tag{18.5.1}$$

and the mean value of x as a function of y

$$\mathcal{M}\{x|y\} \approx X = \bar{x} + b_{x|y}(y - \bar{y}), \quad b_{x|y} = \frac{SPD_{xy}}{SSD_y}, \tag{18.5.2}$$

respectively. In most cases, however, only one of these regression lines is of practical value, as in § 4.7 and Example 18.2.

If one of the two variables, x say, is a *non-stochastic variable* (or a stochastic variable for which the range of variation has been limited at will), only the regression line which has x as the independent variable has a meaning.

In such a case, let us suppose that a regression line has been determined as discussed in the previous paragraphs. Further, a value of y is observed, the corresponding value of x being unknown. The problem then is to estimate the x-value from the observed y and the regression equation.

If, for instance, when manufacturing cement we have determined a regression line giving the fineness of the cement as a function of the production, we may use this relationship to answer the question: Which production may we expect when the cement has a given fineness? In this case we may regard the y-value given as a theoretical quantity.

As another example, where y is an observed quantity, we may mention the dosage-reaction curve of a given preparation, A, which is used to determine the strength of a "similar" preparation, B. In an experiment with preparation B a certain reaction is observed, and then the dosage corresponding to this reaction is read from the regression line for preparation A, in this way finding the strength proportion between A and B.

In what follows, the theory for solving the regression equation is developed, it being assumed that the given value of y is an observed value, whose uncertainty is expressed by the variance $\mathcal{V}\{y\}$. If y is a theoretical value we need merely put $\mathcal{V}\{y\}$ equal to zero.

Let the equation for the regression curve be

$$Y = a+b(x-\bar{x}) .$$

If we substitute an observed value y, independent of the observations used for the determination of a and b, in this equation and solve for x we obtain

$$\boxed{x = x(y) = \bar{x}+\frac{y-a}{b}.} \tag{18.5.3}$$

The estimate $x(y)$ is a function of the three stochastic variables a, b, and y, and is therefore in itself a stochastic variable. Denoting the mean value of y by η, the corresponding theoretical value of x, ξ say, is equal to

$$\xi = \xi(\eta) = \bar{x}+\frac{\eta-a}{\beta}. \tag{18.5.4}$$

Introducing the auxiliary variable

$$\boxed{z = y-a-b(\xi-\bar{x})} \tag{18.5.5}$$

with mean

$$\mathcal{m}\{z\} = \eta-\alpha-\beta(\xi-\bar{x}) = 0 \tag{18.5.6}$$

and variance

$$\boxed{\mathcal{V}\{z\} = \mathcal{V}\{y\}+\mathcal{V}\{a\}+\mathcal{V}\{b\}(\xi-\bar{x})^2} \tag{18.5.7}$$

it follows that the standardized variable $u=z/\sqrt{\mathcal{V}\{z\}}$ is normally distributed with parameters (0, 1). If y is a single observation, we have $\mathcal{V}\{y\} = \sigma^2$, while if y is the mean of n observations for given x, $\mathcal{V}\{y\}$ is equal to σ^2/n. If we insert this expression for $\mathcal{V}\{y\}$ and the corresponding expressions for $\mathcal{V}\{a\}$ and $\mathcal{V}\{b\}$ in $\mathcal{V}\{z\}$, we obtain

$$\mathcal{V}\{z\} = \sigma^2\left(\frac{1}{n}+\frac{1}{\sum_{i=1}^{k} n_i}+\frac{(\xi-\bar{x})^2}{SSD_x}\right). \tag{18.5.8}$$

Replacing the variance σ^2 by the estimate s^2, $f = \Sigma n_i-2$, as determined from the regression analysis, we find that the variable

$$t = \frac{y-a-b(\xi-\bar{x})}{s\sqrt{\dfrac{1}{n}+\dfrac{1}{\Sigma n_i}+\dfrac{(\xi-\bar{x})^2}{SSD_x}}}, \quad f = \Sigma n_i - 2, \tag{18.5.9}$$

has a t-distribution with $\Sigma n_i - 2$ degrees of freedom. If the above expression for t is inserted in the relation

$$P\{t_{P_1} < t < t_{P_2}\} = P_2 - P_1 ,$$

and the inequality is solved for ξ (an equation of the second order), we obtain the following confidence limits for ξ

$$\bar{x}+\frac{y-a}{b_2}-t_{P_2}\frac{s}{b_2}\sqrt{\left(\frac{1}{n}+\frac{1}{\Sigma n_i}\right)\frac{b_2}{b}+\frac{(x(y)-\bar{x})^2}{SSD_x}} < \xi <$$

$$\bar{x}+\frac{y-a}{b_1}-t_{P_1}\frac{s}{b_1}\sqrt{\left(\frac{1}{n}+\frac{1}{\Sigma n_i}\right)\frac{b_1}{b}+\frac{(x(y)-\bar{x})^2}{SSD_x}}, \tag{18.5.10}$$

where

$$b_i = b - \frac{t_{P_i}^2 s^2}{bSSD_x}, \quad i = 1, 2. \tag{18.5.11}$$

If b_i deviates only slightly from b, we have as approximate confidence limits

$$x(y)-t_{P_2}\frac{s}{b}\sqrt{\frac{1}{n}+\frac{1}{\Sigma n_i}+\frac{(x(y)-\bar{x})^2}{SSD_x}} < \xi <$$

$$x(y)-t_{P_1}\frac{s}{b}\sqrt{\frac{1}{n}+\frac{1}{\Sigma n_i}+\frac{(x(y)-\bar{x})^2}{SSD_x}}. \tag{18.5.12}$$

The interpretation of these confidence limits is based on the simultaneous sampling distribution of both the regression line and the observed value of y. Theoretically, therefore, the limits cannot be applied for the calculation of x-values corresponding to a series of given y-values if the *same* regression line is used repeatedly.

18.6. The Variance is Proportional to a Given Function of the Independent Variable.

The analysis in § 18.3 is based on the assumption that the variance of y corresponding to a given value of x, $\mathcal{V}\{y|x\}$, is constant. A similar analysis may, however be carried out if this assumption is generalized, *it now being assumed that the variance is proportional to a given function of* x, i. e.,

$$V\{y|x\} = \sigma^2 h^2(x) , \tag{18.6.1}$$

where $h(x)$ is known and σ^2 is an unknown constant.

As $y_{i\nu}$ is normally distributed about $\eta_i = \alpha+\beta(x_i-\bar{x})$ with variance $\sigma^2 h^2(x_i)$, the variable

$$u_{i\nu} = \frac{y_{i\nu}-\eta_i}{\sigma h(x_i)} \tag{18.6.2}$$

is normally distributed with parameters (0, 1), and from this it follows that

$$\chi^2 = \sum_{i=1}^{k}\sum_{\nu=1}^{n_i} u_{i\nu}^2 = \frac{1}{\sigma^2}\sum_{i=1}^{k}\sum_{\nu=1}^{n_i}\left(\frac{y_{i\nu}-\eta_i}{h(x_i)}\right)^2 \tag{18.6.3}$$

has a χ^2-distribution with $\sum_{i=1}^{k} n_i$ degrees of freedom. If we introduce the notation

$$w_i = \left(\frac{1}{h(x_i)}\right)^2 \tag{18.6.4}$$

and multiply by σ^2, we obtain

$$\sigma^2\chi^2 = \sum_{i=1}^{k}\sum_{\nu=1}^{n_i} w_i(y_{i\nu}-\eta_i)^2 . \tag{18.6.5}$$

This sum of squares may be partitioned in a manner similar to that given in (18.3.8), and by the method of least squares estimates of α and β may be derived by differentiating the sum of squares with respect to α and β and solving the equations corresponding to (18.3.12) and (18.3.13).

We then obtain the empirical regression equation

$$Y = a+b(x-\bar{x}) ,$$

where

$$\bar{x} = \frac{\sum_{i=1}^{k} w_i n_i x_i}{\sum_{i=1}^{k} w_i n_i} , \tag{18.6.6}$$

$$a = \bar{y} = \frac{\sum_{i=1}^{k} w_i n_i \bar{y}_i}{\sum_{i=1}^{k} w_i n_i} \tag{18.6.7}$$

and

$$b = \frac{\sum_{i=1}^{k} w_i n_i (x_i-\bar{x})\bar{y}_i}{\sum_{i=1}^{k} w_i n_i (x_i-\bar{x})^2} . \tag{18.6.8}$$

Table 18.10 shows the partitioning of the variation about the theoretical regression line analogous to the partitioning given in Table 18.2.

TABLE 18.10.

Partitioning of the variation about the theoretical regression line $\eta=\alpha+\beta(x-\bar{x})$. The empirical regression line is denoted by $Y=a+b(x-\bar{x})$.

Variation	Sum of squares	Degrees of freedom	Variance
Deviation of a from α	$(a-\alpha)^2 \sum_{i=1}^{k} w_i n_i$	1	s_4^2
Deviation of b from β	$(b-\beta)^2 \sum_{i=1}^{k} w_i n_i (x_i-\bar{x})^2$	1	s_3^2
About the empirical regression line	$\sum_{i=1}^{k} w_i n_i (\bar{y}_i - Y_i)^2$	$k-2$	s_2^2
Within sets	$\sum_{i=1}^{k} \sum_{\nu=1}^{n_i} w_i (y_{i\nu}-\bar{y}_i)^2$	$\sum_{i=1}^{k} n_i - k$	s_1^2
About the theoretical regression line	$\sum_{i=1}^{k} \sum_{\nu=1}^{n_i} w_i (y_{i\nu}-\eta_i)^2$	$\sum_{i=1}^{k} n_i$	

The linearity of the regression curve is tested by comparing s_1^2 *and* s_2^2. If s_2^2 is not significantly larger than s_1^2, these two variances are combined to the estimate

$$s^2 = \frac{\sum_{i=1}^{k} \sum_{\nu=1}^{n_i} w_i (y_{i\nu}-\bar{y}_i)^2 + \sum_{i=1}^{k} w_i n_i (\bar{y}_i - Y_i)^2}{\left(\sum_{i=1}^{k} n_i - k\right) + (k-2)} = \frac{\sum_{i=1}^{k} \sum_{\nu=1}^{n_i} w_i (y_{i\nu} - Y_i)^2}{\sum_{i=1}^{k} n_i - 2}. \quad (18.6.9)$$

From the two remaining variances it follows that

1. The estimate a is normally distributed about α with variance

$$\mathcal{V}\{a\} = \frac{\sigma^2}{\sum_{i=1}^{k} w_i n_i}. \quad (18.6.10)$$

2. The estimate b is normally distributed about β with variance

$$\mathcal{V}\{b\} = \frac{\sigma^2}{\sum_{i=1}^{k} w_i n_i (x_i-\bar{x})^2}. \quad (18.6.11)$$

3. The estimates a and b are stochastically independent and independent of s^2.

From this it follows that $Y = a+b(x-\bar{x})$ is normally distributed about $\eta = \alpha+\beta(x-\bar{x})$ with variance

$$\mathcal{V}\{Y\} = \mathcal{V}\{a\}+(x-\bar{x})^2\mathcal{V}\{b\} . \tag{18.6.12}$$

The tests of significance corresponding to the above results are analogous to those given in § 18.4.

Consider the two special cases where

$$\mathcal{M}\{y|x\} = \beta_1 x \quad \text{and} \quad \mathcal{V}\{y|x\} = \sigma_1^2 x \tag{18.6.13}$$

and

$$\mathcal{M}\{y|x\} = \beta_2 x \quad \text{and} \quad \mathcal{V}\{y|x\} = \sigma_2^2 x^2 , \tag{18.6.14}$$

respectively. The corresponding estimates of β become

$$b_1 = \frac{\Sigma n_i \bar{y}_i}{\Sigma n_i x_i}, \qquad \mathcal{V}\{b_1\} = \frac{\sigma_1^2}{\Sigma n_i x_i}, \tag{18.6.15}$$

and

$$b_2 = \frac{\Sigma \frac{n_i \bar{y}_i}{x_i}}{\Sigma n_i}, \qquad \mathcal{V}\{b_2\} = \frac{\sigma_2^2}{\Sigma n_i}, \tag{18.6.16}$$

as compared with

$$b = \frac{\Sigma n_i x_i \bar{y}_i}{\Sigma n_i x_i^2}, \qquad \mathcal{V}\{b\} = \frac{\sigma^2}{\Sigma n_i x_i^2}, \tag{18.6.17}$$

in the case where

$$\mathcal{M}\{y|x\} = \beta x \quad \text{and} \quad \mathcal{V}\{y|x\} = \sigma^2 , \tag{18.6.18}$$

see (18.3.38). It will be noted that although the mean value of y in all cases is proportional to x the least squares estimates of the slope differ because the variances of y are different functions of x.

The above results may also be derived from (9.7.11) and (9.7.12), applying these formulas on the variable $z_{iv} = y_{iv}/x_i$. Under the specification (18.6.13), say, we find that $\bar{z}_i = \bar{y}_i/x_i$ is normally distributed about β_1, with variance $\sigma_1^2/x_i n_i$, so that $\bar{z} = \Sigma n_i x_i \bar{z}_i/\Sigma n_i x_i = b_1$ will be normally distributed about β_1 with variance $\sigma_1^2/\Sigma n_i x_i$.

If we use one of these estimates in all cases, independent of the specification of $\mathcal{V}\{y|x\}$, we find that the estimate is unbiased but inefficient, i. e., the estimate has a larger variance than the appropriate least square estimate.

Example 18.3. In an investigation of the relationship between the titanium content of cement and the extinction coefficient of solutions of this same cement, the values given in Table 18.11 were observed. When the extinction co-

TABLE 18.11.

Computation of the means and variances of the extinction coefficients for four values of titanium contents.

Titanium contents	0·1%	0·25%	0·50%	1·00%
Length of cuvette in cm	5	3	2	1
	0·237	0·267	0·308	0·304
	0·235	0·247	0·310	0·300
	0·247	0·260	0·330	0·311
	0·222	0·251	0·304	0·301
	0·245	0·273	0·316	0·322
Computing origin	0·200	0·200	0·300	0·300
n	5	5	5	5
S	0·186	0·298	0·068	0·038
SS	0·007312	0·018228	0·001336	0·000622
S^2/n	0·006919	0·017761	0·000925	0·000289
SSD	0·000393	0·000467	0·000411	0·000333
f	4	4	4	4
	$s_1^2 = \frac{\Sigma SSD}{\Sigma f} = \frac{0{\cdot}001604}{16} = 0{\cdot}0001003$			
S/n	0·0372	0·0596	0·0136	0·0076
Mean	0·2372	0·2596	0·3136	0·3076
$\frac{\text{Mean}}{\text{Length of cuvette}}$	0·0474	0·0865	0·1568	0·3076

efficients were determined the cuvettes were varied in length in order to obtain extinction coefficients in the range 0·2—0·4, the uncertainty being the smallest in this range. The extinction coefficients are referred to the same length of cuvette, e. g., 1 cm, by dividing by the length of the cuvette used. Table 18.11 shows the computation of means and variances for the four sets of measurements. As the variances do not differ significantly, they are combined in the variance $s_1^2 = 0{\cdot}0001003$. If the variance of the observed extinction coefficients is denoted by σ^2, the variances of the extinction coefficients when referred to the same length of cuvette become $\sigma^2/25$, $\sigma^2/9$, $\sigma^2/4$, and $\sigma^2/1$, respectively. When determining the regression line it must therefore be taken into account that the variance is proportional to a given function of the variable.

Table 18.12 shows the regression analysis. The computations are analogous to those of Table 18.7, except that n_i is replaced by $n_i w_i$, but since n_i is

TABLE 18.12. *Regression Analysis.*
Extinction coefficient (y) as a function of titanium contents (x).

x	n	w	$\bar{y}$	$x-\bar{y}$	wx	$w\bar{y}$	$w(x-\bar{y})$	Y	$(\bar{y}-Y)\sqrt{nw}/s_1$
0·10	5	25	0·0474	0·0526	2·50	1·1850	1·3150	0·0465	1·0
0·25	5	9	0·0865	0·1635	2·25	0·7785	1·4715	0·0888	−1·5
0·50	5	4	0·1568	0·3432	2·00	0·6272	1·3728	0·1592	−1·1
1·00	5	1	0·3076	0·6924	1·00	0·3076	0·6924	0·3000	1·7
		39		S	7·75	2·8983	4·8517		
				SS	2·812500	0·316472	1·260322	0·934325 = SP	
				$S^2/39$	1·540064	0·215388	0·603564	0·575944 = $S_x S_{\bar{y}}/39$	
				SSD	1·272436	0·101084	0·656758	0·358381 = SPD	
				b			SPD/SSD_x =	0·28165	
				SPD^2/SSD_x		0·100938			
				$\sum w_i(\bar{y}_i - Y_i)^2$		0·000146			
				$\sum n_i w_i(\bar{y}_i - Y_i)^2$		0·000730			
				f		2			
				s_2^2		0·000365			
				v^2		$s_2^2/s_1^2 = 0{\cdot}000365/0{\cdot}0001003 = 3{\cdot}64$			
				s^2		$\dfrac{0{\cdot}001604+0{\cdot}000730}{16+2} = \dfrac{0{\cdot}002334}{18} = 0{\cdot}0001297$			
				$S/39$	0·19872	0·07432	0·12440		

$$Y = 0{\cdot}07432+0{\cdot}28165(x-0{\cdot}19872) = 0{\cdot}01835+0{\cdot}28165x$$

$$s_a^2 = \frac{s^2}{5\times 39} = 0{\cdot}0^6 6651, \quad s_a = 0{\cdot}00082$$

$$s_b^2 = \frac{s^2}{5\times 1{\cdot}2724} = 0{\cdot}0^4 2039, \quad s_b = 0{\cdot}0045$$

$$s_Y^2 = 0{\cdot}0^6 6651+0{\cdot}0^4 2039(x-0{\cdot}1987)^2$$

constant and equal to 5 this factor may be added after the computation of the sums of squares and products.

The linearity of the regression line is tested by computing the variance ratio $v^2 = 3{\cdot}64$. Since $v^2_{.95}$ (2,16) = 3·63 we are inclined, at first sight, to doubt the hypothesis of linearity. However, this judgment is based on these particular data only, and since other experiments of the same kind suggest a linear relationship, we shall continue to assume that this is true. One v^2-value exactly on the 95% limit does not justify the rejection of previous experience, but only tells us to pay extra attention to future investigations of this type. Table 18.12 gives the equation of the regression curve and the standard errors of the estimated parameters.

If in the estimate of the variance of Y we put $x = 0$, we obtain $s_Y = 0{\cdot}0012$, being the standard error of the "constant" 0·01835, which is thus significantly different from zero. If the linear relation is valid beyond the range of the observations this means that the straight line does not pass through the origin, i. e., the extinction coefficient is not proportional to the concentration.

The regression line is used as a standard curve for computing the titanium contents of cement when the extinction coefficient of the sample has been determined.

Solving the regression equation with regard to the independent variable we find

$$x(y) = 0{\cdot}1987 + \frac{y - 0{\cdot}0743}{0{\cdot}28165} = 3{\cdot}551\,y - 0{\cdot}0651 .$$

If the y-value is determined as the average of n observations with the cuvette length k, we find that $\mathcal{V}\{y\} = \sigma^2/nk^2$, and we obtain as an estimate of the variance of z according to (18.5.5) and (18.5.7)

$$\mathcal{V}\{z\} \approx 0{\cdot}0001297 \left(\frac{1}{nk^2} + \frac{1}{5 \times 39} + \frac{(x - 0{\cdot}1987)^2}{5 \times 1{\cdot}2724} \right)$$

$$= 0{\cdot}0001297 \left(\frac{1}{nk^2} + 0{\cdot}00513 + 0{\cdot}1572 (x - 0{\cdot}1987)^2 \right).$$

If, e. g., we found an extinction coefficient of 0·1729 as the average of 5 determinations with the cuvette length 2 cm, the corresponding titanium percentage would then be

$$x = 0{\cdot}1987 + \frac{0{\cdot}1729 - 0{\cdot}0743}{0{\cdot}28165} = 0{\cdot}1987 + 0{\cdot}3501 = 0{\cdot}549\,\%.$$

The estimate of the variance of z is

$$\mathcal{V}\{z\} \approx 0{\cdot}0001297 (0{\cdot}05 + 0{\cdot}00513 + 0{\cdot}1572 \times 0{\cdot}3501^2) = 0{\cdot}0^5 965 ,$$

and $\sqrt{\mathcal{V}\{z\}} \approx 0{\cdot}00311$. If this estimate is divided by the slope $b = 0{\cdot}282$, we obtain 0·0110, which quantity according to (18.5.12) is the factor of t_P when we are determining the limits of the theoretical value. In this case we thus have as the 95% confidence limits of the required concentration $(0{\cdot}549 \pm 2{\cdot}10 \times 0{\cdot}0110) = (0{\cdot}526,\ 0{\cdot}572)\%$.

18.7. Transforming the Regression Curve to a Straight Line.

From a graph of the observations it will often be obvious that the regression curve cannot be a straight line. In many cases, however, it is possible by simple transformations of the variables to represent the regression curve as *a linear relation between the transformed variables*, e. g., if the regression curve has the equation $y = \gamma x^\beta$ a logarithmic transformation of both variables will lead to a straight line with the equation $\log y = \log \gamma + \beta \log x$, i. e., the points $(\log x, \log y)$ represent a straight line.

In general this principle may be formulated as follows:

If the points (x_i, y_{iv}) *when transformed to* $(g(x_i), f(y_{iv}))$ *are grouped about a straight line in such a manner that* $f(y)$ *is normally distributed with mean value*

$$\boxed{\mathcal{M}\{f(y)|g(x)\} = \alpha+\beta(g(x)-\overline{g(x)})} \tag{18.7.1}$$

and variance

$$\boxed{\mathcal{V}\{f(y)|g(x)\} = \sigma^2} \tag{18.7.2}$$

or

$$\boxed{\mathcal{V}\{f(y)|g(x)\} = \sigma^2 h^2(x),} \tag{18.7.3}$$

then the theory of linear regression may be applied to the transformed observations.

The transformation functions are chosen on the basis of a graphical analysis of the observations, taking all professional knowledge regarding the theoretical behaviour of the regression curve into account.

The same data can often be satisfactorily described by different formulas, that is, described in such a manner that the variation about the regression curve is not significantly larger than the variation within sets, particularly if the data include only a part of the regression curve which is not characteristic of the whole curve. If, however, the theoretical knowledge we have about the behaviour of the curve, e. g., that it passes through the origin and has a horizontal asymptote, is properly utilized, it will often be possible to exclude types of functions which might appear suitable from a purely descriptive point of view. For further illustration, see Example 18.4.

Often we are dealing with several sets of data in which the conditions under which we are studying the interrelation between the variables vary from one set to another. In such cases it is the rule to try to apply *the same type of function to all sets of data*, and let the varying conditions be reflected in the values of the parameters, cf. Example 7.6.

From a purely statistical point of view the regression curve provides a *description* of the interrelation between the two variables *within the limited range of the observations*, and *extrapolations*, i. e., computations of values of one variable corresponding to given values of the other variable *outside* this range are *in principle not justifiable* as perhaps it is not possible to represent the interrelation outside the observed range by the function utilized. It is therefore absolutely necessary that extrapolation be firmly based on professional knowledge concerning the data, cf. the remarks in Example 18.1, p. 545.

The agreement between the observations and the values computed from the empirical regression formula should always be investigated by comparing the deviations with the standard deviation and by examining the variations in the sign and size of the deviations as a function of the independent variable, cf. Chapter 13.

In the following some transformation functions of the type

$$f(y) = \alpha_0 + \beta g(x) \tag{18.7.4}$$

will be discussed from a purely mathematical point of view.

As a simple example of how the formula (18.7.4) may lead to widely differing types of curves by inserting elementary functions for $g(x)$ and $f(y)$, we will put $g(x)$ equal to x, $1/x$, and $\log_e x$, respectively, and $f(y)$ equal to the corresponding functions of y. This leads to the equations given in Table 18.13.

TABLE 18.13.

$f(y) = \alpha_0 + \beta g(x)$

$f(y)$	$g(x)$: x	$1/x$	$\log_e x$
y	$y = \alpha_0 + \beta x$	$y = \alpha_0 + \beta/x$	$y = \alpha_0 + \beta \log_e x$
$1/y$	$1/y = \alpha_0 + \beta x$	$1/y = \alpha_0 + \beta/x$	$1/y = \alpha_0 + \beta \log_e x$
$\log_e y$	$\log_e y = \alpha_0 + \beta x$	$\log_e y = \alpha_0 + \beta/x$	$\log_e y = \alpha_0 + \beta \log_e x$

If these equations are solved with respect to y we obtain the functions entered in Table 18.14, where $\gamma = e^{\alpha_0}$.

When plotted in a $(g(x), f(y))$-coordinate system these functions will represent straight lines. In the following the graphical representation of the most important of these functions will be given in order to make it easier for the reader to identify these types of functions in practice.

TABLE 18.14.

Solution of the equations in Table 18.13 with respect to y.

$f(y) = \alpha_0+\beta g(x)$

$f(y)$	$g(x)$: x	$1/x$	$\log_e x$
y	$y = \alpha_0+\beta x$	$y = \alpha_0+\beta/x$	$y = \alpha_0+\beta \log_e x$
$1/y$	$y = \dfrac{1}{\beta\left(x+\dfrac{\alpha_0}{\beta}\right)}$	$y = \dfrac{1}{\alpha_0}-\dfrac{\beta}{\alpha_0^2}\dfrac{1}{x+\dfrac{\beta}{\alpha_0}}$	$y = \dfrac{1}{\beta\left(\log_e x+\dfrac{\alpha_0}{\beta}\right)}$
$\log_e y$	$y = \gamma e^{\beta x}$	$y = \gamma e^{\beta/x}$	$y = \gamma x^{\beta}$

Hyperbolas with two parameters.

The three functions

$$y = \alpha_0+\beta/x\,, \tag{18.7.5}$$

$$1/y = \alpha_0+\beta x \quad \text{or} \quad y = \frac{1}{\beta\left(x+\dfrac{\alpha_0}{\beta}\right)}, \tag{18.7.6}$$

and

$$1/y = \alpha_0+\beta/x \quad \text{or} \quad y = \frac{1}{\alpha_0}-\frac{\beta}{\alpha_0^2}\frac{1}{x+\dfrac{\beta}{\alpha_0}}, \tag{18.7.7}$$

are special cases of the usual equation of a hyperbola with three parameters

$$(y-\gamma)(x-\lambda) = \varkappa \quad \text{or} \quad y = \gamma+\frac{\varkappa}{x-\lambda}, \tag{18.7.8}$$

which represents a hyperbola with the asymptotes $x = \lambda$ and $y = \gamma$ and curvature $\varkappa$.

TABLE 18.15.

Equation	Vertical asymptote	Horizontal asymptote	Curvature
$y = \alpha_0+\beta/x$	$x = 0$	$y = \alpha_0$	β
$1/y = \alpha_0+\beta x$	$x = -\dfrac{\alpha_0}{\beta}$	$y = 0$	$\dfrac{1}{\beta}$
$1/y = \alpha_0+\beta/x$	$x = -\dfrac{\beta}{\alpha_0}$	$y = \dfrac{1}{\alpha_0}$	$-\dfrac{\beta}{\alpha_0^2}$

Thus, the three hyperbolas (18.7.5)–(18.7.7) may be characterized by the asymptotes and curvatures given in Table 18.15.

Fig. 18.10 illustrates two examples of the behaviour of the function defined by (18.7.7).

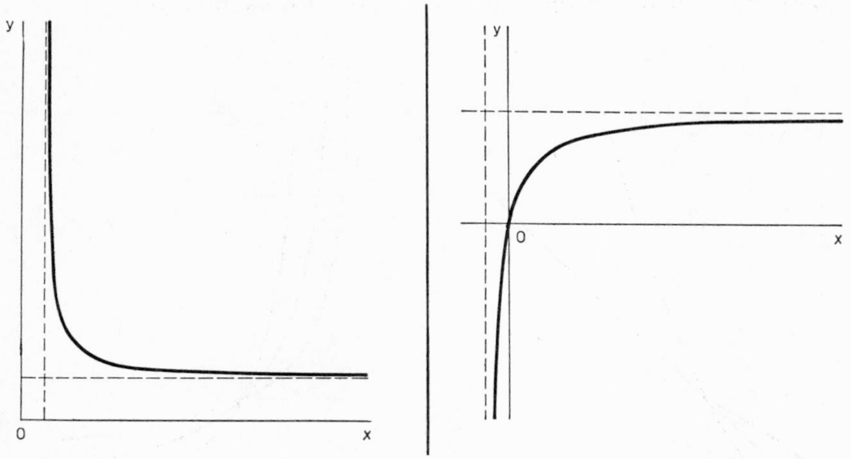

FIG. 18.10 a. Positive curvature. FIG. 18.10 b. Negative curvature.

The hyperbola $\frac{1}{y} = \alpha_0 + \frac{\beta}{x}$.

The exponential function.

Fig. 18.11 illustrates the graphical representation of the exponential function

$$y = \gamma e^{\beta x}, \quad (-\infty < x < \infty). \qquad (18.7.9)$$

The curves pass through the point $(0, \gamma)$ and the x-axis is their asymptote.

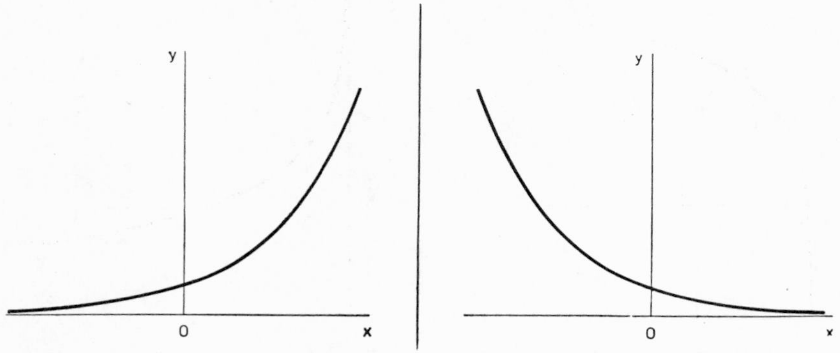

FIG. 18.11 a. $y = \gamma e^{\beta x} (\beta > 0)$. FIG. 18.11 b. $y = \gamma e^{-\beta x} (\beta > 0)$.

Exponential functions.

Power functions.

Fig. 18.12 illustrates the behaviour of the power function

$$y = \gamma x^{\beta} . \tag{18.7.10}$$

All curves pass through the point $(1, \gamma)$. If $\beta > 0$ the curves pass through the point $(0, 0)$; if $\beta < 0$ the coordinate axes are asymptotes.

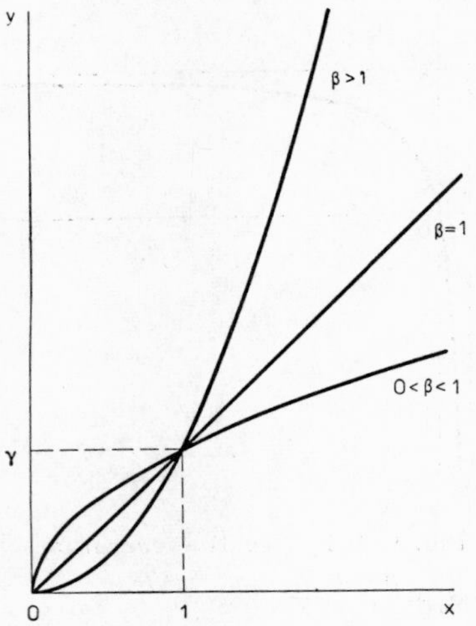

FIG. 18.12a. $y = \gamma x^{\beta} (\beta > 0)$.

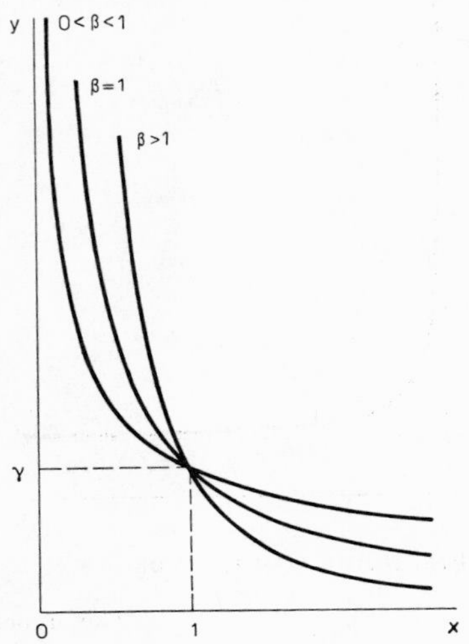

FIG. 18.12b. $y = \gamma x^{-\beta} (\beta > 0)$.

Power functions.

Logarithmic functions.

Fig. 18.13 illustrates the behaviour of the logarithmic curve

$$y = \alpha + \beta \log x, \quad (0 < x < \infty). \tag{18.7.11}$$

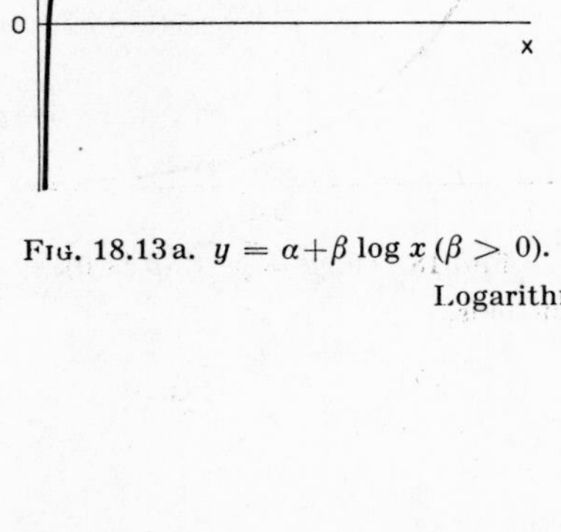

FIG. 18.13a. $y = \alpha + \beta \log x \ (\beta > 0)$.

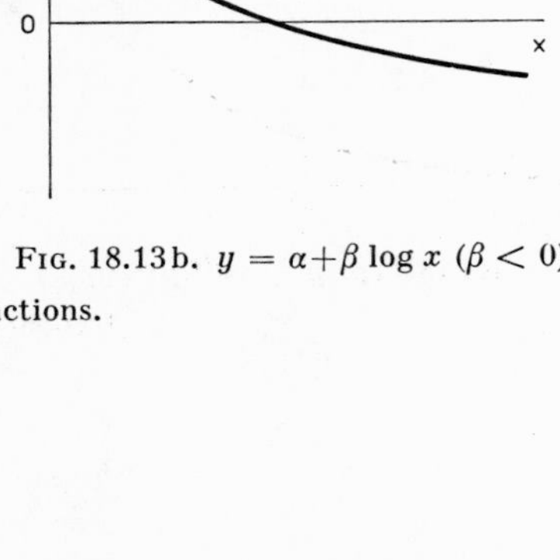

FIG. 18.13b. $y = \alpha + \beta \log x \ (\beta < 0)$.

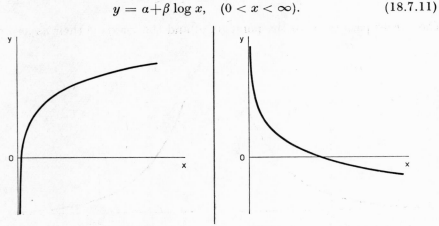

Logarithmic functions.

The y-axis forms the asymptote of the curve which passes through the point $(1, \alpha)$.

The function $y = \gamma e^{\beta/x}$.

Fig. 18.14a gives a graph of the function

$$y = \gamma e^{-\beta/x}, \quad (\beta > 0), \quad (0 < x < \infty) . \tag{18.7.12}$$

The curve passes through the origin, has a point of inflection at $(\beta/2, \gamma/e^2)$ and the asymptote $y = \gamma$.

Fig. 18.14b gives a graph of the function

$$y = \gamma e^{\beta/x}, \quad (\beta > 0), \quad (0 < x < \infty). \tag{18.7.13}$$

The curve has two asymptotes, namely $x = 0$ and $y = \gamma$.

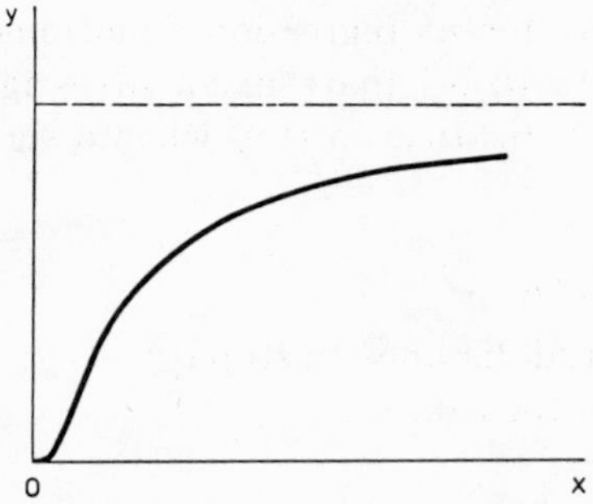

FIG. 18.14a. $y = \gamma e^{-\beta/x}(\beta > 0)$.

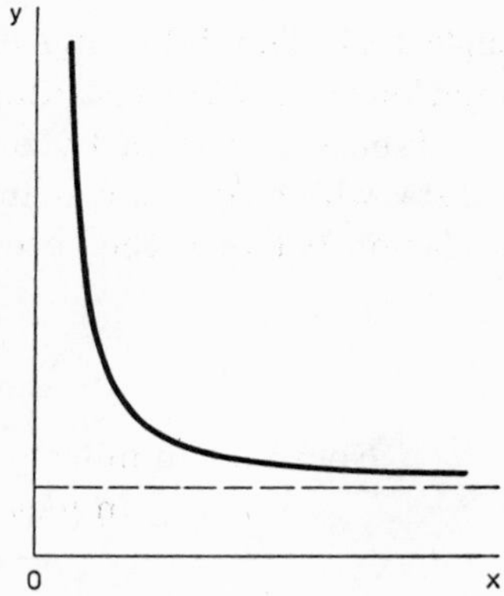

FIG. 18.14b. $y = \gamma e^{\beta/x}(\beta > 0)$.

The formula (18.7.4) includes many other types of functions than those given in Table 18.14. For $f(y) = \dfrac{1}{y}$ and $g(x) = e^{-x}$ we obtain the function

$$y = \frac{1}{\alpha_0 + \beta e^{-x}}, \quad (-\infty < x < \infty) , \tag{18.7.14}$$

being a special case of the "logistic curve" which has the equation

$$y = \frac{1}{\alpha_0 + \beta e^{-\varkappa x}} . \tag{18.7.15}$$

For $\varkappa = 1$, we obtain (18.7.14).

Fig. 18.15 gives a graph of the special logistic curve (18.7.14). The curve has two horizontal asymptotes, $y = 0$ and $y = 1/\alpha_0$, and the point of inflection $\left(\log_e \dfrac{\beta}{\alpha_0}, \dfrac{1}{2\alpha_0}\right)$.

Curves with more than two parameters will be discussed in Chapter 20.

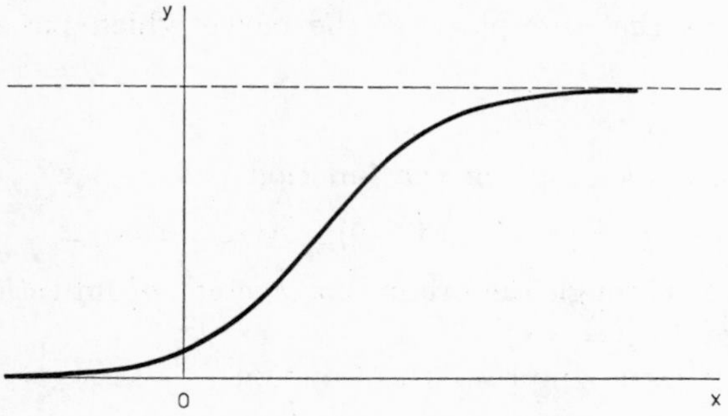

Fig. 18.15. $y = \dfrac{1}{\alpha_0 + \beta e^{-x}}$.

Example 18.4. The following example is discussed in order to illustrate the remarks on p. 558 regarding the fact that often it is possible to describe a given set of data by means of different regression equations.

The data which are shown in Table 18.16 originate in an investigation of the relation between the speed and the distance to stop when a signal is

TABLE 18.16.

Speed (x) in miles per hour and distance to stop (y) in feet for 50 motor-cars.

x	y				
4	2	10			
7	4	22			
8	16				
9	10				
10	18	26	34		
11	17	28			
12	14	20	24	28	
13	26	34	34	46	
14	26	36	60	80	
15	20	26	54		
16	32	40			
17	32	40	50		
18	42	56	76	84	
19	36	46	68		
20	32	48	52	56	64
22	66				
23	54				
24	70	92	93	120	
25	85				

given for 50 motor-cars[1]). Fig. 18.16 illustrates the interrelation between speed and distance to stop. The figure shows that there is a considerable variation in the distance to stop for a given speed, but considering that the observations were made on different cars with different drivers, road surfaces, etc., this was to be expected a priori. For the sake of argument we shall assume that the speed is a predetermined quantity measured without errors.

The way in which the average distance to stop depends on the speed may be determined by a regression analysis. An ordinary linear regression analysis based on the hypothesis

$$\mathcal{M}\{y|x\} = \alpha+\beta(x-\bar{x})$$

and

$$\mathcal{V}\{y|x\} = \sigma^2$$

results in

$$Y = 3{\cdot}93x-17{\cdot}6\,, \quad s^2 = 236{\cdot}5\,,$$

x denoting the speed in miles per hour and y the distance to stop in feet.

Comparison of the variation within sets, $s_1^2 = 218{\cdot}2$, and the variation of the means about the regression line, $s_2^2 = 269{\cdot}9$, demonstrates that the hypothesis regarding linearity cannot be rejected on the basis of this test, as $v^2 = 1{\cdot}24$ is not significant.

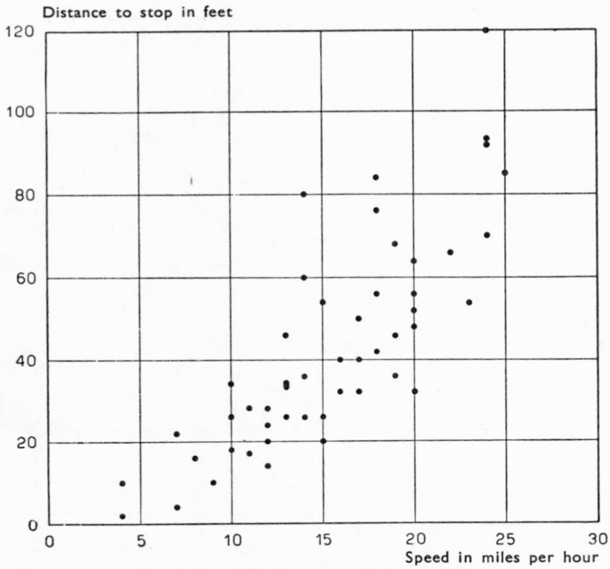

FIG. 18.16. Relation between speed and distance to stop for 50 motor-cars.

---

[1]) The data have been reprinted by permission from Table 10 of M. EZEKIEL: *Methods of Correlation Analysis*, published by John Wiley & Sons, New York, 1930, 2. edt. 1941.

For $x = 0$ we have $y = -17{\cdot}6$, i. e., a nonsensical result if $-17{\cdot}6$ differs significantly from zero. As the corresponding t is equal to $2{\cdot}6$ the deviation must be considered significant.

The regression analysis may also be based on the hypothesis that $\mathcal{M}\{y|x\} = \beta x$, $\mathcal{V}\{y|x\} = \sigma^2$, which leads to

$$Y = 2{\cdot}91x\,, \quad s^2 = 264{\cdot}4\,,$$

see (18.3.38). If the hypothesis is tested by comparing s_1^2 and s_2^2 we obtain $v^2 = 343{\cdot}8/218{\cdot}2 = 1{\cdot}58$, which quantity is not significant.

As a third and fourth hypothesis we will try

$$\mathcal{M}\{y|x\} = \beta x^2, \quad \mathcal{V}\{y|x\} = \sigma^2\,,$$

and

$$\mathcal{M}\{y|x\} = \alpha+\beta(x^2-\overline{x^2})\,, \quad \mathcal{V}\{y|x\} = \sigma^2\,,$$

respectively, x^2 being introduced as the independent variable instead of x, see Fig. 18.17, which gives the relation between the square of the speed and the distance to stop. This leads to the regression equations

$$Y = 0{\cdot}1534x^2\,, \quad s^2 = 243{\cdot}6\,,$$

and

$$Y = 0{\cdot}1290x^2+8{\cdot}86\,, \quad s^2 = 226.5\,,$$

for which the v^2-values are $1{\cdot}32$ and $1{\cdot}11$, respectively.

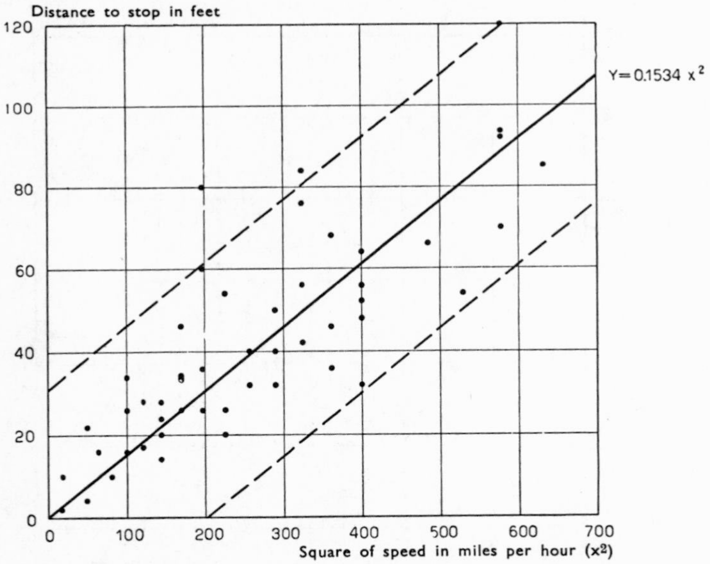

FIG. 18.17. Relation between the square of the speed and the distance to stop for 50 motor-cars.

The above four regression analyses are based on the assumption that $\mathcal{V}\{y|x\} = \sigma^2$. Just as it is possible to suggest several hypotheses regarding the form of the regression curve, it is also possible to suggest various hypotheses as to how the variance depends on the independent variable, and on account of the limitations of the observed data it is not possible from a purely statistical point of view to indicate that any one of them is "best" or "most correct".

In the following we will derive three more regression equations based on the assumption that $\mathcal{V}\{y|x\} = \sigma^2 x^2$ or, what is equivalent, $\mathcal{V}\left\{\frac{y}{x}|x\right\} = \sigma^2$. The assumptions for the three regression analyses are as follows:

$$\mathcal{M}\{y|x\} = \beta x, \qquad \mathcal{V}\{y|x\} = \sigma^2 x^2 ,$$
$$\mathcal{M}\{y|x\} = \beta x^2, \qquad \mathcal{V}\{y|x\} = \sigma^2 x^2 ,$$
$$\mathcal{M}\{y|x\} = \alpha_0 x + \beta x^2, \quad \mathcal{V}\{y|x\} = \sigma^2 x^2 .$$

Introducing the notation $z = y/x$ we find from these equations after dividing by x

$$\mathcal{M}\{z|x\} = \beta, \qquad \mathcal{V}\{z|x\} = \sigma^2 ,$$
$$\mathcal{M}\{z|x\} = \beta x, \qquad \mathcal{V}\{z|x\} = \sigma^2 ,$$
$$\mathcal{M}\{z|x\} = \alpha_0 + \beta x, \quad \mathcal{V}\{z|x\} = \sigma^2 .$$

Transformed in this manner the theory of linear regression may be applied directly with x as independent and z as dependent variable. Fig. 18.18 shows the observed values of (x, z). For the first of the three hypotheses the regression analysis "degenerates" into an analysis of variance,

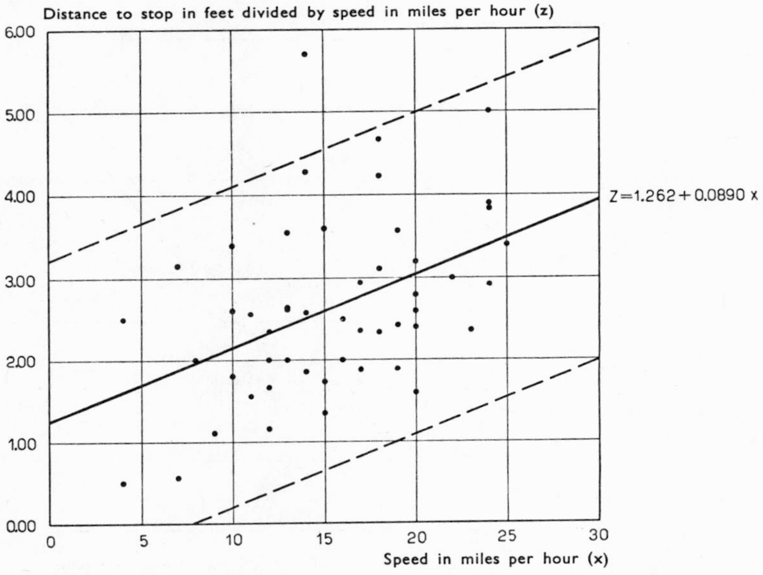

FIG. 18.18. Relation between speed and distance to stop divided by speed for 50 motor-cars.

the mean value of z being assumed to be independent of x, i. e., we suppose that the points of Fig. 18.18 are normally distributed about a horizontal line. The second hypothesis represents the mean value of z as a straight line through the origin, and the third hypothesis represents the mean value of z as a straight line which does not pass through this point, see Fig. 18.18.

Table 18.17 gives the results of the seven regression analyses, and Figs. 18.19 and 18.20 show the corresponding regression curves in cases where we have $\mathcal{V}\{x|y\} = \sigma^2$ and $\mathcal{V}\{y|x\} = \sigma^2 x^2$, respectively.

TABLE 18.17.

The results of various regression analyses of the same data and based on different hypotheses regarding the regression equation and the variance.

Empirical regression equation	Empirical variance	Test for form of regression curve. $v^2 = s_2^2/s_1^2$
$Y = 2{\cdot}91x$	$s^2 = 264{\cdot}4$	$v^2 = 1{\cdot}58$
$Y = 2{\cdot}63x$	$s^2 = 1{\cdot}141x^2$	$v^2 = 1{\cdot}47$
$Y = 3{\cdot}93x - 17{\cdot}6$	$s^2 = 236{\cdot}5$	$v^2 = 1{\cdot}24$
$Y = 0{\cdot}1534x^2$	$s^2 = 243{\cdot}6$	$v^2 = 1{\cdot}32$
$Y = 0{\cdot}1625x^2$	$s^2 = 1{\cdot}088x^2$	$v^2 = 1{\cdot}33$
$Y = 0{\cdot}1290x^2 + 8{\cdot}86$	$s^2 = 226{\cdot}5$	$v^2 = 1{\cdot}11$
$Y = 0{\cdot}0890x^2 + 1{\cdot}26x$	$s^2 = 0{\cdot}938x^2$	$v^2 = 0{\cdot}90$

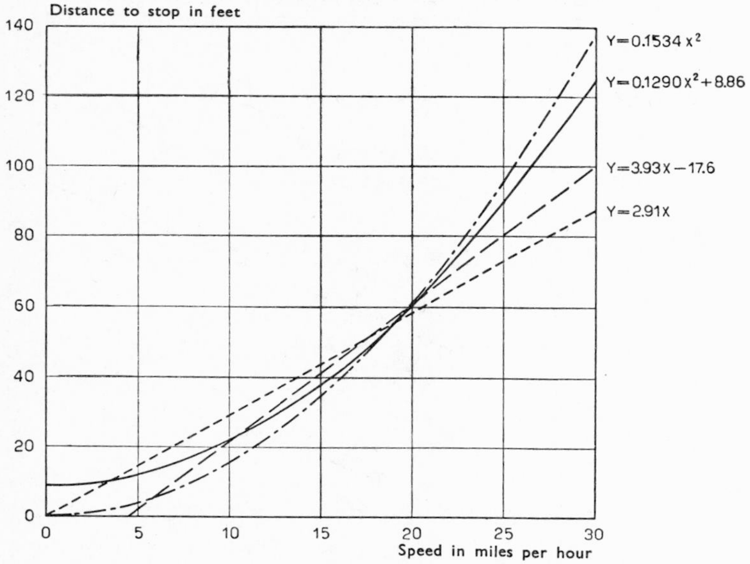

FIG. 18.19. Regression curves derived under the assumption that $\mathcal{V}\{y|x\} = \sigma^2$.

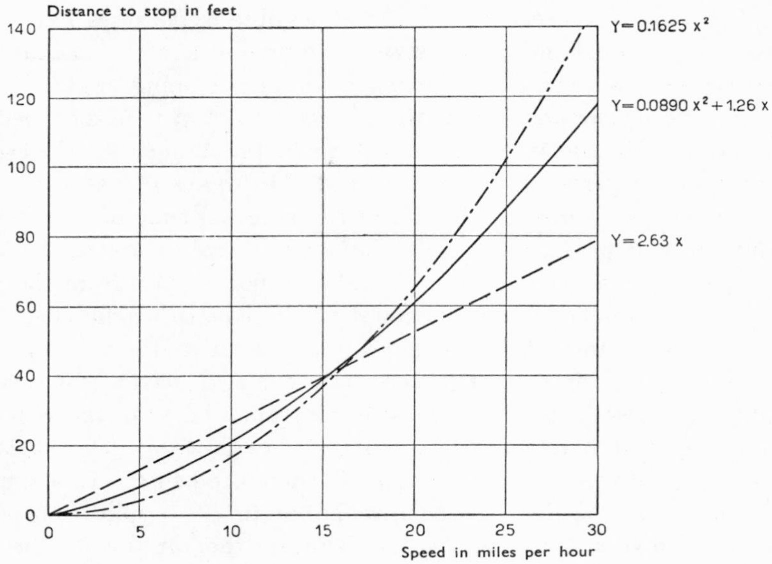

FIG. 18.20. Regression curves derived under the assumption that $\mathcal{V}\{y|x\} = \sigma^2 x^2$.

Besides the above seven regression equations it is possible to derive many others that do not give any significant value of v^2. It may be said about all these equations that in the range between 12 and 20 miles per hour, i. e., in the interval which includes the major part of the observations, they will give distances to stop which deviate slightly compared with the variations observed between the distances to stop corresponding to a given speed.

From a mathematical point of view, however, the equations differ greatly, which results in large deviations between the distances to stop as computed from the equations when the values for speed are outside the range of the observations, see Figs. 18.19 and 18.20 for $x = 30$. This elucidates a characteristic feature of statistical methods which has been mentioned several times before. A test of significance may lead to the rejection of a hypothesis, but never to a proof that a hypothesis is correct, for as it has been shown above, it is always possible to put forth different hypotheses which cannot be rejected by statistical methods alone. It looks tempting to utilize the v^2-criterion to arrange the hypotheses in order of "goodness", the hypothesis leading to the smallest v^2-value being denoted "the best". This is by no means correct, however, as the v^2-value of the "correct" hypothesis is a stochastic variable which may quite possibly possess a comparatively large value for the data in hand; in certain cases the "correct" v^2-value will even be significant, cf. § 9.4.

The choice between the different possible hypotheses must be made by means of professional knowledge of the process generating the observations. In the present case the following mode of reasoning might be used. For a given motor-car and driver the distance to stop is determined mainly by three factors: the speed of the vehicle at the moment of the signal to stop, the reaction time of the driver, and the brakes of the car. If the speed of the car is x miles per hour and the reaction time of the driver α_0 hours the car will proceed $\alpha_0 x$ miles before the brakes are used. According to simple physical laws the theoretical distance to stop from the moment the brakes are applied is proportional to the square of the speed, i. e., equal to βx^2. We must therefore expect a total distance to stop of $\alpha_0 x+\beta x^2$. Repeated experiments with the same car and driver will reveal that the time of reaction varies; it may for instance be that the reaction time is normally distributed about α_0 with the variance σ^2. The variance for the first term of the distance to stop will therefore be $\sigma^2 x^2$. In a similar manner we see that the distance corresponding to the application of the brakes βx^2 will have a variance $\omega^2 x^4$, ω^2 denoting the variance of the distribution of the factor β, the application of the brakes not being exactly the same every time. We must thus expect a variance of the form $\sigma^2 x^2+\omega^2 x^4$.

It is to be expected that for each single car and driver on a given road such a relation holds good. As the observations in our present case include 50 different cars, drivers, and portions of road, each with their value of α_0, β, σ^2 and ω^2, we cannot expect these relationships to show up clearly.

On basis of the above theoretical considerations the hypothesis

$$\mathcal{M}\{y|x\} = \alpha_0 x+\beta x^2$$

must be preferred to the others. Regarding the variance, the form $\mathcal{V}\{y|x\} = \sigma^2 x^2$ has been used in the regression analysis, i. e., it has been assumed that the factor ω^2 is so small that the term $\omega^2 x^4$ may be considered negligible in comparison with $\sigma^2 x^2$.

In the present case it is possible to give a plausible reason for preferring one hypothesis to another. It is often impossible, however, to give a theoretical explanation of an observed relationship. In such cases the choice is determined by the type of the mathematical function, and we try to find a simple mathematical description of the data with as few parameters as possible, taking into account how the results are to be utilized.

The above example gives a clear illustration as to how the mathematical specification affects the result of the regression analysis and warns us to be very careful with the interpretation and the application of formulas that are not firmly based on professional knowledge.

18.8. Comparison of Two Regression Lines.

Let us consider two sets of observational data which have each formed the basis of a regression analysis according to the principles given in the preceding paragraphs. The following quantities which form the basis of the regression analysis are calculated from the two sets of data, see Table 18.4:

N_1	$\bar{x}_1.$	$\bar{y}_1.$	SSD_{x_1}	$SPD_{x_1 y_1}$	b_1	SSD_{y_1}	$SSD_{y_1\|x_1}$	$f_1 = N_1-2$	s_1^2
N_2	$\bar{x}_2.$	$\bar{y}_2.$	SSD_{x_2}	$SPD_{x_2 y_2}$	b_2	SSD_{y_2}	$SSD_{y_2\|x_2}$	$f_2 = N_2-2$	s_2^2

The equations of the two regression lines are:

$$Y^{(1)} = \bar{y}_1.+b_1(x-\bar{x}_1.) \tag{18.8.1}$$

and

$$Y^{(2)} = \bar{y}_2.+b_2(x-\bar{x}_2.) \tag{18.8.2}$$

From each set of observations we thus have three quantities ($\bar{y}_1.$, b_1, s_1^2) and ($\bar{y}_2.$, b_2, s_2^2) which form estimates of the corresponding values (α_1, β_1, σ_1^2) and (α_2, β_2, σ_2^2) in the two populations. The identity of the two populations is tested by comparing the estimates, cf. the corresponding tests in § 15.4 for comparing two (one-dimensional) sets of observational data. First the hypothesis $\sigma_1^2 = \sigma_2^2$ is tested by means of the variance ratio $v^2 = s_1^2/s_2^2$. If this test does not reveal a significant difference between the two variances, the slopes of the two regression lines may be compared by means of a t-test, while there are no exact tests for such a comparison if we must assume that the variances σ_1^2 and σ_2^2 are different, cf. the corresponding problem in § 15.4, case B.

1. *The two variances s_1^2 and s_2^2 do not differ significantly.*

A pooled estimate s^2 of the theoretical variance σ^2 is calculated from the two variances s_1^2 and s_2^2

$$\boxed{s^2 = \frac{(N_1-2)s_1^2+(N_2-2)s_2^2}{N_1+N_2-4}.} \tag{18.8.3}$$

If the test hypothesis $\beta_1 = \beta_2$ is correct, the difference b_1-b_2 is normally distributed about 0 with variance

$$\mathcal{V}\{b_1-b_2\} = \sigma^2\left(\frac{1}{SSD_{x_1}}+\frac{1}{SSD_{x_2}}\right), \tag{18.8.4}$$

the quantity

$$u = \frac{b_1 - b_2}{\sigma\sqrt{\dfrac{1}{SSD_{x_1}} + \dfrac{1}{SSD_{x_2}}}} \tag{18.8.5}$$

being normally distributed with parameters (0, 1). According to (15.1.8) we now have that

$$\boxed{t = \frac{b_1 - b_2}{s\sqrt{\dfrac{1}{SSD_{x_1}} + \dfrac{1}{SSD_{x_2}}}}} \tag{18.8.6}$$

has a t-distribution with $f = N_1 + N_2 - 4$ degrees of freedom. If t exceeds, e. g., the bilateral 99% limit the test hypothesis will be rejected, i. e., we must consider the slopes of the two lines to be really different.

If according to (18.8.6) t is not significant the two regression lines may be considered parallel. A common estimate of the slope $\beta_1 = \beta_2 = \beta$ is obtained by forming the weighted mean of the two slopes b_1 and b_2, using the reciprocal values of the variances as weights, i. e.,

$$\boxed{\bar{b} = \frac{SSD_{x_1} b_1 + SSD_{x_2} b_2}{SSD_{x_1} + SSD_{x_2}} = \frac{SPD_{x_1 y_1} + SPD_{x_2 y_2}}{SSD_{x_1} + SSD_{x_2}},} \tag{18.8.7}$$

cf. (9.7.11). According to the addition theorem for the normal distribution $\bar{b}$ is normally distributed about β with variance

$$\mathcal{V}\{\bar{b}\} = \frac{\sigma^2}{SSD_{x_1} + SSD_{x_2}}. \tag{18.8.8}$$

The quantity

$$\sigma^2 u^2 = (b_1 - b_2)^2 \Big/ \left(\frac{1}{SSD_{x_1}} + \frac{1}{SSD_{x_2}}\right), \tag{18.8.9}$$

see (18.8.5), is in this case distributed as $\sigma^2\chi^2$ with one degree of freedom and may therefore be added to the numerator of s^2, cf. (18.8.3), the one degree of freedom at the same time being added to the denominator, which leads to a new estimate of σ^2 with $N_1 + N_2 - 3$ degrees of freedom.

The equations for the two theoretical regression lines with the same slope are

$$\eta^{(1)} = \alpha_1 + \beta(x - \bar{x}_{1\cdot}) = \alpha_1 - \beta\bar{x}_{1\cdot} + \beta x \tag{18.8.10}$$

and

$$\eta^{(2)} = \alpha_2 + \beta(x - \bar{x}_{2\cdot}) = \alpha_2 - \beta\bar{x}_{2\cdot} + \beta x\,. \tag{18.8.11}$$

If the two constant terms are equal, i. e.,

$$\alpha_1 - \beta \bar{x}_{1\cdot} = \alpha_2 - \beta \bar{x}_{2\cdot} \tag{18.8.12}$$

or

$$\frac{\alpha_1 - \alpha_2}{\bar{x}_{1\cdot} - \bar{x}_{2\cdot}} = \beta, \tag{18.8.13}$$

the lines are identical.

As an estimate of $(\alpha_1 - \alpha_2)/(\bar{x}_{1\cdot} - \bar{x}_{2\cdot})$ we calculate

$$\boxed{\hat{b} = \frac{\bar{y}_{1\cdot} - \bar{y}_{2\cdot}}{\bar{x}_{1\cdot} - \bar{x}_{2\cdot}},} \tag{18.8.14}$$

which is compared with the estimate $\bar{b}$ of β. ($\bar{b}$ has been calculated from the variation within sets and is therefore stochastically independent of $\hat{b}$ which is based on the variation between sets). $\hat{b}$ denotes the slope of the straight line which connects the points $(\bar{x}_{1\cdot}, \bar{y}_{1\cdot})$ and $(\bar{x}_{2\cdot}, \bar{y}_{2\cdot})$. If the two theoretical regression lines are identical, the slope $\hat{b}$ will only deviate at random from the slope $\bar{b}$ of the two empirical regression lines.

As the variance of $\hat{b}$ is equal to

$$\mathcal{V}\{\hat{b}\} = \frac{\sigma^2}{(\bar{x}_{1\cdot} - \bar{x}_{2\cdot})^2}\left(\frac{1}{N_1} + \frac{1}{N_2}\right), \tag{18.8.15}$$

we have

$$\mathcal{V}\{\hat{b} - \bar{b}\} = \sigma^2\left(\frac{1}{(\bar{x}_{1\cdot} - \bar{x}_{2\cdot})^2}\left(\frac{1}{N_1} + \frac{1}{N_2}\right) + \frac{1}{SSD_{x_1} + SSD_{x_2}}\right). \tag{18.8.16}$$

If the two theoretical regression lines are identical, the quantity

$$u = (\hat{b} - \bar{b})/\sqrt{\mathcal{V}\{\hat{b} - \bar{b}\}} \tag{18.8.17}$$

will be normally distributed with parameters (0, 1), and the quantity which appears when σ is replaced by s will have a t-distribution; it will therefore be possible to test the hypothesis by applying a t-test.

2. *The variances s_1^2 and s_2^2 differ significantly.*

If the test hypothesis $\beta_1 = \beta_2$ is correct, the difference $b_1 - b_2$ is normally distributed about 0 with variance

$$\mathcal{V}\{b_1 - b_2\} = \frac{\sigma_1^2}{SSD_{x_1}} + \frac{\sigma_2^2}{SSD_{x_2}}, \tag{18.8.18}$$

and therefore the quantity

$$u = (b_1 - b_2)\bigg/\sqrt{\frac{\sigma_1^2}{SSD_{x_1}} + \frac{\sigma_2^2}{SSD_{x_2}}} \tag{18.8.19}$$

is normally distributed with parameters (0, 1). If the estimates s_1^2 and s_2^2 are substituted for σ_1^2 and σ_2^2, we obtain the quantity

$$t = (b_1 - b_2) \Bigg/ \sqrt{\frac{s_1^2}{SSD_{x_1}} + \frac{s_2^2}{SSD_{x_2}}}, \tag{18.8.20}$$

which is distributed approximately as t with f degrees of freedom, f being determined from (15.4.17) with

$$c = \frac{\dfrac{s_1^2}{SSD_{x_1}}}{\dfrac{s_1^2}{SSD_{x_1}} + \dfrac{s_2^2}{SSD_{x_2}}}. \tag{18.8.21}$$

If the hypothesis $\beta_1 = \beta_2$ is not rejected, a common slope $\bar{b}$ is calculated as the weighted mean of the two slopes b_1 and b_2, using the reciprocal values of the variances as weights, i. e.,

$$\bar{b} = \frac{b_1 \dfrac{SSD_{x_1}}{\sigma_1^2} + b_2 \dfrac{SSD_{x_2}}{\sigma_2^2}}{\dfrac{SSD_{x_1}}{\sigma_1^2} + \dfrac{SSD_{x_2}}{\sigma_2^2}} = \frac{\dfrac{SPD_{x_1y_1}}{\sigma_1^2} + \dfrac{SPD_{x_2y_2}}{\sigma_2^2}}{\dfrac{SSD_{x_1}}{\sigma_1^2} + \dfrac{SSD_{x_2}}{\sigma_2^2}}. \tag{18.8.22}$$

According to the addition theorem for the normal distribution the variance of $\bar{b}$ is equal to

$$v\{\bar{b}\} = 1 \Bigg/ \left(\frac{SSD_{x_1}}{\sigma_1^2} + \frac{SSD_{x_2}}{\sigma_2^2}\right). \tag{18.8.23}$$

As σ_1^2 and σ_2^2 are unknown, they are replaced by the estimates s_1^2 and s_2^2.

In order to test the identity of the regression lines the slope $\bar{b}$ is compared with the slope $\hat{b} = (\bar{y}_{1\cdot} - \bar{y}_{2\cdot})/(\bar{x}_{1\cdot} - \bar{x}_{2\cdot})$. As the variance of $\hat{b}$ is equal to

$$v\{\hat{b}\} = \frac{1}{(\bar{x}_{1\cdot} - \bar{x}_{2\cdot})^2}\left(\frac{\sigma_1^2}{N_1} + \frac{\sigma_2^2}{N_2}\right), \tag{18.8.24}$$

we have

$$v\{\hat{b} - \bar{b}\} = \frac{1}{(\bar{x}_{1\cdot} - \bar{x}_{2\cdot})^2}\left(\frac{\sigma_1^2}{N_1} + \frac{\sigma_2^2}{N_2}\right) + \frac{1}{\dfrac{SSD_{x_1}}{\sigma_1^2} + \dfrac{SSD_{x_2}}{\sigma_2^2}}. \tag{18.8.25}$$

If the test hypothesis is correct, the quantity

$$u = \frac{\hat{b} - \bar{b}}{\sqrt{v\{\hat{b} - b\}}} \tag{18.8.26}$$

is normally distributed with parameters (0, 1). If σ_1^2 and σ_2^2 are replaced by the estimates s_1^2 and s_2^2, (18.8.26) is *approximately* normally distributed *for large values of* f_1 *and* f_2.

Example 18.5. Table 18.18 gives the production figures and the sieve residues for the cement produced in an experiment with two tube mills for grinding clinkers. The sieve residues are determined by taking samples of the cement and using the same sieve for all analyses. The residue gives a (coarse) measure of the fineness of the cement and thus an idea of its quality. As in the range of variation considered here the relation between production and the logarithm of the sieve residue may be regarded as linear, Table 18.18 gives the logarithm of the residue (in %) instead of the residue itself. The question to be answered was whether the experiments in hand showed any difference between the relation between production and sieve residue for the two mills considered.

Table 18.18 gives the two regression analyses, and Fig. 18.21 shows the observations and the two regression lines. The diagram indicates that the slopes of the two lines differ only at random, and that corresponding to a given production mill No. 2 has yielded a finer product than mill No. 1.

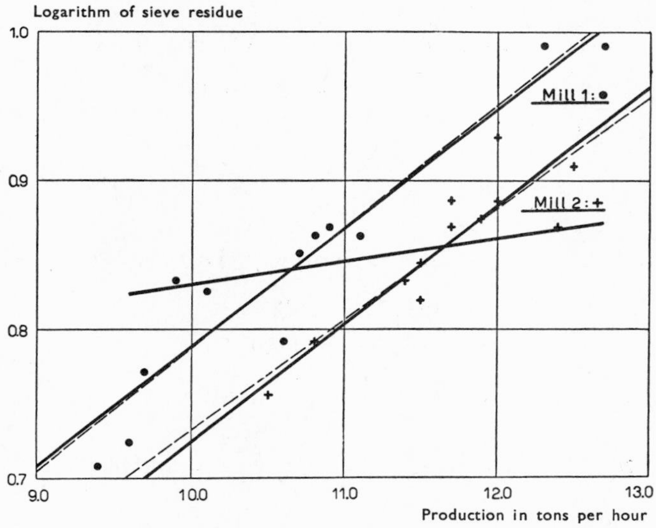

FIG. 18.21. The relationship between production and logarithm of sieve residue for two tube mills for production of cement. The two dashed lines represent the original regression lines, and the two parallel lines are the regression lines with the same (mean) slope. The fifth line passes through the two points corresponding to the means of the two sets of data.

According to the computations in Table 18.18 we find that the equations of the two regression lines are

$$Y^{(1)} = 0{\cdot}8402+0{\cdot}08137(x-10{\cdot}65)$$

and

$$Y^{(2)} = 0{\cdot}8559+0{\cdot}07479(x-11{\cdot}66)\ .$$

The corresponding variances are $s_1^2 = 0{\cdot}000867$ and $s_2^2 = 0{\cdot}000583$.

The variance ratio

$$v^2 = \frac{0{\cdot}000867}{0{\cdot}000583} = 1{\cdot}49$$

demonstrates that the two variances do not differ significantly, as $v^2_{.95}(10,\, 10) = 2{\cdot}98$. We thus find the pooled estimate

$$s^2 = \frac{0{\cdot}008670+0{\cdot}005829}{10+10} = \frac{0{\cdot}014499}{20} = 0{\cdot}000725\ ,$$

see Table 18.18.

The hypothesis $\beta_1 = \beta_2$ is tested by computing the difference between the two slopes

$$b_1-b_2 = 0{\cdot}08137-0{\cdot}07479 = 0{\cdot}00658\ ,$$

the estimate of the variance of this difference,

$$s^2\left(\frac{1}{SSD_{x_1}}+\frac{1}{SSD_{x_2}}\right) = 0{\cdot}000725\left(\frac{1}{11{\cdot}65}+\frac{1}{3{\cdot}749}\right)$$

$$= 0{\cdot}000725\times 0{\cdot}3525 = 0{\cdot}000256\ ,$$

and according to (18.8.6)

$$t = \frac{0{\cdot}00658}{\sqrt{0{\cdot}000256}} = 0{\cdot}41\ .$$

As this value is not significant, we compute the mean slope

$$\bar{b} = \frac{0{\cdot}9480+0{\cdot}2804}{11{\cdot}65+3{\cdot}75} = \frac{1{\cdot}2284}{15{\cdot}40} = 0{\cdot}07977\ ,$$

and the estimate of the variance of this quantity

$$v\{\bar{b}\} \approx \frac{s^2}{SSD_{x_1}+SSD_{x_2}} = \frac{0{\cdot}000725}{15{\cdot}40} = 0{\cdot}0000471\ ,$$

which gives us the standard error $s_{\bar{b}} = 0{\cdot}00686$. (The quantity (18.8.9) is here equal to $0{\cdot}00658^2/0{\cdot}3525 = 0{\cdot}000123$. This quantity may be combined with the above estimate of σ^2 to form the estimate

$$s^2 = \frac{0{\cdot}014499+0{\cdot}000123}{20+1} = 0{\cdot}000696\ .$$

Computation of the pooled estimate of σ^2 is often omitted in practice

TABLE 18.18.

Production (x) in tons per hour and the logarithm (y) of the sieve residue in per cent for two cement mills.

	Mill 1		Mill 2		Cross sums	
	x	y	x	y		
	9·4	0·708	10·5	0·756		
	9·6	0·724	10·8	0·792		
	9·9	0·833	11·7	0·869		
	10·9	0·869	11·5	0·820		
	10·1	0·826	12·0	0·929		
	9·7	0·771	12·5	0·909		
	10·8	0·863	11·7	0·887		
	10·6	0·792	12·4	0·869		
	10·7	0·851	12·0	0·887		
	11·1	0·863	11·9	0·875		
	12·3	0·991	11·4	0·833		
	12·7	0·991	11·5	0·845		
N	12		12		24	
Mean	10·65	0·8402	11·66	0·8559		
SSD	11·65	0·085812	3·749	0·026801	15·40	
SPD	0·9480		0·2804		1·2284	
b	0·08137		0·07479		0·07977	
SPD^2/SSD_x		0·077142		0·020972		
$SSD_{y\|x}$		0·008670		0·005829		0·014499
f		10		10		20
s^2		0·000867		0·000583		0·000725

when the estimate first determined is based on a fairly large number of degrees of freedom).

The two parallel regression lines with the equations

$$Y = 0{\cdot}8402+0{\cdot}07977(x-10{\cdot}65)$$

and

$$Y = 0{\cdot}8559+0{\cdot}07977(x-11{\cdot}66)$$

have been plotted in Fig. 18.21.

The line through the two sets of means has the slope

$$\hat{b} = \frac{0{\cdot}8559-0{\cdot}8402}{11{\cdot}66-10{\cdot}65} = \frac{0{\cdot}0157}{1{\cdot}01} = 0{\cdot}0155\ ,$$

and according to (18.8.15) the variance of this slope is equal to

$$v\{\hat{b}\} \approx \frac{0{\cdot}000725}{1{\cdot}01^2}\left(\frac{1}{12}+\frac{1}{12}\right) = 0{\cdot}000118\ .$$

The identity of the two regression lines is tested by computing the difference

$$\bar{b}-\hat{b} = 0{\cdot}0798-0{\cdot}0155 = 0{\cdot}0643 ,$$

and the variance of this difference

$$v\{\bar{b}-\hat{b}\} = v\{\bar{b}\}+v\{\hat{b}\} \approx 0{\cdot}000165 ,$$

and finally

$$t = \frac{0{\cdot}0643}{\sqrt{0{\cdot}000165}} = 5{\cdot}02 , \quad f = 20 .$$

As this t-value is significant, we cannot consider the two regression lines to be estimates of the same theoretical regression line, which is also indicated from Fig. 18.21.

Thus, the conclusion of this investigation must be that (1) for both mills we find that a change in the production of one ton per hour is followed by a change in the logarithm of the sieve residue of 0·0798, and (2) at a given rate of production mill No. 2 will yield a finer product than mill No. 1. Corresponding to a given rate of production the difference between the logarithms of the sieve residues is

$$\begin{aligned} d_{y|x} &= (\bar{y}_{1\cdot}+\bar{b}(x-\bar{x}_{1\cdot}))-(\bar{y}_{2\cdot}+\bar{b}(x-\bar{x}_{2\cdot})) \\ &= \bar{y}_{1\cdot}-\bar{y}_{2\cdot}-\bar{b}(\bar{x}_{1\cdot}-\bar{x}_{2\cdot}) . \end{aligned}$$

Inserting the values found we get

$$d_{y|x} = 0{\cdot}8402-0{\cdot}8559-0{\cdot}07977(10{\cdot}65-11{\cdot}66) = 0{\cdot}0649 ,$$

i. e., corresponding to the same rate of production the proportion between the sieve residues is antilog 0·0649 = 1·16.

The difference in relationship between production and sieve residue for the two mills may also be expressed as a difference between the productions corresponding to a given sieve residue. This expression corresponds to the *horizontal distance* between the two parallel lines of Fig. 18.21, while the above expression, $d_{y|x}$, denotes the *vertical distance* between the two lines. We may therefore obtain the difference between the two productions corresponding to the same sieve residue as $d_{y|x}$ divided by the slope, i. e.,

$$\frac{d_{y|x}}{\bar{b}} = \frac{\bar{y}_{1\cdot}-\bar{y}_{2\cdot}}{\bar{b}}-(\bar{x}_{1\cdot}-\bar{x}_{2\cdot}) .$$

Inserting the values found we obtain the difference 0·0649/0·07977 = 0·81 t/hour. When producing cement of equal fineness we thus find that mill No. 2 yields 0·81 t/hour more than mill No. 1.

The variance of $d_{y|x}$ is, according to the addition theorem for the normal distribution,

$$\begin{aligned} \mathcal{V}\{d_{y|x}\} &= \mathcal{V}\{\bar{y}_1.\}+\mathcal{V}\{\bar{y}_2.\}+(\bar{x}_1.-\bar{x}_2.)^2\mathcal{V}\{\bar{b}\} \\ &= \sigma^2\left(\frac{1}{N_1}+\frac{1}{N_2}+\frac{(\bar{x}_1.-\bar{x}_2.)^2}{SSD_{x_1}+SSD_{x_2}}\right). \end{aligned}$$

Confidence limits for the theoretical value of $d_{y|x}$ may then be derived by substituting s^2 for σ^2 and applying the t-distribution.

Confidence limits for the theoretical value of $d_{y|x}/\bar{b}$ are derived in analogy with the derivation of the confidence limits for ξ in § 18.5.

18.9. Comparison of Several Regression Lines.

Just as it is possible by application of the analysis of variance to extend the comparison of two sets of (one-dimensional) observations to several sets, see Chapter 16, it is also possible to generalize the principles applied to the comparison of two regression lines so that they may be applied to the comparison of several regression lines.

Let there be m sets of observations from which the following quantities have been computed as a basis of the regression analyses:

TABLE 18.19.

N_1	$\bar{x}_1.$	$\bar{y}_1.$	SSD_{x_1}	$SPD_{x_1y_1}$	b_1	SSD_{y_1}	$SSD_{y_1\|x_1}$	$f_1=N_1-2$	s_{11}^2
N_2	$\bar{x}_2.$	$\bar{y}_2.$	SSD_{x_2}	$SPD_{x_2y_2}$	b_2	SSD_{y_2}	$SSD_{y_2\|x_2}$	$f_2=N_2-2$	s_{21}^2
⋮	⋮	⋮	⋮	⋮	⋮	⋮	⋮	⋮	⋮
N_m	$\bar{x}_m.$	$\bar{y}_m.$	SSD_{x_m}	$SPD_{x_my_m}$	b_m	SSD_{y_m}	$SSD_{y_m\|x_m}$	$f_m=N_m-2$	s_{m1}^2

The μth regression line has the equation

$$Y^{(\mu)} = \bar{y}_{\mu}.+b_{\mu}(x-\bar{x}_{\mu}.)\,,\quad \mu = 1, 2, \ldots, m\,. \tag{18.9.1}$$

We thus compute three quantities from each set of data $(\bar{y}_{\mu}., b_{\mu}, s_{\mu 1}^2)$, $\mu = 1, 2, \ldots, m$, which denote estimates of the corresponding values $(\alpha_{\mu}, \beta_{\mu}, \sigma_{\mu}^2)$, $\mu = 1, 2, \ldots, m$, in the m populations. In the following analysis it is assumed that *the variance about the regression is the same in all populations*, i. e., $\sigma_1^2 = \sigma_2^2 = \ldots = \sigma_m^2 = \sigma^2$. (This assumption may be tested by comparing the variances $s_{11}^2, s_{21}^2, \ldots, s_{m1}^2$ according to BARTLETT'S test.)

As an estimate of σ^2 we calculate

$$s_1^2 = \frac{f_1 s_{11}^2+\ldots+f_m s_{m1}^2}{f_1+\ldots+f_m} = \frac{SSD_{y_1|x_1}+\ldots+SSD_{y_m|x_m}}{N_1+\ldots+N_m-2m}\,. \tag{18.9.2}$$

The variance s_1^2 denotes the variation about the m regression lines, i. e., the variation "within sets".

The parallelism of the regression lines, i. e., the hypothesis $\beta_1 = \beta_2 = \ldots = \beta_m = \beta$, is tested by comparing the variation between the slopes $b_1, b_2, \ldots, b_m$, and the variation within sets. If the hypothesis is correct the slope b_μ is normally distributed with mean value β and variance σ^2/SSD_{x_μ}, so that

$$u_\mu = \frac{b_\mu - \beta}{\sigma/\sqrt{SSD_{x_\mu}}} \tag{18.9.3}$$

is normally distributed with parameters (0, 1). It then follows that

$$\sigma^2 \sum_{\mu=1}^{m} u_\mu^2 = \sum_{\mu=1}^{m} (b_\mu - \beta)^2 SSD_{x_\mu} \tag{18.9.4}$$

is distributed as $\sigma^2\chi^2$ with m degrees of freedom. According to the method of least squares the estimate of β becomes the weighted mean

$$\boxed{\bar{b} = \frac{b_1 SSD_{x_1} + \ldots + b_m SSD_{x_m}}{SSD_{x_1} + \ldots + SSD_{x_m}} = \frac{SPD_{x_1y_1} + \ldots + SPD_{x_my_m}}{SSD_{x_1} + \ldots + SSD_{x_m}}.} \tag{18.9.5}$$

Inserting $b_\mu - \beta = (b_\mu - \bar{b}) + (\bar{b} - \beta)$ in (18.9.4), we obtain

$$\sigma^2\chi^2 = \sum_{\mu=1}^{m} (b_\mu - \beta)^2 SSD_{x_\mu} = \sum_{\mu=1}^{m} (b_\mu - \bar{b})^2 SSD_{x_\mu} + (\bar{b} - \beta)^2 \sum_{\mu=1}^{m} SSD_{x_\mu}. \tag{18.9.6}$$

It now follows from the partition theorem for the χ^2-distribution that

$$\sum_{\mu=1}^{m} (b_\mu - \bar{b})^2 SSD_{x_\mu} \tag{18.9.7}$$

is distributed as $\sigma^2\chi^2$ with $m-1$ degrees of freedom, and that $\bar{b}$ is normally distributed about β with variance $\sigma^2 \Big/ \sum_{\mu=1}^{m} SSD_{x_\mu}$, and that $\bar{b}$ is stochastically independent of the sum of squares (18.9.7).

The parallelism of the regression lines may therefore be tested by calculating the variance

$$s_2^2 = \frac{1}{m-1} \sum_{\mu=1}^{m} (b_\mu - \bar{b})^2 SSD_{x_\mu} \tag{18.9.8}$$

and comparing this variance (the variation between slopes) with the variance s_1^2 (the variation within sets), i. e., by calculating the variance ratio $v^2 = s_2^2/s_1^2$.

If the variation between the slopes is not significantly larger than the variation within sets, the two variances may be combined to a common estimate of σ^2 with $\left(\sum_{\mu=1}^{m} N_\mu - 2m\right) + (m-1) = \sum_{\mu=1}^{m} N_\mu - m - 1$ degrees of freedom.

If the hypothesis $\beta_1 = \beta_2 = \ldots = \beta_m = \beta$ is correct, the m theoretical regression lines may be written

$$\eta^{(\mu)} = \alpha_\mu + \beta(x - \bar{x}_{\mu\cdot}) = \alpha_\mu - \beta\bar{x}_{\mu\cdot} + \beta x, \quad \mu = 1, 2, \ldots, m. \tag{18.9.9}$$

If these lines are identical, the following relation

$$\alpha_1 - \beta\bar{x}_{1\cdot} = \alpha_2 - \beta\bar{x}_{2\cdot} = \ldots = \alpha_m - \beta\bar{x}_{m\cdot} \tag{18.9.10}$$

must hold good, i. e., *the m points $(\bar{x}_{1\cdot}, \alpha_1)$, $(\bar{x}_{2\cdot}, \alpha_2)$, $\ldots$, $(\bar{x}_{m\cdot}, \alpha_m)$ must be situated on a straight line with slope β.* The equation of this line is $\eta = \bar{\alpha} + \beta(x - \bar{x}_{\cdot\cdot})$, where $\bar{\alpha}$ and $\bar{x}_{\cdot\cdot}$ denote the weighted means of $\alpha_1, \alpha_2, \ldots, \alpha_m$ and $\bar{x}_{1\cdot}, \bar{x}_{2\cdot}, \ldots, \bar{x}_{m\cdot}$, respectively.

The identity of the m (parallel) regression lines is therefore tested by calculating the regression line for the m points $(\bar{x}_{1\cdot}, \bar{y}_{1\cdot})$, $(\bar{x}_{2\cdot}, \bar{y}_{2\cdot})$, $\ldots$, $(\bar{x}_{m\cdot}, \bar{y}_{m\cdot})$ and testing its linearity. If the hypothesis of linearity is not rejected the slope of the regression line, $\hat{b}$, is compared with the slope $\bar{b}$.

The equation of the regression line for the means becomes

$$Y = \bar{y}_{\cdot\cdot} + \hat{b}(x - \bar{x}_{\cdot\cdot}), \tag{18.9.11}$$

where $\bar{y}_{\cdot\cdot}$ denotes the weighted mean of $\bar{y}_{1\cdot}, \bar{y}_{2\cdot}, \ldots, \bar{y}_{m\cdot}$, and the slope $\hat{b}$ is calculated from

$$\hat{b} = \frac{SPD_{\bar{x}\bar{y}}}{SSD_{\bar{x}}} = \frac{\sum_{\mu=1}^{m} N_\mu(\bar{x}_{\mu\cdot} - \bar{x}_{\cdot\cdot})\bar{y}_{\mu\cdot}}{\sum_{\mu=1}^{m} N_\mu(\bar{x}_{\mu\cdot} - \bar{x}_{\cdot\cdot})^2}. \tag{18.9.12}$$

The variation about the regression line is expressed by the variance

$$s_3^2 = \frac{1}{m-2} \sum_{\mu=1}^{m} N_\mu(\bar{y}_{\mu\cdot} - Y_\mu)^2, \tag{18.9.13}$$

so that *the linearity of the regression line* may be tested by comparing this variance with the variance s_1^2 or with the above-described common estimate of σ^2 calculated from s_1^2 and s_2^2.

If the hypothesis regarding the linearity of the regression line is not rejected a common estimate of σ^2 may be based on the three variances s_1^2, s_2^2, and s_3^2, the total number of degrees of freedom being $(\Sigma N_\mu - 2m) + (m-1) + (m-2) = \Sigma N_\mu - 3$, cf. (18.9.2), (18.9.8), and (18.9.13).

If *the hypothesis regarding the identity of the m regression lines* is correct, the slope $\hat{b}$ is normally distributed with mean value β and variance $\sigma^2/SSD_{\bar{x}}$, from which it follows that the difference $\hat{b}-\bar{b}$ is normally distributed with mean value 0 and variance

$$v\{\hat{b}-\bar{b}\} = \sigma^2 \left(\frac{1}{SSD_{\bar{x}}} + \frac{1}{\sum\limits_{\mu=1}^{m} SSD_{x_\mu}} \right). \tag{18.9.14}$$

The test hypothesis is therefore examined by calculating the quantity

$$t = \frac{\hat{b}-\bar{b}}{s_0 \sqrt{\dfrac{1}{SSD_{\bar{x}}} + \dfrac{1}{\Sigma\, SSD_{x_\mu}}}}, \tag{18.9.15}$$

s_0^2 denoting the variance calculated from s_1^2, s_2^2, and s_3^2. According to (15.1.8) the above quantity has a t-distribution with $\Sigma N_\mu - 3$ degrees of freedom.

If $\hat{b}$ and $\bar{b}$ do not differ significantly the m empirical regression lines may be considered to be estimates of the same theoretical regression line. We then have the following estimate of this line

$$Y = \bar{y}_{..} + b(x - \bar{x}_{..}), \tag{18.9.16}$$

where b denotes the slope as calculated from the total number of observations, i. e.,

$$b = \frac{SPD_{xy}}{SSD_x} = \frac{\sum\limits_{\mu=1}^{m} SPD_{x_\mu y_\mu} + SPD_{\bar{x}\bar{y}}}{\sum\limits_{\mu=1}^{m} SSD_{x_\mu} + SSD_{\bar{x}}}$$

$$= \frac{\bar{b} \sum\limits_{\mu=1}^{m} SSD_{x_\mu} + \hat{b} SSD_{\bar{x}}}{\sum\limits_{\mu=1}^{m} SSD_{x_\mu} + SSD_{\bar{x}}}, \tag{18.9.17}$$

b thus being the weighted mean of $\bar{b}$ and $\hat{b}$.

The variation about this regression line is expressed by the variance

$$s^2 = \frac{SSD_{y|x}}{\sum\limits_{\mu=1}^{m} N_\mu - 2} = \frac{\left(\sum\limits_{\mu=1}^{m} N_\mu - 2m \right) s_1^2 + (m-1)s_2^2 + (m-2)s_3^2 + s_4^2}{\sum\limits_{\mu=1}^{m} N_\mu - 2}, \tag{18.9.18}$$

where

$$s_4^2 = (\hat{b}-\bar{b})^2 \bigg/ \left(\frac{1}{SSD_{\bar{x}}} + \frac{1}{\sum\limits_{\mu=1}^{m} SSD_{x_\mu}} \right). \tag{18.9.19}$$

The above analysis may be summarized as follows. The following regression lines enter into the analysis:

1. the m "individual" regression lines:

$$Y = \bar{y}_{\mu\cdot} + b_\mu(x - \bar{x}_{\mu\cdot}), \quad \mu = 1, 2, \ldots, m, \tag{18.9.20}$$

2. the m parallel regression lines with the average slope $\bar{b}$:

$$Y = \bar{y}_{\mu\cdot} + \bar{b}(x - \bar{x}_{\mu\cdot}), \quad \mu = 1, 2, \ldots, m, \tag{18.9.21}$$

3. the regression line of the means:

$$Y = \bar{y}_{\cdot\cdot} + \hat{b}(x - \bar{x}_{\cdot\cdot}), \tag{18.9.22}$$

4. the regression line for the total number of observations:

$$Y = \bar{y}_{\cdot\cdot} + b(x - \bar{x}_{\cdot\cdot}). \tag{18.9.23}$$

If the observations are denoted by $(x_{\mu i}, y_{\mu i})$ the deviation between the observation $y_{\mu i}$ and the corresponding point on the regression line for the total number of observations, $\bar{y}_{\cdot\cdot} + b(x_{\mu i} - \bar{x}_{\cdot\cdot})$, may be partitioned as follows:

$$\begin{aligned} y_{\mu i} - \bar{y}_{\cdot\cdot} - b(x_{\mu i} - \bar{x}_{\cdot\cdot}) &= (y_{\mu i} - \bar{y}_{\mu\cdot} - b_\mu(x_{\mu i} - \bar{x}_{\mu\cdot})) \\ &+ (b_\mu - \bar{b})(x_{\mu i} - \bar{x}_{\mu\cdot}) + (\bar{y}_{\mu\cdot} - \bar{y}_{\cdot\cdot} - \hat{b}(\bar{x}_{\mu\cdot} - \bar{x}_{\cdot\cdot})) \\ &+ [(\bar{b} - \hat{b})(x_{\mu i} - \bar{x}_{\mu\cdot}) + (\hat{b} - b)(x_{\mu i} - \bar{x}_{\cdot\cdot})]. \end{aligned} \tag{18.9.24}$$

The first term represents the variation of the observations about the m individual regression lines, the second term the variation between the slopes of these m lines, the third term the variation of the means about the regression line for the means, and the fourth term the difference between the average slope $\bar{b}$ and the slope $\hat{b}$ of the regression line for the means, b representing the weighted mean of $\bar{b}$ and $\hat{b}$, see (18.9.17). Squaring and summing leads to the partitioning of the variation of all observations about the regression line, as given in Table 18.20.

If all observations belong to the same population we know that, according to the partition theorem for the χ^2-distribution, the four variances are distributed as s^2 with mean value σ^2 and the number of degrees of freedom stated in Table 18.20. As s_1^2 is independent of hypotheses regarding the m theoretical regression lines, such hypotheses may be tested by comparing s_2^2, s_3^2, and s_4^2 with s_1^2, as described above.

Regarding the application of the above principles to an example, see: B. H. Wilsdon: *Discrimination by Specification Statistically Considered and Illustrated by the Standard Specification for Portland Cement*, Suppl. Journ. Roy. Stat. Soc., 1, 1934, 152–206, particularly Appendixes I and II, pp. 178–189.

TABLE 18.20.

Partitioning of the variation of all observations about the regression line.

Variation	Sum of squares	Degrees of freedom	Variance
Between $\hat{b}$ and $\bar{b}$	$(\hat{b}-\bar{b})^2 \Big/ \left(\frac{1}{SSD_{\bar{x}}}+\frac{1}{\Sigma SSD_{x\mu}}\right)$	1	s_4^2
About the regresssion line for the means	$\Sigma N_\mu(\bar{y}_{\mu\cdot}-\bar{y}_{\cdot\cdot}-\hat{b}(\bar{x}_{\mu\cdot}-\bar{x}_{\cdot\cdot}))^2$	$m-2$	s_3^2
Between the slopes $b_1, b_2, \ldots, b_m$	$\Sigma(b_\mu-\bar{b})^2 SSD_{x\mu}$	$m-1$	s_2^2
About the m individual regression lines	$\Sigma\Sigma(y_{\mu i}-\bar{y}_{\mu\cdot}-b_\mu(x_{\mu i}-\bar{x}_{\mu\cdot}))^2$	$\Sigma N_\mu-2m$	s_1^2
About the regression line for the total number of observations	$\Sigma\Sigma(y_{\mu i}-\bar{y}_{\cdot\cdot}-b(x_{\mu i}-x_{\cdot\cdot}))^2$	$\Sigma N_\mu-2$	

18.10. Markoff's Theorem.

The regression analysis as developed above is based on the assumption that the y's for given x's are independent and normally distributed. Similar results hold, however, even if the y's are not normally distributed.

Let us suppose that $\mathcal{M}\{y|x\} = \alpha+\beta(x-\bar{x})$, $\mathcal{V}\{y|x\} = \sigma^2h^2(x)$, and that the y's are independent. The MARKOFF theorem then says that *the least square estimate* $Y = \bar{y}+b(x-\bar{x})$ will be an *unbiased* estimate of $\mathcal{M}\{y|x\}$ and further that the variance of Y will be less than the variance of any other *linear* estimate. The least square estimate is therefore also called *the best linear unbiased estimate.* An estimate of σ^2 is obtained by dividing the minimized sum of squares by the number of degrees of freedom, and the variances of $\bar{y}$, b, and Y are obtained from the formulas above.

The proof is analogous to the one given in § 9.7 for proving that the sample mean is the best linear unbiased estimate of the population mean. A proof for the general MARKOFF theorem may be found in F. N. DAVID and J. NEYMAN: *Extension of the Markoff Theorem on Least Squares*, Statistical Research Memoirs, 2, 1938, 105–116.

The distributions of $\bar{y}$, b, and Y will not be normal if the y's are not normally distributed, but these distributions will tend to the normal for $n \to \infty$ according to the central limit theorem.

CHAPTER 19.

THE TWO-DIMENSIONAL NORMAL DISTRIBUTION

19.1. Correlation and Regression.

Consider the conditional distribution $p\{y|x\}$ of y for given x and let us suppose that the mean $\mathcal{M}\{y|x\}$ is a linear function of x and that the variance $\mathcal{V}\{y|x\}$ is a constant, the assumptions underlying the regression analysis of Chapter 18 thus being fulfilled. Assuming further that x *is normally distributed*, the two-dimensional distribution $p\{x, y\}$ can be calculated from the formula

$$p\{x, y\} = p\{x\}p\{y|x\} ,$$

cf. (5.12.6). This distribution function is called *the two-dimensional normal distribution*, and the variables (x, y) are said to be *normally correlated*.

From this distribution function it can be shown that the marginal distribution function of y is normal, and that the conditional distribution of x for given y is normal, the mean being a linear function of y and the variance being a constant. Thus, the properties of the two variables are the same, and they enter the distribution function symmetrically; in what follows they will therefore be denoted by (x_1, x_2) instead of (x, y).

In the following treatment we start with the two-dimensional distribution function and subsequently derive the conditional distribution functions that have been dealt with in Chapter 18.

19.2. The Distribution Function.

Let the two variables x_1 and x_2 be stochastically independent and normally distributed with parameters (ξ_1, σ_1^2) and (ξ_2, σ_2^2). The probability that the variables lie inside the rectangle defined by the intervals (x_1, x_1+dx_1) and (x_2, x_2+dx_2) is, according to the multiplication formula,

$$p\{x_1, x_2\}dx_1dx_2 = p\{x_1\}dx_1p\{x_2\}dx_2 = \frac{1}{2\pi\sigma_1\sigma_2} e^{-\frac{1}{2}\left[\left(\frac{x_1-\xi_1}{\sigma_1}\right)^2+\left(\frac{x_2-\xi_2}{\sigma_2}\right)^2\right]}dx_1dx_2 . \tag{19.2.1}$$

The distribution function $z = p\{x_1, x_2\}$ is represented by a surface in the (x_1, x_2, z) space, and the above probability element is represented by a volume element, the rectangle with area dx_1dx_2 forming the base and the height being z; see Fig. 19.1.

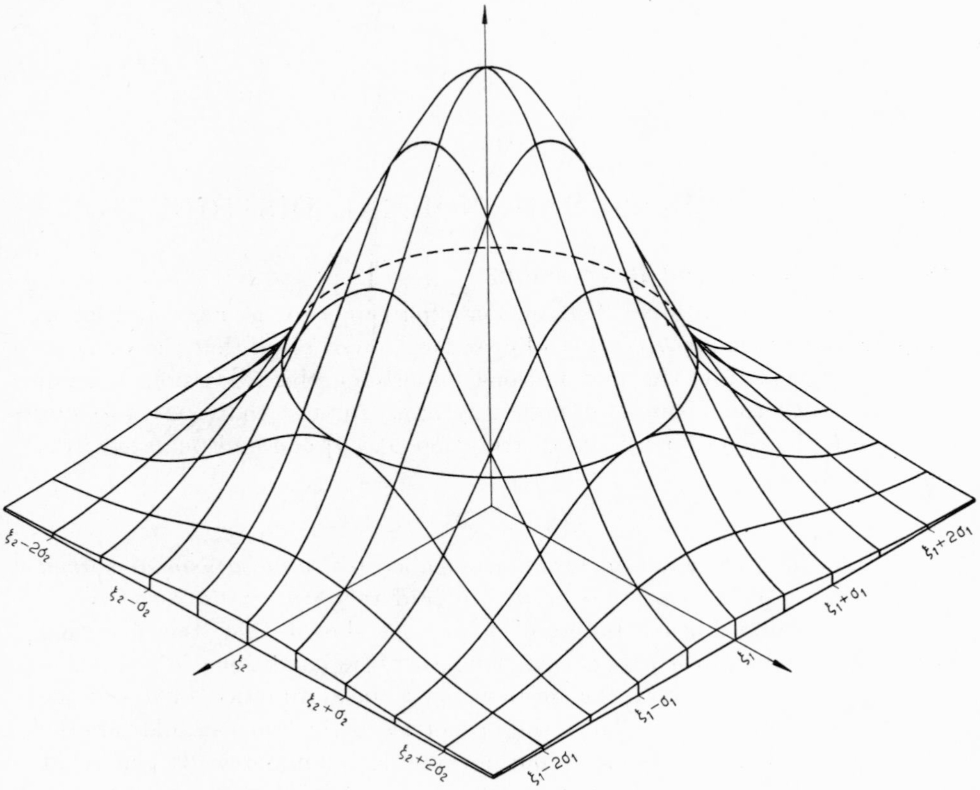

FIG. 19.1. Distribution surface for a two-dimensional normal distribution.

The maximum value of the distribution function is $1/2\,\pi\sigma_1\sigma_2$ for $(x_1, x_2) = (\xi_1, \xi_2)$, and the function tends to zero for $|x_1-\xi_1| \to \infty$ and $|x_2-\xi_2| \to \infty$.

The curve formed by the intersection of the distribution surface and a vertical plane parallel to one of the coordinate planes, e. g., $x_1 = a$, is a curve of the same shape as the normal distribution curve, since the equation

$$z = p\{a, x_2\} = \frac{1}{\sqrt{2\pi}\,\sigma_1} e^{-\frac{(a-\xi_1)^2}{2\sigma_1^2}} \frac{1}{\sqrt{2\pi}\,\sigma_2} e^{-\frac{(x_2-\xi_2)^2}{2\sigma_2^2}} \tag{19.2.2}$$

represents a normal distribution curve with parameters (ξ_2, σ_2^2), the ordinates having been multiplied by the constant

$$\frac{1}{\sqrt{2\pi}\,\sigma_1} e^{-\frac{(a-\xi_1)^2}{2\sigma_1^2}}.$$

The curve formed by the intersection of the distribution surface and a horizontal plane is an ellipse, the equation $p\{x_1, x_2\} = \text{constant}$ being identical with the equation

$$\left(\frac{x_1-\xi_1}{\sigma_1}\right)^2+\left(\frac{x_2-\xi_2}{\sigma_2}\right)^2=c^2, \tag{19.2.3}$$

where c^2 denotes a constant. This equation represents *an ellipse with center in* (ξ_1, ξ_2), *semi-axes of length* $(c\sigma_1, c\sigma_2)$, *and axes that are parallel to the coordinate axes.* An ellipse of this kind is called a *contour ellipse.*

The two types of intersection curves have been illustrated in Fig. 19.1.

According to the addition formula, the probability that (x_1, x_2) lies inside a certain area of the (x_1, x_2)-plane is equal to the integral of $p\{x_1, x_2\}$ over the given area, see also § 5.11. The necessary and sufficient condition that (x_1, x_2) lies inside the ellipse given by (19.2.3) is that the inequality

$$\left(\frac{x_1-\xi_1}{\sigma_1}\right)^2+\left(\frac{x_2-\xi_2}{\sigma_2}\right)^2<c^2 \tag{19.2.4}$$

is satisfied. If we introduce standardized variables

$$u_i=\frac{x_i-\xi_i}{\sigma_i}, \quad i=1, 2, \tag{19.2.5}$$

it follows from (10.2.1) that *the sum of squares*

$$\left(\frac{x_1-\xi_1}{\sigma_1}\right)^2+\left(\frac{x_2-\xi_2}{\sigma_2}\right)^2=u_1^2+u_2^2=\chi^2 \tag{19.2.6}$$

has a χ^2*-distribution with two degrees of freedom.* The probability that the inequality (19.2.4) is satisfied is thus

$$P\{\chi^2<c^2\} \quad \text{for} \quad f=2\,. \tag{19.2.7}$$

Hence, *the probability that* (x_1, x_2) *lies inside the ellipse defined by the equation*

$$\left(\frac{x_1-\xi_1}{\sigma_1}\right)^2+\left(\frac{x_2-\xi_2}{\sigma_2}\right)^2=\chi_P^2, \quad f=2\,, \tag{19.2.8}$$

is equal to P.

Fig. 19.2 shows a system of ellipses corresponding to $P = 0{\cdot}5, 0{\cdot}7, 0{\cdot}9, 0{\cdot}99$, and $0{\cdot}999$.

If x_1 and x_2 are not stochastically independent, we introduce a parameter, *the correlation coefficient* ϱ, which characterizes the dependence between x_1 and x_2; to the square terms in the exponent in (19.2.1) we add a product term, so that *the equation of the two-dimensional normal distribution function takes the form:*

$$p\{x_1, x_2\} = \frac{1}{2\pi\sigma_1\sigma_2\sqrt{1-\varrho^2}} e^{-\frac{1}{2(1-\varrho^2)}\left[\left(\frac{x_1-\xi_1}{\sigma_1}\right)^2 - 2\varrho\frac{x_1-\xi_1}{\sigma_1}\frac{x_2-\xi_2}{\sigma_2} + \left(\frac{x_2-\xi_2}{\sigma_2}\right)^2\right]}. \quad (19.2.9)$$

For $\varrho = 0$ we have (19.2.1).

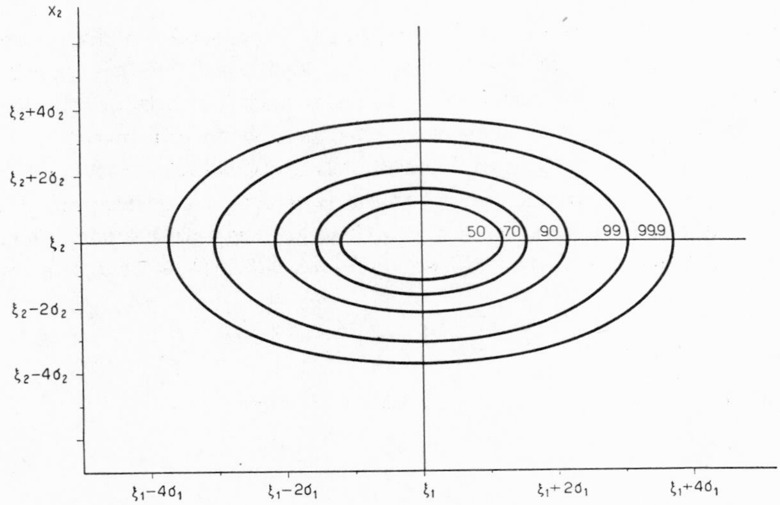

FIG. 19.2. Contour ellipses in a two-dimensional normal distribution with correlation coefficient 0 corresponding to the probabilities 50, 70, 90, 99, and 99·9%.

The following paragraphs will show that the distribution surface corresponding to (19.2.9) is of much *the same shape* as the one illustrated in Fig. 19.1. The curves of intersection between vertical planes parallel to the coordinate planes and the distribution surface—apart from a proportionality factor—form normal distribution curves, and the curves of intersection between horizontal planes and the distribution surface are ellipses. For $\varrho \neq 0$ the axes of the contour ellipses are not parallel to the axes of the coordinate system, and the length of the axes depends both on the standard deviations and on the correlation coefficient.

19.3. The Standardized Distribution.

In analogy with the standardized distribution function $\varphi(u)$ in § 6.3 we introduce the standardized, two-dimensional distribution function defined by

$$\varphi(u_1, u_2) = \frac{1}{2\pi\sqrt{1-\varrho^2}} e^{-\frac{1}{2(1-\varrho^2)}(u_1^2 - 2\varrho u_1 u_2 + u_2^2)} \quad (19.3.1)$$

and the corresponding cumulative distribution function

$$\Phi(u_1, u_2) = \int_{-\infty}^{u_1} \int_{-\infty}^{u_2} \varphi(u_1, u_2) du_1 du_2 . \tag{19.3.2}$$

It is seen that the following relations, in analogy with (6.4.1) and (6.5.1), hold between $p\{x_1, x_2\}$ and $P\{x_1, x_2\}$ and the corresponding standardized functions

$$p\{x_1, x_2\} = \frac{1}{\sigma_1 \sigma_2} \varphi(u_1, u_2) \quad \text{for} \quad u_i = \frac{x_i - \xi_i}{\sigma_i} \tag{19.3.3}$$

and

$$P\{x_1, x_2\} = \Phi(u_1, u_2) \quad \text{for} \quad u_i = \frac{x_i - \xi_i}{\sigma_i} . \tag{19.3.4}$$

On account of these simple relations we shall first prove some theorems for the standardized distribution, and then generalize these results by changing the origin and the scale of the variables.

Several of these proofs are based on the following transformation of the standardized distribution function. The exponent in $\varphi(u_1, u_2)$ is written

$$u_1^2 - 2\varrho u_1 u_2 + u_2^2 = (1-\varrho^2)u_1^2 + (u_2 - \varrho u_1)^2 , \tag{19.3.5}$$

the probability element $\varphi(u_1, u_2)du_1du_2$ then being partitioned into the following product

$$\varphi(u_1, u_2) du_1 du_2 = \frac{1}{\sqrt{2\pi}} e^{-\frac{u_1^2}{2}} \frac{1}{\sqrt{2\pi}\sqrt{1-\varrho^2}} e^{-\frac{(u_2-\varrho u_1)^2}{2(1-\varrho^2)}} du_1 du_2$$

$$= \varphi(u_1) \frac{1}{\sqrt{1-\varrho^2}} \varphi\left(\frac{u_2 - \varrho u_1}{\sqrt{1-\varrho^2}}\right) du_1 du_2 . \tag{19.3.6}$$

Introducing

$$u_{2\cdot 1} = \frac{u_2 - \varrho u_1}{\sqrt{1-\varrho^2}} \tag{19.3.7}$$

as a new variable instead of u_2, we obtain

$$\varphi(u_1, u_2) du_1 du_2 = \varphi(u_1) du_1 \varphi(u_{2\cdot 1}) du_{2\cdot 1} , \tag{19.3.8}$$

i. e., the two variables u_1 and $u_{2\cdot 1}$ are stochastically independent and normally distributed with parameters (0, 1).

In analogy with (19.3.8) we have

$$\boxed{\varphi(u_1, u_2)du_1 du_2 = \varphi(u_2)du_2\varphi(u_{1\cdot 2})du_{1\cdot 2}\,,} \tag{19.3.9}$$

where

$$\boxed{u_{1\cdot 2} = \frac{u_1 - \varrho u_2}{\sqrt{1-\varrho^2}}.} \tag{19.3.10}$$

Further we find from (19.3.8)

$$\int_{-\infty}^{\infty}\int_{-\infty}^{\infty}\varphi(u_1, u_2)du_1 du_2 = \int_{-\infty}^{\infty}\varphi(u_1)du_1\int_{-\infty}^{\infty}\varphi(u_{2\cdot 1})du_{2\cdot 1} = 1\times 1 = 1\,, \tag{19.3.11}$$

i. e., the constant in (19.2.9) is determined as in (5.12.3).

From (19.3.8) it is also seen that the curve formed by intersection of a vertical plane parallel to one of the coordinate planes, e. g., $u_1 = a$, and the distribution surface, is a curve of the same shape as a normal distribution curve, since

$$\varphi(a, u_2) = \varphi(a)\frac{1}{\sqrt{1-\varrho^2}}\varphi\left(\frac{u_2 - \varrho a}{\sqrt{1-\varrho^2}}\right). \tag{19.3.12}$$

The function

$$F(u_1, u_2, \varrho) = \frac{1}{2\pi\sqrt{1-\varrho^2}}\int_{u_1}^{\infty}\int_{u_2}^{\infty} e^{-\frac{1}{2(1-\varrho^2)}(x_1^2 - 2\varrho x_1 x_2 + x_2^2)}\,dx_1\,dx_2 \tag{19.3.13}$$

has been tabulated in K. PEARSON: *Tables for Biometricians and Statisticians*, Part II, London, 1931. With the aid of this table the integral of the two-dimensional normal distribution function over a rectangular area may be calculated. As a special case we have

$$\Phi(u_1, u_2) = \Phi(u_1, u_2, \varrho) = F(-u_1, -u_2, \varrho)\,.$$

The table is rather voluminous since the function tabulated contains three arguments.

A simpler table of a function with two arguments, from which it is comparatively easy to calculate $F(u_1, u_2, \varrho)$, has been given by C. NICHOLSON: *The Probability Integral for Two Variables*, Biometrika, 33, 1943, 59–72.

19.4. The Marginal Distributions.

The marginal distribution function of u_1 is

$$p\{u_1\} = \int_{-\infty}^{\infty}\varphi(u_1, u_2)du_2 = \varphi(u_1)\int_{-\infty}^{\infty}\varphi(u_{2\cdot 1})du_{2\cdot 1} = \varphi(u_1)\,,$$

see (5.12.4). *From this it follows that the marginal distributions of x_1 and x_2 are normal with parameters* (ξ_1, σ_1^2) and (ξ_2, σ_2^2), i. e.,

$$\mathcal{M}\{x_i\} = \xi_i\,, \quad i = 1, 2\,, \tag{19.4.1}$$

and

$$\mathcal{V}\{x_i\} = \sigma_i^2\,, \quad i = 1, 2\,. \tag{19.4.2}$$

19.5. The Correlation Coefficient and the Covariance.

The two-dimensional normal distribution function contains five parameters. Four of them, (ξ_1, σ_1^2) and (ξ_2, σ_2^2), characterize the two marginal distributions, see § 19.4, while the fifth, ϱ, characterizes the relationship between the two variables.

According to (5.16.3) the covariance of (u_1, u_2) is defined by the equation

$$\mathcal{V}\{u_1, u_2\} = \mathcal{M}\{u_1 u_2\} = \int_{-\infty}^{\infty}\int_{-\infty}^{\infty} u_1 u_2 \varphi(u_1, u_2)\,du_1 du_2\,. \tag{19.5.1}$$

If (19.3.8) is introduced into (19.5.1), we obtain

$$\begin{aligned}\mathcal{M}\{u_1 u_2\} &= \int_{-\infty}^{\infty} u_1\varphi(u_1)\int_{-\infty}^{\infty} u_2\varphi(u_{2\cdot 1})\,du_{2\cdot 1}\,du_1 \\ &= \int_{-\infty}^{\infty} u_1\varphi(u_1)\int_{-\infty}^{\infty} (\varrho u_1 + u_{2\cdot 1}\sqrt{1-\varrho^2})\varphi(u_{2\cdot 1})\,du_{2\cdot 1}\,du_1 \\ &= \varrho\int_{-\infty}^{\infty} u_1^2\varphi(u_1)\,du_1\int_{-\infty}^{\infty}\varphi(u_{2\cdot 1})\,du_{2\cdot 1} + \\ &\qquad \sqrt{1-\varrho^2}\int_{-\infty}^{\infty} u_1\varphi(u_1)\,du_1\int_{-\infty}^{\infty} u_{2\cdot 1}\varphi(u_{2\cdot 1})\,du_{2\cdot 1} = \varrho\,,\end{aligned} \tag{19.5.2}$$

since

$$\mathcal{V}\{u_1\} = \mathcal{M}\{u_1^2\} = \int_{-\infty}^{\infty} u_1^2\varphi(u_1)\,du_1 = 1$$

and

$$\int_{-\infty}^{\infty} u_1\varphi(u_1)\,du_1 = \int_{-\infty}^{\infty} u_{2\cdot 1}\varphi(u_{2\cdot 1})\,du_{2\cdot 1} = 0\,.$$

Thus, we have proved that the parameter ϱ is equal to the mean value of the product of the standardized variables

$$\boxed{\varrho = \mathcal{M}\{u_1 u_2\} = \mathcal{M}\left\{\frac{x_1-\xi_1}{\sigma_1}\,\frac{x_2-\xi_2}{\sigma_2}\right\},} \tag{19.5.3}$$

so that ϱ in accordance with the definitions in § 5.16 may be called the correlation coefficient of (x_1, x_2). Since the correlation coefficient depends solely upon the standardized variables, it is a *dimensionless quantity*, i. e., separate linear transformations of x_1 and x_2 will not change ϱ.

From (19.5.3) we find the covariance

$$\mathcal{V}\{x_1, x_2\} = \mathcal{M}\{(x_1-\xi_1)(x_2-\xi_2)\} = \varrho\sigma_1\sigma_2 \,, \tag{19.5.4}$$

so that the correlation coefficient may be expressed as

$$\varrho = \frac{\mathcal{V}\{x_1, x_2\}}{\sqrt{\mathcal{V}\{x_1\}\,\mathcal{V}\{x_2\}}} . \tag{19.5.5}$$

Two variables for which ϱ is positive (negative) are said to be positively (negatively) correlated. If $\varrho = 0$, the distribution function (19.2.9) is partitioned into the product of the marginal distribution functions, i. e., the variables are stochastically independent, see § 5.16.

19.6. The Estimates of the Parameters.

From the two marginal distributions we compute the means and the variances, $(\bar{x}_1, s_1^2)$ and $(\bar{x}_2, s_2^2)$, as estimates of the corresponding theoretical values (ξ_1, σ_1^2) and (ξ_2, σ_2^2).

As an estimate, r, of the correlation coefficient ϱ we compute

$$r = \frac{s_{12}}{s_1 s_2} , \tag{19.6.1}$$

s_{12} denoting the estimate of the covariance. To simplify the arithmetic the latter expression may be written

$$r = \frac{SPD_{x_1x_2}}{\sqrt{SSD_{x_1}\,SSD_{x_2}}} , \tag{19.6.2}$$

see §§ 4.6–4.8 for technique of computations.

The reasons for regarding r as the "best" estimate of the parameter ϱ lie outside the scope of this book. Readers can consult the references given in § 20.9.

Example 19.1. If we assume that the observations in Table 4.7, p. 82, may be regarded as a random sample from a two-dimensional, normally distributed population, the estimates of the five parameters characterizing this population will be as follows:

Mean and standard deviation of the marginal distribution of starch content: 17·546 and 2·221% starch.

Mean and standard deviation of the marginal distribution of specific gravity: 1·0988 and 0·01056 g/cm³.

Correlation coefficient $r = 0{\cdot}949$, see also p. 89.

19.7. The Conditional Distributions.

According to (5.12.7) the distribution function of u_2 for given u_1 is

$$p\{u_2|u_1\} = \frac{\varphi(u_1, u_2)}{\varphi(u_1)} = \varphi(u_{2\cdot 1}) = \frac{1}{\sqrt{2\pi}\sqrt{1-\varrho^2}} e^{-\frac{(u_2-\varrho u_1)^2}{2(1-\varrho^2)}}, \quad (19.7.1)$$

i. e., u_2 is normally distributed with parameters

$$\mathcal{M}\{u_2|u_1\} = \varrho u_1 \quad (19.7.2)$$

and

$$\mathcal{V}\{u_2|u_1\} = 1-\varrho^2 . \quad (19.7.3)$$

By transforming the variables from (u_1, u_2) to (x_1, x_2) we find that, *for a given value of x_1, x_2 is normally distributed with mean*

$$\boxed{\mathcal{M}\{x_2|x_1\} = \xi_2+\varrho\frac{\sigma_2}{\sigma_1}(x_1-\xi_1)} \quad (19.7.4)$$

and variance

$$\boxed{\mathcal{V}\{x_2|x_1\} = \sigma_2^2(1-\varrho^2) .} \quad (19.7.5)$$

Similarly, we find that, *for a given value of x_2, x_1 is normally distributed with mean*

$$\boxed{\mathcal{M}\{x_1|x_2\} = \xi_1+\varrho\frac{\sigma_1}{\sigma_2}(x_2-\xi_2)} \quad (19.7.6)$$

and variance

$$\boxed{\mathcal{V}\{x_1|x_2\} = \sigma_1^2(1-\varrho^2) .} \quad (19.7.7)$$

Thus we see that the means of the conditional distributions are linear functions of the "independent" variable, and the variances are constant, i. e., independent of the "independent" variable.

Further, we see that for a set of observations from a two-dimensional, normally distributed population the assumptions underlying the regression analysis given in Chapter 18 are satisfied for *both* the variables, i. e., both for the regression of x_1 on x_2 and for the regression of x_2 on x_1. However, even though from a purely theoretical, statistical point of view the two regressions are equally important, we will often find as emphasized in § 18.1 that only one of the regressions is of importance for the solution of the problem in hand.

As $\mathcal{V}\{u_2|u_1\} \geq 0$, *it follows from* (19.7.3) *that the correlation coefficient is numerically less than or equal to* 1, *i. e.*,

$$\boxed{-1 \leq \varrho \leq 1.} \tag{19.7.8}$$

The regression lines intersect in the point (ξ_1, ξ_2). The regression coefficients are

$$\beta_{x_2|x_1} = \varrho \frac{\sigma_2}{\sigma_1} \tag{19.7.9}$$

and

$$\beta_{x_1|x_2} = \varrho \frac{\sigma_1}{\sigma_2}, \tag{19.7.10}$$

respectively. Thus, the correlation coefficient is the geometrical mean of the regression coefficients, i. e.,

$$\varrho = \sqrt{\beta_{x_2|x_1} \beta_{x_1|x_2}}. \tag{19.7.11}$$

From the slopes of the regression lines we find that *the angle* θ *between the two regression lines is determined by the equation*

$$\tan \theta = \frac{1-\varrho^2}{\varrho} \frac{\sigma_1 \sigma_2}{\sigma_1^2+\sigma_2^2}. \tag{19.7.12}$$

For $\varrho = 1$, $\tan \theta = 0$, i. e., *the regression lines are identical*. This may also be seen from the equations (19.7.4) and (19.7.6). The equations (19.7.5) and (19.7.7) further show that *the variation about the regression line is zero*, i. e., the two-dimensional distribution is in reality only one-dimensional, *the variables being linearly dependent*.

For $\varrho = -1$ we obtain an analogous result with the modification that for $\varrho = 1$ the slope of the identical regression lines is positive, while for $\varrho = -1$ it is negative.

For $\varrho = 0$ *the variables are stochastically independent, the regression lines being parallel to the coordinate axes and the standard deviations about the regression lines being equal to the marginal standard deviations.*

Between these two extremes of functional (linear) dependence ($|\varrho| = 1$) and stochastic independence ($\varrho = 0$) we have the varying "degrees" of stochastic dependence as expressed by the correlation coefficient lying between 0 and 1 or between 0 and -1. For $0 < \varrho < 1$ the slopes of both regression lines are positive, and for $-1 < \varrho < 0$ both slopes are negative.

As a measure of the "degree" of the dependence we may also utilize *the ratio between the variances in the conditional and the marginal distributions*

$$\boxed{\frac{\mathcal{V}\{x_2|x_1\}}{\mathcal{V}\{x_2\}} = \frac{\mathcal{V}\{x_1|x_2\}}{\mathcal{V}\{x_1\}} = 1-\varrho^2,} \tag{19.7.13}$$

see (19.7.5) and (19.7.7). This equation shows that $|\varrho|$ must be fairly large before the variances of the conditional distributions are conspicuously smaller than the marginal ones.

Thus we may consider the marginal variance σ_2^2 of x_2 as consisting of two parts: the portion $\varrho^2\sigma_2^2$ "is due to" the variation of x_1, while the remainder, $(1-\varrho^2)\sigma_2^2$, is independent of the variation of x_1, and indicates the variance of the distribution of x_2 about the regression line. Similarly, σ_1^2 may be partitioned into $\varrho^2\sigma_1^2$ and $(1-\varrho^2)\sigma_1^2$.

The estimates of the regression lines may be computed either by substitution of the estimates of the parameters into (19.7.4) and (19.7.6) or according to the rules given in § 18.3. Expressed in symbols we have

$$b_{x_2|x_1} = r\frac{s_2}{s_1} = \frac{rs_1s_2}{s_1^2} = \frac{SPD_{x_1x_2}}{SSD_{x_1}} \tag{19.7.14}$$

and

$$b_{x_1|x_2} = r\frac{s_1}{s_2} = \frac{rs_1s_2}{s_2^2} = \frac{SPD_{x_1x_2}}{SSD_{x_2}}, \tag{19.7.15}$$

so that

$$\mathcal{M}\{x_2|x_1\} \approx \bar{x}_2 + b_{x_2|x_1}(x_1-\bar{x}_1) \tag{19.7.16}$$

and

$$\mathcal{M}\{x_1|x_2\} \approx \bar{x}_1 + b_{x_1|x_2}(x_2-\bar{x}_2). \tag{19.7.17}$$

According to (18.3.29) and (18.3.43) the estimates of the variances may be computed from the formulas

$$s_{x_2|x_1}^2 = \frac{1}{n-2}\left(SSD_{x_2} - \frac{SPD_{x_1x_2}^2}{SSD_{x_1}}\right) = \frac{n-1}{n-2}s_2^2(1-r^2) \tag{19.7.18}$$

and

$$s_{x_1|x_2}^2 = \frac{1}{n-2}\left(SSD_{x_1} - \frac{SPD_{x_1x_2}^2}{SSD_{x_2}}\right) = \frac{n-1}{n-2}s_1^2(1-r^2), \tag{19.7.19}$$

cf. (19.7.5) and (19.7.7).

Example 19.2. If we let x_2 denote starch percentage and x_1 the specific gravity, we obtain from Table 4.7, p. 82,

$$b_{x_2|x_1} = \frac{3108{\cdot}8}{3894{\cdot}0}\;\frac{0{\cdot}004 \times 1{\cdot}0}{0{\cdot}004^2} = 199{\cdot}6$$

and

$$b_{x_1|x_2} = \frac{3108{\cdot}8}{2756{\cdot}8}\;\frac{0{\cdot}004 \times 1{\cdot}0}{1{\cdot}0^2} = 0{\cdot}004511.$$

As mentioned before, it is only the regression of the starch percentage on the specific gravity that is of practical interest. According to (19.7.16) we obtain the estimate of this regression line as

$$\mathcal{M}\{x_2|x_1\} \approx 17{\cdot}55 + 199{\cdot}6(x_1 - 1{\cdot}0988)\ .$$

A graph of the regression line has been shown in Fig. 4.4, p. 84.

According to (19.7.18) the conditional variance is computed from the formulas

$$SSD_{x_2|x_1} = 2756{\cdot}8 - \frac{3108{\cdot}8^2}{3894{\cdot}0} = 274{\cdot}9$$

and

$$s^2_{x_2|x_1} = \frac{274{\cdot}9}{558} = 0{\cdot}4927\ ,$$

or from s_2^2 and r using the formula

$$s^2_{x_2|x_1} = \frac{559}{558}\,4{\cdot}9317(1 - 0{\cdot}9488^2) = 0{\cdot}4931\ .$$

(The disagreement between the two results is due to rounding errors in the latter computation).

The standard deviation is computed from the variance giving $s_{x_2|x_1} = 0{\cdot}702\%$ starch. It will be seen that the standard deviation of the conditional distribution is about $^1/_3$ of the standard deviation of the marginal distribution.

Utilizing the line of regression, an estimate of the starch content may be computed from a given specific gravity, in this manner saving a great deal of laboratory work.

19.8. Linear Transformations of the Variables.

In the following section it is shown how by means of a linear (orthogonal) transformation it is possible to find two normally distributed and *stochastically independent* linear functions of the variables (x_1, x_2).

If we introduce two new variables (y_1, y_2) by means of the equations

$$\begin{aligned} y_1 &= (x_1 - \xi_1)\cos\alpha + (x_2 - \xi_2)\sin\alpha \\ y_2 &= -(x_1 - \xi_1)\sin\alpha + (x_2 - \xi_2)\cos\alpha\ , \end{aligned} \qquad (19.8.1)$$

the (y_1, y_2)-coordinate system will have its origin in the point $(x_1, x_2) = (\xi_1, \xi_2)$ and the y-axes will form an angle α with the x-axes, as shown in Fig. 19.3.

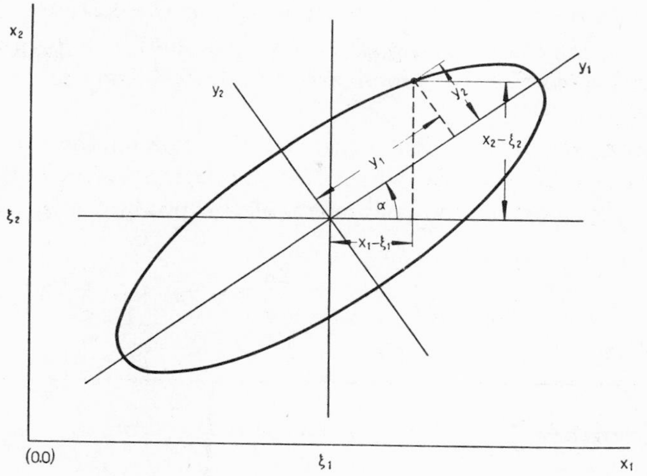

FIG. 19.3. Transformation of the (x_1, x_2)-coordinate system.

By the transformation from (x_1, x_2) to (y_1, y_2) the probability element

$$p\{x_1, x_2\}dx_1dx_2 = \frac{1}{2\pi\sigma_1\sigma_2\sqrt{1-\varrho^2}} e^{-\frac{1}{2}F(x_1, x_2)}dx_1dx_2 , \qquad (19.8.2)$$

where

$$F(x_1, x_2) = \frac{1}{1-\varrho^2}\left[\left(\frac{x_1-\xi_1}{\sigma_1}\right)^2 - 2\varrho\frac{x_1-\xi_1}{\sigma_1}\frac{x_2-\xi_2}{\sigma_2} + \left(\frac{x_2-\xi_2}{\sigma_2}\right)^2\right], \qquad (19.8.3)$$

is changed to the probability element $p\{y_1, y_2\}dy_1dy_2$. As the transformation consists only of a parallel displacement and a rotation of the coordinate axes, the area of the element dx_1dx_2 does not change, i. e., $dx_1dx_2 = dy_1dy_2$.

In order to determine $p\{y_1, y_2\}$ the equations (19.8.1) are solved for (x_1, x_2) giving

$$\begin{aligned} x_1-\xi_1 &= y_1 \cos \alpha - y_2 \sin \alpha \\ x_2-\xi_2 &= y_1 \sin \alpha + y_2 \cos \alpha , \end{aligned} \qquad (19.8.4)$$

which inserted in $p\{x_1, x_2\}$ leads to

$$p\{y_1, y_2\} = \frac{1}{2\pi\sigma_1\sigma_2\sqrt{1-\varrho^2}} e^{-\frac{1}{2}G(y_1, y_2)} , \qquad (19.8.5)$$

where

$$G(y_1, y_2) = \frac{1}{1-\varrho^2}\left[y_1^2\left(\frac{\cos^2\alpha}{\sigma_1^2} + \frac{\sin^2\alpha}{\sigma_2^2} - \frac{\varrho \sin 2\alpha}{\sigma_1\sigma_2}\right)\right.$$

$$\left. + y_2^2\left(\frac{\sin^2\alpha}{\sigma_1^2} + \frac{\cos^2\alpha}{\sigma_2^2} + \frac{\varrho \sin 2\alpha}{\sigma_1\sigma_2}\right) - y_1y_2\left(\frac{\sin 2\alpha}{\sigma_1^2} - \frac{\sin 2\alpha}{\sigma_2^2} + \frac{2\varrho \cos 2\alpha}{\sigma_1\sigma_2}\right)\right] . \qquad (19.8.6)$$

This means that *the distribution function of the variables* (y_1, y_2) *is a two-dimensional normal distribution function* with mean (0, 0) and variances and correlation coefficient which are determined from σ_1^2, σ_2^2, ϱ, and the rotation angle α.

As the correlation coefficient of (y_1, y_2) depends on the angle α, we will try to find the value of α for which the correlation coefficient is zero. This is the same as equating the coefficient of the product $y_1 y_2$ in $G(y_1, y_2)$ to zero, i. e.,

$$\left(\frac{1}{\sigma_1^2}-\frac{1}{\sigma_2^2}\right)\sin 2\alpha+\frac{2\varrho}{\sigma_1\sigma_2}\cos 2\alpha = 0 ,$$

which leads to

$$\boxed{\tan 2\alpha=\frac{2\varrho\sigma_1\sigma_2}{\sigma_1^2-\sigma_2^2} \text{ for } \sigma_1 \neq \sigma_2 \text{ and } \alpha = \frac{\pi}{4} \text{ for } \sigma_1 = \sigma_2 ,} \tag{19.8.7}$$

or

$$\tan \alpha = \frac{1}{2\varrho\sigma_1\sigma_2}\left[\sigma_2^2-\sigma_1^2 \pm \sqrt{(2\varrho\sigma_1\sigma_2)^2+(\sigma_2^2-\sigma_1^2)^2}\right]. \tag{19.8.8}$$

For this value of α, y_1 and y_2 are stochastically independent, so that the distribution function may be written

$$p\{y_1, y_2\} = \frac{1}{2\pi\sigma_{y_1}\sigma_{y_2}} e^{-\frac{1}{2}\left(\frac{y_1^2}{\sigma_{y_1}^2}+\frac{y_2^2}{\sigma_{y_2}^2}\right)} . \tag{19.8.9}$$

By comparing (19.8.5) and (19.8.9) we obtain the following relations for the determination of the two variances:

$$\sigma_{y_1}\sigma_{y_2} = \sigma_1\sigma_2\sqrt{1-\varrho^2} \tag{19.8.10}$$

and

$$\sigma_{y_1}^2+\sigma_{y_2}^2 = \sigma_1^2+\sigma_2^2 . \tag{19.8.11}$$

The first relation is found by equating the constants of the two expressions for $(y_1, y_2) = (0, 0)$, and the second one by applying (19.8.10) to the expression

$$\frac{1}{\sigma_{y_1}^2}+\frac{1}{\sigma_{y_2}^2}=\frac{1}{1-\varrho^2}\left[\frac{1}{\sigma_1^2}+\frac{1}{\sigma_2^2}\right],$$

which is obtained by equating the exponents of (19.8.5) and (19.8.9) for $(y_1, y_2) = (1, 1)$.

From (19.8.10) and (19.8.11) we obtain

$$(\sigma_{y_1}+\sigma_{y_2})^2 = \sigma_1^2+\sigma_2^2+2\sigma_1\sigma_2\sqrt{1-\varrho^2} \tag{19.8.12}$$

and

$$(\sigma_{y_1}-\sigma_{y_2})^2 = \sigma_1^2+\sigma_2^2-2\sigma_1\sigma_2\sqrt{1-\varrho^2}, \tag{19.8.13}$$

from which σ_{y_1} and σ_{y_2} are determined, using $\sigma_{y_1}-\sigma_{y_2} > 0$ for $\varrho > 0$ and $\sigma_{y_1}-\sigma_{y_2} < 0$ for $\varrho < 0$, since σ_{y_1} and σ_{y_2} are proportional to the lengths of the axes of the contour ellipses.

The variables (x_1, x_2) *may thus be converted by a linear transformation into new variables* (y_1, y_2) *which are stochastically independent and each of which is normally distributed.*

From this we may deduce—among other things—that the addition theorem derived in § 9.1 for normally distributed, stochastically independent variables may be extended to deal also with stochastically dependent variables, since

$$z = a_0+a_1x_1+a_2x_2 \tag{19.8.14}$$

may be converted by means of the transformation (19.8.4) into a linear function of the stochastically independent variables (y_1, y_2). It then follows directly from the addition theorem in § 9.1 that *the linear function* z *of the variables* (x_1, x_2) *is normally distributed.* From (5.17.1) and (5.17.2) we obtain

$$\mathcal{M}\{z\} = a_0+a_1\xi_1+a_2\xi_2 \tag{19.8.15}$$

and

$$\mathcal{V}\{z\} = a_1^2\sigma_1^2+a_2^2\sigma_2^2+2a_1a_2\varrho\sigma_1\sigma_2\,. \tag{19.8.16}$$

On p. 265 a special case of the transformation (19.8.1) has been applied in the proof of the partition theorem. If we put $\xi_1 = \xi_2 = 0$, $\sigma_1 = \sigma_2 = 1$ and $\varrho = 0$, so that $(x_1, x_2) = (u_1, u_2)$, we find that the variables defined by the equations

$$y_1 = \frac{1}{\sqrt{2}}(u_1+u_2)$$

and

$$y_2 = -\frac{1}{\sqrt{2}}(u_1-u_2) \tag{19.8.17}$$

are stochastically independent and each of them is normally distributed with parameters (0,1).

19.9. The Contour Ellipses.

The contour ellipses of the distribution surface corresponding to the function (19.2.9) have the equation

$$\frac{1}{1-\varrho^2}\left[\left(\frac{x_1-\xi_1}{\sigma_1}\right)^2-2\varrho\frac{x_1-\xi_1}{\sigma_1}\frac{x_2-\xi_2}{\sigma_2}+\left(\frac{x_2-\xi_2}{\sigma_2}\right)^2\right] = c^2\,, \tag{19.9.1}$$

where c denotes a constant. This equation represents an ellipse in the (x_1, x_2)-plane with center (ξ_1, ξ_2) and axes that are not parallel to the coordinate axes as long as $\varrho \neq 0$.

If the coordinate system is rotated so that the axes of the ellipse are parallel to the coordinate axes, the position of the ellipse corresponds to that of the contour ellipse for two stochastically independent variables, see Fig. 19.2. According to § 19.8 this is equivalent to a linear transformation of (x_1, x_2), introducing (y_1, y_2) as defined by (19.8.1) as new variables. The angle of rotation is determined by (19.8.7).

The quadratic form in (x_1, x_2) *in* (19.9.1) *may thus be written as a sum of squares of two stochastically independent variables, the quantity*

$$\boxed{\frac{1}{1-\varrho^2}\left[\left(\frac{x_1-\xi_1}{\sigma_1}\right)^2 - 2\varrho\frac{x_1-\xi_1}{\sigma_1}\frac{x_2-\xi_2}{\sigma_2} + \left(\frac{x_2-\xi_2}{\sigma_2}\right)^2\right] = \left(\frac{y_1}{\sigma_{y_1}}\right)^2 + \left(\frac{y_2}{\sigma_{y_2}}\right)^2 = \chi^2} \tag{19.9.2}$$

being distributed as χ^2 *with two degrees of freedom. The probability that* (x_1, x_2) *lies inside the ellipse* (19.9.1) *is therefore*

$$P\{\chi^2 < c^2\} \quad \text{for} \quad f = 2\,. \tag{19.9.3}$$

Fig. 19.4 shows some examples of 95% contour ellipses corresponding to various values of ϱ.

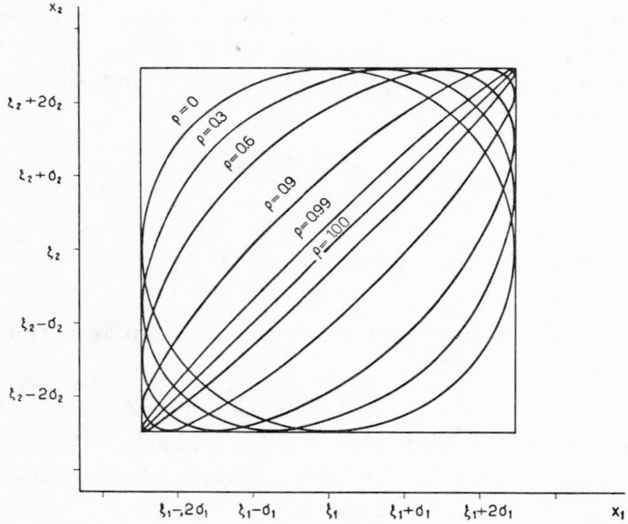

FIG. 19.4. 95% contour ellipses for $\sigma_1 = \sigma_2$ and $\varrho = 0$, 0·3, 0·6, 0·9, 0·99, and 1·00.

The contour ellipse is inscribed in a rectangle with center (ξ_1, ξ_2) and with sides of length $2\sigma_1\chi_P$ and $2\sigma_2\chi_P$, i. e., the length of the sides is independent of ϱ. If we compare contour ellipses from distributions with the same variances but different correlation coefficients, we find that for a given probability all the ellipses are inscribed in the above-mentioned rectangle, and that the eccentricity and the slope of the axes depend on the correlation coefficient; see Fig. 19.4. For $\varrho = 0$ the axes of the ellipse are parallel to the coordinate axes, for $0 < \varrho < 1$ the slope of the major axis is positive, and for $\varrho = 1$ the ellipse degenerates into a straight line, namely the diagonal of the rectangle.

Fig. 19.5 shows the position of the regression lines in relation to the contour ellipses.

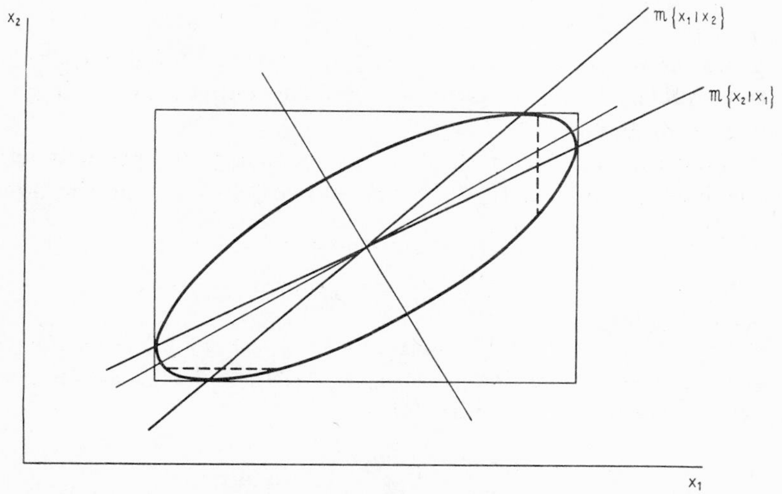

FIG. 19.5. The regression lines and the 95% contour ellipse.

The regression of x_2 on x_1 passes through the centers of the *vertical* chords of the ellipse, while the regression of x_1 on x_2 passes through the centres of the *horizontal* chords. This is due to the fact that the regression lines are derived by minimizing the sums of squares of the vertical and horizontal deviations, respectively, from the regression line.

The major axis of the ellipse, which has the equation

$$x_2 = \xi_2 + \beta(x_1 - \xi_1) , \qquad (19.9.4)$$

where $\beta = \tan \alpha$, see (19.8.8), is also called *the orthogonal regression*, as it may be derived by minimizing the *perpendicular* deviations from the line.

Example 19.3. Given a two-dimensional normal distribution with means (0, 0), standard deviations $\sigma_1 = 1{\cdot}5$ and $\sigma_2 = 1{\cdot}0$, and the correlation coefficient $\varrho = 0{\cdot}75$. Find the regression lines and the 95% contour ellipse.

By substitution in (19.7.4) and (19.7.6) we obtain

$$\mathcal{M}\{x_2|x_1\} = 0{\cdot}75\,\frac{1{\cdot}0}{1{\cdot}5}\,x_1 = 0{\cdot}50x_1$$

and

$$\mathcal{M}\{x_1|x_2\} = 0{\cdot}75\,\frac{1{\cdot}5}{1{\cdot}0}\,x_2 = 1{\cdot}125x_2\,.$$

The slope of the major axis is determined, according to (19.8.7), from the equation

$$\tan 2\alpha = \frac{2\times 0{\cdot}75\times 1{\cdot}5\times 1{\cdot}0}{2{\cdot}25-1{\cdot}00} = 1{\cdot}80\,,$$

which gives $2\alpha = 60{\cdot}95°$, and $\tan\alpha = 0{\cdot}587$, which is the required slope. Thus, the equation for the major axis of the ellipse, the orthogonal regression, is $x_2 = 0{\cdot}587x_1$.

The standard deviations of the two stochastically independent variables are determined, according to (19.8.12) and (19.8.13), from the equations

$$(\sigma_{y_1}+\sigma_{y_2})^2 = 5{\cdot}234$$

and

$$(\sigma_{y_1}-\sigma_{y_2})^2 = 1{\cdot}266\,,$$

since $\sigma_1^2+\sigma_2^2 = 1{\cdot}5^2+1{\cdot}0^2 = 3{\cdot}250$ and $2\sigma_1\sigma_2\sqrt{1-\varrho^2} = 3{\cdot}0\sqrt{1-0{\cdot}75^2} = 1{\cdot}984$. Solution of the equations leads to $\sigma_{y_1} = 1{\cdot}707$ and $\sigma_{y_2} = 0{\cdot}582$, whence the equation of the contour ellipse is

$$\left(\frac{y_1}{1{\cdot}707}\right)^2 + \left(\frac{y_2}{0{\cdot}582}\right)^2 = 5{\cdot}991\,,$$

since $\chi^2_{\cdot 95} = 5{\cdot}991$ for $f = 2$.

The three regression lines and the contour ellipse are shown in Fig. 19.5.

19.10. Examination of an Observed Two-Dimensional Distribution.

a. *Ungrouped observations.*

Given a set of two-dimensional observations, (x_{11}, x_{21}), (x_{12}, x_{22}), ..., (x_{1n}, x_{2n}), the immediate problem arises: what is their joint distribution function $p\{x_1, x_2\}$. We shall give a method to examine whether the distribution could be a two-dimensional normal distribution.

The first way of testing this possibility is to examine the *marginal* distributions. If they are not normal, then $p\{x_1, x_2\}$ cannot be normal. If the observations are not normally distributed, we can try to find functions of the observations such that the transformed values are normally distributed, see Chapter 7.

If the marginal distributions are normal $p\{x_1, x_2\}$ may be normal, but need not necessarily be so. Hence further tests are needed. Quite a simple test is the following.

According to (19.9.2) the quantity

$$\chi^2 = \frac{1}{1-\varrho^2}\left[\left(\frac{x_1-\xi_1}{\sigma_1}\right)^2 - 2\varrho\frac{x_1-\xi_1}{\sigma_1}\frac{x_2-\xi_2}{\sigma_2}+\left(\frac{x_2-\xi_2}{\sigma_2}\right)^2\right] \qquad (19.10.1)$$

is distributed as χ^2 with two degrees of freedom, if (x_1, x_2) are normally correlated. Introducing the n observations in (19.10.1) we obtain n values of χ^2, whose distribution may be compared with the corresponding theoretical distribution. According to (10.2.5) we have

$$P\{\chi^2\} = 1-e^{-\frac{1}{2}\chi^2},$$

and therefore

$$\log_{10}(1-P\{\chi^2\}) = -0{\cdot}217\chi^2. \qquad (19.10.2)$$

If we denote the n χ^2-values, ranked in order of magnitude, $\chi^2_{(1)}, \chi^2_{(2)}, \ldots, \chi^2_{(n)}$, the corresponding estimates of $1-P\{\chi^2\}$ are $(n-1/2)/n$, $(n-3/2)/n$, ..., $1/2n$ and therefore, according to (19.10.2), $\chi^2_{(i)}$ and $(n-i+\frac{1}{2})/n$, when plotted on semi-logarithmic paper, will be distributed at random about a straight line through the point (0, 1) with slope $-0{\cdot}217$. In practical work it is necessary to replace the parameters in (19.10.1) by their estimates, which introduces an uncertainty which cannot be evaluated exactly.

The above method may be simplified if contour ellipses are calculated corresponding to preassigned values of P and the observed and theoretical number of observations which lie inside the ellipses are compared.

b. *Grouped observations.*

If the number of observations is so large that for practical reasons they must be grouped, the examination of the marginal distributions may be supplemented by an examination of the conditional distributions, which should also be normal. Further we may test whether the variances of the conditional distributions differ significantly, see § 11.6, and the linearity of the regression curves may be tested by the method described in § 18.3.

A more detailed investigation involves calculation of the "expected" number of observations in each of the two-dimensional cells, employing the estimates of the parameters, much as in the corresponding one-dimensional case discussed in § 6.8. For this purpose the tables mentioned in § 19.3 may be used. Practical methods of computation are given in the tables.

Example 19.4. In Example 18.2, Table 18.9, the distribution is given of 24 determinations of the percentage dry matter in fresh spinach, and the percentage preserved ascorbic acid after drying at 90° C. An investigation of the two marginal distributions with the aid of probability paper indicates that these distributions do not differ significantly from normal distributions.

The estimates of the five parameters of the two-dimensional distribution are:

$$\bar{x}_1 = 10{\cdot}53, \quad \bar{x}_2 = 73{\cdot}975\,,$$
$$s_1^2 = 2{\cdot}594, \quad s_2^2 = 100{\cdot}39\,,$$
$$r = 0{\cdot}6182.$$

TABLE 19.1.

Investigation of an ungrouped, two-dimensional distribution.

x_{1i}	$\frac{x_{1i}-\bar{x}_1}{s_1}$	x_{2i}	$\frac{x_{2i}-\bar{x}_2}{s_2}$	$\frac{x_{1i}-\bar{x}_1}{s_1}\frac{x_{2i}-\bar{x}_2}{s_2}$	$\chi_i'^2$	$\chi_{(i)}'^2$	$100\times\frac{n-i+\frac{1}{2}}{n}$
10·0	−0·33	70·9	−0·31	0·10	0·13	0·13	97·9
8·9	−1·01	74·0	0·00	0·00	1·65	0·14	93·7
8·9	−1·01	58·6	−1·53	1·55	2·33	0·18	89·6
9·2	−0·83	80·6	0·66	−0·55	2·92	0·28	85·4
7·8	−1·70	69·4	−0·46	0·78	3·46	0·35	81·2
10·1	−0·27	76·0	0·20	−0·05	0·28	0·53	77·1
9·0	−0·95	66·4	−0·76	0·72	0·95	0·54	72·9
8·2	−1·45	50·9	−2·30	3·34	5·26	0·63	68·7
9·5	−0·64	61·9	−1·21	0·77	1·49	0·95	64·6
10·8	0·17	65·2	−0·88	−0·15	1·60	1·02	60·4
11·1	0·35	77·2	0·32	0·11	0·14	1·29	56·2
11·2	0·42	89·6	1·56	0·66	2·90	1·41	52·1
12·5	1·22	74·2	0·02	0·02	2·37	1·49	47·9
12·3	1·10	83·1	0·91	1·00	1·29	1·60	43·7
10·0	−0·33	66·7	−0·72	0·24	0·53	1·65	39·6
10·2	−0·20	77·2	0·32	−0·06	0·35	2·30	35·4
11·2	0·42	83·8	0·98	0·41	1·02	2·33	31·2
11·2	0·42	67·9	−0·61	−0·26	1·41	2·37	27·1
10·0	−0·33	88·9	1·49	−0·49	4·75	2·90	22·9
10·7	0·11	69·0	−0·50	−0·06	0·54	2·92	18·7
10·3	−0·14	69·8	−0·41	0·06	0·18	3·46	14·6
12·9	1·47	86·0	1·20	1·76	2·30	4·75	10·4
11·8	0·79	79·9	0·59	0·47	0·63	5·26	6·2
14·9	2·71	88·2	1·42	3·85	7·43	7·43	2·1
S	−0·01		−0·02	14·22	45·91		
SS	23·01		22·99				

The two-dimensional distribution is examined by computing the 24 values of the quantity

$$\chi_i'^2 = \frac{1}{1-r^2}\left(\left(\frac{x_{1i}-\bar{x}_1}{s_1}\right)^2 - 2r\frac{x_{1i}-\bar{x}_1}{s_1}\frac{x_{2i}-\bar{x}_2}{s_2}+\left(\frac{x_{2i}-\bar{x}_2}{s_2}\right)^2\right).$$

The results are given in Table 19.1.

The columns with $(x_{1i}-\bar{x}_1)/s_1$ and $(x_{2i}-\bar{x}_2)/s_2$ show the "standardized" marginal distributions. As mentioned above, these distributions show good agreement with the normal distribution, and only one value in each distribution is numerically larger than 2. As the sample means and standard deviations have been used when standardizing the observations, the standardized variables are not mutually independent, but satisfy the relations $S = 0$ and $SS = f$, see Table 19.1. (The deviations are due to rounding errors).

The products of the standardized deviations are predominantly positive, i. e., the correlation is positive. Seven out of the 24 products are negative, and their numerical value is comparatively small. The sum of the products is 14·22, which divided by $f = 23$ gives $r = 0{\cdot}618$. Lastly the $\chi_i'^2$ are computed, and as a check $\Sigma\chi_i'^2$ which is 45·91. According to the definition of $\chi_i'^2$ this sum should be $2f = 46$; the difference is due to rounding errors.

The distribution of the 24 values of $\chi_i'^2$ is given in the last two columns of the table, where the values of χ'^2 have been ranked according to order of magnitude, and the corresponding cumulative frequencies stated. Fig. 19.6 shows the relationship between $\chi_{(i)}'^2$ and $\log_{10}\dfrac{n-i+\frac{1}{2}}{n}$. The straight line has the equation (19.10.2). The variation of the points about the

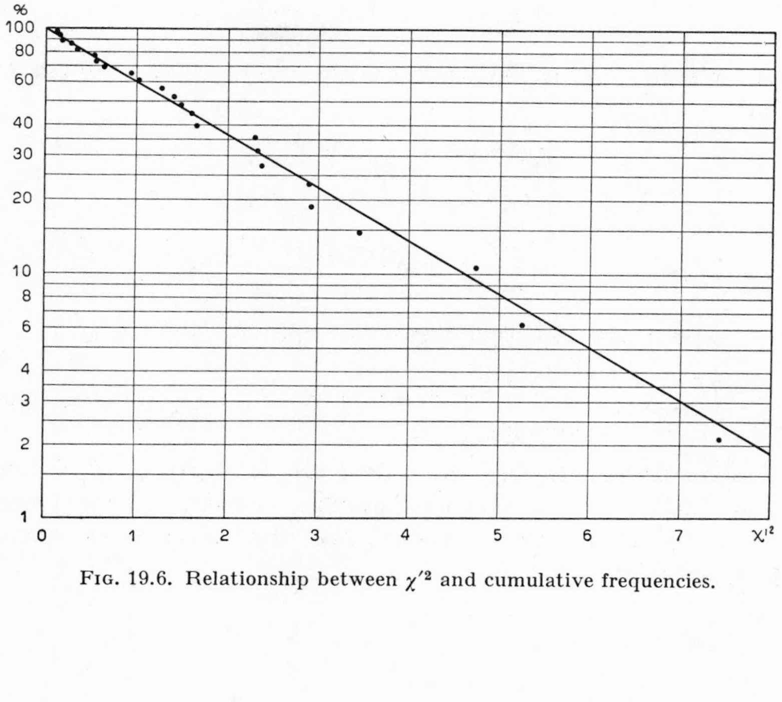

FIG. 19.6. Relationship between χ'^2 and cumulative frequencies.

straight line does not indicate that the distribution examined here differs significantly from the theoretical χ^2-distribution and therefore the hypothesis that the observations belong to a two-dimensional normal population cannot be rejected on the basis of this test.

19.11. Distribution of the Means. Confidence Regions. Tests of Significance.

From n stochastically independent observations (x_{11}, x_{21}), (x_{12}, x_{22}), ..., (x_{1n}, x_{2n}) from a two-dimensional normal population the estimates $(\bar{x}_1, \bar{x}_2, s_1^2, s_2^2, r)$ of the parameters $(\xi_1, \xi_2, \sigma_1^2, \sigma_2^2, \varrho)$ are calculated. As is the case for one-dimensional distributions the determination of confidence regions and various tests of significance are based on *the distributions of the estimates.* The derivation of these distributions, however, is so complicated that it falls outside the scope of this book; in what follows we will therefore deal only with *applications* of these said distributions. For proofs the reader is referred to the bibliography in § 20.9.

The distribution of the sample means $(\bar{x}_1, \bar{x}_2)$ *is a two-dimensional normal distribution with parameters* $\left(\xi_1, \xi_2, \frac{\sigma_1^2}{n}, \frac{\sigma_2^2}{n}, \varrho\right)$, and the sample means are stochastically independent of the estimates (s_1^2, s_2^2, r). Thus, *the correlation coefficient of the sample means is the same as that of the single observations.*

If the parameters of the distribution are given, the probability that $(\bar{x}_1, \bar{x}_2)$ belongs in the region Ω is

$$\iint_\Omega p\{\bar{x}_1, \bar{x}_2\} d\bar{x}_1 d\bar{x}_2 = P\,. \tag{19.11.1}$$

In what follows contour ellipses are employed. According to (19.9.2) the quantity

$$\begin{aligned}\chi^2 &= \frac{1}{1-\varrho^2}\left[\left(\frac{\bar{x}_1-\xi_1}{\sigma_{\bar{x}_1}}\right)^2 - 2\varrho\frac{\bar{x}_1-\xi_1}{\sigma_{\bar{x}_1}}\frac{\bar{x}_2-\xi_2}{\sigma_{\bar{x}_2}} + \left(\frac{\bar{x}_2-\xi_2}{\sigma_{\bar{x}_2}}\right)^2\right]\\ &= \frac{n}{1-\varrho^2}\left[\left(\frac{\bar{x}_1-\xi_1}{\sigma_1}\right)^2 - 2\varrho\frac{\bar{x}_1-\xi_1}{\sigma_1}\frac{\bar{x}_2-\xi_2}{\sigma_2} + \left(\frac{\bar{x}_2-\xi_2}{\sigma_2}\right)^2\right]\end{aligned} \tag{19.11.2}$$

has a χ^2-distribution with two degrees of freedom, and hence

$$P\left\{\frac{n}{1-\varrho^2}\left[\left(\frac{\bar{x}_1-\xi_1}{\sigma_1}\right)^2 - 2\varrho\frac{\bar{x}_1-\xi_1}{\sigma_1}\frac{\bar{x}_2-\xi_2}{\sigma_2} + \left(\frac{\bar{x}_2-\xi_2}{\sigma_2}\right)^2\right] < \chi_P^2\right\} = P\,. \tag{19.11.3}$$

This relation forms the basis for calculating confidence ellipses and for a test of significance, cf. the corresponding one-dimensional problem in § 9.6. For the present we will assume that the values of the parameters $(\sigma_1^2, \sigma_2^2, \varrho)$ are known.

Case I. *Contour ellipses for* $(\bar{x}_1, \bar{x}_2)$. *The population means* (ξ_1, ξ_2) *are known.* When the inequality in (19.11.3) is satisfied, it follows that $(\bar{x}_1, \bar{x}_2)$ lies inside an ellipse with center (ξ_1, ξ_2) and axes and angle of rotation as determined by $(\sigma^2_{\bar{x}_1}, \sigma^2_{\bar{x}_2}, \varrho)$ and χ^2_P, see (19.8.8), (19.8.12), and (19.8.13). The probability that this happens is P.

Case II. *Confidence ellipses for* (ξ_1, ξ_2). *The sample means* $(\bar{x}_1, \bar{x}_2)$ *are known.* When the inequality in (19.11.3) is satisfied it also follows that (ξ_1, ξ_2) lies inside an ellipse with center $(\bar{x}_1, \bar{x}_2)$ and axes and angle of rotation as determined by $(\sigma^2_{\bar{x}_1}, \sigma^2_{\bar{x}_2}, \varrho)$ and χ^2_P. The center of this ellipse is a stochastic variable, and the probability that the ellipse contains the population means is P for every value of (ξ_1, ξ_2). If we have a long series of samples, each containing n observations from the population defined above, we may calculate a corresponding series of ellipses with varying centers, and $100P\%$ of these ellipses must be expected to include the fixed pair of numbers (ξ_1, ξ_2). In practice we use the single pair of sample means to compute such an ellipse which is then called a *confidence ellipse* with a *confidence coefficient* of P.

Case III. *Test of significance. The sample means are known and a pair of population means postulated.* The agreement between a given value of $(\bar{x}_1, \bar{x}_2)$ and a hypothetical value of (ξ_1, ξ_2) is tested by calculating χ^2 according to (19.11.2), and comparing this quantity with χ^2_P.

As mentioned before, the application of the above formulas assumes that the values of the parameters $(\sigma_1^2, \sigma_2^2, \varrho)$ are known. As this is exceptional, the formulas must be modified according to the principles stated by "Student".

In the one-dimensional distribution the variable

$$u = \frac{\bar{x}-\xi}{\sigma_{\bar{x}}} \quad \text{or} \quad u^2 = v^2(1, \infty) = \left(\frac{\bar{x}-\xi}{\sigma_{\bar{x}}}\right)^2$$

changes to

$$t = \frac{\bar{x}-\xi}{s_{\bar{x}}} \quad \text{or} \quad t^2 = v^2(1, f) = \left(\frac{\bar{x}-\xi}{s_{\bar{x}}}\right)^2,$$

$\sigma_{\bar{x}} = \sigma/\sqrt{n}$ being replaced by $s_{\bar{x}} = s/\sqrt{n}$, where s is based on $f = n-1$ degrees of freedom, so that the confidence limits for ξ become independent of σ.

In analogy with the above we change the variable

$$\chi^2 = \frac{1}{1-\varrho^2}\left[\left(\frac{\bar{x}_1-\xi_1}{\sigma_{\bar{x}_1}}\right)^2 - 2\varrho\frac{\bar{x}_1-\xi_1}{\sigma_{\bar{x}_1}}\frac{\bar{x}_2-\xi_2}{\sigma_{\bar{x}_2}} + \left(\frac{\bar{x}_2-\xi_2}{\sigma_{\bar{x}_2}}\right)^2\right] = 2v^2\,(2,\infty) \quad (19.11.4)$$

to

$$T^2 = \frac{1}{1-r^2}\left[\left(\frac{\bar{x}_1-\xi_1}{s_{\bar{x}_1}}\right)^2 - 2r\frac{\bar{x}_1-\xi_1}{s_{\bar{x}_1}}\frac{\bar{x}_2-\xi_2}{s_{\bar{x}_2}} + \left(\frac{\bar{x}_2-\xi_2}{s_{\bar{x}_2}}\right)^2\right] = 2v^2(2, f)\frac{n-1}{n-2}, \quad (19.11.5)$$

where $f = n-2$, cf. (19.7.18) and (19.7.19). The variable

$$v^2(2, n-2) = \frac{n-2}{n-1}\frac{T^2}{2} \tag{19.11.6}$$

thus has a v^2-distribution with $(2, n-2)$ degrees of freedom and may be used in the same manner as χ^2 was used above for calculating confidence ellipses and tests of significance independent of $(\sigma_1^2, \sigma_2^2, \varrho)$[1]).

It has been shown[2]) that the hypothesis $\xi_1 = \xi_2$ may also be tested by the T^2-criterion, which in the present case may be written as

$$t = \frac{\bar{x}_1-\bar{x}_2}{\sqrt{\frac{1}{n}(s_1^2+s_2^2-2s_{12})}}, \quad f = n-1, \tag{19.11.7}$$

which is distributed as t with $n-1$ degrees of freedom. This test is a special case of the test (15.5.7), as $\bar{d} = \bar{x}_1-\bar{x}_2$, and

$$s_d^2 = s_1^2+s_2^2-2s_{12}. \tag{19.11.8}$$

The test in § 15.5, however, did not imply that (x_{1i}, x_{2i}) are random observations from a two-dimensional normal population, but only that the differences $d_i = x_{1i}-x_{2i}$ are normally distributed with constant variance.

19.12. The Distribution of the Correlation Coefficient.

The distribution function of (s_1^2, s_2^2, r) is very complicated, and so is the marginal distribution, $p\{r\}$, of r. As the correlation coefficient is a standardized variable, $p\{r\}$ depends only on the theoretical value ϱ and the number of observations, n. The distribution is symmetrical for $\varrho = 0$. For large values of $|\varrho|$ the distribution is very skew if n is not large. The sample correlation coefficient r *may be considered normally distributed about the population correlation coefficient* ϱ *with standard deviation*

$$\sigma_r = \frac{1-\varrho^2}{\sqrt{n-1}} \tag{19.12.1}$$

only if the value of n *is very large and* ϱ *is near to* 0, so that

$$\boxed{u = \frac{r-\varrho}{1-\varrho^2}\sqrt{n-1} \quad \text{for large values of } n \text{ and small values of } \varrho} \tag{19.12.2}$$

---

[1]) The quantity T^2 is usually denoted HOTELLING's T^2 since it was introduced by H. HOTELLING: *The Generalization of* STUDENT's *Ratio*, Ann. Math. Stat., 2, 1931, 360–378.

[2]) P. L. HSU: *Notes on* HOTELLING's *Generalized T*, Ann. Math. Stat., 9, 1938, 231–243.

is approximately normally distributed with parameters (0, 1). This relation may be used for calculating fractiles for r, confidence limits for ϱ, or for testing a hypothetical value of ϱ according to the principles given for the u-test in §§ 9.3–9.5.

For values of n smaller than 400 the distribution of the correlation coefficient has been tabulated in F. N. DAVID: *Tables of the Ordinates and Probability Integral of the Distribution of the Correlation Coefficient in Small Samples*, London, 1938.

From the distribution of the correlation coefficient it is possible to derive *a simple test for the hypothesis* $\varrho = 0$. If $\varrho = 0$, the quantity

$$t = \frac{r}{\sqrt{1-r^2}}\sqrt{f}, \quad f = n-2, \tag{19.12.3}$$

is distributed as t with $f = n-2$ degrees of freedom. If the t-value corresponding to a given value of r is smaller than, say, the 95% fractile, the data in hand do not justify the conclusion that the two variables are correlated.

As the distribution of the correlation coefficient is cumbersome to deal with in practical work, R. A. FISHER has attempted to transform the correlation coefficient in such a manner that the variance of the new variable is independent of ϱ, see (7.3.4). This leads to the variable

$$z = \tfrac{1}{2}\log_e\frac{1+r}{1-r} = 1{\cdot}1513\log_{10}\frac{1+r}{1-r}, \quad \begin{array}{c} -1 \leqq r \leqq +1 \\ -\infty < z < \infty, \end{array} \tag{19.12.4}$$

which with good approximation is normally distributed—even for small values of n—with mean

$$\mathcal{m}\{z\} = \zeta \simeq \tfrac{1}{2}\log_e\frac{1+\varrho}{1-\varrho} + \frac{\varrho}{2(n-1)} \tag{19.12.5}$$

and variance

$$\mathcal{v}\{z\} \simeq \frac{1}{n-3}, \tag{19.12.6}$$

so that the variable

$$u = (z-\zeta)\sqrt{n-3} \tag{19.12.7}$$

with good approximation is normally distributed with parameters (0, 1). A table of z as a function of r is to be found in R. A. FISHER and F. YATES: *Statistical Tables*.

Usually the quantity $\varrho/2(n-1)$ in ζ may be disregarded, as this quantity is small compared with $1/\sqrt{n-3}$. Only when several z-values are combined in order to calculate an average value it may be of importance to include this term of ζ.

Utilizing the variable z, confidence limits for ζ and thus for ϱ may be determined, the solution of (19.12.4) with respect to r leading to

$$\boxed{r = \frac{e^{2z}-1}{e^{2z}+1}, \quad e^{2z} = \text{antilog}_{10}\frac{z}{1{\cdot}1513}.} \tag{19.12.8}$$

Similarly we can test whether an observed correlation coefficient deviates significantly from a given theoretical value, see § 9.4.

The transformation of r may also be used to examine whether two observed correlation coefficients differ significantly. Let there be, say, n_1 and n_2 pairs of observations from which the correlation coefficients r_1 and r_2 have been calculated. The test hypothesis is $\varrho_1 = \varrho_2 = \varrho$ or $\zeta_1 = \zeta_2 = \zeta$, so that, if the hypothesis is true, $z_1 = \frac{1}{2}(\log_e(1+r_1) - \log_e(1-r_1))$ is normally distributed about ζ with variance $1/(n_1-3)$, and $z_2 = \frac{1}{2}(\log_e(1+r_2) - \log_e(1-r_2))$ is normally distributed about ζ with variance $1/(n_2-3)$. From this it follows that

$$\boxed{u = \frac{z_1 - z_2}{\sqrt{\dfrac{1}{n_1-3} + \dfrac{1}{n_2-3}}}} \tag{19.12.9}$$

is normally distributed with parameters $(0, 1)$. If $|u|$ is smaller than, say, 1·96, the hypothesis cannot be rejected.

For the comparison of several r-values, $r_1, r_2, \ldots, r_k$, the variable

$$\chi^2 = \sum_{i=1}^{k} u_i^2 = \sum_{i=1}^{k} (z_i - \zeta)^2 (n_i - 3), \quad f = k, \tag{19.12.10}$$

may be used, testing the hypothesis $\varrho_1 = \varrho_2 = \ldots = \varrho_k = \varrho$.

As usually it is not possible to specify ζ, this quantity is replaced by the estimate

$$\boxed{\bar{z} = \frac{\sum_{i=1}^{k} (n_i - 3) z_i}{\sum_{i=1}^{k} (n_i - 3)},} \tag{19.12.11}$$

χ^2 being partitioned as follows:

$$\chi^2 = \sum_{i=1}^{k} (n_i-3)(z_i-\zeta)^2 = \sum_{i=1}^{k} (n_i-3)(z_i-\bar{z})^2+(\bar{z}-\zeta)^2 \sum_{i=1}^{k} (n_i-3).$$

According to the partition theorem for the χ^2-distribution

$$\boxed{\chi^2 = \sum_{i=1}^{k} (n_i-3)(z_i-\bar{z})^2, \quad f = k-1,} \tag{19.12.12}$$

is distributed as χ^2 with $k-1$ degrees of freedom, so that this quantity may be employed as a test for our hypothesis. If the hypothesis cannot be rejected, we see that according to (19.12.11) $\bar{z}$ may be utilized as an estimate of ζ with variance

$$\boxed{V\{\bar{z}\} = \frac{1}{\sum_{i=1}^{k} (n_i-3)}.} \tag{19.12.13}$$

If k is large the correction $\varrho/2(n-1)$ in (19.12.5) may become of importance. From (19.12.11) and (19.12.5) we obtain

$$\bar{z} \approx \bar{\zeta} = \tfrac{1}{2} \log_e \frac{1+\varrho}{1-\varrho} + \frac{\varrho}{2n_0}, \tag{19.12.14}$$

where

$$\frac{1}{n_0} = \frac{\sum_{i=1}^{k} \frac{n_i-3}{n_i-1}}{\sum_{i=1}^{k} (n_i-3)}. \tag{19.12.15}$$

Passing from $\bar{z}$ to $\bar{r}$ we solve the equation

$$\bar{z} = \tfrac{1}{2} \log_e \frac{1+\bar{r}}{1-\bar{r}} + \frac{\bar{r}}{2n_0}. \tag{19.12.16}$$

Usually it is sufficient to use an approximation to $\bar{r}$ in the computation of $\bar{z}-\dfrac{\bar{r}}{2n_0}$, and then $\bar{r}$ may be determined from this corrected value of $\bar{z}$ with the aid of (19.12.8).

Example 19.5. In Example 19.4 the correlation coefficient for percentage dry matter and percentage preserved ascorbic acid was determined as $r = 0{\cdot}6182$ on the basis of 24 observations. Does this value differ significantly from zero? This question is answered with the aid of the t-test according to (19.12.3). Employing this formula we obtain

$$t = \frac{0{\cdot}6182}{\sqrt{1-0{\cdot}6182^2}}\sqrt{22} = 3{\cdot}69\ .$$

From Table IV we see that $P\{|t| > 3{\cdot}69\}$ is about 0·1 per cent, which means that the difference must be considered significant, see also Example 18.2.

Example 19.6. Samples were taken from 5 castings of aluminium and the tensile strength and yielding point determined[1]). Each set of observations consists of 12 interdependent values of tensile strength and yielding point. From the 5 sets of observations the following correlation coefficients were found: 0·683, 0·876, 0·714, 0·715, 0·805. Is it possible that this variation in the correlation coefficients during the casting process is random, or has the theoretical correlation coefficient changed? This question is answered by the aid of the χ^2-test, using (19.12.12). As all values of n_i are 12, the weighted averages are replaced by simple arithmetical averages, see Table 19.2, which gives the computations.

TABLE 19.2.
Comparison of 5 Correlation Coefficients.

r	z	$(z-\bar{z})3$
0·683	0·835	−0·55
0·876	1·358	1·02
0·714	0·895	−0·37
0·715	0·897	−0·37
0·805	1·113	0·28
	$\bar{z} = 1{\cdot}020$	
$\Sigma(z_i-\bar{z})^2 = 0{\cdot}1879$,	$\chi^2 = 9\Sigma(z_i-\bar{z})^2 = 1{\cdot}69$,	$f = 4$.

From each r-value a z-value is computed from (19.12.4), and then the mean and the sum of squares are computed, see (19.12.11) and (19.12.12). As the value of χ^2 obtained is not significant, the hypothesis regarding a common theoretical correlation coefficient cannot be rejected. Utilizing $\bar{z}$ and the standard error of $\bar{z}$ it is possible to determine limits for ϱ. From (19.12.13) we obtain

$$\mathcal{V}\{\bar{z}\} \simeq \frac{1}{5\times 9} = 0{\cdot}02222\,,$$

which leads to $\sigma_{\bar{z}} \simeq 0{\cdot}149$ and $1{\cdot}96\times\sigma_{\bar{z}} \simeq 0{\cdot}292$, giving the 95% confidence

[1]) E. S. PEARSON and S. S. WILKS: *Methods of Statistical Analysis Appropriate for k Samples of Two Variables*, Biometrika, 25, 1933, Table I, p. 356.

limits $0{\cdot}728 < \zeta < 1{\cdot}312$. Computing the corresponding limits for ϱ from (19.12.16), we obtain the following equations, n_0 being equal to 11:

$$1{\cdot}1513 \log_{10} \frac{1+r}{1-r} + \frac{r}{22} = \begin{cases} 0{\cdot}728 \\ 1{\cdot}020 \\ 1{\cdot}312. \end{cases}$$

If as a first approximation we put $r = 0{\cdot}76$, we have $r/22 = 0{\cdot}035$, and the equations are reduced as follows

$$1{\cdot}1513 \log_{10} \frac{1+r}{1-r} = \begin{cases} 0{\cdot}693 \\ 0{\cdot}985 \\ 1{\cdot}277 \end{cases}$$

giving $r = 0{\cdot}76$ as an estimate of ϱ and $0{\cdot}60 < \varrho < 0{\cdot}86$ as 95% confidence limits. By application of the three r-values for the determination of new corrections $(r/22)$ a better solution of the three equations may be obtained. The difference from the above values, however, at most amounts to one unit on the second decimal place.

19.13. The Interpretation of the Correlation Coefficient.

From any two-dimensional set of observations we may compute a correlation coefficient r, which is an estimate of the corresponding population coefficient ϱ, but it must be emphasized that *the theory just given holds only when the underlying population is normally distributed.* Either the original observations or transformed values of the original observations should be normally distributed before we attempt to apply the above tests. If the population is not normally distributed the interpretation of the correlation coefficient is very uncertain.

In case of the two-dimensional normal distribution we have seen that there is a certain justification for regarding the correlation coefficient as indicating a measure of the relationship between the two variables, and the limiting cases $\varrho = 0$ and $|\varrho| = 1$ correspond to stochastical independence and linear dependence, respectively. If the distribution is not normal this, however, does not hold good, and in this case $\varrho = 0$ may not indicate stochastical independence. Further, it is even possible to show that certain kinds of functional dependence between the variables may lead to zero correlation, i. e., even though the variables lie on a curve in the (x_1, x_2)-plane without deviating from the curve, the correlation coefficient may be zero.

During the preliminary stage of an investigation on the relationship between two or more factors it is often possible in industrial work to use the routine observations collected under normal working conditions in a plant. By determining the correlation coefficient and then testing the hypothesis of zero correlation it is sometimes possible to prove the existence of a stochastic relationship between the variables. It is, however, necessary to emphasize that a *stochastic* interdependence does not necessarily indicate a *causal* relationship. The correlation coefficient may possibly indicate a stochastic interdependence between x_1 and x_2, but whether x_1 has caused x_2, or x_2 x_1, or their relationship is due to the fact that they are both causally related to other factors, cannot be determined by means of the correlation coefficient. Therefore, *a significant correlation coefficient calls for further work on the problem in order to try to determine the causality present.* In the further analysis, which first and foremost must be planned on the basis of professional knowledge regarding the problem, regression analysis will often play an important part as an aid to the verification of the postulated hypothesis. In experimental work, where it is possible to control the experimental conditions, the experiments usually are planned in such a manner that the values of one variable are chosen a priori, so that the relationship may be determined by a regression analysis.

It is often found that the interpretation of a correlation coefficient is further complicated by the fact that *other factors than the two under investigation have varied* and have thus introduced a spurious correlation. If there is a possibility that this is the case, *the data must be divided into rational sub-groups* in such a manner that the disturbing factors do not vary within the sub-groups, and the correlation coefficient must be determined for each sub-group. It is then possible to test whether the correlation coefficient depends on the disturbing factors by means of (19.12.12). If it is possible to give a quantitative measure of the disturbing factors by introducing new stochastic variables, the influence of these factors may be eliminated by computing partial correlation coefficients according to the principles given in § 20.1.

Spurious correlations may also arise in other ways. For example, suppose we have a set of observations of three stochastically independent variables (x_1, x_2, x_3). The widespread habit of computing index numbers in order to "simplify" the handling of quantitative data may, for instance, lead to the computation of the indices x_1/x_3 and x_2/x_3. These indices are, however, positively correlated even if the three variables are stochastically independent, x_3 being included in both their denominators. Computation of these indices has thus led to a far more complicated description of the observed data than an analysis of the distribution of the original observations. In other cases, it is true, the index numbers may be preferred to the observations themselves, but this must be examined for each set of data.

19.14. The Influence of Errors of Measurement on the Correlation and Regression Coefficients.

In Example 9.2, p. 217, it has been shown how the variance of measures of a certain property, for example a quality characteristic of an industrial product, is composed of the variance of the true property and the variance of the errors of measurement, i. e., $\sigma_T^2 = \sigma_Q^2+\sigma_M^2$.

This result is based on the assumption that the two variables, the "true" value of the property and the error of measurement, are stochastically independent.

When the relationship between two properties denoted by (x_1, x_2) is investigated, *errors of measurement* will similarly influence the two variances and the correlation coefficient. Let the errors of measurement be denoted by (v_1, v_2); the observations are then $(y_1, y_2) = (x_1+v_1, x_2+v_2)$. We will assume that the errors of measurement are stochastically independent and independent of (x_1, x_2), so that the correlation coefficients $\varrho\{v_1, v_2\}$, $\varrho\{x_1, v_1\}$, $\varrho\{x_1, v_2\}$, $\varrho\{x_2, v_1\}$, and $\varrho\{x_2, v_2\}$ are all zero. In order to simplify the derivation of the following formulas, we assume that the means of the four variables are all zero; this assumption, however, is made only for computational purposes. We then find

$$\mathcal{V}\{y_i\} = \mathcal{V}\{x_i\}+\mathcal{V}\{v_i\} = \sigma_{xi}^2+\sigma_{vi}^2 , \quad i = 1, 2 , \tag{19.14.1}$$

and

$$\begin{aligned}\mathcal{V}\{y_1, y_2\} &= \mathcal{M}\{(x_1+v_1)(x_2+v_2)\} \\ &= \mathcal{M}\{x_1x_2\}+\mathcal{M}\{x_1v_2\}+\mathcal{M}\{x_2v_1\}+\mathcal{M}\{v_1v_2\} \\ &= \mathcal{V}\{x_1, x_2\} = \sigma_{x_1}\sigma_{x_2}\varrho\{x_1, x_2\} .\end{aligned} \tag{19.14.2}$$

From the definition of the correlation coefficient it then follows that

$$\varrho\{y_1, y_2\} = \frac{\sigma_{x_1}\sigma_{x_2}\varrho\{x_1, x_2\}}{\sqrt{(\sigma_{x_1}^2+\sigma_{v_1}^2)(\sigma_{x_2}^2+\sigma_{v_2}^2)}} = \frac{\varrho\{x_1, x_2\}}{\sqrt{\left(1+\left(\frac{\sigma_{v1}}{\sigma_{x_1}}\right)^2\right)\left(1+\left(\frac{\sigma_{v2}}{\sigma_{x_2}}\right)^2\right)}}, \tag{19.14.3}$$

i. e., *the correlation coefficient of the properties plus errors of measurement is always less than the correlation coefficient of the properties themselves.* The ratio between the two correlation coefficients depends on the ratios between the variances of the methods of measurement and those of the properties. Thus, the formula (19.14.3) shows—as was to be expected—that relatively large errors of measurement will "disguise" an existing correlation. Estimates of the variances of the methods of measurement, e. g., determined on the basis of duplicate analyses, may enable us to make a correction for this decrease in the correlation coefficient.

Inserting (19.14.3) into (19.7.9) and (19.7.10) we obtain for the regression coefficients

$$\beta_{x_2|x_1} = \beta_{y_2|y_1}\left(1+\frac{\sigma_{v_1}^2}{\sigma_{x_1}^2}\right) \tag{19.14.4}$$

and

$$\beta_{x_1|x_2} = \beta_{y_1|y_2}\left(1+\frac{\sigma_{v_2}^2}{\sigma_{x_2}^2}\right). \tag{19.14.5}$$

If the values of *the independent variable* in a regression analysis are encumbered with random errors the regression coefficient found will thus be smaller than the regression coefficient determined from values of the independent variable without errors.

19.15. Tests of Significance for the Regression Coefficients.

The regression coefficients $b_{x_2|x_1}$ and $b_{x_1|x_2}$ are not normally distributed since both x_1 and x_2 are stochastic variables, so that both the numerators and the denominators of the expressions giving the regression coefficients become stochastic variables. From the results in § 18.4, however, it follows that the quantities

$$t = \frac{b_{x_2|x_1}-\beta_{x_2|x_1}}{s_{x_2|x_1}/\sqrt{\Sigma(x_{1i}-\bar{x}_1)^2}} \tag{19.15.1}$$

and

$$t = \frac{b_{x_1|x_2}-\beta_{x_1|x_2}}{s_{x_1|x_2}/\sqrt{\Sigma(x_{2i}-\bar{x}_2)^2}} \tag{19.15.2}$$

both have t-distributions with $f = n-2$ degrees of freedom.

In particular the hypothesis $\beta_{x_2|x_1} = \beta_{x_1|x_2} = 0$ is tested by calculating

$$t = \frac{b_{x_2|x_1}}{s_{x_2|x_1}/\sqrt{\Sigma(x_{1i}-\bar{x}_1)^2}} = \frac{b_{x_1|x_2}}{s_{x_1|x_2}/\sqrt{\Sigma(x_{2i}-\bar{x}_2)^2}}$$

$$= \frac{r\sqrt{n-2}}{\sqrt{1-r^2}}, \tag{19.15.3}$$

which is identical with the test for the hypothesis $\varrho = 0$, according to (19.12.3), see also Example 18.2 and Example 19.5.

19.16. Comparison of Two Sets of Two-Dimensional Observations.

From the two sets of normally distributed observations we determine estimates of the parameters. Let there be n_1 and n_2 pairs of observations, respectively, giving the estimates $(\bar{x}_1, \bar{x}_2, s_{x_1}^2, s_{x_2}^2, r_{x_1x_2})$ and $(\bar{y}_1, \bar{y}_2, s_{y_1}^2, s_{y_2}^2, r_{y_1y_2})$, say. Comparison of these two sets of parameters may be divided into two tests, since the means are stochastically independent of the variances and the correlation coefficients. The procedure is analogous to that applied in § 15.4.

The variances may be compared two by two by means of the variance ratios, and the correlation coefficients may be compared by means of the test in § 19.12. Application of these three (marginal) tests does not take into account the fact that the variances and the correlation coefficients are stochastically dependent. A combination of these three estimates into one stochastic variable has been made by WILKS, who has introduced *the generalized variance* defined by the expression $s_1^2 s_2^2(1-r^2)$ and has derived the distribution functions of this quantity and the ratio between two generalized variances. As regards this test and its further generalization in order to deal with k sets of two-dimensional observations, we refer to E. S. PEARSON and S. S. WILKS: *Methods of Statistical Analysis Appropriate for k Samples of Two Variables*, Biometrika, 25, 1933, 353–78.

In the following we will derive a test for the hypothesis that two pairs of population means are identical, assuming that the two sets of observations have the same theoretical variances and correlation coefficients, i. e., $\sigma_{x_1}^2 = \sigma_{y_1}^2 = \sigma_1^2$, $\sigma_{x_2}^2 = \sigma_{y_2}^2 = \sigma_2^2$, and $\varrho_{x_1x_2} = \varrho_{y_1y_2} = \varrho$. The test is based on the two differences between the means

$$d_i = \bar{x}_i - \bar{y}_i\,, \quad i = 1, 2\,, \tag{19.16.1}$$

which—if the test hypothesis is true—are normally distributed with mean values (0, 0), variances

$$\mathcal{V}\{d_i\} = \sigma_i^2\left(\frac{1}{n_1}+\frac{1}{n_2}\right), \quad i = 1, 2\,, \tag{19.16.2}$$

and the correlation coefficient ϱ. According to the addition theorem for the s^2-distribution we calculate as an estimate of σ_1^2

$$s_1^2 = \frac{(n_1-1)s_{x_1}^2+(n_2-1)s_{y_1}^2}{n_1+n_2-2} \tag{19.16.3}$$

and as a corresponding estimate of σ_2^2

$$s_2^2 = \frac{(n_1-1)s_{x_2}^2+(n_2-1)s_{y_2}^2}{n_1+n_2-2}\,. \tag{19.16.4}$$

Similarly we obtain by combining the two covariances

$$s_{12} = \frac{(n_1-1)s_{x_1x_2}+(n_2-1)s_{y_1y_2}}{n_1+n_2-2}\,. \tag{19.16.5}$$

Whence the estimate r of the correlation coefficient ϱ becomes $r = s_{12}/s_1s_2$.

The test is now obtained by applying (19.11.5) to (d_1, d_2) which leads to

$$T^2 = \frac{1}{1-r^2}\left[\left(\frac{d_1}{s_{d_1}}\right)^2 - 2r\frac{d_1}{s_{d_1}}\frac{d_2}{s_{d_2}} + \left(\frac{d_2}{s_{d_2}}\right)^2\right] = 2v^2(2, f)\frac{n_1+n_2-2}{n_1+n_2-3}, \tag{19.16.6}$$

where $f = n_1+n_2-3$, and

$$s_{di}^2 = s_i^2\left(\frac{1}{n_1}+\frac{1}{n_2}\right), \quad i = 1, 2. \tag{19.16.7}$$

Thus, the variable

$$v^2(2, n_1+n_2-3) = \frac{n_1+n_2-3}{n_1+n_2-2}\frac{T^2}{2}. \tag{19.16.8}$$

is distributed as v^2 with $(2, n_1+n_2-3)$ degrees of freedom. If v^2 is significant, (d_1, d_2) deviates significantly from (0, 0), i. e., the test hypothesis must be rejected. In this case the differences (d_1, d_2) may be replaced by $(d_1-\delta_1, d_2-\delta_2)$, where (δ_1, δ_2) denote corresponding differences between population means, and by utilizing (19.16.6) it is possible to determine confidence ellipses for (δ_1, δ_2).

The test (19.16.6) is a generalization of the t-test given in (15.4.8). If we apply the t-test to each of the marginal distributions, we obtain

$$t_1 = \frac{\bar{x}_1-\bar{y}_1}{s_1\sqrt{\frac{1}{n_1}+\frac{1}{n_2}}} = \frac{d_1}{s_{d_1}}, \quad f = n_1+n_2-2, \tag{19.16.9}$$

and

$$t_2 = \frac{\bar{x}_2-\bar{y}_2}{s_2\sqrt{\frac{1}{n_1}+\frac{1}{n_2}}} = \frac{d_2}{s_{d_2}}, \quad f = n_1+n_2-2, \tag{19.16.10}$$

so that T^2 may be written

$$T^2 = \frac{1}{1-r^2}(t_1^2-2rt_1t_2+t_2^2). \tag{19.16.11}$$

This relation shows the difference between the application of two marginal tests and a joint test based on the two-dimensional distribution. Let t_0 (> 0) be the significance limit of t corresponding to a given probability, so that the test hypothesis is rejected if $|t| > t_0$. Application of the t-test to the marginal distributions means rejecting the test hypothesis if at least one of the values $|t_1|$ or $|t_2|$ is larger than t_0. Fig. 19.7 shows a (t_1, t_2)-coordinate system in which a square has been drawn with center (0, 0) and sides of length $2t_0$ (the innermost square). If the observed value of (t_1, t_2)

lies inside this square neither of the marginal differences is significant, while at least one of the deviations is significant if (t_1, t_2) lies outside the square.

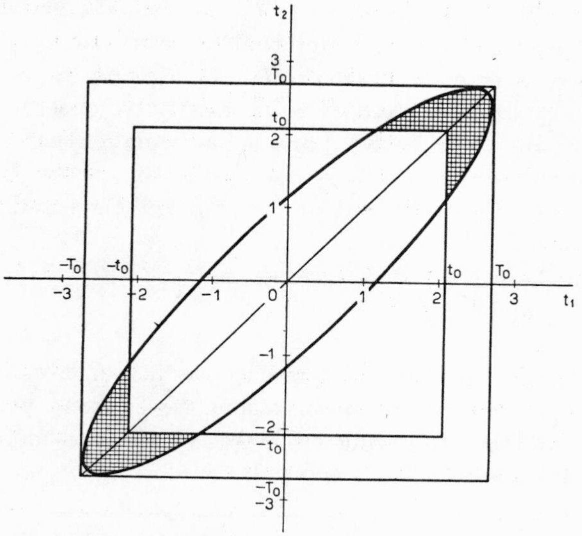

FIG. 19.7. Comparison of two marginal tests with a joint test based on the two-dimensional distribution.

The test derived on the basis of the two-dimensional distribution leads to the rejection of the test hypothesis if $T^2 > T_0^2$, where T_0^2 is the significance limit of T^2 corresponding to the same probability as t_0 for t. The values of (t_1, t_2) which satisfy the equation

$$\frac{1}{1-r^2}(t_1^2-2rt_1t_2+t_2^2) = T_0^2$$

represent an ellipse with center (0, 0), the slope of the major axis equal to 45°, axes of length $(2T_0\sqrt{1+r},\ 2T_0\sqrt{1-r})$, and inscribed in a square with side length $2T_0$, see Fig. 19.7. Since

$$T_0^2 = 2v_P^2(2,\ n_1+n_2-3)\frac{n_1+n_2-2}{n_1+n_2-3}$$

and

$$t_0^2 = v_P^2(1,\ n_1+n_2-2)\ ,$$

we find $T_0^2 > t_0^2$, cf. Table VII. Values of (t_1, t_2) which correspond to points inside the ellipse give us $T^2 < T_0^2$, and points outside the ellipse $T^2 > T_0^2$.

Fig. 19.7 shows that, if at least one of the values $|t_1|$ and $|t_2|$ is larger than T_0, we have $T^2 > T_0^2$, and the test hypothesis may be rejected. If both values are smaller than T_0, but at least one is larger than t_0, the point (t_1, t_2) lies in the area outside the smaller square but inside the larger square. In the hatched part of this area, i. e., the part of the area which lies inside the ellipse, we have $T^2 < T_0^2$, so that the test hypothesis will not be rejected in spite of the fact that at least one of the values—and in the hatched part even both values—of $|t_1|$ and $|t_2|$ is larger than t_0. In the remaining part we have $T^2 > T_0^2$. Lastly we observe the area inside the smaller square where both $|t_1|$ and $|t_2|$ are smaller than t_0. Even though each separate value of t_1 and t_2 is not significant, values lying in the two areas outside the ellipse will lead to $T^2 > T_0^2$, and thus lead to the rejection of the hypothesis.

Fig. 19.7 has been computed from $n_1+n_2 = 24$, $P = 95\%$, i. e., $t_0 = 2{\cdot}07$ and $T_0 = 2{\cdot}70$, and for $r = 0{\cdot}9$.

Example 19.7. In the case of pressure-casting of aluminium described in Example 19.6 two sets of observations were taken, each including 12 observations, and the following estimates of the parameters of the two-dimensional distributions were computed:

Tensile strength (1000 lb. per sq. in.)		Yielding point (Rockwell's E)		Correlation coefficient
Mean	Standard deviation	Mean	Standard deviation	
33·150	3·954	76·12	11·08	0·715
34·269	2·715	69·92	9·88	0·805

Can these two sets of observations be considered as originating from the same two-dimensional, normally distributed population?

First we compute the pooled estimates of the variances and the correlation coefficient, as shown in the following table:

	Tensile strength SSD	Yielding point SSD	SPD	f
	187·618	1473·44	375·91	11
	88·456	1171·73	259·18	11
Sum	276·074	2645·17	635·09	22
Sum/22	$s_1^2 = 12{\cdot}55$	$s_2^2 = 120{\cdot}24$	$s_{12} = 28{\cdot}87$	

From this we obtain $r = s_{12}/s_1 s_2 = 0{\cdot}743$. Further

$$s_{d_1}^2 = s_1^2\left(\frac{1}{12}+\frac{1}{12}\right) = 2{\cdot}092, \quad s_{d_1} = 1{\cdot}446,$$

and

$$s_{d_2}^2 = s_2^2\left(\frac{1}{12}+\frac{1}{12}\right) = 20{\cdot}04\,, \quad s_{d_2} = 4{\cdot}477\,.$$

The differences between the two sets of means are

$$d_1 = 33{\cdot}150-34{\cdot}269 = -1{\cdot}119$$

and

$$d_2 = 76{\cdot}12-69{\cdot}92 = 6{\cdot}20\,.$$

Division by the standard errors of the differences gives the t-values

$$t_1 = -\frac{1{\cdot}119}{1{\cdot}446} = -0{\cdot}774$$

and

$$t_2 = \frac{6{\cdot}20}{4{\cdot}477} = 1{\cdot}385\,.$$

Neither of these t-values is significant, but it is noticeable that they are of opposite sign in spite of the fact that the correlation within sets is positive. If we determine T^2, as given in (19.16.11), we obtain

$$T^2 = \frac{1}{1-0{\cdot}743^2}(0{\cdot}774^2+2\times0{\cdot}743\times0{\cdot}774\times1{\cdot}385+1{\cdot}385^2) = 9{\cdot}17\,,$$

leading to

$$v^2(2,\,21) = \frac{21}{22}\,\frac{9{\cdot}17}{2} = 4{\cdot}38\,,$$

which is larger than $v^2_{.95}(2,21) = 3{\cdot}47$. Thus we see that the two sets of observations differ significantly, which indicates that important factors have changed between one casting and the other.

19.17. Remarks on Some Models of Two-Dimensional Relationships.

The simplest model of a two-dimensional relationship is *the functional relationship*, which in the linear case takes the form $x_1 = \beta_0+\beta_1x_2$ or the equivalent form $x_2 = \alpha_0+\alpha_1x_1$, where $\alpha_0 = -\beta_0/\beta_1$ and $\alpha_1 = 1/\beta_1$.

Functional relationships, however, are never observed directly, because the observed values of (x_1, x_2) are encumbered with errors. Denoting the errors by (v_1, v_2) we observe $(y_1, y_2) = (x_1+v_1, x_2+v_2)$. We assume as usual that the errors are normally distributed with zero means, mutually independent, and independent of (x_1, x_2).

If there are *no errors of measurement of the one variable*, x_2 say, i. e., the errors of this variable are practically negligible as compared with the errors of the other variable, we may estimate the functional relation between x_1 and x_2 by means of regression analysis taking y_1 as dependent and x_2 as independent variable, since

$$\mathcal{M}\{y_1|x_2\} = \mathcal{M}\{x_1+v_1|x_2\} = \beta_0+\beta_1 x_2 \,. \tag{19.17.1}$$

It will be noticed that *the variable without errors has to be used as independent variable* to get an unbiased estimate of the functional relation.

If *both variables are subject to error* regression analysis cannot be used for estimating the functional relationship. This problem of estimation has caused much discussion and a completely satisfactory solution has not yet been given, see the references in § 20.9.

When *the relationship between the variables* (x_1, x_2) *is stochastical*, i. e., the relationship is characterized by a two-dimensional distribution function, it is usually not sufficient for practical purposes just to estimate the parameters of the distribution function. Even when no exact functional relationship exists between two variables we often want to find an equation which gives some sort of "average relationship" between them. Expressed in this way, the problem is indeterminate, not to say meaningless. Whenever such an average relationship is proposed, we must be quite clear as to its exact properties and in what manner it can claim to represent the data.

As examples of such average relationships we may mention the regression lines in a two-dimensional normal population. If we want to predict the value of x_1 for a given value of x_2 with the least possible variance we should use the regression line with x_1 as dependent and x_2 as independent variable.

Observations from a two-dimensional population will usually vary not only on account of the variation inherent in the population but also because errors of measurement are introduced. These errors tend to diminish the regression coefficients as shown in § 19.14.

A model which has been much discussed in econometrics in recent years is based on a system of simultaneous equations[1]). In its simplest form the equation system is

$$x_1 = \beta x_2 + v_1$$

and

$$x_2 = \alpha x_1 + v_2 \tag{19.17.2}$$

where (x_1, x_2) denote *the observable variables*, whereas (v_1, v_2) are *unobservable disturbances* which are assumed to be normally and independently distributed with zero means and variances equal to (σ_1^2, σ_2^2). Also in this case regression analysis does not lead to unbiased estimates of β or α, as might be thought from a consideration of just one equation. The joint distribution of (x_1, x_2) is generated from the distribution of (v_1, v_2) by means of the transformation given by the equation system, which may also be written as

---

[1]) See T. Haavelmo: *The Statistical Implications of a System of Simultaneous Equations*, Econometrica, 11, 1943, 1—12.

and

$$x_1 = (v_1+\beta v_2)/(1-\alpha\beta)$$
$$x_2 = (v_2+\alpha v_1)/(1-\alpha\beta)\ . \qquad (19.17.3)$$

It follows that (x_1, x_2) are normally distributed with zero means and variances and covariance equal to

$$\sigma_{x_1}^2 = (\sigma_1^2+\beta^2\sigma_2^2)/(1-\alpha\beta)^2\ , \qquad (19.17.4)$$

$$\sigma_{x_2}^2 = (\sigma_2^2+\alpha^2\sigma_1^2)/(1-\alpha\beta)^2\ , \qquad (19.17.5)$$

and

$$\sigma_{x_1x_2} = (\alpha\sigma_1^2+\beta\sigma_2^2)/(1-\alpha\beta)^2 \qquad (19.17.6)$$

so that

$$\beta_{x_1|x_2} = \frac{\beta\sigma_2^2+\alpha\sigma_1^2}{\sigma_2^2+\alpha^2\sigma_1^2} = \beta + \frac{\alpha\sigma_1^2(1-\alpha\beta)}{\sigma_2^2+\alpha^2\sigma_1^2}\ . \qquad (19.17.7)$$

Hence, regression analysis with x_1 as dependent and x_2 as independent variable will give an unbiased estimate of $\beta_{x_1|x_2}$, which quantity, however, usually deviates from the required parameter β. We thus see that regression analysis may lead to wrong results when applied to a single equation which forms part of a system of equations. Estimates of the parameters of the equation system may be found by the method of maximum likelihood, see the references in § 20.9.

In connection with systems of equations, and also in other cases, the problem of *identifiability* of parameters comes up. Even if the distribution function of the observable variables is known completely, it may be impossible to determine (identify) all the parameters in the equations which generate the observed variables.

The above remarks indicate a number of problems and difficulties encountered in the analysis of two-dimensional observations. It is always necessary to analyze the generation of the observations very carefully to be able to construct a model of the relation between the variables and choose the adequate statistical technique for handling the data.

CHAPTER 20.

MULTI-DIMENSIONAL CORRELATION AND REGRESSION

20.1. Partial Correlation Coefficients.

The theory of the two-dimensional normal distribution plays a fundamental role in the extension of the theory of normal correlation to several variables, as this is based on the correlation coefficients between variables taken two at a time. We will first consider the distribution of the three variables (x_1, x_2, x_3), which is characterized by a distribution function analogous to (19.2.9). This distribution function includes the following parameters: the mean values $\mathcal{M}\{x_i\}=\xi_i$, $i=1, 2, 3$, the variances $\mathcal{V}\{x_i\} = \mathcal{M}\{(x_i-\xi_i)^2\} = \sigma_i^2$, $i = 1, 2, 3$, and the correlation coefficients

$$\varrho_{ij} = \mathcal{M}\left\{\frac{x_i-\xi_i}{\sigma_i}\frac{x_j-\xi_j}{\sigma_j}\right\}, \quad i \neq j, \quad (i, j) = 1, 2, 3 . \tag{20.1.1}$$

The three marginal distributions, $p\{x_i\}$, $i = 1, 2, 3$, are normal distributions with parameters (ξ_i, σ_i^2), and the three two-dimensional marginal distributions, $p\{x_i, x_j\}$ for $i \neq j$, are two-dimensional normal distributions with parameters $(\xi_i, \xi_j, \sigma_i^2, \sigma_j^2, \varrho_{ij})$. From this it follows that estimates of the parameters in the three-dimensional distribution are determined from the rules given in Chapter 19.

Let there be, say, n stochastically independent, three-dimensional observational results

$$(x_{1\nu}, x_{2\nu}, x_{3\nu}) \quad \text{for} \quad \nu = 1, 2, \ldots, n . \tag{20.1.2}$$

From these we determine the following estimates of the parameters:

$$\bar{x}_i = \frac{1}{n}\sum_{\nu=1}^{n} x_{i\nu} \approx \xi_i, \quad i = 1, 2, 3, \tag{20.1.3}$$

$$s_i^2 = \frac{1}{n-1}\sum_{\nu=1}^{n} (x_{i\nu}-\bar{x}_i)^2 \approx \sigma_i^2, \quad i = 1, 2, 3, \tag{20.1.4}$$

$$s_{ij} = \frac{1}{n-1}\sum_{\nu=1}^{n} (x_{i\nu}-\bar{x}_i)(x_{j\nu}-\bar{x}_j) \approx \varrho_{ij}\sigma_i\sigma_j, \quad i \neq j, \quad (i, j) = 1, 2, 3, \tag{20.1.5}$$

and

$$r_{ij} = \frac{s_{ij}}{s_i s_j} \approx \varrho_{ij},\ i \neq j,\ (i, j) = 1, 2, 3\ . \qquad (20.1.6)$$

As mentioned in § 19.13, p. 613, the total (or marginal) correlation coefficient ϱ_{ij} may be spurious; for example a positive correlation between x_i and x_j may be due to the fact that both these variables are dependent on a third variable x_k, while for every fixed value of x_k the two first variables are stochastically independent. The influence of the variable x_k upon the correlation between x_i and x_j may be studied by examining the conditional distribution of (x_i, x_j) for given x_k. If the three variables are normally correlated this conditional distribution is completely determined by the parameters in the three two-dimensional marginal distributions.

Consider the distribution $p\{x_2, x_3|x_1\}$ which denotes *the distribution of the two variables* (x_2, x_3) *for a given value of the third variable* x_1. It may be shown that this distribution is a two-dimensional normal distribution with correlation coefficient

$$\varrho_{23\cdot 1} = \frac{\varrho_{23} - \varrho_{21}\varrho_{31}}{\sqrt{(1-\varrho_{21}^2)(1-\varrho_{31}^2)}}\ . \qquad (20.1.7)$$

Thus we find that the correlation coefficient $\varrho_{23\cdot 1}$, which is called the *partial* (or conditional) *correlation coefficient of* (x_2, x_3) *after elimination of* x_1, is independent of x_1 and may be expressed solely by means of the three total correlation coefficients ϱ_{12}, ϱ_{13}, and ϱ_{23}.

The elimination of that part of the variation of x_2 and x_3 which may be "ascribed" to x_1 may be demonstrated as follows: According to § 19.7 the means of x_2 and x_3 for given x_1 are

$$\mathcal{M}\{x_2|x_1\} = \xi_2 + \beta_{21}(x_1 - \xi_1) \qquad (20.1.8)$$

and

$$\mathcal{M}\{x_3|x_1\} = \xi_3 + \beta_{31}(x_1 - \xi_1)\ , \qquad (20.1.9)$$

where β_{21} and β_{31} denote the regression coefficients. If we consider the *deviations* between the variables x_2 and x_3 and these means, we obtain

$$x_{2\cdot 1} = x_2 - \mathcal{M}\{x_2|x_1\} = x_2 - \xi_2 - \beta_{21}(x_1 - \xi_1) \qquad (20.1.10)$$

and

$$x_{3\cdot 1} = x_3 - \mathcal{M}\{x_3|x_1\} = x_3 - \xi_3 - \beta_{31}(x_1 - \xi_1)\ . \qquad (20.1.11)$$

The two variables $(x_{2\cdot 1}, x_{3\cdot 1})$ *which denote the variation of* x_2 *and* x_3 *after "eliminating" the variation due to* x_1 *are normally correlated with correlation coefficient* $\varrho_{23\cdot 1}$. (This is easily seen by determining $\mathcal{M}\{x_{2\cdot 1}\, x_{3\cdot 1}\}$).

The expression for the partial correlation coefficient of (x_i, x_j), after eliminating x_k, is

$$\boxed{\varrho_{ij\cdot k} = \frac{\varrho_{ij} - \varrho_{ik}\varrho_{jk}}{\sqrt{(1-\varrho_{ik}^2)(1-\varrho_{jk}^2)}}\ .} \qquad (20.1.12)$$

As mentioned above, it follows from this expression that even though $\varrho_{ij\cdot k} = 0$, i. e., x_i and x_j are stochastically independent for every fixed value of x_k, ϱ_{ij} will differ from zero if both ϱ_{ik} and ϱ_{jk} differ from zero. Conversely we see that, even though $\varrho_{ij} = 0$, $\varrho_{ij\cdot k}$ may differ from zero.

For four normally correlated variables (x_i, x_j, x_k, x_m) we may, besides the partial correlation coefficients of the first order as expressed in (20.1.12), determine partial correlation coefficients of the second order, e. g., $\varrho_{ij\cdot km}$, which denotes the correlation coefficient of (x_i, x_j) for given (x_k, x_m). *Partial correlation coefficients of the second order are calculated by successive elimination of the two variables by means of the formula* (20.1.12), which is first applied to the total correlation coefficients and then to the partial correlation coefficients of the first order. For instance $\varrho_{ij\cdot km}$ is determined by first eliminating x_k, which according to (20.1.12) gives us $\varrho_{ij\cdot k}$, $\varrho_{im\cdot k}$, and $\varrho_{jm\cdot k}$, and then eliminating x_m, whence we obtain

$$\varrho_{ij\cdot km} = \frac{\varrho_{ij\cdot k} - \varrho_{im\cdot k}\varrho_{jm\cdot k}}{\sqrt{(1-\varrho^2_{im\cdot k})(1-\varrho^2_{jm\cdot k})}}. \tag{20.1.13}$$

If elimination is carried out in the opposite order, we obtain

$$\boxed{\varrho_{ij\cdot mk} = \frac{\varrho_{ij\cdot m} - \varrho_{ik\cdot m}\varrho_{jk\cdot m}}{\sqrt{(1-\varrho^2_{ik\cdot m})(1-\varrho^2_{jk\cdot m})}} = \varrho_{ij\cdot km}\,.} \tag{20.1.14}$$

The elimination of several variables follows from the same principles.

From (20.1.14) it is seen that the interpretation of a partial correlation coefficient of the first order involves the same difficulties as the interpretation of a total correlation coefficient, and that we will meet with the same difficulties regarding partial correlation coefficients of any order. The importance which may be attributed to a correlation coefficient—apart from its purely descriptive value—depends on professional knowledge concerning the generation of the relationship between the variables. Correlation coefficients may therefore be dangerous tools for the analysis of observed data, as they may lead to confusion between stochastical and causal interrelationships and thereby to false conclusions.

Estimates of partial correlation coefficients are determined by substituting the estimates r_{ij}, $(i, j) = 1, 2, \ldots,$ for the theoretical values ϱ_{ij}, $(i, j) = 1, 2, \ldots$; e. g., from (20.1.12) we obtain

$$r_{ij\cdot k} = \frac{r_{ij} - r_{ik}r_{jk}}{\sqrt{(1-r^2_{ik})(1-r^2_{jk})}}. \tag{20.1.15}$$

The results regarding the distribution of the estimate r in § 19.12 apply similarly to the partial correlation coefficients, with the modification that

n should be replaced by $n-p$, where p denotes the order of the correlation coefficient in question, i. e., the number of variables eliminated.

Thus we see that, *if the observations are normally correlated*, the correlation between any two of the variables may be studied by successive eliminations of the other variables *without it being necessary to group the observations according to the magnitudes of the variables eliminated*. As mentioned above this is due to the fact that the conditional correlation coefficient of (x_i, x_j) for given (x_k, x_m) is constant, i. e., independent of (x_k, x_m). This will, however, usually not be the case if the variables are not normally correlated, so that for this kind of data the importance of correlation coefficients is extremely dubious.

Regression analysis in connection with normal correlation may be looked upon as a special case of the regression analysis discussed in the following paragraphs, and will therefore not be dealt with separately.

For the generalization of the remaining theorems in Chapter 19: transformations of variables, contour ellipses and the T^2-test, the reader is referred to H. CRAMÉR: *Mathematical Methods of Statistics*, Princeton, 1946, Chapters 24 and 29.

20.2. Linear Regression Analysis with Two Independent Variables.

Linear regression analysis with two independent variables is a generalization of the regression analysis discussed in Chapter 18. The two independent variables will be denoted by (x_1, x_2), and the dependent variable by y. Let the observed data be given as n sets of numbers $(x_{1\nu}, x_{2\nu}, y_\nu)$ for $\nu = 1, 2, \ldots, n$.

The assumptions underlying the regression analysis are as follows:

1. For every fixed value of (x_1, x_2), y is normally distributed.

2. The mean value of y is a linear function of x_1 and x_2. This function may be written

$$\mathcal{M}\{y|x_1, x_2\} = \eta = \alpha+\beta_1(x_1-\bar{x}_1)+\beta_2(x_2-\bar{x}_2)\ , \qquad (20.2.1)$$

where $\bar{x}_1$ and $\bar{x}_2$ denote the means of the n observations of (x_1, x_2).

3. The variance of y is independent of (x_1, x_2),

$$\mathcal{V}\{y|x_1, x_2\} = \sigma^2\ , \qquad (20.2.2)$$

or proportional to a given function of (x_1, x_2), i. e.,

$$\mathcal{V}\{y|x_1, x_2\} = \sigma^2 h^2(x_1, x_2)\ . \qquad (20.2.3)$$

4. The observations are stochastically independent, i. e., the values $(x_{1\nu}, x_{2\nu}, y_\nu)$ are stochastically independent of $(x_{1\mu}, x_{2\mu}, y_\mu)$ for $\nu \neq \mu$.

Estimates of the parameters α, β_1, β_2 and σ^2 are to be determined from the observations.

If the observations are plotted as points in a (x_1, x_2, y)-coordinate system the assumptions mean that the points are distributed about the *plane* represented by (20.2.1), *the vertical distances between the points and the plane,* i. e., $y_\nu - \eta_\nu$, *being normally distributed about zero with variance* σ^2.

The following exposition is based on (20.2.2) while the extension to (20.2.3) will not be dealt with here, as it is analogous with the theory given in § 18.6. Further it is assumed that the observations are ungrouped, whence the variation within sets is not analyzed.

The sum of squares of the deviations of the observations from the theoretical values, $\sum_{\nu=1}^{n} (y_\nu - \eta_\nu)^2$, is distributed as $\sigma^2\chi^2$ with n degrees of freedom. The estimates of the parameters α, β_1, and β_2 are determined by the method of least squares, i. e., the sum of squares of the deviations between the observed values and the empirical regression plane is minimized. If the empirical regression equation is written

$$\boxed{Y = a + b_1(x_1 - \bar{x}_1) + b_2(x_2 - \bar{x}_2)\,,} \tag{20.2.4}$$

the estimates a, b_1, and b_2 are determined so that the sum of squares

$$\sum_{\nu=1}^{n} (y_\nu - Y_\nu)^2 = \sum_{\nu=1}^{n} (y_\nu - a - b_1(x_{1\nu} - \bar{x}_1) - b_2(x_{2\nu} - \bar{x}_2))^2 \tag{20.2.5}$$

takes on the least possible value, i. e., so that the derivatives of this sum of squares with respect to a, b_1, and b_2 are zero. In this manner we obtain the following three equations for the determination of the three unknown quantities:

$$-2 \sum_{\nu=1}^{n} (y_\nu - a - b_1(x_{1\nu} - \bar{x}_1) - b_2(x_{2\nu} - \bar{x}_2)) = 0\,, \tag{20.2.6}$$

$$-2 \sum_{\nu=1}^{n} (y_\nu - a - b_1(x_{1\nu} - \bar{x}_1) - b_2(x_{2\nu} - \bar{x}_2))(x_{1\nu} - \bar{x}_1) = 0\,, \tag{20.2.7}$$

and

$$-2 \sum_{\nu=1}^{n} (y_\nu - a - b_1(x_{1\nu} - \bar{x}_1) - b_2(x_{2\nu} - \bar{x}_2))(x_{2\nu} - \bar{x}_2) = 0\,. \tag{20.2.8}$$

From the first of these equations we obtain, as $\Sigma(x_{1\nu} - \bar{x}_1) = \Sigma(x_{2\nu} - \bar{x}_2) = 0$,

$$\boxed{a = \bar{y} = \frac{1}{n} \sum_{\nu=1}^{n} y_\nu\,.} \tag{20.2.9}$$

Inserting this result in the two last equations we obtain, after a simple reduction:

$$\begin{aligned} b_1\Sigma(x_{1\nu}-\bar{x}_1)^2+b_2\Sigma(x_{1\nu}-\bar{x}_1)(x_{2\nu}-\bar{x}_2) &= \Sigma(x_{1\nu}-\bar{x}_1)(y_\nu-\bar{y}) \\ b_1\Sigma(x_{1\nu}-\bar{x}_1)(x_{2\nu}-\bar{x}_2)+b_2\Sigma(x_{2\nu}-\bar{x}_2)^2 &= \Sigma(x_{2\nu}-\bar{x}_2)(y_\nu-\bar{y}). \end{aligned} \qquad (20.2.10)$$

If we introduce the notation from § 4.4 and § 4.8 the equations may be written

$$\boxed{\begin{aligned} b_1 SSD_{x_1}+b_2 SPD_{x_1x_2} &= SPD_{x_1y} \\ b_1 SPD_{x_1x_2}+b_2 SSD_{x_2} &= SPD_{x_2y}. \end{aligned}} \qquad (20.2.11)$$

The estimates b_1 and b_2 may be determined from these equations if the determinant

$$D_{x_1x_2} = SSD_{x_1}SSD_{x_2}-(SPD_{x_1x_2})^2 \qquad (20.2.12)$$

of the equations differs from zero.

The three estimates a, b_1, and b_2 are linear functions of the y's whence, according to the addition theorem for the normal distribution, they are normally distributed. The more detailed investigation of the distribution of the estimates is best made by applying the partition theorem for the χ^2-distribution. The deviation $y_\nu-\eta_\nu$ is partitioned, by introducing the empirical regression equation, into a sum of four deviations:

$$\begin{aligned} y_\nu-\eta_\nu &= (y_\nu-Y_\nu)+(Y_\nu-\eta_\nu) \\ &= (y_\nu-Y_\nu)+(a-\alpha)+(b_1-\beta_1)(x_{1\nu}-\bar{x}_1)+(b_2-\beta_2)(x_{2\nu}-\bar{x}_2). \end{aligned} \qquad (20.2.13)$$

Squaring and summing gives

$$\begin{aligned} \sum_{\nu=1}^{n}(y_\nu-\eta_\nu)^2 &= \sum_{\nu=1}^{n}(y_\nu-Y_\nu)^2+n(a-\alpha)^2+(b_1-\beta_1)^2\sum_{\nu=1}^{n}(x_{1\nu}-\bar{x}_1)^2 \\ &+2(b_1-\beta_1)(b_2-\beta_2)\sum_{\nu=1}^{n}(x_{1\nu}-\bar{x}_1)(x_{2\nu}-\bar{x}_2)+(b_2-\beta_2)^2\sum_{\nu=1}^{n}(x_{2\nu}-\bar{x}_2)^2, \end{aligned} \qquad (20.2.14)$$

the remaining terms disappearing on account of the three equations (20.2.6)–(20.2.8), which may be written

$$\sum_{\nu=1}^{n}(y_\nu-Y_\nu) = 0, \qquad (20.2.15)$$

$$\sum_{\nu=1}^{n}(y_\nu-Y_\nu)(x_{1\nu}-\bar{x}_1) = 0, \qquad (20.2.16)$$

and

$$\sum_{\nu=1}^{n}(y_\nu-Y_\nu)(x_{2\nu}-\bar{x}_2) = 0, \qquad (20.2.17)$$

and the relations $\Sigma(x_{1\nu}-\bar{x}_1) = \Sigma(x_{2\nu}-\bar{x}_2) = 0$. The number of degrees of freedom for the first sum of squares in (20.2.14) is $n-3$, as the deviations $y_\nu - Y_\nu$ satisfy the three linear relations (20.2.15)–(20.2.17). The number of degrees of freedom for the term $n(a-\alpha)^2$ is one. The last three terms constitute a quadratic form in the two variables (b_1, b_2) of the same type as (19.9.2) for (x_1, x_2) and therefore have 2 degrees of freedom. As the conditions for application of the partition theorem are thus satisfied, (20.2.14) directly leads to the following results:

1. The variance

$$s^2 = \frac{1}{n-3}\sum_{\nu=1}^{n}(y_\nu - Y_\nu)^2, \tag{20.2.18}$$

which is based on the deviations of the observations from the empirical regression values, is distributed as s^2 with parameters $(\sigma^2, n-3)$.

2. The estimate a is normally distributed with mean value α and variance σ^2/n.

3. The estimates (b_1, b_2) are normally correlated with mean values (β_1, β_2), variances

$$\mathcal{V}\{b_1\} = \sigma^2 \frac{SSD_{x_2}}{D_{x_1x_2}} \tag{20.2.19}$$

and

$$\mathcal{V}\{b_2\} = \sigma^2 \frac{SSD_{x_1}}{D_{x_1x_2}} \tag{20.2.20}$$

and the covariance

$$\mathcal{V}\{b_1, b_2\} = -\sigma^2 \frac{SPD_{x_1x_2}}{D_{x_1x_2}}. \tag{20.2.21}$$

This result is seen by comparing the expression

$$\frac{1}{\sigma^2}\left((b_1-\beta_1)^2 SSD_{x_1} + 2(b_1-\beta_1)(b_2-\beta_2)SPD_{x_1x_2} + (b_2-\beta_2)^2 SSD_{x_2}\right), \tag{20.2.22}$$

which according to (20.2.14) is distributed as χ^2 with 2 degrees of freedom, with the analogous expression (19.9.2) written as follows:

$$\varkappa_{11}(x_1-\xi_1)^2 + 2\varkappa_{12}(x_1-\xi_1)(x_2-\xi_2) + \varkappa_{22}(x_2-\xi_2)^2, \tag{20.2.23}$$

where

$$\varkappa_{ii} = \frac{1}{(1-\varrho^2)\sigma_i^2}, \quad i = 1, 2, \tag{20.2.24}$$

and

$$\varkappa_{12} = \frac{-\varrho}{(1-\varrho^2)\sigma_1\sigma_2}. \tag{20.2.25}$$

From (20.2.19)–(20.2.21) it follows that the correlation coefficient for (b_1, b_2) is

$$\varrho_{b_1b_2} = -\frac{SPD_{x_1x_2}}{\sqrt{SSD_{x_1}SSD_{x_2}}}. \tag{20.2.26}$$

4. The estimates s^2, a, and (b_1, b_2) are stochastically independent.

The estimate of the mean value of y

$$Y = a+b_1(x_1-\bar{x}_1)+b_2(x_2-\bar{x}_2) \tag{20.2.27}$$

is a linear function of the estimates a, b_1, and b_2, and consequently this estimate is also normally distributed with mean value $\eta = \alpha+\beta_1(x_1-\bar{x}_1)+\beta_2(x_2-\bar{x}_2)$ and variance

$$\begin{aligned}\mathcal{V}\{Y\} &= \mathcal{V}\{a\}+\mathcal{V}\{b_1\}(x_1-\bar{x}_1)^2+2\mathcal{V}\{b_1, b_2\}(x_1-\bar{x}_1)(x_2-\bar{x}_2)+\mathcal{V}\{b_2\}(x_2-\bar{x}_2)^2\\ &= \sigma^2\left(\frac{1}{n}+\frac{1}{D_{x_1x_2}}(SSD_{x_2}(x_1-\bar{x}_1)^2-2SPD_{x_1x_2}(x_1-\bar{x}_1)(x_2-\bar{x}_2)\right.\\ &\qquad\left.+SSD_{x_1}(x_2-\bar{x}_2)^2\right). \end{aligned} \tag{20.2.28}$$

Tests of significance may be based on these results, which are analogous to the results in § 18.3, in the same manner as in § 18.4. A hypothetical value of β_1 is for instance tested by calculating

$$\boxed{t = \frac{b_1-\beta_1}{s_{b_1}},\quad s_{b_1} = s\sqrt{\frac{SSD_{x_2}}{D_{x_1x_2}}},\quad f = n-3,} \tag{20.2.29}$$

see (18.4.2). This test is a marginal test. A joint test for hypothetical values of (β_1, β_2) where the correlation between b_1 and b_2 is taken into consideration may be obtained by applying (20.2.22), which is stochastically independent of (20.2.18), whence the quantity

$$v^2 = \frac{1}{2s^2}((b_1-\beta_1)^2SSD_{x_1}+2(b_1-\beta_1)(b_2-\beta_2)SPD_{x_1x_2}+(b_2-\beta_2)^2SSD_{x_2}) \tag{20.2.30}$$

is distributed as v^2 with $(2, n-3)$ degrees of freedom.

The hypothesis $(\beta_1, \beta_2) = (0, 0)$, i. e., $\mathcal{M}\{y\} = \alpha$, leads to

$$v^2 = \frac{1}{2s^2}(b_1SPD_{x_1y}+b_2SPD_{x_2y}) \tag{20.2.31}$$

since

$$b_1^2SSD_{x_1}+2b_1b_2SPD_{x_1x_2}+b_2^2SSD_{x_2} = b_1SPD_{x_1y}+b_2SPD_{x_2y}. \tag{20.2.32}$$

The meaning of this test may be elucidated as follows: By means of the regression values Y_ν the deviations $y_\nu-\bar{y}$ are partitioned into a sum of two deviations

$$y_\nu - \bar{y} = (y_\nu - Y_\nu) + (Y_\nu - \bar{y}) \,, \tag{20.2.33}$$

the first of which refers to the deviations of the observations from the regression values, i. e., the part of the variation of y that cannot be determined from the corresponding values of (x_1, x_2), and the second part of which refers to the deviations of the regression values from the total mean. Squaring and adding we obtain

$$\sum_{\nu=1}^{n} (y_\nu - \bar{y})^2 = \sum_{\nu=1}^{n} (y_\nu - Y_\nu)^2 + \sum_{\nu=1}^{n} (Y_\nu - \bar{y})^2$$

$$= \sum_{\nu=1}^{n} (y_\nu - Y_\nu)^2 + b_1 SPD_{x_1y} + b_2 SPD_{x_2y} \,, \tag{20.2.34}$$

see (20.2.11) and (20.2.15)–(20.2.17). These results may be arranged as in Table 20.1.

TABLE 20.1.

Variation	Sum of squares	Degrees of freedom	Variance	Test
Between regression values	$b_1 SPD_{x_1y} + b_2 SPD_{x_2y}$	2	s_2^2	$v^2 = \frac{s_2^2}{s_1^2}$
Residual	$\sum_{\nu=1}^{n} (y_\nu - Y_\nu)^2$	$n-3$	s_1^2	
Total	$\sum_{\nu=1}^{n} (y_\nu - \bar{y})^2$	$n-1$		

We thus see that the test (20.2.31), which is identical with the analysis of variance in Table 20.1, is based on the partitioning of the total variation of the y's into the part which may be determined from the (x_1, x_2)-values, and the residual variation which is due to other factors, the corresponding variances being compared by means of the v^2-test.

The remarks made in Chapter 18 regarding the interpretation and applications of regression analysis, e. g., regarding the determination of confidence limits, repeated application of an empirical regression equation for the prediction of y-values corresponding to given values of x, and extrapolation, apply also to regression with two independent variables. On p. 536 it has been emphasized how important it is that the range of the variation of the independent variable is as large as possible, or in other words that SSD_x is as large as possible, because in this case the regression line is determined with the greatest precision. A similar problem arises when we are dealing with regression analysis with two independent variables, the determinant $D_{x_1x_2}$, see (20.2.12), being of similar importance as SSD_x when dealing with one independent variable. We thus see that it is not sufficient that x_1 and x_2 each varies over a large range, which leads to large values of SSD_{x_1} and SSD_{x_2}; in addition we require that $SPD_{x_1x_2}$ is small in comparison with SSD_{x_1} and SSD_{x_2}, i. e., that as far as possible x_1 and x_2 vary independently. If this condition is not fulfilled but—in order to consider a limiting case—the n values of (x_1, x_2) are linearly dependent

$$x_{2\nu} = \varkappa+\lambda x_{1\nu}\,, \quad \nu = 1, 2, \ldots, n\,, \tag{20.2.35}$$

we obtain

$$x_{2\nu}-\bar{x}_2 = \lambda(x_{1\nu}-\bar{x}_1)\,,$$

so that

$$SSD_{x_2} = \lambda^2 SSD_{x_1}\,,$$

and

$$SPD_{x_1x_2} = \lambda SSD_{x_1}\,,$$

from which it follows that

$$D_{x_1x_2} = SSD_{x_1}SSD_{x_2}-SPD^2_{x_1x_2} = 0\,.$$

Thus the two equations (20.2.11) for the determination of b_1 and b_2 may be reduced to one equation

$$b_1+\lambda b_2 = \frac{SPD_{x_1y}}{SSD_{x_1}}, \tag{20.2.36}$$

which has an infinite number of solutions. This is also obvious if we consider the problem from a geometrical point of view since the observations $(x_{1\nu}, x_{2\nu}, y_\nu)$, $\nu = 1, 2, \ldots, n$, are situated in a plane perpendicular to the (x_1, x_2)-plane and intersecting this plane in the line $x_2 = \varkappa+\lambda x_1$, so that the y-values are normally distributed about the line in space which forms the intersection between the plane $\eta = \alpha+\beta_1(x_1-\bar{x}_1)+\beta_2(x_2-\bar{x}_2)$ and the vertical plane through $x_2 = \varkappa+\lambda x_1$. On account of the relationship between x_1 and x_2 the observations only furnish us with information concerning a *line* in the regression plane, and consequently it is impossible to determine an estimate of the regression *plane*. This also applies if the linear relationship between the n values of (x_1, x_2) is only approximate. Then the determinant $D_{x_1x_2}$ is not exactly equal to zero, and therefore regression analysis with two independent variables will *formally* lead to an equation similar to (20.2.27). The relative values of the coefficients of x_1 and x_2 are, however, of no importance since in reality there is only one independent variable, wherefore the regression plane is unstable.

If the observations originate in experiments it is often possible to plan the experiments so that x_1 and x_2 vary nearly independently. If it is not possible to choose the (x_1, x_2) values at will, but they themselves represent stochastic variables, it is important to have the correlation between x_1 and x_2 in mind, since as shown above the relation between x_1 and x_2 may be so pronounced that determination of the regression plane becomes impossible.

It is often possible to obtain a deeper understanding of the data if regression analysis with two independent variables is carried out as a successive analysis, the first step being to determine the two one-dimensional regression equations:

$$Y_1 = \bar{y}+b_1'(x_1-\bar{x}_1) \tag{20.2.37}$$

and

$$Y_2 = \bar{y}+b_2'(x_2-\bar{x}_2) \tag{20.2.38}$$

and then compare them with

$$Y = \bar{y}+b_1(x_1-\bar{x}_1)+b_2(x_2-\bar{x}_2)\,. \tag{20.2.39}$$

Even if b_1' and b_2' differ significantly from zero the correlation between x_1 and x_2 may cause such uncertainty in the determination of the regression plane that neither b_1 nor b_2 differs significantly from zero. Successive regression analysis may be applied to more than two variables, cf. § 20.3.

When applying regression analysis in order to find the relationship between several variables in cases where, according to our professional knowledge, the variation of the independent variables must be regarded as the *cause* of the variations of the dependent variable it is often of interest to evaluate the relative importance of the different factors as regards the variation of the dependent variable. Comparison of the magnitude of the regression coefficients may be supplemented by an examination of the changes in the variance of y when introducing the independent variables successively. The addition of a variable whose variation greatly influences the variation of the dependent variable will cause a decrease in the variance of y. Such an evaluation of the importance of different factors as regards variations in y is impeded by the correlation between the regression coefficients, and it is further subject to an uncertainty which cannot be expressed statistically. This uncertainty is due to the fact that addition of one more (not observed) variable may completely change the previous conclusions because the b's are altered. It is therefore necessary to have so *much professional knowledge about the subject that all important factors can be included in the investigation.* This state of affairs may be ensured with a considerable degree of certainty in experimentation, but only with great difficulty in other cases.

It is possible to check the question whether the assumptions necessary for a regression analysis have been fulfilled for the data in hand in the following manner: The empirical regression value

$$Y_\nu = a+b_1(x_{1\nu}-\bar{x}_1)+b_2(x_{2\nu}-\bar{x}_2), \quad \nu = 1, 2, \ldots, n\,, \tag{20.2.40}$$

is computed for each of the n observed values of (x_1, x_2), together with the deviation $y_\nu - Y_\nu$ and the deviation divided by the estimated standard deviation $(y_\nu - Y_\nu)/s$. The n points $(x_{1\nu}, x_{2\nu})$, $\nu = 1, 2, \ldots, n$, are plotted in a (x_1, x_2)-coordinate system, and beside each point the corresponding standardized deviation $(y_\nu - Y_\nu)/s$ is entered. From our basic assumptions the standardized deviations should approximately be normally distributed with parameters (0, 1), and this should apply to every sub-space of the (x_1, x_2)-plane. The graphical representation gives us a survey of the variation in magnitude and in signs of the quantities $(y_\nu - Y_\nu)/s$ as a function of (x_1, x_2) from which systematic deviations from the assumptions may be detected.

Example 20.1 shows the computations for a regression analysis with two independent variables. In analogy with (18.3.40) we introduce

$$SSD_{y|x_1, x_2} = \sum_{\nu=1}^{n} (y_\nu - Y_\nu)^2, \; f = n-3. \tag{20.2.41}$$

According to (20.2.34) this sum of squares is most easily computed by means of the formula

$$\boxed{SSD_{y|x_1, x_2} = SSD_y - b_1 SPD_{x_1 y} - b_2 SPD_{x_2 y}.} \tag{20.2.42}$$

As mentioned in § 18.1 and § 19.17 regression analysis is not always the most adequate statistical method for the representation of the relationship between several variables. Regression analysis with two independent variables presupposes that the observations are distributed about a *plane* in space. When this is not the case another model and corresponding statistical technique must be applied, see the references in § 20.9.

Example 20.1. An experimental investigation on the heat evolved during the hardening of Portland cement[1] gave the results shown in Table 20.2. The figures state the relationship between the amounts of tricalcium aluminate, $3CaO \cdot Al_2O_3$, and tricalcium silicate, $3CaO \cdot SiO_2$, contained in the clinkers from which the cement was produced, and the heat evolved. The heat evolved, y, after 180 days' curing is given in calories per gram of cement, and the weights of the two components as percentages of the weight of the clinkers. (Besides the two above-mentioned components, two others, namely $4CaO \cdot Al_2O_3 \cdot Fe_2O_3$ and $2CaO \cdot SiO_2$, will be included in the regression analysis in Example 20.2, p. 645).

---

[1]) H. Woods, H. H. Steinour and H. R. Starke: *Effect of Composition of Portland Cement on Heat Evolved during Hardening*, Industrial and Engineering Chemistry, 24, 1932, 1207–14, Table I. Reprinted by permission.

TABLE 20.2.

Linear Regression Analysis with Two Independent Variables.
Relationship between $3CaO \cdot Al_2O_3$ (x_1) and $3CaO \cdot SiO_2$ (x_2) in % of the weight of the clinkers and heat evolved (y) in calories per gram of cement during the hardening of Portland cement.

	x_1	x_2	y	x_1+x_2+y	Y	$(y-Y)/s$
	7	26	78·5	111·5	80·1	−0·7
	1	29	74·3	104·3	73·3	0·4
	11	56	104·3	171·3	105·8	−0·6
	11	31	87·6	129·6	89·3	−0·7
	7	52	95·9	154·9	97·3	−0·6
	11	55	109·2	175·2	105·1	1·7
	3	71	102·7	176·7	104·0	−0·5
	1	31	72·5	104·5	74·6	−0·9
	2	54	93·1	149·1	91·3	0·7
	21	47	115·9	183·9	114·5	0·6
	1	40	83·8	124·8	80·5	1·4
	11	66	113·3	190·3	112·4	0·4
$n = 13$	10	68	109·4	187·4	112·3	−1·2
S	97	626	1240·5	1963·5	1240·5	0·0
SS	1139	33050	121088·09	309240·69		
S^2/n	723·77	30144·31	118372·33	296564·02		
SSD	415·23	2905·69	2715·76	12676·67		
		x_1x_2	x_1y			
SP		4922	10032·0	Check:		
S_iS_j/n		4670·92	9256·04	12676·66		
SPD		251·08	775·96	=415·23		
				+2905·69		
			x_2y	+2715·76		
SP			62027·8	+2×251·08		
S_iS_j/n			59734·85	+2×775·96		
SPD			2292·95	+2×2292·95		
S/n	7·462	48·154	95·423	151·038		

$$\begin{aligned} 415{\cdot}23b_1 + 251{\cdot}08b_2 &= 775{\cdot}96 \\ 251{\cdot}08b_1 + 2905{\cdot}69b_2 &= 2292{\cdot}95 \end{aligned}, \text{ see (20.2.11).}$$

$$D_{x_1x_2} = 415{\cdot}23 \times 2905{\cdot}69 - 251{\cdot}08^2 = 1143488, \text{ see (20.2.12).}$$

$$b_1 = 1{\cdot}46830. \quad b_2 = 0{\cdot}662249.$$

Substitution in the equations leads to 775·96 and 2292·95, respectively.

$$SSD_{y|x_1,x_2} = 2715{\cdot}76 - 1{\cdot}46830 \times 775{\cdot}96 - 0{\cdot}662249 \times 2292{\cdot}95 = 57{\cdot}91,$$

$$f = 13 - 3 = 10, \text{ see (20.2.42).}$$

$$s^2 = 5{\cdot}791, \quad s = 2{\cdot}41, \text{ see (20.2.18).}$$

$$\begin{aligned} Y &= 95{\cdot}423 + 1{\cdot}4683(x_1 - 7{\cdot}462) + 0{\cdot}66225(x_2 - 48{\cdot}154) \\ &= 52{\cdot}58 + 1{\cdot}468x_1 + 0{\cdot}6622x_2, \text{ see (20.2.4).} \end{aligned}$$

$$s_a^2 = 5{\cdot}791/13 = 0{\cdot}4455, \quad s_a = 0{\cdot}667.$$

$$s_{b_1}^2 = 5{\cdot}791 \times 0{\cdot}002541 = 0{\cdot}01471, \quad s_{b_1} = 0{\cdot}121, \text{ see (20.2.19).}$$

$$s_{b_2}^2 = 5{\cdot}791 \times 0{\cdot}0003631 = 0{\cdot}002103, \quad s_{b_2} = 0{\cdot}0459, \text{ see (20.2.20).}$$

$$s_{b_1b_2} = -5{\cdot}791 \times 0{\cdot}0002196 = -0{\cdot}001272, \text{ see (20.2.21).}$$

Table 20.2 demonstrates the regression analysis in which the heat evolved is represented as a linear function of the two components:

$$Y = 52{\cdot}58 + 1{\cdot}468x_1 + 0{\cdot}6622x_2$$

with a standard deviation about the regression plane equal to $s = 2{\cdot}41$ calories per gram of cement. The table shows the computations, and references to the formulas employed have been added. The column giving x_1+x_2+y has been included for the sake of checking the computations of the sums, sums of squares, and sums of products, see (4.8.7)–(4.8.10).

The total variation of the observed y-values leads to the sum of squares $SSD_y = 2715{\cdot}76$. The variation of the corresponding values of Y as computed from the observed values of (x_1, x_2) leads to a sum of squares which may be computed as $b_1 SPD_{x_1 y} + b_2 SPD_{x_2 y} = 2657{\cdot}85$, and therefore the sum of squares $SSD_{y|x_1, x_2} = 2715{\cdot}76 - 2657{\cdot}85 = 57{\cdot}91$ indicates the part of the variation of y which must be due to factors other than the two independent variables, i. e., other chemical components and experimental errors. The two variances given in Table 20.1 thus become $s_1^2 = 5{\cdot}791$, $f_1 = 10$, and $s_2^2 = 1328{\cdot}93$, $f_2 = 2$, which leads to rejection of the hypothesis $(\beta_1, \beta_2) = (0, 0)$ since the variance ratio $v^2 = 1328{\cdot}93/5{\cdot}791 = 229{\cdot}5$ is highly significant. Each of the two regression coefficients is also significant, since they are 10–20 times their standard errors, see Table 20.2.

Limits for the theoretical value η may be computed from the empirical value Y and the estimate of the variance

$$s_Y^2 = 0{\cdot}4455 + 0{\cdot}01471(x_1 - 7{\cdot}46)^2 + 0{\cdot}002103(x_2 - 48{\cdot}15)^2$$
$$- 0{\cdot}002544(x_1 - 7{\cdot}46)(x_2 - 48{\cdot}15)\ ,$$

the quantity

$$t = \frac{Y - \eta}{s_Y}$$

having a t-distribution with 10 degrees of freedom.

The two last columns of Table 20.2 give the values as computed from the regression equation and the deviations between the observed and computed values divided by the standard deviation. These results give no reason for doubting our assumption that the regression equation is linear and that the theoretical deviations are normally distributed.

The regression analysis shows that it is possible to calculate the amount of heat evolved fairly precisely from the amounts of the two chemical components in the clinkers. The analysis will be extended in Example 20.2.

20.3. Linear Regression Analysis with m Independent Variables.

Regression analysis with several independent variables is similar to the analysis with two independent variables, but the estimation of the variances and covariances in the distribution of the regression coefficients requires further elucidation.

Let the observations be written as, say, $(x_{1\nu}, x_{2\nu}, \ldots, x_{m\nu}, y_\nu)$, $\nu = 1, 2, \ldots, n$. It is assumed that the dependent variable is normally distributed with mean value

$$\eta = \mathcal{M}\{y|x_1, x_2, \ldots, x_m\} = \alpha+\beta_1(x_1-\bar{x}_1)$$
$$+\beta_2(x_2-\bar{x}_2)+\ldots+\beta_m(x_m-\bar{x}_m) \quad (20.3.1)$$

and variance

$$\mathcal{V}\{y|x_1, x_2, \ldots, x_m\} = \sigma^2. \quad (20.3.2)$$

The estimate of the regression equation is written

$$\boxed{Y = a+b_1(x_1-\bar{x}_1)+b_2(x_2-\bar{x}_2)+\ldots+b_m(x_m-\bar{x}_m).} \quad (20.3.3)$$

According to the method of least squares the estimates of the parameters are determined from the equations

$$\boxed{a = \bar{y}} \quad (20.3.4)$$

and the "normal equations"

$$\boxed{\begin{aligned} b_1 SSD_{x_1} &+b_2 SPD_{x_1x_2} +\ldots+b_m SPD_{x_1x_m} = SPD_{x_1y} \\ b_1 SPD_{x_2x_1} &+b_2 SSD_{x_2} +\ldots+b_m SPD_{x_2x_m} = SPD_{x_2y} \\ &\vdots \\ b_1 SPD_{x_mx_1} &+b_2 SPD_{x_mx_2}+\ldots+b_m SSD_{x_m} = SPD_{x_my}. \end{aligned}} \quad (20.3.5)$$

In analogy with (20.2.13)–(20.2.14) the sum of squares $\Sigma(y_\nu-\eta_\nu)^2$ is partitioned as follows:

$$\sum_{\nu=1}^{n}(y_\nu-\eta_\nu)^2 = \sum_{\nu=1}^{n}(y_\nu-Y_\nu)^2+n(a-\alpha)^2$$
$$+\sum_{i=1}^{m}(b_i-\beta_i)^2 SSD_{x_i}+2\sum_{i=1}^{m-1}\sum_{j=i+1}^{m}(b_i-\beta_i)(b_j-\beta_j)SPD_{x_ix_j}. \quad (20.3.6)$$

Application of the partition theorem for the χ^2-distribution leads to the following results:

1. The variance

$$s^2 = \frac{1}{n-m-1} \sum_{\nu=1}^{n} (y_\nu - Y_\nu)^2 \tag{20.3.7}$$

is distributed as s^2 with parameters $(\sigma^2, n-m-1)$.

2. The estimate a is normally distributed with mean α and variance σ^2/n.

3. The estimates $(b_1, b_2, \ldots, b_m)$ are normally correlated with means $(\beta_1, \beta_2, \ldots, \beta_m)$, variances

$$\mathcal{V}\{b_i\} = c_{ii}\sigma^2, \quad i = 1, 2, \ldots, m, \tag{20.3.8}$$

and covariances

$$\mathcal{V}\{b_i, b_j\} = c_{ij}\sigma^2, \quad i \neq j, \quad (i, j) = 1, 2, \ldots, m, \tag{20.3.9}$$

where the quantities c_{ii} and c_{ij}, the computation of which will be described below, depend solely on the coefficients of the normal equations.

4. The estimates s^2, a, and $(b_1, b_2, \ldots, b_m)$ are stochastically independent. From this it follows, e. g., that

$$\mathcal{V}\{Y\} = \mathcal{V}\{a\} + \sum_{i=1}^{m} \mathcal{V}\{b_i\}(x_i - \bar{x}_i)^2 + 2 \sum_{i=1}^{m-1} \sum_{j=i+1}^{m} \mathcal{V}\{b_i, b_j\}(x_i - \bar{x}_i)(x_j - \bar{x}_j)$$

$$= \sigma^2 \left(\frac{1}{n} + \sum_{i=1}^{m} c_{ii}(x_i - \bar{x}_i)^2 + 2 \sum_{i=1}^{m-1} \sum_{j=i+1}^{m} c_{ij}(x_i - \bar{x}_i)(x_j - \bar{x}_j) \right), \tag{20.3.10}$$

which may be used for the determination of confidence limits for η, since $(Y-\eta)/s_Y$ is distributed as t with $n-m-1$ degrees of freedom, s_Y^2 being determined from (20.3.10) by substituting s^2 for σ^2.

Further, we obtain as a marginal test for any specified hypothetical value of β_i

$$t = \frac{b_i - \beta_i}{s_{b_i}}, \quad s_{b_i} = s\sqrt{c_{ii}}, \quad f = n-m-1, \tag{20.3.11}$$

which is distributed as t with $n-m-1$ degrees of freedom, and as a test for a specified set of values $(\beta_1, \beta_2, \ldots, \beta_m)$ the quantity

$$v^2 = \frac{1}{ms^2} \left[\sum_{i=1}^{m} (b_i - \beta_i)^2 SSD_{x_i} + 2 \sum_{i=1}^{m-1} \sum_{j=i+1}^{m} (b_i - \beta_i)(b_j - \beta_j) SPD_{x_i x_j} \right], \tag{20.3.12}$$

which is distributed as v^2 with $(m, n-m-1)$ degrees of freedom.

Computation of the numerator of s^2 defined by (20.3.7) is made from the formula

$$\begin{aligned} SSD_{y|x_1, x_2, \ldots, x_m} &= \sum_{\nu=1}^{n} (y_\nu - Y_\nu)^2 \\ &= SSD_y - b_1 SPD_{x_1y} - b_2 SPD_{x_2y} - \ldots - b_m SPD_{x_my}, \end{aligned} \tag{20.3.13}$$

which is analogous with (20.2.42).

Definition of c_{ij}.

The quantities $c_{11}, c_{12}, \ldots, c_{1m}$ are defined as follows: In the normal equations (20.3.5) we wish to eliminate the last $(m-1)$ b's so that there remains one equation containing only the unknown b_1. This is done by multiplying the first equation by c_{11}, the second by c_{12}, etc., and adding the m equations, the m quantities, $c_{11}, c_{12}, \ldots, c_{1m}$, being determined in such a manner that in the equation obtained in this way the coefficient of b_1 is equal to 1 and the coefficients of $b_2, \ldots, b_m$ are equal to zero. The equation obtained thus becomes

$$b_1 = c_{11} SPD_{x_1y} + c_{12} SPD_{x_2y} + \ldots + c_{1m} SPD_{x_my}, \tag{20.3.14}$$

and the c's must satisfy the following conditions:

$$\begin{aligned} c_{11} SSD_{x_1} + c_{12} SPD_{x_2x_1} + \ldots + c_{1m} SPD_{x_mx_1} &= 1, \\ c_{11} SPD_{x_1x_2} + c_{12} SSD_{x_2} + \ldots + c_{1m} SPD_{x_mx_2} &= 0, \\ &\vdots \\ c_{11} SPD_{x_1x_m} + c_{12} SPD_{x_2x_m} + \ldots + c_{1m} SSD_{x_m} &= 0, \end{aligned} \tag{20.3.15}$$

i. e., the c's satisfy a system of equations similar to the normal equations for the b's, with the only modification that the quantities on the right-hand side of the equations have been replaced by 1, 0, ..., 0.

If the elimination is continued in this manner we obtain

$$b_i = c_{i1} SPD_{x_1y} + c_{i2} SPD_{x_2y} + \ldots + c_{im} SPD_{x_my}, \tag{20.3.16}$$

where the c's are determined from a modification of the normal equations (20.3.5), the b's being replaced by the c's and the quantities on the right-hand side by 0, 0, ..., 0, 1, 0, ..., 0, the figure 1 occurring in the ith equation.

The quantities c_{ij} are defined in this manner, $(i, j) = 1, 2, \ldots, m$, and it is easily seen that $c_{ij} = c_{ji}$.

Proofs of (20.3.8) *and* (20.3.9).

It follows from (20.3.16) that the b's are linear functions of the y's, since

$$SPD_{x_\mu y} = \sum_{\nu=1}^{n} (x_{\mu\nu}-\bar{x}_\mu)(y_\nu-\bar{y}) = \sum_{\nu=1}^{n} (x_{\mu\nu}-\bar{x}_\mu)y_\nu, \tag{20.3.17}$$

and the c's are independent of the y's. For b_i we obtain

$$b_i = \sum_{\nu=1}^{n} y_\nu(c_{i1}(x_{1\nu}-\bar{x}_1)+c_{i2}(x_{2\nu}-\bar{x}_2)+\ldots+c_{im}(x_{m\nu}-\bar{x}_m)) = \sum_{\nu=1}^{n} y_\nu k_{i\nu}, \tag{20.3.18}$$

where

$$k_{i\nu} = c_{i1}(x_{1\nu}-\bar{x}_1)+c_{i2}(x_{2\nu}-\bar{x}_2)+\ldots+c_{im}(x_{m\nu}-\bar{x}_m)\,. \tag{20.3.19}$$

According to the addition theorem for the normal distribution b_i is normally distributed with variance

$$\mathfrak{V}\{b_i\} = \sum_{\nu=1}^{n} k_{i\nu}^2 \mathfrak{V}\{y_\nu\} = \sigma^2 \sum_{\nu=1}^{n} k_{i\nu}^2\,. \tag{20.3.20}$$

From (20.3.19) we obtain

$$\sum_{\nu=1}^{n} k_{i\nu}^2 = c_{i1} \sum_{\nu=1}^{n} k_{i\nu}(x_{1\nu}-\bar{x}_1)+c_{i2} \sum_{\nu=1}^{n} k_{i\nu}(x_{2\nu}-\bar{x}_2)+\ldots+c_{im} \sum_{\nu=1}^{n} k_{i\nu}(x_{m\nu}-\bar{x}_m)\,. \tag{20.3.21}$$

If we consider a single term of this sum, e. g., the μth term, (20.3.19) leads to

$$\sum_{\nu=1}^{n} k_{i\nu}(x_{\mu\nu}-\bar{x}_\mu) = c_{i1}SPD_{x_1x_\mu}+c_{i2}SPD_{x_2x_\mu}+\ldots$$
$$+c_{im}SPD_{x_mx_\mu} = \begin{cases} 0 & \text{for} \quad \mu \neq i \\ 1 & \text{for} \quad \mu = i \end{cases}, \tag{20.3.22}$$

according to the definition of $c_{i1}, c_{i2}, \ldots, c_{im}$. Inserting this result in (20.3.21) we obtain

$$\sum_{\nu=1}^{n} k_{i\nu}^2 = c_{ii} \sum_{\nu=1}^{n} k_{i\nu}(x_{i\nu}-\bar{x}_i) = c_{ii}\,, \tag{20.3.23}$$

whence—as stated in (20.3.8)—the variance of b_i is equal to

$$\mathfrak{V}\{b_i\} = \sigma^2 c_{ii}\,.$$

The result,

$$\mathfrak{V}\{b_i, b_j\} = \sigma^2 c_{ij}\,,$$

may be proved in a similar manner since by means of (20.3.19) and (20.3.22) we find that

$$\sum_{\nu=1}^{n} k_{i\nu}k_{j\nu} = c_{ij}\,. \tag{20.3.24}$$

If we put $m = 2$ in (20.3.15) we obtain the following equations for the determination of c_{11} and c_{12}:

$$c_{11}SSD_{x_1} + c_{12}SPD_{x_2x_1} = 1$$
$$c_{11}SPD_{x_1x_2} + c_{12}SSD_{x_2} = 0,$$

from which we obtain

$$c_{11} = \frac{SSD_{x_2}}{SSD_{x_1}SSD_{x_2} - (SPD_{x_1x_2})^2}$$

and

$$c_{12} = \frac{-SPD_{x_1x_2}}{SSD_{x_1}SSD_{x_2} - (SPD_{x_1x_2})^2},$$

which is in agreement with (20.2.19) and (20.2.21). Similarly we may determine c_{22} (and c_{21}) from the equations

$$c_{21}SSD_{x_1} + c_{22}SPD_{x_2x_1} = 0$$
$$c_{21}SPD_{x_1x_2} + c_{22}SSD_{x_2} = 1,$$

which gives us $c_{21} = c_{12}$ and

$$c_{22} = \frac{SSD_{x_1}}{SSD_{x_1}SSD_{x_2} - (SPD_{x_1x_2})^2}$$

in agreement with (20.2.20).

Solving the normal equations.

The normal equations can be solved in various manners. The simplest method, which has been stated by GAUSS, is based on successive elimination of the unknowns, the first equation being used for the elimination of the first of the unknowns, b_1, in the remaining equations, so that the original set of equations is reduced to one equation (the first) with m unknowns and a set of $(m-1)$ equations with $(m-1)$ unknowns. In the latter set of equations the first equation is used for the elimination of b_2, and the process is repeated until we have one equation with one unknown. By solving this equation and successive substitution in the reduced equations we obtain the final solution.

As an example we will consider a set of equations with three unknowns

$$a_{11}b_1 + a_{12}b_2 + a_{13}b_3 = a_{10} \qquad (1)$$
$$a_{21}b_1 + a_{22}b_2 + a_{23}b_3 = a_{20} \qquad (2) \qquad (20.3.25)$$
$$a_{31}b_1 + a_{32}b_2 + a_{33}b_3 = a_{30}, \qquad (3)$$

where $a_{ij} = a_{ji}$. The equation (1) is multiplied by $-a_{12}/a_{11}$ and $-a_{13}/a_{11}$, respectively, which leads to

$$-a_{12}b_1 - \frac{a_{12}^2}{a_{11}}b_2 - \frac{a_{12}a_{13}}{a_{11}}b_3 = -\frac{a_{12}a_{10}}{a_{11}} \qquad (1_2) \qquad (20.3.26)$$

and

$$-a_{13}b_1 - \frac{a_{12}a_{13}}{a_{11}}b_2 - \frac{a_{13}^2}{a_{11}}b_3 = -\frac{a_{13}a_{10}}{a_{11}}. \qquad (1_3) \qquad (20.3.27)$$

By adding these equations to equations (2) and (3), respectively, we obtain

$$\left(a_{22} - \frac{a_{12}^2}{a_{11}}\right)b_2 + \left(a_{23} - \frac{a_{12}a_{13}}{a_{11}}\right)b_3 = a_{20} - \frac{a_{12}a_{10}}{a_{11}} \qquad (2')$$

and (20.3.28)

$$\left(a_{32} - \frac{a_{12}a_{13}}{a_{11}}\right)b_2 + \left(a_{33} - \frac{a_{13}^2}{a_{11}}\right)b_3 = a_{30} - \frac{a_{13}a_{10}}{a_{11}}. \qquad (3')$$

This set of equations may be written

$$a_{22\cdot 1}b_2 + a_{23\cdot 1}b_3 = a_{20\cdot 1} \qquad (2')$$

and (20.3.29)

$$a_{32\cdot 1}b_2 + a_{33\cdot 1}b_3 = a_{30\cdot 1}, \qquad (3')$$

$a_{ij\cdot 1}$ denoting that the equations have been obtained by eliminating the unknown b_1 from the original set of equations.

Multiplication of equation (2′) by $-a_{23\cdot 1}/a_{22\cdot 1}$ gives

$$-a_{23\cdot 1}b_2 - \frac{a_{23\cdot 1}^2}{a_{22\cdot 1}}b_3 = -\frac{a_{23\cdot 1}a_{20\cdot 1}}{a_{22\cdot 1}}, \qquad (2_3') \qquad (20.3.30)$$

which added to (3′) leads to

$$\left(a_{33\cdot 1} - \frac{a_{23\cdot 1}^2}{a_{22\cdot 1}}\right)b_3 = a_{30\cdot 1} - \frac{a_{23\cdot 1}a_{20\cdot 1}}{a_{22\cdot 1}}, \qquad (3'') \qquad (20.3.31)$$

which may be written

$$a_{33\cdot 12}b_3 = a_{30\cdot 12}\,. \qquad (3'') \qquad (20.3.32)$$

By solving this equation with respect to b_3 and introducing b_3 in the equation (2′) we obtain b_2, whence b_1 is determined from equation (1) by introducing b_2 and b_3.

These computations may conveniently be arranged as in Table 20.3, which is due originally to DOOLITTLE.

In this scheme the unknowns have been omitted, and use has been made of the symmetry of the set of coefficients. The order of the computations is slightly different from that just considered. First equation (2′) is determined (the equation (5) of the scheme), and then equation (3″) (equation (8) of the scheme) is found as the sum of (3), (1_3), and $(2_3')$, so that the equation (3′) is not found explicitly in the computation scheme.

The determination of the c's may be fitted into this scheme, columns being added which include (1, 0, 0) corresponding to (a_{10}, a_{20}, a_{30}), etc., see Table 20.5.

In order to check the computations during the course of the eliminations, the "transverse sums"

$$\begin{aligned} a_1 &= a_{11}+a_{12}+a_{13}+a_{10}\,, \\ a_2 &= a_{12}+a_{22}+a_{23}+a_{20}\,, \\ a_3 &= a_{13}+a_{23}+a_{33}+a_{30}\,, \end{aligned} \tag{20.3.33}$$

are determined.

TABLE 20.3.

Computational Scheme for Solving the Normal Equations.

(A numerical example is given in Table 20.5, p. 648).

Explanation of computations	b_1	b_2	b_3	
(1)	a_{11}	a_{12}	a_{13}	a_{10}
(2)		a_{22}	a_{23}	a_{20}
(3)			a_{33}	a_{30}
$(4) = (1)\times\left(-\dfrac{a_{12}}{a_{11}}\right)$		$-\dfrac{a_{12}^2}{a_{11}}$	$-\dfrac{a_{12}a_{13}}{a_{11}}$	$-\dfrac{a_{12}a_{10}}{a_{11}}$
$(5) = (2)+(4)$		$a_{22\cdot 1}$	$a_{23\cdot 1}$	$a_{20\cdot 1}$
$(6) = (1)\times\left(-\dfrac{a_{13}}{a_{11}}\right)$			$-\dfrac{a_{13}^2}{a_{11}}$	$-\dfrac{a_{13}a_{10}}{a_{11}}$
$(7) = (5)\times\left(-\dfrac{a_{23\cdot 1}}{a_{22\cdot 1}}\right)$			$-\dfrac{a_{23\cdot 1}^2}{a_{22\cdot 1}}$	$-\dfrac{a_{23\cdot 1}a_{20\cdot 1}}{a_{22\cdot 1}}$
$(8) = (3)+(6)+(7)$			$a_{33\cdot 12}$	$a_{30\cdot 12}$
$(9) =$ Solution of (8)	$b_3 = \dfrac{a_{30\cdot 12}}{a_{33\cdot 12}}$			
$(10) =$ Substitution in (5)	$b_2 = \dfrac{a_{20\cdot 1}-a_{23\cdot 1}b_3}{a_{22\cdot 1}}$			
$(11) =$ Substitution in (1)	$b_1 = \dfrac{a_{10}-a_{13}b_3-a_{12}b_2}{a_{11}}$			
$(12) =$ Substitution in (3) as a check	$a_{13}b_1+a_{23}b_2+a_{33}b_3 = a_{30}$			

These check figures are included in the computations in the same manner as the coefficients and the right-hand sides of the equations, so that for the reduced equations we have the following checks

$$a_{2\cdot1} = a_{22\cdot1}+a_{23\cdot1}+a_{20\cdot1} \tag{20.3.34}$$

and

$$a_{3\cdot12} = a_{33\cdot12}+a_{30\cdot12}\,. \tag{20.3.35}$$

A numerical example with four unknowns has been given in Table 20.5

A discussion of different methods for solving the normal equations may be found in H. JENSEN: *An Attempt at a Systematic Classification of Some Methods for the Solution of Normal Equations*, Geodætisk Institut, Meddelelse No. 18, Copenhagen, 1944, and P. S. DWYER: *Linear Computations*, John Wiley, New York, 1951.

Omission or addition of an independent variable changes the regression coefficients and the c's. A simple method for computing these changes is found in R. A. FISHER: *Statistical Methods*, § 29.1, and in W. G. COCHRAN: *The Omission or Addition of an Independent Variate in Multiple Linear Regression*, Suppl. Journ. Roy. Stat. Soc., 5, 1938, 171–176.

Example 20.2. In the investigation described in Example 20.1 on the heat evolved during the hardening of Portland cement, considered as a function of the chemical composition of the cement, the results entered in Table 20.4 were obtained. The table shows the computation of the sums of squares and products which belong to the normal equations. Table 20.5 includes the normal equations and 4 columns corresponding to the definition of the c's, see (20.3.15); for computational reasons 1 has been replaced by 100, so that all figures are of the same order of magnitude. Finally, the last column gives the sums used for checking, e. g.,

$$-372{\cdot}62-166{\cdot}54+492{\cdot}31+38{\cdot}00-618{\cdot}23+100{\cdot}00 = -527{\cdot}08.$$

The factors given in the first column have been computed as follows:

$$-\frac{a_{12}}{a_{11}} = -\frac{251{\cdot}08}{415{\cdot}23} = -\,0{\cdot}6046769$$

$$-\frac{a_{13}}{a_{11}} = \frac{372{\cdot}62}{415{\cdot}23} = 0{\cdot}8973822$$

$$-\frac{a_{23\cdot1}}{a_{22\cdot1}} = -\frac{58{\cdot}775}{2753{\cdot}868} = -\,0{\cdot}0213427$$

$$-\frac{a_{14}}{a_{11}} = \frac{290{\cdot}00}{415{\cdot}23} = 0{\cdot}6984081$$

$$-\frac{a_{24\cdot 1}}{a_{22\cdot 1}} = \frac{2865\cdot 644}{2753\cdot 868} = 1\cdot 0405887$$

$$-\frac{a_{34\cdot 12}}{a_{33\cdot 12}} = \frac{161\cdot 080}{156\cdot 673} = 1\cdot 028129$$

$$\frac{1}{a_{44\cdot 123}} = \frac{1}{11\cdot 894} = 0\cdot 0840760\,.$$

The remaining computations will be understood from the table and Table 20.3.

As a further check the b's may be computed from the c's and $SPD_{x_i y}$, see (20.3.16) (the disagreement is due to rounding errors). It is usually necessary to carry out the solution of the normal equations with a large number of significant figures, for instance two figures more than wanted in the final results, because rounding errors may accumulate and make the last figures in the b's unreliable.

Table 20.5 shows that the variation between the regression values, $s_2^2 = \frac{1}{4} \times 2667\cdot 84 = 666\cdot 96$, is significantly larger than the residual variation, $s^2 = 5\cdot 99$. A comparison of the four regression coefficients with their standard errors given in the table shows that none of the regression coefficients differ significantly from zero when the t-test is applied, see (20.3.11). This result, that the regression equation on the whole gives a very considerable reduction in the variation of y while the influence of each of the independent variables is not significant, calls for a further investigation of the data.

An examination of the relationship between the variations of the independent variables shows that x_1 and x_3 as well as x_2 and x_4 are practically linearly related except for random variations. This explains the indefiniteness of the regression coefficients, see p. 633. Table 20.6 demonstrates a more detailed analysis of the observations showing the results of a successive regression analysis, starting with one independent variable and ending with the results of Table 20.5.

The first row of the table gives SSD_y and the corresponding variance. The following four rows give the results of the four regression analyses with one independent variable. It is seen that one independent variable causes a comparatively small reduction in the variation of y, i. e., information concerning one of the four components is of no great value as far as computation of the heat evolved is concerned. The next part of the table gives the results of the six regression analyses with two independent variables. The combination (x_1, x_2) is the best, since it causes a very con-

TABLE 20.4.

Computation of sums of squares and products.
Observations: $x_1 = 3CaO \cdot Al_2O_3$, $x_2 = 3CaO \cdot SiO_2$, $x_3 = 4CaO \cdot Al_2O_3 \cdot Fe_2O_3$, and $x_4 = 2CaO \cdot SiO_2$ in % of weight of clinkers.
y = Heat evolved in calories per gram of cement.

	x_1	x_2	x_3	x_4	y	$\Sigma x_i + y$
	7	26	6	60	78·5	177·5
	1	29	15	52	74·3	171·3
	11	56	8	20	104·3	199·3
	11	31	8	47	87·6	184·6
	7	52	6	33	95·9	193·9
	11	55	9	22	109·2	206·2
	3	71	17	6	102·7	199·7
	1	31	22	44	72·5	170·5
	2	54	18	22	93·1	189·1
	21	47	4	26	115·9	213·9
	1	40	23	34	83·8	181·8
	11	66	9	12	113·3	211·3
$n = 13$	10	68	8	12	109·4	207·4
S	97	626	153	390	1240·5	2506·5
SS	1139	33050	2293	15062	121088·09	485939·29
S^2/n	723·77	30144·31	1800·69	11700·00	118372·33	483272·48
SSD	415·23	2905·69	492·31	3362·00	2715·76	2666·81
$SP_{x_1x_i}$		4922	769	2620	10032·00	Check sum:
$S_{x_1}S_{x_i}/n$		4670·92	1141·62	2910·00	9256·04	2666·79
$SPD_{x_1x_i}$		251·08	−372·62	−290·00	775·96	
$SP_{x_2x_i}$			7201	15739	62027·80	
$S_{x_2}S_{x_i}/n$			7367·54	18780·00	59734·85	
$SPD_{x_2x_i}$			−166·54	−3041·00	2292·95	
$SP_{x_3x_i}$				4628	13981·50	
$S_{x_3}S_{x_i}/n$				4590·00	14599·73	
$SPD_{x_3x_i}$				38·00	−618·23	
SP_{x_4y}					34733·30	
$S_{x_4}S_y/n$					37215·00	
SPD_{x_4y}					−2481·70	
S/n	7·46	48·15	11·77	30·00	95·42	192·81

siderable reduction of the variance (from 226·31 to 5·79). Thus, information regarding these two components allows of a very accurate determination

TABLE 20.5. Linear regression analysis with four in

Explanation of computations	b_1	b_2	b_3	b_4
(1)	415·23	251·08	−372·62	−290·00
(2)		2905·69	−166·54	−3041·00
(3)			492·31	38·00
(4)				3362·00
(5) = (1) × (−0·6046769)		−151·822	225·315	175·356
(6) = (2) + (5)		2753·868	58·775	−2865·644
(7) = (1) × 0·8973822			−334·383	−260·241
(8) = (6) × (−0·0213427)			−1·254	61·161
(9) = (3) + (7) + (8)			156·673	−161·080
(10) = (1) × 0·6984081				−202·538
(11) = (6) × 1·0405887				−2981·957
(12) = (9) × 1·028129				−165·611
(13) = (4) + (10) + (11) + (12)				11·894
(14) = (13) × 0·0840760				$b_4 = -0·1441$
Substitution in (9)			$b_3 = 0·1019$	
Substitution in (6)		$b_2 = 0·5101$		
Substitution in (1)	$b_1 = 1·5511$			
Substitution in (4)				
Computation of b from (20.3.16)				
$\Sigma b_i \times SPD_{x_i y}$	2667·84			
$SSD_{y\|x_1, x_2, x_3, x_4}$	2715·76 − 2667·84 = 47·92, $f = 13 - 5 = 8$.			
s^2	47·92/8 = 5·99, $s = 2·45$.			
$s_{b_i} = \sqrt{c_{ii} s^2}$	0·745	0·724	0·755	0·710

of the heat evolved, see Example 20.1. The table further shows that (x_1, x_3) as independent variables give practically the same variance as x_1 alone, i. e., the addition of x_3 is of no value, because x_3 is linearly related to x_1, apart from random variations. The relation between the two regression coefficients is therefore indefinite. In the same manner we see that when we have utilized x_2 the addition of x_4 as an independent variable is of no consequence. The two last parts of the table show that regressions with three or four independent variables are of little more value than the regression which employ only (x_1, x_2) as independent variables.

Thus we see that the data only allow a determination of the heat evolved as a function of two out of the four components given. The best description ndent variables. Solution of the normal equations.

SPD_{xiy}	$100c_{i1}$	$100c_{i2}$	$100c_{i3}$	$100c_{i4}$	Check
775·96	100·00	0·00	0·00	0·00	879·65
2292·95	0·00	100·00	0·00	0·00	2342·18
−618·23	0·00	0·00	100·00	0·00	−527·08
−2481·70	0·00	0·00	0·00	100·00	−2312·70
−469·205	−60·468	0·000	0·000	0·000	−531·904
1823·745	−60·468	100·000	0·000	0·000	1810·276
696·333	89·738	0·000	0·000	0·000	789·382
−38·924	1·291	−2·134	0·000	0·000	−38·636
39·179	91·029	−2·134	100·000	0·000	223·666
541·937	69·841	0·000	0·000	0·000	614·355
1897·768	−62·922	104·059	0·000	0·000	1883·753
40·281	93·590	−2·194	102·813	0·000	229·958
−1·714	100·509	101·865	102·813	100·000	415·366
	$100c_{41}$ = 8·4504	$100c_{42}$ = 8·5644	$100c_{43}$ = 8·6441	$100c_{44}$ = 8·4076	
	$100c_{31}$ = 9·2691	$100c_{32}$ = 8·7917	$100c_{33}$ = 9·5255	$100c_{34}$ = 8·6441	
	$100c_{21}$ = 8·5736	$100c_{22}$ = 8·7607	$100c_{23}$ = 8·7917	$100c_{24}$ = 8·5644	
	$100c_{11}$ = 9·2763	$100c_{12}$ = 8·5736	$100c_{13}$ = 9·2690	$100c_{14}$ = 8·4503	
−2481·63	0·03	−0·04	−0·14	99·90	
	1·5508	0·5105	0·1029	−0·1435	

of the observed y-values is obtained when (x_1, x_2) are used as independent variables.

20.4. Non-Linear Regression with One Independent Variable.

In Chapter 18 we have dealt with the cases of regression analysis with one independent variable where it is possible to write the regression equation in the form

$$\mathcal{M}\{f(y)|g(x)\} = \alpha+\beta(g(x)-\overline{g(x)}), \tag{20.4.1}$$

i. e., the mean value of $f(y)$ is a linear function of $g(x)$. If it is not possible to describe the relationship between the variables in a satisfactory manner

TABLE 20.6.
Successive regression analysis of the observations in Table 20.4.
Regression coefficients and variances.

b_1	b_2	b_3	b_4	SSD	f	s^2
				2715·76	12	226·31
1·87				1265·69	11	115·06
	0·79			906·34	11	82·39
		−1·26		1939·40	11	176·31
			−0·74	883·86	11	80·35
1·47	0·66			57·91	10	5·79
2·31		0·49		1227·06	10	122·71
1·44			−0·61	74·76	10	7·48
	0·73	−1·01		415·45	10	41·54
	0·31		−0·46	868·88	10	86·89
		−1·20	−0·72	175·74	10	17·57
1·70	0·66	0·25		48·12	9	5·35
1·45	0·42		−0·24	47·97	9	5·33
1·05		−0·41	−0·64	50·84	9	5·65
	−0·92	−1·45	−1·56	73·81	9	8·20
1·55	0·51	0·10	−0·14	47·92	8	5·99

by an equation of this type, (20.4.1) may be generalized as follows:

$$\mathcal{M}\{f(y)|x\} = \alpha+\beta_1\big(g_1(x)-\overline{g_1(x)}\big)+\ldots+\beta_m\big(g_m(x)-\overline{g_m(x)}\big)\,, \quad (20.4.2)$$

where

$$\overline{g_i(x)} = \frac{1}{n}\sum_{\nu=1}^{n} g_i(x_\nu)\,. \quad (20.4.3)$$

It is seen that *the determination of the estimates of $\alpha, \beta_1, \ldots, \beta_m$ from the method of least squares leads to computations as given in § 20.3, the m functions $g_1(x), g_2(x), \ldots, g_m(x)$ being interpreted as m variables $x_1, x_2, \ldots, x_m$*. Thus, the coefficients of the normal equations are

$$SPD_{x_ix_j} = \sum_{\nu=1}^{n}\big(g_i(x_\nu)-\overline{g_i(x)}\big)\big(g_j(x_\nu)-\overline{g_j(x)}\big)\,. \quad (20.4.4)$$

If we introduce

$$\alpha_0 = \alpha-\beta_1\overline{g_1(x)}-\ldots-\beta_m\overline{g_m(x)}\,, \quad (20.4.5)$$

(20.4.2) may be written

$$\mathcal{M}\{f(y)|x\} = \alpha_0+\beta_1 g_1(x)+\ldots+\beta_m g_m(x)\,. \quad (20.4.6)$$

As a simple special case we may take $g_i(x) = x^i$, which leads to

$$\mathcal{M}\{f(y)|x\} = \alpha_0+\beta_1 x+\beta_2 x^2+\ldots+\beta_m x^m\,, \quad (20.4.7)$$

i. e., the mean value of y or a function of y is represented by a polynomial of the m*th order in x.* If the observed values of x are equidistant, and exactly one (or the same number of) y-values are observed for each value of x, the polynomial may be built up successively by means of *orthogonal* polynomials of increasing order, which has both computational and theoretical advantages, see R. A. FISHER: *Statistical Methods*, § 27, CH. JORDAN: *Approximation and Graduation According to the Principle of Least Squares by Orthogonal Polynomials*, Ann. Math. Stat., 3, 1932, 257–357, and P. LORENZ: *Der Trend*, Vierteljahrshefte zur Konjunkturforschung, Sonderheft 21, Berlin, 1931.

Among other simple examples of functions of the type

$$f(y) = \alpha_0+\beta_1 g_1(x)+\beta_2 g_2(x) \tag{20.4.8}$$

we may mention

$$y = \alpha_0+\beta_1 x+\beta_2 \log x\,, \quad (x > 0)\,, \tag{20.4.9}$$

and

$$y = \alpha_0+\beta_1 x+\beta_2 e^x \tag{20.4.10}$$

and further

$$\log_e y = \alpha_0+\beta_1 x+\beta_2 \log_e x\,, \quad (x > 0,\ y > 0)\,. \tag{20.4.11}$$

Examples of the two first-mentioned functions have been given in Fig. 7.16 and Fig. 7.17, p. 183. Thus, the estimates of the parameters in the transformation function $u = \alpha \log x+\beta x+\gamma$ stated in Example 7.6 may be determined by means of a regression analysis with $(x, \log x)$ as independent variables.

If the equation (20.4.11) is solved with respect to y, we obtain

$$y = \gamma e^{\beta_1 x} x^{\beta_2}\,, \quad \gamma = e^{\alpha_0}\,, \quad (x > 0, y > 0)\,. \tag{20.4.12}$$

For $\beta_1 < 0$ and $\beta_2 > 0$ this function increases in the interval $(0 < x < -\beta_2/\beta_1)$ from $(0, 0)$ to a maximum value for $x = -\beta_2/\beta_1$, and then decreases towards zero for $x \to \infty$, see Example 20.3.

The above representation of the mean value of y or of $f(y)$ as a non-linear function of x presupposes that *the parameters* $(\alpha, \beta_1, \ldots, \beta_m)$ *enter the function linearly*. If this is not the case but the required mean value is represented by the formula

$$\mathcal{M}\{f(y)|x\} = g(x; \beta_1, \beta_2, \ldots, \beta_m)\,, \tag{20.4.13}$$

direct application of the method of least squares will usually lead to equations that are very difficult to solve. According to the method of least squares the estimates $(b_1, \ldots, b_m)$ of $(\beta_1, \ldots, \beta_m)$ are determined by minimizing the sum of squares

$$\sum_{\nu=1}^{n} (f(y_\nu)-g(x_\nu; b_1, \ldots, b_m))^2. \tag{20.4.14}$$

Putting the derivatives of this sum of squares with respect to the b's equal to zero, we obtain the following system of equations for the determination of the b's:

$$\sum_{\nu=1}^{n}(f(y_\nu)-g(x_\nu; b_1,\ldots, b_m))g_i(x_\nu; b_1,\ldots, b_m) = 0\,, \; i = 1,\ldots, m\,, \quad (20.4.15)$$

where

$$g_i(x_\nu; b_1,\ldots, b_m) = \frac{\partial}{\partial b_i} g(x_\nu; b_1,\ldots, b_m)\,. \quad (20.4.16)$$

It is seen that this system of equations takes the simple form (20.3.5) only if the b's enter linearly into $g(x; b_1, \ldots, b_m)$.

General rules cannot be stated regarding the solution of the above m non-linear equations. It is, however, possible to state a general method for *the determination of* $b_1, \ldots, b_m$ *by successive approximation*. Let there be given a set of approximate values, say, $(b_{10}, \ldots, b_{m0})$, which have, e. g., been determined graphically. By expanding $g(x; b_1, \ldots, b_m)$ in a Taylor's series with $(b_{10}, \ldots, b_{m0})$ as origin and discarding terms of the second and higher orders we obtain:

$$g(x; b_1,\ldots, b_m) \simeq g(x; b_{10},\ldots, b_{m0}) + (b_1-b_{10})g_{10}+\ldots+(b_m-b_{m0})g_{m0}\,, \quad (20.4.17)$$

where

$$g_{i0} = g_i(x; b_{10},\ldots, b_{m0})\,. \quad (20.4.18)$$

Through this transcription $g(x; b_1, \ldots, b_m)$ *becomes a linear function of* $((b_1-b_{10}), \ldots, (b_m-b_{m0}))$, so that these quantities may be determined from the normal equations (20.3.5), b_i in this system being replaced by (b_i-b_{i0}), $SPD_{x_ix_j}$ by $SPD_{g_{i0}g_{j0}}$, and SPD_{x_iy} by

$$SPD_{g_{i0}f(y)} = \sum_{\nu=1}^{n}(g_i(x_\nu; b_{10},\ldots, b_{m0})-\bar{g}_{i0})(f(y_\nu)-g(x_\nu; b_{10},\ldots, b_{m0}))\,. \quad (20.4.19)$$

By solving the equations we determine the corrections $(b_1-b_{10}), \ldots, (b_m-b_{m0})$, and then the whole procedure is repeated with the corrected b-values as starting point. If the original values have been fortunately chosen so that the corrections are small, it is unnecessary to repeat the procedure.

As simple examples of functions of the above type we may mention

$$y = \beta_1+\beta_2 x+\beta_3 x^{\beta_4} \quad (20.4.20)$$

and

$$y = \beta_1+\beta_2 x+\beta_3 e^{\beta_4 x}\,. \quad (20.4.21)$$

Example 20.3. Table 20.7 shows the relation between the assimilation of nitrogen and the production of dry matter from oats in experiments where

Chile saltpetre was administered. These experiments were carried out at the Copenhagen Agricultural School in 1924[1]). The table also contains the values computed from the regression equation.

TABLE 20.7.
Assimilation of nitrogen and production of dry matter by oats in experiments with Chile saltpetre.

Assimilation of nitrogen g/vessel	Production of dry matter	
	Observed g/vessel	Calculated g/vessel
0·09	15·1	15·3
0·32	57·3	54·9
0·69	103·3	107·4
1·51	174·6	174·1
2·29	191·5	192·6
3·06	193·2	186·5
3·39	178·7	179·3
3·63	172·3	173·4
3·77	167·5	169·5

A graphical representation of the observations, see Fig. 20.1, indicates that the relation between the two variables may be expressed by the equation

$$\text{Production of dry matter} = \gamma e^{\beta_1 x} x^{\beta_2} ,$$

where x denotes the amount of nitrogen. From this it follows that the logarithm (y) of the production of dry matter is equal to

$$y = \log \gamma + \beta_1 x + \beta_2 \log_e x ,$$

see (20.4.11) and (20.4.12).

Table 20.8 shows the regression analysis with $x_1 = 100x$ and $x_2 = \log_{10} x_1$ as independent variables. From the regression equation

$$Y = 2{\cdot}3429 - 0{\cdot}196908x + 1{\cdot}09112 \log_{10} x$$

we obtain, after transferring to natural logarithms,

$$\text{Production of dry matter} = 220{\cdot}2e^{-0{\cdot}4534x} x^{1{\cdot}0911}$$

The corresponding curve has been plotted in Fig. 20.1. Both the figure and the two last columns of Table 20.8 show that there is a good agreement between the observed and the calculated values.

---

[1]) K. A. BONDORFF: *Landbrugets Jorddyrkning* II, København, 1939, p. 243.
H. C. PLESSING: *Et matematisk Udtryk for Udbyttekurven for Landbrug*, Nordisk Tidsskrift for teknisk Økonomi, 1942, 123–134, Table III.

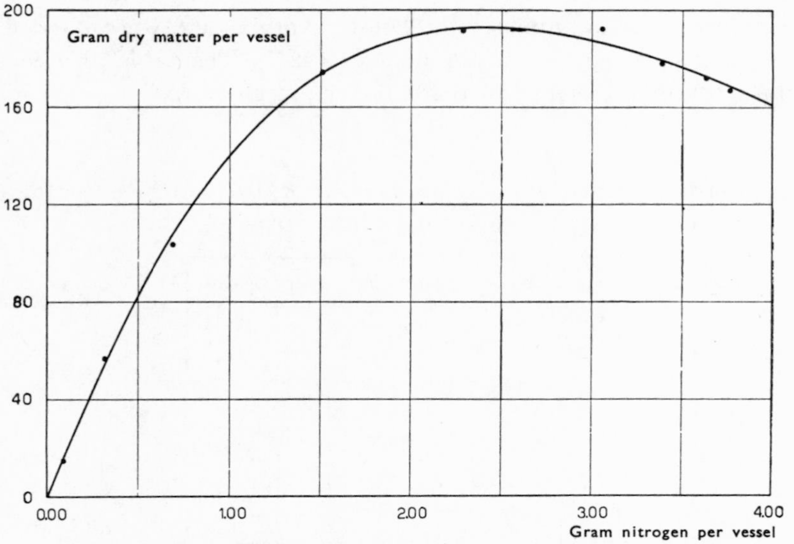

FIG. 20.1. Relation between assimilation of nitrogen and production of dry matter from oats in experiments with Chile saltpetre.

From the regression equation it is possible to compute an estimate of the nitrogen assimilation which results in the greatest production of dry matter. Differentiating the regression equation with respect to x, we obtain

$$\frac{dY}{dx} = b_1 + b_2 \frac{M}{x} = 0 \,, \quad M = 0{\cdot}43429 \,,$$

which leads to

$$x = -\frac{b_2}{b_1} M = \frac{1{\cdot}09112}{0{\cdot}196908} 0{\cdot}43429 = 2{\cdot}407 \quad \text{g/vessel} \,.$$

As the standard errors of b_1 and b_2 are very small in comparison with b_1 and b_2 themselves, the standard error of x may be determined from the approximation formula (9.9.11). This gives us

$$\left(\frac{s_x}{x}\right)^2 \simeq \left(\frac{s_{b_1}}{b_1}\right)^2 + \left(\frac{s_{b_2}}{b_2}\right)^2 - 2\frac{s_{b_1b_2}}{b_1b_2} = 0{\cdot}000557, \quad \frac{s_x}{x} \simeq 0{\cdot}0236 \,,$$

the standard error of x being $s_x \simeq 0{\cdot}0236 \times 2{\cdot}407 = 0{\cdot}0568$ g/vessel. From this we compute the 95% confidence limits for the theoretical value as $2{\cdot}407 - 2{\cdot}447 \times 0{\cdot}0568 = 2{\cdot}27$ g/vessel and $2{\cdot}407 + 2{\cdot}447 \times 0{\cdot}0568 = 2.55$ g/vessel, 2·447 indicating the bilateral 95% limit in the t-distribution with 6 degrees of freedom, see Table IV.

The above approximate determination of the confidence limits for the theoretical "maximum abscissa" ξ may be replaced by an exact determina-

TABLE 20.8.

Regression analysis for the logarithm (y) of the production of dry matter as a function of the nitrogen assimilated (x).

	$x_1 = 100x$	$x_2 = \log(100x)$	y	$0{\cdot}001x_1+x_2+y$	Y	$(y-Y)/s$
	9	0·954	1·179	2·142	1·1838	−0·38
	32	1·505	1·758	3·295	1·7397	1·46
	69	1·839	2·014	3·922	2·0313	−1·38
	151	2·179	2·242	4·572	2·2408	0·10
	229	2·360	2·282	4·871	2·2847	−0·22
	306	2·486	2·286	5·078	2·2706	1·23
	339	2·530	2·252	5·121	2·2536	−0·13
	363	2·560	2·236	5·159	2·2391	−0·25
$n = 9$	377	2·576	2·224	5·177	2·2290	−0·40
S	1875	18·989	18·473	39·337	18·4726	
SS	563563	42·645175	39·014061	180·884433		
S^2/n	390625	40·064680	37·916859	171·933285		
SSD	172938	2·580495	1·097202	8·951148		
		x_1x_2	x_1y			
SP		4571·924	4180·013	Check sum:		
S_iS_j/n		3956·042	3848·542	8·951147		
SPD		615·882	331·471			
			x_2y			
SP			40·578880			
S_iS_j/n			38·975977			
SPD			1·602903			
S/n	208·3	2·1099	2·0526	4·3708		

$$\begin{array}{l} 172938b_1 + 615{\cdot}882b_2 = 331{\cdot}471 \\ 615{\cdot}882b_1 + 2{\cdot}580495b_2 = 1{\cdot}602903 \end{array}, \quad \text{see (20.2.11).}$$

$$D_{x_1x_2} = 172938 \times 2{\cdot}580495 - 615{\cdot}882^2 = 66955{\cdot}01, \quad \text{see (20.2.12).}$$

$$b_1 = -0{\cdot}00196908. \quad b_2 = 1{\cdot}09112.$$

Substitution in the equations gives 331·472 and 1·602909, respectively.

$$SSD_{y|x_1,x_2} = 1{\cdot}097202 + 0{\cdot}00196908 \times 331{\cdot}471 - 1{\cdot}09112 \times 1{\cdot}602903 = 0{\cdot}000935,$$

$$f = 9 - 3 = 6, \quad \text{see (20.2.42).}$$

$$s^2 = 0{\cdot}000156, \quad s = 0{\cdot}0125, \quad \text{see (20.2.18).}$$

$$\begin{aligned} Y &= 2{\cdot}0526 - 0{\cdot}00196908(x_1 - 208{\cdot}3) + 1{\cdot}09112(x_2 - 2{\cdot}1099) \\ &= 2{\cdot}0526 - 0{\cdot}00196908(100x - 208{\cdot}3) + 1{\cdot}09112(\log x + 2 - 2{\cdot}1099) \\ &= 2{\cdot}3429 - 0{\cdot}196908x + 1{\cdot}09112 \log x. \end{aligned}$$

$$\begin{aligned} s_a^2 &= 0{\cdot}000156/9 = 0{\cdot}0000173, \quad s_a = 0{\cdot}00416. \\ s_{b_1}^2 &= 0{\cdot}000156 \times 0{\cdot}00003854 = 0{\cdot}0^8601, \quad s_{b_1} = 0{\cdot}0^4775, \quad \text{see (20.2.19).} \\ s_{b_2}^2 &= 0{\cdot}000156 \times 2{\cdot}583 = 0{\cdot}000403, \quad s_{b_2} = 0{\cdot}0201, \quad \text{see (20.2.20).} \\ s_{b_1b_2} &= -0{\cdot}000156 \times 0{\cdot}009198 = -0{\cdot}0^5143, \quad \text{see (20.2.21).} \end{aligned}$$

tion, the quantity $z = b_1+b_2\frac{M}{\xi}$ being normally distributed with zero mean and variance

$$s_z^2 = s_{b_1}^2+s_{b_2}^2\frac{M^2}{\xi^2}+2s_{b_1b_2}\frac{M}{\xi},$$

so that z/s_z is distributed as t with 6 degrees of freedom. Solution of the inequality $|z|/s_z < 2{\cdot}447$, which leads to an equation of the second degree in ξ, gives us the confidence limits for ξ as 2·28 and 2·56 g/vessel, respectively. In the present case where the coefficients of variation for b_1 and b_2 are small the approximation formula leads to practically the same results as the exact formula.

Above it has been demonstrated how an estimate may be determined of a "maximum abscissa" and its confidence limits for a regression equation of a given form. The uncertainty of the "maximum abscissa", however, besides being dependent on the experimental error also depends on which function is chosen, and it is difficult to evaluate how much the choice of a special function influences the position of the maximum abscissa.

20.5. Non-Linear Regression with *m* Independent Variables.

In analogy with the generalization of the linear regression analysis in § 18.7 and § 20.4 the linear regression with m independent variables may also be generalized, the variables $(x_1, x_2, \ldots, x_m, y)$ being replaced by the new variables $(g_1, g_2, \ldots, g_k, f)$ where

$$g_i = g_i(x_1, x_2, \ldots, x_m)\,, \quad i = 1, 2, \ldots, k\,, \tag{20.5.1}$$

and $f = f(y)$ are *given functions of the directly observed variables.*

Hence, the assumptions for the regression analysis are that for every given value of the independent variables the stochastic variable $f(y)$ is normally distributed with mean value

$$\mathcal{M}\{f(y)|x_1, x_2, \ldots, x_m\} = \alpha+\beta_1(g_1-\bar{g}_1)+\ldots+\beta_k(g_k-\bar{g}_k)\,, \tag{20.5.2}$$

where

$$\bar{g}_i = \frac{1}{n}\sum_{\nu=1}^{n} g_i(x_{1\nu}, x_{2\nu}, \ldots, x_{m\nu})\,,$$

and variance

$$\mathcal{V}\{f(y)|x_1, x_2, \ldots, x_m\} = \sigma^2 h^2(x_1, x_2, \ldots, x_m)\,, \tag{20.5.3}$$

where $h(x_1, x_2, \ldots, x_m)$ is a given function.

The estimates of the β's are determined from k normal equations analogous with (20.3.5), the coefficients of the unknowns taking the form

$$SPD_{g_ig_j} = \sum_{\nu=1}^{n}(g_i(x_{1\nu}, \ldots, x_{m\nu})-\bar{g}_i)(g_j(x_{1\nu}, \ldots, x_{m\nu})-\bar{g}_j)\,,$$

$$(i, j) = 1, 2, \ldots, k\,, \tag{20.5.4}$$

and the quantities on the right-hand side of the equation being

$$SPD_{g_if} = \sum_{\nu=1}^{n} (g_i(x_{1\nu}, \ldots, x_{m\nu}) - \bar{g}_i)(f(y_\nu) - \bar{f}). \qquad (20.5.5)$$

As a special case of (20.5.2) we have for $k = m$ and $g_i = g_i(x_i)$ the following function

$$\mathcal{M}\{f(y)|x_1, \ldots, x_m) = \alpha + \beta_1(g_1(x_1) - \bar{g}_1) + \ldots + \beta_m(g_m(x_m) - \bar{g}_m), \qquad (20.5.6)$$

i. e., each of the m variables in (20.3.1) is replaced by a function of this variable.

Simple examples of functions of the type

$$f(y) = \alpha_0 + \beta_1 g_1(x_1) + \beta_2 g_2(x_2) \qquad (20.5.7)$$

may be found in the same manner as in § 18.7 by substituting x, $1/x$, and $\log x$ for the $g(x)$'s. We obtain, for instance,

$$\log_e y = \alpha_0 + \beta_1 x_1 + \beta_2 \log_e x_2 \qquad (20.5.8)$$

and

$$\log_e y = \alpha_0 + \beta_1 \log_e x_1 + \beta_2 \log_e x_2 \,. \qquad (20.5.9)$$

Solving these equations for y, we find

$$y = \gamma e^{\beta_1 x_1} x_2^{\beta_2}, \quad \gamma = e^{\alpha_0}, \qquad (20.5.10)$$

and

$$y = \gamma x_1^{\beta_1} x_2^{\beta_2}, \quad \gamma = e^{\alpha_0}. \qquad (20.5.11)$$

As a simple example of the application of the general formula (20.5.2) we may mention

$$y = \alpha_0 + \beta_1 x_1 + \beta_2 x_2 + \beta_3 x_1 x_2 \,, \qquad (20.5.12)$$

where $g_1(x_1, x_2) = x_1$, $g_2(x_1, x_2) = x_2$, and $g_3(x_1, x_2) = x_1 x_2$.

If the mean is not a linear function of the parameters, but

$$\mathcal{M}\{f(y)|x_1, \ldots, x_m\} = g(x_1, \ldots x_m; \beta_1, \ldots \beta_k),$$

estimates of the parameters may be determined by successive approximations by a method similar to that described in § 20.4.

20.6. Regression Analysis and the Analysis of Variance.

A comparison between the analysis of variance in Chapter 16 and the regression analysis in Chapters 18 and 20 shows that the assumptions underlying the two methods are partly the same. In the discussion of the analysis of variance in § 16.4 and § 16.7 it was assumed that the observations (which now are denoted by y instead of by x as in Chapter 16) can be written

$$y_{i\nu} = \xi + \eta_i + z_{i\nu} \,, \qquad (20.6.1)$$

and

$$y_{ij\nu} = \xi + \eta_i + \zeta_j + \vartheta_{ij} + z_{ij\nu} , \tag{20.6.2}$$

respectively, i and j indicating the two criteria used for classification, and z being normally distributed with parameters $(0, \sigma^2)$. Similarly it is assumed that the observations in a regression analysis can be written

$$y_\nu = \alpha + \beta\big(g(x_\nu) - \overline{g(x)}\big) + z_\nu , \tag{20.6.3}$$

and

$$y_\nu = \alpha + \beta_1\big(g_1(x_{1\nu}) - \bar{g}_1\big) + \beta_2\big(g_2(x_{2\nu}) - \bar{g}_2\big) + \beta_3\big(g_3(x_{1\nu}, x_{2\nu}) - \bar{g}_3\big) + z_\nu , \tag{20.6.4}$$

respectively, x, x_1, and x_2 indicating independent (quantitative) variables. Thus the parameter ξ in the analysis of variance corresponds to the parameter α in regression analysis, and the parameters η_i, $i = 1, 2, \ldots, k$, correspond to the expression $\beta\big(g(x) - \overline{g(x)}\big)$, which includes only one unknown parameter β, as the function $g(x)$ is supposed to be given. Likewise the interaction of the analysis of variance, the parameters ϑ_{ij}, $i=1, 2, \ldots, k$, and $j = 1, 2, \ldots, m$, correspond to the expression $\beta_3(g_3(x_1, x_2) - \bar{g}_3)$ which consists of the given function of *two* variables, $g_3(x_1, x_2)$, multiplied by the unknown parameter β_3. Thus, contrary to the analysis of variance, the regression analysis demands that the criteria used for classification can be quantitatively specified, so that the dependent variable can be expressed as a known function of the independent variables, which diminishes the number of unknown parameters. The test of the hypothesis $\eta_1 = \eta_2 = \ldots = \eta_k = 0$ which is applied in the analysis of variance is in regression analysis partitioned into two tests, one regarding the form of the regression equation, the other a test of the hypothesis $\beta = 0$, see Table 18.3.

Regression analysis and analysis of variance may both be considered as special cases of the analysis of covariance where the dependent variable is a function of both qualitative classification criteria and quantitative variables. As an example we may mention the specification

$$y_{ij\nu} = \xi + \eta_i + \zeta_j + \beta(x_{ij\nu} - \bar{x}) + z_{ij\nu} . \tag{20.6.5}$$

Estimates of the parameters may be found by the method of least squares.

20.7. Growth Curves.

In the above chapters it has repeatedly been emphasized that the function used to represent the relationship between two variables should as far as possible be chosen on the basis of professional knowledge about the problem under discussion and that the reasons advanced for this choice are of fundamental importance as regards confidence in extrapolations.

As sometimes it is possible to characterize a process by a differential equation we shall here sketch some examples of simple differential equations which have been applied to population statistics and to biological statistics in order to characterize the phenomena of growth. The results may be employed in the discussion of similar processes in chemistry, physics, and economics.

Let x denote time or the magnitude of a growth factor which influences the size y of the phenomenon observed. Then the differential coefficient dy/dx denotes the rate of growth, i. e., the increase per unit of time. It is assumed that the growth process may be characterized by the differential equation

$$\frac{dy}{dx} = f(x, y) ,$$

which indicates that the growth rate depends both on the time (x), and on the size obtained (y). In the following we will deal only with special cases of the type

$$\frac{dy}{dx} = f(y)g(x) , \qquad (20.7.1)$$

which may be written as

$$\frac{dy}{f(y)} = g(x)dx .$$

By integration we obtain

$$F(y) = G(x) .$$

By means of this equation y is determined as a function of x.

Introducing $f(y) = 1, y, \lambda - y$, and $y(\lambda - y)$ in (20.7.1), we obtain the following four differential equations

$$\frac{dy}{dx} = \begin{cases} g(x) , \\ yg(x) , \\ (\lambda - y)g(x) , \quad (0 < y < \lambda) , \\ y(\lambda - y)g(x) , \quad (0 < y < \lambda) . \end{cases} \qquad (20.7.2)$$

In the last two equations, which include λ, the growth is limited, as λ denotes the maximum value of y. The four equations, respectively, indicate that at a given time the growth rate (1) depends on the time, but is independent of the size reached, (2) is proportional to the size reached and to a function of the time, (3) is proportional to the "remaining size", i. e., the maximum size minus the size reached, and a function of the time, (4) is proportional to both the size reached and the remaining size as well as a function of the time.

If the "logarithmic differential coefficient" is introduced

$$\frac{d \log_e y}{dx} = \frac{1}{y}\frac{dy}{dx},$$

which denotes the relative growth rate, i. e., the proportional increase per unit of time, the last three of the above four equations may be written

$$\frac{d \log_e y}{dx} = g(x) ,$$

$$-\frac{d \log_e (\lambda - y)}{dx} = g(x) ,$$

and

$$\frac{d \log_e y}{dx} - \frac{d \log_e (\lambda - y)}{dx} = \lambda g(x) .$$

This gives us the following solution of the four differential equations

$$\begin{aligned} y &= G(x) , \\ \log_e y &= G(x) , \\ \log_e (\lambda - y) &= \log_e \lambda - G(x) , \\ \log_e (\lambda - y) - \log_e y &= -\lambda G(x) , \end{aligned}$$

or if solved for y:

$$\left.\begin{aligned} y &= G(x) , \\ y &= e^{G(x)} , \\ y &= \lambda(1 - e^{-G(x)}) , \\ y &= \frac{\lambda}{1 + e^{-\lambda G(x)}} . \end{aligned}\right\} \quad (20.7.3)$$

These equations include an constant of integration which may be determined from a given value of (x, y).

By introducing special functions for $g(x)$, e. g.,

$$g_1(x) = \beta_0 + \beta_1 x + \beta_2 x^2$$

and

$$g_2(x) = \beta_0 + \frac{\beta_1}{x} + \frac{\beta_2}{x^2} ,$$

we obtain a number of examples of frequently applied growth curves. In the simplest case, $g(x) = \beta$, we obtain the following four functions:

$$\left.\begin{aligned} y &= \alpha + \beta x , \\ y &= \gamma e^{\beta x} , \quad \gamma = e^{\alpha} , \\ y &= \lambda(1 - \gamma e^{-\beta x}) , \quad \gamma = e^{-\alpha} , \\ y &= \frac{\lambda}{1 + \gamma e^{-\lambda \beta x}} , \quad \gamma = e^{-\lambda \alpha} . \end{aligned}\right\} \quad (20.7.4)$$

For $\beta > 0$ these four functions are increasing functions of x, the last two having the asymptote λ.

The growth curve with the equation

$$y = \frac{\lambda}{1+\gamma e^{-\varkappa x}} \tag{20.7.5}$$

is called the *logistic* curve. The course of the curve between the two asymptotes $y = 0$ and $y = \lambda$ is seen in Fig. 18.15, p. 564, which illustrates a curve corresponding to $\varkappa = 1$. The logistic curve has been frequently used to illustrate the growth of "populations" (cells, human populations, telephone subscribers, etc.), the development of business transactions between different countries, education of persons in various manual and mental accomplishments, etc. Regarding most of these applications it may be said that the *theoretical* analysis of the growth process in hand is so uncertain that it is doubtful whether or not the process is governed by a differential equation as (20.7.2), wherefore application of the logistic curve is mainly based on its descriptive properties. The results of the extrapolations regarding population figures, production, etc., which have been carried out on this basis should therefore be regarded with great scepticism.

A special problem arises in the estimation of the parameters of the equations for the growth curves because *the observations usually are not stochastically independent*. The values of, say, successive annual, or five-yearly, population figures will not be stochastically independent. Thus, the assumptions underlying regression analysis usually do not hold. If regression analysis is applied notwithstanding this fact in order to obtain an objective estimate of the parameters, it must be kept in mind that the assumptions for applying the usual standard errors of the estimates and the corresponding tests of significance have not been fulfilled either.

The statistical methods to be used for the handling of observations from growth processes have as yet not been clarified. In certain cases it may be justifiable to consider successive rates of growth or relative rates of growth as stochastically independent so that the constants of the differential equation may be determined by means of regression analysis. For the logistic curve we have

$$\frac{1}{y}\frac{dy}{dx} = \frac{d \log_e y}{dx} = (\lambda - y)\beta\,, \tag{20.7.6}$$

i. e., the relative rate of growth is a linear function of y. (This relation may be utilized to examine whether a given growth curve may be represented by the logistic curve, the observed y values being plotted as abscissas and the corresponding values of $\Delta \log y/\Delta x$ as ordinates). In this case the constant of integration must be determined in some other manner.

For a more detailed discussion of growth curves the reader is referred to the following publications.

Literature on Growth Curves.

R. PEARL: *Studies in Human Biology*, Baltimore, 1927.

— *The Biology of Population Growth*, New York, 1928.

A. J. LOTKA: *Elements of Physical Biology*, Baltimore, 1925.

G. U. YULE: *The Growth of Population and the Factors Which Control It*, Journ. Roy. Stat. Soc., 87, 1925, 1–58.

H. HOTELLING: *Differential Equations Subject to Error, and Population Estimates*, Jour. Amer. Stat. Ass., 22, 1927, 283–314.

E. B. WILSON and R. R. PUFFER: *Least Squares and Laws of Population Growth*, Proc. Amer. Acad. Arts and Sciences, 68, 1933, 285–382.

H. C. PLESSING: *Træk af Telefonens Udvikling*, Nordisk Tidsskrift for teknisk Økonomi, 1942, 81–122.

S. S. KUZNETS: *Secular Movements in Production and Prices*, New York, 1930.

H. S. WILL: *On a General Solution for the Parameters of Any Function with Application to the Theory of Organic Growth*, Ann. Math. Stat., 7, 1936, 165–190.

20.8. Time Series.

By a time series we mean a set of observations, $y(x_1), y(x_2), \ldots, y(x_n), \ldots$, where $x_1 < x_2 < \ldots < x_n < \ldots$ denote time, and $y(x)$ the corresponding observations. As examples of time series we may mention the recording of temperature, barometric pressure, humidity, etc., at regular intervals, recording of the values of various factors in industrial plants, the daily production and sales from the works, electricity used per hour in a given district, daily prices of certain goods, prices on the Stock Exchange, etc.

In the analysis of time series the behaviour of the observations is studied as a function of time. If the observations vary at random with time, see Chapter 13, the observations may be analyzed according to the methods given in the previous chapters. If this is not the case, the observations are stochastically dependent, see § 8.10, and special methods must be employed for the analysis.

The following remarks are only meant to draw attention to some of the methods usually applied to such problems. The existing methods are, however, far from satisfactory, and the whole problem of time series is at present in the melting pot.

The classical analysis of time series interprets a time series $y(x)$ as the sum of several components: a trend $t(x)$ (progressive changes over a long period), a cyclic movement $c(x)$ (an irregular undulating movement, the "business cycles"), a seasonal variation $s(x)$ (a periodical movement with a fixed period), and a random component $z(x)$, i. e.,

$$y(x) = t(x)+c(x)+s(x)+z(x) \,. \tag{20.8.1}$$

The behaviour of these four components is supposed to be determined by different groups of causes, and the time series analysis aims at *decomposing* the given series into its four components as an introduction to a deeper investigation of the causes of the observed variations.

The usual procedure is based on a successive determination of estimates of the seasonal component $s(x)$ and the trend $t(x)$, followed by a determination of the cyclic and random components from the differences

$$c(x)+z(x) = y(x)-t(x)-s(x) . \tag{20.8.2}$$

The seasonal variation is eliminated from a given time series by means of an appropriate moving average. An estimate of the seasonal variation is now determined from the differences between the observed time series and the moving average. After eliminating the seasonal variation an estimate of the trend is made, e. g., by the means of regression analysis, expressing the progressive changes by means of a polynomial in x. The assumptions underlying the regression analysis are, however, seldom fulfilled, since—on account of the cyclic component—the observations are not normally distributed about the trend, and furthermore the observations are not stochastically independent. As stated above, $c(x)+z(x)$ is estimated by means of the "estimates" of $s(x)$ and $t(x)$. This "stationary" time series may then be analyzed by several methods, each of which represents an "explanation" of the cyclic movement. Classical harmonic analysis interprets the cyclic movement as a sum of periodic variations. This point of view has in later years been replaced by a more stochastic point of view, according to which each element of the time series is represented by a function of a certain number of previous elements plus a stochastic variable (the autoregression equation). A cyclic variation may also be generated by the moving average of a random variable (the SLUTZKY-YULE effect). Once again it must be remembered that the model and statistical technique must be chosen in agreement with professional knowledge regarding the "mechanism" which causes the variations. Lastly there exists a method, "*the variate-difference method*", for the determination of the variance in the distribution of the random component.

Investigations of the correlation between two time series also meet with great difficulties and in several cases the results obtained have been misleading (spurious correlation).

Literature on Time Series.

The number of articles published in the journals regarding the problem of time series—particularly concerning economic data—is very large indeed. In the following only some few articles of particular importance will be mentioned together with some modern monographs and text-books which include surveys of the subject and references to the literature.

General Expositions of Methods Applied to Time Series.

H. T. DAVIS: *The Analysis of Economic Time Series*, The Cowles Commission for Research in Economics, Monograph No. 6, Bloomington, Indiana, 1941, 620 pp.

M. G. KENDALL: *The Advanced Theory of Statistics*, London, 1946, Vol. II, Chap. 29–30.

On the Determination of the Trend.

P. LORENZ: *Der Trend*, Vierteljahrshefte zur Konjunkturforschung, Sonderheft 21, Berlin, 1931.

M. SASULY: *Trend Analysis of Statistics*, Washington, 1934.

P. LORENZ: *Annual Survey of Statistical Technique: Trends and Seasonal Variations*, Econometrica, 3, 1935, 456–71.

Methods for the Handling of Stationary Time Series (the cyclic component):

E. SLUTZKY: *The Summation of Random Causes as the Source of Cyclic Processes*, Econometrica, 5, 1937, 105–46.

K. STUMPFF: *Grundlagen und Methoden der Periodenforschung*, Berlin, 1937, 332 pp.

H. WOLD: *A Study in the Analysis of Stationary Time Series*, Uppsala, 1938, 214 pp.

M. G. KENDALL: *Contributions to the Study of Oscillatory Time Series*, Cambridge, 1946, 76 pp.

N. WIENER: *Extrapolation, Interpolation and Smoothing of Stationary Time Series*, New York, 1949, 163 pp.

P. WHITTLE: *Hypothesis Testing in Time Series Analysis*, Uppsala, 1951, 120 pp.

Methods for the Determination of the Seasonal Fluctuations.

O. DONNER: *Die Saisonschwankungen als Problem der Konjunkturforschung*, Vierteljahrshefte zur Konjunkturforschung, Sonderheft 6, Berlin, 1928.

A. WALD: *Berechnung und Ausschaltung von Saisonschwankungen*, Beiträge zur Konjunkturforschung, herausgegeben vom Österreichischen Institut für Konjunkturforschung, Nr. 9, Wien, 1936, 140 pp.

H. MENDERSHAUSEN: *Annual Survey of Statistical Technique: Methods of Computing and Eliminating Changing Seasonal Fluctuations*, Econometrica, 5, 1937, 234–62.

A. HALD: *The Decomposition of a Series of Observations Composed of a Trend, a Periodic Movement and a Stochastic Variable*, Copenhagen, 1948, 134 pp.

Methods for the Determination of the Variance of the Random Component.
O. ANDERSON: *Die Korrelationsrechnung in der Konjunkturforschung*, Frankfurter Gesellschaft für Konjunkturforschung, Heft 4, Bonn, 1929.
G. TINTNER: *The Variate Difference Method*, The Cowles Commission for Research in Economics, Monograph No. 5, Bloomington, Indiana, 1940, 175 pp.

Literature on the Correlation between Time Series:
G. U. YULE: *Why Do We sometimes Get Nonsense-Correlations between Time Series?* Jour. Roy. Stat. Soc., 89, 1926, 1–64.
C. F. ROOS: *Annual Survey of Statistical Technique: The Correlation and Analysis of Time Series*, Econometrica, 4, 1936, 368–81.
G. H. ORCUTT and S. F. JAMES: *Testing the Significance of Correlation between Time Series*, Biometrika, 35, 1948, 397–413.
Furthermore, reference is made to the literature in § 20.9 on confluence analysis and simultaneous equations.

20.9. Literature on Regression, Correlation and Related Subjects.

The method of least squares, which forms the theoretical basis for the determination of estimates of the parameters in the regression equation, was formulated and applied in connection with the solution of statistical problems in astronomy and geodesy by A. M. LEGENDRE, P. S. LAPLACE, and C. F. GAUSS during the first part of the 19th century. An abundant literature on the method of least squares and its application in various fields followed upon these fundamental works. In the latter part of the 19th century F. GALTON applied regression and correlation analysis to the problems of heredity, and as a continuation of this work there appeared the writings of K. PEARSON and G. U. YULE, which have formed the basis of later expositions of regression and correlation analysis. Lastly, R. A. FISHER and his school have derived the distribution functions for the estimates of the parameters and the corresponding tests of significance.

In the following the titles of some important original treatises on the subject are given and reference is made to some more advanced text-book expositions.

YULE's exposition of regression and correlation analysis is found in:
G. U. YULE and M. G. KENDALL: *Theory of Statistics*, Ch. Griffin, London, 1940, Chap. 11–16.

The following are the most important of R. A. FISHER's publications on the subject:

Frequency Distribution of the Values of the Correlation Coefficient in Samples from an Indefinitely Large Population, Biometrika, 10, 1915, 507–21.

On the "Probable Error" of a Coefficient of Correlation Deduced from a Small Sample, Metron, 1, No. 4, 1921, 1–32.

The Goodness of Fit of Regression Formulæ, and the Distribution of Regression Coefficients, Jour. Roy. Stat. Soc., 85, 1922, 597–612.

The Distribution of the Partial Correlation Coefficient, Metron, 3, 1923, 329–32.

Applications of "Student's" Distribution, Metron, 5, 1925, 90–104.

The General Sampling Distribution of the Multiple Correlation Coefficient, Proc. Roy. Soc. London, A, 121, 1928, 654–73.

As an introduction to the more advanced theoretical problems connected with the multi-dimensional normal distribution the reader is referred to: S. S. WILKS: *Mathematical Statistics*, Princeton, 1944, Chap. 11, and M. G. KENDALL: *The Advanced Theory of Statistics*, II, Ch. Griffin, London, 1946, Chap. 28.

An extensive, elementary exposition of regression and correlation analysis with many examples is given in M. EZEKIEL: *Methods of Correlation Analysis*, J. Wiley, New York, 1941.

Some of the most important recent papers on confluence analysis and the theory of simultaneous equations are mentioned below.

R. FRISCH: *Statistical Confluence Analysis by Means of Complete Regression Systems*, Oslo, 1934.

T. KOOPMANS: *Linear Regression Analysis of Economic Time Series*, Netherlands Economic Institute, Haarlem, 1937.

A. WALD: *The Fitting of Straight Lines if Both Variables are Subject to Error*, Ann. Math. Stat., 11, 1940, 284–300.

T. HAAVELMO: *The Statistical Implications of a System of Simultaneous Equations*, Econometrica, 11, 1943, 1–12.

H. B. MANN and A. WALD: *On the Statistical Treatment of Linear Stochastic Difference Equations*, Econometrica, 11, 1943, 173–220.

T. KOOPMANS: *Statistical Estimation of Simultaneous Economic Relations*, Jour. Amer. Stat. Ass., 40, 1945, 448–466.

O. REIERSÖL: *Confluence Analysis by Means of Instrumental Sets of Variables*, Arkiv för Matematik, Astronomi och Fysik, Stockholm, 1945.

G. TINTNER: *Some Applications of Multivariate Analysis to Economic Data*, Jour. Amer. Stat. Ass., 41, 1946, 472–500.

M. S. BARTLETT: *Multivariate Analysis*, Suppl. Jour. Roy. Stat. Soc., 9, 1947, 176–190.

D. V. LINDLEY: *Regression Lines and the Linear Functional Relationship*, Suppl. Jour. Roy. Stat. Soc., 9, 1947, 218–244.

M. A. GIRSCHICK and T. HAAVELMO: *Statistical Analysis of the Demand for Food: Examples of Simultaneous Estimation of Structural Equations*, Econometrica, 15, 1947, 79–110.

R. C. GEARY: *Determination of Linear Relations Between Systematic Parts of Variables with Errors of Observation the Variances of Which are Unknown*, Econometrica, 17, 1949, 30–58.

T. KOOPMANS: *Identification Problems in Economic Model Construction*, Econometrica, 17, 1949, 125–144.

G. H. ORCUTT and D. COCHRANE: *A Sampling Study of the Merits of Autoregressive and Reduced Form Transformations in Regression Analysis*, Jour. Amer. Stat. Ass., 44, 1949, 356–372.

T. W. ANDERSON and H. RUBIN: *Estimation of the Parameters of a Single Equation in a Complete System of Stochastic Equations*, Ann. Math. Stat. 20, 1949, 46–62, and 21, 1950, 570–582.

O. REIERSÖL: *Identifiability of a Linear Relation Between Variables Which are Subject to Error*, Econometrica, 18, 1950, 375–389.

Statistical Inference in Dynamic Economic Models, Edited by T. KOOPMANS, COWLES COMMISSION, Monograph No. 10, J. Wiley, New York, 1950.

Numerous examples of the application of regression, correlation and confluence analysis have been published. Here only some expositions of theoretical importance will be mentioned.

Applications to *economics*: H. SCHULTZ: *The Theory and Measurement of Demand*, Chicago, 1938; J. TINBERGEN: *A Method and Its Application to Investment Activity*, League of Nations, Geneva, 1939; L. R. KLEIN: *Economic Fluctuations in the United States 1921–1941*, Cowles Commission, Monograph No. 11, J. Wiley, New York, 1950.

Applications to *biological standardization*: J. IPSEN: *Contribution to the Theory of Biological Standardization*, Copenhagen, 1941; D. J. FINNEY: *Probit Analysis*, Cambridge, 1947; J. H. BURN: *Biological Standardization*, Oxford, 1950.

Application to the analysis of *costs*: PH. LYLE: *Regression Analysis of Production Costs and Factory Operations*, London, 1946.

A discussion of non-normal, two-dimensional distributions has been given in S. J. PRETORIUS: *Skew Bivariate Frequency Surfaces Examined in the Light of Numerical Illustrations*, Biometrika, 22, 1930, 109–223, and M. J. VAN UVEN: *Extension of Pearson's Probability Distribution to Two Variables*, Koninklijke Nederlandsche Akademie van Wetenschappen, 50, 1947, 477–84 and 578–90, and further 51, 1948, 12–23 and 62–67.

CHAPTER 21.

THE BINOMIAL DISTRIBUTION

21.1. The Distribution Function.

Let the probability of the event U be $P\{U\} = \theta$. If the probability of the event is θ at each observation, irrespective of the outcome of previous observations, then the probability that the event will occur exactly x times in n observations is

$$p_n\{x\} = \binom{n}{x} \theta^x (1-\theta)^{n-x}, \quad x = 0, 1, \ldots, n. \tag{21.1.1}$$

This distribution, the binomial distribution, has been derived in § 2.1 and some elementary examples of its application have been given. In the following a systematic exposition will be given of the properties of the binomial distribution and the corresponding tests of significance.

The distribution function $p\{x\}$ is discontinuous, as it is only defined for $x = 0, 1, \ldots, n$. A survey of the course of the distribution function is obtained by examining the ratio of two successive terms

$$f(x) = \frac{p\{x+1\}}{p\{x\}} = \frac{n-x}{x+1}\frac{\theta}{1-\theta}, \quad x = 0, 1, \ldots, n-1. \tag{21.1.2}$$

Hence $f(x)$ is a decreasing function,

$$f(0) > f(1) > \ldots > f(n-1), \tag{21.1.3}$$

wherefore the course of the distribution function may be characterized as follows:

1. The distribution function is steadily decreasing if $f(x) < 1$ for all x, which leads to

$$f(0) = \frac{n\theta}{1-\theta} < 1 \quad \text{or} \quad \theta < \frac{1}{n+1}. \tag{21.1.4}$$

2. The distribution function is steadily increasing if $f(x) > 1$ for all x, i. e.,

$$f(n-1) = \frac{\theta}{n(1-\theta)} > 1 \quad \text{or} \quad \theta > \frac{n}{n+1}. \tag{21.1.5}$$

3. The distribution function first increases and then decreases if $f(0) > 1 > f(n-1)$, i. e.,

$$\frac{1}{n+1} < \theta < \frac{n}{n+1}. \tag{21.1.6}$$

In the third case the mode, $x = r$, of the distribution may be found from the inequality

$$f(r-1) \geq 1 > f(r), \tag{21.1.7}$$

which is identical with the inequalities $p\{r\} \geq p\{r-1\}$ and $p\{r\} > p\{r+1\}$. Substitution of (21.1.2) into (21.1.7) leads to

$$\frac{n-r+1}{r}\,\frac{\theta}{1-\theta} \geq 1 > \frac{n-r}{r+1}\,\frac{\theta}{1-\theta},$$

which gives

$$(n+1)\theta - 1 < r \leq (n+1)\theta, \tag{21.1.8}$$

i. e., the mode is equal to the largest integer smaller than or equal to $(n+1)\theta$. From this it further follows that $r = n\theta$ if $n\theta$ is an integer, and that if $(n+1)\theta$ is an integer $p\{r-1\} = p\{r\}$. Comparison with cases 1 and 2 shows that the above result is generally valid, as in these cases the mode is 0 and n, respectively. Figs. 2.1 and 2.2, pp. 29 and 31, illustrate two distribution functions of the third type.

As it is simple to tabulate $f(x)$, the distribution function may be tabulated by computing a single value of $p\{x\}$, and thence the other values may be obtained by successive multiplications and divisions by $f(x)$ according to the formulas:

$$p\{x+1\} = p\{x\}f(x) \tag{21.1.9}$$

and

$$p\{x-1\} = \frac{p\{x\}}{f(x-1)}. \tag{21.1.10}$$

The single value of $p\{x\}$ may, e. g., be calculated by means of logarithms, as

$$\log p\{x\} = \log \binom{n}{x} + x \log \theta + (n-x) \log (1-\theta). \tag{21.1.11}$$

For $n \leq 100$ $\log_{10} \binom{n}{x}$ has been tabulated in Table XIV, and for $n > 100$ $\log \binom{n}{x}$ may be computed from

$$\log \binom{n}{x} = \log n! - \log x! - \log (n-x)! \,, \qquad (21.1.12)$$

using Table XIII, which gives $\log_{10} n!$ for $n \leq 1000$. It is usually practical to choose the starting value of $p\{x\}$ near the maximum of the distribution. If the series of values to be computed is long, it is advisable to compute, e. g., every tenth value directly from the formula (21.1.11).

A table of the distribution function to seven decimal places for $\theta = 0{\cdot}01\ (0{\cdot}01)\ 0{\cdot}50$ and $n = 2\ (1)\ 49$ may be found in *Tables of the Binomial Probability Distribution*, National Bureau of Standards, Applied Mathematics Series, 6, Washington, 1950.

Example 21.1. As an example we will compute the binomial distribution for $n = 100$ and $\theta = 0{\cdot}1$, i. e.,

$$p\{x\} = \binom{100}{x} 0{\cdot}1^x \ 0{\cdot}9^{100-x} \,, \quad x = 0,1, \ldots, 100 \,.$$

The mode is $r = n\theta = 10$, and the maximum value is

$$p\{10\} = \binom{100}{10} 0{\cdot}1^{10} \ 0{\cdot}9^{90} \,.$$

From Table XIV we obtain $\log \binom{100}{10} = 13{\cdot}2383$, so that

$$\log p\{10\} = 13{\cdot}2383 - 10{\cdot}0000 + 0{\cdot}8818 - 5 = 0{\cdot}1201 - 1 \,,$$

which gives us $p\{10\} = 0{\cdot}1319$. As we later on want to compare this binomial distribution with a corresponding normal distribution, to four correct decimal places, we will utilize the more accurate value $p\{10\} = 0{\cdot}13187$ instead of $0{\cdot}1319$ as our starting value.

Formula (21.1.2) leads to

$$f(x) = \frac{100-x}{x+1} \frac{1}{9} \,, \quad x = 0, 1, \ldots, 99 \,.$$

Table 21.1 illustrates the computation of $f(x)$ and $p\{x\}$ according to the formulas (21.1.9) and (21.1.10), the starting value being $p\{10\}$. The last column of the table gives the cumulative distribution function $P\{x\} = p\{0\} + p\{1\} + \ldots + p\{x\}$. On account of rounding errors the sum of the probabilities amounts to $1{\cdot}00003$ instead of $1{\cdot}00000$.

Example 21.2. By sampling inspection of the production process mentioned in Example 2.2, p. 28, where the normal proportion defective was $\theta = 2\%$, the number of defective items found in 50 samples each of 100

TABLE 21.1.

Computation of the distribution function $p\{x\}$ and the cumulative distribution function $P\{x\}$ for the binomial distribution with $n=100$ and $\theta=0{\cdot}1$.

x	$f(x)$	$p\{x\}$	$P\{x\}$
0	11·1	·00003	·00003
1	5·500	·00029	·00032
2	3·630	·00162	·00194
3	2·6944	·00589	·00783
4	2·1333	·01588	·02371
5	1·7593	·03387	·05758
6	1·4921	·05958	·11716
7	1·29167	·08890	·20606
8	1·13580	·11483	·32089
9	1·01111	·13042	·45131
10	0·909091	·13187	·58318
11	0·824074	·11988	·70306
12	0·75214	·09879	·80185
13	0·69048	·07430	·87615
14	0·63704	·05130	·92745
15	0·59028	·03268	·96013
16	0·54902	·01929	·97942
17	0·51235	·01059	·99001
18	0·4795	·00543	·99544
19	0·4500	·00260	·99804
20	0·4233	·00117	·99921
21	0·399	·00050	·99971
22	0·377	·00020	·99991
23	0·36	·00008	·99999
24	0·34	·00003	1·00002
25	0·32	·00001	1·00003
26		·00000	1·00003

TABLE 21.2.

Number of defective items in 50 samples of 100 items each.

Sample No.	No. of defectives	Sample No.	No. of defectives	Sample No.	No. of defectives	Sample No.	No. of defectives	Sample No.	No. of defect ves
1	2	11	1	21	2	31	4	41	1
2	0	12	6	22	2	32	0	42	1
3	3	13	1	23	4	33	2	43	3
4	2	14	1	24	2	34	3	44	1
5	2	15	2	25	3	35	1	45	3
6	0	16	0	26	0	36	3	46	4
7	4	17	4	27	1	37	0	47	2
8	3	18	0	28	3	38	1	48	2
9	2	19	1	29	2	39	2	49	0
10	3	20	1	30	4	40	4	50	1

TABLE 21.3.
Distribution of 50 samples according to number of defective items.

Number of defective items	Number of samples	"Expected" number of samples $= 50\,p\{x\}$
0	8	6·6
1	12	13·5
2	13	13·7
3	9	9·1
4	7	4·5
5	0	1·8
6	1	0·6
7	0	0·2
	50	50·0

items was that given in Table 21.2. Enumeration of the number of samples with 0, 1, 2, ... defective items leads to the distribution given in Table 21.3, where the corresponding results according to the binomial distribution for $n = 100$ and $\theta = 0{\cdot}02$ have also been entered for the sake of comparison, see Table 2.2, p. 29, for the values of $p\{x\}$. The agreement between the observed and the theoretical numbers (considered at face value) is quite good. The total percentage of defective items among the 5000 observed items is $100 \times 99/5000 = 1{\cdot}98$.

21.2. The Cumulative Distribution Function and the Fractiles.

The cumulative distribution function $P\{x\}$ is found by summation of the distribution function, the probability of a result smaller than or equal to x being

$$P\{x\} = p\{0\}+p\{1\}+\ldots+p\{x\} = \sum_{\nu=0}^{x} \binom{n}{\nu} \theta^{\nu}(1-\theta)^{n-\nu}, \quad x = 0, 1, \ldots, n. \tag{21.2.1}$$

Table 21.1 shows the cumulative distribution function for $n = 100$ and $\theta = 0{\cdot}1$. As $P\{x\}$ is discontinuous, the equation

$$P\{x\} = P, \quad \text{or} \quad x = x_P, \tag{21.2.2}$$

can be solved only for $P = P\{0\}, P\{1\}, \ldots, P\{n\}$, and for these values of P the solutions are $x_P = 0, 1, \ldots, n$. From Table 21.1 we see, e. g., that $x_{0\cdot451} = 9$, and $x_{0\cdot583} = 10$, while $x_{0\cdot50}$, the 50%-fractile, does not exist.

On account of this discontinuity we see that—contrary to what applies to continuous distributions—the probabilities $P\{x_{P_1} \leq x \leq x_{P_2}\}$ and $P\{x_{P_1} < x \leq x_{P_2}\}$ are not equal since

$$P\{x_{P_1} < x \leq x_{P_2}\} = P\{x_{P_1}+1 \leq x \leq x_{P_2}\} = P\{x_{P_2}\}-P\{x_{P_1}\} = P_2-P_1 . \quad (21.2.3)$$

while

$$P\{x_{P_1} \leq x \leq x_{P_2}\} = P\{x_{P_1} < x \leq x_{P_2}\}+p\{x_{P_1}\} . \quad (21.2.4)$$

Table 21.1 shows, e. g., that

$$P\{5 < x \leq 10\} = P\{10\}-P\{5\} = 0{\cdot}58318-0{\cdot}05758 = 0{\cdot}52560 ,$$

while

$$P\{5 \leq x \leq 10\} = P\{10\}-P\{4\} = 0{\cdot}58318-0{\cdot}02371 = 0{\cdot}55947 .$$

Dividing by n and putting $h = x/n$ we obtain from (21.2.3)

$$P\{h_{P_1} < h \leq h_{P_2}\} = P\left\{h_{P_1}+\frac{1}{n} \leq h \leq h_{P_2}\right\} = P\{h_{P_2}\}-P\{h_{P_1}\} = P_2-P_1 . \quad (21.2.5)$$

In § 21.3 it will be proved that the fractiles x_P of the binomial distribution may be expressed by the fractiles of the v^2-distribution, as

$$P\{x\} = 1-P\left\{v^2 < \frac{n-x}{x+1}\frac{\theta}{1-\theta}\right\} , \quad (21.2.6)$$

where the degrees of freedom for v^2 are $f_1 = 2(x+1)$ and $f_2 = 2(n-x)$. Thus, the equation $P\{x_P\} = P$ is identical with

$$P\left\{v^2 < \frac{n-x_P}{x_P+1}\frac{\theta}{1-\theta}\right\} = 1-P ,$$

and as $P\{v^2 < v^2_{1-P}\} = 1-P$, we directly obtain

$$\frac{n-x_P}{x_P+1}\frac{\theta}{1-\theta} = v^2_{1-P} , \quad \begin{array}{l} f_1 = 2(x_P+1) , \\ f_2 = 2(n-x_P) , \end{array} \quad (21.2.7)$$

or by "solving" the equation with respect to x_P

$$x_P = \frac{n\theta-(1-\theta)v^2_{1-P}}{\theta+(1-\theta)v^2_{1-P}} , \quad \begin{array}{l} f_1 = 2(x_P+1) , \\ f_2 = 2(n-x_P) . \end{array} \quad (21.2.8)$$

Division by n gives

$$h_P = \frac{\theta-(1-\theta)\dfrac{v^2_{1-P}}{n}}{\theta+(1-\theta)v^2_{1-P}} , \quad \begin{array}{l} f_1 = 2(x_P+1) , \\ f_2 = 2(n-x_P) . \end{array} \quad (21.2.9)$$

In § 21.11 these results are utilized for the determination of confidence limits for θ.

In § 21.3 it is further shown that

$$P\{x\} = 1-I_\theta(x+1, n-x)\ ,$$

where $I_\theta(s, t)$ denotes the ratio between the incomplete and the complete Beta-function, which for $s \leq 50$ and $t \leq 50$ have been tabulated in K. PEARSON: *Tables of the Incomplete Beta-function*, Biometrika Office, London, 1934. For $n \leq 49$ and $\theta = 0{\cdot}00\ (0{\cdot}01)\ 1{\cdot}00$ all values of $P\{x\}$ may be directly read from this table, which forms the basis for *Tables of the Binomial Probability Distribution*, National Bureau of Standards, Applied Mathematics Series, 6, Washington, 1950, containing $1-P\{x\}$ to seven decimal places for $\theta = 0{\cdot}01\ (0{\cdot}01)\ 0{\cdot}50$ and $n = 2\ (1)\ 49$.

21.3. The Binomial Distribution and the v^2-Distribution.

When deriving (21.2.6) we consider the integral

$$\int_\theta^1 y^x(1-y)^{n-x-1}dy\ ,$$

which by integration by parts gives

$$\int_\theta^1 y^x(1-y)^{n-x-1}dy = \frac{1}{n-x}\,\theta^x(1-\theta)^{n-x} + \frac{x}{n-x}\int_\theta^1 y^{x-1}(1-y)^{n-x}dy\ . \tag{21.3.1}$$

If reduction of the integral is continued by successive applications of this formula, we obtain, after multiplication by $n\binom{n-1}{x}$,

$$n\binom{n-1}{x}\int_\theta^1 y^x(1-y)^{n-x-1}dy = \sum_{\nu=0}^{x}\binom{n}{\nu}\theta^\nu(1-\theta)^{n-\nu} = P\{x\}\ . \tag{21.3.2}$$

This result may be expressed by the incomplete Beta-function, which according to (14.4.15) may be written as

$$\mathrm{B}_\theta(x+1, n-x) = \int_0^\theta y^x(1-y)^{n-x-1}dy\ .$$

Substitution into (21.3.2) leads to

$$P\{x\} = \frac{\mathrm{B}(x+1, n-x)-\mathrm{B}_\theta(x+1, n-x)}{\mathrm{B}(x+1, n-x)} = 1-I_\theta(x+1, n-x)\ , \tag{21.3.3}$$

see (14.4.14) and (14.4.16).

Furthermore, according to (14.4.17) the cumulative distribution function of the variable

$$y = \frac{f_1 v^2}{f_2+f_1 v^2}$$

is

$$P\{y\} = I_y\left(\frac{f_1}{2}, \frac{f_2}{2}\right),$$

from which we—after comparison with (21.3.3)—see that

$$P\{x\} = 1 - P\{y\} \tag{21.3.4}$$

for

$$y = \frac{f_1 v^2}{f_2 + f_1 v^2} = \theta \tag{21.3.5}$$

and

$$f_1 = 2(x+1)\ , f_2 = 2(n-x)\ . \tag{21.3.6}$$

This result may also be written

$$P\{x\} = 1 - P\left\{\frac{(x+1)v^2}{(n-x)+(x+1)v^2} < \theta\right\} \tag{21.3.7}$$

or as in (21.2.6)

$$P\{x\} = 1 - P\left\{v^2 < \frac{n-x}{x+1}\frac{\theta}{1-\theta}\right\}.$$

21.4. The Mean and the Variance.

The binomial distribution is completely determined by the parameters n and θ, wherefore the mean and the variance are functions of these parameters. According to (5.7.4) the mean of x is

$$\boxed{\mathcal{M}\{x\} = n\theta\ ,} \tag{21.4.1}$$

i. e., the number of observations multiplied by the probability of the occurrence of the event at a single observation. According to (5.9.14) the variance of x is

$$\boxed{\mathcal{V}\{x\} = n\theta(1-\theta)\,.} \tag{21.4.2}$$

For a given value of n the variance takes its maximum value $n/4$ for $\theta = 1-\theta = \frac{1}{2}$ and decreases symmetrically towards zero as θ departs from $^1/_2$ towards 0 or 1.

The mean is proportional to n, while the standard deviation $\sqrt{n\theta(1-\theta)}$ is proportional to $\sqrt{n}$.

The frequency $h = x/n$ denotes the frequency of the event in n observations. From (21.4.1) and (21.4.2) we have

$$\boxed{\mathcal{M}\{h\} = \theta\ ,} \tag{21.4.3}$$

and

$$\boxed{v\{h\} = \frac{\theta(1-\theta)}{n},} \tag{21.4.4}$$

as

$$v\{h\} = v\left\{\frac{x}{n}\right\} = \frac{1}{n^2}v\{x\}.$$

Thus, the mean of the frequency in repeated samples is equal to the probability, and the standard error of the frequency is inversely proportional to the square root of the number of observations.

21.5. The Binomial Distribution and the Normal Distribution.

Table 21.4 contains the values of the distribution function $p\{x\}$ for $\theta = 0{\cdot}1$ and $n = 5$, 10, 20, 50, and 100, and for comparison with the last of these functions a normal distribution with mean $\xi = n\theta = 10$ and standard deviation $\sigma = \sqrt{n\theta(1-\theta)} = \sqrt{9} = 3$. (The computation of this normal distribution will be further explained in the following). It follows from this table that the skewness of the binomial distribution for a given value of θ diminishes as n increases, and that for $n = 100$ and $\theta = 0{\cdot}1$ the normal distribution forms a good approximation to the binomial distribution, see Fig. 21.1 which shows the cumulative distribution func-

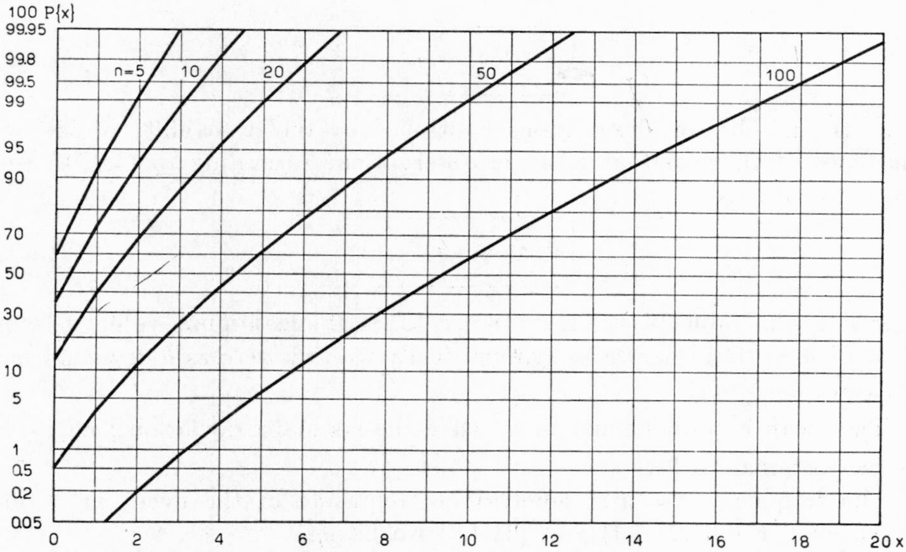

FIG. 21.1. Cumulative distribution functions for the binomial distribution for $\theta = 0{\cdot}1$ and $n = 5, 10, 20, 50$, and 100 plotted on probability paper. The points corresponding to the values of the cumulative distribution function have been connected by straight lines.

TABLE 21.4.

The binomial distribution $p_n\{x\}$ for $\theta = 0{\cdot}1$ and $n = 5, 10, 20, 50$, and 100 and the normal distribution for $\xi = 10$ and $\sigma = 3$.

n $n\theta$ $n\theta(1-\theta)$ $\sqrt{n\theta(1-\theta)}$	5 0·5 0·45 0·67	10 1·0 0·90 0·95	20 2·0 1·80 1·34	50 5·0 4·50 2·12	100 10·0 9·00 3·00	$\xi = 10{\cdot}0$ $\sigma^2 = 9{\cdot}00$ $\sigma = 3{\cdot}00$	Difference
x	$p_5\{x\}$	$p_{10}\{x\}$	$p_{20}\{x\}$	$p_{50}\{x\}$	$p_{100}\{x\}$	$p_N\{x\}$	$p_{100}\{x\}-p_N\{x\}$
0	·5905	·3487	·1216	·0052	·0000	·0005	−·0005
1	·3281	·3874	·2702	·0286	·0003	·0015	−·0012
2	·0729	·1937	·2852	·0779	·0016	·0039	−·0023
3	·0081	·0574	·1901	·1386	·0059	·0089	−·0030
4	·0005	·0112	·0898	·1809	·0159	·0183	−·0024
5		·0015	·0319	·1849	·0339	·0334	·0005
6		·0001	·0089	·1541	·0596	·0549	·0047
7			·0020	·1076	·0889	·0807	·0082
8			·0004	·0643	·1148	·1062	·0086
9			·0001	·0333	·1304	·1253	·0051
10				·0152	·1319	·1324	−·0005
11				·0061	·1199	·1253	−·0054
12				·0022	·0988	·1062	−·0074
13				·0007	·0743	·0807	−·0064
14				·0002	·0513	·0549	−·0036
15				·0001	·0327	·0334	−·0007
16					·0193	·0183	·0010
17					·0106	·0089	·0017
18					·0054	·0039	·0015
19					·0026	·0015	·0011
20					·0012	·0005	·0007
21					·0005	·0002	·0003
22					·0002	·0000	·0002
23					·0001	·0000	·0001
24					·0000	·0000	·0000

tions plotted on normal probability paper. Fig. 21.2 further shows the distribution function for $n = 100$ and $\theta = 0{\cdot}1$ together with the corresponding normal distribution function, $p\{x\}$ for the binomial distribution being represented by rectangles with base 1 and height $p\{x\}$, i. e., area $p\{x\}$.

It may be shown, see p. 681, that *the binomial distribution for a given value of θ and increasing values of n converges to the normal distribution*, wherefore for large values of n the binomial distribution may be approximated by the normal distribution with mean $\xi = n\theta$ and variance $\sigma^2 = n\theta(1-\theta)$, i. e.,

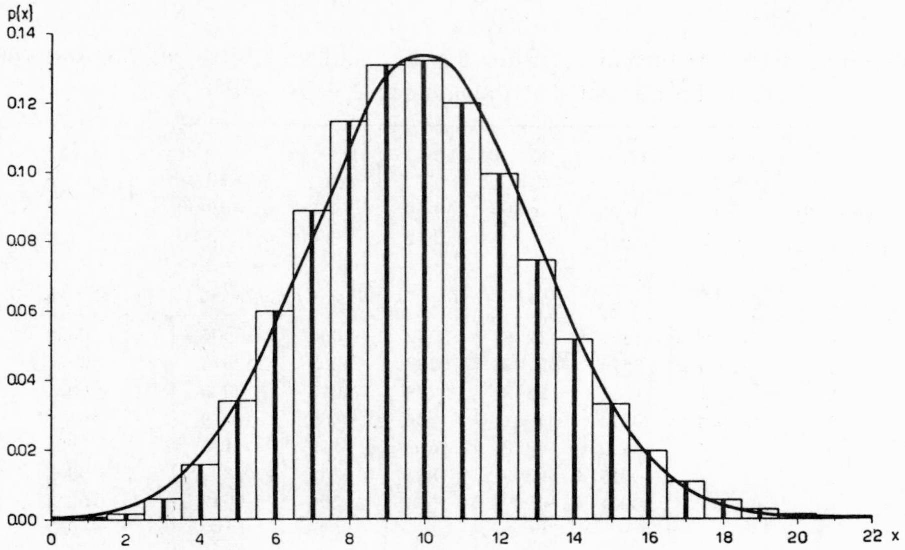

FIG. 21.2. Graphs of the binomial distribution for $n = 100$ and $\theta = 0{\cdot}1$ and the normal distribution for $\xi = 10$ and $\sigma = 3$.

$$p\{x\} = \binom{n}{x}\theta^x(1-\theta)^{n-x} \simeq \frac{1}{\sqrt{2\pi n\theta(1-\theta)}}\, e^{-\frac{(x-n\theta)^2}{2n\theta(1-\theta)}}. \tag{21.5.1}$$

Thus, for large values of n, x is approximately normally distributed with parameters $(n\theta, n\theta(1-\theta))$, wherefore

$$u = \frac{x-n\theta}{\sqrt{n\theta(1-\theta)}} \tag{21.5.2}$$

is approximately normally distributed with parameters (0, 1). Dividing both numerator and denominator by n we obtain

$$u = \frac{h-\theta}{\sqrt{\dfrac{\theta(1-\theta)}{n}}}, \tag{21.5.3}$$

i. e., the frequency h is approximately normally distributed with mean θ and variance $\theta(1-\theta)/n$.

In most applications of the binomial distribution it is not a single value of $p\{x\}$ which is required, but sums of $p\{x\}$ of the form

$$P\{x_1 \le x \le x_2\} = \sum_{x=x_1}^{x_2} p\{x\}.$$

Fig. 21.2 shows that the areas of the rectangles corresponding to this sum may be approximated by the area of the normal distribution from $x_1-\frac{1}{2}$ to $x_2+\frac{1}{2}$, so that

$$P\{x_1 \leq x \leq x_2\} = \sum_{x=x_1}^{x_2} p\{x\} \simeq \frac{1}{\sqrt{2\pi}\,\sigma} \int_{x_1-\frac{1}{2}}^{x_2+\frac{1}{2}} e^{-\frac{(x-\xi)^2}{2\sigma^2}} dx$$

$$= \Phi\left(\frac{x_2+\frac{1}{2}-\xi}{\sigma}\right) - \Phi\left(\frac{x_1-\frac{1}{2}-\xi}{\sigma}\right), \tag{21.5.4}$$

where $\xi = n\theta$ and $\sigma = \sqrt{n\theta(1-\theta)}$. Thus, when we pass from the discontinuous to the continuous distribution the limits of summation (x_1, x_2) must be changed to the limits of integration $(x_1-\frac{1}{2}, x_2+\frac{1}{2})$. This correction, which always appears when we are dealing with approximations of this kind, is denoted *the correction for continuity*. For $x = x_1 = x_2$ we obtain an approximation to $p\{x\}$ slightly different from (21.5.1), viz.

$$p\{x\} \simeq \Phi\left(\frac{x+\frac{1}{2}-\xi}{\sigma}\right) - \Phi\left(\frac{x-\frac{1}{2}-\xi}{\sigma}\right). \tag{21.5.5}$$

The figures in the last but one column of Table 21.4 have been computed from (21.5.5).

For the cumulative distribution function $P\{x\}$ we have

$$\boxed{P\{x\} \simeq \Phi\left(\frac{x+\frac{1}{2}-n\theta}{\sqrt{n\theta(1-\theta)}}\right),} \tag{21.5.6}$$

which leads to

$$\frac{x_P+\frac{1}{2}-n\theta}{\sqrt{n\theta(1-\theta)}} \simeq u_P\,, \tag{21.5.7}$$

or

$$\boxed{x_P \simeq n\theta-\tfrac{1}{2}+u_P\sqrt{n\theta(1-\theta)}\,,} \tag{21.5.8}$$

and by division by n

$$\boxed{h_P \simeq \theta-\frac{1}{2n}+u_P\sqrt{\frac{\theta(1-\theta)}{n}}.} \tag{21.5.9}$$

Thus, also for the determination of fractiles it is necessary to take the correction for continuity into account, the effect of the correction being to diminish the fractile.

An exhaustive examination of the accuracy of the approximation formulas has not yet been made, and we can therefore only give rough rules for the applicability of the formulas.

In cases where the binomial distribution is very skew, as for $\theta < \dfrac{1}{n+1}$ or $\theta > \dfrac{n}{n+1}$ where the distribution function is steadily decreasing or increasing, respectively, the approximation formulas cannot be applied.

As, for $\theta \neq 0{\cdot}5$, the binomial distribution is skew, while the normal distribution is always symmetrical, the approximation (21.5.1) or (21.5.5) has an error which usually carries opposite signs for values of x lying symmetrically about $n\theta$, see the last column of Table 21.4. Even though the approximation to the distribution function on account of this skewness is rather bad, *the approximation* (21.5.4) *to the sum of the probabilities will often be good if summation takes place over an interval which has* $n\theta$ *as mid-point*, as in this case the errors will counteract one another.

In the following the accuracy of the approximation to the cumulative distribution function is examined when *the range of application is limited by the inequality* $n\theta(1-\theta) > 9$. Thus, for a given value of θ we can according to this rule determine the minimum number of observations $n = 9/\theta(1-\theta)$ which allows the application of the normal distribution, see Table 21.5. For values of θ between 0·30 and 0·70 this minimum number of observations amounts to about 40. Outside this interval the number increases as θ approaches 0 or 1, and in the intervals $0{\cdot}0 < \theta < 0{\cdot}1$ and $0{\cdot}9 < \theta < 1{\cdot}0$,

TABLE 21.5.
Values of n and θ computed from the formula $n\theta(1-\theta) = 9$ $(\sqrt{n\theta(1-\theta)} = 3)$.

θ	$1-\theta$	n	$n\theta$	$n(1-\theta)$
0·50	0·50	36	18	18
0·40	0·60	38	15	23
0·30	0·70	43	13	30
0·20	0·80	56	11	45
0·15	0·85	71	11	60
0·10	0·90	100	10	90
0·05	0·95	189	9	180
0·02	0·98	459	9	450
0·01	0·99	909	9	900

respectively, the number is practically inversely proportional to θ and $1-\theta$. Table 21.6 elucidates the accuracy for 5 sets of values of n and θ, the table showing the deviations between the exact and the approximate values of the cumulative distribution function. (For $\theta = 0{\cdot}2$ and $\theta = 0{\cdot}01$ we have used $n = 55$ and $n = 900$ instead of $n = 56$ and $n = 909$, see

Table 21.5, as hereby $n\theta$ becomes an integer, and $n\theta(1-\theta)$ differs only slightly from 9. For $n \to \infty$ and $\theta = 9/n \to 0$ the binomial distribution tends to the POISSON-distribution, see § 21.7). Besides the deviations between the cumulative distribution functions the deviations, Δ, obtained by bilateral summation and bilateral integration, respectively, with $n\theta$ as midpoint have been given. In the last column of the table the probabilities derived by bilateral summation of the POISSON-distribution have further been given; for the other distributions the corresponding probabilities are of the same order of magnitude as for the POISSON-distribution.

Table 21.6 demonstrates that delimitation of the range of applicability by means of the inequality $n\theta(1-\theta) > 9$ does not result in the same accuracy for different corresponding values of n and θ determined from the formula $n\theta(1-\theta) = 9$. For $\theta = 0{\cdot}5$ the deviations are very small, so that the approximation may also be applied for values of n smaller than 36. The deviations increase with increasing values of $|\theta-\frac{1}{2}|$ and converge to the deviations given for the POISSON-distribution. For example we have, for $n = 100$ and $\theta = 0{\cdot}1$, that $P\{x \leq 4\} = 2{\cdot}37\%$, while the approximation formula gives $3{\cdot}34\%$, i. e., a deviation of $0{\cdot}97\%$. For values of $\theta < 0{\cdot}2$ (or larger than 0·8) the deviations between the cumulative distribution functions in the neighbourhood of the values 2·5% and 97·5% are of the magnitude 1%, and in the neighbourhood of 0·5% and 99·5% the deviations are of the magnitude 0·2—0·9%. On account of the skewness, however, there is a considerable difference between the accuracy at the two ends of the distribution, whence the approximate formulas must be applied with care to unilateral summations. In bilateral sums the error for $\theta \neq \frac{1}{2}$ will usually be considerably smaller, which is also seen in Table 21.6. For $n = 100$ and $\theta = 0{\cdot}1$ we find, e. g., that $P\{|x-n\theta| \geq 6\} = P\{x-n\theta \leq -6\} + P\{x-n\theta \geq 6\} = P\{x-n\theta \leq -6\} + 1 - P\{x-n\theta \leq 5\} = 0{\cdot}0237 + 0{\cdot}0399 = 0{\cdot}0636$, while the approximation formula gives 0·0668, i. e., a deviation of $-0{\cdot}0032$. The corresponding figures for the POISSON-distribution are 0·0627, 0·0668 and $-0{\cdot}041$, see the two last columns of Table 21.6. For values of n and θ for which $n\theta(1-\theta)$ is larger than 9, the deviations will be smaller than stated in Table 21.6.

The approximation formula for $P\{x\}$ and $1-P\{x\}$, respectively, can only be applied in the interval $\pm$ 2–3 times the standard deviation about the mean, i. e., for $n\theta - 3\sqrt{n\theta(1-\theta)} < x < n\theta + 3\sqrt{n\theta(1-\theta)}$, as outside this interval the relative error is usually considerable.

Proof of (21.5.1).

The basis of the transcription of the binomial distribution is STIRLING's formula

TABLE

Comparison of the cumulative distribution functions for the

$\Delta = P_B\{|x-n\theta| \geq d\} - P_N\{|x-n\theta| \geq d\}$ where d

n	36	55	100
θ	0·5	0·2	0·1
$n\theta$	18·0	11·0	10·0
$n\theta(1-\theta)$	9·00	8·80	9·00
$\sqrt{n\theta(1-\theta)}$	3·00	2·9665	3·00

$x-n\theta$	$P_B\{x\}$	$P_N\{x\}$	P_B-P_N	Δ	$P_B\{x\}$	$P_N\{x\}$	P_B-P_N	Δ	$P_B\{x\}$	$P_N\{x\}$
−12	·0000	·0001	−·0001							
−11	·0002	·0002	·0000		·0000	·0002	−·0002			
−10	·0006	·0008	−·0002		·0001	·0007	−·0006		·0000	·0008
−9	·0020	·0023	−·0003		·0005	·0021	−·0016		·0003	·0023
−8	·0057	·0062	−·0005		·0024	·0057	−·0033		·0019	·0062
−7	·0144	·0151	−·0007		·0086	·0142	−·0056		·0078	·0151
−6	·0326	·0334	−·0008		·0245	·0319	−·0074		·0237	·0334
−5	·0663	·0668	−·0005		·0576	·0646	−·0070		·0576	·0668
−4	·1215	·1217	−·0002		·1156	·1190	−·0034		·1172	·1217
−3	·2025	·2023	·0002		·2025	·1997	·0028		·2061	·2023
−2	·3089	·3085	·0004		·3159	·3066	·0093		·3209	·3085
−1	·4340	·4338	·0002		·4463	·4331	·0132		·4513	·4338
0	·5660	·5662	−·0002		·5798	·5669	·0129		·5832	·5662
1	·6911	·6915	−·0004	·0004	·7021	·6934	·0087	·0003	·7031	·6915
2	·7975	·7977	−·0002	·0008	·8032	·8003	·0029	·0006	·8019	·7977
3	·8785	·8783	·0002	·0004	·8790	·8810	−·0020	−·0001	·8762	·8783
4	·9337	·9332	·0005	−·0004	·9309	·9354	−·0045	−·0014	·9275	·9332
5	·9674	·9666	·0008	−·0010	·9633	·9681	−·0048	−·0025	·9601	·9666
6	·9856	·9849	·0007	−·0016	·9818	·9858	−·0040	−·0026	·9794	·9849
7	·9943	·9938	·0005	−·0014	·9917	·9943	−·0026	−·0016	·9900	·9938
8	·9980	·9977	·0003	−·0010	·9964	·9979	−·0015	−·0007	·9954	·9977
9	·9994	·9992	·0002	−·0006	·9986	·9993	−·0007	−·0001	·9980	·9992
10	·9998	·9998	·0000	−·0004	·9998	·9998	·0000	·0001	·9992	·9998
11	1·0000	·9999	·0001	·0000	·9998	·9999	−·0001	−·0002	·9997	·9999
12		1·0000	·0000	·0002	·9999	1·0000	−·0001		·9999	1·0000
13					1·0000	1·0000	·0000		1·0000	1·0000
14										

$$n! \simeq n^{n+\frac{1}{2}}e^{-n}\sqrt{2\pi}\,. \tag{21.5.10}$$

As an illustration of the accuracy of this formula we see that, for $n = 10$, $10! = 3\,628\,800$, while $10^{10\cdot5}e^{-10}\sqrt{2\pi} = 3\,598\,699$. The difference between the exact and the approximate value is 30 101, which is 0·8% of the exact value. For larger values of n the relative error is smaller.

If we introduce

$$y = x - n\theta\,,$$

21.6.

binomial distribution $P_B\{x\}$ and the normal distribution $P_N\{x\}$.
denotes the positive figures given in the first column.

100		900				∞				
0·1		0·01				$9/n \to 0$				
10·0		9·00				9				(POISSON)
9·00		8·9100				9				
3·00		2·9850				3				
P_B-P_N	Δ	$P_B\{x\}$	$P_N\{x\}$	P_B-P_N	Δ	$P_B\{x\}$	$P_N\{x\}$	P_B-P_N	Δ	$P_B\{\lvert x-n\theta\rvert\geq d\}$
−·0008										
−·0020		·0001	·0022	−·0021		0001	·0023	−·0022		
−·0043		0012	·0060	−·0048		·0012	·0062	−·0050		
−·0073		·0061	·0147	−·0086		·0062	·0151	−·0089		
−·0097		·0208	·0327	−·0119		·0212	·0334	−·0122		
−·0092		·0541	·0658	−·0117		·0550	·0668	−·0118		
−·0045		·1145	·1205	−·0060		·1157	·1217	−·0060		
·0038		·2054	·2012	·0042		·2068	·2023	·0045		
·0124		·3227	·3077	·0150		·3239	·3085	·0154		
·0175		·4550	·4335	·0215		·4557	·4338	·0219		
·0170		·5874	·5665	·0209		·5874	·5662	·0212		
·0116	·0005	·7066	·6923	·0143	·0006	·7060	·6915	·0145	·0007	·8683
·0042	·0008	·8040	·7988	·0052	·0007	·8030	·7977	·0053	·0009	·6179
−·0021	−·0004	·8769	·8795	−·0026	−·0010	·8758	·8783	−·0025	−·0008	·4038
−·0057	−·0024	·9272	·9342	−·0070	−·0034	·9261	·9332	−·0071	−·0035	·2399
−·0065	−·0035	·9593	·9673	−·0080	−·0047	·9585	·9666	−·0081	−·0047	·1289
−·0055	−·0032	·9785	·9853	−·0068	−·0039	·9780	·9849	−·0069	−·0041	·0627
−·0038	−·0018	·9893	·9940	−·0047	−·0018	·9889	·9938	−·0049	−·0020	·0282
−·0023	−·0005	·9949	·9978	−·0029	−·0001	·9947	·9977	−·0030	−·0001	·0123
−·0012	·0003	·9977	·9993	−·0016	·0008	·9976	·9992	−·0016	·0008	·0054
−·0006	·0004	·9990	·9998	−·0008		·9989	·9998	−·0009		
−·0002		·9996	·9999	−·0003		·9996	·9999	−·0003		
−·0001		·9998	1·0000	−·0002		·9998	1·0000	−·0002		
·0000		·9999	1·0000	−·0001		·9999	1·0000	−·0001		
		1·0000	1·0000	·0000		1·0000	1·0000	·0000		

the distribution function may be written

$$p\{x\} = \frac{n!}{x!(n-x)!}\,\theta^x(1-\theta)^{n-x} = \frac{n!}{(n\theta+y)!(n(1-\theta)-y)!}\,\theta^{n\theta+y}(1-\theta)^{n(1-\theta)-y}.$$

Application of STIRLING's formula to $n!$, $(n\theta+y)!$, and $(n(1-\theta)-y)!$ leads to

$$p\{x\} \simeq \frac{n^{n+\frac{1}{2}} e^{-n} \sqrt{2\pi}\, \theta^{n\theta+y} (1-\theta)^{n(1-\theta)-y}}{(n\theta+y)^{n\theta+y+\frac{1}{2}} e^{-n\theta-y} \sqrt{2\pi} (n(1-\theta)-y)^{n(1-\theta)-y+\frac{1}{2}} e^{-n(1-\theta)+y} \sqrt{2\pi}}$$

$$= \frac{1}{\sqrt{2\pi n\theta(1-\theta)}} \left(\frac{n\theta}{n\theta+y}\right)^{n\theta+y+\frac{1}{2}} \left(\frac{n(1-\theta)}{n(1-\theta)-y}\right)^{n(1-\theta)-y+\frac{1}{2}}$$

$$= \frac{1}{\sqrt{2\pi n\theta(1-\theta)}} \left(1+\frac{y}{n\theta}\right)^{-(n\theta+y+\frac{1}{2})} \left(1-\frac{y}{n(1-\theta)}\right)^{-(n(1-\theta)-y+\frac{1}{2})}. \quad (21.5.11)$$

The further development of this expression is based on the two expansions

$$\log_e \left(1+\frac{y}{n\theta}\right) = \frac{y}{n\theta} - \frac{y^2}{2n^2\theta^2} + \frac{y^3}{3n^3\theta^3} - \ldots$$

and

$$\log_e \left(1-\frac{y}{n(1-\theta)}\right) = -\frac{y}{n(1-\theta)} - \frac{y^2}{2n^2(1-\theta)^2} - \frac{y^3}{3n^3(1-\theta)^3} - \ldots.$$

which are convergent provided

$$\left|\frac{y}{n\theta}\right| = \left|\frac{x-n\theta}{n\theta}\right| < 1 \quad \text{and} \quad \left|\frac{y}{n(1-\theta)}\right| = \left|\frac{x-n\theta}{n(1-\theta)}\right| < 1 .$$

If we only consider values of x for which $|x-n\theta|$ is of the order of magnitude $\sqrt{n\theta(1-\theta)}$, e. g., $|x-n\theta| < 3\sqrt{n\theta(1-\theta)}$, the above inequalities will always be satisfied for large values of n, as $|y/n\theta|$ then is of the order of magnitude $1/\sqrt{n}$.

Introduction of the two serial expansions in

$$\log_e p\{x\} \simeq -\log_e \sqrt{2\pi n\theta(1-\theta)} - (n\theta+y+\tfrac{1}{2}) \log_e \left(1+\frac{y}{n\theta}\right)$$

$$-(n(1-\theta)-y+\tfrac{1}{2}) \log_e \left(1-\frac{y}{n(1-\theta)}\right)$$

and rejection of terms of the order of magnitude $1/\sqrt{n}$ leads to

$$\log_e p\{x\} \simeq -\log_e \sqrt{2\pi n\theta(1-\theta)} - \frac{y^2}{2n\theta(1-\theta)}, \quad (21.5.12)$$

which gives

$$p\{x\} \simeq \frac{1}{\sqrt{2\pi n\theta(1-\theta)}} e^{-\frac{(x-n\theta)^2}{2n\theta(1-\theta)}} .$$

This result was first derived by A. DE MOIVRE in 1733, see p. 189.

In the above proof remainder terms for the determination of the accuracy of the approximation formula have not been considered. Regarding this problem the reader is referred to J. V. USPENSKY: *Introduction to Mathematical Probability*, McGraw-Hill, New York, 1937, p. 129, and W. FELLER: *On the Normal Approximation to the Binomial Distribution*, Ann. Math. Stat., 16, 1945, 319–29.

21.6. The Transformation 2 arcsin $\sqrt{h}$.

On the basis of the relation between the mean θ and the variance $\theta(1-\theta)/n$ for the frequency $h = x/n$ we may determine a function $y = g(h)$ in such a manner that *the variance of the transformed variable y is independent of θ*, see the remarks on p. 176. Application of (7.3.4) leads to the transformation function

$$y = 2 \arcsin \sqrt{h} \tag{21.6.1}$$

with variance

$$\mathcal{V}\{y\} \simeq \frac{1}{n}. \tag{21.6.2}$$

If h is approximately normally distributed with mean θ and variance $\theta(1-\theta)/n$, then $y = 2 \arcsin \sqrt{h}$ is also approximately normally distributed with mean $\eta = 2 \arcsin \sqrt{\theta}$ and variance $1/n$, so that

$$u = (2 \arcsin \sqrt{h} - 2 \arcsin \sqrt{\theta})\sqrt{n} \tag{21.6.3}$$

is approximately normally distributed with parameters (0, 1). This transformation has been given by R. A. FISHER and has been particularly useful in analyses of variance[1]) and regression analyses involving frequencies. Table XII gives the values of the function $y = 2 \arcsin \sqrt{x}$ for $0 \leq x \leq 1$, and Fig. 21.3 shows a graph of this function.

In analogy with (21.5.6) we obtain

$$P\{h\} \simeq \Phi\left(\left(2 \arcsin \sqrt{h + \frac{1}{2n}} - 2 \arcsin \sqrt{\theta}\right)\sqrt{n}\right), \tag{21.6.4}$$

wherefore the fractile h_P is determined by the relation

[1]) M. S. BARTLETT: *The Square Root Transformation in Analysis of Variance*, Suppl. Journ. Roy. Stat. Soc., 3, 1936, 68–78.

W. G. COCHRAN: *Some Difficulties in the Statistical Analysis of Replicated Experiments*, Empire Journ. Experimental Agric., 6, 1938, 157–175.

W. G. COCHRAN: *The Analysis of Variance when Experimental Errors Follow the* POISSON *or Binomial Laws*, Ann. Math. Stat., 11, 1940, 335–347.

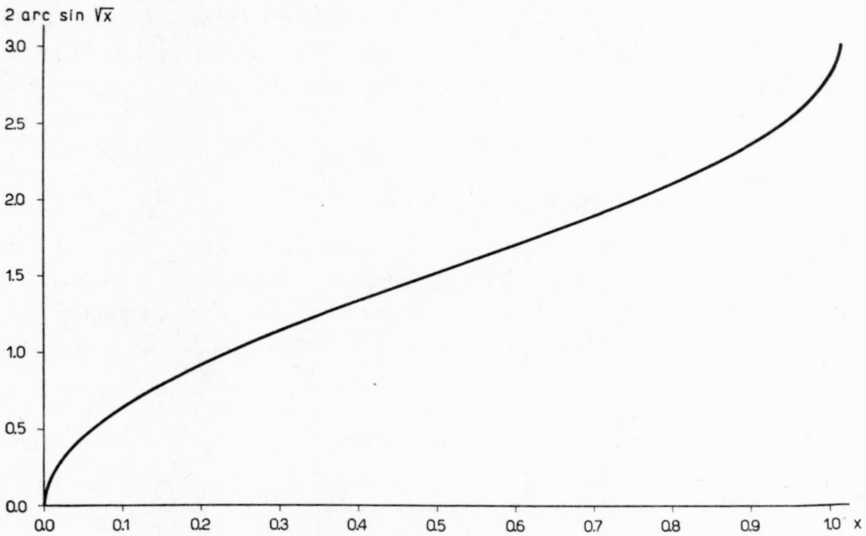

FIG. 21.3. Graph of $y = 2 \arcsin \sqrt{x}$.

$$\boxed{2 \arcsin \sqrt{h_P + \frac{1}{2n}} \simeq 2 \arcsin \sqrt{\theta} + \frac{u_P}{\sqrt{n}}.} \tag{21.6.5}$$

Systematic examinations of the accuracy of this approximation formula have not been made, but the scattered experience we have indicates that the two approximation formulas (21.5.6) and (21.6.4) usually lead to deviations from the exact values of the same order of magnitude, but with opposite signs, so that the *mean of the two approximations usually gives a considerably better approximation to the exact value than either of the single values*. Our experience, however, covers only the commonly applied range for the tests of significance, i. e., unilateral and bilateral 5% and 1% limits. Thus, as an approximation to $P\{h\}$ for $0{\cdot}005 < P < 0{\cdot}05$ and $0{\cdot}95 < P < 0{\cdot}995$ we may apply $\Phi(u)$ for

$$u = \tfrac{1}{2}(u_1 + u_2)\,, \tag{21.6.6}$$

where

$$u_1 = \frac{h + \frac{1}{2n} - \theta}{\sqrt{\frac{\theta(1-\theta)}{n}}}\,, \tag{21.6.7}$$

and

$$u_2 = \left(2 \arcsin \sqrt{h + \frac{1}{2n}} - 2 \arcsin \sqrt{\theta}\right)\sqrt{n}\,. \tag{21.6.8}$$

The use of the arc sine transformation has been greatly facilitated and many new applications have been developed by F. MOSTELLER and J. W. TUKEY in *The Uses and Usefulness of Binomial Probability Paper*, Journ. Amer. Stat. Ass., 44, 1949, 174–212, to which the reader is referred. The basic idea of the "binomial probability paper" is to plot the "sample point" $(n-x, x)$ on double-square-root paper, i. e., in reality we plot $(\sqrt{n-x}, \sqrt{x})$. This point will lie somewhere on a quarter-circle with radius $\sqrt{n}$. The angle made by the line connecting the origin with the sample point and the horizontal axis is arcsin $\sqrt{h}$. Repeated random samples of size n will give sample points varying at random on the circle, their standard error measured on the circle being $\frac{1}{2}$. This simple property makes tests of significance involving one or more frequencies comparatively easy when carried out graphically.

Example 21.3. As an example of the application of the above formulas we shall calculate the approximations to $P\{h \leq 0{\cdot}15\}$ for $n = 100$ and $\theta = 0{\cdot}1$. From (21.6.7) we obtain

$$u_1 = \frac{0{\cdot}150 + 0{\cdot}005 - 0{\cdot}100}{0{\cdot}03} = 1{\cdot}833\,,$$

and from (21.6.8)

$$u_2 = (2 \arcsin \sqrt{0{\cdot}155} - 2 \arcsin \sqrt{0{\cdot}1}) \times 10 = 1{\cdot}658\,,$$

whence $u = \frac{1}{2}(1{\cdot}833 + 1{\cdot}658) = 1{\cdot}746$. From Table II or III we have $\Phi(u) = 0{\cdot}9596$ as approximation to the exact value $P\{h \leq 0{\cdot}15\} = 0{\cdot}9601$, see Table 21.6. For comparison we give the approximations $\Phi(u_1) = 0{\cdot}9666$ and $\Phi(u_2) = 0{\cdot}9513$. Thus, the approximation $\Phi(u)$ is considerably better than $\Phi(u_1)$ and $\Phi(u_2)$. Instead of $\Phi(u)$ we may use the mean $\frac{1}{2}(\Phi(u_1) + \Phi(u_2)) = 0{\cdot}9590$, but the difference between these two values is usually very small, and therefore we prefer $\Phi(u)$, which only requires the looking up of one tabular value.

Table 21.6 shows that for $n = 100$ and $\theta = 0{\cdot}1$ we have the fractile $h_{.9601} = 0{\cdot}15$. From the approximation formula (21.5.9) we obtain

$$h_{.9601} \simeq 0{\cdot}1000 - 0{\cdot}0050 + 1{\cdot}752 \sqrt{\frac{0{\cdot}1 \times 0{\cdot}9}{100}} = 0{\cdot}1476\,,$$

as $u_{.9601} = 1{\cdot}752$, see Table III. From (21.6.5) we find

$$2 \arcsin \sqrt{h_{.9601} + 0{\cdot}0050} \simeq 0{\cdot}6435 + \frac{1{\cdot}752}{\sqrt{100}} = 0{\cdot}8187\,.$$

Interpolation in Table XII gives $h_{.9601} + 0{\cdot}0050 \simeq 0{\cdot}1584$, i. e., $h_{.9601} \simeq 0{\cdot}1534$. The mean of the two results is $\frac{1}{2}(0{\cdot}1476 + 0{\cdot}1534) = 0{\cdot}1505$, which is a better approximation to the exact value 0·1500 than either of the single values. As h_P can take only the values 0·00, 0·01, ..., all the approximate values on rounding off to two decimals give the right figure 0·15.

21.7. The Binomial Distribution and the Poisson-Distribution.

For fixed θ and increasing n the binomial distribution will, as shown in § 21.5, converge to the normal distribution. But if θ varies in inverse proportion to n, i. e., $\theta = \xi/n$, so that $\theta \to 0$ for $n \to \infty$, the binomial distribution converges to the Poisson-distribution

$$\boxed{p\{x\} = e^{-\xi}\frac{\xi^x}{x!}, \quad x = 0, 1, \ldots .} \tag{21.7.1}$$

This result is derived by the following transformation of the binomial distribution

$$p\{x\} = \binom{n}{x}\theta^x(1-\theta)^{n-x} = \frac{n(n-1)\ldots(n-x+1)}{x!}\frac{\xi^x}{n^x}\left(1-\frac{\xi}{n}\right)^{n-x}$$

$$= \left[1\left(1-\frac{1}{n}\right)\ldots\left(1-\frac{x-1}{n}\right)\left(1-\frac{\xi}{n}\right)^{-x}\right]\frac{\xi^x}{x!}\left(1-\frac{\xi}{n}\right)^n$$

$$\to \frac{\xi^x}{x!}e^{-\xi} \text{ for } n \to \infty , \tag{21.7.2}$$

the factors in the square brackets converging to 1, so that the product

TABLE 21.7.

The binomial distribution $p_n\{x\}$ for $n = 5, 10, 20, 50$, and 100, and $\theta = 1/n$ $(n\theta = 1)$ together with the Poisson-distribution.

n	5	10	20	50	100	∞	Difference
θ	0·2	0·1	0·05	0·02	0·01	0	
$n\theta$	1	1	1	1	1	1	
$n\theta(1-\theta)$	0·8	0·9	0·95	0·98	0·99	1	
$\sqrt{n\theta(1-\theta)}$	0·894	0·949	0·975	0·990	0·995	1	
x	$p_5\{x\}$	$p_{10}\{x\}$	$p_{20}\{x\}$	$p_{50}\{x\}$	$p_{100}\{x\}$	$p_\infty\{x\}$	$p_{100}\{x\}-p_\infty\{x\}$
0	·3277	·3487	·3585	·3642	·3660	·3679	−·0019
1	·4096	·3874	·3774	·3716	·3697	·3679	·0018
2	·2048	·1937	·1887	·1858	·1849	·1839	·0010
3	·0512	·0574	·0596	·0607	·0610	·0613	−·0003
4	·0064	·0112	·0133	·0146	·0149	·0153	−·0004
5	·0003	·0015	·0023	·0027	·0029	·0031	−·0002
6		·0001	·0003	·0004	·0005	·0005	·0000
7			·0000	·0001	·0001	·0001	·0000

TABLE 21.8.

Comparison of the cumulative distribution functions of the binomial distribution $P_B\{x\}$ and the POISSON-distribution $P_P\{x\}$ for $\theta = 0{\cdot}1$ and $n = 5, 10, 20, 50,$ and 100.

n	5			10			20			50			100		
θ	0·1			0·1			0·1			0·1			0·1		
$n\theta$	0·5	0·5		1·0	1·0		2·0	2·0		5·0	5·0		10·0	10·0	
$n\theta(1-\theta)$	0·45	0·50		0·90	1·00		1·80	2·00		4·50	5·00		9·00	10·00	
$\sqrt{n\theta(1-\theta)}$	0·67	0·71		0·95	1·00		1·34	1·41		2·12	2·24		3·00	3·16	
x	$P_B\{x\}$	$P_P\{x\}$	P_B-P_P	$P_B\{x\}$	$P_P\{x\}$	P_B-P_P	$P_B\{x\}$	$P_P\{x\}$	P_B-P_P	$P_B\{x\}$	$P_P\{x\}$	P_B-P_P	$P_B\{x\}$	$P_P\{x\}$	P_B-P_P
0	·5905	·6065	—·0160	·3487	·3679	—·0192	·1216	·1353	—·0137	·0052	·0067	—·0015	·0000	·0000	·0000
1	·9185	·9098	·0087	·7361	·7358	·0003	·3918	·4060	—·0142	·0338	·0404	—·0066	·0003	·0005	—·0002
2	·9914	·9856	·0058	·9298	·9197	·0101	·6769	·6767	·0002	·1117	·1247	—·0130	·0019	·0028	—·0009
3	·9995	·9982	·0013	·9872	·9810	·0062	·8671	·8571	·0100	·2503	·2650	—·0147	·0078	·0103	—·0025
4	·9999	·9998	·0001	·9984	·9963	·0021	·9568	·9473	·0095	·4312	·4405	—·0093	·0237	·0293	—·0056
5	1·0000	1·0000	·0000	·9999	·9994	·0005	·9888	·9834	·0054	·6161	·6160	·0001	·0576	·0671	—·0095
6				1·0000	·9999	·0001	·9976	·9955	·0021	·7702	·7622	·0080	·1172	·1301	—·0129
7				1·0000	1·0000	·0000	·9996	·9989	·0007	·8779	·8666	·0113	·2061	·2202	—·0141
8							·9999	·9998	·0001	·9421	·9319	·0102	·3209	·3328	—·0119
9							1·0000	1·0000	·0000	·9755	·9682	·0073	·4513	·4579	—·0066
10										·9907	·9863	·0044	·5832	·5830	·0002
11										·9968	·9945	·0023	·7031	·6968	·0063
12										·9990	·9980	·0010	·8019	·7916	·0103
13										·9997	·9993	·0004	·8762	·8645	·0117
14										·9999	·9998	·0001	·9275	·9165	·0110
15										1·0000	·9999	·0001	·9601	·9513	·0088
16										1·0000	1·0000	·0000	·9794	·9730	·0064
17													·9900	·9857	·0043
18													·9954	·9928	·0026
19													·9980	·9965	·0015
20													·9992	·9984	·0008
21													·9997	·9993	·0004
22													·9999	·9997	·0002
23													1·0000	·9999	·0001
24													1·0000	1·0000	·0000

also converges to 1 as the number of factors is finite, while $\left(1-\frac{\xi}{n}\right)^n \to e^{-\xi}$.

Thus, the mean in the binomial distribution $\xi = n\theta$ is fixed and the variance $n\theta(1-\theta) = \xi(1-\theta)$ converges to ξ, as $n \to \infty$ and $\theta \to 0$, wherefore *the mean and the variance in the* POISSON-*distribution are the same and equal to* ξ. The distribution was derived by S. D. POISSON in 1837. A table of the distribution function and the cumulative distribution function to six decimal places for values of ξ between 0 and 100 has been published by E. C. MOLINA: POISSON's *Exponential Binomial Limit*, Van Nostrand, New York, 1945.

In Table 21.7 an example is given of the convergence of the binomial to the POISSON-distribution for $\xi = n\theta = 1$. It is seen that the POISSON-distribution is a good approximation to the binomial distribution, also for small values of n, when θ is sufficiently small. *It is generally considered justifiable to apply the* POISSON-*distribution as approximation to the binomial distribution when* $\theta < 0{\cdot}1$.

Table 21.8 shows a number of examples of the accuracy of the approximation formula in the limiting case $\theta = 0{\cdot}1$. When evaluating the magnitude of the relative error it must be borne in mind that in unilateral tests we apply either $P\{x\}$ or $1-P\{x\}$, and that the absolute error of these two quantities is the same.

Introducing $\theta = \xi/n$ and letting $n \to \infty$ in (21.2.6) we find for the POISSON-distribution

$$P\{x\} = 1-P\{\chi^2 < 2\xi\} \text{ for } f = 2(x+1) \tag{21.7.3}$$

since $v^2(f_1, \infty) = \chi^2/f_1$ according to (14.5.1). Thus the cumulative POISSON-distribution can be expressed by the χ^2-distribution just as the cumulative binomial distribution is expressed by the v^2-distribution.

A more detailed discussion of the POISSON-distribution is given in Chapter 22.

21.8. The Binomial Distribution and the Hypergeometric Distribution.

Let there be given a finite population of N elements, exactly M of which belong to the category U. The probability that a random sample of n elements includes exactly x elements of category U is

$$\boxed{p\{x\} = \frac{\binom{M}{x}\binom{N-M}{n-x}}{\binom{N}{n}}.} \tag{21.8.1}$$

This distribution, the hypergeometric distribution, has been derived in § 2.4 and some elementary examples of its application have been given.

Further it has been shown in § 17.2 that the mean and the variance in the hypergeometric distribution are

$$\boxed{\mathcal{m}\{x\} = n\theta} \tag{21.8.2}$$

and

$$\boxed{\mathcal{v}\{x\} = n\theta(1-\theta)\left(1-\frac{n-1}{N-1}\right) \simeq n\theta(1-\theta)\left(1-\frac{n}{N}\right),} \tag{21.8.3}$$

where

$$\boxed{\theta = \frac{M}{N}} \tag{21.8.4}$$

denotes the fraction of elements belonging to the category U in the population.

Like the binomial distribution, the hypergeometric distribution may be approximated to by the normal distribution, as

$$p\{x\} \simeq \frac{1}{\sqrt{2\pi}\sigma} e^{-\frac{(x-\xi)^2}{2\sigma^2}}, \tag{21.8.5}$$

for $\xi = n\theta$ and $\sigma^2 = n\theta(1-\theta)\left(1-\frac{n}{N}\right)$. The range of applicability is limited as in the case of the binomial distribution by the inequality

$$\sigma^2 = n\theta(1-\theta)\left(1-\frac{n}{N}\right) > 9\,. \tag{21.8.6}$$

For $N \to \infty$ and a fixed value of $\theta = \frac{M}{N}$ the hypergeometric distribution converges to the binomial distribution, as proved in (2.4.2). It is usually considered justifiable to use the binomial distribution as an approximation to the hypergeometric distribution when the sample drawn constitutes less than 10% of the population, i. e.,

$$p\{x\} = \frac{\binom{M}{x}\binom{N-M}{n-x}}{\binom{N}{n}} \simeq \binom{n}{x}\theta^x(1-\theta)^{n-x},\ \theta = \frac{M}{N},\ \text{for } \frac{n}{N} < 0{\cdot}1\,. \tag{21.8.7}$$

Tables 2.2 and 2.9 show an example for $n = 0{\cdot}1N$ and $\theta = 0{\cdot}02$.

If both $\theta < 0{\cdot}1$ and $n < 0{\cdot}1N$ the Poisson-distribution may be used as an approximation.

21.9. The Control Chart for Proportion Defective.

In many cases the only way to characterize the quality of a product is to classify the items produced as conforming or nonconforming to given specifications and then calculate the proportion of nonconforming (defective) items. Even if measurements actually are possible it may be more economical to use the proportion defective as quality characteristic, because the cost of the detailed measurements may be considerably larger than the cost of classifying the items into two groups, for example by go and not-go gauges. For a given number of observations the control chart based on measurements is more efficient than the chart for proportion defective, but the opposite may well be true for given cost.

For a production process in a state of statistical control with a standard proportion defective of θ the 95% and 99·8% control limits for the proportion defective in samples of size n are

$$\theta \pm 1{\cdot}96 \sqrt{\frac{\theta(1-\theta)}{n}} \quad \text{and} \quad \theta \pm 3{\cdot}09 \sqrt{\frac{\theta(1-\theta)}{n}}, \tag{21.9.1}$$

respectively, if the normal distribution applies, i. e., if $n > 9/\theta(1-\theta)$. By plotting the proportion defective found in successive samples on the control chart, lack of control may be discovered as discussed in § 9.3 and § 13.6.

If sample size varies considerably from one sample to another control limits must be computed anew for each sample. In many cases, however, variations in sample size are so small that an average sample size may be used.

For constant sample size the control chart for proportion defective may be replaced by a control chart for the number of defective items in each sample, the control limits being obtained from (21.9.1) by multiplication by n.

When the process has not been analyzed previously so that we do not know whether or not it is in a state of statistical control the observations are analyzed according to the principles given in § 13.7, i. e., a trial value of θ is computed from the total number of observations, whereafter the proportion defective for each rational subgroup of observations is plotted on a control chart with limits calculated from the trial value of θ.

If more accurate limits than the above are needed we proceed as follows. According to (21.5.9) and (21.6.5) approximations to the fractile h_P are obtained from the formulas

$$h_P + \frac{1}{2n} \simeq \theta + u_P \sqrt{\frac{\theta(1-\theta)}{n}} \tag{21.9.2}$$

and

$$2 \arcsin \sqrt{h_P + \frac{1}{2n}} \simeq 2 \arcsin \sqrt{\theta} + \frac{u_P}{\sqrt{n}}, \qquad (21.9.3)$$

the mean of the two h_P's usually being a better approximation than either single value. However, since h can take on only the values 0, $1/n$, $2/n$, . . ., 1, the approximate value of the lower limit has to be "rounded off" downwards and the approximate value of the upper limit "rounded off" upwards, the effect being that slightly more than the stipulated percentage of the distribution will be included between these limits. (According to (21.2.5) $1/n$ has to be added to the lower fractile if the region of acceptance shall include *both* limits). The probability corresponding to the limits finally found may be determined from (21.6.6)–(21.6.8), see Example 21.4.

If the normal distribution does not apply the fractiles must be computed from the binomial distribution as shown in § 21.2.

Example 21.4. As an example we will compute the 95% control limits for $n = 100$ and $\theta = 0{\cdot}1$. From (21.9.2) we obtain

$$h_{.975} + 0{\cdot}0050 \simeq 0{\cdot}1000 + 1{\cdot}960 \sqrt{\frac{0{\cdot}1 \times 0{\cdot}9}{100}} = 0{\cdot}1588$$

and

$$h_{.025} + 0{\cdot}0050 \simeq 0{\cdot}1000 - 1{\cdot}960 \sqrt{\frac{0{\cdot}1 \times 0{\cdot}9}{100}} = 0{\cdot}0412\,,$$

which leads to $h_{.975} \simeq 0{\cdot}1538$ and $h_{.025} + 0{\cdot}0100 \simeq 0{\cdot}0462$.

From (21.9.3) we obtain

$$2 \arcsin \sqrt{h_{.975} + 0{\cdot}0050} \simeq 0{\cdot}6435 + 0{\cdot}1960 = 0{\cdot}8395$$

and

$$2 \arcsin \sqrt{h_{.025} + 0{\cdot}0050} \simeq 0{\cdot}6435 - 0{\cdot}1960 = 0{\cdot}4475\,,$$

interpolation in Table XII giving $h_{.975} \simeq 0{\cdot}1661 - 0{\cdot}0050 = 0{\cdot}1611$ and $h_{.025} + 0{\cdot}0100 \simeq 0{\cdot}0492 + 0{\cdot}0050 = 0{\cdot}0542$.

The mean of the two sets of results gives the lower control limit 0·0502 and the upper control limit 0·1575. By "rounding off" downwards and upwards, respectively, to multiples of $1/n = 0{\cdot}01$ we obtain the limits 0·05 and 0·16.

An approximation to the probability $P\{0{\cdot}05 \leq h \leq 0{\cdot}16\}$ can be determined from (21.6.6) which leads to

$$P\{0{\cdot}05 \leq h \leq 0{\cdot}16\} \simeq \Phi(2{\cdot}049) - \Phi(-1{\cdot}997) = 0{\cdot}9798 - 0{\cdot}0229 = 0{\cdot}9569\,.$$

Table 21.6 which includes the binomial distribution for $n = 100$ and $\theta = 0{\cdot}1$ shows that the exact value is 0·9557, since

$$P\{h \leq 0{\cdot}16\} = 0{\cdot}9794 \quad \text{and} \quad P\{h \leq 0{\cdot}04\} = 0{\cdot}0237\,.$$

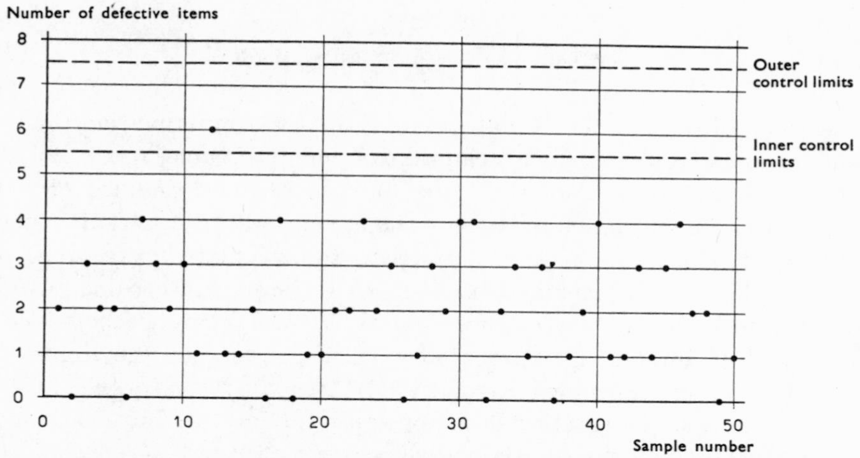

FIG. 21.4. Control chart for the number of defective items in samples of 100 items from a production process in a state of statistical control with a standard proportion defective of 2%.

Example 21.5. Fig. 21.4 shows the control chart for the 50 samples in Table 21.2. The number of defective items in each sample has been plotted as ordinate and the sample number as abscissa. From Table 2.2, p. 29, it follows that only the upper control limits have any importance since the probability of getting no defectives is 13·26%. An exact 97·5% limit does not exist, but $P\{x \leq 5\} = 98{\cdot}5\%$; therefore the inner control limit has been drawn between $x = 5$ and $x = 6$. Similarly, the outer control limit has been drawn between $x = 7$ and $x = 8$, since $P\{x \leq 7\} = 0{\cdot}999$. The control chart does not show any sign of lack of control.

A more sensitive control chart showing possible deviations in both directions requires a sample size of about 500.

21.10. Comparing an Observed Frequency with a Hypothetical Probability.

Assume first that the normal distribution applies. If the test hypothesis $\theta = \theta_0$ is true, the standardized variable

$$u = \frac{h-\theta_0}{\sigma_0}, \quad \sigma_0 = \sqrt{\frac{\theta_0(1-\theta_0)}{n}}, \tag{21.10.1}$$

will be normally distributed with parameters (0, 1), so that the rules of § 9.4 for the u-test apply. For example, in the case of a two-sided alternative we use $h-\frac{1}{2n}-\theta_0 > 1{\cdot}96\sigma_0$ and $h+\frac{1}{2n}-\theta_0 < -1{\cdot}96\sigma_0$ as the critical region at the 5% level of significance.

Since the standard deviation

$$\sigma = \sqrt{\frac{\theta(1-\theta)}{n}} \tag{21.10.2}$$

depends on θ, the power function becomes slightly more complicated than (9.4.9), viz.

$$\pi(\theta) = P\left\{\left|\frac{h-\theta_0}{\sigma_0}\right| > u_{1-\frac{\alpha}{2}};\ \mathcal{M}\{h\} = \theta\right\}$$

$$= \Phi\left((u_{\frac{\alpha}{2}}-\lambda)\frac{\sigma_0}{\sigma}\right)+\Phi\left((u_{\frac{\alpha}{2}}+\lambda)\frac{\sigma_0}{\sigma}\right) \text{ for } \lambda = \frac{\theta-\theta_0}{\sigma_0}. \quad (21.10.3)$$

This complication may be avoided by using the arc sine transformation, the standardized variable being

$$u = (y-\eta_0)\sqrt{n} \quad (21.10.4)$$

where $y = 2 \arcsin \sqrt{h}$ and $\eta_0 = 2 \arcsin\sqrt{\theta_0}$, see (21.6.3). The power function then becomes

$$\pi(\theta) = P\{|(y-\eta_0)\sqrt{n}| > u_{1-\frac{\alpha}{2}};\ \mathcal{M}\{h\} = \theta\}$$

$$= \Phi(u_{\frac{\alpha}{2}}-\lambda)+\Phi(u_{\frac{\alpha}{2}}+\lambda) \quad \text{for} \quad \lambda = (\eta-\eta_0)\sqrt{n}, \quad (21.10.5)$$

which is analogous to (9.4.9).

As in § 9.4 the power function may be used in the planning of experiments to determine the number of observations necessary to obtain a given discriminating power (or to determine the sensitivity of a control chart to changes in the standard proportion defective). The outcome of the experiment is supposed to be a frequency $h = x/n$ which forms the basis for choosing between the two hypotheses $\theta \leq \theta_0$ and $\theta \geq \theta_1$ ($\theta_0 < \theta_1$). If we require that the probability of rejecting the hypothesis $\theta \leq \theta_0$ when it is true should be equal to or less than α, and similarly that the probability of rejecting the hypothesis $\theta \geq \theta_1$ when it is true should be equal to or less than β, we find that the minimum number of observations is determined from the equation

$$\sqrt{n} = \frac{u_{1-\beta}+u_{1-\alpha}}{2 \arcsin \sqrt{\theta_1}-2 \arcsin \sqrt{\theta_0}}, \quad (21.10.6)$$

which is analogous to (9.4.18). The critical region becomes

$$2 \arcsin \sqrt{h} > \frac{u_{1-\beta}\, 2 \arcsin \sqrt{\theta_0}+u_{1-\alpha}\, 2 \arcsin \sqrt{\theta_1}}{u_{1-\beta}+u_{1-\alpha}}. \quad (21.10.7)$$

The results given in Table 21.9 indicate the number of observations necessary to discriminate between two probabilities having a difference of 0·05 or a ratio of 2 for $\alpha = \beta = 0{\cdot}05$.

TABLE 21.9.

The number of observations necessary to discriminate between θ_0 and θ_1 by means of an observed frequency. The probabilities of errors of the first and second kind are both 0·05. $n = 10{\cdot}82/(\eta_1-\eta_0)^2$.

Difference = 0·05			Ratio = 2		
θ_0	θ_1	n	θ_0	θ_1	n
0·01	0·06	125	0·01	0·02	1550
0·05	0·10	292	0·05	0·10	292
0·10	0·15	469	0·10	0·20	134
0·20	0·25	752	0·20	0·40	55
0·50	0·55	1078	0·30	0·60	29

If a more accurate approximation is wanted the correction for continuity must be taken into account, cf. (21.5.6) and (21.6.4).

A test based on the binomial distribution itself may be formulated as follows. Assuming that the alternative to the test hypothesis $\theta = \theta_0$ is $\theta > \theta_0$ we want to determine a critical region $x \geq c$ for the absolute frequency so that $P\{x \geq c;\ \theta_0\} = \alpha$. Since c has to be an integer the equality above must be replaced by the inequality

$$P\{x \geq c;\ \theta_0\} \leq \alpha < P\{x \geq c-1;\ \theta_0\}\,. \tag{21.10.8}$$

To decide whether or not an observed frequency $h_0 = x_0/n$ deviates significantly from θ_0 we have to compare the observed absolute frequency x_0 with the "rejection number" c. If $x_0 \geq c$, then it follows that $P\{x \geq x_0;\ \theta_0\} \leq \alpha$, whence we may replace the first criterion by the second. Consequently, *the test may be carried out by computing the probability* $P\{x \geq x_0;\ \theta_0\}$ *and comparing this probability with* α. From (21.2.6) we find

$$P\{x \geq x_0\} = P\left\{v^2 < \frac{n-x_0+1}{x_0}\,\frac{\theta_0}{1-\theta_0}\right\}, \quad \begin{aligned} f_1 &= 2x_0\,, \\ f_2 &= 2(n-x_0+1)\,, \end{aligned} \tag{21.10.9}$$

which probability may be found by interpolating in Table VII.

If the alternative hypothesis is $\theta < \theta_0$ the critical region is $x \leq c$ where

$$P\{x \leq c;\ \theta_0\} \leq \alpha < P\{x \leq c+1;\ \theta_0\}\,. \tag{21.10.10}$$

Hence we compute

$$P\{x \leq x_0\} = P\left\{v^2 > \frac{n-x_0}{x_0+1}\,\frac{\theta_0}{1-\theta_0}\right\}, \quad \begin{aligned} f_1 &= 2(x_0+1)\,, \\ f_2 &= 2(n-x_0)\,. \end{aligned} \tag{21.10.11}$$

In the case of the two-sided alternative $\theta \neq \theta_0$ a difficulty arises in defining the critical region $x \leq c_1$ and $x \geq c_2$. The simplest solution is to use a critical region based on equal tail areas, i. e., to determine (c_1, c_2) as in the one-sided test replacing α by $\alpha/2$. If, for example, the observed $h_0 = \frac{x_0}{n} > \theta_0$ we compute $P\{x \geq x_0;\ \theta_0\}$ and compare this probability with $\alpha/2$.

Table XI shows the two-sided 95% and 99% confidence limits for θ corresponding to an observed frequency and computed from the v^2-distribution, see § 21.11. Thus, the two-sided test at the 5% and 1% levels of significance (or the one-sided test at the 2·5% and 0·5% levels of significance) may be carried out by comparing the hypothetical θ_0 with the confidence limits corresponding to h_0.

Another solution consists in computing the probability

$$P\{|h-\theta_0| \geq |h_0-\theta_0|;\ \theta_0\} \qquad (21.10.12)$$

and comparing this probability with α. In this case the region of acceptance is chosen to be symmetrical around θ_0.

Example 21.6. Examination of a random sample of 50 items revealed that 6 items (12%) were defective. Is this result consistent with the hypothesis $\theta = 0{\cdot}05$, i. e., a theoretical proportion defective of 5%?

According to (21.10.9) the probability of obtaining 6 or more defective items is

$$P\{x \geq 6\} = P\left\{v^2 < \frac{50-6+1}{6}\,\frac{0{\cdot}05}{0{\cdot}95}\right\} = P\{v^2(12,90) < 0{\cdot}395\}\,.$$

Application of (14.1.9) leads to

$$P\{v^2(12,90) < 0{\cdot}395\} = P\{v^2(90,12) > 2{\cdot}53\}\,.$$

From Table VII we find $v^2_{.95}(90,\ 12) = 2{\cdot}35$ and $v^2_{.975}(90,\ 12) = 2{\cdot}81$, and thus we have $0{\cdot}025 < P\{x \geq 6\} < 0{\cdot}05$. Therefore, the said hypothesis must be rejected at the 5% level of significance for the alternative $\theta > 0{\cdot}05$. If, however, the alternative is double-sided, the hypothesis cannot be rejected at the 5% level of significance. This may also be seen immediately from Table XI which shows that the double-sided 95% confidence limits are (0·045, 0·244) for $x = 6$ and $n-x = 44$.

21.11. Confidence Limits.

From an observed frequency, $h_0 = \frac{x_0}{n}$, we may determine confidence limits $\underline{\theta}(h_0)$ and $\bar{\theta}(h_0)$ for the probability θ in such a manner that

$$P\{\underline{\theta}(h_0) \leq \theta \leq \bar{\theta}(h_0)\} = P_2-P_1 = P\,, \qquad (21.11.1)$$

for any value of θ[1]). The limits are functions of x_0, n, P_1, and P_2. The upper limit $\bar{\theta}$ is determined so that the probability of frequencies smaller than or equal to the one observed is P_1 for $\theta = \bar{\theta}$, i. e., $\bar{\theta}$ is determined from the equation

$$P\{h \leq h_0;\, \theta = \bar{\theta}\} = \sum_{x=0}^{x_0} \binom{n}{x} \bar{\theta}^x (1-\bar{\theta})^{n-x} = P_1\,, \qquad (21.11.2)$$

and similarly the lower limit $\underline{\theta}$ is determined from the equation

$$P\{h \geq h_0;\, \theta = \underline{\theta}\} = \sum_{x=x_0}^{n} \binom{n}{x} \underline{\theta}^x (1-\underline{\theta})^{n-x} = 1-P_2; \qquad (21.11.3)$$

see the corresponding definitions in Fig. 9.7, p. 241. These equations may be solved directly by means of (21.2.6) from which we obtain

$$\boxed{\bar{\theta} = \frac{(x_0+1)v_{1-P_1}^2}{n-x_0+(x_0+1)v_{1-P_1}^2}\,, \quad \begin{array}{l} f_1 = 2(x_0+1)\,, \\ f_2 = 2(n-x_0)\,, \end{array}} \qquad (21.11.4)$$

and

$$\boxed{\underline{\theta} = \frac{x_0}{x_0+(n-x_0+1)v_{P_2}^2}\,, \quad \begin{array}{l} f_1 = 2(n-x_0+1)\,, \\ f_2 = 2x_0\,. \end{array}} \qquad (21.11.5)$$

Bilateral 95% and 99% limits for θ have been tabulated in Table XI.

For large values of $n\theta(1-\theta)$ the approximation formulas in § 21.5 and § 21.6 may be applied. Application of (21.5.6) leads to the following equations for the determination of $\bar{\theta}$ and $\underline{\theta}$

$$\frac{h_0+\dfrac{1}{2n}-\bar{\theta}}{\sqrt{\dfrac{\bar{\theta}(1-\bar{\theta})}{n}}} \simeq u_{P_1} \qquad (21.11.6)$$

and

$$\frac{h_0-\dfrac{1}{2n}-\underline{\theta}}{\sqrt{\dfrac{\underline{\theta}(1-\underline{\theta})}{n}}} \simeq u_{P_2}, \qquad (21.11.7)$$

which have the solutions

$$\boxed{\left.\begin{array}{l}\bar{\theta} \\ \underline{\theta}\end{array}\right\} \simeq \frac{1}{n+u_P^2}\left[x_0 \pm \frac{1}{2} + \frac{u_P^2}{2} - u_P\sqrt{\frac{(x_0\pm\frac{1}{2})(n-x_0\mp\frac{1}{2})}{n}+\frac{u_P^2}{4}}\right],} \qquad (21.11.8)$$

[1]) See C. J. Clopper and E. S. Pearson: *The Use of Confidence or Fiducial Limits Illustrated in the Case of the Binomial*, Biometrika, 26, 1934, 404–413.

where the upper sign corresponds to $\bar{\theta}$ and the lower one to $\underline{\theta}$, and P is replaced by P_1 and P_2, respectively[1]).

In analogy with these results we obtain by applying (21.6.4)

$$\left.\begin{array}{l} 2 \arcsin \sqrt{\bar{\theta}} \\ 2 \arcsin \sqrt{\underline{\theta}} \end{array}\right\} \simeq 2 \arcsin \sqrt{\frac{x_0 \pm \frac{1}{2}}{n}} - \frac{u_P}{\sqrt{n}}. \tag{21.11.9}$$

The means of the results according to (21.11.8) and (21.11.9) will usually give good approximations to the exact values[2]).

As a less accurate approximation we may utilize the fact that h is approximately normally distributed about θ with variance $\theta(1-\theta)/n$, so that

$$P\left\{ u_{P_1} < \frac{h_0 - \theta}{\sqrt{\frac{\theta(1-\theta)}{n}}} < u_{P_2} \right\} \simeq P_2 - P_1 . \tag{21.11.10}$$

Solution of the inequality with respect to θ leads to confidence limits for θ.

Example 21.7. In an experiment with a toxic drug 8 out of 30 mice died after injection with the drug. Within which limits should we expect the corresponding lethal probability?[2])

The given quantities are $x_0 = 8$ and $n = 30$ and the probabilities P_1 and P_2, which are chosen as $P_1 = 0{\cdot}025$ and $P_2 = 0{\cdot}975$, say. Table VII shows that $v^2_{\cdot 975}(18,44) = 2{\cdot}07$ and $v^2_{\cdot 975}(46,16) = 2{\cdot}49$, and then substitution in (21.11.4) and (21.11.5) gives the limits

$$\bar{\theta} = \frac{9 \times 2{\cdot}07}{22 + 9 \times 2{\cdot}07} = 0{\cdot}459$$

and

$$\underline{\theta} = \frac{8}{8 + 23 \times 2{\cdot}49} = 0{\cdot}123 .$$

Thus, the 95% confidence limits for the lethal probability determined from the frequency $8/30 = 0{\cdot}267$ are (0.123, 0·459). This result may also be read directly from Table XI for $x = 8$ and $n - x = 22$.

From the approximation formula (21.11.8) we obtain for $u_{P_1} = -1{\cdot}960$ and $u_{P_2} = 1{\cdot}960$,

$$\bar{\theta} \simeq \frac{1}{33{\cdot}84}\left[8{\cdot}50 + 1{\cdot}92 + 1{\cdot}96\sqrt{\frac{8{\cdot}5 \times 21{\cdot}5}{30} + 0{\cdot}96}\right] = 0{\cdot}462$$

---

[1]) S. Millot: *Théorie nouvelle de la probabilité des causes*, Paris, 1925.

O. N. Anderson: *Einführung in die mathematische Statistik*, Chap. 1, § 11, J. Springer, Vienna, 1935.

[2]) M. S. Bartlett: *Sub-sampling for Attributes*, Suppl. Journ. Roy. Stat. Soc., 4, 1937, 131–135.

and

$$\underline{\theta} \simeq \frac{1}{33{\cdot}84}\left[7{\cdot}50+1{\cdot}92-1{\cdot}96\sqrt{\frac{7{\cdot}5\times 22{\cdot}5}{30}+0{\cdot}96}\right] = 0{\cdot}129\ .$$

From (21.11.9) we further find

$$2 \arcsin \sqrt{\bar{\theta}} \simeq 2 \arcsin \sqrt{\frac{8{\cdot}5}{30}}+\frac{1{\cdot}960}{\sqrt{30}} = 1{\cdot}4805,\ \bar{\theta} \simeq 0{\cdot}455\ ,$$

and

$$2 \arcsin \sqrt{\underline{\theta}} \simeq 2 \arcsin \sqrt{\frac{7{\cdot}5}{30}}-\frac{1{\cdot}960}{\sqrt{30}} = 0{\cdot}6893,\ \underline{\theta} \simeq 0{\cdot}114\ .$$

Thus, the means $\frac{1}{2}(0{\cdot}462+0{\cdot}455) = 0{\cdot}459$ and $\frac{1}{2}(0{\cdot}129+0{\cdot}114) = 0{\cdot}122$ furnish us with excellent approximations to the exact values.

21.12. Sampling Inspection by Attributes.

In this section we shall outline some basic principles for sampling inspection by attributes. Detailed descriptions of both theoretical and practical aspects of sampling inspection together with a large number of sampling inspection plans and tables may be found in H. F. Dodge and H. G. Romig: *Sampling Inspection Tables*, J. Wiley, New York, 1944, *Sampling Inspection* by the Statistical Research Group, *Columbia University*, McGraw-Hill, New York, 1948, and three papers by H. C. Hamaker: *Lot Inspection by Sampling, The Theory of Sampling Inspection Plans*, and *The Practical Application of Sampling Inspection Plans and Tables*, Philips Technical Review, Vol. 11, No. 6, 176–182, No. 9, 260–270, and No. 12, 362–370, 1949–50.

Single Sampling Inspection.

Suppose that a lot of N items is submitted for inspection and that each item can be classified as defective or nondefective according to given specifications. The unknown proportion of defective items in the lot is denoted by $\theta = M/N$. Let us investigate the consequences of applying the following sampling inspection procedure: *Accept the lot if a random sample of n items contains c or less defective items; otherwise reject the lot.* The number c is called the *acceptance number*. The probability of rejection is found from the hypergeometric distribution function as

$$P\{x \geq c+1\} = \sum_{x\geq c+1} \binom{M}{x}\binom{N-M}{n-x}\Big/\binom{N}{n}\ . \qquad (21.12.1)$$

For a given sampling plan characterized by lot size, sample size, and acceptance number, (N, n, c), this probability, considered as a function of θ, will be called *the power function* $\pi(\theta)$ *of the sampling plan*. The probability of acceptance, $1-\pi(\theta)$, is also commonly called *the operating characteristic* of the plan.

Theoretically, the power function depends on all three parameters. Under normal factory conditions, however, the proportion defective will usually be less than 10% and the sample will contain less than 10% of the lot, so that the hypergeometric distribution may be replaced by the POISSON-distribution with sufficient accuracy, see § 21.8. We thus find

$$\pi(\theta) = P\{x \geq c+1\} \simeq e^{-n\theta} \sum_{x\geq c+1} \frac{(n\theta)^x}{x!}, \tag{21.12.2}$$

which may be written as

$$\pi(\theta) \simeq P\{\chi^2 < 2n\theta\} \quad \text{for} \quad f = 2(c+1) \tag{21.12.3}$$

according to (21.7.3). The probability of rejection considered as a function of the proportion of defective items in the lot may thus be obtained directly from Table V.

This approximation to the power function depends only on the two quantities n and c. Conversely, n and c may be determined from two given values of the power function. Usually the producer requires that at most $100\alpha\%$ of submitted lots with a proportion defective less than θ_1 shall be rejected and the consumer requires that at most $100\beta\%$ of submitted lots with a proportion defective larger than θ_2 shall be accepted. These requirements lead to the equations $\pi(\theta_1) = \alpha$ and $\pi(\theta_2) = 1-\beta$ which have the solutions $\chi^2_\alpha = 2n\theta_1$ and $\chi^2_{1-\beta} = 2n\theta_2$ for $f = 2(c+1)$. Therefore, the acceptance number c may be found as $c = \frac{f}{2} - 1$ where f is determined as the number of degrees of freedom satisfying the equation

$$\frac{\chi^2_{1-\beta}}{\chi^2_\alpha} = \frac{\theta_2}{\theta_1} \quad \text{for} \quad f = 2(c+1), \tag{21.12.4}$$

whence the sample size is found as

$$n = \frac{\chi^2_\alpha}{2\theta_1} = \frac{\chi^2_{1-\beta}}{2\theta_2}. \tag{21.12.5}$$

Let us now construct a sampling plan with the same properties as the plan described on p. 311, i. e., we require that at most 5% of submitted lots with a proportion defective less than $\theta_1 = 0{\cdot}03$ shall be rejected and that at most 10% of submitted lots with a proportion defective larger than $\theta_2 = 0{\cdot}08$ shall be accepted. Inserting these values into (21.12.4) we find

$$\chi^2_{\cdot 90}/\chi^2_{\cdot 05} = 2{\cdot}67 \quad \text{for} \quad f = 2(c+1).$$

From Table V we find $\chi^2_{\cdot 90}/\chi^2_{\cdot 05} = 2{\cdot}61$ for $f = 20$ so that $c \doteq 9$. Further (21.12.5) gives $n = 180$. The sampling inspection is thus carried out by drawing a random sample of 180 items from each lot, rejecting all lots for which the sample contains more than 9 defectives. Comparing with the results on p. 311 it is seen that sampling inspection by attributes requires a sample size of 180 whereas "inspection by variables" gives practically the same quality assurance with only 87 items. However, the costs of the two sampling plans may be such that it pays to use the sampling inspection by attributes even if a larger sample is required to obtain the same discriminating power.

It will be noted that when the Poisson-approximation applies and the quality assurance wanted has been specified by the two values of the power function then the sample size and the acceptance number follow, i. e., the sample size necessary to give the required quality assurance is independent of the lot size. It is a common misapprehension that the sample size has to be a fixed proportion of the lot size to give the same quality assurance.

Also, when the normal distribution applies as an approximation to the hypergeometric distribution and $n < 0{\cdot}1N$, it follows from (21.8.3) that σ_x and consequently the power function is practically independent of N.

If n/N is large, the hypergeometric distribution must be used to find the power function. The Poisson-distribution will, however, always give a lower bound to the exact power function.

In setting up sampling plans it is usual to require a better protection against misclassifications for large lots than for small lots. As proved above a constant sample size will give about the same power curve for small and large lots, whereas a constant percentage of inspected items will give a discriminating power which is relatively poor for small lots and good for large lots.

In practice the choice of the steepness of the power curve depends on the average quality of the product submitted, the nature of the product, the costs of inspection, and similar considerations.

The Statistical Research Group (SRG) has worked out a set of single sampling plans with a fixed relation between sample size and lot size, so that sample size increases with lot size but less than proportionally with lot size. The location of the power curve is specified by the "acceptable-quality level", i. e., the proportion defective satisfying the equation $\pi(\theta) = 0{\cdot}05$.

In the Philips *Standard Sampling System* (*Philips SSS*) the location of the power curve is specified by the "point of control", also called the "indifference quality", i. e., the proportion defective satisfying the equation $\pi(\theta) = 0{\cdot}50$. The steepness of the power curve is characterized by its relative slope at the point of control, and the amount of inspection as measured by the relative slope is made a function of both lot size and point of control.

The DODGE-ROMIG tables have been constructed under the assumption that the lots submitted for inspection originate from a production process in a state of statistical control. Further it is assumed that *all the items of rejected lots are inspected* and that *all defective items found are corrected or replaced by nondefective items*. Under these assumptions the average number of items inspected per lot and the average proportion defective in the product after inspection may be calculated as functions of θ. Two kinds of tables have been published. In the *lot quality protection tables* the location of the power curve is given by the value of θ satisfying the equation $\pi(\theta) = 0{\cdot}90$. (This value of θ is called the lot tolerance per cent defective). In the *average quality protection tables* the location of the power curve is characterized by the maximum average outgoing quality. In both cases a specified degree of protection leads to a relation between n and c, whereafter the tabulated values are determined so that *the average number of items inspected per lot of average quality is a minimum.*

Double Sampling Inspection.

Double sampling inspection is defined by the following set of rules:

1. A random sample of size n_1 is drawn from a lot of size N.
2. If the sample contains c_1 or less defective items, the lot is accepted.
3. If the sample contains more than c_2 defective items, the lot is rejected.
4. If the number of defective items exceeds c_1 but not c_2, a second sample of n_2 items is randomly drawn from the remainder of the lot.
5. Accept the lot if the total number of defectives in both samples does not exceed c_3; otherwise reject the lot.

A double sampling inspection plan is thus characterized by the six numbers $(N, n_1, n_2, c_1, c_2, c_3)$, i. e., lot size, the two sample sizes, and the three "acceptance numbers".

As HAMAKER has pointed out, the principles of double sampling may be clearly brought out by means of a "random walk" diagram as shown in Fig. 21.5, using the number of items inspected as abscissa and the number of defective items successively found as ordinate.

For each lot the results of inspection are shown by a series of points in the diagram, it being assumed that the results of inspecting the items of the sample have been recorded successively.

The criteria used have been represented by two screens, the lower part of each screen giving the region of acceptance (A) and the upper part the region of rejection (R). Further the opening of the first screen represents the region where final decision is deferred until a second sample has been taken. Very good and very bad lots will usually result in random walks ending in A_1 and R_1, respectively, so that the final decision is taken from the evidence of the first sample. Only the doubtful cases are passed on to further sampling inspection. This procedure naturally results in a saving in the average number of items inspected as compared with a single sampling plan giving the same quality assurance.

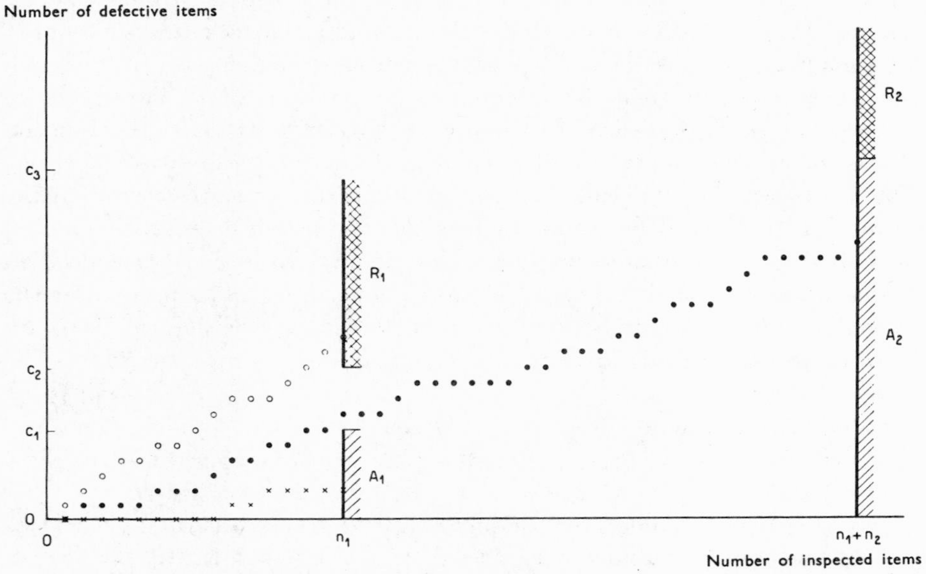

FIG. 21.5. A random walk diagram illustrating the principles of double sampling inspection.

The saving naturally depends on the quality of the product submitted. For very bad and very good inspection lots the average saving is larger than for lots of intermediate quality. Under usual conditions in practice the average saving amounts to about 30%. In choosing between a single and a double sampling plan with the same power function this saving has to be compared with the larger administrative difficulties in applying the double sampling plan, which perhaps more than outweigh the saving in the number of observations.

Denoting the number of defective items found in the first and second sample by x_1 and x_2, respectively, the probability of rejection (the power function) may be found as

$$P\{x_1 \geq c_2+1 \text{ or } x_1+x_2 \geq c_3+1\} = P\{x_1 \geq c_2+1\}+ \sum_{x_1=c_1+1}^{c_2} p\{x_1\}P\{x_2 \geq c_3+1-x_1\} \quad (21.12.6)$$

where the probabilities $p\{x_1\}$ and $p\{x_2\}$ are given as hypergeometric distributions.

HAMAKER has pointed out that the criteria of a double sampling plan must satisfy the relationship

$$c_1 < \frac{n_1}{n_1+n_2} c_3 < c_2 , \qquad (21.12.7)$$

see Fig. 21.5, to give a reasonable efficiency.

The number of parameters of the double sampling plans published has been restricted to less than the six above.

DODGE and ROMIG use the same rejection numbers for the first sample and the combined samples, i. e., they choose $c_2 = c_3$, whence the efficiency of the first sample in sorting out bad lots will not be as high as possible. This principle has been followed by the SRG and in the PHILIPS SSS.

The STATISTICAL RESEARCH GROUP uses as mentioned above a fixed relation between sample size and lot size and the second sample is always twice as large as the first, i. e., $n_2 = 2n_1$.

In the PHILIPS SSS the second sample is twice as large as the first and in most cases $c_3 = 5c_1$.

In the DODGE-ROMIG double sampling tables the average number of items inspected per lot of average quality is made a minimum under similar assumptions as in the corresponding single sampling tables.

Multiple and Sequential Sampling Inspection.

The principles of double sampling may be extended to multiple sampling and to sequential (item-by-item) sampling.

The tables of the STATISTICAL RESEARCH GROUP contain a system of multiple sampling inspection plans corresponding to the single and double sampling plans.

Sequential sampling plans may be found in *Sequential Analysis of Statistical Data: Applications* by the STATISTICAL RESEARCH GROUP, Columbia University Press, 1945; see also Chapter 24.

21.13. Comparison of Two Frequencies.

Suppose that the probability of the event U in one (infinite) population (experiment) is θ_1 and in another θ_2. From each population a random sample of size n_1 and n_2, respectively, is drawn giving the frequencies $h_1 = \frac{x_1}{n_1} \approx \theta_1$ and $h_2 = \frac{x_2}{n_2} \approx \theta_2$ of the event U. We want to test the equality of the two probabilities θ_1 and θ_2, i. e., $\theta_1 = \theta_2 = \theta$.

If the normal distribution applies to both frequencies, and the test hypothesis is true, the difference $h_1 - h_2$ will be normally distributed with zero mean and variance

$$\mathcal{V}\{h_1 - h_2\} = \theta(1-\theta)\left(\frac{1}{n_1}+\frac{1}{n_2}\right). \tag{21.13.1}$$

As an estimate of θ we introduce the total frequency of U in both samples

$$h = \frac{x_1+x_2}{n_1+n_2} = \frac{x}{n} \approx \theta\,. \tag{21.13.2}$$

Thus the standardized variable

$$u = \frac{h_1-h_2}{\sqrt{h(1-h)\left(\frac{1}{n_1}+\frac{1}{n_2}\right)}} \tag{21.13.3}$$

will be approximately normally distributed with parameters (0, 1). Taking the correction for continuity into account the critical region for testing the hypothesis $\theta_1 = \theta_2$ against the two-sided alternative $\theta_1 \neq \theta_2$ at the α level of significance becomes

$$\frac{\left(h_1-\frac{1}{2n_1}\right)-\left(h_2+\frac{1}{2n_2}\right)}{\sqrt{h(1-h)\left(\frac{1}{n_1}+\frac{1}{n_2}\right)}} > u_{1-\frac{\alpha}{2}},\ (h_1 > h_2)\,. \tag{21.13.4}$$

For the one-sided alternative $\theta_1 > \theta_2$, $u_{1-\frac{\alpha}{2}}$ has to be replaced by $u_{1-\alpha}$.

Using the arc sine transformation we see that the standardized variable

$$u = \frac{2\arcsin\sqrt{h_1} - 2\arcsin\sqrt{h_2}}{\sqrt{\frac{1}{n_1}+\frac{1}{n_2}}} \tag{21.13.5}$$

is approximately normally distributed with parameters (0, 1). The critical region becomes

$$\frac{2\arcsin\sqrt{h_1-\frac{1}{2n_1}} - 2\arcsin\sqrt{h_2+\frac{1}{2n_2}}}{\sqrt{\frac{1}{n_1}+\frac{1}{n_2}}} > u_{1-\frac{\alpha}{2}},\ (h_1 > h_2)\,. \tag{21.13.6}$$

The mean of the two u-values will usually be a better approximation than either single value.

The calculations according to the approximation formulas may be carried out as shown in Table 21.10.

TABLE 21.10.

Calculations for testing the hypothesis $\theta_1 = \theta_2$ on the basis of two frequencies h_1 and h_2, $(h_1 > h_2)$.

n_1	x_1	$h_1 = \dfrac{x_1}{n_1}$	$\dfrac{1}{n_1}$	$\dfrac{1}{2n_1}$	$h_1' = h_1 - \dfrac{1}{2n_1}$	$y_1' = 2 \arcsin \sqrt{h_1'}$
n_2	x_2	$h_2 = \dfrac{x_2}{n_2}$	$\dfrac{1}{n_2}$	$\dfrac{1}{2n_2}$	$h_2' = h_2 + \dfrac{1}{2n_2}$	$y_2' = 2 \arcsin \sqrt{h_2'}$
n_1+n_2	x_1+x_2	$h = \dfrac{x_1+x_2}{n_1+n_2}$	$\dfrac{1}{n_1}+\dfrac{1}{n_2}$		$h_1'-h_2'$	$y_1'-y_2'$

$$\sqrt{h(1-h)} = \ldots \qquad \sqrt{\frac{1}{n_1}+\frac{1}{n_2}} = \ldots$$

$$u_1 = \frac{h_1'-h_2'}{\sqrt{h(1-h)}\sqrt{\dfrac{1}{n_1}+\dfrac{1}{n_2}}}. \quad u_2 = \frac{y_1'-y_2'}{\sqrt{\dfrac{1}{n_1}+\dfrac{1}{n_2}}}. \quad u = \tfrac{1}{2}(u_1+u_2).$$

According to (9.4.13) the power function of the one-sided test based on the arc sine transformation will be

$$\pi(\delta) = \Phi(u_\alpha + \lambda) \tag{21.13.7}$$

where

$$\lambda = \frac{|\delta|}{\sigma} = \frac{|2 \arcsin \sqrt{\theta_1} - 2 \arcsin \sqrt{\theta_2}|}{\sqrt{\dfrac{1}{n_1}+\dfrac{1}{n_2}}}. \tag{21.13.8}$$

The number of observations necessary to discriminate between θ_1 and θ_2 so that $\pi(0) = \alpha$ and $\pi(\delta) = 1-\beta$ may be found from the equation

$$\sqrt{\frac{1}{n_1}+\frac{1}{n_2}} = \frac{|2 \arcsin \sqrt{\theta_1} - 2 \arcsin \sqrt{\theta_2}|}{u_{1-\beta}+u_{1-\alpha}}. \tag{21.13.9}$$

For $n_1 = n_2 = m$ we find

$$\sqrt{\frac{m}{2}} = \frac{u_{1-\beta}+u_{1-\alpha}}{|2 \arcsin \sqrt{\theta_1} - 2 \arcsin \sqrt{\theta_2}|} \tag{21.13.10}$$

analogous to (21.10.6). Thus Table 21.9 also gives half the number of observations from each population ($n = m/2$) necessary to discriminate between the two given probabilities with the power stated.

When the alternative is two-sided an approximation to the number of observations may be found by substituting $\alpha/2$ for α in the above formulas, cf. (9.4.24).

A test based on the binomial distribution may be developed as follows. Consider all possible outcomes of n_1 and n_2 observations, denoting the absolute frequencies of the event U by ν_1 and ν_2, respectively. (We introduce this notation because we want to distinguish between the theoretical variables (ν_1, ν_2) and an actually observed pair of frequencies (x_1, x_2)). If the test hypothesis is true the probability of getting the result (ν_1, ν_2) is

$$p\{\nu_1, \nu_2\} = p\{\nu_1\}p\{\nu_2\} = \binom{n_1}{\nu_1}\theta^{\nu_1}(1-\theta)^{n_1-\nu_1}\binom{n_2}{\nu_2}\theta^{\nu_2}(1-\theta)^{n_2-\nu_2}$$

$$= \frac{\binom{n_1}{\nu_1}\binom{n_2}{\nu_2}}{\binom{n_1+n_2}{\nu_1+\nu_2}} \times \binom{n_1+n_2}{\nu_1+\nu_2}\theta^{\nu_1+\nu_2}(1-\theta)^{n_1+n_2-\nu_1-\nu_2}. \tag{21.13.11}$$

As this probability depends on the unknown parameter θ it cannot be determined from the observations, and we therefore attempt to find an upper bound which does not include θ.

Introducing the sum $\nu_1+\nu_2 = x$ as a new variable we have

$$p\{x\} = \binom{n_1+n_2}{\nu_1+\nu_2}\theta^{\nu_1+\nu_2}(1-\theta)^{n_1+n_2-\nu_1-\nu_2}, \tag{21.13.12}$$

since x represents the absolute frequency of U in n_1+n_2 trials with the probability θ. We thus see that $p\{\nu_1, \nu_2\}$ may be written as the product of $p\{x\}$ and $p\{\nu_1|x\}$, the conditional probability of ν_1 for given x

$$p\{\nu_1|x\} = \frac{\binom{n_1}{\nu_1}\binom{n_2}{\nu_2}}{\binom{n_1+n_2}{\nu_1+\nu_2}} = \frac{\binom{n_1}{\nu_1}\binom{n_2}{x-\nu_1}}{\binom{n_1+n_2}{x}}. \tag{21.13.13}$$

This conditional probability may also be derived directly from the hypergeometric distribution. Suppose we have a finite population of n_1+n_2 elements, x elements having the attribute U. Selecting a random sample of n_1 elements it follows from § 21.8 that the probability of getting ν_1 elements with attribute U is given by (21.13.13).

Thus we have

$$p\{\nu_1, \nu_2\} = p\{x\}p\{\nu_1|x\}. \tag{21.13.14}$$

Consider first the alternative $\theta_1 > \theta_2$. For each x we may determine a critical region for ν_1, $\nu_1 \geq c_x$ say, independent of θ, so that

$$P\{\nu_1 \geq c_x|x\} \leq \alpha < P\{\nu_1 \geq c_x - 1|x\}\,. \tag{21.13.15}$$

Defining the critical region for (ν_1, ν_2) as the domain consisting of all the above critical regions, i. e., the point set satisfying the relations

$$\nu_1 \geq c_x \quad \text{and} \quad \nu_2 = x - \nu_1 \quad \text{for} \quad 0 \leq x \leq n_1 + n_2\,, \tag{21.13.16}$$

we see that the probability of a result (ν_1, ν_2) inside this domain will be

$$P = \sum\sum p\{\nu_1, \nu_2\} = \sum_{x=0}^{n_1+n_2} p\{x\} \sum_{\nu_1 \geq c_x} p\{\nu_1|x\} \leq \sum_{x=0}^{n_1+n_2} p\{x\} \times \alpha = \alpha\,. \tag{21.13.17}$$

Hence by using this critical region we are working at a level of significance which does not exceed α. The true level of significance depends on the unknown θ and may in some cases for small (n_1, n_2) be considerably less than α. The test based on $p\{\nu_1|x\}$ is called *a conditional test* because it only uses the conditional distribution of ν_1 which is independent of θ.

In practice we do not determine the critical region but proceed as in § 21.10. To test the hypothesis $\theta_1 = \theta_2$ against the alternative hypothesis $\theta_1 > \theta_2$ on the basis of the observed frequencies $h_1 = \dfrac{x_1}{n_1}$ and $h_2 = \dfrac{x_2}{n_2}$ we should investigate whether $x_1 \geq c_x$ for $x_1 + x_2 = x$. However, if $x_1 \geq c_x$ then $P\{\nu_1 \geq x_1|x\} \leq \alpha$ so that *the test may be carried out by computing* $P\{\nu_1 \geq x_1|x_1+x_2\}$ *and comparing this probability with* α.

In the case of a two-sided alternative $\theta_1 \neq \theta_2$, where $h_1 > h_2$, say, we compare $P\{\nu_1 \geq x_1|x_1+x_2\}$ with $\alpha/2$, using a critical region based on equal tail areas.

Another possibility is to use a critical region for the difference between the two frequencies which is symmetrical about zero.

The computation of

$$P\{\nu_1 \geq x_1|x_1+x_2\} = \sum_{\nu_1 \geq x_1} \binom{n_1}{\nu_1}\binom{n_2}{x_1+x_2-\nu_1} \Big/ \binom{n_1+n_2}{x_1+x_2} \tag{21.13.18}$$

may be carried out by means of Tables XIII and XIV. For $n \leq 100$ all values of $\log \dbinom{n}{x}$ are to be found in Table XIV. For $n_1 + n_2 > 100$ Table XIII may be used to determine a single term of (21.13.18) and then the following terms may be computed successively by the recursion formula

$$f(\nu_1) = \frac{p\{\nu_1+1|x_1+x_2\}}{p\{\nu_1|x_1+x_2\}} = \frac{n_1-\nu_1}{\nu_1+1} \times \frac{\nu_2}{n_2-\nu_2+1}, \quad \nu_2 = x_1+x_2-\nu_1\,. \tag{21.13.19}$$

If (21.13.18) contains more than three terms the approximation given in Table 21.10 will usually be sufficiently accurate.

For both n_1 and n_2 less than or equal to 15 the 0·05 and 0·01 significance limits have been tabulated in D. J. Finney: *The* Fisher-Yates *Test of Significance in* 2×2 *Contingency Tables*, Biometrika, 35, 1948, 145–156.

The conditional test and the correction for continuity, which is also called Yates' correction, have been investigated in F. Yates: *Contingency Tables Involving Small Numbers and the* χ^2*-Test*, Suppl. Jour. Roy. Stat. Soc., 1, 1934, 217–235.

The principles underlying the conditional test have been discussed by G. A. Barnard: *Significance Tests for* 2×2 *Tables*, Biometrika, 34, 1947, 123–38, and E. S. Pearson: *The Choice of Statistical Tests Illustrated on the Interpretation of Data Classed in a* 2×2 *Table*, Biometrika, 34, 1947, 139–167.

Example 21.8. Two experiments including 24 and 60 trials, respectively, have given the frequencies $h_1 = 11/24 = 0{\cdot}458$ and $h_2 = 14/60 = 0{\cdot}233$ of the event U. Is this sufficient evidence for concluding that the probabilities of U in the two experiments differ?

The conditional test is based on the probability

$$P = \sum_{\nu_1=11}^{24} \binom{24}{\nu_1}\binom{60}{25-\nu_1} \Big/ \binom{84}{25},$$

see (21.13.18) for $n_1 = 24$, $n_2 = 60$, $x_1 = 11$, and $x_2 = 14$.

By means of Table XIV we compute each term of P (for $\nu_1 \geq 16$ the terms are negligible) as follows:

ν_1	$\log\binom{24}{\nu_1}$	$\log\binom{60}{25-\nu_1}$	$-\log\binom{84}{25}$	Sum	Antilog
11	6·3973	13·2392	−21·1877	$\bar{2}{\cdot}4488$	0·0281
12	6·4320	12·7132	-	$\bar{3}{\cdot}9575$	0·0091
13	6·3973	12·1459	-	$\bar{3}{\cdot}3555$	0·0023
14	6·2925	11·5349	-	$\bar{4}{\cdot}6397$	0·0004
15	6·1164	10·8773	-	$\bar{5}{\cdot}8060$	0·0001
					0·0400

Hence the required probability is 0·0400 so that the difference observed is not significant at the 5% level, using a two-sided test.

In the present case the approximation formulas give a very accurate result, viz. $P = 0{\cdot}401$, as will be seen from the computations below, which have been arranged as in Table 21.10.

n	x	h	$\frac{1}{n}$	$\frac{1}{2n}$	h'	$2\arcsin\sqrt{h'}$
24	11	0·458	0·04167	0·021	0·437	1·4445
60	14	0·233	0·01667	0·008	0·241	1·0263
84	25	0·298	0·05834		0·196	0·4182

$$\sqrt{0{\cdot}298\times 0{\cdot}702} = 0{\cdot}457\,. \qquad \sqrt{0{\cdot}05834} = 0{\cdot}242\,.$$

$$u_1 = \frac{0{\cdot}196}{0{\cdot}457\times 0{\cdot}242} = 1{\cdot}77\,.\ u_2 = \frac{0{\cdot}4182}{0{\cdot}242} = 1{\cdot}73.\ u = 1{\cdot}75.\ \Phi(-1{\cdot}75) = {\cdot}0401$$

21.14. Comparison of k Frequencies.

The problem of § 21.13 may be generalized to the comparison of k frequencies resulting from k independent samples. In order to test the hypothesis $\theta_1 = \theta_2 = \ldots = \theta_k = \theta$ the k frequencies

$$h_1 = \frac{x_1}{n_1},\ h_2 = \frac{x_2}{n_2},\ \ldots,\ h_k = \frac{x_k}{n_k},$$

may be compared according to the principles of the analysis of variance, see § 16.4, it being assumed that the conditions for applying the approximation formulas in § 21.5 and § 21.6 have been fulfilled. If the hypothesis is true, the k standardized variables

$$u_i = \frac{h_i - \theta}{\sqrt{\dfrac{\theta(1-\theta)}{n_i}}},\ i = 1, 2, \ldots, k, \tag{21.14.1}$$

will be approximately normally distributed with parameters (0, 1), so that the sum of squares

$$\chi^2 = \sum_{i=1}^{k} u_i^2 = \frac{1}{\theta(1-\theta)} \sum_{i=1}^{k} n_i (h_i - \theta)^2\,,\quad f = k\,, \tag{21.14.2}$$

will be approximately distributed as χ^2 with k degrees of freedom, see (10.2.1). By minimizing the sum of squares with respect to θ we obtain the estimate

$$\bar{h} = \frac{\sum_{i=1}^{k} n_i h_i}{\sum_{i=1}^{k} n_i} = \frac{\sum_{i=1}^{k} x_i}{\sum_{i=1}^{k} n_i} \approx \theta\,. \tag{21.14.3}$$

From this it follows—by analogy with the results in § 16.3—that the sum of squares

$$\boxed{\chi^2 = \frac{1}{\bar{h}(1-\bar{h})} \sum_{i=1}^{k} n_i(h_i-\bar{h})^2\ , \ f = k-1\ ,} \tag{21.14.4}$$

is approximately distributed as χ^2 with $k-1$ degrees of freedom, and that $\bar{h}$ is normally distributed with mean θ and variance $\theta(1-\theta)/\sum n_i$. The hypothesis is tested by calculating χ^2 according to (21.14.4) and comparing this value with the appropriate fractile of the χ^2-distribution. In order to evaluate how much each of the frequencies contributes to the χ^2-value it is advisable to calculate the k terms

$$\frac{h_i-\bar{h}}{\sqrt{\dfrac{\bar{h}(1-\bar{h})}{n_i}}} = \frac{x_i-n_i\bar{h}}{\sqrt{n_i\bar{h}(1-\bar{h})}}\ . \tag{21.14.5}$$

Similarly the hypothesis may be tested by applying the arc sine transformation

$$y_i = 2 \arcsin \sqrt{h_i}\ . \tag{21.14.6}$$

Introducing the mean

$$\bar{y} = \frac{\sum n_i y_i}{\sum n_i} \tag{21.14.7}$$

we find the sum of squares

$$\boxed{\chi^2 = \sum_{i=1}^{k} n_i(y_i-\bar{y})^2\ , \quad f = k-1\ ,} \tag{21.14.8}$$

which is approximately distributed as χ^2 with $k-1$ degrees of freedom.

More advanced problems are treated in a similar manner according to the principles of the analysis of variance, see Chapter 16 and the references on p. 685.

Example 21.9. In a mass production the number of items, n, produced each day and the corresponding number of defective items, x, were registered as shown in the following table. The table shows the comparison of the seven fraction defectives to test the hypothesis that the manufacturing process has been in a state of statistical control during the week in question. The fraction defective for the whole week taken in one is 1·192 %. We compute $\bar{h}(1-\bar{h}) = 0{\cdot}01178$ and the quantities

$$\frac{h_i - \bar{h}}{\sqrt{\dfrac{\bar{h}(1-\bar{h})}{n_i}}}$$

given in the last column of the table. The sum of squares of these quantities leads to $\chi^2 = 5{\cdot}74$, $f = 6$, so that on the basis of this test there is no reason to doubt the stability of the process.

n	x	$h = \dfrac{x}{n}$	$h-\bar{h}$	$\sqrt{\dfrac{\bar{h}(1-\bar{h})}{n}}$	
2688	27	0·01004	−0·00188	0·00209	−0·90
2937	43	0·01464	0·00272	0·00200	1·36
2979	36	0·01208	0·00016	0·00199	0·08
2822	34	0·01205	0·00013	0·00204	0·06
2447	23	0·00940	−0·00252	0·00219	−1·15
3372	47	0·01394	0·00202	0·00187	1·08
2887	30	0·01039	−0·00153	0·00202	−0·76
20132	240	0·01192			

The hypothesis may also be tested by constructing a control chart for fraction defective, see § 21.9.

CHAPTER 22.

THE POISSON DISTRIBUTION

22.1. The Distribution Function and the Cumulative Distribution Function.

In § 21.7 the POISSON-distribution was derived as the limit of the binomial distribution for $n \to \infty$ and $\theta \to 0$, $n\theta = \xi$ being fixed. The distribution function is

$$p\{x\} = e^{-n\theta}\frac{(n\theta)^x}{x!} = e^{-\xi}\frac{\xi^x}{x!}, \quad x = 0, 1, \ldots . \tag{22.1.1}$$

Thus, we have for the POISSON-distribution a number of theorems analogous to the theorems for the binomial distribution.

From the ratio $f(x)$ of two successive terms of the distribution function

$$f(x) = \frac{p\{x+1\}}{p\{x\}} = \frac{\xi}{x+1}, \quad x = 0, 1, \ldots , \tag{22.1.2}$$

we see that the distribution function is steadily decreasing for $\xi < 1$, while for $\xi > 1$ it is first increasing and then decreasing. The mode is found as the largest integer less than or equal to ξ. By tabulating $f(x)$ and computing a single value of $p\{x\}$, the distribution function may be tabulated by successive multiplications and divisions as in Example 21.1.

The mean and the variance are identical and equal to the parameter ξ, i. e.,

$$\mathcal{m}\{x\} = \mathcal{v}\{x\} = \xi . \tag{22.1.3}$$

The sample mean $\bar{x}$ is a sufficient estimate of ξ.

As shown in § 21.7

$$P\{x\} = 1 - P\{\chi^2 < 2\xi\} \text{ for } f = 2(x+1) . \tag{22.1.4}$$

In analogy with the proof given in § 21.3 this result may also be obtained by successive partial integrations of

$$\frac{1}{x!}\int_{\xi}^{\infty} t^x e^{-t} dt \,. \qquad (22.1.5)$$

Thus, for given values of ξ and x we may determine $P\{x\}$ by interpolation in Table V. From (22.1.4) it follows that

$$\boxed{\chi^2_{1-P} = 2\xi \quad \text{for } f = 2(x_P+1)\,,} \qquad (22.1.6)$$

a relation which in § 22.5 is utilized for the determination of confidence limits for ξ.

In § 22.10 the POISSON-distribution will be derived from a slightly different point of view from the one in § 21.7.

In E. C. MOLINA: POISSON'S *Exponential Binomial Limit*, Van Nostrand, New York, 1945, a table is given of $p\{x\}$ and $1-P\{x\}$ to 6 decimal places for about 300 values of ξ in the interval $0{\cdot}001 \leq \xi \leq 100$.

Table 21.8 of this book includes five examples of the POISSON-distribution for values of ξ in the interval (0·5, 10·0).

The POISSON-distribution was derived by S. D. POISSON in 1837. The following publications include discussions on this distribution and examples of various modes of application:

L. v. BORTKIEWICZ: *Das Gesetz der kleinen Zahlen*, Leipzig, 1898.

"STUDENT": *On the Error of Counting with a Haemacytometer*, Biometrika, 5, 1907, 351–60.

L. WHITAKER: *On the* POISSON *Law of Small Numbers*, Biometrika, 10, 1915, 36–71.

L. v. BORTKIEWICZ: *Realismus und Formalismus in der mathematischen Statistik*, Allgemeines statistischen Archiv, 9, 1915, 225–56.

"STUDENT": *An Explanation of Deviations from* POISSON'S *Law in Practice*, Biometrika, 12, 1919, 211–15.

F. THORNDIKE: *Applications of* POISSON'S *Probability Summation*, The Bell System Technical Journal, 5, 1926, 604–24.

Example 22.1. As a classic example of the application of POISSON'S distribution we may here describe "STUDENT'S" publication: *On the Error of Counting with a Haemacytometer*, Biometrika, 5, 1907, 351–60. This investigation dealt with the distribution of yeast cells per unit volume of a suspension placed in a haemacytometer.

It is first imagined that the suspension is divided into a large number, N, of volume units, and that the cells are distributed at random and mutually independently in the units, so that the probability that a cell belongs to a given unit is $1/N$. Further, let the average number of cells per unit volume be ξ, and therefore the total number of cells in the suspension $N\xi = n$; then the probability that a cell belongs in a given unit volume is $1/N = \xi/n$. Thus for $N \to \infty$ the probability that a given unit volume contains x cells will be given by (22.1.1).

When we want to determine an estimate of the number of cells in a suspension, the suspension is well mixed and then spread out in a thin layer in a haemacytometer. Through the microscope it is now possible to count the number of cells per unit volume. Table 22.1 gives the results of counting the cells in 400 such units, 20 units having 1 cell each, 43 units having 2 cells each, etc. The total number of cells found is 1872, the average number per unit thus being $1872/400 = 4{\cdot}68$. If our assumption regarding the random and mutually independent distribution of the cells holds good, we expect that the observed distribution with good approximation may be described by a Poisson-distribution with mean 4·68. Column 3 of Table 22.1 gives the calculated values of a Poisson-distribution with mean 4·68, and column 4 these values multiplied by 400. It is seen that the deviations between the theoretical and the observed distribution are small, and the signs of the deviations seem to vary at random, so that the agreement is apparently good. (The agreement may be tested according to the rules given in § 23.2, see Example 23.2).

Table 22.1.

Distribution of 400 volume units according to number of yeast cells per unit. (Student's data).

Number of yeast cells (x)	Number of volume units	Poisson $p\{x\}$	$400\ p\{x\}$
0	0	0·00928	3·71
1	20	0·04343	17·37
2	43	0·10162	40·65
3	53	0·15852	63·41
4	86	0·18547	74·19
5	70	0·17360	69·44
6	54	0·13541	54·16
7	37	0·09053	36·21
8	18	0·05296	21·18
9	10	0·02754	11·02
10	5	0·01289	5·16
11	2	0·00548	2·19
12	2	0·00214	0·86
≥ 13	0	0·00113	0·45
Total	400	1·00000	400·00

When counting other particles, e. g., blood corpuscles and bacteria, other investigators have found systematic deviations from the POISSON-distribution, see J. BERKSON, T. B. MARGATH and M. HURN: *Laboratory Standards in Relation to Chance Fluctuations of the Erythrocyte Count as Estimated with the Hemacytometer*, Journ. Amer. Stat. Ass., 30, 1935, 414–26, and E. JACOBSEN, C. M. PLUM and G. RASCH: *On the Accuracy of Reticulocyte Counts*, Acta path., 24, 1947, 554–66.

22.2. The POISSON-Distribution and the Normal Distribution.

In § 21.5 it was shown that the normal distribution may be used as an approximation to the binomial distribution when $n\theta(1-\theta) > 9$. From this it follows directly that, *when the value of ξ is large, the* POISSON-*distribution may be approximated by the normal distribution with mean ξ and variance ξ*, see Table 21.6, p. 683, which gives a comparison between the two distributions for $\xi = 9$. This means that the cumulative distribution function may be written

$$\boxed{P\{x\} = e^{-\xi}\sum_{\nu=0}^{x}\frac{\xi^\nu}{\nu!} \simeq \Phi\left(\frac{x+\frac{1}{2}-\xi}{\sqrt{\xi}}\right),} \tag{22.2.1}$$

so that

$$\frac{x_P+\frac{1}{2}-\xi}{\sqrt{\xi}} \simeq u_P \tag{22.2.2}$$

or

$$x_P \simeq \xi-\tfrac{1}{2}+u_P\sqrt{\xi}\,. \tag{22.2.3}$$

The transformation $y = 2 \arcsin \sqrt{h}$, given in § 21.6, becomes for the POISSON-distribution $y = 2\sqrt{x}$. *If x is approximately normally distributed with mean ξ and variance ξ, $y = 2\sqrt{x}$ is also approximately normally distributed with mean $2\sqrt{\xi}$ and variance* 1, see (7.3.4). From this it follows that the cumulative distribution function may also be approximated by

$$\boxed{P\{x\} = e^{-\xi}\sum_{\nu=0}^{x}\frac{\xi^\nu}{\nu!} \simeq \Phi(2(\sqrt{x+\tfrac{1}{2}}-\sqrt{\xi}))\,,} \tag{22.2.4}$$

so that

$$2(\sqrt{x_P+\tfrac{1}{2}}-\sqrt{\xi}) \simeq u_P \tag{22.2.5}$$

or

$$x_P \simeq \xi-\tfrac{1}{2}+u_P\sqrt{\xi}+\tfrac{1}{4}u_P^2\,. \tag{22.2.6}$$

In variance and regression analyses involving variables distributed according to POISSON's law it is often practical to apply the square root transformation, as the variance then becomes constant, which simplifies the calculations considerably, see the references given on p. 685.

Systematic investigations have not been made on the accuracy of the approximation formulas, but the scattered experience published so far seems to indicate that the mean of the two values (22.2.1) and (22.2.4) usually gives a better approximation than either of the single values in the range generally used for tests of significance. The mean of (22.2.3) and (22.2.6) becomes

$$\boxed{x_P \simeq \xi - \tfrac{1}{2} + u_P \sqrt{\xi} + \tfrac{1}{8} u_P^2 \, .} \tag{22.2.7}$$

As an example of the application of these formulas we will determine the approximations to $P\{x \leq 15\}$ for $\xi = 9$. From (22.2.1) we find

$$u_1 = \frac{15{\cdot}5 - 9}{3} = 2{\cdot}167 \, ,$$

which gives $P\{x \leq 15\} \simeq \Phi(2.167) = 0{\cdot}9849$. (22.2.4) leads to

$$u_2 = 2(\sqrt{15{\cdot}5} - \sqrt{9}) = 1{\cdot}874 \, ,$$

which gives $P\{x \leq 15\} \simeq \Phi(1{\cdot}874) = 0{\cdot}9695$. The exact value is $P\{x \leq 15\} = 0{\cdot}9780$, see the last part of Table 21.6. The mean of the two approximations is $0{\cdot}9772$. The approximate value may also be calculated as $\Phi(u)$, where $u = \frac{1}{2}(u_1 + u_2)$, which leads to $\Phi(2{\cdot}021) = 0{\cdot}9784$.

22.3. The Control Charts for the Number of Defectives and Defects.

As mentioned on p. 692 the control chart for proportion defective may be replaced by a control chart for the number of defectives *if the sample size is constant*. If further the sample size is large and the average number of defectives per sample is small, the POISSON-distribution may be used instead of the binomial for computing the control limits.

If the production process is in a state of statistical control and *the expected number of defectives per sample of given size* is ξ, then the samples will be distributed according to number of defectives as given by the POISSON-distribution with mean ξ.

In some cases the *number of defects* will be a more convenient quality characteristic than the number of defectives. For example in controlling the quality of plates of building material the number of surface defects per square meter may be used as a quality characteristic. Similarly in the production of spectacle glasses sheets of glass of an appropriate size are controlled by counting the number of bubbles ("seeds") on each sheet. If the defects are *distributed at random and mutually independent* over each item examined, cf. Example 22.1, the number of items will be distributed according to number of defects as given by the POISSON-distribution, see also the discussion in § 22.10.

If ξ is large the 95% and 99·8% control limits for the number of defectives (or defects) may be computed as

$$\xi \pm 1{\cdot}96\sqrt{\xi} \quad \text{and} \quad \xi \pm 3{\cdot}09\sqrt{\xi}\,. \tag{22.3.1}$$

If more accurate limits are needed the fractiles may be computed from (22.2.7) or from the POISSON-distribution itself using

$$\chi^2_{1-P} = 2\xi \quad \text{for} \quad f = 2(x_P + 1)\,,$$

since ξ and P are given.

It should be remembered that when we are to determine, e. g., the 95% limits it is necessary, on account of the discontinuity of the distribution, to consider $P_1 \leq 0{\cdot}025$ and $P_2 \geq 0{\cdot}0975$. For instance for $\xi = 9$ and $P = 2{\cdot}5\%$ we have $\chi^2_{\cdot 975} = 18$ which corresponds to a number of degrees of freedom between 8 and 9, as Table V gives $\chi^2_{\cdot 975} = 17{\cdot}5$ for $f = 8$ and $\chi^2_{\cdot 975} = 19{\cdot}0$ for $f = 9$. From the relation $f = 2(x_P + 1)$ and $P_1 \leq 0{\cdot}025$ it follows that $x_{P_1} = 3$, as x_{P_1} is an integer. By analogy we have for $P = 97{\cdot}5\%$ that $\chi^2_{\cdot 025} = 18$, which corresponds to a number of degrees of freedom between 31 and 32, i. e., $x_{P_2} = 15$, as $P_2 \geq 0{\cdot}975$. The result for $\xi = 9$ is therefore that

$$P\{4 \leq x \leq 15\} \geq 0{\cdot}950$$

as $P\{3\} \leq 0{\cdot}025 < P\{4\}$ and $P\{14\} < 0{\cdot}975 \leq P\{15\}$.

From the approximation formula (22.2.7) we obtain for $\xi = 9$

$$x_{\cdot 025} \simeq 9 - 0{\cdot}5 - 1{\cdot}96 \times 3 + 0{\cdot}125 \times 1{\cdot}96^2 = 3{\cdot}1$$

and

$$x_{\cdot 975} \simeq 9 - 0{\cdot}5 + 1{\cdot}96 \times 3 + 0{\cdot}125 \times 1{\cdot}96^2 = 14{\cdot}9\,.$$

By rounding off downwards ($P_1 \leq 0{\cdot}025$) and upwards ($P_2 \geq 0{\cdot}975$) we find $x_{P_1} = 3$ and $x_{P_2} = 15$.

Example 22.2. Consider a production process in a state of statistical control where each workman fairly constantly produces 600 units in every shift. The expected number of defectives is 3·6 for each workman in every shift. The variation in the number of defectives may then be controlled by plotting the number of defectives per workman per shift on a control chart, the limits being computed from a POISSON-distribution with mean 3·6.

Table 22·2 gives the number of defectives in 52 shifts, and Fig. 22.1 shows the corresponding control chart.

Since the probability of finding no defectives is 2·7% no lower control limit has been drawn. The upper control limits are drawn between 8 and 9, and 11 and 12 defectives, respectively, since $P\{7\} < 0{\cdot}975 \leq P\{8\}$ and $P\{10\} < 0{\cdot}999 \leq P\{11\}$.

TABLE 22.2.
Number of defectives in 52 shifts.

3	4	2	2	2	5	6	4	3	3	6
1	4	3	2	1	1	5	4	4	3	1
0	4	3	5	3	3	9	6	4	1	
7	5	5	2	2	0	4	0	1	2	
3	3	1	5	3	2	4	7	2	3	

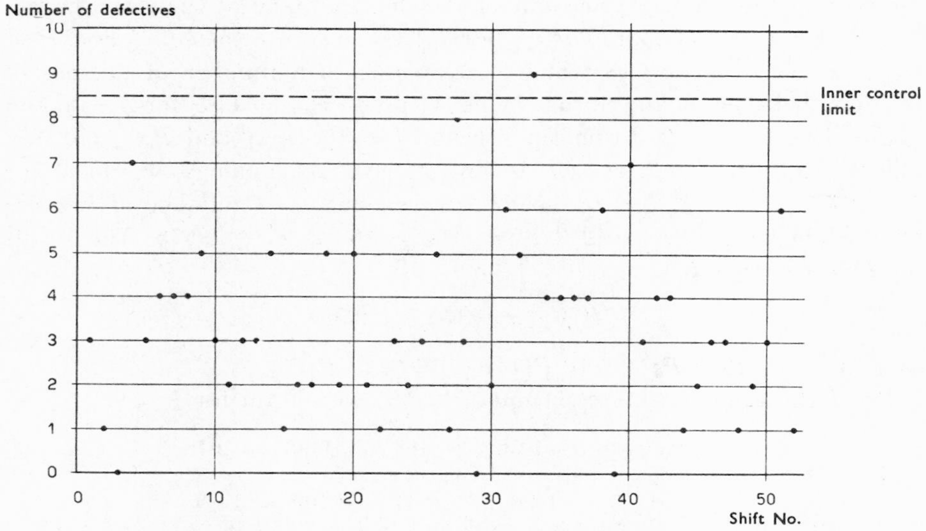

FIG. 22.1. Control chart for number of defectives per shift.

The control chart does not show any sign of lack of control during the period considered.

The distribution of the 52 shifts according to number of defectives has been given in Table 22.3, which also shows the theoretical distribution. As was the case for the control chart, a comparison between the observed and the calculated distribution indicates that during the period considered the number of defectives did not show "abnormal" deviations from the expected number. The average number of defectives per shift was $\bar{x} = 168/52 = 3{\cdot}23$.

TABLE 22.3.
Distribution of 52 shifts according to number of defectives per shift.

Number of defectives (x)	Number of shifts	POISSON $p\{x\}$	Calculated number of shifts: $52\ p\{x\}$
0	3	0·0273	1·42
1	7	0·0984	5·12
2	9	0·1771	9·21
3	12	0·2125	11·05
4	9	0·1912	9·94
5	6	0·1377	7·16
6	3	0·0826	4·30
7	2	0·0425	2·21
8	0	0·0191	0·99
9	1	0·0076	0·40
≥ 10	0	0·0040	0·21
	52	1·0000	52·01

22.4. A Test of Significance for the Difference between an Observed and an Expected Number.

Consider the problem of testing the hypothesis $\xi = \xi_0$ against the alternative $\xi > \xi_0$ by means of an observation x_0 which is known to originate from a POISSON-distributed population with unknown mean. As usual we define a critical region $x \geq c$ by the relation

$$P\{x \geq c;\ \xi_0\} \leq a < P\{x \geq c-1;\ \xi_0\}\,, \tag{22.4.1}$$

rejecting the test hypothesis if $x_0 \geq c$. Instead of calculating c *we calculate* $P\{x \geq x_0;\ \xi_0\}$ *and reject the hypothesis if this probability is less than or equal to* a. This comparison is easily carried out by means of Table V since

$$P\{x \geq x_0;\ \xi_0\} = P\{\chi^2 < 2\xi_0\} \quad \text{for} \quad f = 2x_0\,. \tag{22.4.2}$$

Hence *the test hypothesis will be rejected at the* a *level of significance if* $2\xi_0 < \chi^2_a$ *for* $f = 2x_0$, see also (22.5.3).

If the alternative is $\xi < \xi_0$ we use

$$P\{x \leq x_0;\ \xi_0\} = 1 - P\{\chi^2 < 2\xi_0\} \quad \text{for} \quad f = 2(x_0+1)\,, \tag{22.4.3}$$

leading to $2\xi_0 > \chi^2_{1-a}$, $f = 2(x_0+1)$ as criterion for rejection.

For the two-sided alternative we replace a by $a/2$.

For large values of ξ_0 a u-test may be applied, see § 22.2.

Example 22.3. In Example 22.2 registration of the number of defectives in 52 shifts led to an average of 3·23 defectives per shift. Does this figure differ significantly from the "expected" value 3·6?

According to the addition theorem for the POISSON-distribution stated in § 22.6 the sum of k POISSON-distributed variables is in itself POISSON-distributed with a parameter equal to the sum of the k parameters. Thus, the total number of defectives in 52 shifts will be POISSON-distributed with parameter 52ξ if in each shift the number of defectives is POISSON-distributed with parameter ξ. The above question may therefore be expressed: Does the total number of defectives, 168 in all, differ significantly from the "expected" number which is $52\times 3{\cdot}6 = 187{\cdot}2$? As a POISSON-distribution with parameter 187·2 may with good approximation be replaced by a normal distribution with mean 187·2 and standard deviation $\sqrt{187{\cdot}2} = 13{\cdot}7$ we will first provisionally calculate the value

$$u \simeq \frac{168{\cdot}0-187{\cdot}2}{13{\cdot}7} = -1{\cdot}40\,,$$

which immediately shows that the difference is not significant.

A more accurate evaluation requires the determination of the probability $P\{x \leq 168\}$ which according to (22.2.1) is

$$P\{x \leq 168\} \simeq \Phi\left(\frac{168{\cdot}5-187{\cdot}2}{13{\cdot}7}\right) = \Phi(-1{\cdot}36) = 0{\cdot}087\,.$$

22.5. Confidence Limits.

From the observation $x = x_0$, confidence limits, $\underline{\xi}(x_0)$ and $\overline{\xi}(x_0)$, for the parameter ξ may be determined so that

$$P\{\underline{\xi}(x_0) \leq \xi \leq \overline{\xi}(x_0)\} = P_2 - P_1 = P\,. \tag{22.5.1}$$

From (22.1.4) and (22.1.6) we obtain in analogy with the results of § 21.11

$$\boxed{\overline{\xi} = \tfrac{1}{2}\chi^2_{1-P_1} \text{ for } f = 2(x_0+1)} \tag{22.5.2}$$

and

$$\boxed{\underline{\xi} = \tfrac{1}{2}\chi^2_{1-P_2} \quad \text{for} \quad f = 2x_0\,.} \tag{22.5.3}$$

[1]

[1]) A discussion with examples of the applications of these limits has been given in J. PRZYBOROWSKI and H. WILEŃSKI: *Statistical Principles of Routine Work in Testing Clover Seeds for Dodder*, Biometrika, 27, 1935, 273–92, and F. GARWOOD: *Fiducial Limits for the* POISSON *Distribution*, Biometrika, 28, 1936, 437–42.

For large values of ξ the approximation formulas in § 22.2 may be applied leading to

$$\left.\begin{matrix}\overline{\xi}\\ \underline{\xi}\end{matrix}\right\} \simeq x_0 \pm \tfrac{1}{2} + \frac{u_P^2}{2} - u_P \sqrt{x_0 \pm \tfrac{1}{2} + \frac{u_P^2}{4}} \tag{22.5.4}$$

and

$$\left.\begin{matrix}\overline{\xi}\\ \underline{\xi}\end{matrix}\right\} \simeq x_0 \pm \tfrac{1}{2} + \frac{u_P^2}{4} - u_P \sqrt{x_0 \pm \tfrac{1}{2}}\,, \tag{22.5.5}$$

the uppermost sign corresponding to $\overline{\xi}$, the lower one to $\underline{\xi}$, and P being replaced by P_1 and P_2, respectively.

Example 22.4. When the purity of seeds is to be tested, e. g., clover seed is examined for its content of dodder, a random sample including a large number of seeds is usually taken, and the number of dodder seeds in the sample counted. If the whole lot of seeds has been well mixed the drawing of one large sample will correspond to the drawing of a large number of small samples with a very small probability of finding any dodder in each sample. Instead of drawing a sample of, e. g., 50,000 seeds, the practical experiment is made by taking a corresponding weight, e. g., 100 grammes. Under the above-described circumstances it is to be expected that the number of dodder seeds found in 100 grammes of seeds is distributed according to POISSON's law, see J. PRZYBOROWSKI and H. WILÉNSKI, cited above.

Let us imagine that a sample of 100 grammes contained no dodder. What is the upper 99% confidence limit for the mean number of dodder seeds in a sample of 100 grammes?

Substituting $x_0 = 0$ and $P_1 = 0{\cdot}01$ in (22.5.2), we get $f = 2$ and

$$\overline{\xi} = \frac{1}{2}\chi^2_{\cdot 99} = \frac{9{\cdot}21}{2} = 4{\cdot}6, \text{ see Table V.}$$

If the inspector concludes that, each time a random sample of 100 grammes of clover seed contains no dodder, the corresponding consignment will at most contain 4·6 dodder seeds per 100 grammes clover (46 dodder seeds per kg), 1% of these conclusions will in the long run be false, the average number of dodder seeds exceeding 4·6.

If the sample taken had been 5 kgs instead of 100 grammes, the conclusion would be (with 99% confidence) that the consignment at most contained 4·6 dodder seeds per 5 kgs, i. e., 0·92 per kg.

If, instead of 0 dodder seeds, 3 are found in a sample of 100 grammes, (22.5.2) for $x_0 = 3$ and $P_1 = 0{\cdot}01$ leads to

$$\overline{\xi} = \frac{1}{2}\chi^2_{\cdot 99} = \frac{20{\cdot}1}{2} = 10{\cdot}1,\ f = 8\,, \text{ see Table V.}$$

22.6. The Addition Theorem for the POISSON-Distribution.

In analogy with the addition theorem for the χ^2-distribution, see § 10.5, the following addition theorem holds for the POISSON-distribution:

If $x_1, x_2, \ldots, x_k$ *are stochastically independent and* POISSON-*distributed with parameters* $\xi_1, \xi_2, \ldots, \xi_k$, *then the sum* $x = x_1+x_2+\ldots+x_k$ *will be* POISSON-*distributed with parameter* $\xi = \xi_1+\xi_2+\ldots+\xi_k$.

For $k = 2$ we have according to the addition and multiplication formulas

$$p\{x_1+x_2 = x\} = \sum_{x_1=0}^{x} p\{x_1\}p\{x_2 = x-x_1\} . \tag{22.6.1}$$

Introducing

$$p\{x_1\} = e^{-\xi_1}\frac{\xi_1^{x_1}}{x_1!}$$

and

$$p\{x_2\} = e^{-\xi_2}\frac{\xi_2^{x_2}}{x_2!},$$

we obtain

$$p\{x_1+x_2 = x\} = e^{-\xi_1-\xi_2}\,\xi_2^{x} \sum_{x_1=0}^{x} \frac{1}{x_1!(x-x_1)!}\left(\frac{\xi_1}{\xi_2}\right)^{x_1} . \tag{22.6.2}$$

As

$$\left(1+\frac{\xi_1}{\xi_2}\right)^{x} = \sum_{x_1=0}^{x} \frac{x!}{x_1!(x-x_1)!}\left(\frac{\xi_1}{\xi_2}\right)^{x_1},$$

substitution in (22.6.2) leads to

$$p\{x_1+x_2 = x\} = e^{-(\xi_1+\xi_2)}\frac{(\xi_1+\xi_2)^{x}}{x!}, \tag{22.6.3}$$

i. e., $x = x_1+x_2$ is POISSON-distributed with parameter $\xi = \xi_1+\xi_2$. From this the general theorem may be derived by induction.

Application of the addition theorem to n observations from a given POISSON-distributed population with parameter ξ shows that the sum $\sum x_i$ is also POISSON-distributed but with parameter $n\xi$. If $n\xi$ is larger than 9 we thus have that $\sum x_i$ is approximately normally distributed with mean $n\xi$ and variance $n\xi$, i. e., the mean $\bar{x}$ is approximately normally distributed with mean ξ and variance ξ/n. Hence the standardized variable

$$u = \frac{\sum_{i=1}^{n} x_i - n\xi}{\sqrt{n\xi}} = \frac{\bar{x}-\xi}{\sqrt{\frac{\xi}{n}}} \tag{22.6.4}$$

is approximately normally distributed with parameters (0, 1) for $n\xi > 9$. If ξ is unknown, the condition $n\xi > 9$ may be replaced by the condition $\sum x_i > 15$, as $15 = 9+2\sqrt{9}$.

22.7. Comparison of Two Poisson-Distributed Observations.

Consider two observations x_1 and x_2 originating from two Poisson-distributed populations with unknown means. We want to test the equality of these means, i. e., $\xi_1 = \xi_2 = \xi$.

If the normal distribution applies to both populations, and the test hypothesis is true, the difference $x_1 - x_2$ will be approximately normally distributed with zero mean and variance 2ξ. Introducing the sum $x_1 + x_2$ as an estimate of 2ξ the standardized variable $u = (x_1 - x_2)/\sqrt{x_1 + x_2}$ will be approximately normally distributed with parameters (0, 1). Taking the correction for continuity into account the critical region for testing the hypothesis $\xi_1 = \xi_2$ against the two-sided alternative $\xi_1 \neq \xi_2$ at the α level of significance becomes

$$\frac{(x_1 - \frac{1}{2}) - (x_2 + \frac{1}{2})}{\sqrt{x_1 + x_2}} = \frac{x_1 - x_2 - 1}{\sqrt{x_1 + x_2}} > u_{1-\frac{\alpha}{2}} \text{ for } x_1 > x_2 . \tag{22.7.1}$$

The corresponding test based on the square root transformation is

$$(\sqrt{x_1 - \tfrac{1}{2}} - \sqrt{x_2 + \tfrac{1}{2}})\sqrt{2} > u_{1-\frac{\alpha}{2}} \text{ for } x_1 > x_2 . \tag{22.7.2}$$

A test based on the Poisson-distribution, analogous to the conditional test given in § 21.13, may be developed as follows. If the test hypothesis is true, the probability of the result (ν_1, ν_2) is

$$p\{\nu_1, \nu_2\} = p\{\nu_1\} p\{\nu_2\} = e^{-\xi} \frac{\xi^{\nu_1}}{\nu_1!} \times e^{-\xi} \frac{\xi^{\nu_2}}{\nu_2!} = e^{-2\xi} \frac{\xi^{\nu_1 + \nu_2}}{\nu_1! \, \nu_2!} . \tag{22.7.3}$$

Introducing the sum $\nu_1 + \nu_2 = x$ as a new variable we find according to the addition theorem in § 22.6 that the distribution function of x is

$$p\{x\} = e^{-2\xi} \frac{(2\xi)^x}{x!} . \tag{22.7.4}$$

Hence the required conditional probability becomes

$$p\{\nu_1 | x\} = \frac{p\{\nu_1, \nu_2\}}{p\{x\}} = \frac{x!}{\nu_1! \, \nu_2!} \left(\frac{1}{2}\right)^x = \binom{x}{\nu_1} \left(\frac{1}{2}\right)^x , \tag{22.7.5}$$

i. e., the $(\nu_1 + 1)$th term of the binomial distribution for $n = x$ and $\theta = \frac{1}{2}$.

Testing the hypothesis $\xi_1 = \xi_2$ against the alternative $\xi_1 > \xi_2$ we therefore compute

$$\begin{aligned} P\{\nu_1 \geq x_1 | x_1 + x_2\} &= \left(\frac{1}{2}\right)^{x_1 + x_2} \sum_{\nu_1 = x_1}^{x_1 + x_2} \binom{x_1 + x_2}{\nu_1} \\ &= 1 - P\left\{v^2 < \frac{x_1}{x_2 + 1}\right\}, \quad \begin{aligned} f_1 &= 2(x_2 + 1), \\ f_2 &= 2x_1 , \end{aligned} \end{aligned} \tag{22.7.6}$$

according to (21.2.6). If this probability is less than or equal to α, the level of significance chosen, the test hypothesis is rejected. Since the inequality

$$P\{v_1 \geq x_1 | x_1+x_2\} \leq \alpha \qquad (22.7.7)$$

may be transformed to

$$\frac{x_1}{x_2+1} \geq v^2_{1-\alpha}(2(x_2+1),\, 2x_1)\,, \qquad (22.7.8)$$

the test is simply carried out by computing the ratio $x_1/(x_2+1)$ and comparing this quantity with the appropriate fractile of the v^2-distribution given in Table VII.

For the two-sided alternative α has to be replaced by $\alpha/2$.

A table of the critical region for x_1 for given $x_1+x_2 = 1\ (1)\ 80$ and $\alpha = 0{\cdot}01, 0{\cdot}05, 0{\cdot}10$, and $0{\cdot}20$ together with an investigation of the power of the test with corresponding tables and diagrams has been given in J. Przyborowski and H. Wiléński: *Homogeneity of Results in Testing Samples From* Poisson *Series*, Biometrika, 31, 1940, 313–323.

Example 22.5. In controlling a production process two items produced by different operators were examined for defects, the number of defects found being 9 and 2, respectively. Is this sufficient evidence for concluding that items produced by the one operator are of poorer quality than items produced by the other?

We assume that the items observed are randomly chosen from a population of items which are Poisson-distributed according to number of defects.

According to (22.7.8) we compute the ratio $x_1/(x_2+1) = 9/3 = 3$. In Table VII we find for $f_1 = 6$ and $f_2 = 18$ that $v^2_{\cdot 975} = 3{\cdot}22$ so that the ratio is not significant at the 5% level.

As a rough test we may use (22.7.1) which gives $u = (9-2-1)/\sqrt{11} = 1{\cdot}81$.

22.8. Comparison of *k* Poisson-Distributed Observations.

Let there be given k Poisson-distributed observations $x_1, x_2, \ldots, x_k$, for which we wish to test the hypothesis $\xi_1 = \xi_2 = \ldots = \xi_k = \xi$. *If k is large*, the distribution of the observations may be tabulated as shown in Example 22.1, where $k = 400$, and the hypothesis tested by comparing this distribution with the theoretical Poisson-distribution as calculated from the sample mean, see § 23.2.

For *small values of k*, when the *shape* of the distribution cannot be studied, the test referred to above must be replaced by a comparison between the sample mean and variance, which—if the test hypothesis is true—should be equal apart from random variations. The test is derived as follows: For large values of ξ, x_i is approximately normally distributed with parameters (ξ, ξ), so that

$$u_i \simeq \frac{x_i - \xi}{\sqrt{\xi}} \qquad (22.8.1)$$

and therefore

$$\chi^2 = \sum_{i=1}^{k} u_i^2 \simeq \frac{\sum_{i=1}^{k}(x_i-\xi)^2}{\xi}, \quad f = k. \qquad (22.8.2)$$

Introducing the mean $\bar{x}$ as an estimate of ξ, we obtain

$$\boxed{\chi^2 \simeq \frac{\sum_{i=1}^{k}(x_i - \bar{x})^2}{\bar{x}} = \frac{k\sum_{i=1}^{k} x_i^2}{\sum_{i=1}^{k} x_i} - \sum_{i=1}^{k} x_i, \quad f = k-1,} \qquad (22.8.3)$$

which is approximately distributed as χ^2 with $k-1$ degrees of freedom. An investigation by P. V. SUKHATME: *On the Distribution of χ^2 in Samples of the* POISSON *Series*, Suppl. Journ. Roy. Stat. Soc., 5, 1938, 75–79, indicates that the approximation to the χ^2-distribution is good even for small values of ξ, $(1 < \xi < 5)$, and k, $(5 < k < 15)$. For smaller values of ξ, i. e., ξ about 1, the values of k must, however, be somewhat larger, i. e., $k > 15$, in order to apply (22.8.3).

22.9. The Compound POISSON-Distribution.

When the analysis of variance is applied to k sets of normally distributed observations and a significant variation between sets has been proved, the analysis may sometimes be continued on the basis of the hypothesis that the variation between sets is a random variation, i. e., the k means $\xi_1, \ldots, \xi_k$ are considered as a random sample from a normally distributed population of means, see § 16.5. It is thus assumed that the variation of the observations is brought about from two sources, which act successively, the first causing the variation between sets, and the second the variation within sets.

The POISSON-distribution may be generalized in a similar manner as we may imagine that a significant variation between k POISSON-distributed observations has been generated by random variation of the parameters of the k POISSON-distributions, the k parameters representing a random sample from a population of parameters. The distribution of the observations will then be dependent on the distribution of this population. This hypothesis was first suggested by M. GREENWOOD and G. U. YULE: *An Inquiry into the Nature of Frequency Distributions Representative of Multiple Happenings*, Journ. Roy. Stat. Soc., 83, 1920, 255–79. In this publication the authors found that the distribution of a number of workmen according to the number of accidents that befell each workman during a certain time interval did not follow POISSON's law. If all workers have the same probability of being subject to an accident, i. e., they are exposed to the same risk and they present equal skill and attention, and the accidents occur independently, it is a priori to be expected that the observations are distributed according to POISSON's law. The assumption regarding the independency of the accidents means that if an accident befalls one workman it must not influence the accidents befalling the other workmen, nor must the said workman later have become more accident prone. In field practice circumstances differ from these assumptions on several points, but GREENWOOD and YULE assume that the main disagreement lies in the fact that as far as accidents are concerned all workers do not present equal skill and attention. Instead they assume that this factor varies from one workman to another. Thus, each workman has his own "accident intensity", ξ, indicating the parameter in the corresponding POISSON-distribution, which gives the probability that this workman will be subject to x accidents.

Hence, the hypothesis is characterized by the following probabilities:

1. $p\{\xi\}d\xi$ = the probability that the accident intensity of a workman, chosen at random, lies between ξ and $\xi+d\xi$.
2. $p\{x;\ \xi\}$ = the probability that a workman with accident intensity ξ is subject to x accidents in the given time interval.

From the multiplication theorem we obtain the probability that a workman chosen at random will present the accident intensity ξ and be subject to x accidents as $p\{\xi\}d\xi\ p\{x;\ \xi\}$. According to the addition theorem integration of this expression leads to the probability that a workman chosen at random (with unknown accident intensity) is subject to x accidents

$$p\{x\} = \int_0^\infty p\{x;\ \xi\}\ p\{\xi\}d\xi\ . \tag{22.9.1}$$

According to our assumptions we have

$$p\{x;\ \xi\} = e^{-\xi}\frac{\xi^x}{x!}\ , \quad x = 0,\ 1,\ \ldots, \tag{22.9.2}$$

and GREENWOOD and YULE have chosen as working hypothesis

$$p\{\xi\} = \frac{\gamma^\alpha}{(\alpha-1)!}e^{-\gamma\xi}\,\xi^{\alpha-1}\ , \quad \xi \geq 0\ . \tag{22.9.3}$$

Insertion in (22.9.1) gives

$$p\{x\} = \frac{\gamma^\alpha}{(\alpha-1)!\,x!}\int_0^\infty e^{-(\gamma+1)\xi}\,\xi^{x+\alpha-1}d\xi$$

$$= \left(\frac{\gamma}{\gamma+1}\right)^\alpha \frac{\alpha(\alpha+1)\ldots(\alpha+x-1)}{x!}\left(\frac{1}{\gamma+1}\right)^x$$

$$= \left(\frac{\gamma}{\gamma+1}\right)^\alpha \binom{-\alpha}{x}(-1)^x\left(\frac{1}{\gamma+1}\right)^x, \tag{22.9.4}$$

where

$$\binom{-\alpha}{x} = (-1)^x \frac{\alpha(\alpha+1)\ldots(\alpha+x-1)}{x!}. \tag{22.9.5}$$

This distribution is called *the compound* Poisson-*distribution* or *the binomial distribution with negative index*, $p\{x\}$ being equal to the $(x+1)$st term in the expansion of

$$\left(\frac{\gamma+1}{\gamma}-\frac{1}{\gamma}\right)^{-\alpha} = \left(\frac{\gamma}{\gamma+1}\right)^\alpha\left(1-\frac{1}{\gamma+1}\right)^{-\alpha}. \tag{22.9.6}$$

(The notation "the compound Poisson-distribution" is not quite adequate, as Poisson-distributions may be combined in many other ways than that given by the special "weight-function" (22.9.3)).

Utilizing (5.7.2) and (5.9.5) we find the mean and variance as

$$\mathcal{M}\{x\} = \frac{\alpha}{\gamma} \tag{22.9.7}$$

and

$$\mathcal{V}\{x\} = \frac{\alpha}{\gamma}\frac{\gamma+1}{\gamma}. \tag{22.9.8}$$

As the distribution has two parameters, α and γ, it is more flexible than the Poisson-distribution. Estimates of the parameters may be determined by substituting the mean $\bar{x}$ and the variance s^2 for $\mathcal{M}\{x\}$ and $\mathcal{V}\{x\}$ in (22.9.7) and (22.9.8). Solution of the equations leads to

$$\alpha \approx \frac{\bar{x}^2}{s^2-\bar{x}} \tag{22.9.9}$$

and

$$\gamma \approx \frac{\bar{x}}{s^2-\bar{x}}. \tag{22.9.10}$$

The distribution function may be calculated by successive multiplications from

$$p\{0\} = \left(\frac{\gamma}{\gamma+1}\right)^\alpha, \tag{22.9.11}$$

$f(x)$ first being tabulated from the formula

$$f(x) = \frac{p\{x+1\}}{p\{x\}} = \frac{1}{\gamma+1}\frac{x+\alpha}{x+1}, \quad x = 0, 1, \ldots . \qquad (22.9.12)$$

The principal publication on the compound POISSON-distribution is the one cited above by GREENWOOD and YULE. A more extensive analysis of the problems set forth in this article has been made in E. NEWBOLD: *Practical Applications of the Statistics of Repeated Events, Particularly to Industrial Accidents*, Journ. Roy. Stat. Soc., 90, 1927, 487–547, and E. G. CHAMBERS and G. U. YULE: *Theory and Observation in the Investigation of Accident Causation*, Suppl. Journ. Roy. Stat. Soc., 7, 1941, 89–109.

The estimates of α and γ calculated from (22.9.9) and (22.9.10) are not efficient, as we can determine estimates with a smaller standard error. The determination of the maximum likelihood estimates requires more complicated computations than (22.9.9) and (22.9.10), cf. J. B. S. HALDANE: *The Fitting of Binomial Distributions*, Annals of Eugenics, 11, 1941, 179–81, R. A. FISHER: *The Negative Binomial Distribution*, Annals of Eugenics, 11, 1941, 182–87, and M. E. WISE: *The Use of the Negative Binomial Distribution in an Industrial Sampling Problem*, Suppl. Journ. Roy. Stat. Soc., 8, 1946, 202–11.

In § 22.10 the compound POISSON-distribution is derived in another manner than that employed by GREENWOOD and YULE.

Example 22.6. Table 22.4 shows the distribution of 647 women according to the accidents that befell them in a shell factory during a period of 5 weeks, see GREENWOOD and YULE, cited above.

TABLE 22.4.

Distribution of 647 women according to number of accidents.

Number of accidents (x)	Number of women	POISSON 647 $p\{x\}$	Compound POISSON 647 $p\{x\}$
0	447	406·3	442·9
1	132	189·0	138·5
2	42	44·0	44·4
3	21	6·8	14·3
4	3	0·8	4·6
5	2	0·1	1·5
≥ 6	0	0·0	0·7
	647	647·0	646·9

Thus we see that among the 647 women 2 were subject to 5 accidents, 3 to 4 accidents, 21 to 3 accidents, etc., during the 5 weeks.

First we examine whether the distribution can be interpreted as a Poisson-distribution. The mean is 0·4652 and the variance 0·6919. This leads to the ratio

$$\frac{s^2}{\bar{x}} = \frac{0{\cdot}6919}{0{\cdot}4652} = 1{\cdot}49 ,$$

which according to (22.8.3) is approximately distributed as χ^2/f with $f = 646$, if the observations are Poisson-distributed. As $\chi^2_{\cdot 9995}/f = 1{\cdot}19$, see Table VI, the hypothesis must be rejected. For comparison with the observed distribution the Poisson-distribution with mean 0·4652 has been entered in the third column of Table 22.4. This shows that the number of women who, according to the Poisson-distribution, were expected to have 0 accidents is considerably smaller than the observed number.

Inserting $\bar{x} = 0{\cdot}4652$ and $s^2 = 0{\cdot}6919$ in (22.9.9) and (22.9.10) we obtain $\alpha \approx 0{\cdot}9546$ and $\gamma \approx 2{\cdot}052$, which when substituted in (22.9.4) give us an estimate of $p\{x\}$. The numbers computed in this manner have been entered in the last column of Table 22.4, which shows a considerably closer agreement with the observations than the Poisson-distribution.

22.10. Discontinuous Stochastic Processes.

By a discontinuous stochastic process we understand a process which includes a stochastic variable with a discontinuous distribution function depending on a parameter that varies continuously or discontinuously.

As a simple example of a discontinuous stochastic process we may consider a process where the stochastic variable is defined as the number of times a certain event (the event U) occurs during a period of length t (t = the continuously varying parameter). We assume that the probability of the occurrence of U in a period of length Δt is asymptotically proportional to Δt and independent of the position of this period and of the number of times the event has previously occurred, i. e., the probability that the event will occur in the period $(t, t+\Delta t)$ is supposed to be equal to $\xi \Delta t + o(\Delta t)$, $o(\Delta t)$ denoting an infinitesimal of higher order than Δt, i. e., $o(\Delta t)/\Delta t$ converges to 0 for $\Delta t \to 0$.

The course of the process in the period $(t, t+\Delta t)$ may be characterized by the following probabilities:

1. The probability that the event will not occur is $1-\xi\Delta t$.
2. The probability that the event will occur exactly once is $\xi\Delta t$.
3. The probability that the event will occur twice or more often is $o(\Delta t)$.

Using these probabilities we can obtain a formula which gives the probability $P_x(t+\Delta t)$ that the event U will occur exactly x times during the period $(0, t+\Delta t)$. This period is divided by the time t into two intervals $(0, t)$ and $(t, t+\Delta t)$, so that the condition for x occurrences of the event in the interval $(0, t+\Delta t)$ may be partitioned into the following three complementary conditions:

1. The event will occur exactly x times in $(0, t)$ and not in $(t, t+\Delta t)$.
2. The event will occur $x-1$ times in $(0, t)$ and once in $(t, t+\Delta t)$.
3. The event will occur $x-\nu$ times in $(0, t)$ and ν times in $(t, t+\Delta t)$, where $\nu = 2, 3, \ldots, x$.

According to the multiplication theorem the three corresponding probabilities are:

1. $P_x(t)(1-\xi\Delta t)$.
2. $P_{x-1}(t)\xi\Delta t$.
3. $o(\Delta t)$.

According to the addition theorem the required probability may then be found as

$$\boxed{P_x(t+\Delta t) = P_x(t)(1-\xi\Delta t)+P_{x-1}(t)\xi\Delta t+o(\Delta t)\ , \quad x = 1, 2, \ldots\ ,} \tag{22.10.1}$$

and

$$P_0(t+\Delta t) = P_0(t)(1-\xi\Delta t)+o(\Delta t)\ , \tag{22.10.2}$$

which may be transformed into

$$\frac{P_x(t+\Delta t)-P_x(t)}{\Delta t} = \xi\,(P_{x-1}(t)-P_x(t)) + \frac{o(\Delta t)}{\Delta t}\ , \quad x = 1, 2, \ldots\ ,$$

and

$$\frac{P_0(t+\Delta t)-P_0(t)}{\Delta t} = -\xi P_0(t)+\frac{o(\Delta t)}{\Delta t}\ .$$

For $\Delta t \to 0$ we find the differential equations

$$\frac{dP_x(t)}{dt} = \xi(P_{x-1}(t)-P_x(t))\ , \quad x = 1, 2, \ldots\ , \tag{22.10.3}$$

and

$$\frac{dP_0(t)}{dt} = -\xi P_0(t)\ . \tag{22.10.4}$$

If these equations are supplemented by the equations $P_0(0) = 1$ and $P_x(0) = 0$ for $x \geq 1$ as expressing the initial stage, we obtain the solution

$$\boxed{P_x(t) = e^{-\xi t}\frac{(\xi t)^x}{x!}\ , \quad x = 0, 1, \ldots\ ,} \tag{22.10.5}$$

i. e., the POISSON-distribution with parameter ξt. (This result may also be obtained directly from the binomial distribution if we introduce a "duration" of each observation and assume that this "duration" is inversely proportional to the number, n, of observations for $n \to \infty$).

The above derivation of the POISSON-distribution shows that we may expect *the number of occurrences of an event during a given period* to be distributed according to POISSON's law if *the probability that the event occurs in an (infinitesimally) short period Δt is proportional to Δt, and the events occur independently of one another.* In this mode of reasoning time may be replaced by some other continuous variable, e. g., length, area, or volume, so that the number of occurrences of the event within a given area must be expected to be distributed according to POISSON's law if the probability that the event will occur within an (infinitesimally) small area Δt is proportional to Δt, and the events occur independently of one another.

As examples of the application of the POISSON-distribution from the above viewpoint we may state that (22.10.5) has been used to describe the distribution of the number of calls to a telephone exchange during a time interval of length t, or the number of cars passing a given street in the time interval t, the number of radioactive atoms decaying in a given time period, and the number of particles (blood corpuscles, bacteria, grains of cement) within a given area of the microscopic field.

The mode of reasoning which led to the POISSON process may be applied to the derivation of more general processes, the "intensity" ξ being replaced by a function of x and t, i. e., it is presumed that *the probability that the event U will occur in the interval $(t, t+\Delta t)$ is proportional to Δt and depends on t and on the number of times, x, the event has occurred before t.* The probability that the event U will occur exactly once in $(t, t+\Delta t)$ may thus be written as $p_x(t)\Delta t$, from which it follows that the probability that the event will not occur is $1-p_x(t)\Delta t$. It is further assumed that the probability that the event will occur twice or more times is $o(\Delta t)$. We then have, by analogy with (22.10.1),

$$\boxed{P_x(t+\Delta t) = P_x(t)(1-p_x(t)\Delta t)+P_{x-1}(t)p_{x-1}(t)\Delta t+o(\Delta t)\,, \quad x = 1, 2, \ldots,} \tag{22.10.6}$$

which leads to the differential equations

$$\frac{dP_x(t)}{dt} = p_{x-1}(t)P_{x-1}(t)-p_x(t)P_x(t)\,, \quad x = 1, 2, \ldots\,, \tag{22.10.7}$$

supplemented by

$$\frac{dP_0(t)}{dt} = -p_0(t)P_0(t) \tag{22.10.8}$$

and

$$P_0(0) = 1 \quad \text{and} \quad P_x(0) = 0 \quad \text{for} \quad x \geq 1 \, .$$

As a special case of (22.10.7) we will consider a process with intensity

$$p_x(t) = \frac{\alpha+x}{\beta+t}, \quad \alpha \geq 0 \text{ and } \beta \geq 0 \, , \qquad (22.10.9)$$

i. e., at a given time t the probability that the event U will occur in the interval $(t, t+\Delta t)$ increases linearly with the number, x, of the events which have occurred in $(0, t)$, and for a given value of x the probability decreases with increasing values of t. This process is called the PÓLYA-process, and amongst other applications the corresponding distribution has been used to describe the number of deaths from epidemic disease which occur during each month for a certain number of years. (Several other "contagious-distributions" have been suggested corresponding to other intensities of contagiousness than (22.10.9), see the references on p. 735).

For the sake of analogy with the POISSON-process, (22.10.9) is transcribed as follows

$$p_x(t) = \frac{1+\frac{x}{\alpha}}{1+\frac{t}{\beta}} \frac{\alpha}{\beta} = \frac{1+\frac{x}{\alpha}}{1+\frac{\xi t}{\alpha}} \xi \, , \quad \alpha \geq 0 \text{ and } \xi \geq 0 \, , \qquad (22.10.10)$$

where $\xi = \alpha/\beta$. Introducing (22.10.10) in the differential equations (22.10.7) and (22.10.8), and solving these equations with respect to $P_x(t)$, we obtain the PÓLYA-distribution

$$P_x(t) = \left(\frac{\alpha}{\alpha+\xi t}\right)^{\alpha} \frac{\alpha(\alpha+1)\ldots(\alpha+x-1)}{x!} \left(\frac{\xi t}{\alpha+\xi t}\right)^{x}$$

$$= \left(\frac{\alpha}{\alpha+\xi t}\right)^{\alpha} \binom{-\alpha}{x} (-1)^x \left(\frac{\xi t}{\alpha+\xi t}\right)^{x} \, , \quad x = 0, 1, \ldots \qquad (22.10.11)$$

This distribution is identical with GREENWOOD-YULE's compound POISSON-distribution, see (22.9.4), for $\gamma = \alpha/\xi t$. Thus, the mean is ξt, and the variance $\xi t \left(1+\frac{\xi t}{\alpha}\right)$. For $1/\alpha \to 0$ we obtain the POISSON-distribution (22.10.5). For $\alpha \to 1$ we obtain a PASCAL-distribution, cf. (2.3.1).

It is noteworthy that the GREENWOOD-YULE-PÓLYA-distribution may be considered as resulting from two very different hypotheses; GREENWOOD and YULE assume that the events are mutually independent, and that the intensities vary from individual to individual, while PÓLYA assumes that the events are stochastically dependent, the occurrence of an event increasing the probability that further events will occur. Thus, a good agreement between an observed distribution and the distribution (22.10.11) may be interpreted at any rate in two ways, so that a further analysis is necessary in order to choose the model suitable for the explanation of the generation of the observations.

The POISSON and PÓLYA processes are simple examples of stochastic processes. The theory of stochastic processes and their application are as yet in their first stages of development, but it looks as if in the years to come further development will take place, and we shall probably find that we here have one of the most valuable tools of statistics.

References regarding stochastic processes.

A. KOLMOGOROFF: *Über die analytischen Methoden in der Wahrscheinlichkeitsrechnung*, Math. Annalen, 104, 1931, 415–58.

W. FELLER: *Zur Theorie der stochastischen Prozesse*, Math. Annalen, 113, 1936, 113–60.

W. FELLER: *Die Grundlagen der Volterraschen Theorie des Kampfes um Dasein in wahrscheinlichkeitstheoretischer Behandlung*, Acta Biotheoretica, Leiden, 1939, 11–40.

M. FRÉCHET: *Recherches théoriques modernes*, Livre II, Traité du calcul des probabilités et de ses applications, tome I, fasc. 3, Paris, 1938.

O. LUNDBERG: *On Random Processes and Their Application to Sickness and Accident Statistics*, Uppsala, 1940 (Thesis).

N. ARLEY: *On the Theory of Stochastic Processes and Their Application to the Theory of Cosmic Radiation*, Copenhagen, 1943 (Thesis).

C. PALM: *Intensitätsschwankungen in Fernsprechverkehr. Untersuchungen über die Darstellung auf Fernsprechverkehrsprobleme anwendbarer stochastischer Processe*, Ericsson Technics, No. 44, 1943, Stockholm (Thesis).

D. G. KENDALL: *On the Generalized "Birth and Death" Process*, Ann. Math. Stat., 19, 1948, 1–15.

W. FELLER: *On the Theory of Stochastic Processes, with Particular Reference to Applications*, Berkeley Symposium on Mathematical Statistics, 403–432, Univ. of Cal. Press, 1949.

Symposium on Stochastic Processes, Journ. Roy. Stat. Soc., B, 11, 1949, 150–282.

W. FELLER: *Probability Theory and Its Applications*, J. Wiley, New York, 1950.

LUNDBERG'S and ARLEY'S theses among other things include an exposition of the POISSON and PÓLYA processes. These subjects and related matters have also been discussed in:

M. GREENWOOD and G. U. YULE: *An Inquiry into the Nature of Frequency Distributions Representative of Multiple Happenings*, Journ. Roy. Stat. Soc., 83, 1920, 255–79.

E. Newbold: *Practical Applications of the Statistics of Repeated Events, Particularly to Industrial Accidents*, Journ. Roy. Stat. Soc., 90, 1927, 487–547.

E. G. Chambers and G. U. Yule: *Theory and Observation in the Investigation of Accident Causation*, Suppl. Journ. Roy. Stat. Soc., 7, 1941, 89–109.

F. Eggenberger and G. Pólya: *Über die Statistik verketteter Vorgänge*, Zeitschrift für angewandte Mathematik und Mechanik, 1, 1923, 279–89.

F. Eggenberger: *Die Wahrscheinlichkeitsansteckung*, Mitteilungen der Vereinigung schweizerischer Versicherungs-Mathematiker, 1924, 31–144.

G. Pólya: *Sur quelques points de la théorie des probabilités*, Annales de l'Institut Henri Poincaré, 1, 1930, 117–61.

W. Feller: *On a General Class of "Contagious Distributions"*, Ann. Math. Stat., 14, 1943, 389–400.

As an example of the application of the theory of stochastic processes in the technical world we may mention:

H. A. Einstein: *Der Geschiebetrieb als Wahrscheinlichkeitsproblem*, Mitteilung der Versuchsanstalt für Wasserbau an der Eidg. Technischen Hochschule, Zürich, 1937.

G. Pólya: *Zur Kinematik der Geschiebebewegung*, Mitteilung der Versuchsanstalt für Wasserbau an der Eidg. Technischen Hochschule, Zürich, 1937.

Applications to telephone problems will be indicated in § 22.11.

22.11. Statistical Equilibrium.

Even before the general theory was developed in the 1930's discontinuous stochastic processes had been applied to practical work, for instance in the theory of telephone systems. In 1909 A. K. Erlang derived the Poisson-distribution in a manner similar to that described in § 22.10 for the description of the distribution of telephone calls. By similar methods A. K. Erlang has solved several important problems related to the application of the probability calculus to the theory of telephone systems; usually, however, he regarded only the process for $t \to \infty$, a condition denoted as "statistical equilibrium". As a typical example of Erlang's method we shall below derive Erlang's B-formula for the probability that a call will be blocked (lost).

Let us consider a simple cable, i. e., a group of lines which co-operate in such a manner that any call may seek any of the free lines. The charge on the cable depends not only on the number of calls, but also on the time, "the holding time", during which each line is occupied. *The traffic intensity ξ is defined as the average number of calls per average holding time*, i. e., the mean holding time is used as unit of time. It is assumed that the calls arrive independently with intensity ξ. Further, it is assumed that the probability that a call which has lasted t time units will end in the time interval $(t, t+\Delta t)$ is proportional to Δt and independent of t, i. e., equal to Δt, as the time unit is equal to the average holding time. If the cable contains a limited number of lines, k, some of the calls will occur at a time when all lines are occupied, so that these calls are lost. ERLANG's formula gives the probability, P_x, $x = 0, 1, \ldots, k$, that x lines are occupied, and in particular P_k denotes the probability of losing a call (blocking).

According to the addition theorem the probability is $x\Delta t$ that a call will end in the time interval Δt, when exactly x calls are taking place at the beginning of Δt. Further, the probability that more than one event (arrival or departure of a call) will occur in the time interval Δt is $o(\Delta t)$, and the probability of the combination "one arrival and no departures" during Δt is equal to $\xi\Delta t(1-x\Delta t) = \xi\Delta t+o(\Delta t)$.

The probability that x lines are occupied at the time $t+\Delta t$ may then be represented by the formula

$$P_x(t+\Delta t) = P_{x-1}(t)\xi\Delta t+P_x(t)(1-\xi\Delta t-x\Delta t)+P_{x+1}(t)(x+1)\Delta t+o(\Delta t)\,, \tag{22.11.1}$$

where $\xi\Delta t$ denotes the probability of a call during Δt, $1-\xi\Delta t-x\Delta t$ the probability of no new arrivals or departures of any of the x calls, and $(x+1)\Delta t$ the probability that one of $(x+1)$ calls will depart during Δt. As we are only considering the equilibrium state where the probabilities are independent of t, we obtain from (22.11.1), as $P_x(t+\Delta t) = P_x(t) = P_x$, the following difference equations:

$$(x+1)P_{x+1}-(\xi+x)P_x+\xi P_{x-1} = 0\,, \quad x = 1, 2, \ldots, k-1\,, \tag{22.11.2}$$

and

$$P_1-\xi P_0 = 0\,. \tag{22.11.3}$$

This leads to

$$P_x = \frac{\xi^x}{x!}\Big/\sum_{x=0}^{k}\frac{\xi^x}{x!}\,, \quad x = 0, 1, \ldots, k\,. \tag{22.11.4}$$

Thus, the distribution may be called *a truncated* POISSON-*distribution*, as it arises by "normalization" of the first $k+1$ terms in a POISSON-distribution with parameter ξ. For $k \to \infty$ we obtain the POISSON-distribution.

The probability, P_k, that all k lines are occupied plays an important rôle when it is to be determined how many lines are to be included in a cable.

A discussion of A. K. ERLANG's results has been given by E. BROCKMEYER, H. L. HALSTROM and ARNE JENSEN: *The Life and Works of* A. K. ERLANG, Trans. Danish Acad. Technical Sci., No. 3, 1948. A thorough discussion of the theory of probability as applied to problems of congestion may be found in Chapter 10 of T. C. FRY: *Probability and Its Engineering*

Uses, Van Nostrand, New York, 1929, see also the thesis by C. PALM quoted in § 22.10 and L. KOSTEN, J. R. MANNING and F. GARWOOD: *On the Accuracy of Measurements of Probabilities of Loss in Telephone Systems*, Journ. Roy. Stat. Soc., B, 1949, 11, 54–67.

The methods and results from the application of the probability calculus to telephone systems may be transferred to other fields of work, see for example F. GARWOOD: *An Application of the Theory of Probability to the Operation of Vehicular-Controlled Traffic Signals*, Suppl. Journ. Roy. Stat. Soc., 7, 1940–41, 65–77.

CHAPTER 23.

THE MULTINOMIAL DISTRIBUTION AND THE χ^2-TEST

23.1. The Multinomial Distribution and the χ^2-Distribution.

In § 2.2 we derived the formula

$$p\{x_1, x_2, \ldots, x_k\} = \frac{n!}{x_1!\,x_2!\ldots x_k!}\,\theta_1^{x_1}\theta_2^{x_2}\ldots\theta_k^{x_k}, \tag{23.1.1}$$

which denotes the probability that the event U_1 will occur x_1 times, the event U_2 x_2 times, ..., and the event U_k x_k times in n observations, $x_1+x_2\ldots+x_k = n$, assuming (1) that the observations may have k different results, $U_1, U_2, \ldots, U_k$, which are mutually exclusive, (2) the probability of the result U_i is θ_i, $\theta_1+\theta_2+\ldots+\theta_k = 1$, and (3) the observations are stochastically independent. As a special case we have the binomial distribution for $k = 2$.

It follows from § 21.5 that the standardized variable

$$\frac{x_i - n\theta_i}{\sqrt{n\theta_i(1-\theta_i)}} \tag{23.1.2}$$

for large values of $n\theta(1-\theta)$ will be approximately normally distributed with parameters (0, 1), so that

$$\sum_{i=1}^{k}\frac{(x_i - n\theta_i)^2}{n\theta_i(1-\theta_i)} \tag{23.1.3}$$

would be approximately distributed as χ^2 with k degrees of freedom *if the x's were mutually independent*. On account of the correlation between the x's and the relation $\sum x_i = n$, (23.1.3) must be replaced by

$$\boxed{\chi^2 = \sum_{i=1}^{k}\frac{(x_i - n\theta_i)^2}{n\theta_i}, \quad f = k-1,} \tag{23.1.4}$$

which is approximately distributed as χ^2 with $k-1$ degrees of freedom, if all $n\theta > 5$, cf. e. g. H. CRAMÉR: *Mathematical Methods of Statistics*, Princeton, 1946, Chap. 30.1.

The sum of squares (23.1.4) thus includes k terms

$$y_i = \frac{x_i - n\theta_i}{\sqrt{n\theta_i}}, \quad i = 1, 2, \ldots, k, \tag{23.1.5}$$

satisfying the linear relation

$$\sum_{i=1}^{k} y_i \sqrt{n\theta_i} = \sum_{i=1}^{k} x_i - n \sum_{i=1}^{k} \theta_i = 0, \tag{23.1.6}$$

wherefore the number of degrees of freedom is $k-1$.

The k y-values denote the deviations between the observed and the theoretical ("expected") numbers divided by the square root of the theoretical numbers.

For $k = 2$ we obtain from (23.1.4) for $\theta_1 = \theta$, $\theta_2 = 1-\theta$, $x_1 = x$ and $x_2 = n-x$:

$$\chi^2 = \frac{(x_1 - n\theta_1)^2}{n\theta_1} + \frac{(x_2 - n\theta_2)^2}{n\theta_2} = \frac{(x-n\theta)^2}{n\theta} + \frac{(n\theta - x)^2}{n(1-\theta)} = \frac{(x-n\theta)^2}{n\theta(1-\theta)},$$

which quantity is approximately distributed as χ^2 with one degree of freedom in accordance with the result obtained in § 21.5 for the binomial distribution.

The above result may for instance be applied to the comparison of an observed grouped distribution with a theoretical distribution, see Table 23.1, in which the notation used in Table 3.3, p. 47, has again been applied. The classes must be chosen in such a manner that all $n\theta_i > 5$. The probabilities θ_i are determined from the distribution function by summation or integration over the class interval in question. If the value obtained for χ^2 exceeds, e. g., $\chi^2_{.95}$, the θ's are rejected as theoretical values of the observed frequencies.

TABLE 23.1.

Comparison between an observed grouped distribution and a theoretical distribution.

Class midpoint t_i	Class length Δt_i	Number of observations a_i	Theoretical number $n\theta_i$	Deviations $a_i - n\theta_i$	$y_i^2 = \frac{(a_i - n\theta_i)^2}{n\theta_i}$
t_1	Δt_1	a_1	$n\theta_1$	$a_1 - n\theta_1$	y_1^2
t_2	Δt_2	a_2	$n\theta_2$	$a_2 - n\theta_2$	y_2^2
$\vdots$	$\vdots$	$\vdots$	$\vdots$	$\vdots$	$\vdots$
t_k	Δt_k	a_k	$n\theta_k$	$a_k - n\theta_k$	y_k^2
Total		n	n	0	χ^2

The above results are due to K. PEARSON: *On a Criterion that a Given System of Deviations from the Probable in the Case of a Correlated System of Variables is such that It Can be Reasonably Supposed to Have Arisen from Random Sampling*, Phil. Mag., Series V, 50, 1900, 157–75.

Example 23.1. When reading a scale, e. g., on a balance or a slide rule, where the last figure is estimated, it is often seen that the observer prefers certain figures to others. Table 23.2 gives the distribution of 200 randomly chosen routine readings by a certain observer according to the magnitude of the estimated figure. The table shows that the figures 0 and 8 appear somewhat more frequently than the other figures. The question to be decided is whether the observer is making a systematic error. This question is answered by comparing the observed distribution with a polynomial distribution in which

$$\theta_i = \frac{1}{10} \text{ for } i = 1, 2, \ldots, 10.$$

TABLE 23.2.
Distribution of 200 readings according to magnitude of last figure.

Figure	Observed number a	Theoretical number $n\theta$	Deviations $a-n\theta$	$\frac{(a-n\theta)^2}{n\theta}$
0	35	20	15	11·25
1	16	20	−4	0·80
2	15	20	−5	1·25
3	17	20	−3	0·45
4	17	20	−3	0·45
5	19	20	−1	0·05
6	11	20	−9	4·05
7	16	20	−4	0·80
8	30	20	10	5·00
9	24	20	4	0·80
	200	200	0	24·90

Table 23.2 gives the computation of the deviations between the observed and the theoretical numbers together with the calculated value of χ^2 according to (23.1.4). As $\chi^2 = 24{\cdot}90$ with 9 degrees of freedom exceeds the 99·5% fractile of the χ^2-distribution, see Table V, the test hypothesis must be rejected.

23.2. Comparison of an Observed Grouped Distribution and a Theoretical Distribution.

The method described in § 23.1 for comparing an observed and a theoretical distribution is based on the presumption that the theoretical distribution is known, so that it is possible to calculate the probabilities θ_i. Usually, however, *estimates* of the parameters must be determined from the observed distribution, whence it is necessary to modify the test given in § 23.1. The probabilities θ_i are replaced by the estimates $\hat{\theta}_i$, which are determined in the same manner as θ_i, the theoretical values of the parameters of the distribution being replaced by their estimates, see § 6.8, Example 7.1, and Example 22.1. If the estimates of c parameters are used, the k quantities

$$\frac{a_i - n\hat{\theta}_i}{\sqrt{n\hat{\theta}_i}}, \quad i = 1, 2, \ldots, k\,, \tag{23.2.1}$$

will be subject to c linear constraints, so that

$$\boxed{\chi^2 = \sum_{i=1}^{k} \frac{(a_i - n\hat{\theta}_i)^2}{n\hat{\theta}_i}, \quad f = k - c - 1\,,} \tag{23.2.2}$$

is approximately distributed as χ^2 with $k-c-1$ degrees of freedom, if all $n\hat{\theta}_i > 5$. (We here assume that the estimates used are the "best" estimates, cf. H. Cramér, loc. cit., Chap. 30.3). Hence, the agreement between an observed and a hypothetical distribution is tested by calculating χ^2 according to (23.2.2) and comparing this with the fractiles of the χ^2-distribution. If χ^2 exceeds for instance $\chi^2_{.95}$, the hypothetical distribution is rejected. R. A. Fisher has emphasized that a small value of χ^2, e. g., $\chi^2 < \chi^2_{.05}$, should lead to rejection of the hypothesis, the agreement being "too good to be true". Too good an agreement may be due to the fact that the hypothetical distribution includes too large a number of parameters, so that it should be possible to use a smaller number of parameters and still obtain an adequate description of the observed distribution.

The χ^2-test should be applied with caution, partly because it is based on an approximation and depends on the grouping used, and partly because the signs of the deviations are not taken into account. The test should therefore be supplemented by an investigation of the runs in the series of deviations.

The above modification of K. Pearson's χ^2-test has been given by R. A. Fisher: *The Conditions under Which χ^2 Measures the Discrepancy between Observation and Hypothesis*, Journ. Roy. Stat. Soc., 87, 1924, 442–49.

Example 23.2. Table 23.3 gives a comparison between the observed distribution in Table 22.1 and the POISSON-distribution with the same mean. The extreme groups have been combined, so that $n\theta_i > 5$ for all values of i.

TABLE 23.3.
Comparison between an observed distribution and a calculated POISSON-distribution. ("Student's" data).

Number of yeast cells x	Number of volume units a	Calculated number $n\hat{\theta}$	Deviations $a-n\hat{\theta}$	$\frac{(a-n\hat{\theta})^2}{n\hat{\theta}}$
≤ 1	20	21·08	−1·08	0·06
2	43	40·65	2·35	0·14
3	53	63·41	−10·41	1·71
4	86	74·19	11·81	1·88
5	70	69·44	0·56	0·00
6	54	54·16	−0·16	0·00
7	37	36·21	0·79	0·02
8	18	21·18	−3·18	0·48
9	10	11·02	−1·02	0·09
≥ 10	9	8·66	0·34	0·01
Total	400	400·00	0·00	4·39

The observed and the calculated distributions are in agreement on two points: they both include 400 results and they have the same mean. The number of degrees of freedom is therefore 8 (the number of classes minus 2). As $\chi^2 = 4{\cdot}39$ is not significant, see Table V, the observed distribution does not differ significantly from the calculated POISSON-distribution.

Example 23.3. Table 23.4 gives a comparison between the distribution in Table 7.2 and the calculated normal distribution with the same mean, $\bar{t} = 2{\cdot}7959$, and standard deviation, $s = 0{\cdot}2421$. The method used for calculating the normal distribution follows from § 6.8.

The extreme groups have been combined, so that $n\hat{\theta}_i > 5$ for all i. The distribution includes 13 classes, which makes the number of degrees of freedom equal to 10, as $c = 2$ parameters, the mean and the variance, are estimated. The value of χ^2 found is somewhat larger than the 95% fractile, 18·3. It is thus seen that the agreement between the observed distribution and the normal distribution is not particularly good, and it must be considered doubtful whether the theoretical distribution corresponding to the observed distribution is normal.

TABLE 23.4.

Comparison between the distribution of 750 consumers of electricity according to the logarithms of the consumption time in hours and the corresponding normal distribution.

Class midpoint t	Number of consumers a	Calculated number $n\hat{\theta}$	Deviations $a-n\hat{\theta}$	$\frac{(a-n\hat{\theta})^2}{n\hat{\theta}}$
$\leq 2{\cdot}15$	9	5·2	3·8	2·8
2·25	8	10·0	−2·0	0·4
2·35	15	23·1	−8·1	2·8
2·45	39	44·8	−5·8	0·8
2·55	73	73·8	−0·8	0·0
2·65	109	102·6	6·4	0·4
2·75	141	120·6	20·4	3·5
2·85	123	119·7	3·3	0·1
2·95	86	100·5	−14·5	2·1
3·05	65	71·3	−6·3	0·6
3·15	52	42·7	9·3	2·0
3·25	14	21·7	−7·7	2·7
$\geq 3{\cdot}35$	16	14·0	2·0	0·3
	750	750·0	0·0	18·5

23.3. A Test of Stochastical Independence.

Let n observations be classified according to two attributes, U and V, as shown in Table 23.5, a_{ij} denoting the number of results with attributes U_iV_j. The attributes may be qualitative as in Table 1.3 or quantitative as in Table 4.7. The probability of obtaining the result U_iV_j is denoted by θ_{ij}.

If a set of hypothetical probabilities θ_{ij} is given, the agreement between the observed and the hypothetical distribution may be tested by calculating

TABLE 23.5.

Classification of n observations according to two attributes, U and V.

	V_1	V_2	$\dots$	V_k	Total
U_1	a_{11}	a_{12}	$\dots$	a_{1k}	a_{10}
U_2	a_{21}	a_{22}	$\dots$	a_{2k}	a_{20}
$\vdots$	$\vdots$	$\vdots$		$\vdots$	$\vdots$
U_m	a_{m1}	a_{m2}	$\dots$	a_{mk}	a_{m0}
Total	a_{01}	a_{02}		a_{0k}	n

$$\chi^2 = \sum_{i=1}^{m} \sum_{j=1}^{k} \frac{(a_{ij}-n\theta_{ij})^2}{n\theta_{ij}}\,, \quad f = mk-1\,, \tag{23.3.1}$$

which is approximately distributed as χ^2 with $mk-1$ degrees of freedom according to (23.1.4).

Generally the probabilities θ_{ij} are unknown, and estimates of θ_{ij} are found according to a hypothesis with a number of parameters less than mk.

The hypothesis of stochastical independence means that

$$P\{U_iV_j\} = P\{U_i\}P\{V_j\}$$

or

$$\theta_{ij} = \theta_{i0}\theta_{0j}\,, \tag{23.3.2}$$

where we have $P\{U_i\} = \theta_{i0}$ and $P\{V_j\} = \theta_{0j}$, see § 1.5. As

$$\sum_{i=1}^{m} \theta_{i0} = \sum_{j=1}^{k} \theta_{0j} = 1\,, \tag{23.3.3}$$

the hypothetical distribution includes $(m-1)+(k-1) = m+k-2$ parameters. The estimates of these parameters are given by the marginal frequencies

$$\frac{a_{i0}}{n} \approx \theta_{i0} \quad \text{and} \quad \frac{a_{0j}}{n} \approx \theta_{0j}\,, \tag{23.3.4}$$

so that the estimate of the "expected" number of observations with attribute U_iV_j is

$$\frac{a_{i0}a_{0j}}{n} \approx n\theta_{ij}\,. \tag{23.3.5}$$

Inserting this estimate in (23.3.1) instead of $n\theta_{ij}$, we obtain

$$\boxed{\chi^2 = \sum_{i=1}^{m} \sum_{j=1}^{k} \frac{\left(a_{ij}-\dfrac{a_{i0}a_{0j}}{n}\right)^2}{\dfrac{a_{i0}a_{0j}}{n}} = n\left(\sum_{i=1}^{m} \sum_{j=1}^{k} \frac{a_{ij}^2}{a_{i0}a_{0j}}-1\right),} \tag{23.3.6}$$

which, in analogy with (23.2.2), is approximately distributed as χ^2 with $(m-1)\times(k-1)$ degrees of freedom, as $c = m+k-2$. It may also be seen directly by vertical and horizontal summations of the mk deviations,

$$a_{ij}-\frac{a_{i0}a_{0j}}{n}\,,$$

that they satisfy $m+k-1$ linear relations.

For $m = k = 2$ we obtain from (23.3.6)

$$\chi^2 = \frac{(a_{11}a_{22}-a_{12}a_{21})^2 n}{a_{10}a_{20}a_{01}a_{02}}\,, \quad f = 1\,. \tag{23.3.7}$$

Introducing the correction for continuity (YATES' correction), we obtain

$$\chi^2 = \frac{(|a_{11}a_{22}-a_{12}a_{21}|-\frac{1}{2}n)^2 n}{a_{10}a_{20}a_{01}a_{02}}, \quad f=1, \tag{23.3.8}$$

which should be applied for small values of n.

Example 23.4. Table 1.3 gives the distribution of 6805 moulded pieces of vulcanite classified according to two reasons for rejection. If defective dimensions and porosity occur independently the expected number of pieces rejected for both reasons should be $473 \times 1375/6805 = 95{\cdot}6$, which is a considerably smaller number than that observed. According to (23.3.7) the hypothesis of independence is tested by computing

$$a_{11}a_{22}-a_{12}a_{21} = 142\times 5099-331\times 1233 = 315935$$

and

$$\chi^2 = \frac{315935^2\times 6805}{473\times 6332\times 1375\times 5430} = 30{\cdot}4, \quad f=1.$$

As the value of χ^2 found is significant, see Table V, the hypothesis of independence must be rejected. Thus, the two types of defects are positively correlated.

23.4. Comparison of k Observed Grouped Distributions.

Consider k observed grouped distributions including $n_1, n_2, \ldots, n_k$ observations, respectively, as shown in Table 23.6.

TABLE 23.6.
Comparison of k observed grouped distributions.

Group No.	Distribution No. 1	2	...	k	Total
	Number of Observations				
1	a_{11}	a_{12}	...	a_{1k}	a_{10}
2	a_{21}	a_{22}	...	a_{2k}	a_{20}
⋮	⋮	⋮		⋮	⋮
m	a_{m1}	a_{m2}	...	a_{mk}	a_{m0}
Total	n_1	n_2	...	n_k	n_0

The attribute used for classification is assumed to be the same for all distributions and may be either qualitative or quantitative. (Formally Table 23.6 has the same form as Table 23.5).

As test hypothesis we postulate that the k observed distributions are random samples from the same population, i. e., the probability that an observation will fall in the ith group is θ_i, which is assumed to take the same value in all distributions. Thus, the theoretical number of observations in the ith group of the jth distribution is $\theta_i n_j$, so that

$$\chi_j^2 = \sum_{i=1}^{m} \frac{(a_{ij}-\theta_i n_j)^2}{\theta_i n_j} \tag{23.4.1}$$

according to (23.1.4) is distributed as χ^2 with $m-1$ degrees of freedom. According to the addition theorem for the χ^2-distribution the sum

$$\chi^2 = \sum_{j=1}^{k} \chi_j^2 = \sum_{i=1}^{m} \sum_{j=1}^{k} \frac{(a_{ij}-\theta_i n_j)^2}{\theta_i n_j} \tag{23.4.2}$$

will also be distributed as χ^2, the degrees of freedom being $k(m-1)$.

As the probabilities θ_i are usually unknown, we introduce the estimates a_{i0}/n_0, so that (23.4.2) is replaced by

$$\chi^2 = \sum_{i=1}^{m} \sum_{j=1}^{k} \frac{\left(a_{ij}-\dfrac{a_{i0}n_j}{n_0}\right)^2}{\dfrac{a_{i0}n_j}{n_0}} = n_0\left(\sum_{i=1}^{m} \sum_{j=1}^{k} \frac{a_{ij}^2}{a_{i0}n_j}-1\right), \tag{23.4.3}$$

which is approximately distributed as χ^2 with $(m-1)(k-1)$ degrees of freedom ($f = k(m-1)-c = (m-1)(k-1)$, as $c = m-1$, because $\sum \theta_i = 1$). Formally this test is identical with (23.3.6).

For $m = 2$, (23.4.3) is equal to (21.14.4), and for $m = k = 2$, (23.4.3) is equal to the square of (21.13.3).

Example 23.5. At an examination of the balls in a ball mill, samples of 300 balls were taken from five different localities in the mill and the

TABLE 23.7.
Distributions of 5 samples of 300 balls according to diameter.

Diameter in mm	Sample No. 1	2	3	4	5	Total	Average number
30	133	155	145	150	152	735	147·0
40	98	101	80	88	90	457	91·4
60	21	22	17	24	19	103	20·6
70	42	19	43	26	31	161	32·2
90	6	3	15	12	8	44	8·8
Total	300	300	300	300	300	1500	300·0

diameters of the balls measured, see Table 23.7. The question is: Do these 5 distributions of diameters differ significantly so that we must conclude that the balls are sorted during the milling process?

In this case it is particularly easy to calculate the "expected" numbers, $a_{i0}n_j/n_0$, as $n_j/n_0 = 1/5$, wherefore the "expected" number is equal to the average number, see the last column of Table 23.7. Table 23.8 shows the deviations.

TABLE 23.8.

Deviations between observed and "expected" numbers, $a_{ij}-a_{i0}n_j/n_0$.

Diameter in mm	Sample No. 1	2	3	4	5	Total
30	−14·0	8·0	−2·0	3·0	5·0	0·0
40	6·6	9·6	−11·4	−3·4	−1·4	0·0
60	0·4	1·4	−3·6	3·4	−1·6	0·0
70	9·8	−13·2	10·8	−6·2	−1·2	0·0
90	−2·8	−5·8	6·2	3·2	−0·8	0·0
Total	0·0	0·0	0·0	0·0	0·0	0·0

Table 23.9 shows each single term of χ^2 according to (23.4.3). As the value of χ^2, 30·06, exceeds $\chi^2_{\cdot 975}$ we conclude that the distribution according to diameter of the balls is not the same at the 5 different localities in the mill. Table 23.9 shows that the largest contribution to χ^2 is given

TABLE 23.9.

Computation of $\dfrac{n_0}{a_{i0}n_j}\left(a_{ij}-\dfrac{a_{i0}n_j}{n_0}\right)^2$.

Diameter in mm	Sample No. 1	2	3	4	5	Total
30	1·33	0·44	0·03	0·06	0·17	2·03
40	0·48	1·01	1·42	0·13	0·02	3·06
60	0·01	0·10	0·63	0·56	0·12	1·42
70	2·98	5·41	3·62	1·19	0·04	13·24
90	0·89	3·82	4·37	1·16	0·07	10·31
Total	5·69	10·78	10·07	3·10	0·42	30·06

by samples 2 and 3 for the diameters 70 and 90 millimetres. The deviations in Table 23.8 show that the number of large balls in sample 2 is smaller and in sample 3 larger than "expected".

CHAPTER 24.

SEQUENTIAL ANALYSIS

24.1. Some Principles of Sequential Tests.

As explained in § 9.4 the u-test

$$\frac{\bar{x}-\xi_0}{\sigma/\sqrt{n}} > u_{1-\alpha} \tag{24.1.1}$$

is uniformly most powerful for testing the hypothesis $\xi = \xi_0$ against the one-sided alternative $\xi > \xi_0$, i. e., for a given number of observations no other test can be found with a greater power than the test above. If we want to discriminate between the hypotheses $\xi \leq \xi_0$ and $\xi \geq \xi_1$, where $\xi_1 > \xi_0$, so that the probability of rejecting the test hypothesis $\xi \leq \xi_0$ when it is true is at most α, and the probability of rejecting the alternative hypothesis $\xi \geq \xi_1$ when it is true is at most β, that is $\pi(\xi_0) = \alpha$ and $\pi(\xi_1) = 1-\beta$, then the number of observations and the critical region may be found from (9.4.18) and (9.4.19).

Like the u-test above, all other tests described in this book—with the exception of the double sampling inspection plans—are based on the assumption that the number of observations is independent of the observational results. For a sequential test, however, the number of observations depends on the outcome of the observations and is therefore a random variable. A sequential test is based on a criterion (a decision function) for making one of the following three decisions for each of the successive observations:

1. Accept the test hypothesis.
2. Reject the test hypothesis.
3. Continue observation by making one further observation.

In the classical tests we use only two regions: the region of acceptance and the region of rejection, and these are functions of the predetermined number of observations. In the sequential tests one further region is introduced: the region where no final decision is taken because the observations made so far justify neither acceptance nor rejection of the test hypothesis and therefore a further observation must be made. The process of observation is continued as far as the observations fall inside this region, whence the number of observations becomes a random variable.

The sequential test requires on the average only about half as many observations as the classical most powerful test to discriminate between two hypotheses $\xi \leq \xi_0$ and $\xi \geq \xi_1$ with a given power (expressed by α and β). The average sample number depends on ξ, intermediate cases ($\xi_0 < \xi < \xi_1$) requiring a larger sample size than extreme cases where a decision may be readily obtained.

Only one type of sequential tests, namely *the probability ratio test* devised by A. WALD, will be considered here.

To illustrate the principles of this test consider a sequence of observations $x_1, x_2, \ldots, x_m, \ldots$ from a normally distributed population with known standard deviation σ and unknown mean. The test hypothesis is $\xi = \xi_0$ and the alternative $\xi = \xi_1$. For these values of ξ the probability density of the observations $x_1, x_2, \ldots, x_m$ takes on the values

$$p_{0m} = \left(\frac{1}{\sqrt{2\pi}\sigma}\right)^m e^{-\frac{1}{2\sigma^2}\sum_{i=1}^{m}(x_i-\xi_0)^2} \tag{24.1.2}$$

and

$$p_{1m} = \left(\frac{1}{\sqrt{2\pi}\sigma}\right)^m e^{-\frac{1}{2\sigma^2}\sum_{i=1}^{m}(x_i-\xi_1)^2} . \tag{24.1.3}$$

The test is based on a comparison of p_{0m} and p_{1m} for $m = 1, 2, \ldots$. As long as p_{0m} and p_{1m} do not deviate considerably a final decision cannot be taken, but as soon as for example p_{1m} becomes considerably larger than p_{0m}, ξ_1 is preferred to ξ_0.

Thus, the test takes the following form: For each value of m the probability ratio p_{1m}/p_{0m} is calculated. As long as the ratio does not deviate considerably from 1, i. e., as long as

$$B < \frac{p_{1m}}{p_{0m}} < A , \tag{24.1.4}$$

one more observation is made, but as soon as

$$\frac{p_{1m}}{p_{0m}} \geq A , \tag{24.1.5}$$

or

$$\frac{p_{1m}}{p_{0m}} \leq B , \tag{24.1.6}$$

the process is ended by accepting $\xi = \xi_1$ or $\xi = \xi_0$, respectively. The constants A and B are determined so that the power function takes the given values α and $1-\beta$ for $\xi = \xi_0$ and $\xi = \xi_1$, respectively. WALD has shown that this is obtained when

$$A \leq \frac{1-\beta}{\alpha} \tag{24.1.7}$$

and

$$B \geq \frac{\beta}{1-\alpha}, \tag{24.1.8}$$

and that the error introduced by substituting the above limits for A and B is of no practical consequence.

Thus, the basis of the sequential test is the inequality

$$\boxed{\frac{\beta}{1-\alpha} < \frac{p_{1m}}{p_{0m}} < \frac{1-\beta}{\alpha}.} \tag{24.1.9}$$

As long as this inequality is satisfied, the process of observation is continued. WALD has further shown that the probability that the sequential test will lead to a termination (the choice of either ξ_0 or ξ_1) is equal to 1.

Inserting (24.1.2) and (24.1.3) into (24.1.9), we obtain

$$\frac{\beta}{1-\alpha} < e^{-\frac{1}{2\sigma^2}\sum\limits_{i=1}^{m}[(x_i-\xi_1)^2-(x_i-\xi_0)^2]} < \frac{1-\beta}{\alpha}, \tag{24.1.10}$$

or

$$\log_e \frac{\beta}{1-\alpha} < -\frac{1}{2\sigma^2}\sum_{i=1}^{m}\left[(x_i-\xi_1)^2-(x_i-\xi_0)^2\right] < \log_e \frac{1-\beta}{\alpha}. \tag{24.1.11}$$

Further reduction leads to

$$\boxed{\frac{\sigma^2}{\xi_1-\xi_0}\log_e\frac{\beta}{1-\alpha}+m\frac{\xi_0+\xi_1}{2} < \sum_{i=1}^{m} x_i < \frac{\sigma^2}{\xi_1-\xi_0}\log_e\frac{1-\beta}{\alpha}+m\frac{\xi_0+\xi_1}{2}.} \tag{24.1.12}$$

Introducing the notation

$$\bar{x}_m = \frac{1}{m}\sum_{i=1}^{m} x_i\,, \quad m = 1, 2, \ldots\,, \tag{24.1.13}$$

and

$$\bar{\xi} = \tfrac{1}{2}(\xi_0+\xi_1)\,, \tag{24.1.14}$$

(24.1.12) may be written

$$\bar{\xi}+\frac{\sigma^2}{m}\frac{\log_e\dfrac{\beta}{1-\alpha}}{\xi_1-\xi_0} < \bar{x}_m < \bar{\xi}+\frac{\sigma^2}{m}\frac{\log_e\dfrac{1-\beta}{\alpha}}{\xi_1-\xi_0}. \tag{24.1.15}$$

Hence according to the sequential test we cannot choose between ξ_0 and ξ_1 as long as the successive means lie in the "neighbourhood" of $\bar{\xi}$. For increasing values of m, however, the limits converge to $\bar{\xi}$.

As an example we put $\xi_1 - \xi_0 = 1$, $\sigma = 3{\cdot}5$, and $\alpha = \beta = 0{\cdot}05$, (24.1.15) leading to

$$\bar{\xi} - \frac{36{\cdot}06}{m} < \bar{x}_m < \bar{\xi} + \frac{36{\cdot}06}{m}. \qquad (24.1.16)$$

As long as the successive means satisfy (24.1.16) new observations must be made.

The corresponding number of observations according to the classical u-test is $n = 132$, see (9.4.15). According to whether the mean of these 132 observations is smaller or larger than $\bar{\xi}$, the test hypothesis is accepted or rejected.

In the sequential test the number of observations is not given a priori. However, the number of observations which must be made before a decision is reached will usually be smaller than 132. It has been shown that the average number of observations necessary when the sequential test is repeatedly applied is only about half the number which is necessary to obtain the same discriminating power with the classical test. Fig. 24.1 gives a typical illustration of this fact. The limits in this figure have been determined from (24.1.16). At the 48th observation the mean is smaller than the lower limit and the hypothesis $\xi = \xi_0$ is accepted. In the classical test only one point on the curve of the successive means is utilized, namely the point corresponding to $\bar{x}_{132}$ (which is not seen on the figure), and this

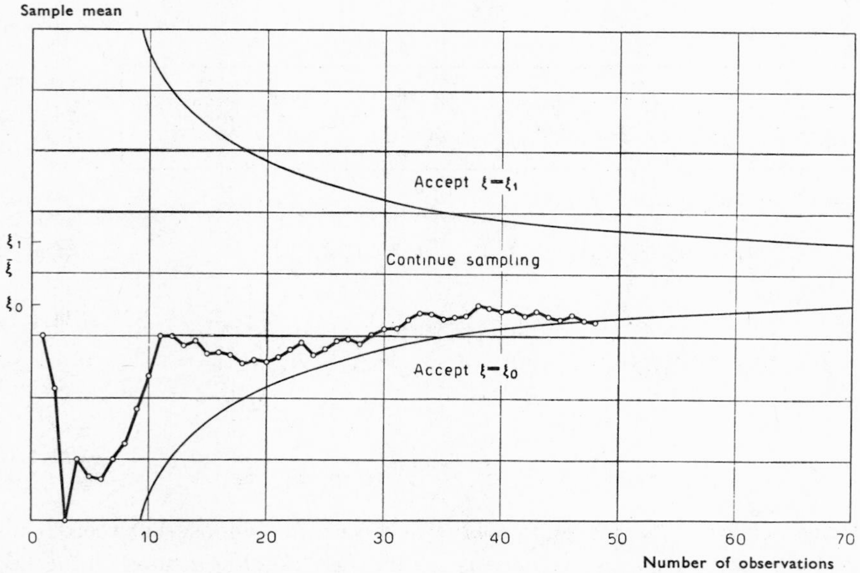

FIG. 24.1. Graphical representation of a sequential test corresponding to (24.1.16).

point is compared with the horizontal line through $\bar{\xi}$. Thus, the more detailed utilization of the observations which takes place in the sequential test will on the average lead to a considerable saving in the number of observations.

When the sequential test is applied in practice, (24.1.12) is used instead of (24.1.15). According to (24.1.12) the sums of the successive observations are compared with limits that increase linearly with the number of observations (the proportionality factor is $\bar{\xi}$), see Fig. 24.2.

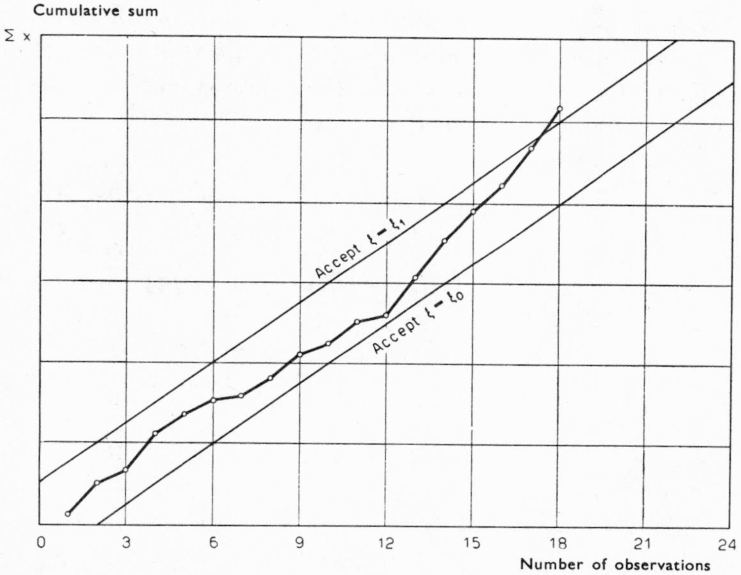

FIG. 24.2. Graphical representation of a sequential test corresponding to (24.1.12).

The general basis for the sequential test is given by the formula (24.1.9), which assumes that

1. The mathematical form of the distribution function is known.
2. A test hypothesis and an alternative (ξ_0 and ξ_1) regarding the values of an unknown parameter ξ in the distribution function is specified.
3. A choice of the values of the power function, α and $1-\beta$, for $\xi = \xi_0$ and $\xi = \xi_1$ is made.

The probability ratio test was developed by A. WALD in 1943, see A. WALD: *Sequential Analysis*, J. Wiley, New York, 1947.

In *Sequential Analysis of Statistical Data: Applications* by the STATISTICAL RESEARCH GROUP, Columbia University, 1945, formulas and tables have been given to facilitate the applications of sequential tests.

These two publications include formulas analogous with (24.1.12) for testing the usual hypotheses regarding the parameters in the normal distribution, the binomial distribution and the POISSON-distribution.

24.2. Some Special Cases of Sequential Tests.

In this section formulas analogous with (24.1.12) are given for some important special cases. For other problems, the reader is referred to the above-mentioned publications.

1. Let the observations be normally distributed with mean ξ and unknown standard deviation. *The test hypothesis is* $\sigma = \sigma_0$, *and the alternative* $\sigma = \sigma_1$, i. e., for $\sigma \leq \sigma_0$ it is of practical importance that the test hypothesis is accepted and for $\sigma \geq \sigma_1$ it is of practical importance that the test hypothesis is rejected. It is assumed that $\sigma_1 > \sigma_0$.

This leads to

$$p_m = \left(\frac{1}{\sqrt{2\pi}\,\sigma}\right)^m e^{-\frac{1}{2\sigma^2}\sum_{i=1}^{m}(x_i-\xi)^2},$$

which for $\sigma = \sigma_0$ and $\sigma = \sigma_1$ after introduction in (24.1.9) and some reduction gives

$$\frac{2\log_e\dfrac{\beta}{1-\alpha}+m\log_e\dfrac{\sigma_1^2}{\sigma_0^2}}{\dfrac{1}{\sigma_0^2}-\dfrac{1}{\sigma_1^2}} < \sum_{i=1}^{m}(x_i-\xi)^2 < \frac{2\log_e\dfrac{1-\beta}{\alpha}+m\log_e\dfrac{\sigma_1^2}{\sigma_0^2}}{\dfrac{1}{\sigma_0^2}-\dfrac{1}{\sigma_1^2}}. \tag{24.2.1}$$

The process of observation is continued as long as the sum of squares lies within these limits.

If ξ is unknown, the sum of squares in (24.2.1) is replaced by the sum of squares of the deviations from the sample mean

$$\sum_{i=1}^{m}(x_i-\bar{x}_m)^2 = \sum_{i=1}^{m}x_i^2 - \frac{\left(\sum_{i=1}^{m}x_i\right)^2}{m}, \tag{24.2.2}$$

and in the limits m is replaced by $m-1$ (the number of degrees of freedom).

2. When the proportion defective for a lot of produced units is to be examined by means of sampling inspection, sequential analysis represents a continuation of the ideas which form the basis of the double sampling inspection discussed in § 21.12. The test hypothesis is $\theta = \theta_0$ and the alternative $\theta = \theta_1$, and we want to be reasonably sure to accept lots with a proportion defective less than θ_0 and reject lots with a proportion defective larger than θ_1. It is assumed that the lots are so large that the probability

of finding a defective item is practically constant during the drawing of the sample.

From a lot with a proportion defective of θ the probability of obtaining the results $x_1, x_2, \ldots, x_m$, where x_i denotes defective or nondefective, is

$$p_m = \theta^{d_m}(1-\theta)^{m-d_m}, \tag{24.2.3}$$

d_m denoting the number of defective items among the m items examined. Substituting θ_0 and θ_1 for θ and inserting these expressions in (24.1.9) we obtain after reduction

$$\boxed{\frac{\log \dfrac{\beta}{1-\alpha}+m \log \dfrac{1-\theta_0}{1-\theta_1}}{\log \dfrac{\theta_1}{\theta_0}-\log \dfrac{1-\theta_1}{1-\theta_0}} < d_m < \frac{\log \dfrac{1-\beta}{\alpha}+m \log \dfrac{1-\theta_0}{1-\theta_1}}{\log \dfrac{\theta_1}{\theta_0}-\log \dfrac{1-\theta_1}{1-\theta_0}}.} \tag{24.2.4}$$

3. Let $x_1, x_2, \ldots, x_m$ be Poisson-distributed with the same mean. The test hypothesis specifies the mean as $\xi = \xi_0$, and the alternative is $\xi = \xi_1$, $\xi_1 > \xi_0$.

With mean ξ we obtain

$$p_m = \frac{\xi^{\Sigma x_i}}{x_1! \ldots x_m!} e^{-m\xi}. \tag{24.2.5}$$

Substituting ξ_0 and ξ_1 for ξ and inserting these expressions in (24.1.9) we obtain after reduction

$$\boxed{\frac{\log_e \dfrac{\beta}{1-\alpha}+m(\xi_1-\xi_0)}{\log_e \xi_1-\log_e \xi_0} < \sum_{i=1}^{m} x_i < \frac{\log_e \dfrac{1-\beta}{\alpha}+m(\xi_1-\xi_0)}{\log_e \xi_1-\log_e \xi_0}.} \tag{24.2.6}$$

CHAPTER 25.

THE MAIN POINTS OF A STATISTICAL ANALYSIS

25.1. Formulation of the Problem.

The exposition of statistics in this book has admittedly many shortcomings, especially regarded as a general theory of statistics. Amongst other things this is due to the fact that the aim of this book is first and foremost to furnish the reader with simple and practical methods—which can be understood and applied by non-statisticians—for the handling of the majority of the problems which occur in everyday work. A general exposition of the theory of statistics lies outside the scope of the present publication. Consequently, certain fundamental questions such as, e. g., the theory of estimation have been discussed only superficially.

As we go through the examples we get the impression that there is quite good agreement between theory and practical experience, which may lead to an optimism concerning the usefulness of the methods described which is not quite justified. In practical work we often find that it is difficult to obtain satisfactory agreement between for instance the distribution of the observations and a simple theoretical distribution. If nevertheless standard statistical methods are applied to such cases it must be kept in mind that possible errors in the conclusions drawn may be due to the fact that the assumptions underlying the theory in question have not exactly been fulfilled; this should not lead to a general rejection of the theory, but to greater caution in applying it before having properly examined the premises, and to an understanding of the fact that *standard methods will not always lead to the best solution of a problem;* the investigator must be prepared, on occasion, to develop his own methods.

According to the viewpoint of R. A. Fisher as given in the preceding chapters *a statistical analysis can usually be divided into the following five steps*:

1. The *planning* of the investigation, in particular the sampling method and the number of observations.
2. The *specification*, i. e., the formulation of a mathematical-statistical description, a model, of the observations.

3. The *estimation* of the parameters pertaining to the model and derivation of *the sampling distributions* of these estimates.
4. An *examination of the agreement* between the model and the observations.
5. The real solution of the problem by the aid of the estimates of the parameters and tests of significance.

These steps are not mutually independent, e. g., the number of observations necessary to obtain a certain accuracy in estimating a parameter depends on which estimate is computed from the observations, see for instance the comparison between the standard deviation and the range as estimates of the population standard deviation given in § 12.3.

In the following a few comments on the above five steps in a statistical analysis will be given.

25.2. The Sampling Method and the Number of Observations.

It is a basic condition for the application of statistical methods as developed here that the observations are the results of a *random operation*, see §§ 1.1 and 13.2. This may be obtained in sampling from a finite population by using random numbers, see § 17.2, and similarly in experimentation by randomizing the experimental units, see § 17.6. In other cases in practice it may be extremely difficult to ensure randomness. It is difficult to state general rules for the sampling, but it is at any rate always possible to adhere to two rules: The choice of a unit must not depend (1) on its properties, or (2) on the personal discretion of the investigator. As far as possible the series of observations should be tested for randomness using the tests given in Chapter 13.

The *number of observations* necessary to obtain a given accuracy of the estimate of the population mean depends on the population variance, whether the population includes a finite or an infinite number of elements, see § 17.2. Usually it will pay to *divide the population into a number of homogeneous sub-populations*, drawing a random sample from each sub-population instead of a single random sample from the total population. By handling the data according to the principles of the *analysis of variance*, the variation between the sub-populations may be eliminated, see Chapter 16 and §§ 17.3 and 17.4. For example, if the output of a plant consists of the output of k machines, it is usually best to choose the units for investigation by random sampling from the output of each machine, instead of investigating the total output as a whole.

The basic principles pertaining to the *design of experiments* in connection with the analysis of variance have been discussed in Chapter 17.

If the investigation aims at elucidating *the connection between two or more factors, some of which are of a non-stochastic nature*, the principle is to vary the non-stochastic factors over as wide a range as possible, see for example the remarks on regression analysis p. 536 and p. 632.

In the planning of investigations it is often possible from previous experience or by intelligent guesswork to indicate the magnitude of some of the population parameters. Combining these "guesstimated" parameter values with the values following from the hypothesis to be tested and the alternative hypothesis, *the number of observations necessary to obtain a reliable decision follows from the power function of the test to be applied*, see for example §§ 9.4, 11.3, 14.2, 16.4, 16.5, and 21.13.

25.3. The Specification.

The second step in the statistical analysis: the formulation of *a mathematical-statistical model which gives a satisfactory description of the data*, is not in principle a statistical task, but belongs within the professional subject from which the observations have been derived. In practical work, however, we often find that professional knowledge is so small that it is not possible to formulate a proper (theoretical) model, i. e., a description based on general laws regarding the process which has generated the observations. In such cases the specification becomes merely a *phenomenological description*, i. e., a purely empirical description of the observed phenomenon without any attempt at linking this description up with theoretical reasoning based on professional knowledge.

The theory of statistics includes *a collection of standard models*, e. g., the various systems of distribution functions and regression and correlation analysis, which are well suited for application to such a phenomenological description. It is usual to choose the mathematical form of the distribution function and regression equation on the basis of a *graphical analysis* of the observed data. To a certain degree the choice of model is arbitrary as long as the only requirement is a satisfactory agreement between the observed data and the model, cf. § 7.5, § 18.7, and Example 18.4, p. 564. (Usually this requirement is supplemented by a demand for the greatest possible "simplicity" in describing the process—either that the description is mathematically simple or that it leads to a simple test of significance). It should, however, be borne in mind that in the long run it does not pay to be satisfied with a phenomenological description; this should be resorted to only when all attempts at giving a theoretical description have proved impracticable.

Basing the specification solely on an analysis of the data is dangerous because an effective test of the model cannot be carried out by using the same data both for constructing and for testing the model.

The construction of a *theoretical model* of the phenomena is difficult not only because professional knowledge is lacking but also because statistics does not yet possess the necessary tools. In this respect, the further development of the theory of stochastic processes will be very important, see § 22.10 and § 22.11.

According to the above remarks we see that the basis of a statistical analysis of a set of data is the choice of a distribution function for the population from which the observations have been randomly drawn. This leads to the following two questions: (1) To what degree do the conclusions resulting from the statistical analysis depend on the specification? (2) What conclusions is it possible to draw directly from the observations without making assumptions regarding the mathematical form of the distribution? So far these questions have not been fully answered.

As an example of the type of answers which may be given to question 2 we refer to TCHEBYCHEFF's theorem, see § 5.10, illustrating that conclusions drawn by "distribution-free" methods are usually weaker than conclusions based on a specified distribution.

25.4. Estimating the Parameters.

When the mathematical form of the distribution of the population has been specified the next step in the statistical analysis is the derivation of estimates of the parameters from the observations.

In the theory of estimation a classification of estimates is attempted on the basis of properties derived from their sampling distributions. As examples of such classes of estimates we may mention the consistent, efficient, sufficient, and unbiased estimates defined in § 8.8 and § 8.9. Further, methods for deriving estimates with specified properties have been developed, for example the method of maximum likelihood, see § 8.8, and the method of least squares, see § 18.10.

Most of the estimates used in this book are sufficient or, if sufficient estimates do not exist, efficient. In the greater part of the problems treated, it is assumed that the observations are randomly drawn from a normally distributed population, and the estimates of the parameters have been derived using the method of least squares which in this case leads to sufficient estimates.

25.5. Investigation of the Agreement between the Model and the Observations.

As mentioned in § 25.3, a *graphical* analysis is usually made of the observations in connection with the specification, so that often we obtain a good idea of the agreement between model and observations at the same time that we are constructing the model. Following the computation of estimates of the parameters the graphical analysis ought to be supplemented by *a numerical test*. An observed grouped distribution may, e. g., first be examined on probability paper, which leads to the formulation of a distribution function. After computing the estimates of the parameters the "expected" distribution may be compared with the observed distribution by means of the χ^2-test in § 23.2. As another example we may mention the v^2-test for the linearity of the regression curve, given on p. 535.

As a general rule it may be stated that the *deviations* between the observations and the estimates of the corresponding theoretical values should be computed, so that the size and signs of these deviations may be studied.

25.6. Tests of Significance.

Some of the principles of tests of significance have been developed in § 9.4, and further references have been given in § 9.10. Here the author merely wants once again to emphasize two of the fundamental conditions for the tests described in this book.

Everywhere it has been assumed that the observations are *stochastically independent*. The importance of this assumption has been only partly elucidated, but the theoretical investigations published so far indicate that the tests may be highly misleading if the observations are even slightly correlated.

Further, it must be kept in mind that a satisfactory agreement between the observations and the test hypothesis is not equivalent to a proof of the hypothesis. The agreement only means that *the hypothesis has not been disproved*, i. e., that so far the hypothesis can be accepted as a working hypothesis—on line with many other hypotheses—and that the collection of more observational data may perhaps lead to modification or rejection of the hypothesis.

NOTATION

As far as possible the *parameters* of distribution functions are denoted by *Greek* letters and the *estimates* of the parameters by *Latin* letters. Examples: ξ and $\bar{x}$ (population and sample mean), σ^2 and s^2 (population and sample variance), ϱ and r (population and sample correlation coefficient), and β and b (population and sample regression coefficient).

Further, the letters t and χ^2 have been used in the same manner as in most other texts. The notation u and v^2 are, however, not universally applied in the same manner as here; the variance ratio v^2 is sometimes denoted by e^{2z} or F (F = FISHER).

The operator $\mathcal{M}\{x\}$ is in other textbooks often denoted by $E(x)$ (E = expectation).

The book of tables and formulas includes a list of the symbols used.

In order to make it easier for the reader to read many of the formulas, the Greek alphabet is given below:

THE GREEK ALPHABET.

A	α	Alpha	N	ν	Nu
B	β	Beta	Ξ	ξ	Xi
Γ	γ	Gamma	O	o	Omĭcron
Δ	δ	Delta	Π	π	Pi
E	ε	Epsilon	P	ϱ	Rho
Z	ζ	Zeta	Σ	σ	Sigma
H	η	Eta	T	τ	Tau
Θ	θ, ϑ	Theta	Y	υ	Upsīlon
I	ι	Iota	Φ	φ	Phi
K	$\varkappa$	Kappa	X	χ	Chi
Λ	λ	Lambda	Ψ	ψ	Psi
M	μ	Mu	Ω	ω	Omĕga

LIST OF AUTHOR REFERENCES

INDEX OF SYMBOLS AND FORMULAS

INDEX OF SUBJECTS